CANADIAN CITIES IN TRANSITION

SIXTH EDITION

CANADIAN CITIES IN TRANSITION

Understanding Contemporary Urbanism

Edited by Markus Moos | Tara Vinodrai | Ryan Walker

Oxford University Press is a department of the University of Oxford.
It furthers the University's objective of excellence in research, scholarship,
and education by publishing worldwide. Oxford is a registered trade mark of
Oxford University Press in the UK and in certain other countries.

Published in Canada by
Oxford University Press
8 Sampson Mews, Suite 204,
Don Mills, Ontario M3C 0H5 Canada

www.oupcanada.com

First Edition published in 1991
Second Edition published in 2000
Third Edition published in 2006
Fourth Edition published in 2010
Fifth Edition published in 2015

Library and Archives Canada Cataloguing in Publication
Title: Canadian cities in transition : understanding contemporary urbanism / edited by Markus Moos, Tara Vinodrai and Ryan Walker.
Names: Moos, Markus, 1980- editor. | Vinodrai, Tara, 1974- editor. | Walker, Ryan, 1975- editor.
Description: Sixth edition. | Includes index.
Identifiers: Canadiana (print) 20190169842 | Canadiana (ebook) 20190169850 | ISBN 9780199032693 (softcover) | ISBN 9780199038695 (looseleaf) | ISBN 9780199032709 (EPUB)
Subjects: LCSH: Cities and towns—Canada—Textbooks. | LCSH: Urbanization—Canada—Textbooks. | LCGFT: Textbooks. Classification: LCC HT127 .C32 2020 | DDC 307.760971—dc23

Cover image: © Paul McKinnon/Shutterstock.com
Cover design: Laurie McGregor
Interior design: Sherill Chapman

Oxford University Press is committed to our environment.
This book is printed on Forest Stewardship Council® certified paper
and comes from responsible sources.

Printed and bound in the United States of America

1 2 3 4 — 23 22 21 20

Contents

Publisher's Preface vii
Preface: A Guide to the Text viii
Contributors xii
Tribute to Trudi Bunting (1944–2017) xiv

I • City Building Blocks 1

1 Fundamentals of Cities 2
Pierre Filion

2 Urban Transitions: Historical, Present, and Future Perspectives on Canadian Urban Development 15
Pierre Filion

3 Governing Canadian Cities 33
Zack Taylor and Neil Bradford

4 Indigenous Peoples, Indigenous Cities 51
R. Ben Fawcett and Ryan Walker

5 Digital Cities: Contemporary Issues in Urban Policy and Planning 70
Shauna Brail and Betsy Donald

II • Demography, Identity, and Home 87

6 Age and Generational Change in the City 89
Markus Moos

7 Urban Change through the Lens of Demography, Life Course, and Lifestyles 110
Ivan Townshend and Ryan Walker

8 International Migration and Immigration: Remaking the Multicultural Canadian City 134
Audrey Kobayashi and Valerie Preston

9 Gender, Sexuality, and the City 155
Damaris Rose

10 Urban Divisions: Inequality, Neighbourhood Poverty, and Homelessness in the Canadian City 175
Alan Walks

11 Housing: Dreams and Nightmares 195
Richard Harris

III • Urban Form, Structure, and Design 213

12 The Market, Planning, and Emerging Urban Forms 215
Jill L. Grant and Pierre Filion

13 Public Space in the City Centre: Design, Activity, and Measurement 232
Ryan Walker and Jill Blakley

14 Integrated Transport and Land-Use Planning: Connecting Individuals to Opportunities 255
Ahmed El-Geneidy and Emily Grisé

15 Re-Imaging, Re-Elevating, and Re-Placing the Urban: Gentrification and the Transformation of Canadian Inner Cities 277
Alison Bain and Bryan Mark

16 Cities on the Edge: Suburban Constellations in Canada 292
Jean-Paul D. Addie, Robert S. Fiedler, and Roger Keil

IV • Economies of Cities 311

17 The New Economy of Canadian Cities? Employment, Creativity, and Industrial Change 313
Tara Vinodrai

18 Innovation and Canadian Cities: Exploring Different Viewpoints 331
Richard Shearmur

19 The Economic Value of Urban Land 347
Andrejs Skaburskis and Markus Moos

20 Slow Growth and Decline in Canadian Cities 364
Heather M. Hall and Betsy Donald

V • The Environmental Imperative 379

21 Getting Serious about Urban Sustainability: Eco-Footprints and the Vulnerability of Twenty-First-Century Cities 381
William E. Rees

22 The Ups and Downs of a Sustainable and Climate Resilient Development Path in Canadian Cities 397
Meg Holden and Robin Chang

23 Food Systems and the City 417
Alison Blay-Palmer and Karen Landman

24 Climate Change Threats and Urban Design Responses 432
Maged Senbel, Simon Liem, and Alexandra Lesnikowski

Glossary 446
Index 453

Publisher's Preface

Introducing *Canadian Cities in Transition*

Oxford University Press is proud to introduce the sixth edition of *Canadian Cities in Transition: Understanding Contemporary Urbanism,* edited by Markus Moos, Tara Vinodrai, and Ryan Walker.

The new edition retains its unique Canadian focus on urban geography and urban planning while providing a relevant and comprehensive survey of urbanization from both modern and traditional perspectives. From founding principles to the current trends shaping the discipline today, *Canadian Cities in Transition* explores the recent and ongoing transformational change to our urban environment while building on the strength of previous editions and their authoritative contributions.

Highlights of the New Edition

- **New voices.** A blend of new and seasoned scholars brings a lively mix of perspectives in the field of urban geography and planning to the text, and ensures balance in the interpretation of trends, both past and present, that are shaping Canadian cities in the twenty-first century.
- **New and expanded topical coverage.** Coverage of issues including gentrification, climate resilient design, and digital innovation draw students into the current and relevant social and physical factors that have an impact on the nation's urban environments.
- **Canadian perspective.** *Canadian Cities in Transition* remains the only book in urban geography and urban planning to focus on Canadian urban areas and the uniquely Canadian forces that shape our cities.
- **Updated data and visuals.** The most recent statistics and new additions to the art program help to contextualize key issues and locales.

Preface: A Guide to the Text

This sixth edition of *Canadian Cities in Transition: Understanding Contemporary Urbanism* is designed to serve a number of purposes. It is an introduction for university students to the Canadian urban phenomenon, presenting different facets of the city: its historical evolution, economic dynamics, environmental impacts, dependence on natural systems, urban lifestyles, cultural makeup, social structure, infrastructures, governance, planning, and design. The volume also aims to assist the next generation of citizens, consumers, experts, business people, and politicians in their efforts to solve the urban problems—traffic congestion, different forms of environmental damage, crime, social segregation, inequality, housing affordability, governance—that they are inheriting. Canadian cities are not simply a collection of problems to be solved, however, and this book helps to articulate the promise of Canada's urban age, where people and public space are re-centred for economic, environmental, and social reasons, and where "quality" instead of simply "growth" becomes a unifying hallmark of urbanism. The book offers the most current knowledge and perspectives on the Canadian city. The contributors review the recent literature and research on different aspects of the city, and provide their expert opinion on how to focus our examination of contemporary urban issues. Finally, the volume provides an update on urban Canada by identifying the main characteristics of the contemporary Canadian urban phenomenon, its problems, achievements, and opportunities. In this regard, the text will help students and other citizens make sense of the vast flow of information on cities circulated by the media. Because quality information is a condition for judicious decisions, knowledge of the city is vital to effective planning, private and public development, consumer choice, and functioning of democratic processes.

The text is informed by different disciplines with an urban dimension: mainly geography and urban planning, but also economics, political science, sociology, ecology, and history. It focuses on different urban themes and draws on all the disciplines relevant to their exploration. It also considers cities belonging to all size categories, as well as to different Canadian regions. Contributors who represent all parts of the country are able to highlight cross-country differences, as well as similarities by drawing on examples from their own regions.

The 24 chapters of this edition are organized into five parts. The five chapters in Part I "City Building Blocks" serve an introductory role by setting the context for the inquiries that occupy the following chapters, and also a formative role, building for readers a foundation upon which to shape their interpretation of the content in subsequent parts of the book. In Chapter 1, Pierre Filion lays out seven universal properties common to all cities. These properties explain the existence of the urban phenomenon as well as its different manifestations over time. In Chapter 2, Filion provides historical background by exploring the themes of transition and transformation that are at the centre of the book. It describes three different epochs of urban development in Canada: the pre–World War II city, the rise of suburbanization in the postwar period, and the urban development patterns—both new and inherited—that have prevailed since the post-industrial structural shifts of the 1970s. The chapter closes with an exploration of possible future trajectories of urban development. Chapter 3, by Zack Taylor and Neil Bradford, begins by discussing the role of local governments across Canada in shaping urban policy. They identify examples of how Canadian governments and civic leaders at all levels have engaged in collaborative multi-level urban governance to address today's pressing urban challenges. In Chapter 4, Ben Fawcett and Ryan Walker discuss cities as Indigenous places, the concept of Indigenous urbanism, and key foundations to understanding the future of Canadian cities where shared sovereignty, governance, cultural resurgence, and guiding principles for truth-seeking and reconciliation need to be understood by all who would make the Canadian city a subject of analysis, and not just the few who in the past have taken a special interest in the topic. The final building block, Chapter 5 by Shauna Brail and Betsy Donald, addresses how cities are being shaped by new—and potentially disruptive—technologies. They argue that we have entered a period where rapid digitization and the fast-paced adoption of new technologies are disrupting almost every aspect of urban life.

The six chapters in Part II, "Demography, Identity, and Home," look at how the social and cultural space of cities is shaped by demographic trends, and is responsive to changing societal and global contexts, and how the range of urban experiences is brought into sharp contrast by the extent to which people are able to meet the need for housing and home. In Chapter 6, Markus Moos considers the age and generational dimensions of urban restructuring. He tracks the changing age compositions of our cities and the changing location patterns of young adults and seniors. Moos points to growing segregation of young adults in central areas of our cities and discusses the deteriorating economic prospects of those just entering labour markets. He describes current policies aimed at planning for an aging population and explores exclusion and inequalities arising from ageism. Chapter 7, by Ivan Townshend and Ryan Walker, is about social changes affecting cities: demography, life course, and lifestyles. Among other things, the chapter highlights the effects of aging, the extension of youth, and the co-existence of numerous lifestyles, as well as the impact of these trends on the built environment and on community dynamics within our cities. In Chapter 8, Audrey Kobayashi and Valerie Preston focus on immigration and the resulting social diversity. They chart the geography of immigration in Canada—the urban areas that especially attract immigrants and where immigrants concentrate in these cities. They also describe new urban phenomena associated with immigration, such as the emergence of "ethnoburbs," and the broader policy frameworks through which immigration is managed. Damaris Rose explores the nature of gender and sexuality in the city in Chapter 9. She adopts a perspective that underscores the freedoms and constraints confronting gender differences and the expression of sexual orientations in cities. The chapter discusses the impact of values and of the spatial organization of cities on the lives of women. It also looks at how LGBTQ+ people (lesbian, gay, bisexual, transgender, queer, and others) negotiate the city. The subject of Chapter 10 is social polarization. Alan Walks identifies a range of factors accounting for growing income polarization among households over the last decades. He also paints a picture of the urban consequences of polarization—an urban social geography that is increasingly characterized by unevenness—and the policy responses that could address social inequality. Chapter 11, by Richard Harris, is devoted to housing, one of the largest uses of space in Canadian cities. The chapter reviews the socio-economic and geographic landscape of housing in Canada and the policy environment in which housing is managed, and ends with a discussion of some of the pressing issues facing the Canadian housing system as a whole.

Part III, "Urban Form, Structure, and Design," is about responses to contemporary patterns and challenges that structure our cities, as well as specific design and policy approaches seeking to enhance Canadian urbanism. In Chapter 12, Jill L. Grant and Pierre Filion pursue the planning transition theme introduced in Chapter 2 through an exploration of the newer urban forms emerging in the Canadian city. Chapter 12 is about the loud call within the planning profession for a change in urban development trends. It describes and evaluates attempts at intensifying the urban environment while acknowledging the counter-effect of many new automobile-dependent urban forms, such as power malls. Ryan Walker and Jill Blakley examine the movement, market, and meeting functions of public space in Chapter 13, focusing on streets and squares in our city centres. Topics range from the conceptual to measurement and metrics, and to lessons learned over multiple generations of scholarship and observation to enhance the design and programming of public spaces. The connections between activity and urban form are discussed. The authors give examples of how public life is being re-ignited in the shared spaces of our cities by place-specific public art and performance, by recalibrating transportation infrastructure to induce more of the movement we want in our urban age, and by decades- or centuries-old lessons for good physical design of streets and squares. In Chapter 14, Ahmed El-Geneidy and Emily Grisé use the lens of accessibility to examine the relationship between transportation and land use. Following a history of urban transport, congestion levels, and the distribution of transport mode share between private automobiles, active transport, and public transit in Canadian cities are compared with urban areas internationally. The authors discuss how the modelling of travel demand is changing in pursuit of a multimodal sustainable approach that blends transport options with land-use decisions. Chapter 15, by Alison Bain and Bryan Mark, deals with emerging places in Canada's inner or central cities: gentrified neighbourhoods, high-rise condominiums, and the like. It relates the conditions that have led to their development and the impact of these places on the social structure and the functioning of cities. Given the increasingly suburban nature of urban life in Canada, it is fitting for this section of the book to close with Chapter 16, by Jean-Paul D. Addie, Robert S. Fiedler, and Roger Keil, an exploration of how understandings of the Canadian urban form are broadening to include the conceptual study of suburbanisms, the "cities on the edge."

The four chapters in Part IV, "Economies of Cities," undertake an examination of urban economics and the dynamics of growth and decline in our Canadian urban system. Chapter 17, by Tara Vinodrai, focuses on how cities are affected by present and recent economic trends and how they have responded. She traces the shift from the industrial to the post-industrial city and the growing emphasis placed on innovation, creativity, and knowledge-based work. She sees this trend as having both positive and negative impacts on cities. In Chapter 18, Richard Shearmur explores the different viewpoints on the relationship between cities and innovation, and examines how the debate applies to Canadian cities. He presses us to understand what is specific about innovation in an urban context and why innovation is more likely to occur in cities. Chapter 19, by Andrejs Skaburskis and Markus Moos, examines the economics of urban land use. The chapter introduces the structuring parameters of urban land use and describes the origin and operation of urban land markets as well as their outcomes. Skaburskis and Moos end by introducing the dimensions of timing and strategy to help explain development decisions that challenge our conventional views of when, where, and how intensively land is capitalized. Chapter 20, by Heather M. Hall and Betsy Donald, zeroes in on polarization *among* urban areas. This last contribution of the book's fourth part describes the multiple challenges that declining urban areas face and policy responses that could mitigate the consequences of decline. This issue is increasingly relevant given the present concentration of demographic and economic growth in a few large metropolitan regions and their surroundings. Hall and Donald convey the opportunities missed by urban decision-makers too narrowly focused on growth rather than on qualitative development.

Part V, "The Environmental Imperative," is about the most pressing structuring parameters of Canadian urbanism in the twenty-first century, environment and sustainability. Chapter 21, by William E. Rees, is about the environmental impact of cities. It pictures cities as an important contributor to global environmental damage. The chapter also explores the vulnerability of cities to environmental deterioration and the need for them to deploy long-term sustainability strategies. In Chapter 22, by Meg Holden and Robin Chang, climate change, sustainability, and resilience are examined in relation to the ups and downs of Canada's battle with urban sprawl, its costs, and the challenges of enhancing the density of our built environment in relation to its ecology. Model sustainable neighbourhoods are discussed, and the chapter ends with critical reflections on where the powerful concepts of sustainability and resilience join to direct a pathway for Canadian urban development. Chapter 23, by Alison Blay-Palmer and Karen Landman, looks at the new-found interest in the geography of food, as evidenced in movements such as the 100-mile diet. This chapter is about how food is procured and distributed within Canadian cities. Issues include accessibility to different forms of food outlets and the problem of food deserts, efforts to increase reliance on food grown nearby, and food production within cities themselves. The book closes with Chapter 24, an urgent discussion of how climate change threats must prompt responses to how we design our cities. In this chapter, Maged Senbel, Simon Liem, and Alexandra Lesnikowski explore the challenges and opportunities inherent in various approaches to sustainable urban design intended to create low-carbon, resilient, and healthy neighbourhoods.

Contributors

Jean-Paul D. Addie
Urban Studies Institute, Andrew Young School of Policy Studies
Georgia State University

Alison Bain
Department of Geography
York University

Jill Blakley
Department of Geography and Planning
University of Saskatchewan

Alison Blay-Palmer
Department of Geography and Environmental Studies
Wilfrid Laurier University

Neil Bradford
Department of Political Science
Huron University College, Western University

Shauna Brail
Urban Studies Program, Innis College
University of Toronto

Robin Chang
Department of European Planning Cultures
Technical University of Dortmund

Betsy Donald
Department of Geography and Planning
Queen's University

Ahmed El-Geneidy
School of Urban Planning
McGill University

R. Ben Fawcett
Department of Geography and Planning
University of Saskatchewan

Robert S. Fiedler
Department of Geography
York University

Pierre Filion
School of Planning
University of Waterloo

Jill L. Grant
School of Planning
Dalhousie University

Emily Grisé
School of Urban and Regional Planning
University of Alberta

Heather M. Hall
School of Environment, Enterprise and Development
University of Waterloo

Richard Harris
School of Geography and Earth Sciences
McMaster University

Meg Holden
Urban Studies Program and Department of Geography
Simon Fraser University

Roger Keil
Faculty of Environmental Studies
York University

Audrey Kobayashi
Department of Geography and Planning
Queen's University

Karen Landman
School of Environmental Design and Rural Development
University of Guelph

Alexandra Lesnikowski
Department of Geography
McGill University

Simon Liem
School of Community and Regional Planning
University of British Columbia

Bryan Mark
Department of Geography
York University

Markus Moos
School of Planning
University of Waterloo

Valerie Preston
Department of Geography
York University

William E. Rees
School of Community and Regional Planning
University of British Columbia

Damaris Rose
Centre urbanisation culture société
Institut national de la recherche scientifique (INRS)

Maged Senbel
School of Community and Regional Planning
University of British Columbia

Richard Shearmur
School of Urban Planning
McGill University

Andrejs Skaburskis
School of Urban and Regional Planning
Queen's University

Zack Taylor
Department of Political Science
Western University

Ivan Townshend
Department of Geography
University of Lethbridge

Tara Vinodrai
Department of Geography and Institute for Management & Innovation (UTM)
University of Toronto

Ryan Walker
Department of Geography and Planning
University of Saskatchewan

Alan Walks
Department of Geography and Planning
University of Toronto

Tribute to Trudi Bunting (1944–2017)

Co-editor of the First Four Editions of Canadian Cities in Transition

Six editions of *Canadian Cities in Transition* ago, in the late 1980s, we were experiencing a surge of interest for Canadian cities as a distinct area of study—different enough from their US counterparts to be an object of investigation in their own right. *The Myth of the North American City* by Michael Goldberg and John Mercer, published in 1986, highlighted differences between Canadian and US cities. These researchers demonstrated that Canadian cities registered higher density and public transit use, were much less racially segmented, and benefited from a greater involvement of the public sector. It is in this context that Trudi Bunting and I decided to proceed with the first edition of *Canadian Cities in Transition* (we did not know at the time that there would be a second, let alone a sixth, edition). From the start, we adopted principles that have defined the book through all its editions: chapters had to communicate original research, they were to be written by the leading experts on the aspects of the city they addressed, their material was to be accessible to undergraduate students without being dumbed down, and the book had to advance knowledge of the Canadian urban phenomenon.

Trudi Bunting both made the book possible and was responsible for the orientation and quality of its first four editions. We came up with the idea of such a book together, but I hardly knew anyone within the Canadian urban research community. I had just finished my doctorate in England a few years earlier, so my Canadian contacts were limited. Trudi Bunting, on the other hand, knew nearly all Canadian urban researchers. After we identified the themes to be covered by the different chapters, she quickly matched them with potential contributors. And then came the Trudi Bunting personal touch. It was important to her that we meet face-to-face with as many of the contributors as possible in order to discuss together a detailed plan for their chapter. She had a vision of each edition of the book she co-edited and wanted to make sure that all chapters fit this vision. Then she would carefully edit the chapters, not hesitating to restructure them to clarify their message and ease their reading. Trudi Bunting had an exceptional ability to perceive problems with a text and correct them. The four editions of the book benefited greatly from this talent.

The work of Trudi Bunting on *Canadian Cities in Transition* was an extension of her dedication to the education of undergraduate and graduate students. She was exceptionally supportive of her students, but at the same time highly rigorous about the quality of their work. Their papers and theses would be given the same detailed editing treatment as the book chapters. The book also reflected the breadth of her interest in urban geography. She began her career focusing on children in the city from a behavioural psychology perspective. Then her research work shifted to other aspects of cities, particularly mid-size cities, the revitalization of the inner city, and the dynamics of suburban areas. Later in life, she left urban research to investigate the characteristics of rural places that attract artists.

At the beginning, I disagreed with Trudi Bunting about the title "Canadian Cities in Transition." My point was that we did not know if this system of cities was actually undergoing a transition. She, on the other hand, was adamant that Canadian cities were indeed in a transitionary phase. Thirty years later, with the revival of core areas, widespread densification, ethnic and racial diversification, and renewed interest in public transit, walking, and cycling, it is obvious that she was right.

—Pierre Filion, co-editor of the first five editions of *Canadian Cities in Transition*.

I City Building Blocks

The chapters in the first part of the book introduce readers to foundational concepts needed to understand and study contemporary Canadian cities. The first two chapters outline fundamental properties of cities. They outline why cities form and the factors that shape them. Considered are the internal structures of Canadian cities in different time periods, and the reasons for the changes over time. It is important to consider the past because it demonstrates the long-lasting elements of historical urban development that continue to influence urban life today. Understanding of historical trends also helps us see more evidently where new changes are occurring in cities. Finally, a historical lens highlights the existence of some universal features of cities that remain consistent over time.

The remaining three chapters in this section deal with more specific dimensions of urban Canada that are also context setting. Chapter 3 considers how Canadian cities are governed. Governance, or how we organize ourselves as a society and make decisions, influences all aspects of urban life. Readers will learn about the historical context that continues to shape municipal governance today, the role of municipalities, the involvement of various levels of government in urban policy-making, and important changes underway that influence how cities are governed.

Chapter 4 outlines another foundational dimension to understanding Canadian urban life. We are reminded that Canada and its cities exist on Indigenous land. Readers will learn about the history of settler colonialism and the experiences of Indigenous peoples with contemporary urbanism. The chapter also calls on readers to work toward a decolonized co-existence between Indigenous and non-Indigenous society. Examples of attempts at reconciliation are provided.

Chapter 5 considers the relationship between technology and the city. Technology has shaped human settlements from the first cities of Mesopotamia until today. However, the fast-paced adoption of technology and increasing digitization means that we simply cannot understand urban transformations without thinking about what has become known as the "digital" or "smart" city. The chapter defines and outlines the growth of the platform economy, the generation and use of "big data" by city governments, and the role of technology in shaping urban economies. Readers will learn about what urbanism might hold when people can increasingly interact, work, and shop in digital spaces.

After reading the chapters in Part I, readers will have understanding of

- the factors shaping cities;
- the historic and current forms and structure of Canadian cities;
- how cities are governed;
- the history of settler colonialism upon which Canadian cities are built; and
- how digitization is shaping contemporary urbanism.

1 Fundamentals of Cities

Pierre Filion[1]

Introduction

This volume is about understanding and dealing with change and transition in twenty-first-century cities. At the outset, however, we need sound knowledge of how cities operate and how they relate to broader societal trends in order to address contemporary urban issues. More often than not, past mistakes can be seen in hindsight to be the result of deficient understanding of **urban dynamics**. Yet cities resist understanding because they are such complex systems. We can argue that, along with language, the large city is the most intricate of human creations. In both cases, complexity stems from the presence of a relatively stable structure upon which interchangeable elements can be affixed in different fashions. In the case of language, the structure is syntax, which supports nearly unlimited combinations of words; cities owe their structure to major infrastructure networks, which provide connections between different assemblages of buildings and other land uses. The degree of complexity further increases when we consider economic and value systems behind the creation of the urban built environment, the multiple ways people use this environment, and the perceptions and interpretations of this environment and of the activities that take place therein.

This chapter introduces universal properties that define the city and represent universal features of the urban phenomenon. While Canadian cities have their own distinctiveness—a central theme in this book—they share many similarities with their counterparts developed at different times in different parts of the globe. This chapter, then, establishes some basic principles inherent to urban areas from the very beginning of urban settlement.

Seven properties are fundamental to understanding the urban phenomenon, regardless of time or place: *production*, *proximity*, *reproduction*, *capitalization*, *place*, *governance*, and *environment* (Table 1.1). These properties account for the existence of cities, their diverse configurations, and the way they operate, along with the need for specialized knowledge and interventions to deal with urban issues, and have been inherent in the city from its beginnings some 4000 years ago in China and Mesopotamia. As expected, their manifestations vary across periods. The seven properties that we set out here provide a framework to discuss problems confronting the contemporary Canadian city and possible solutions.

Table 1.1 Urban Properties and Their Effects on the Contemporary City

Properties	Definition	Manifestation in the contemporary city
Production	Need for cities to produce goods and services for their own residents and to be exported beyond their territories to ensure the purchase of the goods and services that cannot be procured within their territories. Production attracts people to cities.	Links between cities and their hinterlands are replaced by economic interconnections between cities across the world. Transition from an industrialized to a service economy along with a renewed emphasis on resources.
Proximity	Cities are made of numerous overlapping markets of frequently repeated exchanges, with a predominant structuring role taken by the labour market. Proximity makes these exchanges possible; otherwise they would be ruled out by excessive travel time and cost.	Reliance on the car has greatly extended spatial range whereby repeated exchanges can be carried out. However, decentralization tendencies are in part countered by the stress placed on culture, entertainment, and, generally, face-to-face contacts by new economic tendencies.
Reproduction	Reproduction refers to the different conditions needed for the availability of a labour force that is well suited to the needs of the production sector of an urban area. A narrow definition of reproduction relates to the replacement of generations and the presence of conditions needed to maintain health. A broader definition includes education and much of household consumption, including even entertainment.	With the vast majority of working-age adults in the labour force, Canada faces below-reproduction birth rates, and households have difficulties in providing reproduction-related services to their members. The problem is that increasing demand for state reproduction-related services happens at a time when public sector willingness and capacity to intervene is limited by insufficient resources.
Capitalization	Refers to all investments in the built environment of cities, as well as to this built environment itself. The capitalization property of cities derives from its dense urban environment. Capitalization is a factor of stability and durability for cities, and can be an obstacle to implementing innovations.	Over recent decades capitalization in cities has taken two forms: suburban development and inner-city intensification. Urban capitalization is conducive to speculation as demonstrated in the recent property bubble and associated adverse economic consequences.
Place	Place is about feelings, either positive or negative, associated with different locales in the urban environment. It refers to subjective reactions to these aspects of the city. Efforts are made by different professions involved in urban development to associate positive meaning with their projects. The types of urban places that are most valued vary over time.	Renewed attention given to place characteristics coincides with the growing importance of services and leisure in the post-industrial city. Quality places are seen as a way of attracting the creative class, which has the potential to propel an urban area's economy. Meanwhile, standard and poor-quality places are still being produced, especially in the suburban retailing sector.
Governance	Cities require interventions that are suited to their reality and, thus, specialized forms of administrations to formulate and deliver these interventions. They also rely on the knowledge that is essential to these interventions. Cities need to deal with issues related to concentration of activities and urban infrastructures.	Expansion of administrations with responsibility for urban interventions in response to growing demand for such interventions. This expansion is followed by cutbacks in tight budgetary circumstances. Under pressure from the municipal lobby, the federal government has directed an important share of its economic stimulus budget to urban infrastructures.
Environment	Historically, to survive, cities had to respect their environment. Cities that did not do so were unable to draw natural resources essential to their survival and vanished over time.	Cities are able to draw resources from ever-longer distances. They are thus less dependent on their immediate environment. At the same time, environmental awareness becomes global, and concern about different planetary impacts of cities (especially on climate change) is on the rise.

Production

Throughout history the foremost raison d'être for urban settlement has been accommodation of the need for specialized production activities that could not survive in isolation in rural settings—for example, markets, production of fine crafts, centralized governance. Indeed, if asked why they live in a city, most people will respond to the effect that "I work there or a member of my family does." Economic production creates jobs and brings people into the city and is thus the main reason for urban growth. Economic production, too, is the property most often associated with transformative change in **urban form** and structure. Thus, for example, Canada's first mercantile settlements were established to export "staple" products, such as fish, furs, and timber, to European colonial "mother" countries.

A further impetus for urban growth was the administration of resource industries and transportation systems required for their export. These conditions were conducive to the growth of only a small number of cities, however. But with agricultural development, centres grew to service rural areas, and then with wide-scale industrialization urban growth took off in Canada as it did in other developed countries (Innis, 1995a [1931], 1995b [1938]).

Many economic activities, of course, are aimed at the consumption needs of a city's own residents (Watkins, 1980). What most sets urban settlements apart from traditional self-reliant rural economies, however, is the city's historic inability to satisfy all its consumption requirements. As a result, it becomes imperative that cities export so as to generate the revenues needed to acquire products that can only be obtained outside their territory. Inside the urban envelope, for example, the need for specialized activities to be close to one another and to their workers (the proximity property, discussed below) rules out any possibility of devoting large surfaces to agriculture as would be needed to feed a large resident population (at least this has been the case until recent attempts at urban-based agriculture) (Christensen, 2007; Lawson, 2005). A city also must reach beyond its territory to secure other products and resources essential for sustaining its population and economic activity, including different forms of staples and energy and, often, water. Research shows that cities depend for their natural resources on a territory (or **ecological footprint**; see Rees, Chapter 21) that far exceeds the urbanized perimeter.

For a city to exist, it must be in a position to export sufficient goods and services to counterbalance its imports. But exports need not be tied to products. Capital cities, for example, export decisions and derive their monetary returns from tax revenues; likewise, in medieval times, it was not unusual for cities to draw taxes, often in kind, from a hinterland to which they extended military protection. Cities that fail to export decline and may disappear altogether. Throughout the Canadian periphery, resource communities whose staple has run out must either find alternative economic activities or perish (Lorch, Johnston, and Challen, 2004; Mitchell and O'Neill, 2016; see Hall and Donald, Chapter 20).

Over recent decades, goods and resources (including labour) that cities draw from outside their territory have increasingly originated from foreign countries and continents. Canadian producers have also moved toward new international markets. This change in the reach of economic exchange is loosely referred to as **globalization**. In the global period, the interdependence that existed in the past between the city and a well-defined hinterland has lost much of its importance (Harrison and Hoyler, 2015; Kloosterman, Mamadouh, and Terhost, 2018). The tendency today is for cities to transform and consume goods from around the world and sell their products on international markets. Yet, there are signs that present and future change, driven by environmental awareness, the possibility of escalating energy costs, and the imposition by some countries of tariffs may bring about a renewed emphasis on nearness in economic exchanges (Rubin, 2009). Thus, Chapter 23 discusses responses to the new emphasis on "local" procurement of goods and services, particularly foodstuffs (e.g., the **100-mile diet**—see Harris, 2009; *Time*, 2006), that accompanies attempts to decrease the urban "footprint" and so make Canadian cities more environmentally sustainable. But it remains the case that cities have to trade with areas outside their territory, either near or afar.

Though the nature of export-based specialized economic production has changed dramatically in the last hundred years in Canadian cities, economic production still is imperative for urban growth and development. With the decline of routine production, developed countries such as Canada must rely on innovation and knowledge-intensive activities in order to compete on the world stage (the **new economy** is sometimes spoken about as a "knowledge" economy) (see Vinodrai, Chapter 17). This explains the emphasis Florida (2002, 2005) places on measures cities can take to attract the **creative class** that he sees as key to the knowledge economy. Above all, the growing importance of the service sector relative to the manufacturing sector has characterized the last decades (Daniels and Bryson, 2002). This trend was propelled by accelerated loss of industrial jobs in Canada over the last decades, which in late 2018 involved the decision by General Motors (GM) to close its Oshawa automobile assembly plant. The departure of GM marked the end of a century of car production in Oshawa. The transition is highly visible, moreover, in the urban landscape. On the one hand, we see abandoned industrial premises (or their twenty-first-century "makeover" as lofts), and, on the other, we are confronted with an explosion of restaurants and personal services, and places of entertainment and cultural activities (Jones, Comfort, and Hillier, 2005) (see Bain and Mark, Chapter 15). The shift toward the service economy is also felt, albeit with less intensity, within cities' export sphere. Among services Canadian cities export are financial services, engineering and development expertise, and culture.

If **deindustrialization** persists, Canadian cities may become focused on services and resources, with major centres being responsible for the production, administration, and export of services and resources, and smaller ones for the extraction and early-stage transformation of commodities. A renewed emphasis on staples would bring us back to early phases of Canadian economic development, even if urban distribution and forms have clearly evolved over the centuries (see Chapter 2). While their common resource orientation ties eighteenth-century fur-trading depots along the St. Lawrence or eighteenth- and nineteenth-century fishing communities along Canada's Atlantic coast with fast-growing Alberta oil-patch communities, their configuration obviously differs sharply.

It is important, finally, to point out that production impacts all aspects of urban life, which is particularly notable in periods of transformative change such as the present. Thus, for example, Chapter 10 speaks to increasing socio-economic **polarization** in the Canadian city in the twenty-first century. The chapter attributes this change primarily to a shift in the prevailing economic regime, to deindustrialization and the rise of the hourglass income distribution characteristic of the growing service sector.

Proximity

Individuals and activities have always congregated in cities to facilitate communication and minimize the cost (in terms of time, effort, and money) of interaction. If we probed reasons why people live in cities, most would place the need to be close to work at or near the top of their list of answers. Other explanations would include proximity to educational establishments, to shopping opportunities, to cultural activities, to entertainment and recreation, to family and friends, and to medical facilities. People opt for urban living because of their need for frequent interaction outside the home. Likewise, businesses and institutions locate in cities so they can be close to their market, labour force, and the establishments with which they maintain linkages (in other words, to enjoy agglomerative economies). By concentrating activities and people and thus creating proximity, the city makes frequent interactions affordable in terms of cost and time. In a rural setting, in contrast, many recurring contacts that are routine in the city would involve prohibitive transportation times and/or costs due to long distances. A by-product of interaction is innovation (Hall, 1999); ease of interaction is why cities have been catalysts for change in the social, economic, technological, and cultural realms. The renewed attraction of central city living that so distinguishes cities of the twenty-first century from the previous era is closely related to widespread recognition that the urban environment promotes and welcomes all kinds of low-cost and/or spontaneous interactions.

The city can be perceived as comprising numerous overlapping markets of frequently repeated exchanges. The fact that cities are fundamentally places of economic enterprise makes daily commuting between residences and workplaces of unparalleled importance in explaining urban structure. Also important from an economic perspective are linkage networks between enterprises, which benefit from proximity, especially in the case of just-in-time delivery. Other frequent exchanges that affect the size and spatial organization of cities include the connections of retail facilities to their market and of public services to their clients. In addition, markets connect cultural and recreational activities to their public—the archetypical attraction of the city's "bright lights."

Proximity is a relative condition, largely determined by prevailing transportation systems and activity distribution patterns (see Chapters 14 and 19). In the 1960s the urban planner Richard Meier (1962) pointed out how innovation in both communication and transportation could trigger new transformative eras of urban development. Urbanist Lewis Mumford in *The City in History* (1961) provides details that lend credence to Meier's hypothesis. Mumford points out that in the pre-industrial city, which depended on non-mechanized forms of transportation (primarily walking and horse-powered transportation), the principle of proximity dictated that important activities be centralized; likewise, the principle of proximity meant that the outer expanse of the built-up, urbanized perimeter was largely dictated by the prevailing mode of transportation.

The car- and truck-dominated contemporary city, on the other hand, takes on a highly decentralized form. In the decentralized or dispersed metropolis, adequate accessibility levels can be maintained over large territories so that residents and activities can consume far more land than in the past (Bottles, 1987). As rush-hour gridlock across an extensive section of Highway 401 attests, effective early twenty-first-century boundaries of the Greater Toronto Area run from Kitchener–Waterloo–Cambridge in the west to beyond Clarington in the east (a distance of 165 km). Similarly, heavy commuter traffic extends from the City of Toronto to points south toward Niagara Falls and north toward Orillia (a distance of 180 km). This extended Toronto region is referred to as the "Greater Golden Horseshoe" in recent planning documents (Ontario, 2017). But the proximity principle remains influential even in these more dispersed circumstances, as evidenced by the enduring existence of higher densities in cities than in the countryside, and at accessibility peaks within the city itself (at rail transit stations and junctions of major arterials and/or expressways). Chapter 2 in this volume can be read as a testimonial to the changing role that accessibility has played since the early twentieth century in promoting, first, centralization and high-density styles of urban development; second, dispersion and low density; and recently, a return to high residential density, at least in central parts of the metropolitan envelope (also see Chapter 19).

Today, debate is ongoing as to whether our ability to substitute telecommunications for actual movement holds the potential for an even more dispersed urban form. But forecasts predicting the death of the city as we know it have proven to be wrong because they did not anticipate the impact of a changing mode of production on urban form. The "new" urban economy, with its focus on services, entertainment, and culture, has witnessed a renewed centralization of activities and has considerably elevated the importance of face-to-face contact and spontaneous connecting. In large metropolitan regions this is reflected in the increase in inner-city living, where access to all kinds of people and activities is within a short walk.

Reproduction

As properties of the urban phenomenon, production and reproduction are intimately tied to each other. Reproduction, as understood in the Marxian sense, centres on the conditions essential to the continued provision of an ample labour force, that is, the literal re-production of workers. These requisites include birth and child-rearing but also other conditions that relate most directly to the well-being of family/household units within the city: health care, education, social services, family and community support facilities, immigration policies, and so on (Castells, 1977; Jarvis, Pratt, and Wu, 2001; Jessop, 2002: 47, 77).

Before disease control, engineered infrastructure, public health, food security, and general welfare improved rapidly after the Industrial Revolution, living conditions were unhealthy in cities. Afflicted by successive epidemics, for long periods cities were unable to maintain their populations without a constant inflow from the country (Howard, 1968; Russel, 1972). Only after the introduction of water treatment and sewer systems and the advent of immunization did city living cease to be a worse threat to health than living in rural areas. Indeed, historians have traced the origins of urban planning to early efforts at alleviating adverse health effects associated with the crowding and pollution of the early industrial city (Hodge and Gordon, 2014). Still, we should avoid being smug about the improved health conditions of contemporary cities. Possible deadly flu epidemics, antibiotic-resistant bacterial infections, adverse health effects associated with poor air quality and other sources of pollution, and increasing obesity due to insufficient exercise in an auto-centric culture all point to health problems associated with the contemporary city (Nieuwenhuijsen and Khreis, 2019; Sallis et al., 2016).

At the most fundamental level, Canada's low birth rate—and consequent inability to reproduce its own population—is the fundamental reason why rates of foreign immigration have soared over recent decades. Immigration policies are thus central to the city's reproduction property, especially since, in Canada, large cities are the destination of virtually all immigrants (Filion, 2010; Vézina and Houle, 2017; see Kobayashi and Preston, Chapter 8).

Beyond demographic growth, examples of reproduction-related urban sites include homes, schools, hospitals, and water treatment and distribution systems, as well as parks and other recreational facilities that promote health and reduce stress. A smooth operation of the reproduction system allows employers to find an abundant workforce that is healthy, qualified, and possesses a work ethic compatible with types of employment present in a given city. Reproductive activities are usually centred on the home but are increasingly supplemented with services provided by outside agencies in the public and private sectors. Today, in fact, reproduction-related consumption of both services (e.g., fast food, child-care) and goods (e.g., dishwashers, microwave ovens, health and hygiene and home maintenance products) represents an important outlet for the production sector. In reality, virtually all household consumption can be seen as having a reproductive aspect. Meanwhile, resources needed for reproduction-related goods and services are derived from the production sector in the form of household expenditure and tax revenues.

A clear feminist dimension to our present understanding of reproduction concerns movement away from the traditional role of women. In the past, and still to a large degree, women have assumed the major burden of reproductive work without any payment. While "equal pay for equal work" has yet to be achieved in most sectors of the economy, the majority of Canadian women (ages 15–64) now participate in the labour force (Beaudry and Lemieux, 2000). Women's roles in both the productive and reproductive spheres have considerable impact on the way we live (see Townshend and Walker, Chapter 7) and on how essential services, such as health, education, and child-care, are delivered. Increased participation of women in productive sectors of the economy, for example, appears to be an important driver of our consumerist lifestyles as well as of demographic stagnation. Had it not been for the massive entry of women into the job market over the past 50 years, household incomes and standards of living would have declined considerably. Labour force participation of all adult household members, however, also can be a source of tension, as in the case of more health care being off-loaded to the home at a time when there is unlikely to be a stay-at-home caregiver to assist with ill family members (Allan and Crow, 1989; Charles et al., 2017; Wakabayshi and Donata, 2005).

Today, we witness growing public sector difficulties in providing essential conditions for reproduction. From the Great Depression until roughly the early 1980s, governments expanded the **welfare state** and thereby their role in the reproduction sphere. In recent decades, however, the opening of international markets has allowed producers to seek low-tax and low-wage jurisdictions offshore, thus cutting corporate-based fiscal revenues; at the same time, public resentment about high tax levels has made increasing income tax a politically unrealistic option (Campeau, 2005; Finkel, 2006; Graham, Swift, and Delaney, 2009; Myles, 2015). As a result, governments have faced reduced spending capacity. Paradoxically, the subsequent cutbacks in public sector reproduction services have coincided with rising expectations and demand for such services. At the very time when near-total engagement of working-age adults in the labour force makes it difficult for them to attend to the reproduction needs of society (having and rearing children) and of their dependants (old parents and sick or disabled family members), government support in these matters is either stagnant or declining.

In this same vein, the knowledge-intensive economy places a growing burden on the post-secondary education system. The response has been a gradual shift from public to private funding, evidenced by inflating community college and university fees. Yet, the persistent shift toward private funding is a major source of social inequality (see Walks, Chapter 10). Underfunding of reproduction-related public infrastructures and services can have consequences whose costs, in terms of suffering and monetary expense, far exceed initial savings. A revealing example is the plight of those who are homeless, which can be associated to reduced government involvement in the funding of assisted housing for the very poor. Not only do people who are homeless suffer personally in terms of sleeping in rough or overcrowded shelters and being exposed to cold, violence, and hunger, but the important medical expenses and their lack of integration into society related to their homeless condition may well represent costs that exceed government subsidies that could provide them with supportive housing (Evans, Collins, and Anderson, 2016; Ly and Latimer, 2015). The same logic pertains, in a less dramatic fashion, to the consequences of reduced services in sectors such as social services, education, and health care. For example, deferred treatment of health problems can bring about enormous suffering and productivity losses, often with cumulative costs far beyond that of the treatment.

Capitalization

The capitalization property of cities derives from their compact spatial form. Because urban land is scarce, it becomes the object of substantial capital investment so its use can be maximized. *Capitalization* refers to the vast resources invested to accommodate agglomerations of residents, businesses, and services. The nature of capitalization and, hence, the form cities take are largely influenced by the engineering possibilities of the time. Over the centuries, improving technologies have promoted larger city size and, until the relatively recent predilection for suburban forms, higher densities.

Once built-up urban environments are capitalized and populated, they become highly durable and thus contribute a considerable degree of continuity to the urban landscape. As a result, they become a factor in **path dependence** (favouring the perpetuation of existing patterns at the expense of innovation) (Pierson, 2000; see Filion, Chapter 2). However, as technology and lifestyles tend to change faster than urban form, capitalization of urban land also engenders obsolescence. Change, especially concerning modes of production or transportation technologies, demands adjustments of the built environment to new conditions. But a city is not easily retrofitted. Typically, costs of redevelopment on **brownfield** or **greyfield sites** are higher than those on undeveloped land, often called greenfield sites, at the urban edge. While financial constraints can play a critical role, they are not the only impediment to altering the urban environment. One obstacle to urban environment adaptability is the symbiosis that binds patterns of behaviour to built environments. For example, high-capacity road systems encourage reliance on the automobile and the truck, and high rates of car and truck use generate a continued demand for improved and expanded roads (Noland, 2000; Parthasarathi, Levinson, and Karamalaputi, 2003). Another obstacle to changing the way the built environment is capitalized comes in the form of citizen resistance that occurs when proposals for redevelopment of previously built-up areas clash with residents' strong emotional attachment to their homes and neighbourhoods. An important challenge facing planners and politicians today, then, is how to reconcile citizens' attachment to their home "places," where they desire to maintain the status quo (**NIMBY** [not in my back yard]), with the need for change—particularly in the face of looming environmental crisis (Inhaber, 1998).

Capitalization trends over the past decades have promoted two very different urban forms. First, across Canada most urban development still occurs in suburban-like settings. Large investments are targeted, therefore, at the conditions required for suburban growth: for example, peripheral expressways, arterials, local road systems, water and sewer systems, and public services, such as schools and hospitals. At the same time, a sizeable private industry is dedicated to suburban residential and commercial development, which caters to an enduring strong demand for suburban living. Interest groups that presently oppose measures to contain peripheral urban development mostly stem from these industries. Meanwhile, inner cities of large metropolitan regions are experiencing residential **intensification**, largely in the form of high-rise condominiums (Filion, 2018).

We cannot underestimate the importance on the economy of investments, both from the private and the public sectors, targeted at the urban environment. The home is indeed the main asset for a majority of households. The presence of a speculative dimension to urban-related investments, which mirrors the dynamics of the stock market, is thus not surprising. This is especially the case since space is limited within metropolitan regions and the development process can take years. Such conditions can result in an imbalance between vigorous demand and lagging supply, which translates into escalating property prices. Especially in Toronto and Vancouver, strong housing demand, buttressed by a prosperous economy and low interest rates, has fuelled over recent decades accelerated house value inflation. Past experience, however, teaches us that such ascending house value trends can give way to precipitous market corrections (Head and Lloyd-Ellis, 2016).

Place

Sense of place is the least tangible among urban properties discussed in this chapter. This does not mean, however, that sense of place is any less important. Indeed, a renewed sensitivity to place most distinguishes current styles of urban development from earlier, modern growth (see **Fordism**). Enhanced interest in place is consistent with a shift in economic priority from the city as a centre of industrial production to the city as primarily an agglomeration of services, including culture and entertainment. Urbanites always attach meaning to space, whether it conforms to developers' intent or not. Recently, however, developers have been paying more attention to the messages conveyed by place, either in an effort to reap financial benefits or simply to enhance users' well-being (also see Chapter 13).

The noted geographer Yi-Fu Tuan (1974) coined the term **topophilia** to denote the personal identity with, and love of, a place. Thus, whereas "space" relations in cities are mostly about objective attributes of proximity and access, "place" is all about subjective attachment. Design-oriented professionals—such as Jane Jacobs (1961), Jon Lang (1994), planner Kevin Lynch (1964, 1984), and architect Christopher Alexander (1979; Alexander, Ishikawa, and Silverstein, 1977)—believe that fundamental "place" principles can provide guidelines to good urban form. They argue that, applied to urban development, such principles will lead to higher quality of life (the topophilia factor) as well as to more efficient use of urban space. Scholars such as Relph (1976, 1987) use the term "sense of place" in a related fashion to speak about subjective and emotional feelings associated with different parts of the urban environment. Relph argues that in modern times the perpetuation of monotonous landscapes in the suburbs and the lack of concern about imaging and good urban design left a vacuum in the urban entity. Others have been concerned that most people who live in car-oriented cities are missing out by virtue of being "detached" from their surroundings, and that the quality of individual and collective life, as well as the quality of the physical environment, has deteriorated as a result (e.g., Kunstler, 1993, 1996).

The diminishment of topophilic places as intrinsic urban attributes is considered characteristic of the industrial city. Meanwhile the elevation, indeed celebration, of positive features of strong places is one of the foremost differences distinguishing the postmodern city from the modern, industrial city (Ellin, 1999; Lance, Dixon, and Gillham, 2008; see Chapter 13). Nonetheless, even in the largest of Canadian cities, detachment from landscape remains largely characteristic of the suburban environments, where most current residents of Canadian cities spend much of their time. In suburban settings, place attachment is mostly associated with the home and the neighbourhood. Other places are considered to be, at best, highly standardized (such as shopping malls and **power centres**) or, at worst, "junkscapes" (as in the case of haphazard car-oriented retail strips).

Place, then, is the attribute of urbanity that engenders strong connections, positive or negative, between urban dwellers and the locales that surround them. Specific characteristics of place that are favoured will vary in terms of the particular style or symbolic meaning conveyed through urban developments at any point in time. Radical or transformative change, as occurred during the Industrial Revolution, can cause a shift in the extent to which topophilia is elevated as an urban property. Sense of place is thus highly fluid. Former industrial areas provide a good example of this fluidity—these were avoided by most urbanites until their rebirth as loft developments toward the end of the twentieth century (Zukin, 1982). If the industrial transformation led to a relative demise of principles of good urban design, we seem to stand today at another crossroads of transformative shift in urban development wherein the "selling" or marketing of place has increasingly become an intended goal of municipal land-use policies. Aspects of cities that are highlighted for publicity purposes, such as economic development or tourism, reflect the evolution in the types of places that people value. If smokestacks were considered to be iconic of progressive cities at the turn of the last century, in the 1960s city promotions featured sleek modernist office towers, thus reflecting a shift in the economy from a predominance of blue- to white-collar employment. The emphasis is now on cultural and festival places, heritage buildings and neighbourhoods, and well-designed public spaces.

In sum, place is the intangible that makes some locales feel good while others do not, and that invites or repels visitors. Place interfaces with all the other properties discussed here. Under conditions of the "new" economy, the manipulation of place properties has become a marketing device used to attract global interest and bring in outside investment (see Zukin, 1991). Thus, place features can be seen as a lure for the creative class, and are associated with the benefits of economic development believed to derive from this class (Florida, 2002, 2005; see Vinodrai, Chapter 17).

Governance

As understood here, **governance** consists of administrative structures and political processes aimed at generating policies suited to the specific circumstances confronting cities. The proximity and capitalization properties central to urban as compared to non-urban settlements require distinct management measures for the urban community (e.g., Booth and Jouve, 2005; Lightbody, 2006). Proximity requires collective control and co-operation between nearby neighbours over communal space. The smooth functioning of cities relies on shared infrastructures (e.g., transportation, communication, electricity, water mains, and sewers) and services (e.g., policing and garbage collection) and on a battery of legal measures (e.g., property rights, payment for shared facilities, bylaws) intended to assure the orderly cohabitation of a wide variety of land uses. Haphazard development decisions can plunge a city into a state of chaos. For example, without planning controls, noisy and polluting industries and high-traffic generators could locate in residential areas. Likewise, new developments could proceed without heeding infrastructural capacity, thus provoking all sorts of bottlenecks. A pure laissez-faire approach is clearly not suited to the city.

Various types of administrative arrangements have developed to provide urban infrastructures, services, and controls. These administrations have been local or regional or have been lodged in senior governments, as is the case with provincial ministries of municipal affairs or federal housing programs. Issues of governance generally belong to the public sector, but some urban management responsibilities can be vested in community-based or private sector organizations. Over time, as cities grew, as buildings became bigger and required more infrastructure (roads, water, sewage), as reliance on mechanized forms of transportation (particularly the automobile) increased, and as the public demanded more and better services, administrations responded by becoming larger and more complex.

The need for interventions specific to the urban context has spawned disciplines and university programs that generate and impart the knowledge essential to these interventions. Relevant disciplines include urban planning, urban geography, local public administration, urban sociology, urban economics, and subfields within civil engineering.

Today, coincident with the growing realization of the importance of local governance in matters such as environmental protection, economic development, the equitable provision of services, and the promotion of health and quality of life, we face the harsh reality that municipal administrations are confronting severe financial restrictions. In the absence of a reliable funding stream from these higher levels, municipalities presently rely largely on property tax revenues, which are deemed inadequate to deal with their considerable police and public transit expenditures, or to update infrastructures and address housing affordability problems (Bird and Slack, 2017; Kitchen and Slack, 2016).

Environment

To survive in the long term, cities must respect their natural environment. In recent years, human life everywhere has been threatened by an environment where soil, air, and water quality is severely degraded. As a consequence, increased dangers to health arise from by-products of our industrial and consumption processes. In the past, cities were generally situated at favourable locales; the immediate environment would usually be chosen for reasons of "site" (e.g., Quebec City's easily defendable position or Halifax's deep and sheltered harbour) or "situation" (good connectivity by water in the case of Montreal or by rail in the case of western Canadian cities—Winnipeg is a prime example). Usually, however, the condition of the regional environment had not been thought of in active terms.

As demonstrated in Chapters 21 and 22, even today cities tend to ignore to a large extent the environmental damage they inflict by exporting their pollution. Thus, for example, by sending its garbage to a Michigan landfill, Toronto does not have to deal directly with the consequences of its consumption. In a similar vein, the Montreal sewer system discharges its partially treated effluents downstream from Montreal Island in the St. Lawrence River, and Victoria still pumps raw sewage in the Strait of Juan de Fuca. The core argument of Chapter 21 is that cities have an environmental footprint that exceeds, manifold, their built area (Rees, 2008; Wackernagel and Rees, 1996). With time, environmental awareness has extended its scope from the very local to the global (Carr, 2005; Malhi, 2017; Krooth, 2009). Prior to the twentieth century, the well-off classes would leave cities during times when disease spread most quickly; the urban residences of the wealthy would also command sites that were deemed to be most attractive and relatively "risk-free" according to the dictates of the time. In the 1950s, the nascent environmental movement targeted local consequences of pollution, such as high bacteria counts preventing swimming and air pollution caused by specific close-by industries (Crenson, 1971). Later, as air pollution worsened due mostly to rising automobile use, environmental awareness became metropolitan in scale. Residents of metropolitan regions soon became conscious of the fact that retreating to distant leafy suburbs offered little relief from many forms of air pollution. Today, environmental awareness is decidedly global. There is increasing realization that with a world population approaching eight billion, much of which is moving toward a consumerist lifestyle, the effects of human behaviour on the entire biosphere are inevitable (Friedman, 2008). Above all, global environmental awareness is driven by concern over climate change. And as hubs of industrial production and consumption containing over 50 per cent of the world population, cities are major contributors to global warming. It is thus inevitable that any attempt to control **greenhouse gas emissions** will have major urban repercussions.

In his book *Collapse: How Societies Choose to Fail or Succeed*, Jared Diamond has documented the disappearance of civilizations over the ages due to circumstances affecting their immediate natural environment (Diamond, 2005; see also Whyte, 2008). In some instances, climate change (then mostly the outcome of natural circumstances) caused droughts; in other cases, communities and entire civilizations carelessly depleted the natural resources on which they depended. The lesson from Diamond's book is that the survival of any human group, and of humanity itself, depends on its ability to achieve sustainable forms of development in harmony with the natural environment. Otherwise, that group is doomed. The same goes for cities, except for the fact that they have long been successful in exporting, and thus overlooking, their environmental harm. But with the increasingly global impact of human activity, it is becoming difficult for cities to ignore their environmental effects. The environmentally related collapse of individual civilizations can now be transposed to a worldwide scale. It is now the fate of human civilization in its entirety (and thus the global urban system) that is at stake. Scenarios of doom in a warmer planet proliferate, with stark implications for cities: the flooding of low-lying coastal urban areas; unprecedented heat waves that will take a heavy toll on vulnerable urban populations; and violent conflicts over remaining water sources and fertile areas as deserts expand (Dalby, 2017; Dyer, 2008; Smil, 2008). To be sure, cities have engaged in initiatives to alleviate their environmental impact (Register, 2006; Satterthwaite, 1999; Tomalty and Mallach, 2015). Yet, while these measures have had positive results at local and regional levels (cleaner air and water, soil decontamination, preservation of natural areas), it remains to be seen if their scope will be sufficient to abate global trends such as climate change (Bulkeley, 2012).

Conclusion

In an effort to conceptualize the urban phenomenon, this chapter has considered its essentials. Seven properties—production, proximity, reproduction, capitalization, place, governance, and environment—are present in all cities across history (see Table 1.1). These properties help to explain the reasons for the existence of cities and identify the main principles that drive their functioning. While this chapter has been about shared features of cities, the remainder of the book is about ramifications of the urban phenomenon. The book investigates diverse facets of cities and differences in how these facets manifest themselves according to their place within a metropolitan region and the position of a city in Canada's urban system. It also looks at how different aspects of society are mirrored in cities. Chapter 1 has been about common characteristics of urban areas; the remainder of the book is about their multiple dimensions.

Review Questions

1. In your opinion, do the seven properties described in the chapter cover all aspects of the urban phenomenon? If not, which would you add?
2. Which properties, in your estimation, are the most important in the present context? How is this different from previous periods?

Note

1. An earlier version of this chapter was written with Trudi Bunting (1944–2017).

References

Alexander, C. 1979. *The Timeless Way of Building*. New York: Oxford University Press.

———, S. Ishikawa, and M. Silverstein. 1977. *A Pattern Language: Towns, Buildings, Construction*. New York: Oxford University Press.

Allan, G., and G. Crow, eds. 1989. *Home and Family: Creating the Domestic Sphere*. Basingstoke: Macmillan.

Beaudry, P., and T. Lemieux. 2000. *Evolution of the Female Labour Force Participation Rate in Canada, 1976–1994: A Cohort Analysis*. Hull, QC: Human Resources Development Canada, Applied Research Branch.

Bird, R.M., and E. Slack, eds. 2017. *Financing Infrastructures: Who Should Pay?* Montreal and Kingston: McGill-Queen's University Press.

Booth, P., and B. Jouve, eds. 2005. *Metropolitan Democracies: Transformations of the State and Urban Policy in Canada, France and Great Britain*. Aldershot, Hants: Ashgate.

Bottles, S. 1987. *Los Angeles and the Automobile: The Making of a Modern City*. Berkeley: University of California Press.

Bulkeley, H. 2012. *Cities and Climate Change*. London: Routledge.

Campeau, G. 2005. *From* UI *to* EI: *Waging War Again on the Welfare State*. Vancouver: University of British Columbia Press.

Carr, M. 2005. *Bioregionalism and Civil Society: Democratic Challenges to Corporate Globalism*. Vancouver: University of British Columbia Press.

Castells, M. 1977. *The Urban Question: A Marxist Approach*. London: Edward Arnold.

Charles, L., S. Brémault-Phillips, J. Parmar, M. Johnson, and L-A. Sacrey. 2017. "Understanding how to support family caregivers of seniors with complex needs," *Canadian Geriatric Journal* 20(2): 75–84.

Christensen, R. 2017. "Spin-Farming: Advancing urban agriculture from pipe dream to populist movement," *Sustainability: Science, Practice and Policy* 3(2): 57–60.

Crenson, M.A. 1971. *The Un-politics of Air Pollution: A Study of Non-Decision-Making in the Cities*. Baltimore: Johns Hopkins University Press.

Dalby, S. 2017. "Anthropocene formations: Environmental security, geopolitics and disaster," *Theory, Culture and Society* 34(2–3): 233–252.

Daniels, P.N., and J.R. Bryson. 2002. "Manufacturing services and servicing manufacturing: Knowledge-based cities and changing forms of production," *Urban Studies* 39: 977–91.

Diamond, J. 2005. *Collapse: How Societies Choose to Fail or Succeed*. New York: Viking.

Dyer, G. 2008. *Climate Wars*. Toronto: Random House Canada.

Ellin, N. 1999. *Postmodern Urbanism*, rev. edn. Princeton, NJ: Princeton University Press.

Evans, J., D. Collins, and J. Anderson. 2016. "Homelessness, bedspace and the case for Housing First in Canada," *Social Science and Medicine* 168: 249–56.

Filion, P. 2010. "Growth and decline in the Canadian urban system: The impact of emerging economic, policy and demographic trends," *Geojournal* 75: 517–38.

———. 2018. "Time scales and planning history: Medium- and long-term interpretations of downtown Toronto planning and development," *Planning Perspectives*. At: https://www.tandfonline.com/doi/full/10.1080/02665433.2018.1554451

Finkel, A. 2006. *Social Policy and Practice in Canada: A History*. Waterloo, ON: Wilfrid Laurier University Press.

Florida, R.I. 2002. *The Rise of the Creative Class: And How It's Transforming Work, Leisure, Community and Everyday Life*. New York: Basic Books.

———. 2005. *Cities and the Creative Class*. New York: Routledge.

Friedman, T.L. 2008. *Hot, Flat, and Crowded: Why the World Needs a Green Revolution, and How We Can Renew Our Global Future*. London: Allen Lane.

Graham, J.R., K. Swift, and R. Delaney. 2009. *Canadian Social Policy: An Introduction*. Toronto: Pearson Prentice-Hall.

Hall, P. 1999. *Cities in Civilization: Culture, Innovation, and Urban Order*. London: Phoenix.

Harris, E. 2009. "Neoliberal subjectivities or a politics of the possible? Reading for difference in alternative food networks," *Area* 41: 55–63.

Harrison, J., and M. Hoyler, eds. 2015. *Megaregions: Globalization's New Urban Form?* Cheltenham, UK: Edward Elgar.

Head, A., and H. Lloyd-Ellis. 2016. "Has Canadian house price growth been excessive?," *Canadian Journal of Economics* 49(4): 1367–1400.

Hodge, G., and D.L.A. Gordon. 2014. *Planning Canadian Communities: An Introduction to the Principles, Practice, and Participants*, 6th edn. Toronto: Nelson Education.

Howard, S. 1968. *Medieval Cities*. New York: Braziller.

Inhaber, H. 1998. Slaying the NIMBY Dragon. New Brunswick, NJ: Transaction.

Innis, H.A. 1995a [1931]. "Transportation as a factor in Canadian economic history," in D. Drache, ed., *Staples, Markets and Cultural Change: Selected Essays, Harold A. Innis*. Montreal and Kingston: McGill-Queen's University Press.

———. 1995b [1938]. "The penetrative process of the price system on new world states," in D. Drache, ed., *Staples, Markets and Cultural Change: Selected Essays, Harold A. Innis*. Montreal and Kingston: McGill-Queen's University Press.

Jacobs, J. 1961. *The Death and Life of Great American Cities*. New York: Random House.

Jarvis, H., A.C. Pratt, and P. C-C. Wu. 2001. *The Secret Lives of Cities: Social Reproduction of Everyday Life*. London: Routledge.

Jessop, B. 2002. *The Future of the Capitalist State*. Cambridge: Polity Press.

Jones, P., D. Comfort, and D. Hillier. 2005. "Regeneration through culture," *Geography Review* 18(4): 21–3.

Kitchen, H., and E. Slack. 2016. *More Tax Sources for Canada's Largest Cities: Why, What and How?* Toronto: University of Toronto, Institute on Municipal Finance and Governance, No. 27.

Kloosterman, R.C., V. Manadouh, and P. Terhorst, eds. 2018. *Handbook on the Geographies of Globalization*. Cheltenham, UK: Edward Elgar.

Krooth, R. 2009. *Gaia and the Fate of Midas: Wrenching Planet Earth*. Lanham, MD: University Press of America.

Kunstler, J.H. 1993. *The Geography of Nowhere: The Rise and Decline of America's Man-Made Landscape*. New York: Simon & Shuster.

———. 1996. *Home from Nowhere: Remaking Our Everyday World for the Twenty-First Century*. New York: Simon & Shuster.

Lance, J.B., D. Dixon, and O. Graham. 2008. *Urban Design for an Urban Century: Placemaking for People*. New York: Wiley.

Lang, J. 1994. *Urban Design: The American Experience*. New York: Van Nostrand Reinhold.

Lawson, L.J. 2005. *City Bountiful: A Century of Community Gardening in America*. Berkeley: University of California Press.

Lightbody, J.M.A. 2006. *City Politics, Canada*. Peterborough, ON: Broadview Press.

Lorch, B., M. Johnson, and D. Challen. 2004. "Views of community sustainability after a mine closure: A case study of Manitouwadge, Ontario," *Environments* 32: 15–29.

Ly, A., and E. Latimer. 2015. "Housing First impact on costs and associated cost offsets: A review of the literature," *The Canadian Journal of Psychiatry* 60(11): 475–87.

Lynch, K. 1964. *The Image of the City*. Cambridge, MA: MIT Press.

———. 1984. *Good City Form*. Cambridge, MA: MIT Press.

Malhi, Y. 2017. "The concept of the anthropocene," *Annual Review of Environment and Resources* 42: 77–104.

Meier, R.L. 1962. *A Communications Theory of Urban Growth*. Cambridge, MA: MIT Press.

Mitchell, C.J.A., and K. O'Neill. 2016. "Tracing economic transition in the mine towns of northern Ontario: An application of the 'resource-depending' model," *The Canadian Geographer* 60(1): 91–106.

Mumford, L. 1961. *The City in History: Its Origins, Its Transformations, Its Prospects*. New York: Harcourt, Brace and World.

Myles, J. 2015. "Canadian Sociological Association Outstanding Contribution Lecture: The fading of redistributive politics in Canada," *Canadian Review of Sociology* 52(1): 1–21.

Nieuwenhuijsen, M., and H. Khreis, eds. 2019. *Integrating Human Health into Urban and Transport Planning: A Framework*. Cham, Switzerland: Springer.

Noland, R.B. 2000. "Relationship between highway capacity and induced vehicle travel," *Transportation Research Part A, Policy and Practice* 35: 47–72.

Ontario. 2017. *Growth Plan for the Greater Golden Horseshoe*. Toronto: Government of Ontario, Ministry of Municipal Affairs.

Parthasarathi, P., D.M. Levinson, and R. Karamalaputi. 2003. "Induced demand: A microscopic perspective," *Urban Studies* 40: 1335–51.

Pierson, P. 2000. "Increasing returns, path dependence, and the study of politics," *American Political Science Review* 94: 251–67.

Rees, W.E. 2008. "Human nature, eco-footprints and environmental injustice," *Local Environment: The International Journal of Justice and Sustainability* 13: 685–701.

Register, R. 2006. *EcoCities: Rebuilding Cities in Balance with Nature*, rev. edn. Gabriola Island, BC: New Society Publishers.

Relph, E. 1976. *Place and Placelessness*. London: Pion.

———. 1987. *The Modern Urban Landscape*. Baltimore: Johns Hopkins University Press.

Rubin, J. 2009. *Why Your World Is About to Get a Whole Lot Smaller: Oil and the End of Globalization*. New York: Random House.

Russel, J.C. 1972. *Medieval Regions and Their Cities*. Newton Abbot: David and Charles Press.

Sallis, J.F., F. Bull, R. Burdett, L.D. Frank, P. Griffiths, B. Giles-Corti, and M. Stevenson. 2016. "Use of science to guide city planning policy and planning: How to achieve healthy and sustainable future cities," *The Lancet* 388(10062): 2936–47.

Satterthwaite, D., ed. 1999. *Sustainable Cities*. London: Earthscan.

Smil, V. 2008. *Global Catastrophes and Trends: The Next 50 Years*. Cambridge, MA: MIT Press.

Time. 2006. "The lure of the 100-mile diet," 11 June. At: www.time.com/time/magazine/article/0,9171,1200783,00.html

Tomalty, R., and A. Mallach. 2015. *America's Urban Future: Lessons from North of the Border*. Washington, DC: Island Press.

Tuan, Y. 1974. "Space and place: A humanistic perspective," *Progress in Geography* 6: 233–46.

Vézina, M., and R. Houle. 2017. *Settlement Patterns and Social Integration of the Population with an Immigrant Background in the Montreal, Toronto and Vancouver Metropolitan Areas*. Ottawa, ON: Statistics Canada. At: www.statcan.gc.ca/pub/89-657-x2016002-eng.pdf

Wackernagel, M., and W.E. Rees. 1996. *Our Ecological Footprint: Reducing Human Impact on the Earth*. Gabriola Island, BC: New Society Publishers.

Wakabayshi, C., and K.M. Donata. 2005. "The consequences of caregiving: Effects on women's employment and earnings," *Population Research and Policy Review* 24: 467–88.

Watkins, A.J. 1980. *The Practice of Urban Economics*. Beverly Hills, CA: Sage.

Whyte, I. 2008. *World without End? Environmental Disaster and the Collapse of Empires*. London: I.B. Tauris.

Zukin, S. 1982. *Loft Living: Culture and Capital in Urban Change*. Baltimore: Johns Hopkins University Press.

———. 1991. *Landscapes of Power: From Detroit to Disney World*. Berkeley: University of California Press.

2 Urban Transitions

Historical, Present, and Future Perspectives on Canadian Urban Development

Pierre Filion[1]

Introduction

The present chapter discusses the evolution of cities, how they change over time. The chapter concentrates on the factors that bring about urban transitions in Canada after colonization. What form do these factors take? What circumstances favour their emergence? And in what conditions are they most apt to alter urban development trajectories and thus launch new urban patterns? Our attempts at answering these questions will examine tensions between urban inertia and forces of change. The urban phenomenon entails major sources of inertia, the most obvious being the inherent durability of the built environment. At the same time, however, it is under pressure to adapt to society-wide changes. Clearly, those trends that result in profound and lasting transformations of society will have more impact than those that are more fleeting and superficial. But in all instances, factors inducing urban change will need to overcome urban inertia.

The chapter first considers the tension between urban inertia and change; it then goes on to describe factors of urban transformation. The remainder of the chapter is given to a description of three periods of urban development in Canada: the pre-1945 city; the 1945–1975 period when **suburbanization** prevailed; and post-1975 development patterns, characterized by a critique of, as well as ongoing prevalence of, forms inherited from the previous period. The chapter closes with an exploration of possible future urban transitions.

Inertia and Change

If you happen to look at pictures taken over time of a given urban location, you will note a discrepancy between changes affecting their foreground and background. You will observe in the foreground variations in the modes of transportation: an early predominance of streetcars, which were gradually replaced by automobiles with their changing size and design. People, too, will change. You will see either more or fewer pedestrians, according to fluctuations in the appeal of the area, changes in their clothing and hairstyles, shifts in the prevalence of age groups in response to variations in the nature of close-by activities and, in the Canadian context, increasing racial diversity over time. In addition, the signs in front of the stores, the appearance of their facades, and, above all, the content of shop windows will undergo repeated transformations. In comparison, the background of these pictures, comprising the buildings themselves, will appear to be much more static. It is not that transformations affecting the built environment do not take place but, rather, that such changes happen much more slowly. It is as though the photos of the foreground and background were taken by two different time-lapse cameras: one showing transformations at a rapid pace; and the other presenting changes that unfold much more slowly.

You would experience this phenomenon more directly were you to return to the neighbourhood of your childhood after a 15-year absence. You could not escape feeling the contrast between how much you have changed in 15 years and how closely the neighbourhood conforms to your memories, provided it was not the object of major development or redevelopment as happens at the edge of urban areas and in certain central districts. True, some things will be different. Neighbours will have been replaced and some homes will have been painted different colours. Here and there new additions will have been built. But, overall, the neighbourhood will have changed very little. Meanwhile, 15 years in a young person's life bring many important transformations.

Cities do not offer a clean canvas, they always come with an inheritance from the past. There is naturally the persistence of the built environment. The large sums required in construction deter a systematic replacement of old by new **urban forms**. The redevelopment of a site involves assembling the land, possibly at a cost exceeding market value due to the presence of owners who are not necessarily motivated to sell. Redevelopment also entails acquiring the site's structures and then incurring the cost of demolishing them—hence the predilection for greenfield development over re-urbanization. However, when the value of the land (determined largely by its accessibility and the prestige of its location) substantially exceeds that of the buildings occupying it, a site is ripe for redevelopment. Such a tendency is seen in the succession of buildings in downtowns of growing cities. Indeed, it is not unusual for three generations of buildings or more to have occupied given downtown sites in Toronto or Montreal (Gad, 1991).

The endurance of the built environment is not, however, the sole source of urban inertia. The urban phenomenon provides fertile ground for spawning powerful **path dependencies** (Atkinson and Oleson, 1996; Pierson, 2000; Sorensen, 2015). A city's form (land-use patterns) and dynamics (people's behaviour, including their transportation choices) are kept in place by interactions between different components of the urban reality. Transforming these interactions, and thus urban form and dynamics, is made difficult by the fact that doing so must involve all interconnected aspects of the city, engaged in any given interaction.

The most fundamental and influential of the **urban dynamics** that give rise to path dependencies are those that evolve around the mutual relationship between transportation and land use. Different modes of transportation foster different accessibility patterns, which in turn influence the nature of the built environment (e.g., compact vs. sprawling) (Boarnet and Crane, 2001; Cervero, 2016; Ewing and Cervero, 2010; Frank, Kavage, and Appleyard, 2007; Millward and Xue, 2007; van de Coevering and Schwanen, 2006). At the same time, urban forms shaped by a given mode of transportation result in additional demand for this mode. Once such a relationship is embedded, it becomes difficult to modify modes of transportation without simultaneously transforming land use, and vice versa.

In addition, there are other powerful urban path dependencies at work. One example is the mutual expectation that prevails between developers and their clients. As in the case of the transportation–land use relationship, a profound transformation of the city requires a concurrent modification of the expectations of both developers and clients. There is also the dependence of municipal administrations on property tax, which causes them to stick to conventional development in order to avoid financial risks that come with innovation. A final example of a path dependency is attachment on the part of residents to existing urban settings and their suspicion of any change to their living environment. These sentiments explain the prevalence of **NIMBY (not in my back yard)** reactions (Brown and Glanz, 2018; Curic and Bunting, 2006; Schively, 2007). But urban path dependencies are not immutable, as visible transformations of cities over time attest.

Factors of Change

We now consider seven interrelated factors of change. While most factors unfold at the scale of a society as a whole, and beyond in the case of global trends, some emanate at the local level. The factors discussed are the economy, demography, technology, **governance**, values, urban models, and the consequences of urban growth and decline. While we consider the impact of these factors on cities, we must remember that cities themselves also shape these factors.

The Economy

Economic trends affect the city in different ways. Most obviously, as economic prosperity fuels urban growth, development models popular in periods of economic expansion come to dominate the urban environment (Frenken and Boschma, 2007; Glaeser, 2000). In this sense, prosperity acts as a catalyst for urban development models. Which is not to say, however, that periods of economic growth necessarily bring changes in urban development. Economic expansion may simply lead to a further extension of established urban forms into new areas.

Economic changes also lead to adaptive efforts to ensure that urban environments can accommodate these transformations. Governance (discussed below) generally plays a key role in this matter. Reliant on the economic performance of their jurisdiction for their fiscal entries and the satisfaction of their electorate, governments indeed show deep interest in the economic competitiveness of the urban areas under their mandate (Harvey, 1990). It follows that a foremost motivation for policies targeting urban areas is to ensure their compatibility with the requirements of economic development, as interpreted by public sector agencies.

Spaces of consumption bear the mark of economic tendencies: economic performance and shifts in the importance of different economic sectors have an impact on the type and amount of housing that is built, as well as on the quantity and format of retailing and services. The same goes for spaces of production. Moreover, the consequences of the distribution of resources within society, which varies with the evolution of the economy, are visible in the urban environment: homelessness, overcrowding, and poor-quality housing contrast with opulent high-rise condo towers and large single-family-home subdivisions.

Demography

Demography is a measure of urban growth or decline, as cities are generally ranked according to their population. In Canada, a country where the labour force is mobile and reliance on immigration is high, the demographic growth of cities tends to be closely tied to their economic performance (Bourne and Rose, 2001; Filion, 2010a). People are attracted to urban areas that offer well-paying jobs and leave those parts of the country where the economy is depressed.

The impact of demography on cities takes different forms. For example, variations in the size of different age **cohorts** leave their imprint on cities. The presence of children, moreover, is associated with outdoor space, hence the predilection of families for ground-related housing (single-family and semi-detached homes as well as townhouses). Meanwhile, young and old adults are more likely to opt for high-density living. From a demographic point of view, the Canadian urban history of the past 60 years has largely been shaped by the life cycle of the baby boom generation, although immigration and the millennials are gradually overtaking its influence (Foot with Stoffman, 1996; Moos, Pfeiffer, and Vinodrai, 2018; Trovato, 2009).

Technology

Technology is at the heart of the urban phenomenon. In the past, construction techniques were a prerequisite for permanent human settlement, as were the technologies required to carry essential resources to cities: baskets, jugs, bottles, carts, and, eventually, mechanical means of transportation now carrying containers. Perhaps most remarkable was the expertise the Roman Empire developed to bring clean water to major cities. Some of its aqueducts still stand. Not only did technologies enable the creation of cities, they also provided the conditions for their growth. Before the advent of public health measures (largely reliant on technologies), cities were victims of repeated epidemics, to the extent that it was difficult for them to maintain their population level; life was often short in cities. Then, from the second half of the nineteenth century onwards, knowledge of the conditions required for healthy urban living led to a number of innovations: water treatment, vaccination, and antibiotics (Rosen, 1993 [1958]). Just as basic construction techniques were conditions for the launching of the urban phenomenon, health-related "technologies" were essential to public health in urban areas.

Successive transportation technologies—trains, streetcars, subways, and automobiles—have most transformed the urban form over the past two centuries. Different types of rail transportation shaped a distinctive urban form, which contrasts with the previous pattern fashioned by walking and animal traction. In a similar fashion, the automobile fostered an urban morphology that superseded the one inherited from the railway era (Graham and Marvin, 2001; Muller, 2004). The most recent technological innovations concern information technology: personal computers, the Internet and smart phones. While these devices have a profound impact on our lives, their effect on the city is not as clear (Audirac, 2005; Sohn, Tschangho, and Hewings, 2005). As yet, there is no evidence of significant reduction in commuting or a change in urban form resulting from the widespread adoption of these communication technologies, with the possible exception of longer residential distances from places of employment for people who can work from home several days a week thanks to electronic communication.

Governance

The organizational structure of public agencies has a determining effect on the priorities they adopt and their capacity to carry them out. Across Canada, municipal administrations have undergone considerable organizational change over the last decades. These include annexations, amalgamations, and the creation of regional administrations. The purpose of these reorganizations is to design governments appropriate to the size of the geographic areas they cover, and to more effectively govern metropolitan areas that have grown beyond political boundaries (Collin, Léveillée, and Poitras, 2002; LeBlanc, 2006; Meligrana, 2004; Vojnovic, 2000). The organizational architecture of public institutions also determines which groups of interests are best able to influence their decision-making. Some institutional structures make ample room for public participation while others are responsive primarily to dominant economic interests. In addition, changes in the priorities of the different levels of government reverberate on cities by causing fluctuations in the funding earmarked for infrastructures and shifts in the types of projects that will benefit from public sector resources (see Taylor and Bradford, Chapter 3).

Values

Beliefs or values that are widely held in a society can have multiple repercussions on cities. They affect consumer choice and, thus, the type of housing that is built, along with the forms that retailing takes and the nature of available services. A society's values also have an effect on the political process and thereby influence policy-making, including policies targeting urban issues. Values can have a unifying or divisive effect. At times, there may be near consensus regarding dominant values, which then contribute to the zeitgeist (the spirit of the times) of a given society. In other instances, disagreement over values may be a source of social divisions (Harvey, 1990; Hutcheon, 2002). As they change over time, values are a factor in social and urban transformation (see Bain and Mark, Chapter 15).

Urban Planning Models

Since the end of World War II, urban planning has had a major influence on the form urban development takes. Over the 70 years since the war, urban planning has mostly been concerned with the specialization of land use, relying on zoning to prevent the co-existence of incompatible activities. For example, land uses that can depress quality of life and property values are kept away from residential areas. Planning has also strived, not always successfully, to balance land use and the capacity of infrastructures in order to prevent bottlenecks, for instance by expanding road capacity in growing metropolitan areas. In its effort to organize cities, planning draws inspiration from models proposing different patterns of urban development. The last 100 years or so have witnessed the influence of the Garden City (new, fully planned, and self-contained low-density communities with plentiful green space), **Tower-in-the-Park** housing (high-rise buildings erected in a park-like setting), transit-oriented development (medium- and high-density development located adjacent to public transit stations), and **New Urbanism** models (attempts at replicating pedestrian-oriented forms as they existed before World War II). Some of these—in particular the Garden City and the Tower in the Park—have captured the imagination of planners, developers, and the public and have contributed to shaping the urban environment (Fishman, 1977; Howard, 2003 [1902]; Le Corbusier, 1973 [1933]). The impact of the other models has been more modest (at least until now). Note that planning models are rarely implemented in a pure fashion but are, rather, the object of compromises and adjustments as they confront the realities of the policy-making and development process. Still, they can inspire those who look for alternatives to prevailing urban forms.

Changes in the Size of Urban Regions

One of the factors that account for differences in how urban areas function and are organized is their size. Small urban areas offer a more limited range of options (employment, retailing, public and private services, modes of transportation, and types of neighbourhoods) than large metropolitan regions. In Canada, large metropolitan regions (Toronto, Montreal, and Vancouver) have strong downtowns and extensive public transit systems, and they register heavy traffic congestion and long commutes. In contrast, small urban areas are often decentralized and automobile oriented. Access by car in these metropolitan regions is made easy by their modest size and relative absence of congestion.

These size-related differences suggest that as urban areas grow (or shrink) they experience transformations in their structure and dynamics (Bessey, 2002). Calgary illustrates changes associated with the growing size of an urban area. When the Calgary metropolitan region reached a population of 593,000 in 1981, it inaugurated a light-rail transit system (the CTrain). Before this, it was a near fully automobile-dependent and largely decentralized metropolitan region. Over the following years, Calgary's downtown area grew considerably, while the CTrain system was extended on an ongoing basis. Today, with the population of metropolitan Calgary at 1,392,609 (2016 census), it boasts a light-rail network that extends 59.9 kilometres; a strong downtown, especially in terms of employment; and a growing core population as traffic congestion is on the rise and the advantages of core area living become more apparent.

Factors of urban change are interconnected. For example, economic growth provides a fertile terrain for the spread of innovations, whose accommodation in the urban environment requires government interventions that involve, in certain cases, setting up new infrastructure networks. Fiscal rewards generated by periods of prosperity provide the means for governments to engage in such interventions. Public support for the idea of a major urban transition is also helpful, as is a planning model that portrays and justifies the proposed transformations (Filion, 2010b).

As indicated, factors of change must overcome deeply rooted path dependencies. It follows that large-scale urban transitions will occur only when path dependencies are weakened and factors of change are both aligned and powerful. These circumstances explain the periodic toppling of long periods of stability by transformative epochs when conditions are conducive to urban transitions. We now turn to the history of Canadian cities in order to identify the major transitions that have marked their evolution.

Canadian Urban Transformations

The focus of this chapter is on relatively recent urban transformations, that is, those that have unfolded since 1945. However, we begin by briefly looking at the pre-1945 Canadian urban history since colonization, and describe the urban form this period bequeathed to the post–World War II city. While the characteristics of cities vary in any period, some generalities may be observed in each time period.

From Urban Origins to 1945: The Development of the Railroad

Canadian cities were built on Indigenous land, following colonization largely by European settlers from the United Kingdom and France as well as a host of other nations. Until the second half of the nineteenth century, walking was the main mode of transportation in Canadian cities. Their relatively small size made it possible to access most destinations on foot. However, major transformations unfolded in the latter part of that century that had considerable effects on transportation within Canadian urban areas, as well as on their structure and size. From the 1860s, for example, horse-drawn streetcars serviced the largest Canadian cities. In 1866, the first electric streetcar began operation in Windsor, Ontario. Quickly thereafter, systems spread across the country, so much so that by World War I streetcars were present in 48 Canadian urban areas.

Unlike prior animal traction, electric streetcars were able to provide relatively cheap mass transportation. The presence of streetcars allowed cities to expand rapidly at a time of accelerated population growth, primarily the result of industrialization. The streetcars propelled the first generation of suburbs. These suburbs were generally constituted of a commercial street along the streetcar line, with stores on the ground floor and apartments above (Jacobs, 1961). The remainder of the neighbourhood was residential, often with an industrial presence, and within walking distance of streetcar stops. From the late nineteenth century, in Montreal and Toronto railway companies operated commuter services on their main lines. These train services shaped peripheral urban development, which took the form of commuter suburbs centred on railway stations (Hoyt, 1939; Warner, 1962). In their original configuration, these commuter suburbs were still within walking distance from train stations even though the built form was generally low density.

Together, streetcar and commuter rail systems made it possible to focus much office, retail, and service development on a central point—the downtown. It was the capacity of these systems to draw passengers from the entire metropolitan region and deliver them downtown that propelled the growth of this sector and the appearance of its early twentieth-century hallmarks, the department store and the office skyscraper (Colby, 1933).

Another profound urban transformation that took place from the late nineteenth century to 1945 was the proliferation of large industrial areas along railway lines, which stimulated urban development. Industrialization came later in Canada than in the United Kingdom and the United States. The rapid urban development that industrialization triggered in these two countries in the early (in the UK) and mid- (in the US) nineteenth century reached Canada only in the latter part of that century (Naylor, 2006; Smucker, 1980).

The late nineteenth century, then, marked the passage from small walking cities to much larger rail-oriented cities. The influence of rail on urban development resulted in the rapid construction of streetcar lines, the introduction of commuter rail, the resulting commuter suburbs, as well as the development of large industrial districts along railway lines. (See Figure 2.1 for a depiction of some of the key features of the pre-1945 city.)

From 1945 to 1975: "Urban Dispersion" in the Fordist City

The 15 years or so that followed the end of World War II, often called the Fordist period (see **Fordism**), witnessed the formulation and implementation of a new model of urban development, which involved an all-out adaptation of the city to the automobile (Hardwick, 2004; Sewell, 2009). We label this model "urban dispersion" to highlight, along with its land-use specialization, the dependence on the automobile and the city's low density, as well as its radical departure from multi-functional centralization (Filion, Bunting,

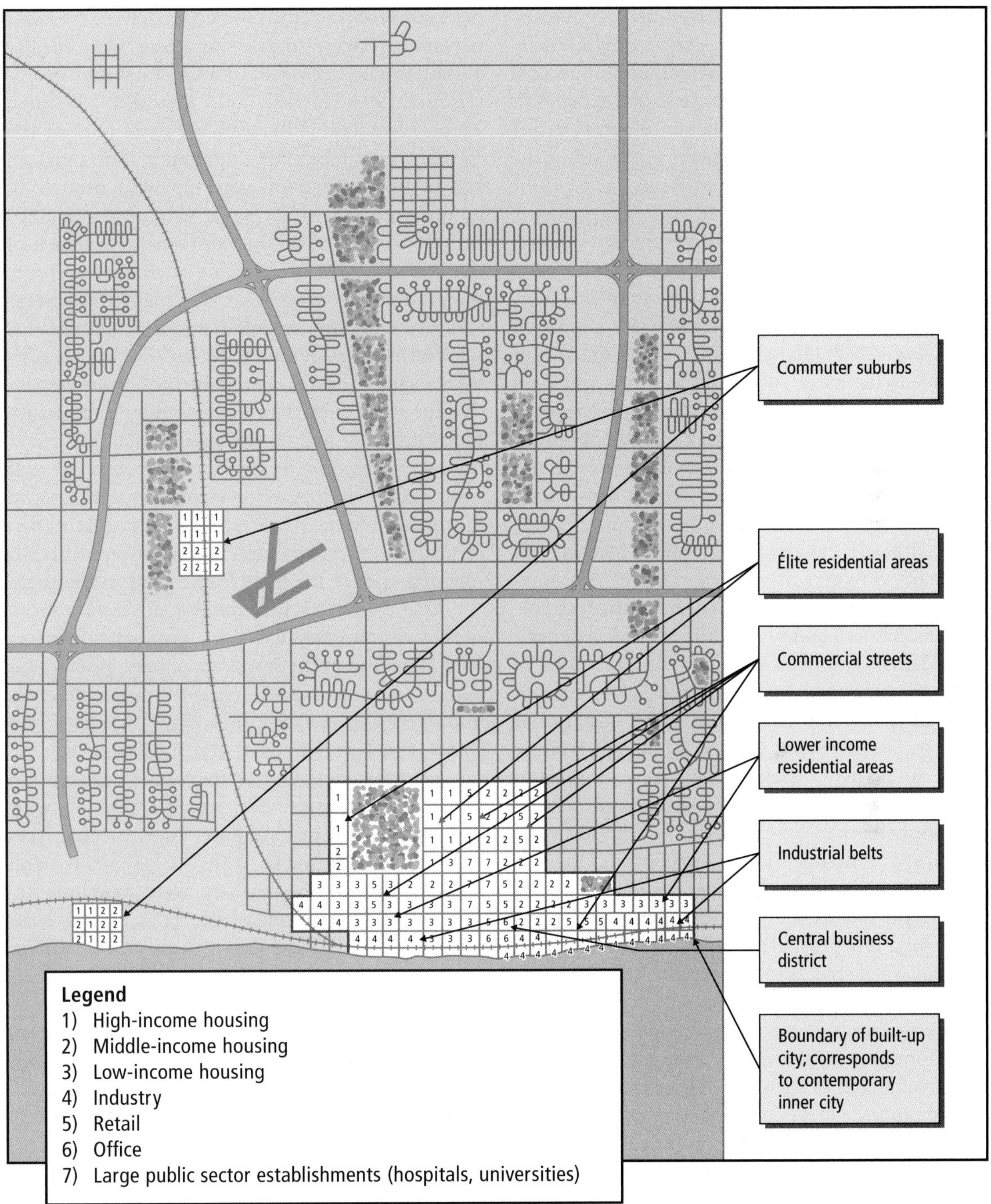

Figure 2.1 Urban Structure before 1945

and Warriner, 1999). In dispersed urban environments, employment, retailing, services, and institutions are mostly scattered along numerous axes of automobile accessibility. Once in place, the path dependencies generated by urban dispersion ensured this model's entrenchment and a strengthening of the relationship between generalized automobile reliance and land use. The postwar urban transition may well represent the deepest change cities have ever experienced in a relatively short period.

A near-perfect alignment of the identified factors of urban change made such a transition possible. (The next sub-section, which explores the obstacles to a large-scale transition in the present context, will highlight the exceptional nature of the transformative circumstances that materialized between 55 and 70 years ago.) A prosperous economy fanned mass consumption of automobiles. Although cars had been around by this time for nearly 50 years, it was the postwar economy that made a generalized reliance on this technology possible as a ballooning middle class acquired vehicles. The two decades following World War II were characterized by rapid population growth driven by the birth of the baby boom generation and, to a lesser extent, the arrival of immigrants. With so many households engaged in rearing children, the popularity over these years of single-family homes and of neighbourhoods with plentiful green space is not surprising. Governments were, moreover, enablers of this postwar urban transformation. Healthy public finances, buttressed by prosperity, allowed governments to build infrastructure, especially highways, required for the accommodation of transformative trends.

Dominant values of the time also promoted urban change. Wide-scale adherence to modernism and the attendant rejection of traditions, for example, explains the popularity of new car-dominated suburban-type environments at the expense of older inner-city areas (Sewell, 1993). What is more, after nearly two decades of privation due to the Great Depression and wartime restrictions, there was a pent-up aspiration for a middle-class consumerist lifestyle. New suburban subdivisions, with their abundant private interior and exterior space, were ideally suited to the accumulation of consumer goods (Hamel, 1993; Hayden, 2003; Miron, 1988; Spurr, 1976). Over the 1930s and the early part of the 1940s, while urban development was at a virtual standstill, visionaries came up with models of automobile-reliant forms of urbanization. Representations of these models were mainstays of the world's fairs of the time (Leinberger, 2008; Rydell and Schiavo, 2010). The planning profession drew on two models from prior decades for inspiration: the Garden City for low-density suburban areas and the Tower-in-the-Park perspective for some suburban development, but mostly for inner-city **urban renewal** projects. Finally, the rapid urban growth of the time meant that urban areas were moving beyond population and economic activity thresholds where existing infrastructure systems and demands were in equilibrium—hence the need for new infrastructure.

One cannot exclude the role of serendipity in such an apparently flawless alignment of factors of change. But at the same time, there is clear evidence of intentional strategies to transform cities into agents of consumption. The postwar urban transformation happened at a time when several mechanisms contributed to swell consumption and thereby ensure that demand kept up with rising industrial productivity. These mechanisms included government transfer payments and trade unions, both of which were instrumental in creating a blue- and white-collar middle class (Aglietta, 1979; Jessop and Sum, 2006). There were also **Keynesian** policies, government subsidies common during Fordism, meant to stimulate demand (Frazer, 1994; Jones, 2008). Some of these took the form of infrastructure development and others of guaranteed mortgages for single-family home purchasers (Dennis and Fish, 1972). Policy-makers of the time were aware of the sizeable economic spinoffs the construction and sale of a new home generate: building materials, construction workforce, new appliances, and furniture (Harris, 2004). By encouraging the purchase and use of automobiles, the building of urban highways had similar effects.

Over this period, while Canada took part in the continental trend toward urban dispersion, it did so with more moderation than the United States. In Canada, unlike in many US metropolitan regions where inner-city blight was common, inner-city neighbourhoods remained for the most part vibrant. Moreover, while in most US urban areas urban transportation policies were entirely focused on the car, large Canadian cities adopted a more balanced approach. Toronto inaugurated its first subway line in 1954, and the Montreal metro began operation in 1966.

In the present age of **deindustrialization** and social **polarization**, it is easy to look back longingly at this period as an economic utopia. We must, however, avoid glorifying this time. If it was true that the middle class was expanding in a climate of job security and rising standards of living, it was also true that a large segment of the population was excluded from the rewards of this economic system. White men dominated the workplace; the prevailing conformity made it difficult for people with alternative lifestyles. As well, by today's standards, goods were expensive; they also lacked the diversity the present global economy provides. Indeed, despite all its downsides, **globalization** can be credited for the fact that—despite stagnating incomes—more of us can afford a wide range of goods, notably electronics products. Finally, as cities at that time became cauldrons of consumption in a car-oriented environment, they contributed to a narrowing of transportation options and urban ways of life (public transit services are a poor alternative to the automobile in dispersed urban settings and virtually all journeys exceed walkable distances), depleted quality of life due to time spent in traffic, and a deterioration in air quality (Huang et al., 2018; Miller and Shalaby, 2003).

Figure 2.2 identifies some of the more salient properties of suburban areas developed in the postwar period, such as the separation of unlike activities (e.g., housing and industry). With rising car use and the presence of expressways and arterials, land values and density gradients became flatter, creating numerous points of equivalent importance. Major expressway interchanges and arterial intersections became sites for regional malls or power malls (Jones, 2006; Jones and Simmons, 1990). Other locations offering good accessibility along, off, or close to high-capacity roads attracted industrial and business parks, small retail malls, and self-standing retail and service establishments (e.g., fast-food outlets, car dealerships, gas stations, and so on). Land-hungry activities such as university campuses also opted for suburban locations. Suburbs distinguished themselves from the inner city by adopting an inwardly focused system of curvilinear streets within super blocks.

Meanwhile, quality of life in inner-city neighbourhoods suffered considerably from the construction of highways intended to improve suburban commuters' accessibility to the downtown (Nowlan and Nowlan, 1970). A perceived decline of the older housing stock was a major incentive for urban renewal schemes that demolished existing structures to make way for public housing. More frequently, though, "slum clearance" was the outcome of private development (Birch, 1971; Bourne, 1967; Hoover and Vernon, 1962; Miron, 1993; Smith, 1964; White, 2016). While everywhere the share of metropolitan jobs and retailing found in the **central business district (cbd)** declined, in large metropolitan regions these districts enjoyed considerable absolute office and retailing growth. The situation was different in smaller urban areas where the CBD shrunk in both relative and absolute terms (Filion and Hammond, 2009).

Post 1975: Dissatisfaction with Modern Urban Expansion

From the early 1970s, attitudes toward prevailing patterns of urban development became progressively less supportive. Residents were becoming more aware and less tolerant of the downsides of existing urban patterns, giving rise to political reactions. These mostly took the form of citizen protest movements, whose reverberations were felt within city councils. These reactions had a profound impact on the form that city growth took thereafter. The most important achievement

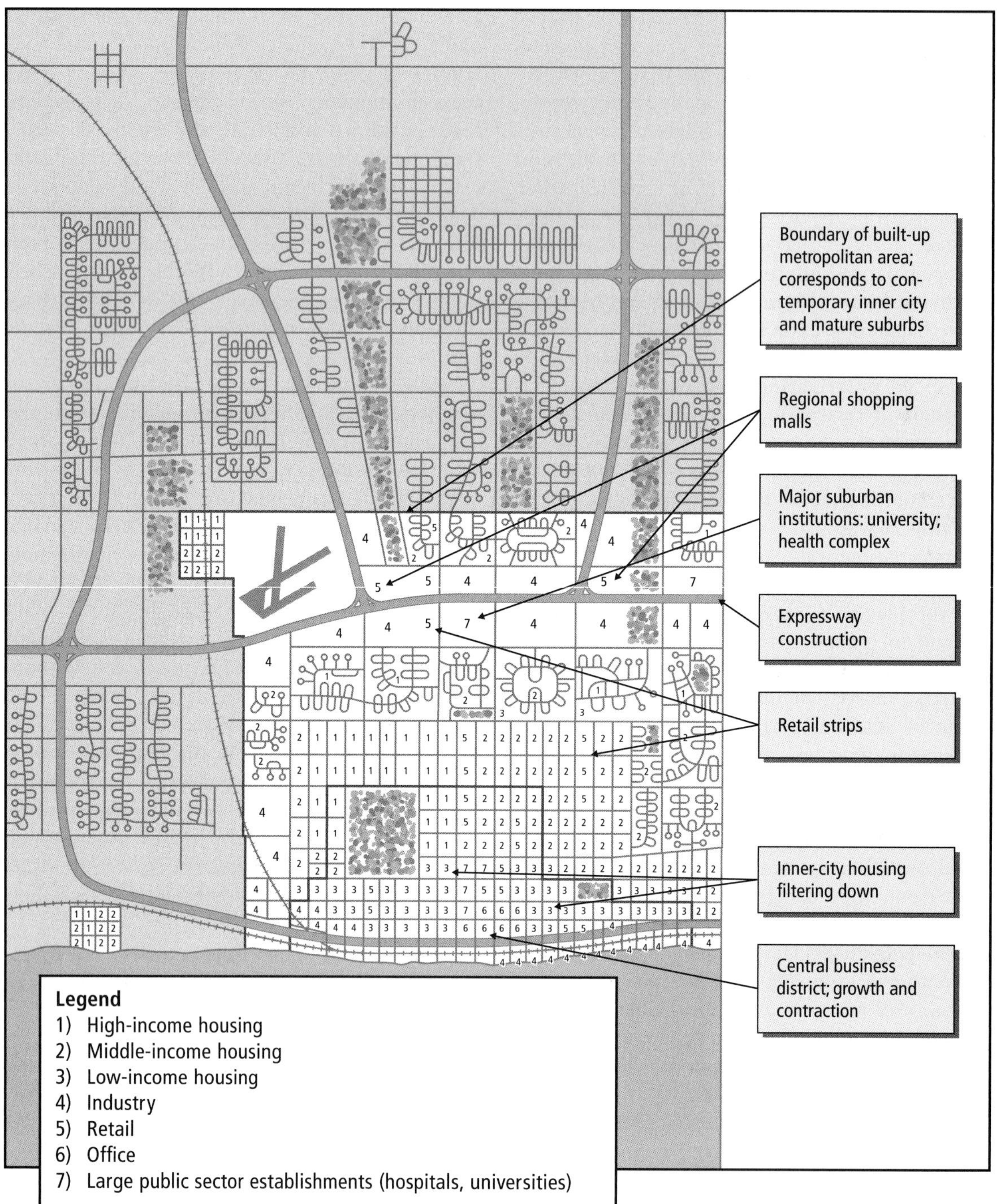

Figure 2.2 Urban Structure, 1945–1975

of such protest movements was the halting of public sector-sponsored urban renewal and the construction of urban expressways. Indeed, everywhere in Canada (with the exception of the Ville-Marie Expressway in Montreal) these movements were successful in ending the construction of urban expressways. In contrast, suburban expressways, whose right-of-way had generally been safeguarded from the time individual suburbs were originally planned and therefore did not require expropriations and demolitions, went ahead. This change in attitude toward dispersed patterns also caused some middle-class households, which would have previously routinely opted for suburban living, to choose inner-city locations (Lees, Slater, and Wyly, 2008; Ley, 1996). These were the households that fuelled inner-city redevelopment processes, which have been gathering momentum ever since.

Other circumstances caused dispersed urbanization to lose some of its gloss. Economic growth had slowed, with recessions becoming deeper and more frequent. A sputtering economy made it difficult for depleted government budgets to bankroll the heavy infrastructure expenses of dispersed urbanization, especially their transportation networks. As expected, one outcome was a deterioration of traffic conditions. It is not irrelevant, given the energy-hungry nature of dispersed urbanization, that global oil supply disruptions often played a role in triggering recessions. As environmental concerns gathered momentum, the dispersed model was cast in an increasingly negative light because of its automobile reliance. It was first blamed for worsening air pollution and more recently for high emissions of greenhouse gases. All of this was taking place as income stagnation and polarization were impairing the housing consumption potential of the middle class.

In the face of rising critiques of urban dispersion and increasing discordance between this model and society-wide economic conditions, one might expect that we would be at the cusp of another profound urban transition. Urban reality suggests otherwise. The dispersed model appears to be deeply entrenched and, if anything, the object of an ever-advancing mutual adaptation between driving and land use. For example, recent retail and entertainment formats, such as new-generation supermarkets, power malls, big-box stores, and multi-screen cinemas, draw from large catchment areas of motorists.

Meanwhile, the factors of urban change that propelled the postwar urban transformation are not likely to be as effective in the present context. We have already seen that economic growth has slowed considerably since the 1945–1975 period, as has demographic expansion. With a birth rate below replacement since the early 1970s, Canada relies on immigration to maintain a slow overall population growth. These economic and demographic conditions account for the fact that across the country the proportion of the urban built environment that is added annually is modest compared to the situation that prevailed over the previous period. From a technological perspective, the extraordinary wave of information technology innovation has had, to date, limited impact on **urban form** and dynamics. For example, although information technology makes it possible for many people to work from home, workplace regulations, peer pressure, and a desire to fully fit in with co-workers account for the fact that these innovations have had negligible effects on commuting and residential location patterns (Janelle, 2004; Moos and Skaburskis, 2007). Add to this an absence of new modes of urban transportation and we have an explanation for the limited effects of emerging technologies on the way cities operate, in sharp contrast with the impact of the widespread acquisition of cars in the 1950s and 1960s. Leaner economic times make it difficult for governments to invest the large sums needed to create extensive and efficient public transit systems, which could spawn alternative forms of urban development.

Any transformative initiative on the part of governments must also confront the present currency of the neo-liberal ideology, which gives precedence to market trends over government intervention (Brenner and Theodore, 2002; Hackworth, 2006; Harvey, 2005; Springer, Birch and MacLeavy, 2016). Fragmentation of values also impedes governments' urban transformation capacity by ruling out consensus around an alternative urban model. The vanishing common ground is not unrelated to the economic polarization of society as the size of the middle-class is reduced due to deindustrialization and a bifurcation of the economy between low- and high-wage service sector jobs. Whereas in the postwar period the dispersed model of urbanization garnered widespread public support, there is no comparable adherence to a dominant model at the present time. Alternative urban forms, such as those represented by New Urbanism and transit-oriented development, have failed to capture the public's imagination to an extent comparable to the postwar enthusiasm for dispersed patterns (Duany, Plater-Zyberk, and Speck, 2000; Grant and Bohdanow, 2008; Talen, 2000). Meanwhile, attachment to prevailing suburban forms remains strong, and suburban residents use their political power to influence development at the local level and determine urban (and other) policies at the provincial and federal levels. Finally, with demographic growth concentrated in a few metropolitan regions, these regions must adjust their transportation and urban form to their expanding size. Such pressures are absent from other, slow- or no-growth, urban areas.

There is thus much doubt about the capacity of achieving a major urban transition at the present time (Table 2.1). Factors of urban transformation, and concentration, currently muster insufficient power relative to the entrenchment of urban dispersion. It is difficult to see how cities can overcome 70 years of massive public and private sector investment in urban dispersion, along with the path dependencies such urban development generates (White, 2003). There has been no replication of the alignment of factors of urban change that triggered the previous great urban transformation. All of this leaves cities in a predicament. Changes of attitudes on the part of many and discordance between urban dispersion and present trends affecting society do not translate into different ways of building cities. To be sure, such transformations do happen in certain urban sectors (e.g., the high-rise condominium boom in downtown Toronto and Vancouver), but despite some examples of urban and suburban infill, much current urban growth still conforms to the dispersed model (Berelowitz, 2005; Punter, 2003).

Figure 2.3 confirms that while the post-1975 period has not ushered in a profound urban transition, it has resulted in a number of innovations (also see Table 2.1). With the decentralization of offices from downtowns, the period has witnessed the creation of suburban business (or office) parks, and, more recently, the emergence of so-called suburban downtowns that combine office, retail, and high-rise residential concentrations with civic centres (Cervero, 1986; Filion, 2007; Garreau, 1991). On the retail scene, the appearance of big-box stores and **power centres**, and mostly online shopping, has disrupted retail distributions inherited from previous periods (Jones and Doucet, 2000; Jones and Simmons, 1990; Visser, Nemoto, and Browne, 2014). Meanwhile, although the density of suburban employment and retail space has either remained steady or declined, that of residential areas is on the rise. Homes are generally larger than they were in the previous period, but occupy comparatively smaller lots. Nonetheless, auto-based configurations of residential developments continue. One major difference between the two generations of suburban residential areas concerns the nature of green space. Whereas suburban green areas in the previous period came mostly in the form of large private lots and public playgrounds, more space is now allocated to the preservation of natural features, such as woods, creeks and their riparian zones, and marshes and ponds (Hough, 2004; Manuel, 2003).

Another suburban innovation over the post-1975 period has been the introduction of the New Urbanism model of development. However, this urban formula has not caught on as much as originally anticipated. In large metropolitan regions, the previous period was a time of downtown and inner-city large-scale commercial and residential redevelopment (mostly high-rise rental apartments); the focus during the present epoch has been on **gentrification** and residential **intensification**, mostly in the form of high-rise condominiums. The inner city and the core have become a highly appealing place to live for a large segment

Table 2.1 Conditions for Urban Transitions: The Post–World War II Period and the Present Time

Conditions for post–World War II urban transitions	Insufficient conditions for urban transitions at the present time
The Economy	
Prolonged period of prosperity; expansion of the middle class; consumerist lifestyle	Faltering economy; decline of the middle class; income polarization; globalization
Demography	
Rapid population growth, baby boom; many families with children; popularity of single-family homes and suburbs with abundant green space	Slow demographic growth; fewer households with children
Technology	
The automobile had been around for many decades; economic conditions were present over the period for the generalization of its use	Explosion of information technology innovations but they have limited impact on urban form and transportation; no new transportation technology
Governance	
Governments are enablers of post-WWII urban transformations; healthy public sector budgets and adoption of measures to sustain the economy: suburban-type development as well as expressways and arterials	Depleted government coffers prevent major public sector investments in new infrastructure; impedes large-scale public transit schemes
Values	
Conformity, modernist values; rejection of traditions; broad-scale support for the suburban model; pent-up demand for a consumerist lifestyle	Fragmentation of values in parallel with income polarization; absence of consensus around any urban model
Planning Models	
Pre-WWII visions of car-oriented cities; draws from Garden City and Tower-in-the-Park models	Recentralization model dominates in planning documents but its popularity is mostly confined to planning circles
Changes in the Size of Urban Regions	
With urban growth there is a need to redefine urban infrastructures	Fast growth in some large metropolitan regions, where there is a need to adapt infrastructures; much slower growth or decline in other cases

of the population sensitive to their urban amenities: entertainment, culture, proximity to workplaces, and walking-hospitable environments.

Conclusion: Possible Future Transitions

With the emergence of suburban office and employment nodes—areas of high employment concentrations—and the resurgence of the downtown, some have speculated that we may be moving toward a new urban form characterized by several nodes (not just one downtown), which are connected by higher-density transportation corridors, and surrounded by low-density expanses of residential or commercial development (Moos and Mendez, 2013). In fact, a survey of North American planning documents with a metropolitan-wide focus indicates a near consensus among planning agencies around an alternative pattern of urban development involving the further development of downtowns, the creation of new secondary urban centres in the suburbs, and the interconnection of these different centres by public transit (either rail transit or bus rapid

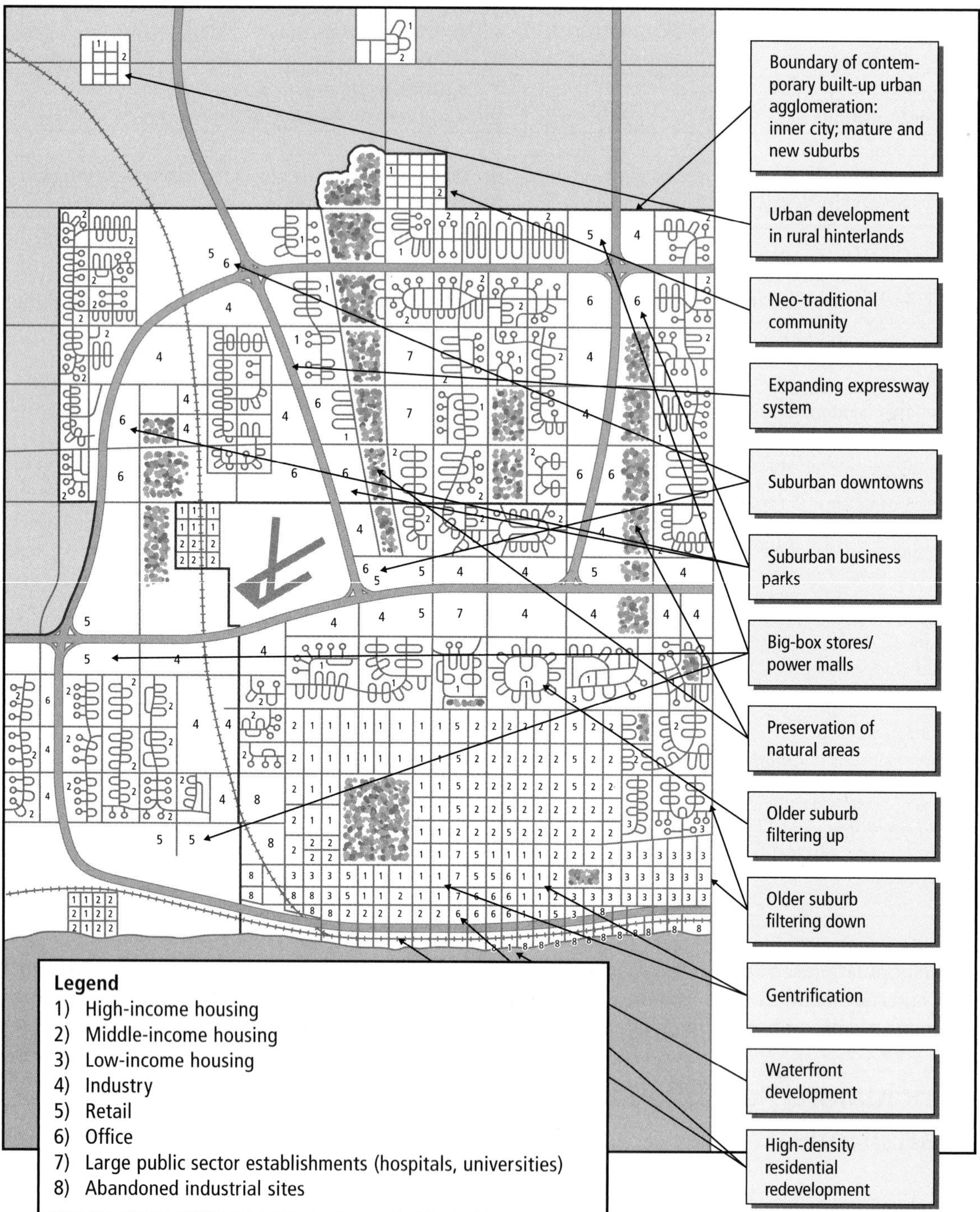

Figure 2.3 Urban Structure, 1975 to Present

transit) (Filion, Kramer, and Sands, 2016). The recentralization model purports to abate land consumption, car reliance, and the need for additional highway infrastructures associated with urban growth. It is thus in accord with environmental values, economic concerns, and the quest by many residents for locations with handy urban amenities. If implemented on a grand scale, we anticipate that this model will have a profound transformative effect on urban form and transportation. But we can also expect barriers to implementing this strategy. First, there is no large-scale public mobilization around the recentralization concept. It is still mostly the domain of planners. Second, even if the land-use dimensions of recentralization can be achieved largely through planning regulations, and thus avoid straining already tight public sector budgets, major funding will still be required for the public transit investments essential to the achievement of this type of urbanization (Filion and Kramer, 2011). It is a challenge for governments to fund new transit projects in a time of fiscal conservatism, hence the search for creative funding formulas for such projects. For example, nearly half the funding for the $6.3 billion Montreal Réseau express métropolitain, a 67 kilometre rail public transit system, will originate from the Quebec pension fund (Caisse de dépôt) (REM, n.d.).

A recentralization strategy may, however, cause serious shifts in the social geography of urban areas. In a fashion that mirrors the supersession of social by environmental issues in the urban planning discourse of the last two decades, plans emphasize the environmental benefits of recentralization but give scant attention to its possible impact on different social groups (Gunder, 2006). In its ideal version, this strategy would bring centrality and accessibility to everyone by saturating a metropolitan region with centres and different types of rapid transit. A more realistic but also more problematic outcome would be the creation of a limited number of centres and new rapid transit lines. These centres themselves, along with the areas surrounding them and the bordering rapid transit lines, would be perceived as attractive residential locations. And because areas offering such urban amenities and accessibility potential would be limited, they would likely be objects of gentrification resulting from competition for location in these sectors. Over time, and under pressure from higher-income groups, incumbent residents would move out and therefore be deprived of the new advantages offered by these areas—this has been referred to as eco-gentrification (Moos et al., 2018; Quastel, Moos, and Lynch, 2012).

We cannot exclude the possibility of technical innovations having a transformative impact on cities. But such innovations would have to fulfill a number of conditions to be widely adopted and to contribute to a transformation of the urban environment. In the present economic climate, it is difficult to foresee the setting up of expensive new infrastructures. In consequence, to be successful new technologies would have to rely on cheaper infrastructures than prior technologies, as in the case of cellular phones in comparison to landlines, or accommodate themselves of existing infrastructures. Perhaps the most probable urban transportation innovation on the horizon is the driverless, and perhaps eventually fully electric, car. A key advantage of this innovation would be its suitability to existing transportation infrastructures. In fact, allowing cars to travel safely faster and closer to each other would raise the efficiency of highway networks. But as this technology would accentuate the urban impacts of the car, without however its adverse safety, energy, and pollution consequences, it would be more likely to reinforce, rather than cause a departure from, the dispersed model.

Review Questions

1. What were the main factors influencing the structure of Canadian cities during different periods of urban development?
2. How do you think the structure of Canadian cities will change in the next 20 to 30 years? Explain what factors you think would bring about these changes.
3. Considering the city that you grew up in, what factors were important in shaping the physical urban structure? How so?

Note

1. An earlier version of this chapter was written in collaboration with Trudi Bunting (1944–2017).

References

Aglietta, M. 1979. *A Theory of Capitalist Regulation: The U.S. Experience*. London: New Left Books.

Atkinson, G., and T. Oleson. 1996. "Urban sprawl as a path dependent process," *Journal of Economic Issues* 30: 609–15.

Audirac, I. 2005. "Information technology and urban form: Challenges to smart growth," *International Regional Science Review* 28: 119–45.

Berelowitz, L. 2005. *Dream City: Vancouver and the Global Imagination*. Vancouver: Douglas & McIntyre.

Bessey, K.M. 2002. "Structure and dynamics in an urban landscape: Toward a multiscale view," *Ecosystems* 5: 360–75.

Birch, D. 1971. "Toward a stage model of urban growth," *Journal of the American Institute of Planners* 37: 78–87.

Boarnet, M.G., and R. Crane. 2001. *Travel by Design: The Influence of Urban Form on Travel*. Oxford: Oxford University Press.

Bourne, L.S. 1967. *Private Redevelopment of the Central City*. Chicago: University of Chicago, Department of Geography Research Paper No. 112.

———, and D. Rose. 2001. "The changing face of Canada: The uneven geographies of population and social change," *The Canadian Geographer* 45: 105–19.

Brenner, N., and N. Theodore, eds. 2002. *Spaces of Neoliberalism: Urban Restructuring in North America and Western Europe*. Malden, MA: Blackwell.

Brown, G., and H. Glanz. 2018. "Identifying potential NIMBY and YIMBY effects in general land use planning and zoning," *Applied Geography* 99: 1–11.

Cevero, R. 1986. "Urban transit in Canada: Integration and innovation at its best," *Transportation Quarterly* 40: 293–316.

———. 2016. "Public transit and sustainable urbanism: Global lessons," in C. Curtis, J.L. Renne, and L. Bertolini, eds., *Transit Oriented Development: Making It Happen*. London: Routledge.

Colby, C. 1933. "Centripetal and centrifugal forces in urban geography," *Annals, Association of American Geographers* 23: 1–20.

Collin, J-P., J. Léveillée, and C. Poitras. 2002. "New challenges and old solutions: Metropolitan reorganization in Canadian and US city regions," *Journal of Urban Affairs* 24: 317–32.

Curic, T., and T. Bunting. 2006. "Does compatible mean same as? Lessons learned from the residential intensification of surplus hydro lands in four older suburban neighbourhoods in the City of Toronto," *Canadian Journal of Urban Research* 15: 202–24.

Dennis, M., and S. Fish. 1972. *Programs in Search of a Policy*. Toronto: Hakkert.

Duany, A., E. Plater-Zyberk, and J. Speck. 2000. *Suburban Nation: The Rise of Sprawl and the Decline of the American Dream*. New York: North Point Press.

Ewing, R., and R. Cervero. 2010. "Travel and the built environment: A meta-analysis," *Journal of the American Planning Association* 76: 265–94.

Filion, P. 2007. *The Urban Growth Centre Strategy in the Greater Golden Horseshoe: Lessons from Downtowns, Nodes and Corridors*. Toronto: Neptis Foundation.

———. 2010a. "Growth and decline in the Canadian urban system: The impact of emerging economic, policy and demographic trends," *Geojournal* 75: 517–38.

———. 2010b. "Reorienting urban development? Structural obstruction to new urban forms," *International Journal of Urban and Regional Research* 34: 1–19.

———, T. Bunting, and K. Warriner. 1999. "The entrenchment of urban dispersion: Residential preferences and location patterns in the dispersed city," *Urban Studies* 36: 1317–47.

———, and K. Hammond. 2009. "When planning fails: Downtown malls in mid-size cities," *Canadian Journal of Urban Research* 17, 2: 1–27.

———, and A. Kramer. 2011. "Metropolitan-scale planning in neo-liberal times: Financial and political obstacles to urban form transition," *Space and Polity* 15: 197–212.

———, ———, and G. Sands. 2016. "Recentralization as an alternative to urban dispersion: Transformative planning in a neoliberal societal context," *International Journal of Urban and Regional Research* 40: 658–78.

Fishman, R. 1977. *Urban Utopias in the Twentieth Century: Ebenezer Howard, Frank Lloyd Wright, and Le Corbusier.* New York: Basic Books.

Foot, D.K., with D. Stoffman. 1996. *Boom, Bust and Echo: Profiting from the Demographic Shift in the New Millennium.* Toronto: Macfarlane Walter and Ross.

Frank, L.D., S. Kavage, and S. Appleyard. 2007. "The urban form and climate change gamble," *Planning* 73: 18–23.

Frazer, W.J. 1994. *The Legacy of Keynes and Friedman: Economic Analysis, Money, and Ideology.* Westport, CT: Praeger.

Frenken, K., and R.A. Boschma. 2007. "A theoretical framework for evolutionary economic geography: Industrial dynamics and urban growth as a branching process," *Journal of Economic Geography* 7: 635–49.

Gad, G. 1991. "Toronto's financial district," *The Canadian Geographer* 35: 203–7.

Garreau, J. 1991. *Edge City: Life on New Frontier.* New York: Doubleday.

Glaeser, E.L. 2000. "The new economics of urban and regional growth," in G.L. Clark, M.P. Feldman, and M.S. Gertler, eds., *The Oxford Handbook of Economic Geography.* Oxford: Oxford University Press.

Graham, S., and S. Marvin. 2001. *Splintering Urbanism: Networked Infrastructures, Technological Mobilities and the Urban Condition.* New York: Routledge.

Grant, J.L., and S. Bohdanow. 2008. "New urbanism developments in Canada: A survey," *Journal of Urbanism* 1: 109–27.

Gunder, M. 2006. "Sustainability: Planning's saving grace on road to perdition?" *Journal of Planning Education and Research* 26: 208–21.

Hackworth, J. 2006. *The Neoliberal City: Governance, Ideology, and Development in American Urbanism.* Ithaca, NY: Cornell University Press.

Hamel, P. 1993. "Modernity and postmodernity: The crisis of urban planning," *Canadian Journal of Urban Research* 2: 16–29.

Hardwick, M.J. 2004. *Mall Maker: Victor Gruen, Architect of an American Dream.* Philadelphia, PA: University of Pennsylvania Press.

Harris, R. 2004. *Creeping Conformity: How Canada Became Suburban, 1900–1960.* Toronto: University of Toronto Press.

Harvey, D. 1990. *The Conditions of Postmodernity: An Enquiry into the Origins of Cultural Change.* Oxford: Blackwell.

Harvey, D. 2005. *A Brief History of Neoliberalism.* Oxford: Oxford University Press.

Hayden, D. 2003. *Building Suburbia: Green Fields and Urban Growth, 1820–2000.* New York: Pantheon Books.

Hoover, E.M., and R. Vernon. 1962. *Anatomy of a Metropolis.* New York: Doubleday Anchor.

Hough, M. 2004. *Cities and Natural Processes: A Basis for Sustainability.* London: Routledge.

Howard, E. 2003 [1902]. *To-morrow: A Peaceful Path to Real Reform.* London: Routledge.

Hoyt, H. 1939. *The Structure and Growth of Residential Neighborhoods in American Cities.* Washington: Federal Housing Administration.

Huang, W., A. Xu, Y. Yan, and A. Zipf. 2018. "An exploration of the interaction between urban human activities and daily traffic conditions: A case study of Toronto, Canada," *Cities.* At: https://doi.org/10.106/j/cities.2018.07.001

Hutcheon, L. 2002. *The Politics of Postmodernism,* 2nd edn. London: Routledge.

Jacobs, J. 1961. *The Death and Life of Great American Cities.* New York: Random House.

Janelle, D.G. 2004. "Impact of information technologies," in S. Hanson and G. Giuliano, eds., *The Geography of Urban Transportation.* New York: Guilford.

Jessop, B., and N.-L. Sum. 2006. *Beyond the Regulation Approach: Putting Capitalist Economies in their Place.* Cheltenham: Edward Elgar.

Jones, J.P. 2008. *Keynes's Vision: Why the Great Depression Did Not Return.* London: Routledge.

Jones, K. 2006. "The urban retail landscape," in T. Bunting and P. Filion, eds., *Canadian Cities in Transition: Local through Global Perspectives,* 3rd edn. Toronto: Oxford University Press.

———, and M. Doucet. 2000. "Big-box retailing and the urban retail structure: The case of the Toronto area," *Journal of Retailing and Consumer Services* 7: 233–47.

———, and J. Simmons. 1990. *The Retail Environment.* London: Routledge.

LeBlanc, M-F. 2006. "Two tales of municipal reorganization: Toronto's and Montreal's diverging paths towards regional governance and social sustainability," *Canadian Journal of Political Science* 39: 571–90.

Le Corbusier. 1973 [1933]. *The Athens Charter.* New York: Grossman.

Lees, L., T. Slater, and E.K. Wyly. 2008. *Gentrification.* New York: Routledge.

Leinberger, C.B. 2008. *The Option of Urbanism: Investing in a New American Dream.* Washington: Island Press.

Ley, D. 1996. *The New Middle Class and the Remaking of the Central City.* Oxford: Oxford University Press.

Manuel, P.M. 2003. "Cultural perceptions of small urban wetlands: Cases from the Halifax Regional Municipality, Nova Scotia, Canada," *Wetlands* 23: 921–40.

Meligrana, J., ed. 2004. *Redrawing Local Government Boundaries: An International Study of Politics, Procedures, and Decisions.* Vancouver, BC: UBC Press.

Miller, E.J., and A. Shalaby. 2003. "Evolution of personal travel in the Toronto area and policy implication," *Journal of Urban Planning and Development* 129: 1–26.

Millward, H., and G. Xue. 2007. "Local urban form measures related to land-use and development period: A case study for Halifax, Nova Scotia," *Canadian Journal of Urban Research* 16: 53–72.

Miron, J. 1988. *Housing in Postwar Canada: Demographic Change, Household Formation and Housing Markets.* Montreal and Kingston: McGill-Queen's University Press.

———, ed. 1993. *House, Homes, and Community: Progress in Housing Canadians, 1945–1986.* Montreal and Kingston: McGill-Queen's University Press.

Moos, M., D. Pfeiffer, and T. Vinodrai, eds. 2018. *The Millennial City: Trends, Implications, and Prospects for Urban Planning and Policy.* Oxford: Routledge.

———, and P. Mendez. 2013. "Suburbanization and the remaking of metropolitan Canada," in R. Keil, ed., *Suburban Constellations: Governance, Land and Infrastructure in the 21st Century.* Berlin: Jovis Publishers.

———, and A. Skaburskis. 2007. "The characteristics and location of home workers in Montreal, Toronto and Vancouver," *Urban Studies* 44: 1781–1808.

———, T. Vinodrai, N. Revington, and M. Seasons. 2018. "Planning for mixed use: Affordable for whom," *Journal of the American Planning Association* 84: 7–20.

Muller, P.O. 2004. "Transportation and urban form: Stages in the spatial evolution of the American metropolis," in S. Hanson and G. Guiliano, eds., *The Geography of Urban Transportation.* New York: Guilford.

Naylor, R.T. 2006. *The History of Canadian Business, 1867–1914.* Montreal and Kingston: McGill-Queen's University Press.

Nowlan, D., and N. Nowlan. 1970. *The Bad Trip: The Untold Story of Spadina Highway.* Toronto: Anansi Press.

Pierson, P. 2000. "Increasing returns, path dependence, and the study of politics," *American Political Science Review* 94: 251–67.

Punter, J. 2003. *The Vancouver Achievement: Urban Planning and Design.* Vancouver: University of British Columbia Press.

Quastel, N., M. Moos, and N. Lynch, N. 2012. "Sustainability as density and the return of the social: The case of Vancouver, British Columbia," *Urban Geography 33*: 1055–84.

REM, n.d. *Réseau express métropolitain: A New Line that Transports Greater Montreal.* At: https://www.cdpqinfra.com/en/reseau_electrique_metropolitain

Rosen, G. 1993 [1958]. *A History of Public Health.* Baltimore, MD: The Johns Hopkins University Press.

Rydell, R.W., and L.B. Schiavo, ed. 2010. *Designing Tomorrow: America's World's Fairs of the 1930s.* New Haven, CT: Yale University Press.

Schively, C. 2007. "Understanding the NIMBY and LULU phenomena: Reassessing our knowledge base and informing future research," *Journal of Planning Literature* 21: 255–66.

Sewell, J. 1993. *The Shape of the City: Toronto Struggles with Modern Planning.* Toronto: University of Toronto Press.

———. 2009. *The Shape of the Suburbs: Understanding Toronto's Sprawl.* Toronto: University of Toronto Press.

Smith, W. 1964. *Filtering and Neighborhood Change.* Berkeley: University of California, Center for Real State and Urban Economics, Report No. 24.

Smucker, J. 1980. *Industrialization in Canada.* Scarborough, ON: Prentice-Hall.

Sohn, J., J.K. Tschangho, and G.J.D. Hewings. 2005. "Information technology and urban spatial structure: A comparative analysis of the Chicago and Seoul regions," in H.W. Richardson and C-H. C. Bae, eds., *Globalization and Urban Development.* Berlin: Springer.

Sorensen, A. 2015. "Taking path dependence seriously: An historical institutionalist research agenda in planning history," *Planning Perspectives* 30: 17–38.

Springer, S., K. Birch, and J. MacLeavy, eds. 2016. *The Handbook of Neoliberalism.* New York: Routledge.

Spurr, P. 1976. *Land and Urban Development: A Preliminary Study.* Toronto: James Lorimer.

Talen, E. 2000. "New urbanism and the culture of criticism," *Urban Geography* 21: 318–41.

Trovato, F. 2009. *Canada's Population in a Global Context: An Introduction to Social Demography.* Toronto: Oxford University Press.

van de Coevering, P., and T. Schwanen. 2006. "Re-evaluating the impact of urban form on travel patterns in Europe and North America," *Transport Policy* 13: 229–39.

Visser, J., T. Nemeto, and M. Browne. 2014. "Home delivery and the impacts on urban freight transport: A review," *Procedia–Social and Behavioral Sciences* 125: 15–27.

Vojnovic, I. 2000. "The transitional impacts of municipal consolidations," *Journal of Urban Affairs* 22: 385–417.

Warner, B.J. 1962. *Streetcar Suburbs: The Process of Growth in Boston.* Cambridge, MA: Harvard University Press.

White, R. 2003. *Urban Infrastructure and Urban Growth in the Toronto Region: 1950s to the 1990s.* Toronto: Neptis Foundation.

———. 2016. Planning Toronto: The Planners, the Plans, their Legacies, 1940-1980. Vancouver, BC: UBC Press.

3 Governing Canadian Cities

Zack Taylor and Neil Bradford

Introduction

In a way that would have been unthinkable a generation ago, some of the most vibrant debates in Canadian politics and policy now revolve around how cities ought to be governed and what place municipalities should occupy in Canada's federal system. These debates have been driven by accelerating economic, social, and environmental change. Many of today's most pressing policy dilemmas are to be found in cities. As discussed elsewhere in this volume, **globalization**'s most important flows—of people, capital, and ideas—intersect primarily in cities, making urban centres focal points for major public policy challenges. Urban economies are now understood to be the drivers of national prosperity, where creativity and innovation flourish (see Vinodrai, Chapter 17; Shearmur, Chapter 18). At the same time, there is a widening opportunity gap between "haves" and "have-nots" in Canadian cities, with many residents struggling to find stable employment or affordable housing (see Walks, Chapter 10). Demographic changes—the uneven greying and "youthification" of cities, the concentration of immigrants in only a few metropolitan areas, and the urbanization of Canada's Indigenous peoples—pose unprecedented policy challenges (see Fawcett and Walker, Chapter 4; Moos, Chapter 6; Kobayashi and Preston, Chapter 8). And low-density suburbanization in rapidly growing cities puts pressure on agriculture and the natural environment, undermining food security and ecosystem resilience (see Filion, Chapter 2; Rees, Chapter 21).

As challenges of national consequence increasingly play out at the urban scale, how cities are governed matters more than ever. This chapter takes stock of urban **governance** in Canada—how it has changed and where it may be headed in the future. The policy challenges described above are complex and interconnected. They do not fit comfortably within the constitutional jurisdiction or legal authority of any one level of government, and so cannot be solved by any single level of government alone. For this reason, this chapter does not equate *urban* governance with *local* government, although the quality of democratic political debate and decision-making at the local level is nonetheless important to efficient and equitable urban governance. Urban governance is necessarily multi-level, in the sense that responses to our most difficult policy problems involve federal-provincial-municipal collaboration of some kind or another.

In this chapter, we chart growing experimentation by Canadian governments and civic leaders at all levels with new forms of what we call *collaborative multi-level urban governance*. We ask whether and how Canadian local governments are positioned for growing policy responsibility and governance innovation. We conclude that while Canadian municipalities remain constrained by a historical legacy of legal and fiscal dependence on senior governments, new forms of multi-level urban governance may yield a fruitful balance between local autonomy and national standards as governments seek to address pressing policy problems in cities. First, however, we begin with some context. What do Canadian local governments do, what powers do they have, and how might their autonomy—their independent capacity to make and implement policies—be constrained?

Canadian Local Government: A Bit of Context

Canada's federal government, ten provinces, and three territories co-exist with thousands of local governments—not only **general-purpose municipal governments** that provide a broad range of services to their residents, but also **special-purpose bodies** such as school boards, conservation authorities, health boards, and utilities. There are so many different types of local governments, and so much variation within and between provinces and territories, that it is difficult to arrive at a precise count. Canada lacks an up-to-date national inventory of local governments, their responsibilities, and their finances. (Unlike in the United States, where the federal government conducts a census of governments every five years, the Canadian government does not collect information about local governments.)

We should note that Indigenous governments, including the band councils that administer reserves under the Indian Act, and also territories administered through self-government agreements, are important and complex topics in their own right. While some may be similar to municipalities in so far as they have elected councils, collect taxes, and provide local services to residents, they are constitutionally and historically distinct in that they are creations of federal law or treaties between the Crown and Indigenous nations. This chapter focuses most directly on municipal governments that derive their authority from provincial law, and urban ones in particular. We discuss evolving federal-local policy frameworks for urban Indigenous communities later in this chapter; for a discussion of urban Indigenous communities' struggles for recognition and services, see Fawcett and Walker, Chapter 4.

Municipal Government and the Canadian Urban System

There are approximately 3700 municipalities in Canada, with considerable variation across provinces and territories in terms of their size and responsibilities. Table 3.1 shows the number of municipalities in each province. Saskatchewan's one million residents live in almost 800 municipalities, while Ontario's 13 million reside in only 444. Most municipalities have directly elected councils, although British Columbia, Ontario, New Brunswick, and Quebec have two-tier systems of local government in which upper-tier regional or county councils, boards, or commissions are composed of delegates from their constituent lower-tier municipalities. Upper-tier bodies (called regional districts in British Columbia, regional service commissions in New Brunswick, counties and regional municipalities in Ontario, and regional county municipalities in Quebec) are generally involved in regional planning and coordinating or operating major infrastructure systems. Lower-tier municipalities generally provide property-related services such as solid waste management, land-use planning, fire protection, and parks and recreation. There is no consistency across the country as to which level runs policing and transit systems. In single-tier systems, all municipal services are offered by a single level of local government.

The average population of municipalities is large in Canada compared to other countries, including the United States and Australia. This matters because larger municipalities have greater administrative and fiscal capacity to make more meaningful policy decisions and provide a broader range of services. Approximately two-thirds of Canadians live in municipalities with more than 50,000 residents (and over half in municipalities with more than 100,000), and one-third in municipalities with more than half a million people. Canadians are also concentrated in only a few metropolitan areas. Two-thirds of Canadians live in the 33 **census metropolitan areas (CMAs),** half of them in the five largest (see Table 3.2). Metropolitan-area

Table 3.1 Municipalities by Province and Territory, 2018

	Lower-tier	Upper-tier	Single-tier	Total	Provincial population (2016)	Average population of lower- and single-tier municipalities
Newfoundland & Labrador			275[a]	275	519,716	1,693
Prince Edward Island			72	72	142,907	1,173
Nova Scotia			49	49	923,598	14,900
New Brunswick	104[a]	12		116	747,101	4,936
Quebec	1,108	89[b]		1,197	8,164,361	7,227
Ontario	241	30	173	444	13,448,494	32,264
Manitoba			137	137	1,278,365	8,790
Saskatchewan			776	776	1,098,352	1,341
Alberta		75[c]	341	416	4,067,175	11,759
British Columbia	162	27		189	4,648,055	25,601
Yukon			8	8	35,874	3,629
Northwest Territories			33	33	41,786	1,245
Nunavut			25	25	35,944	1,435
Canada	1,615	233	1,889	3,737	35,151,728	9,608

Note: Most provinces have single-tier local government systems. British Columbia, New Brunswick, Ontario, and Quebec have two-tier systems. In some provinces, particularly New Brunswick and Newfoundland and Labrador, a high proportion of the population lives outside of incorporated municipalities. In these mostly rural places residents receive services from upper-tier general- or special-purpose governments, or directly from the provincial government. Totals do not include Indigenous governing bodies even if incorporated under provincial law.

[a] Does not include Local Service Districts.

[b] Includes 87 regional county municipalities (RCMs) and 2 metropolitan communities that overlap with RCMs.

[c] Includes 73 regional service commissions and 2 growth management boards

Source: Municipality counts are from provincial government websites. Populations are from the 2016 Census.

municipalities are more populous than rural ones. Fully 40 per cent of the national population lives in the **central cities** of metropolitan areas.

This national pattern of population concentration in metropolitan areas and metropolitan central cities, and relatively **consolidated local government** within them, suggests the potential for urban municipalities, and large central cities in particular, to play a leading role in multi-level urban governance.

What Do Municipalities Do?

Municipalities' most visible role is providing services. In all provinces, municipalities are primarily concerned with property-related services, including garbage pickup, policing, and fire protection. They also build, manage, and maintain physical infrastructure, including roads and bridges, transit systems, and water and sewer systems, as well as public facilities such as libraries, parks, and arenas. In fact, the total value of municipal public assets is greater than those owned by the federal and provincial governments (Sancton, 2015: 291–92). In some provinces, municipalities also deliver human services, including housing, immigrant settlement, child-care, and, in the case of Ontario, social assistance. While some of these activities may seem mundane, they are central to urban quality of life. How, and how equitably and efficiently, these services are delivered and infrastructure systems are maintained and expanded is an important determinant of cities' economic, environmental, and social health.

Less visible is the regulatory role played by municipalities. Unlike nation-states, municipalities

Table 3.2 The Metropolitan Concentration of the Canadian Population, 2016 Census

Rank	Number of single- and lower-tier municipalities	Population in incorporated municipalities	Region	% of national population	Cumulative % of national population	Average municipal population size
1	58	8,588,686	Greater Golden Horseshoe	24.4	24.4	148,081
2	91	4,098,927	Montreal	11.7	36.1	45,043
3	23	2,618,475	BC Lower Mainland	7.4	43.5	113,847
4	8	1,390,966	Calgary	4	47.5	173,871
5	19	1,323,783	Ottawa– Gatineau	3.8	51.3	69,673
6	31	1,317,015	Edmonton	3.7	55.0	42,484
7	28	798,162	Quebec	2.3	57.3	28,506
8	11	777,973	Winnipeg	2.2	59.5	70,725
9	8	494,069	London	1.4	60.9	61,759
10	1	403,131	Halifax	1.1	62.0	403,131
	129	3,011,155	Other 25 CMAs	8.6	70.6	23,342
		84,794	Unincorporated places within CMAs	0.2	70.9	
	3,097	9,226,103	Incorporated municipalities outside of CMAs	26.2	97.1	2,979
		1,018,489	Unincorporated places outside CMAs	2.9	100.0	

Most Canadians live in large municipalities in only a few metropolitan areas. The Greater Golden Horseshoe region combines the contiguous Toronto, Hamilton, Kitchener, St. Catharines–Niagara, Oshawa, Barrie, Guelph, Brantford, and Peterborough CMAs. British Columbia's Lower Mainland region includes the Vancouver and Abbotsford–Mission CMAs.
Source: Municipality and population counts are from the 2016 Census.

have virtually no control over flows of money and people. They cannot check passports or collect tariffs when people and goods cross city limits. They can, however, regulate the private use of land and buildings, which are immobile by their very nature. Municipalities can use their control over infrastructure systems and land-use regulation to strategically shape the urban **built environment** in ways that yield social, economic, and environmental benefits.

Finally, municipal governments are much closer to their communities, and therefore potentially more democratically accountable, than other levels of government. Unlike other levels of government, decisions are made in public, in open sessions of council. Ordinary people have a level of access to local politicians and public servants that is impossible at the provincial and federal levels. Local governments are also important leadership training grounds: federal and provincial political careers often begin with municipal office, and leaders of national and provincial advocacy groups often emerge from locally oriented organizations. Being "close to the ground" enables local leaders to identify local policy problems that might be invisible to provincial or federal decision-makers. Municipalities can also much more easily access the experiential knowledge of local residents, businesses, and interest groups as they make policy and advocate for policy changes at other levels. As discussed later in this chapter, municipalities' deep knowledge of local conditions and problems can be leveraged when tailoring provincial and federal programs to local needs.

How Much Autonomy Do Municipalities Have?

While municipalities do many things, they cannot do whatever they want. That is, **municipal autonomy** is limited (Goldsmith, 1995). As already suggested, their autonomy is constrained by their limited influence over mobile capital and labour; if a company or resident no longer wants to live a particular municipality, they can very easily relocate to another. Indeed, economists have long emphasized how "voting with your feet" is likely a more powerful influence on municipal policy than voting at the ballot box (Tiebout, 1956; Fischel, 2005). A municipality with poor-quality services, decaying infrastructure, and high tax rates cannot easily compete with one with high-quality services and infrastructure and modest tax rates.

A second type of constraint is that imposed by higher levels of government. While there is considerable variation in the number, size, and responsibilities of municipalities within and between provinces, all Canadian municipalities have something in common: under the Canadian constitution, local government falls under provincial jurisdiction. Although municipal councils are elected—and therefore are as democratically legitimate as provincial or federal legislatures—the scope of municipal autonomy has always been defined by provincial legal frameworks and fiscal arrangements (see Box 3.1). The degree to which these have constrained and enabled municipalities has changed over time. Since the Great Depression of the 1930s, provincial governments have responded to social and economic change and, after World War II, rapid urban growth, by actively regulating what local governments can and cannot do, "uploading" and "downloading" responsibilities and resources, mandating particular activities and functions, and reorganizing municipal boundaries. Municipalities and school boards effectively became administrative extensions of provincial governments during the postwar period, their actions reflecting provincial policy priorities. As discussed below, changes in the provincial-municipal relationship since the early 2000s have increased the scope of municipal autonomy in several provinces, although limits remain.

As a result of these constraints, municipalities have traditionally sat on the sidelines, either as the implementers of provincial policies or as silent partners in shifting federal priorities. This has shaped public perceptions of the purpose of local governments. Especially in large cities, municipal politics increasingly features debates between "expansionists" who embrace broader expectations of local democracy and policy participation by municipalities in confronting social, economic, and environmental challenges, and "traditionalists" who prefer local officials stick to their narrow historical role as providers of

Box 3.1 "Creatures of the Provinces"

Local governments have been traditionally been characterized as "creatures of the provinces" because the Canadian constitution assigns them to provincial jurisdiction:

- Provincial governments may unilaterally create, dissolve, amalgamate, and otherwise alter the boundaries of local governments, without consulting with local residents.
- Municipalities derive all of their authority from provincial law, and so can only perform functions authorized in provincial law.
- Provincial laws specify municipalities' access to revenue, including what taxes they can levy. For example, no Canadian municipality may levy an income or general sales tax, and so Canadian local governments are highly reliant on property taxes, user fees (such as transit fares and facility admission charges), and licensing fees.
- Provincial governments can, and do, dictate or influence local government activities by mandating that they perform specific functions and by making financial support conditional on following provincial rules.

property-related services, leaving big-picture policies to other levels of government (Sancton, 2015; Horak, 2012: 350–1). This conflict speaks not only to growing expectations about local government and its interactions with community organizations and citizens, but equally to views about what different levels of government should do and how they should interact with one another. With this context, we now discuss Canada's uneven engagement in national urban policy-making and the recent evolution of Canadian multi-level urban governance.

Toward Place-Based National Urban Policies

National and subnational governments—provinces and states—have traditionally developed uniform, "one-size-fits-all" policies that applied equally across their territories. Since the 1990s, however, a different approach has emerged in Europe and North America that recognizes that particular geographic spaces and community contexts generate distinct policy problems that are best addressed by aligning federal and provincial resources with municipalities' local knowledge and community networks (Dreier, Mollenkopf, and Swanstrom, 2014; Graham and Andrew, 2014).

This **place-based public policy** approach builds on research demonstrating that today's most significant economic, social, and environmental challenges are complex—interconnected in their causes and localized in their manifestations—and therefore not solved unilaterally by any single actor or agency (Klowdawsky, Siltanen, and Andrew, 2017; Wolfe and Gertler, 2016). Poverty is now seen through a wider lens of **social exclusion** shaped by an array of contextual factors, not only a lack of income support for individuals and households. Economic geographers have shown that national prosperity is driven by **urban innovation systems** that attract **"clusters"** of talent and investment for knowledge-intensive production. In Canada, "superstar cities" such as Toronto and Vancouver are leaders of the **cognitive-creative economy**, powered by well-educated millennial workers and rapid technological change (see Vinodrai, Chapter 17; Shearmur, Chapter 18). At the same time, mid-sized and smaller cities struggle to retain population and restructure their traditional economies. Integrated **regional planning** of infrastructure and land development have the potential to promote social equity by better distributing economic opportunity, facilitating mobility, preserving fragile ecosystems, and protecting scarce farmland (Jones, Lord, and Shields, 2015; Knaap, Nedovic-Budic, and Carbonell, 2015).

Think tanks and research networks close to governments have paid attention to these academic perspectives on how "cities matter." The Organisation for Economic Co-operation and Development's (OECD) local research program aims to "get cities right" based on recognition that they are the places where "policies and people meet" (OECD, 2014). The influential American-based Brookings Institution advocates a globally focused "new localism" as the optimal public policy route to sustainable and inclusive national economic growth (Katz and Nowack, 2017). The European Union features cities in its regional programming, putting the "urban dimension at the very heart of Cohesion policy" (European Commission, 2018). These transnational networks underscore how urban policy formulation is increasingly shaped through learning processes at the global scale. As cities around the world confront similar challenges of sustainability, inclusion, and livability, urban leaders look to one another for lessons and innovations. In many countries, including Canada, participation in global networks such as the C40 Cities Climate Leadership Group or the Commonwealth Local Government Forum adds another layer to collaborative multi-level urban governance.

Most recently the OECD and UN-Habitat coalesced these ideas and networks in a joint declaration calling on all nation-states to implement a "National Urban Policy," which they defined as "a coherent set of decisions derived through a deliberate government-led process of coordinating and rallying various actors for a common vision and goal that will promote more transformative, productive, inclusive and resilient urban development for the long term" (UN-Habitat-OECD, 2017; Van den Berg, Braun, and Van der Meer 2007). This place-based public policy perspective offers national governments a robust framework to coordinate urban interventions across policy sectors and between government scales (Scruggs, 2017).

Cities on Canada's Governing Agenda

Despite the growing popularity of these ideas, a national, place-based urban policy has proved elusive in Canada. Generations of Canadian urbanists have catalogued the obstacles to such an integrated approach to cities (Eidelman and Taylor, 2010; Lithwick 1970). Not surprisingly, the OECD concluded its 2002 territorial review of Canada with the observation that the federation's "disjointed approach" resulted in "a failure to draw up an integrated urban policy" (OECD, 2002: 159). Nevertheless, urban affairs in Canada is an active and evolving inter-governmental file. As understanding deepens of the strategic importance of cities to national (and individual) well-being, Canadian policy analysts have explored how municipal, provincial, and federal governments can relate to each other and engage with communities to resolve pressing public problems (Friendly, 2016; Canadian Global Citizens Council, 2018). Over the past 15 years, the federal government and several provincial governments have implemented various "New Deals" for cities, all involving devolution of legal responsibilities and revenue-raising capacity while also testing multi-level policy partnerships (Bradford, 2007). The Gas Tax Fund, introduced with great fanfare by a Liberal government in 2005 and quietly made permanent by the Conservatives in 2008, is a prime example of the principle of "flexible conditionality," whereby federal grants finance diverse municipal infrastructures within a common national sustainability framework (Adams and Maslove, 2014).

Buffeted by shifting partisan winds amid ongoing inter-governmental tensions, Canada's decade of "New Deal" experimentation after 2004 has not become entrenched in Canadian federalism. Nonetheless, these policy legacies continue to shape multi-level urban governance along two jurisdictional tracks:

- In the realm of provincial-municipal relations, a *local autonomy* track focuses on giving municipalities greater recognition, authority, and fiscal resources to address local problems and become more effective partners in multi-level governance.
- In parallel, a *federally orchestrated multi-level governance* track focuses on mobilizing all levels of government in relation to identified urban policy dilemmas: infrastructure, affordable housing, economic development, immigrant settlement, Indigenous peoples, and environmental sustainability.

The following sections examine these tracks, suggesting that they collectively represent the development of an implicit, rather than explicit, national urban policy. Unlike the highly institutionalized "City Deals" or "Big-City Policies" adopted in other OECD countries, including the United Kingdom and France, the Canadian approach relies more on institutional experimentation and informal collaborative leadership than legislated mandates, formal transfers of legal jurisdiction, or the creation of substantial new governing institutions (Graham, 2010; Van den Berg et al., 2007).

The Provincial-Municipal Local Autonomy Track

After the rancorous provincial–municipal conflicts over amalgamations and disentanglement in the 1990s, the new political leaders who gained office at both levels in the early 2000s sought to make amends. This was driven in part by political calculation—voters who had opposed unilateral provincial interventions in local affairs were attracted to opposition parties that promised respect for community identities and preferences. Liberal governments elected in British Columbia in 2001, and in Ontario and Quebec in 2003, for example, promised more constructive relationships with municipalities than they had had with their predecessors. Looking across the country, we can identify three shifts in the provincial-municipal relationship during the first decade of the 2000s: provincial recognition of municipalities as a democratic and accountable order of government, the empowerment of municipalities through legislative change, and the expansion of fiscal resources available to municipalities.

Political Recognition

Several provinces have officially recognized local governments as more than subordinates—as an "order of government." Quebec, for example, passed a law in 2017 whose preamble "recognizes that municipalities are . . . an integral part of the Québec State" yet "municipal officers have the necessary legitimacy, from a representative democracy perspective, to govern according to their powers and responsibilities" (Government of Quebec, 2017). The Ontario, Quebec, British Columbia, and Nova Scotia governments have signed agreements with municipal associations that recognize municipalities as democratically constituted governments that legitimately represent their citizens. These agreements establish a duty for the province to consult municipalities, or seek their approval, before making interventions that affect them. While municipalities remain "creatures of the provinces," embedding political recognition of municipal government in agreements and legislation signals a departure from the postwar command-and-control pattern. This new, collaborative tone may serve as the foundation for future provincial-municipal partnerships.

Legal Empowerment

As noted earlier, municipalities have no independent status in Canada's constitution. Their authority is delegated to them by the provinces. In recent years, however, provincial governments have amended general municipal legislation to enlarge the scope of municipal authority (Lidstone, 2004; Tindal et al., 2013: 206). In several provinces, this has entailed shift in legal doctrine from **"express powers"** to broadly defined **"spheres of jurisdiction."** Within the express powers framework, introduced by British authorities before Confederation, municipalities could only perform narrowly defined tasks that were specifically listed in provincial legislation. Over the years, the courts repeatedly found that municipalities would frequently have to ask the provincial legislature to pass special laws when confronted with unanticipated situations (Levi and Valverde, 2006). Under the "spheres of jurisdiction" framework, municipalities are authorized to act with broad discretion within more permissively construed areas of jurisdiction. Ontario's recently amended Municipal Act, for example, states that municipalities may pass bylaws respecting "the economic, social and environmental well-being of the municipality, including respecting climate change" (Government of Ontario, 2001: s. 10(2)(5)). Most municipal acts now also contain a "broad authority" clause stating that the courts shall interpret the scope of municipal powers generously rather than restrictively. Ontario, Alberta, and other provinces have also granted municipalities "natural person powers," which means that they are free to enter into certain kinds of contractual arrangements and to establish public corporations such as economic development agencies without provincial approval.

In recognition of their distinctive characteristics and problems, some large cities, including Toronto, Montreal, Quebec City, and Halifax, have recently joined Winnipeg, Vancouver, Saint John, Lloydminster, Corner Brook, and St. John's by acquiring their own **city charter**—special legislation, pertaining only to them, that sometimes grants them powers beyond those available to other municipalities (Kitchen, 2016). In 2018, Alberta established special regulations for Calgary and Edmonton that amount to much the same thing. The impact of these city charters may be more symbolic than anything else. Detaching big-city governance from general municipal legislation—in other words, introducing asymmetry into the general local government legal framework—is justified as necessary for cities to solve their own problems. It appears, however, that this merely stimulates demand from other municipalities for the same powers. Indeed, after the City of Toronto Act was adopted in 2006, the Ontario government amended the Municipal Act to bring many of its provisions into alignment with Toronto's. In British Columbia, the reverse dynamic has occurred but with much the same effect. After the general municipal law was modernized, Vancouver's charter has come to appear inflexible and outdated. In early 2019, the BC government introduced a bill to align some aspects of the Vancouver Charter with the province's Local Government Act. One thing is clear: stand-alone charters do not in themselves increase the autonomy of municipalities in the sense of providing protection from unilateral provincial interventions in their affairs. This was made very apparent by the government of Ontario's reorganization of Toronto's ward system in the middle of the fall 2018 municipal election over the opposition of city council.

Overall, it is difficult to assess whether more permissive general legislation and big-city charters have made much of a difference in practice. Nevertheless, even if it has not yet been fully exercised—a theme to which we will return later in the chapter—the general trend toward greater municipal policy autonomy is clear.

More Fiscal Resources

Finally, while there is variation among municipalities within and between provincial and territorial boundaries, there has been a generalized increase in the fiscal resources available to municipalities. This has occurred in two ways. Statistics Canada data suggest that, after adjusting for inflation and population, the municipal sector as a whole raises and spends more of its own money—**"own-source revenues"**—than in the past, and also receive more grants from provincial and federal grants. Figure 3.1 shows that from 1988 through to about 1997, municipalities spent more than they raised, with grants from other levels of government filling the gap. The increase in municipal spending and grants reflected the

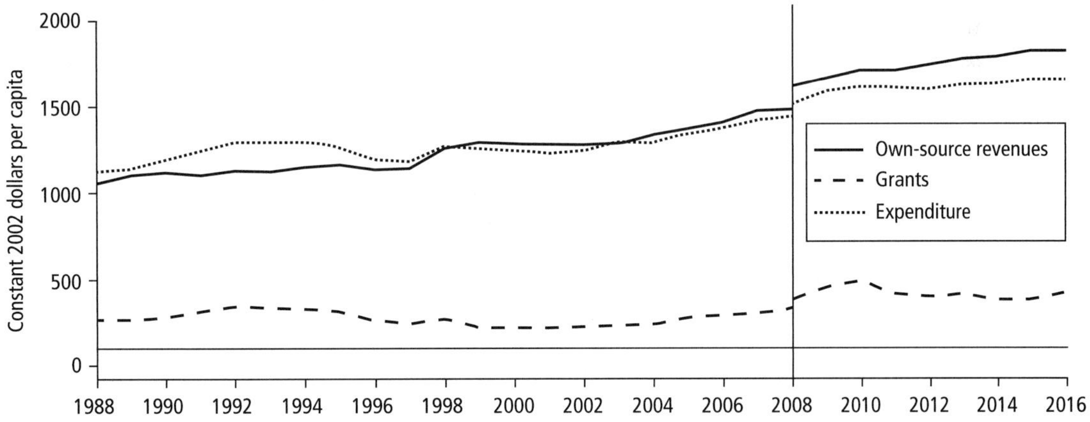

Figure 3.1 Municipal Own-Source Revenues, Grants from Other Levels of Government, and Expenditures, Inflation-Adjusted Per Capita Amounts, 1988–2016

Note: Values for the 1988–2008 and 2008–2016 periods are not directly comparable due to changes in public sector accounting rules. Still, the general trends hold.

Source: Calculated from Statistics Canada, Local General Government Own-Source Revenues and Expenditures (Current Account), cansim Table 3850024 (1988–2008); Canadian government finance statistics, statement of operations and balance sheet for municipalities and other local public administrations, annual, cansim Table 385-0037 (2008–2016); National Population Estimates, CANSIM Table 510001; and Consumer Price Index (all items, 2009 Basket), CANSIM Table 3260021.

difficult economic recession of the early 1990s, particularly in Ontario, where municipalities administer provincially funded social assistance. This pattern inverted between the mid-1990s and the mid-2000s. While spending levels remained about the same (roughly $1250 per capita in 2002 dollars), municipalities became less dependent on grants and more reliant on own-source revenues. Municipal own-source revenues have climbed in step with rising expenditures since the mid-2000s, likely reflecting the devolution of additional responsibilities to municipalities in several provinces. Importantly, grants from upper-level governments have also increased, peaking in the aftermath of the 2008 recession and remaining high compared to the trough of the mid-1990s.

The upshot: while there is considerable variation within and between provinces, municipalities are, on the whole, more reliant on their own revenues than in the past *and* benefit from greater support from the provincial and federal governments. More fiscal resources, coupled with political recognition and greater and more flexible legal authority, has increased the impact of local governments on people and their communities, and created new opportunities for creative municipal policy-making and involvement in **multi-level governance**.

The Federally Orchestrated Multi-Level Governance Track

The 2015 election put cities and urban issues back on the federal public policy agenda. In the months leading up to the campaign, Justin Trudeau cited the example of former Liberal prime minister Paul Martin's 2004–2006 **New Deal for Cities and Communities** when offering "a new spirit of cooperation" and acknowledgement of "local needs and priorities" (Trudeau, 2015). Once in office, Prime Minister Trudeau announced a $180 billion, 12-year investment plan for "inclusive growth." Expanding traditional conceptions of physical infrastructure to include economic, social, and digital dimensions, federal budgets from 2016 to 2018 have featured significant investments across a host of urban priorities including transportation and transit, housing and homelessness, economic development and immigrant settlement, Indigenous communities, and ecological resilience. At the same time, the Trudeau government fully engaged with the UN-Habitat negotiations on the international urban agenda, observing that the principles "closely align" with federal economic, social, and environmental plans and launching a public consultation on "key challenges, opportunities and trends in urbanization" (Government of Canada, 2016; Scruggs, 2016).

Moving from principles to practice, the Trudeau government is effectively using an "urban lens" to coordinate multi-pronged investments in cities (Sgro, 2002; Gurria, 2016). The urban lens is a policy tool that has been bandied about by federal governments dating back to the 1970s. Applied to planning and implementation across government departments with an interest or impact in cities, the urban lens enables horizontal and vertical policy coordination. Horizontal coordination takes place at the upper governmental levels as departments align their respective sectoral interventions to eliminate duplication or conflicting program criteria. Vertical coordination pays close attention to implementation details—securing municipal partners, assisting with local governance, and tailoring policies to community contexts.

Seeking to simultaneously address "local needs and national priorities" (Infrastructure Canada, 2018), the Trudeau government has used the urban lens to design or redesign place-based policies and programs that, in the words of the Federation of Canadian Municipalities (FCM), rely on "municipalities to deliver local solutions to national challenges—from economic growth to climate change to social inclusion" (FCM, 2018). We identify two types of multi-level urban governance, each of which is defined by distinct inter-governmental relationships and mechanisms for sharing authority and responsibility.

- *Federal-provincial/territorial agreement with municipal participation.* A form of tri-level engagement, these agreements tailor pan-Canadian sectoral policies to varied local circumstances. The federal-provincial/territorial negotiations incorporate municipal interests and priorities in program implementation replacing one-size-fits-all policies with differentiated interventions reflecting Canada's increasingly urbanized economic and social geography.
- *Federal-local partnerships with municipalities, communities, and Indigenous peoples.* A direct form of federal-local engagement, these partnerships combine the respective policy resources of the federal government with varying local actors. Federal financial support and technical assistance leverages local knowledge through place-based collaborations that mobilize multi-sectoral strategies for investments in local innovations.

Federal-Provincial/Territorial Agreements with Municipal Participation: Infrastructure and Immigration

A driving force in all of Canada's New Deal debates has been the country's **"infrastructure deficit"** estimated to be in the range of $120 billion. The Trudeau government's "Investing in Canada" plan proposes to invest over $180 billion over 12 years in public transit, social infrastructure, green infrastructure, trade and transportation infrastructure, and rural and northern communities (Infrastructure Canada, 2018). For implementation, the federal government is negotiating "Integrated Bilateral Agreements" (IBAs) with provinces and territories. The Federation of Canadian Municipalities and the Big City Mayors' Caucus welcomed "historic federal investments for local priorities" (FCM, 2017). Four aspects of approach are notable:

- increased federal financial contribution to cost-shared municipal projects;
- federal requirements that infrastructure investments balance provincial/territorial and municipal priorities;
- federally defined outcome targets for inclusiveness, sustainability, and accessibility; and
- federal investment in the capacity of municipalities to collect data and manage their assets.

The result is a promising national framework that aligns federal goals with municipal priorities. Moving forward, as the FCM has also observed, turning "historic investments into historic outcomes will require the full partnership of all orders of government."

In this context, other sector-specific federal initiatives illustrate the workings of federal multilevel urban governance. For example, the 2017 National Housing Strategy (NHS) is a 10-year, $40 billion plan to support the 1.7 million Canadians struggling with housing or homelessness, residing overwhelmingly in cities (Falvo, 2017). The federal government has set specific policy targets, such as the removal of 530,000 households from housing need and the reduction of chronic homelessness by 50 per cent over 10 years. The federal-provincial/territorial investment framework permits flexibility for municipalities in how they contribute to the national housing targets—for example, municipalities can provide public land, streamline approvals processes, offer financial incentives, or make regulatory changes such as inclusionary zoning. Similarly, the NHS includes a $2.2 billion "Reaching Home" program that increases local shelter funding and allows greater municipal planning flexibility in exchange for more accountability in meeting federal targets. This responds to concerns that earlier federal homelessness programs blocked locally responsive solutions because they offered insufficient money and imposed onerous reporting requirements (Klodawsky and Evans, 2014).

Another field that demonstrates how federal-provincial/territorial agreements include municipal representation is immigrant settlement (Andrew et al., 2013). Urban economic restructuring and the increasing diversity of newcomers arriving in Canadian cities over the past two decades have complicated the successful social and economic integration of immigrants. In response, federal officials and their provincial and territorial counterparts have invited municipalities, especially big-city governments, to the policy table. For example, the pioneering Canada-Ontario Immigration Agreement (COIA) signed in 2005 and renewed in 2018 leverages municipal and community-based planning for city-specific coordination of federal and provincial settlement services. Recognizing municipalities as policy partners through the Association of Municipalities of Ontario and the City of Toronto, the COIA empowered nearly 50 multi-sectoral Local Immigration Partnership Councils (LIPs) across the province, each tasked with identifying pressure points and gaps in local settlement systems. Moving well beyond federally funded language training, their goal is to align health, education, housing, and labour market services with the specific needs of newcomers as these are revealed in community contexts. This federally driven place-based approach has been extended to cities in five other provinces.

Federal-Local Partnerships: Economic Development, Indigenous Communities, and Smart Cities

Another model, evident in economic development and Indigenous affairs, is the federal-local partnership. The Department of Innovation, Science and Economic Development (ISED) administers two pan-Canadian policy frameworks that involve local economic development officials. The first is designed to consolidate leading-edge economic clusters in Canada's most globally connected cities. The Innovation Superclusters Initiative (ISI) makes available $950 million over five years to industry-led consortia of firms, academia, and community associations in five urban regions: Halifax (oceans technology), Quebec City-Montreal (data science/machine learning), Greater Toronto Area (advanced manufacturing), Saskatoon (plant proteins), and Vancouver (digital technology) (Wolfe, 2018).

At the same time, ISED also invests in smaller cities and rural communities through partnerships with municipal-community economic development networks (Bradford, 2017). Federal Regional Development Agencies (RDAs) serve the six major regions, including the north. They support nearly 300 Community Futures Organizations (CFOs) with volunteer boards that deliver business services and community economic development for local projects and regional infrastructure. Through the CFOs, the RDAs communicate evolving federal policy priorities such as support for Indigenous entrepreneurs, social enterprises, and youth employment opportunities. Individual CFOs have considerable latitude within federal parameters to develop strategies customized to local assets and needs.

Another example of direct, place-based federal-local engagement concerns the provision of services to Indigenous peoples in cities, a growing population that the federal government has struggled to reach since the late 1990s. Launched in 12 cities in 1998, the Urban Aboriginal Strategy (UAS) was intended to better integrate the activities of nearly a dozen federal departments and agencies, and to develop partnerships with Indigenous communities and municipalities. While enhancing services, the UAS was perceived as overly top-down and insensitive to legacies of colonization and exploitation (Bradford and Chouinard, 2011). The Trudeau government replaced the UAS with the Urban Programming for Indigenous Peoples (UPIP) in 2017, making available $53 million over five years to better understand the urban Indigenous experience, research innovations, and design services. Based on principles of "self-determination, reconciliation, respect, and cooperation," the UPIP dedicates funding to Inuit, Métis, and First Nations organizations, and requires non-Indigenous partners such as municipalities to demonstrate support from Indigenous groups in order to receive federal funds. The National Association of Friendship Centres in cities are recognized as hubs to coordinate programs and partners. Among the UPIP's four funding streams, two are notable from a collaborative urban multi-level governance perspective: one supports the formation of "local coalitions" among Indigenous representatives and all orders of government; the other builds the "organizational capacity" of the Friendship Centre to "maintain a stable base" for delivering services and managing partnerships (National Association of Friendship Centres, 2018).

A final example of federal-local engagement is the 10-year, $300 million Smart Cities Challenge (SCC), launched in 2017, which links public infrastructure, economic development, and Indigenous communities (see Brail and Donald, Chapter 5). Municipalities and Indigenous communities compete for funding for projects that mobilize data and technology to create smarter and more connected places. The SCC has two interesting design features: four tiers of competition based on comparable city populations, and the prioritization of applications that yield urban place-based social and environmental benefits, such as real-time information on homeless shelters or environmental sensors for monitoring areas at risk of flooding.

To sum up, while the mechanisms differ, the federal-provincial/territorial agreements and federal-local partnerships that have taken shape since the Trudeau government won the 2015 election share two common features. First, they signal the federal government's renewed interest in tailoring national social, economic, and environmental policies to local conditions, with a particular emphasis on cities. Second, they exemplify a departure from the traditional command-and-control approach in which Ottawa set stringent and inflexible standards. Instead, flexible policy frameworks are designed and implemented in collaboration with provincial, territorial, and municipal governments, as well as relevant stakeholders: Indigenous peoples, businesses and business associations, academic institutions, social enterprises, and groups in society that are affected by policy decisions.

Conclusion: Sunny, But with a Chance of Clouds

This chapter has taken stock of recent experimentation and innovation in Canadian urban multilevel governance. We have argued that, at least potentially, municipalities are emerging as effective policy-makers and policy partners with the federal and provincial governments. Over the past 20 years, most provincial governments have given municipalities new and more flexible authority as well as political recognition. Municipalities also have more financial resources—both own-source and transfers from other governments. Municipalities now sit at the table as inter-governmental social and economic policies are made and implemented. Nonetheless, it would be premature to characterize the last two decades' experiments with collaborative multi-level urban governance and place-based public policy-making, and municipal participation in them, as an unvarnished success or as finished business. We conclude this chapter by discussing some of the political and institutional constraints that inhibit creative and sustained urban governance.

The Local Politics of Municipal Empowerment: Too Much Democracy?

Asking for more authority is easy; facing the political consequences of exercising it is much harder. Toronto's case is illustrative. In the late 1990s and early 2000s, the city faced an ongoing budget crisis that was resolved every year with provincial bailouts. As part of Toronto's new 2006 charter, the city was given the authority to impose new taxes so it could reduce its dependence on commercial and residential property taxes and be democratically accountable for its actions. Two taxes were introduced: one levied on residential property sales (the land transfer tax), the other on motor vehicle registrations. While they soon brought in hundreds of millions of dollars a year, bridging the budget gap, they were unpopular. On taking office in 2010, Mayor Rob Ford, backed by city council, cancelled the vehicle tax and pledged to eliminate the land transfer tax, effectively turning the city's back on the decade-long revenue diversification strategy. The land transfer tax remains in effect, but only because Toronto's hot housing market generates so much revenue that there is no way the city could do without it. Whether this case exemplifies a democratic response to unpopular taxes or the irresponsible avoidance of political costs associated with pursuing a long-term fiscal agenda depends on one's political perspective. Ultimately, local government's relative responsiveness to the community is a double-edged sword. On the one hand, it permits a level of face-to-face interaction between politicians and residents that does not exist at the provincial or federal levels. On the other, the lack of insulation from local interests may inhibit municipal politicians' ability to make and sustain sometimes-unpopular policy positions in pursuit of long-term gains.

Institutional Constraints on Local Policy Innovation: Do We Need Strong Mayors?

The institutional structure of Canadian municipal government stacks the odds against creative policy-making and decisive action. Featuring "weak mayors" and, in most provinces, no political parties, Canadian municipal institutions widely disperse authority and influence. While "strong mayors" in some large American cities can veto council decisions, directly control the budgetary process, hire and fire senior city staff, and direct the bureaucracy without council approval, Canadian mayors have but one vote on council, have few resources of their own to develop policy proposals, and must rely on persuasion to implement their vision and platform. Since coalitions must be built issue by issue, one councillor at a time, politicians must work together to come up with responsive and mutually acceptable solutions. Sancton (2015) argues that the **weak-mayor system** inhibits policy innovation while rewarding "brokers" who avoid conflict and big ideas. Whether the consensual and incremental, rather than unilateral and innovative, nature of decision-making on Canadian councils is a virtue is a matter of opinion. Several European countries have sought to emulate the American strong-mayor model in the belief that stronger executive accountability, unilateral authority, and resources will make cities more innovative and responsive (Magre and Bertana, 2007). While strong-mayor reform proposals come up from time to time—indeed, Doug Ford endorsed it before becoming Ontario premier, and Toronto mayor John Tory has recently called for it—few, if any, Canadian cities are likely to be run by strong mayors anytime soon.

The Provincial Politics of Local Empowerment: How Much Autonomy Will Provinces Allow?

Even as provincial governments have expanded municipal authority and resources and afforded them greater political recognition, they remain reluctant to cede independent decision-making authority when their electoral self-interest is at stake.

A first example is the 2015 transit funding referendum in Greater Vancouver. In the 1990s, the BC government empowered regional districts (federations of local governments) to make binding long-term land-use plans and created TransLink, a regional transit agency for Greater Vancouver with dedicated revenues, to plan and operate an integrated regional transportation system, including bus and rail transit, major roads, and bridges. While on the face of it these actions were consistent with the trend toward expanded municipal self-determination, the provincial government has repeatedly overturned local long-term planning efforts. Recognizing that TransLink's existing revenues were insufficient to build new lines and improve service, the region's mayors agreed to a 10-year, $7.7-billion transportation plan funded by a 0.5 per cent sales tax increase. After the provincial Liberal government required a referendum on any new transportation taxes in Greater Vancouver—an unprecedented move in the contemporary Canadian context—area voters defeated the proposal in 2015. Perhaps reflecting their suburban electoral base, the same government's replacements of the Port Mann Bridge, which opened in 2012, and the Massey Tunnel, which broke ground in 2017, required no referendum, despite costing about $6 billion—almost as much as the mayors' transit plan. Both were also opposed by the Metro Vancouver regional district and TransLink, which argued that they undermined the policy of concentrating growth north of the Fraser River. The NDP provincial government elected in 2017 has been a more sympathetic partner. In March 2018, the federal, provincial, regional district, and municipal governments agreed on a funding plan that includes increases to local taxes and fees, but not a sales tax increase.

A second example is Toronto's battle with the provincial government over the tolling of municipally owned highways to fund infrastructure repairs. After city council voted to request that the province allow the city to impose the tolls, and the mayor received quiet assurances that the request would be approved, the premier publicly denied the request without first informing the mayor. Mayor John Tory condemned the province for its paternalism: "It is time that we stop being treated . . . as a little boy going up to Queen's Park in short pants" (Reitl, 2017). Instead of enhancing municipal own-source revenues—and letting municipal politicians bear the political cost of imposing the fee on residents—the provincial government instead promised to double the municipal share of the provincial gasoline tax by 2022. The Liberal government of the day was clearly responding to the many voters in suburban "905" ridings who use those highways to commute to jobs in Toronto. At the time of this writing, it is not clear whether the Progressive Conservative government elected in June 2018, which campaigned on reducing provincial gas taxes, will honour the previous government's promise.

A third example, also from Ontario, is the new provincial government's interest in restructuring and reorganizing local governments, whether they want it or not. One element mentioned earlier was the province's unilateral reduction of the number of Toronto's wards from 47 to 25 in the midst of the city's elections. Potentially more disruptive is the province's review, announced in January 2019, of two-tier regional government systems in the Greater Toronto Area, Niagara, Muskoka, and Oxford and Simcoe Counties. Whether this will result in single-tier consolidations, the dissolution of upper-tier governments, or the merger of lower-tier governments, is not yet known. Whatever ultimately happens, the impact on local democratic accountability, administrative efficiency, and ultimately the exercise of local autonomy, will be profound. It is no surprise that local politicians across the province are nervously staking out positions on the issue.

The Federal Politics of Multi-Level Management: What Happens When the Going Gets Rough?

The new Liberal government that took office in 2015 has taken a different direction than the Conservative government that had been in power since 2006. From infrastructure and housing agreements to economic development and immigrant settlement partnerships, the Trudeau government has established new forms of collaborative multilevel urban governance that leverage the growing capacity and authority of municipalities. These are, however, early days in what is planned to be a decade-long policy rollout. It remains to be seen whether and how these collaborative multi-level partnerships and agreements will be able to adapt to future political shocks and stresses.

The episodic history of federal interest in cities and urban problems is a reminder of the uncertainties ahead. The abrupt end of the previous Liberal government's New Deal following the Conservatives' return to power in 2006 underscores how changing political leadership can have sharp policy consequences. As the initiator and major funder of many of these agendas, the federal government is the essential actor. If federal political or financial support is withdrawn due to electoral change or fiscal restraint, they are unlikely to survive in their current form. A further source of risk is the vagueness of several aspects of the new federal agenda—for example, the meaning of the human-rights-based approach to housing and how partners will be held accountable to federally defined targets. While vagueness may be the "grease" that permits the successful negotiation of complex agreements among numerous stakeholders across multiple files, it may also render the achievement of concrete long-term results more elusive.

Much also depends on the federal government's capacity to deftly manage the diverse interests of its partners. Even with strong commitment to collaboration and flexibility, federally orchestrated multi-level urban governance will inevitably run up against deep-seated historical political positions: Indigenous communities' desire for self-determination, Quebec's desire for greater autonomy, rural areas' distrust of urban-focused policies, and provincial governments' resistance to federal intervention in municipal affairs. The support of mayors and municipal associations for the urban policy lens and local representation at inter-governmental tables will likely be an important source of political support for these arrangements.

Canada's Implicit National Urban Policy

Canada does not have an explicit national urban policy and it is unlikely to develop one. Instead, the governance of Canadian cities is characterized by creative experiments that crosscut the traditional jurisdictions of federal, provincial, and municipal governments. These experiments have occurred in a political atmosphere that is favourable to the empowerment of local governments and open to tailoring of policies to the needs of distinct places.

While these debates and policy experiments have ebbed and flowed over the past 20 years, mediated by political and institutional constraints, it would be inaccurate to conclude that they have not altered Canada's urban governance and policy landscape in interesting ways. While modest in scope and scale, these governance innovations constitute Canada's *implicit* urban policy. They allow municipalities with strong leadership and strategic purpose to assume a more prominent role in Canadian federalism and local governance of public policy. It may be that Canada will successfully chart a middle path to the "new localism" between an overemphasis on either municipal autonomy or non-governmental actors to make and deliver public policies. In the United States, for example, autonomous local government has led to undesirable inequities both within metropolitan areas and between more dynamic and struggling cities. At the same time, those European jurisdictions where multi-level governance and stakeholder-led policy-making are most advanced reportedly suffer from muddled accountability and a lack of fiscal transparency. Canada's recent experience embedding **place-based public policy-making** within traditional inter-governmental relations may ease these tensions while also reconciling this country's enduring political challenge of accommodating national standards to community particularity.

Review Questions

1. What are the advantages and disadvantages of place-based policy-making through multi-level governance as opposed to traditional federal policy-making?
2. Why have cities only recently come onto Canada's governing agenda?
3. How has the jurisdictional and fiscal autonomy of municipalities changed over time?
4. What are the political and institutional barriers to effective multi-level urban governance?
5. In the Canadian context, what are the advantages and disadvantages of switching to a strong-mayor system of local government?

References

Adams, E., and A. Maslove. 2014. The federal gas tax cession: From advocacy efforts to thirteen signed agreements," in K.A.H. Graham and C. Andrew, eds., *Canada in Cities: The Politics and Policy of Federal-Local Governance*. Montreal and Kingston: McGill-Queen's University Press.

Andrew, C., J. Biles, M. Burstein, V. Esses, and E. Tolley. 2013. *Immigration, Integration and Inclusion in Ontario*

Cities. Montreal and Kingston: McGill-Queens University Press.

Bradford, N. 2007. *Whither the Federal Urban Agenda? A New Deal in Transition*. Ottawa: Canadian Policy Research Networks.

Bradford, N. 2017. *Canadian Regional Development Policy: Flexible governance and adaptive implementation*. OECD-European Commission. Paris: OECD. At: www.oecd.org/cfe/regional-policy/Bradford_Canadian-Regional-Development-Policy.pdf

———, and J.A. Chouinard. 2010. "Learning through evaluation? Reflections on two federal community-building initiatives," *Canadian Journal of Program Evaluation* 24: 51–77.

Canadian Global Citizens Council. 2018. "Planning for an Urban Future: Our Call for a National Urban Strategy for Canada." At: https://globalcitiescouncil.ca/wp-content/uploads/2018/02/CGCC-National-Urban-Strategy-Report.pdf

Dreier, P., J. Mollenkopf, and T. Swanstrom. 2014. *Place Matters: Metropolitics for the Twenty-first Century*, 3rd edn. Kansas City, KS: University Press of Kansas.

Eidelman, G., and Z. Taylor. 2010. "Canadian urban politics: Another 'black hole'?" *Journal of Urban Affairs* 32: 305–20.

European Commission. 2018. "EU Urban and Regional Development." At: http://ec.europa.eu/regional_policy/en/policy/themes/urban-development/

Falvo, N. 2017. "Ten things to know about Canada's newly-unveiled National Housing Strategy." At: http://ec.europa.eu/regional_policy/en/policy/themes/urban-development/

Federation of Canadian Municipalities (FCM). 2017. "FCM President and Big City Mayors' Caucus Chair respond to launch of infrastructure negotiations." Retrieved from https://fcm.ca/home/media/news-and-commentary/2017/fcm-president-and-big-city-mayors%E2%80%99-caucus-chair-respond-to-launch-of-infrastructure-negotiations.htm

———. 2018. "From opportunities to outcomes: How federal budget 2018 can empower municipalities to deliver for Canadians." At: https://fcm.ca/documents/issues/2018-PreBudget-Submission-EN.pdf

Fischel, W.A. 2005. *The Homevoter Hypothesis: How Home Values Influence Local Government Taxation, School Finance, and Land-Use Policies*. Cambridge, MA: Harvard University Press.

Friendly, A. 2016. *National Urban Policy: A Roadmap for Canadian Cities*. IMFG Perspectives No. 14/2016.

Goldsmith, M. 1995. "Autonomy and city limits," in D. Judge, G. Stoker, and H. Wolman, eds., *Theories of Urban Politics*, 1st edn., London: Sage.

Government of Canada. 2016. "Government of Canada engages Canadians leading up to Global Summit on Housing and Sustainable Urban Development." At: https://www.newswire.ca/news-releases/government-of-canada-engages-canadians-leading-up-to-global-summit-on-housing-and-sustainable-urban-development-583991021.html

Government of Ontario. Municipal Act. S.O. 2001, c. 25.

Government of Quebec. An Act mainly to recognize that municipalities are local governments and to increase their autonomy and powers (Bill 122). S.Q. 2017, c. 13.

Graham, K. 2010. "No joke! Local government and intergovernmental relations in Canada," in E. Brunet-Jailly and J.F. Martin, eds., *Local Government in a Global World: Australia and Canada in Comparative Perspective*. Toronto: University of Toronto Press.

Graham, K.H., and C. Andrew, eds., 2014. *Canada in Cities: The Politics and Policy of Federal-Local Governance*. Montreal and Kingston: McGill-Queen's University Press.

Gurria, Á. 2016. "Implementing the New Urban Agenda through National Urban Policy: Ministerial Perspectives." Paris: OECD. At: www.oecd.org/regional/implementing-the-new-urban-agenda-through-national-urban-policy-ministerial-perspectives.htm

Horak, M. 2012. "Conclusion: Understanding multilevel governance in Canada's cities," in M. Horak and R. Young, eds., *Sites of Governance: Multilevel Governance and Policy Making in Canada's Big Cities*. Montreal and Kingston: McGill-Queen's University Press.

Infrastructure Canada. 2018. *Investing in Canada Plan: Canada's Long-term Infrastructure Plan*. At: www.infrastructure.gc.ca/plan/about-invest-apropos-eng.html?pedisable=true

Jones, K.E., A. Lord, and R. Shields, eds. 2015. *City-Regions in Prospect? Exploring Points between Place and Practice*. Montreal and Kingston: McGill-Queen's University Press.

Katz, B., and J. Nowack. 2017. *The New Localism: How Cities Can Thrive in the Age of Populism*. Washington, DC: Brookings Institution Press.

Kitchen, H. 2016. *Is "Charter-City Status" A Solution for Financing City Services in Canada—Or Is That a Myth?* SPP Research Papers 9:2. Calgary: School of Public Policy, University of Calgary.

Klodawsky, F., and L. Evans. 2014. "Homelessness on the federal agenda: Progressive architecture but no solution in sight" in C. Andrew and K.A.H. Graham, eds., *Canada in Cities: The Politics and Policy of Federal-Local Governance*. Montreal and Kingston: McGill-Queen's University Press.

———, J. Siltanen, and C. Andrew, eds. 2017. *Toward Equity and Inclusion in Canadian Cities: Lessons from Critical Praxis-Oriented Research*. Montreal and Kingston: McGill-Queen's University Press.

Knaap, G.-J., Z. Nedovic-Budic, and A. Carbonell, eds. 2015. *Planning for States and Nation-States in the U.S. and Europe*. Cambridge, MA: Lincoln Institute of Land Policy.

Levi, R., and M. Valverde. 2006. "Freedom of the city: Canadian cities and the quest for governmental status," *Osgoode Hall Law Journal* 44(3): 409–59.

Lidstone, D. 2004. *Assessment of the Municipal Acts of the Provinces and Territories*. Ottawa: Federation of Canadian Municipalities.

Lithwick, N.H. 1970. *Urban Canada: Problems and Prospects*. Ottawa: Central Mortgage and Housing Corporation.

Magre, J. and X. Bertana. 2007. "Exploring the limits of institutional change: The direct election of mayors in Western Europe," *Local Government Studies* 33(2): 181–94.

National Association of Friendship Centres. 2018. "Programs and Initiatives." At: http://nafc.ca/en/programs-and-services/

Organisation for Economic Co-operation and Development (OECD). 2002. *Territorial Review: Canada*. Paris: OECD.

———. 2014. OECD *Regional Outlook 2014: Where Policies and People Meet*. Paris: OECD.

Reitl, J. 2017. "Mayor decries 'short-sighted' road-toll rejection by province." CBC News. At: www.cbc.ca/news/canada/toronto/john-tory-road-tolls-1.3954882

Sancton, A. 2015. *Canadian Local Government: An Urban Perspective*, 2nd edn. Toronto: Oxford University Press.

Scruggs, G. 2016. "Canada has emerged as one of Habitat III's strongest advocates for vulnerable groups." *Citiscope*. At: http://archive.citiscope.org/habitatIII/news/2016/09/canada-has-emerged-one-habitat-ii-is-strongest-advocates-vulnerable-groups

———. 2017. "Since Habitat III, an uptick in interest around national urban policies." *Citiscope*. At: http://archive.citiscope.org/story/2017/habitat-iii-uptick-interest-around-national-urban-policies

Sgro, J. 2002. *Canada's Urban Strategy: A Blueprint for Action*. Final Report Prime Minister's Caucus Task Force on Urban Issues. Ottawa.

Tiebout, C.M. 1956. "A pure theory of local expenditures," *Journal of Political Economy* 64: 416–24.

Tindal, C.R., and S.N. Tindal, K. Stewart, and P.J. Smith. 2013. *Local Government in Canada*, 8th edn. Toronto: Nelson.

Trudeau, J. 2015. "Justin Trudeau's Speech to the Federation of Canadian Municipalities Conference." At: https://www.liberal.ca/justin-trudeaus-speech-to-the-federation-of-canadian-municipalities-conference

UN-Habitat– OECD. 2017. The State of National Urban Policy in OECD Countries: A special report prepared for the United States Conference on Housing and Sustainable Urban Development (Habitat III). At: www.oecd.org/cfe/regional-policy/the-state-of-national-urban-policy-in-OECD-countries.pdf

Van den Berg, L., E. Braun, and J. Van der Meer. 2007. *National Responses to Urban Challenges in Europe*. Hampshire: Ashgate Press.

Wolfe, D. 2018. *Creating Digital Opportunity for Canada*. Toronto: Brookfield Institute for Innovation and Entrepreneurship and Innovation Policy Lab, Munk School of Global Affairs.

Wolfe, D., and M. Gertler, eds. 2016. Growing Urban Economies: Innovation, Creativity, and Governance in Canadian City-Regions. Toronto: University of Toronto Press.

4 Indigenous Peoples, Indigenous Cities

R. Ben Fawcett and Ryan Walker

Introduction

Canadian cities are Indigenous places, and Indigenous peoples are progressively shaping urban areas to reflect this truth. These efforts have been complicated by the Canadian state and non-Indigenous **settler** society's reluctance to yield its own privileging power. Indigenous people have struggled to build urban community infrastructures through which to live good lives emanating from their personal senses of **indigeneity** and place. This chapter offers a critical overview of Canadian cities as geographical sites and scales where Indigenous-settler relations are magnified, and wherein Canada's colonial entanglements may be reconfigured to forge a more just and equitable co-existence (Porter and Barry, 2016). As practitioners of spatial production and instruments of urban change, readers are encouraged to situate themselves as bearers of responsibility and creative potential to "turn the corner" toward this pursuit. As authors, we also position ourselves as non-Indigenous white men residing in Treaty Six territory and the homeland of the Métis nation, in Saskatoon, Saskatchewan. Our observations are inseparable from our lived experiences as beneficiaries of settler colonial power in and of Canada. Our perspectives are also informed by our research and relationships with Indigenous community members in our shared endeavour to unsettle the privileging power of whiteness in geography and planning across the prairies.

To begin to comprehend the enormous complexity of Indigenous peoples' experiences in Canadian cities requires an understanding of settler colonialism, Indigenous urbanism, rights and resistance, cultural identity and innovation, and forces of social change. According to Canada's most recent census, nearly 52 per cent (867,415) of Aboriginal people who live within Canada's borders now reside in metropolitan areas with populations over 30,000 (Statistics Canada, 2017a; see Table 4.1). The number of Aboriginal people residing in large cities has grown by roughly 60 per cent between 2006 and 2016 (Statistics Canada, 2017a). While this figure indicates that indigeneity increasingly encapsulates urban attachment and experience, it is only a snapshot in time, which says nothing of the dynamic configurations of social networks and mobility, political organization and action, socio-economic circumstances, nor daily acts of cultural resurgence among Indigenous peoples in cities. Further, as Glenn (2017) has argued, it is important to think of Indigenous peoples in the context of how many centuries they have been in their home territory, and not exclusively as a "population size," growing, shrinking, or whatever the case may be in a given city.

According to section 35.1 of the Constitution Act, 1982, the Canadian state recognizes "Aboriginal" peoples as the "Indian, Inuit, and Métis." Statistics Canada uses the term "Indian" in reference to both "North American Indians"—descendants of the continent's original inhabitants—and those

Table 4.1 Aboriginal Identity Populations in Selected Census Metropolitan Areas (CMAs), 2016

	Total Aboriginal	% of CMA	% First Nations*	% Métis*	% Inuit*
Victoria	17,240	4.8	57.7	37.9	0.8
Vancouver	61,455	2.5	58.2	38.1	0.7
Edmonton	76,205	5.9	44.5	51.7	1.5
Calgary	41,645	3.0	43.1	53.4	1.1
Saskatoon	31,350	10.9	50.3	47.5	0.3
Regina	21,650	9.3	60.7	36.8	0.4
Winnipeg	92,810	12.2	41.7	56.2	0.3
Thunder Bay	15,070	13.0	75.2	22.8	0.2
Toronto	46,315	0.8	60.0	32.9	1.5
Ottawa-Gatineau	38,115	2.9	46.7	45.0	3.4
Montreal	34,745	0.9	46.4	44.5	2.8
Quebec	11,515	1.5	54.1	40.1	1.3
Halifax	15,810	4.0	48.6	43.7	2.6

* Totals do not add up to 100 per cent because of rounding, because some individuals identified with more than one category, and because some individuals identified themselves as Aboriginal but did not identify as First Nations, Métis, or Inuit.
Source: Statistics Canada, 2017b.

who are "registered" or "Treaty Indians" under the federal Indian Act. While the term "First Nations" has commonly replaced "Indians" in modern commentary, many individuals identify with specific confederated nations such as Algonquin, Mi'kmaq, Cree, and Salish, as well as many sub-cultural tribal and linguistic groups. "Inuit" also contains nine distinct Indigenous groups from northern and arctic regions, while "Métis" has often been used to describe people of mixed European and Indigenous ancestry but principally represents the cultural group descending from the Red River settlement in what is now Manitoba, a province created by the Métis. Some people also claim general urban Indigenous identities; others identify with multiple affiliations. In this chapter we refer to "Indigenous" peoples and communities collectively as the descendants of this continent's original inhabitants who formed complex societies prior to Canadian Confederation. We use "Aboriginal" where it applies to the Canadian state's constitutional and policy relationship with Indigenous peoples, and we use "Indian" where we refer to "registered Indian" status under the federal Indian Act.

Every Canadian city contains a unique assemblage of Indigenous identity groups, as well as place-specific agglomerations of social, economic, political, and cultural geographies and institutions that have historically impacted upon and been shaped by each urban region's Indigenous community presence. Canadian cities were established in areas that have long been important gathering places or settlements for Indigenous peoples. They are often located on historically significant trade routes. Many cities even bear names derived from local Indigenous words such as *tkaronto* (Toronto), *odawa* (Ottawa), *winipek* (Winnipeg), and *misâskwatômina* (Saskatoon). City life is now a core feature of indigeneity, but as the next section explains, urban development has also played a central role in the consolidation of settler colonialism across Canada. The chapter then examines the concept of Indigenous urbanism, relating some of its dynamics through a discussion of case examples from Saskatoon and Calgary. This is followed by an examination of Indigenous rights and the resurgence of indigeneity in urban areas, prior to the chapter's conclusion.

Settler Colonialism and Urbanization in Canada

"Settler colonialism" is a term that is used to describe the structures and processes through which a colonizing society claims and reproduces its governing authority over a pre-existing society's customary territory. According to Bonds and Inwood (2016: 716), settler colonialism historically signifies the "*permanent* occupation of a territory and removal of Indigenous peoples with the express purpose of building an ethnically distinct national community." Although the ethnocultural fabric of Canadian society has shifted since Confederation in 1867 to include a diverse array of peoples emigrating from all over the world, the institutions, practices, and processes of settler colonialism's persistent structure are, and always have been, dominated by western, European (normatively white) systems of knowledge, authority, and assumptions of cultural superiority (Denis, 1997; Wolfe, 2006).

Prior to Canadian Confederation, Britain's legal justification for its colonization of North America historically relied on the misguided and racist concept of *terra nullius* in international law, which maintained that Indigenous peoples are uncivilized, unproductive, and therefore their lands could be deemed lawfully empty and destined for European settlement (Henderson, 2000). The Supreme Court of Canada and some legal scholars have argued that *terra nullius* did not apply in Canada because Britain's *Royal Proclamation, 1763* recognized Indigenous title and bestowed upon the Crown the sole legal right and obligation to negotiate treaties for the cession or purchase of Indigenous land (see Banner, 2005). Britain's formal acknowledgement of Indigenous peoples' prior occupancy has influenced Canadian court decisions recognizing Indigenous title such as *Tsilhqot'in Nation v. British Columbia*. Yet Borrows (2015: 702) asserts that "Canadian law still has *terra nullius* written all over it" as the state continues to claim absolute sovereignty and radical underlying title over all land within its territorial boundaries, even in regions with no prior treaties such as mainland British Columbia. Crown title and Canadian sovereignty are enacted through practices of delegated jurisdiction on the ground, which have overwhelmingly prioritized the wealth creation potential of land through private "fee simple" ownership over collective Indigenous interests or distinctive forms of land use and occupancy. In other words, Indigenous land continues to be considered "alienable" in Canadian law and policy. This "frontier" ethos continues to hold powerful weight in Canadian cities where land and space are mobilized as productive forces in highly competitive and exclusionary property markets (Blomley, 2004; Granzow and Dean, 2007; Tomiak, 2017).

The governing authority of Canada is predicated on the physical colonization of Indigenous land, which has relied on treaty making. Across most of western Canada, the numbered treaties promised First Nations reserve land and farming implements to transition to agricultural economies, annuity payments to band members, the construction of schools, and emergency medical and food provisions in exchange for peaceful co-existence. Tribal leaders believed they would retain their relationships and responsibilities over the stewardship of land, water, and human and non-human life through traditional governance practices (Henderson, 2008). Treaties did not relinquish Indigenous nations' nor individuals' self-determining autonomy, which is defined in the United Nations Declaration on the Rights of Indigenous Peoples (UNDRIP) as the ability to "freely determine their political status and freely pursue their economic, social and cultural development" (United Nations General Assembly, 2007: 8). But as Schmidt (2018) describes, the Canadian state deceptively negotiated treaties to peaceably consolidate its own jurisdictional authority over Indigenous territory. In other words, Canada's written versions of the numbered treaties assert that First Nations agreed to cede, release, and surrender their rights and title to their land, which contradicts Indigenous oral versions emphasizing the sharing of land and respect for mutual sovereignty (Henderson, 2008).

In 1876 Canada passed the Indian Act, a legislative mechanism to not only administer the Crown's fiduciary relationship with First Nations, but to also fulfill the state's social and economic development ambitions by asserting control over Indigenous lives and land. Through the Indian Act, reserves became policed enclosures, which allowed for colonization and urban growth across an expanding Canada (Harris, 2002). Canadian cities developed as distribution, industrial, and commercial centres of regional economies (see Filion, Chapter 2) through which western values became embedded in property relations, civic governance, labour relations, and the built form. Urbanization accompanied the Canadian state's dispossession of Indigenous land and displacement of Indigenous communities from their traditional territories (Edmonds, 2010). Blomley (2003) describes how surveying, mapping, and the grid system of land allotment implanted a legal order of property on top of already existing geographies of Indigenous territorial governance. The Métis, for example, were denied land title over their settlements in Manitoba and the North-West (Saskatchewan and Alberta), and were instead offered a scrip system of parcel allotment that extinguished their Aboriginal rights and displaced families, many of whom relocated to the margins of then-burgeoning prairie cities.

For a century and a half, Canada's consolidation of private and Crown property has relied on the Indian Act and what Wolfe (2006) calls "elimination" or "replacement" policies such as isolation on reserves; dissolution of traditional governance structures; state regulation and gendered erasure of women's Indian status and band membership if, for example, they married a non-status man; criminalization of ceremony; restriction of traditional economies and denial of agricultural technologies; violent indoctrination of Indigenous children into Western knowledge and value systems through Canada's genocidal residential schools; and the systematic removal of Indigenous children from their families during the "Sixties Scoop" (Warnock, 2004). These policies have wrought immense intergenerational harm onto Indigenous peoples in ways that are scarcely understood by non-Indigenous Canadian society.

It is against this settler colonial backdrop that Indigenous urbanization must be located. Prior to 1951, the Indigenous presence in Canadian cities was small, attributable in part to urban policies aimed at removing Indigenous people and their "memory (i.e., recall of experience, even existence)" and "materiality (i.e., physical quality, presence, and structure)" from cities (Matunga, 2013: 8; Wilson and Peters, 2005). In his account of Vancouver's city planning strategies from 1928 to the 1950s, Stanger-Ross (2008) reveals a legacy of "municipal colonialism" through which City Hall expropriated reserve lands belonging to the Squamish and Musqueam First Nations. Edmonds (2010) similarly explains how a 1911 amendment to the Indian Act allowed the provinces and their municipalities to seize reserves in Canadian cities for public use and benefit, which empowered British Columbia to purchase Lekwammen and Esquimalt reserve land around Victoria's harbour without consent. Band members who remained in the city were deemed squatters on their home territories.

It has been estimated that in 1951 only 6.7 per cent of Canada's Indigenous population resided in urban areas (Norris et al., 2013), though the authors note that official data on Indigenous urbanization is skewed prior to 1996 due to Statistics Canada's failure to account for Indigenous identities, omitting the Métis entirely from the 1951, 1961, and 1971 censuses (Norris et al., 2013). Since the 1980s, the Indigenous population across Canada has increased substantially (see Table 4.2), while the proportion of Indigenous people residing in Canadian cities has also grown acutely due to rural-to-urban movement, natural fertility rates, increased self-identification of Aboriginal identities, and improved statistical accounting methods (Peters, 2015; see Townshend and Walker, Chapter 7).

Rapid Indigenous urbanization since the 1980s has many causal factors. While urban policies in the first half of the twentieth century aimed to keep Indigenous people outside of settler cities,

Table 4.2 Population Growth of Aboriginal Identity Groups in Canada, 1996-2016

	Total Aboriginal Identity*	Métis**	First Nations	Inuit	Registered Indian
Total population, 1996	799,010	210,190	554,290	41,080	N/A
Total population, 2006	1,172,790	389,780	698,025	50,480	623,780
Total population, 2011	1,400,690	451,795	851,560	59,440	697,510
Total population, 2016	1,673,785	587,545	977,235	65,030	820,120

* The total Aboriginal identity population includes persons who reported more than one Aboriginal identity group and those who reported being a registered Indian and/or band member without reporting an Aboriginal identity.
** The counts for Métis, First Nations, and Inuit were based on single responses to census questions about Aboriginal identity. Some individuals identifying as Aboriginal claimed more than one Aboriginal identity.
Sources: Statistics Canada, 1998, 2007, 2013, 2017a.

Indian Act legislation and federal policies entrenched conditions of poverty on reserves in part to coerce Indigenous assimilation into the Canadian polity (Cairns, 2000). When people chose or were forced to leave their reserve communities and move into cities, they were determined by Canada to have relinquished their Indigenous cultures, their Indian status, and therefore their Aboriginal rights (Newhouse and Peters, 2003). As urban Indigenous populations grew immensely in the second half of the twentieth century—particularly in prairie cities—urban policy became dictated by the goal of integration into mainstream society rather than support to live good urban lives *as* Indigenous people. As Andersen (2013a: 267) explains,

> Canadian governing rationalities were anchored in the assumption that Aboriginal society was not evolutionarily equipped to compete with the complexities of civilization (now terminologically evolved into apparently less odious euphemism "modern life"), and as such, Aboriginal policy was predicated either on absorbing "the Indian problem" into the Canadian body politic in the assimilationist era, or to correct the problems that existed in Aboriginal communities to "bring them up to speed," in the integrationist era. Such assumptions were rooted in the similar assertion that contemporaneous Aboriginal ways of life were developmentally delayed—the only difference in policy rationality was related to how to correct it.

This narrative of Indigenous peoples' incompatibility with modern life influenced academic thinking as well, contributing to assumptions about "authentic" Indigenous cultures as primitive and essentially rural (Peters and Andersen, 2013).

As we will discuss in the next two sections, urban Indigenous communities have not simply adjusted to life in Canadian cities; rather, they have continuously worked at adapting institutions and spaces to support their cultural identities, needs, and ambitions. This is not to say that Indigenous peoples' urban experiences are not also struggles for survival. Many Indigenous people disproportionately live in the poorest neighbourhoods in Canadian cities (Peters, 2011). Indigenous experiences with urban poverty are prevalent and represent the material conditions of colonialism and economic exclusion under capitalism, which can significantly impact individuals' quality of life and daily choices (Silver, 2008). On average, urban Indigenous residents experience higher rates of unemployment, residential mobility, household crowding, proportion of income spent on rent, and homelessness, and lower levels of income, home ownership, and education attainment than non-Indigenous residents (Andersen 2013a; Silver et al., 2008; see Table 4.3).

Silver (2006: 17) argues that in Winnipeg, the disproportionately high concentration of Indigenous residents in poor inner-city areas represents the reproduction of economic and spatial marginalization experienced on reserves and in other rural areas, resulting in "the relative absence

Table 4.3 Socio-Economic Characteristics of Aboriginal and Non-Aboriginal Identity Populations in Selected CMAs, 2016

	Unemployment rates (%)		Median after-tax individual income ($)		Population with university bachelor's degree or above (%)	
	Aboriginal	**Non-Aboriginal**	**Aboriginal**	**Non-Aboriginal**	**Aboriginal**	**Non-Aboriginal**
Victoria	9.9	5.4	24,658	33,813	10.6	26.0
Vancouver	9.6	5.7	24,649	29,504	9.3	26.6
Edmonton	14.6	8.1	31,117	39,309	6.2	20.5
Calgary	13.8	9.2	32,830	39,134	10.5	26.5
Saskatoon	13.0	6.0	29,553	37,030	9.5	22.3
Regina	14.0	5.2	27,994	39,245	7.0	21.4
Winnipeg	11.8	5.7	25,753	32,176	7.6	22.8
Thunder Bay	15.0	6.9	22,115	34,159	8.0	15.6
Toronto	11.1	7.7	25,698	29,280	13.4	27.9
Ottawa-Gatineau	8.6	6.9	31,430	36,411	14.8	27.9
Montreal	10.5	7.4	25,781	29,712	12.2	21.2
Quebec	6.8	4.6	28,403	33,357	13.9	20.8
Halifax	10.8	7.2	27,910	31,832	14.9	25.1

Source: Statistics Canada, 2018.

from the labour market and the core institutions of society … of a large proportion of the urban Aboriginal population." Research conducted by Peters and Lafond (2013) in Saskatoon demonstrates that many Indigenous people have experienced social interactions that mark them as "out of place" (Cresswell, 1996). Urban residents face reminders of social exclusion when they are harassed by police, when their presence is surveilled in wealthier and predominantly white suburban neighbourhoods, when their identities are associated with acts of theft or not having financial means to purchase goods and services in retail spaces, and through discriminatory practices impeding access to meaningful employment and quality housing (Peters and Lafond, 2013).

Although most Indigenous people have experienced some form of racism and marginalization in Canadian cities, many have also flourished. Indeed, some of the rapid growth of Indigenous populations in Canadian cities stems from people seeking employment, education, and other opportunities to be found or created in urban areas. Between 1981 and 2001, Indigenous people made gains in education, employment, and wages (Siggner and Costa, 2005). There is also evidence in all Canadian cities of a growing "Aboriginal middle class," which Newhouse and Peters (2003) have linked to an expanding Indigenous civil service. What ties most urban Indigenous people together, however, is a sense of pride in their distinctive cultural identities and a desire to express or actualize indigeneity in daily life (Environics Institute, 2010). We turn now to how this expression has begun influencing urban development.

Indigenous Urbanism

Urbanism is a concept that is used to discuss how cities are designed and built to accommodate their inhabitants in pursuit of a good life. Most of the built environment, civic institutions, public space design, and programming that we see around us in cities is the result of decisions made by citizens, developers, and governments emanating mostly from Western, European-derived worldviews, practices, and institutions. Though cities are situated in Indigenous territories, the social and physical constructs that we experience as "urban life" are—materially (i.e., relating to built form) and discursively (i.e., relating to governance)—for the most part products of non-Indigenous city plans and decision-making processes. **Indigenous urbanism** is a concept referring to the adjustment of our cities, so that Indigenous approaches to the production of urban space gain greater priority and currency. This should lead to better lives for urban Indigenous peoples, and for non-Indigenous peoples that would see the depth of shared civic identity and a strengthened attachment to community and place as a result (Walker, 2013).

Self-governance and cultural density (Andersen, 2013a) are important to shaping "Bimaadiziwin" in the city, an Anishinaabemowin concept meaning the good life or pursuit of the good life (Newhouse, 2014). In Cree a term used similarly is "Pimatisiwin" (Settee, 2013). Indigenous communities working to build good lives in the city may find that decades of negative **symbolic capital** (Bourdieu, 2000) has been generated, led by a non-Indigenous mainstream society that has produced an enveloping perception that Indigenous culture does not belong in the city (Nejad and Walker, 2018). Associated with this negative symbolic capital has been the dominant social discourse centred on what is "different" or "lacking" in Indigenous communities and cultures, relative to the markers of positive symbolic capital attached to mainstream urban society as a whole (Newhouse, 2011). The project of Indigenous urbanism redirects attention from whitestream frames of reference, such as statistical "development gaps" that situate Indigenous "difference" as deficient or in need of remedy, toward an appreciation for **Indigenous density**. Density accounts for the nuanced complexities of indigeneity and respects Indigenous people's distinctive right and desire to evaluate, respond to, and transform urban conditions *separately* from settler society (Denis, 1997). Density also embodies Indigenous self-governance over urban affairs and community development through structures and processes that may or may not align with non-Indigenous priorities (Andersen, 2013a). Indigenous urbanism will be important to "unsettling" the colonial urbanism in Canadian cities, leading to an urban co-existence between Indigenous and non-Indigenous peoples where power relations account for Indigenous sovereignty at the urban scale, as within the broader traditional territory (Tomiak, 2016; Porter and Barry, 2016; Heritz, 2018).

A network of Indigenous urban organizations has grown in cities across Canada, in sectors ranging widely from housing, education, health, human resources, child and family services, culture and spirituality, legal services, art, design, and real estate development, to name a few (Peters, 2005). This network of organizations owes much, directly or indirectly, to the **Friendship Centre** movement that developed out of cities like Winnipeg, Toronto, and Vancouver in the 1950s, and spread to cities and towns across Canada to meet the needs of Indigenous peoples who were urbanizing (i.e., moving from reserve and rural communities to settler cities) in large numbers in the second half of the last century, and who needed help getting established and finding some sense of community in the city. *Urbanization*, and the challenge of adaptation to urban life, which was of central importance in the 1900s, is joined to a much larger degree in the twenty-first century by a focus on *urbanism*—namely, adapting cities themselves so that they are positively cast Indigenous places in which the large and relatively young urban Indigenous communities can participate in and pursue a good life (Walker et al., 2017). Urbanism encompasses the rich density of Indigenous cultural identities, experiences, and knowledge that can inform valuable insights about living well together—not just differently. We turn now to three examples from Prairie cities where aspects of Indigenous urbanism are emergent.

First Nation Urban Reserves

Urban reserves are parcels of land, acquired on a willing-buyer-from-willing-seller basis in the city's real estate market, converted to reserve status on application to the Minister of Crown-Indigenous Relations under the federal government's Additions to Reserves Policy, and are satellite holdings at some distance from a First Nation community's principal reserve. The first urban reserve in a large city was created in 1988 by the Muskeg Lake Cree Nation in the City of Saskatoon (Figure 4.1). Saskatoon is the largest urban centre in the Saskatchewan portion of Treaty 6 territory (1876). The urban reserve houses a commercial complex, with a variety of shops, services, restaurant, and offices in the Sutherland neighbourhood, though urban reserves can contain other uses such as housing, educational institutions, gas stations, and high-rise buildings, among any number of other city land uses (Figure 4.2).

Prior to a land parcel's conversion to reserve status by the federal government, the First Nation government and local municipal government must work together to create an agreement dealing with four key issues: (1) compensation paid to the municipality in place of property taxes once the land is converted to reserve, (2) the types and payment arrangements for municipal services to be extended to the urban reserve parcel, (3) land-use bylaw compatibility between the urban reserve and surrounding municipal land uses, and (4) a mechanism for dispute resolution and periodic consultation between the municipality and First Nation (Sully et al., 2008).

Urban reserves are becoming quite common in Saskatchewan and Manitoba, in large part because of the financial resources allocated to First Nations in the Treaty Land Entitlement Framework Agreements in both provinces starting in the 1990s, used then to acquire urban land holdings. These agreements were between the federal, provincial, and many First Nations governments to settle specific land claims where the Crown owed land to First Nations to settle outstanding land debts under the numbered treaties. There are around 50 urban reserves in Saskatchewan and many in Manitoba. They are rare in other regions of Canada, though interest in creating

Leanne Mathieson

Figure 4.1 Muskeg Lake Cree Nation urban reserve in Sutherland neighbourhood, Saskatoon

Leanne Mathieson

Figure 4.2 Yellow Quill First Nation urban reserve in downtown Saskatoon

urban reserves may increase over time, given that they can be a good way for First Nations to create spaces in cities that can be useful as commercial enterprises, with exchange value, and as places with community use value, a base for any number of Indigenous urban land uses where people can connect with one another in the city. The presence of First Nation urban reserves is one of the ways that we see the erroneous distinction promulgated through generations of Canadians that Indigenous places are outside the city (e.g., on reserves or rural areas). Instead, urban reserves remind us that the urban scale is just as much a part of Indigenous territories as rural and remote reserve parcels, which themselves, incidentally, were often rather arbitrarily designated by the federal government for First Nations. Indigenous territory follows Indigenous peoples and communities, in urban and rural settings.

Round Prairie Métis and the Saskatoon Public Library

Saskatoon is the largest city in Saskatchewan's Métis nation homeland. The most common local history of how the city of Saskatoon was settled, however, centres on the arrival of the Temperance colonists from Ontario in the early 1880s. Mention is made of Chief Whitecap, who is said to have suggested to the Temperance colonists that the present location of Saskatoon would be a good place to erect their settlement. Rarely discussed are the roughly 30 buffalo-hunting Métis families that settled Round Prairie (i.e., La Prairie Ronde) beginning in the 1850s roughly 40 kilometres south of the present-day city of Saskatoon (Troupe, 2009), decades prior to the Temperance colonists, whose migration into the city in subsequent decades was an important part of Saskatoon's urban evolution (and for a related and fascinating story of Métis settlement at Rooster Town in Winnipeg, another Prairie city, see Peters, Stock, and Werner, 2018).

Round Prairie was first settled as a wintering site for the families, part of their seasonal pattern of living and hunting between Round Prairie and Red River, the main centre in the Métis nation's homeland at present-day Winnipeg. In 1870 they loaded hundreds of Red River carts with their belongings and left Red River to settle permanently at Round Prairie (Schilling, 1983), making it one of the largest Métis settlements in what would become Saskatchewan. Kinship ties with Gabriel Dumont and the Métis of the Batoche community to the northeast led to them fighting alongside one another at the Battle of Fish Creek as part of the Northwest Resistance in 1885, led by Dumont and Louis Riel, to try and protect Métis rights in the territory against the encroaching Canadian government.

When the resistance ended, many Métis from Round Prairie relocated to Montana to escape Canadian government reprisal, but by the early 1900s they were moving back to Round Prairie. And by the end of the 1930s, the Round Prairie Métis had migrated to the built-up and fringe areas of Saskatoon, in the Holiday Park and King George neighbourhoods on the west side of the city, and on the east side near Clarence, Lansdowne, William, and Dufferin Avenues and First, Second, and Taylor Streets, including a prominent community garden at the site of today's Aden Bowman Collegiate (Nejad and Walker, 2018). Many of the Métis in Saskatoon descend from the Round Prairie community. Much of the network of urban Métis civic organizations that helped create the foundation for an emergent Indigenous urbanism grew from the Round Prairie urban community that settled in Saskatoon during the early decades of the last century (e.g., Saskatoon Indian and Métis Friendship Centre, Central Urban Métis Federation Inc., Métis Nation-Saskatchewan, SaskNative Rentals and Camponi Housing).

The Round Prairie Métis have played a significant role in Saskatoon's urban settlement and evolution of its urban indigeneity. Yet it was only in 2016 with the opening of the newest branch of the Saskatoon Public Library (SPL) that this principal community in the interpretation of the depth of Saskatoon's civic identity was commemorated by Saskatoon city officials. In consultations hosted by Saskatchewan's Office of the Treaty Commissioner between the SPL and Métis and First Nations Elders, the new library branch in one of Saskatoon's newest neighbourhoods was named to commemorate the Round Prairie Métis (Figure 4.3). The area where the branch is located, in the Stonebridge neighbourhood, was a popular site used by Round Prairie community members for community purposes such as berry picking and hunting (Troupe, 2009). In addition to the library branch's name, it also contains interpretive material recounting the history and influence of the Round Prairie Métis in the city and region (Figure 4.4).

Ryan Walker

Figure 4.3 Round Prairie Branch, Saskatoon Public Library

Figure 4.4 Interpretive Materials, Round Prairie Branch, Saskatoon Public Library

Bear in mind that thousands of colonial actions and experiments attempting to erase indigeneity from the Canadian landscape have led us to where we are today. Initiatives like the naming process used for this prominent Saskatoon civic institution serve as an example of one of the thousands of initiatives that will collectively be required over time to build a stronger platform for Indigenous urbanism in Canadian cities and repair our capacity for constructive co-existence. The example here of naming the Round Prairie Branch of the SPL, both in process and outcome, brings attention to this Métis urban settlement story and takes the important step of shifting it from the urban fringes to the centre of Saskatoon's civic identity.

Medicine Hill, Moh'kins Tsis, Calgary

Calgary is the largest city in Treaty 7 (1877) territory and was a major centre where two Blackfoot trading routes intersect. Its Blackfoot name is Moh'kins tsis (Crowshoe, 2015). East of the Canada Olympic Park, site of the 1988 Winter Olympics, an archaeological inventory done in the late-1990s indicated around 40 camp and buffalo kill sites in an area referred to as the Paskapoo Slopes (Figure 4.5). The area is described by Blackfoot Elders as having a large cultural footprint, left behind by their ancestors, and the Elders expressed the importance of protecting the area, its archaeological sites, and the functioning of the natural landscape and processes there to

City of Calgary

Figure 4.5 Medicine Hill (Paskapoo Slopes), Calgary

help teach and sustain Blackfoot knowledge and worldview (Crowshoe, 2015).

In 2013 the Trinity Development Group acquired 105 hectares of land at the Paskapoo Slopes area and submitted a concept plan to the City of Calgary shortly thereafter called the Trinity Hills Project, a mixed-use development of office, residential, retail and entertainment on 40 hectares of the lower part of the site. This development proposal set in motion a stakeholder engagement process. Blackfoot people were consulted as stakeholders, given the site's importance within their traditional territory. This was the first time in Calgary's land development approval processes that they were invited to do so in the context of recognizing traditional territory (Crowshoe, 2015). Since the Blackfoot reserves are located a considerable distance from the city itself, this was significant in signalling the city's recognition of First Nations traditional territory, as opposed to simply recognizing the comparatively small pockets of reserve land within the much larger Blackfoot territory (Nejad and Walker, 2018). On the other hand, being recognized as a stakeholder by City Hall is not the same as being engaged as a territorial partner, co-existing at Moh'kins Tsis/Calgary on a balanced field of power relations, with the Indigenous right to co-produce planning objectives and implement development results alongside the municipal authority in shared territory (i.e., Blackfoot and municipal) (Belanger et al., 2019).

Blackfoot Elders walked through the site with city officials and the developer, identifying waterways and natural springs requiring continued care and stewardship. They also conducted a blessing ceremony at the Trinity Hills site and undertook efforts to help the public understand the significance of these cultural protocols and offerings on the land, and to understand the natural system, local plant species, and patterns of the animals living at Paskapoo Slopes. The Elders invited city officials and staff from Trinity Development Group to the Blood Reserve, hosting them at some of the sacred and cultural sites to explain the connections between practices like the Sundance, medicine wheel, the buffalo pound, and the natural cycles of life in Blackfoot culture (Crowshoe, 2015). Over the course of 18 months, several meetings between Blackfoot Elders, Trinity Development staff, and City of Calgary officials led to refinements to the development proposal to protect waterways and natural springs, a commitment to continue working together to protect and interpret the area's archaeology, including hiring Blackfoot workers. City council approved the development plan in 2015. The Elders provided the name for the site, Medicine Hill in English, and a list of other names in the Blackfoot language to use for streets and parks in the new development. In 2016 further discussions got underway—though paused due to funding limitations—to create a master plan for a regional park on the upper (and largest) portion of the site, where hopefully future efforts will be undertaken to create space for Blackfoot traditional practices, interpretation, and (natural and cultural) landscape protection.

There was mutual learning and the Indigenous sense of place was elevated rather than erased, to a greater extent than in comparable city development processes in Calgary's past, presumably setting a new standard to be followed in future development processes around the city (Nejad and Walker, 2018). Promising though it may be, it should not be seen as a comfortable resting place for City Hall's approach to working with the Blackfoot in the territory they share, given that much of the process and the approval authority continues to reside solely at City Hall. One simple example to illustrate pertains to naming the site. City council voted against adopting the Blackfoot language name (i.e., *Aiss ka pooma*) proposed by its Calgary Planning Commission for the site, and instead adopted the English translation of the term (i.e., Medicine Hill) (Canadian Broadcasting Corporation, 2016). It is not that this was a bad outcome or a good one, but simply that it was a decision made by City council after receiving a recommendation from its Planning Commission, and not a decision jointly made with the Blackfoot people. In effect, this demonstrated City council's ultimate belief in its sole authority for decision-making. Canadian cities will need to progress further still in creating processes that reflect a high standard of co-production of urban space, where authority from the point of agenda-setting through to implementation is shared between the city and Indigenous leaders (Belanger and Walker, 2009; Fawcett, Walker, and Greene, 2015).

Urban Indigenous Rights and Resurgence

Many Indigenous people expect their cultural distinctiveness and political separateness to afford respect and influence in urban life, and much of the recent academic literature about the urban Indigenous experience has focused on public policy and service provision in this regard (see Peters, 2011). Historically, the diverse needs of urban Indigenous populations have been circumvented by both federal and provincial governments that often play a perverse game of constitutional avoidance over Indigenous-specific urban policy and program funding (Hanselmann and Gibbins, 2005). While the federal government has tended to restrict its fiduciary responsibilities and scope of service delivery to on-reserve registered Indians, provincial governments have also been reluctant to develop Indigenous approaches to, funding for, and control over urban services (Graham and Peters, 2002). This "jurisdictional quagmire" (Andersen, 2013b) has produced a "patchwork of short-term, overlapping, and inefficient urban Aboriginal programs and policies" (Andersen and Strachan, 2011: 127). In the chronic absence of effective government support, Indigenous people have developed intricate networks of self-administered organizations that have anchored community development in Canadian cities (Newhouse, 2003).

Municipal governments have recently become more aware of the important role that this "invisible infrastructure" (Newhouse, 2003) plays in the lives of Indigenous residents, and they have started to develop strategic partnerships to generate more Indigenous input into the planning and development of urban programs and service delivery (Walker and Belanger, 2013). Since 2015, in the wake of the Truth and Reconciliation Commission (TRC) and its 94 calls to action, some municipalities are also allocating resources to address issues related to the intergenerational trauma of residential schools. The TRC report is the culmination of a seven-year process gathering testimonies from Indigenous survivors and conducting research linking this legacy to Canadian colonialism. The TRC's 10 guiding principles (Table 4.4), as well as many of the calls to action, provide direction for strengthening urban co-existence through principled relations that value Indigenous density and transform colonial mentalities. In efforts to develop reconciliatory partnerships, De Costa and Clark (2016) argue that non-Indigenous people must be active, willing listeners and participants, and that the existing configuration of local Indigenous networks will largely determine the breadth and depth of actions taken in each city.

The focus upon reconciliation, however, has also been attributed to a liberal politics of recognition that continues to omit urban Indigenous communities from constitutional claims to the restitution of land, self-governance, and nationhood, with all the attendant rights and responsibilities that should flow from this special status (Coulthard, 2014). The courts have long been the primary arbiter of Aboriginal rights in Canada, and decisions affecting urban Indigenous peoples have challenged federal and provincial governments to broaden their limited accommodation of Aboriginal rights. The *Corbiere* decision of 1999 ruled that members of First Nations living off-reserve are entitled to vote in band elections, and the *Esquega* decision of 2007 ruled that First Nation members could run for band council elections even if they do not live on reserve (Walker and Belanger, 2013). In *Misquadis v Canada*, 2002, the Supreme Court ruled that Indigenous organizations can represent urban Indigenous interests, and that Human Resources and Skills Development Canada had discriminated against urban Indigenous communities by refusing to fund the infrastructure required to establish urban service

Table 4.4 Truth and Reconciliation Commission 10 Guiding Principles

1. The *United Nations Declaration on the Rights of Indigenous Peoples* is the framework for reconciliation at all levels and across all sectors of Canadian society.
2. First Nations, Inuit, and Métis peoples, as the original peoples of this country and as self-determining peoples, have Treaty, constitutional, and human rights that must be recognized and respected.
3. Reconciliation is a process of healing relationships that requires public truth sharing, apology, and commemoration that acknowledge and redress past harms.
4. Reconciliation requires constructive action on addressing the ongoing legacies of colonialism that have had destructive impacts on Aboriginal peoples' education, cultures and languages, health, child welfare, the administration of justice, and economic opportunities and prosperity.
5. Reconciliation must create a more equitable and inclusive society by closing the gaps in social, health, and economic outcomes that exist between Aboriginal and non-Aboriginal Canadians.
6. All Canadians, as Treaty peoples, share responsibility for establishing and maintaining mutually respectful relationships.
7. The perspectives and understandings of Aboriginal Elders and Traditional Knowledge Keepers of the ethics, concepts, and practices of reconciliation are vital to long-term reconciliation.
8. Supporting Aboriginal peoples' cultural revitalization and integrating Indigenous knowledge systems, oral histories, laws, protocols, and connections to the land into the reconciliation process are essential.
9. Reconciliation requires political will, joint leadership, trust building, accountability, and transparency, as well as a substantial investment of resources.
10. Reconciliation requires sustained public education and dialogue, including youth engagement, about the history and legacy of residential schools, Treaties, and Aboriginal rights, as well as the historical and contemporary contributions of Aboriginal peoples to Canadian society.

Source: Truth and Reconciliation Commission of Canada, 2015.

delivery and representative governance. Walker and Belanger (2013: 199) further explain that

> the three cases proclaimed urban Aboriginal communities to be political communities and in the process established a legal framework to guide Aboriginal community leaders; Canadian policy-makers; and federal, provincial, and municipal officials to better determine what an urban Aboriginal community is and what it represents to the various agencies drawn into their orbit. These are not absolute conventions, but rather represent contours that inform an evolutionary process that began several decades ago, and one that continues to offer perspectives that assist in defining "urban Aboriginal community" within complex socio-political and socio-economic matrices.

Most recently, the 2016 *Daniels* decision ruled that Canada has a constitutional responsibility for roughly 200,000 Métis and 400,000 non-status Indians under section 91.24 of the Constitution Act, 1982, many of whom live in urban areas.

As was previously mentioned, Canada is also a late signatory to the UNDRIP, an international framework for reconstituting Indigenous-state relations. Of the many rights outlined in UNDRIP, self-determination by Indigenous peoples through freely determined strategies for economic, social, cultural, and political organization and development, as well as unrestricted access to, use of, and control over lands and resources in traditional territories is a central theme. UNDRIP provides an important framework for addressing injustices committed against Indigenous peoples and their territories, and the declaration is often cited by Indigenous leaders. It remains to be seen, however, if and how UNDRIP will be implemented in complex urban areas (Belanger, 2011). As with reconciliation, the application of UNDRIP will pose significant challenges and opportunities for practitioners of urban governance, planning, design, and programming into the future.

Unless radically transformative changes that empower Indigenous self-determination are accepted by the Canadian state and settler society, we should expect that cities will continue to be key sites of Indigenous resistance. Urban protest movements and spatial occupations have been frequently undertaken by Indigenous communities to protect (land, natural resources, each other) and advance (political demands, rights, cultural identities) their interests. Such actions might also be viewed through the concept of Indigenous resurgence, which Corntassel (2012) defines as everyday choices that undermine the legitimacy of colonialism, simultaneously affirming traditional ways of being and adding to the cultural density of urban Indigenous communities.

One form of Indigenous resurgence is "affective resistance" (Barker 2015: 46), which reinforces social relations and attachment to place through "spontaneous, creative action or contention." For example, Culhane (2003: 593–94) describes how an annual Valentine's Day Women's Memorial March in Vancouver's Downtown Eastside "gives political expression to a complex process through which Aboriginal women here are struggling to change the language, metaphors, and images through which they come to be (re) known as they emerge into public visibility." Since the 1970s, dozens of Indigenous women have been murdered or gone missing from Vancouver's Downtown Eastside, many of whom were coerced into sex work and chronically exploited, criminalized, and dehumanized by police and mainstream society (Dean, 2015). A flyer that was distributed at the 2001 march proclaims that "We are Aboriginal women. Givers of life. We are mothers, sisters, daughters, aunties and grandmothers. Not just prostitutes and drug addicts. Not welfare cheats. We stand on our mother earth and we demand respect. We are not there to be beaten, abused, murdered, ignored" (Culhane, 2003: 593). The annual women's march weaves through some of the most affluent commercial districts of downtown Vancouver, demanding an empowered visibility of Indigenous women's presence and resistance against the racialized and gendered violence that many have experienced in Vancouver and across Canada.

Countless other urban actions have successfully publicized Indigenous demands due to the high visibility and potential for disruption that are unique to cities. For example, the Idle No More movement was kindled through public teach-ins, round dances in shopping malls, and street marches in the wake of Canada's omnibus Bill C-45, the Jobs and Growth Act, 2012, which unilaterally proposed amendments to the Indian Act and the removal of thousands of bodies of water from federal protections under the Navigable Waters Protection Act. Urban Indigenous communities have continued to organize around the Idle No More vision, which "calls on all people to join in a peaceful revolution, to honour Indigenous sovereignty, and to protect the land and water" through grassroots resurgences of Indigenous nationhood ("Idle No More: The Vision"). The Idle No More movement has connected Indigenous peoples, non-Indigenous allies, and their resistance efforts across local to global scales. Through its diffuse networks it continues to support countless actions and community projects.

With support from Idle No More, Chief Theresa Spence of Attawapiskat First Nation mounted a hunger strike in 2012–2013 on Victoria Island near Parliament Hill in Ottawa to protest her northern community's deplorable housing conditions. In 2013–2014 many Indigenous communities and non-Indigenous allies in major cities rallied in solidarity with Elsipogtog First Nation's barricade against natural gas "fracking" exploration and RCMP enforcement in New Brunswick. These strategies involved marches, round dances, and sacred fires in public urban spaces. In 2018, a "Justice for our Stolen Children" tipi camp was established for 197 days on the Saskatchewan provincial legislature grounds in Regina. The camp initially consisted of two tipis that were raised to advocate for the overhaul of provincial and federal child welfare and justice systems in response to the violent deaths of 15-year-old Tina Fontaine and 22-year-old Colten Boushie, and the acquittals of the accused in both cases. Regina police responded to provincial government concerns about the camp's potential disruption of Canada Day celebrations by forcefully dismantling the tipis for violating municipal bylaws on the 111th day of the camp's existence. Police arrested six people without charge, but the camp was re-established three days later and eventually grew to 15 tipis (Baxter, 2018).

These examples of organized urban action should be understood as a continuation of over three centuries of Indigenous resistance against colonization and the violent, dispossessory legacy of settler colonialism. Urban forms of resistance contribute to a broader decolonization movement, and the radical pursuit of Indigenous self-determination at multiple scales, from the embodied "self" to the nation and beyond (Alfred, 1999). Decolonization will fundamentally dismantle or reconfigure the oppressive and dispossessory frameworks, functions, and ideologies that colonial governance, property ownership, capitalism, and state-determining authority have imposed upon Indigenous peoples (Tuck and Yang, 2012). Such a transformation will require that the ascendancy of decision-making authority practised by the Canadian state and mainstream institutions be curtailed, in the process forging material, symbolic, and functional space for indigeneity to thrive and co-exist with non-Indigenous society.

Conclusion

In this chapter we have positioned Canadian cities as Indigenous cities, or urban scales that continue to encompass, and be encompassed by, Indigenous territory. Indigenous territory follows Indigenous people in cities embodying a rich density of experience, identity, knowledge, and attachment to place, both rural and urban. The social and physical constructs that shape urban life have, for the most part, been produced through non-Indigenous planning and decision-making processes dominated by Western worldviews and institutions. Urban development is also entangled with Canada's legacy of settler colonial governance, which has sought to replace Indigenous society via colonization. Indigenous peoples have always resisted oppressive incursions on their lives and land, and their struggles for self-determination are increasingly tied to resurgent actions in urban spaces. We have encouraged readers to see themselves as creative actors in this complex urban fabric contributing to decolonized co-existence between Indigenous and non-Indigenous society. This radical change will require future urbanists and the non-Indigenous citizenry to help advance Indigenous urbanism through respectful adherence to Indigenous sovereignty, and solidarity with political movements seeking to reconstitute a more just relationship with the Canadian state.

Review Questions

1. What is meant by Canada's legacy of settler colonialism, and how has it influenced the development of cities in relation to Indigenous peoples' self-determination?
2. How might a shift in focus from Indigenous "urbanization and difference" to "urbanism and density" enhance our ability to conceive and experience cities as Indigenous places?
3. How do Indigenous resistance, resurgence, and decolonization relate to the ideal of co-existence, and in what ways have these pursuits been performed in urban contexts?

References

Alfred, T. 1999. *Peace, Power, Righteousness: An Indigenous Manifesto*. Don Mills: Oxford University Press.

Andersen, C. 2013a. "Urban Aboriginal planning: Towards a transformative statistical praxis," in R. Walker, T. Jojola, and D. Natcher, eds., *Reclaiming Indigenous Planning*. Montreal and Kingston: McGill-Queen's University Press.

———. 2013b. "Urban Aboriginality as a distinctive identity, in twelve parts," in E. Peters and C. Andersen, eds., *Indigenous in the City: Contemporary Identities and Cultural Innovation*. Vancouver: UBC Press.

———, and J. Strachan. 2011. "Urban Aboriginal policy in a coordination vacuum: The Alberta (dis)advantage," in E. Peters, ed., *Urban Aboriginal Policy Making in Canadian Municipalities*. Montreal and Kingston: McGill-Queen's University Press.

Banner, S. 2005. "Why *Terra Nullius?* Anthropology and property law in early Australia," *Law and History Review* 23: 95–131.

Barker, A. 2015. "A direct act of resurgence, a direct act of sovereignty: Reflections on Idle No More, Indigenous activism, and Canadian settler colonialism," *Globalizations* 12: 43–65.

Baxter, D. 2018. "Justice for our stolen children camp to hold 'sendoff' potluck and round dance," *Global News*. At: https://globalnews.ca/news/4440604/justice-for-our-stolen-children-camp-to-hold-sendoff-potluck-and-round-dance/

Belanger, Y. 2011. "The United Nations Declaration on the Rights of Indigenous Peoples and urban Aboriginal self-determination in Canada: A preliminary assessment," *Aboriginal Policy Studies* 1: 132–61.

———, K. Dekruyf, and R. Walker. 2019. "Calgary, Canada: Policy co-production and Indigenous development in urban settings," in S. Darchen and G. Searle, eds., *Global Planning Innovations for Urban Sustainability*. New York: Routledge.

———, R. Walker. 2009. "Interest convergence & co-production of plans: An examination of Winnipeg's 'Aboriginal Pathways,'" *Canadian Journal of Urban Research* 18(1) Supplement: 118–39.

Blomley, N. 2003. "Law, property, and the geography of violence: The frontier, the survey, and the grid," *Annals of the Association of American Geographers* 93: 121–41.

———. 2004. *Unsettling the City: Urban Land and the Politics of Property*. New York: Routledge.

Bonds, A., and J. Inwood. 2016. "Beyond white privilege: Geographies of white supremacy and settler colonialism," *Progress in Human Geography* 40: 715–33.

Borrows, J. 2015. "The durability of *Terra Nullius: Tsilhqot'in Nation v. British Columbia*," *University of British Columbia Law Review* 48: 701–42.

Bourdieu, P. 2000. *Pascalian Meditations*. Cambridge: Polity Press.

Cairns, A. 2000. *Citizens Plus: Aboriginal Peoples and the Canadian State*. Vancouver: UBC Press.

Canadian Broadcasting Corporation (CBC). 2016. "Medicine Hill, not Aiss ka pooma, selected as new community name." At: www.cbc.ca/news/canada/calgary/paskapoo-slopes-medicine-hill-aiss-ka-pooma-1.3479084

Corntassel, J. 2012. "Re-envisioning resurgence: Indigenous pathways to decolonization and sustainable self-determination," *Decolonization: Indigeneity, Education and Society* 1: 86–101.

Coulthard, G. 2014. *Red Skin, White Masks: Rejecting the Colonial Politics of Recognition*. Vancouver: UBC Press.

Cresswell, T. 1996. *In Place/Out of Place: Geography, Ideology, and Transgression*. Minneapolis: University of Minnesota Press.

Crowshoe, L. 2015. "Paskapoo Slopes: Accommodating Blackfoot cultural heritage," *Heritage* 18(3): 16–19.

Culhane, D. 2003. "Their spirits live within us: Aboriginal women in Downtown Eastside Vancouver emerging into visibility," *American Indian Quarterly* 27: 593–606.

Dean, A. 2015. *Remembering Vancouver's Disappeared Women: Settler Colonialism and the Difficulty of Inheritance*. Toronto: Toronto University Press.

De Costa, R., and T. Clark. 2016. "On the responsibility to engage: Non-Indigenous peoples in settler cities," *Settler Colonial Studies* 6: 191–208.

Denis, C. 1997. *We Are Not You: First Nations and Canadian Modernity*. Peterborough: Broadview Press.

Edmonds, P. 2010. "Unpacking settler colonialism's urban strategies: Indigenous peoples in Victoria, British Columbia, and the transition to a settler-colonial city," *Urban History Review* 38: 4–20.

Environics Institute. 2010. *Urban Aboriginal Peoples Study: Main Report*. Toronto: Environics Institute.

Fawcett, R., R. Walker, and J. Greene. 2015. "Indigenizing city planning processes in Saskatoon, Canada," *Canadian Journal of Urban Research* 24: 158–75.

Glenn, D. 2017. Contribution to plenary discussion at Architecture Canada's *International Indigenous Architecture and Design Symposium: Reconciliation, Place-Making, and Identity*. Ottawa, May 27.

Graham, K., and E. Peters. 2002. *Aboriginal Communities and Urban Sustainability*. Ottawa: Policy Research Network.

Granzow, K., and A. Dean. 2007. "Revanchism in the Canadian West: Gentrification and resettlement in a prairie city," *Canadian Journal of Cultural Studies (TOPIA)* 18: 89–106.

Hanselmann, C., and R. Gibbins. 2005. "Another voice is needed: Intergovernmentalism in the urban Aboriginal context," in M. Murphy, ed., *Reconfiguring Aboriginal-State Relations. Canada. The State of the Federation*. Montreal and Kingston: McGill-Queen's University Press.

Harris, C. 2002. *Making Native Space: Colonialism, Resistance, and Reserves in British Columbia*. Vancouver: UBC Press.

Henderson, J. 2000. "Postcolonial ghost dancing: Diagnosing European colonialism," in M. Battiste, ed., *Reclaiming Indigenous Voice and Vision*. Vancouver: UBC Press.

———. 2008. "Treaty governance," in Y. Belanger, ed., *Aboriginal Self-Government in Canada: Current Trends and Issues*. Saskatoon: Purich.

Heritz, J. 2018. "Municipal-Indigenous relations in Saskatchewan: Getting started in Regina, Saskatoon and Prince Albert," *Canadian Public Administration* 61(4): 616–40.

Idle No More: The Vision. At: www.idlenomore.ca/vision

Matunga, H. 2013. "Theorizing Indigenous planning," in R. Walker, T. Jojola, and D. Natcher, eds., *Reclaiming Indigenous Planning*. Montreal and Kingston: McGill-Queen's University Press.

Nejad, S., and R. Walker. 2018. "Contemporary urban Indigenous placemaking in Canada," in E. Grant, K. Greenop, A. Refiti, and D. Glenn, eds., *The Handbook of Contemporary Indigenous Architecture*. Singapore: Springer.

Newhouse, D. 2003. "The invisible infrastructure: Urban Aboriginal institutions and organizations,", in D. Newhouse and E. Peters, eds., *Not Strangers in These Parts: Urban Aboriginal Peoples*. Ottawa: Policy Research Initiative.

———. 2011. "Urban life: Reflections of a middle-class Indian," in H. Howard and C. Proulx, eds., *Aboriginal Peoples in Canadian Cities: Transformations and Continuities*, Waterloo: Wilfrid Laurier University Press.

———. 2014. "Fostering Bimaadiziwin: The city as home for urban Aboriginal peoples," *Canadian Diversity* 11(4): 43–48.

———, and E. Peters, eds. 2003. *Not Strangers in These Parts: Urban Aboriginal Peoples*. Ottawa: Policy Research Initiative.

Norris, J., S. Clatworthy, and E. Peters. 2013. "Urbanization of Aboriginal populations in Canada: A half century in review," in C. Andersen and E. Peters, eds., *Indigenous in the City: Contemporary and Cultural Innovation*. Vancouver: UBC Press.

Peters, E. 2005. "Indigeneity and marginalization: Planning for and with urban Aboriginal communities in Canada," *Progress in Planning* 63: 327–404.

———, ed. 2011. *Urban Aboriginal Policy Making in Canadian Municipalities*. Kingston and Montreal: McGill-Queen's University Press.

———. 2015. "Aboriginal people in Canadian cities," in P. Filion, M. Moos, T. Vinodrai, and R. Walker, eds., *Canadian Cities in Transition: Perspectives for an Urban Age*, 5th edn. Toronto: Oxford University Press.

———, and C. Andersen. 2013. "Introduction," in E. Peters and C. Andersen, eds., *Indigenous in the City: Contemporary Identities and Cultural Innovation*. Vancouver: UBC Press.

———, and C. Lafond. 2013. "I basically mostly stick with my own kind: First Nations appropriation of urban space in Saskatoon, Saskatchewan," in E. Peters and C. Andersen, eds., *Indigenous in the City: Contemporary Identities and Cultural Innovation*. Vancouver: UBC Press. 88–109.

———, M. Stock, and A. Werner. 2018. *Rooster Town: The History of an Urban Métis Community, 1901–1961*. Winnipeg: University of Manitoba Press.

Porter, L., and J. Barry. 2016. *Planning for Coexistence? Recognizing Indigenous Rights through Land-use Planning in Canada and Australia*. New York: Routledge.

Schilling, R. 1983. *Gabriel's Children*. Local 11, Saskatoon: Saskatoon Métis Society.

Schmidt, J. 2018. "Bureaucratic territory: First Nations, private property, and "turn-key" colonialism in Canada,"

Annals of the American Association of Geographers, 1-16.

Settee, P. 2013. *Pimatisiwin: The Good Life, Global Indigenous Knowledge Systems*. Vernon: JCharlton Publishing Ltd.

Siggner, A., and R. Costa. 2005. *Aboriginal Conditions in Census Metropolitan Areas, 1981–2001*. Catalogue no. 89-613-MIE. Ottawa: Statistics Canada.

Silver, J. 2006. *In Their Own Voices: Building Urban Aboriginal Communities*. Halifax: Fernwood.

———. 2008. *The Inner Cities of Saskatoon and Winnipeg: A New and Distinctive Form of Development*. Ottawa: Canadian Centre for Policy Alternatives.

———, P. Ghorayshi, J. Hay, and D. Klyne, eds. 2008. "Sharing, community, and decolonization: Urban Aboriginal community development," in M. Cannon and L. Sunseri, eds., *Racism, Colonialism, and Indigeneity in Canada*. Toronto: Oxford University Press.

Stanger-Ross, J. 2008. "Municipal colonialism in Vancouver: City planning and the conflict over Indian reserves, 1928–1950s," *The Canadian Historical Review* 89: 541–80.

Statistics Canada. 1998. "1996 Census: Aboriginal Data," *The Daily*. January 13. Statistics Canada Catalogue no. 93F0025XDB96000.

———. 2007. *Aboriginal Population Profile. 2006 Census*. Statistics Canada Catalogue no. 92-594-WXE.

———. 2013. "2011 National Household Survey: Aboriginal Peoples in Canada: First Nations People, Métis and Inuit," *The Daily*. May 8. Statistics Canada Catalogue no. 99-011-X2011001.

———. 2017a. "Aboriginal Peoples in Canada: Key Results from the 2016 Census," *The Daily*. October 25. Statistics Canada Catalogue no. 11-001-X.

———. 2017b. *Aboriginal Identity Population by Both Sexes, Total—Age, 2016 Counts, Canada and Census Metropolitan Areas and Census Agglomerations, 2016 Census*. Aboriginal Peoples Highlight Tables. Statistics Canada Catalogue no. 98-402-X2016009.

———. 2018. *Aboriginal Identity, Employment Income Statistics, Highest Certificate, Diploma, or Degree, and Labour Force Status Status of Canada, Provinces and Territories, Census Metropolitan Areas and Census Agglomeration, 2016 Census*. Statistics Canada Catalogue nos. 98-400-X2016178 and 98-400-X2016176.

Sully, L., L. Kellett, J. Garcea, and R. Walker, 2008. "First Nations urban reserves in Saskatoon: Partnerships for positive development," *Plan Canada* 48(1): 39–42.

Tomiak, J. 2016. "Unsettling Ottawa: Settler colonialism, Indigenous resistance, and the politics of scale," *Canadian Journal of Urban Research* 25(1): 8–21.

———. 2017. "Contesting the settler city: Indigenous self-determination, new urban reserves, and the neoliberalization of colonialism," *Antipode* 49(4): 1-18.

Troupe, C. 2009. *Métis Women: Social Structure, Urbanization and Political Activism, 1850–1980*. M.A. thesis, University of Saskatchewan, Saskatoon.

Truth and Reconciliation Commission of Canada. 2015. *What We Have Learned: Principles of Truth and Reconciliation*. Ottawa: Truth and Reconciliation Commission of Canada.

Tuck, E., and W. Yang. "Decolonization is not a metaphor," *Decolonization: Indigeneity, Education and Society* 1: 1–40.

United Nations General Assembly. 2007. *Declaration on the Rights of Indigenous Peoples*.

Walker, R. 2013. "Increasing the depth of our civic identity: Future seeking and place making with Aboriginal communities," in E. Peters and C. Andersen, eds., *Indigenous in the City: Contemporary Identities and Cultural Innovation*. Vancouver: UBC Press.

———, and Y. Belanger. 2013. "Aboriginality and planning in Canada's large Prairie cities," in R. Walker, T. Jojola, and D. Natcher, eds., *Reclaiming Indigenous Planning*. Montreal and Kingston: McGill-Queen's University Press.

———, L. Berdahl, E. Lashta, D. Newhouse, and Y. Belanger. 2017. "Public attitudes towards Indigeneity in Canadian Prairie urbanism," *The Canadian Geographer* 61(2): 212–23.

Warnock, J. 2004. *Saskatchewan: The Roots of Discontent and Protest*. Montreal: Black Rose Books.

Wilson, K., and E. Peters. 2005. "You can make a place for it: Remapping urban First Nations spaces of identity," *Environment and Planning D: Society and Space* 23: 395–413.

Wolfe, P. 2006. "Settler colonialism and the elimination of the Native," *Journal of Genocide Research* 8: 387–409.

5 Digital Cities

Contemporary Issues in Urban Policy and Planning

Shauna Brail and Betsy Donald

Introduction

Technological innovation and its impact on urban settlement patterns and city life are not new. The relationship between technology and the city dates back centuries—from the introduction of agriculture in Mesopotamia, to steam engines in Britain, to the automobile in the United States—technology has had a hand in shaping how we live, work, and plan our cities and settlements. Today, intense digitization with the widespread use of computers, smart phones, and wireless data services is once again changing our relationship to the city.

The purpose of this chapter is to introduce the concept of the **digital city** and explore what it means for contemporary issues in urban policy and planning. The first part of the chapter will review concepts and definitions associated with the digital city, smart cities, the platform economy, and other terms that have taken on new meaning in contemporary digital life. We have entered a period where rapid digitization and the fast-paced adoption of new technologies are disrupting almost every aspect of urban life—including public services like transportation and private services like short-term accommodations. These shifts have an impact on how infrastructure is provided and laid out in the city. The next section details the ways in which urban scholars understand and define the digital city, and how they contribute to discussions about the value and challenges associated with digitization. We then explore debates about the ways in which efforts to exploit technological change influence local economies. Following this, the chapter provides current case studies on digital cities in the Canadian context. The final section examines the implications of the digital city for Canadian policy and planning.

Concepts, Definitions, and Debates: Digital Cities

There are many terms used to capture the adoption of ubiquitous technology and computing systems in our everyday urban lives. The terms **digital city** and **smart city** are often used interchangeably to capture mobile technology adoption in the field of urban services. Governments and corporations like the terms "digital city" and "smart city" because they imply that cities can dramatically improve city-service delivery performance by adopting new technologies to make services run more efficiently. Often government or business reports on the smart city speak about cities that have embraced technological innovation as making those cities "smarter, more innovative, or entrepreneurial." For example, the City of Montreal has a Smart and Digital City Action Plan that emphasizes the ways in which a smart city strategy can lead to improvements in economic development, mobility, access to services, digital divides, and access to democracy. But what does this really mean? Below, we expand on the adoption of digital technologies in Canada, discuss the types of data being generated, and address the ways in which digital or smart city applications are implemented at three different scales.

Cities, Digital Technologies, and Big Data

Overall, Canadians have adopted digital technologies, although patterns of usage vary by geography. The latest statistics indicate that nearly 90 per cent of Canadians now use the Internet, with 86 per cent of Canadians having a broadband Internet connection at home (Canadian Internet Registration Authority [CIRA], 2018). Seventy-six per cent of Canadians owned a smart phone in 2016, with 94 per cent of 15- to 34-year-olds reporting that they own a cellphone (Statistics Canada, 2016). According to the Canadian Radio-television and Telecommunications Commission (CRTC), broadband is now seen as a "necessary service to the quality of life for Canadians." But there is a growing digital divide between urban and rural Canada: many places in rural and remote areas of Canada do not have the same access to the Internet in terms of speed, capacity, quality, and price. Most Canadian census metropolitan areas (CMAs), however, are well covered, except for remote or rural parts within some of those CMAs.

Hand in hand with the widespread use of digital technologies is the generation of large volumes of data, often referred to as "big data" (Kitchin, 2014). Big data are everywhere. They are huge in volume, high in velocity, and growing every year. The former chief economist at Google, for example, estimates that more data are being produced every two days at present than in all of history prior to 2003. Kitchin (2014) notes that explosive growth in data is due to a number of different enabling technologies, infrastructures, techniques, and processes. He divides the categories of big data into three areas: (1) directed data; (2) automated data; and (3) volunteered data. It is worth reviewing these categories to provide context for understanding how data are collected in the digital city.

Directed data are usually generated by traditional forms of surveillance. In these cases, the technology is focused on a person or place by a human operator. For example, when you travel through immigration passport control at an airport, passenger details are collected and checked against various databases and this information is linked with real-time data in terms of closed-circuit television cameras, photographs, and sometimes through iris scans or fingerprints.

The second category of data is automated data, which are generated automatically and collected in multiple ways. Automated data are produced via capture systems, which means that by performing a task, data are captured about that task. For example, every time you scan an item through a checkout at a grocery store (and use either a loyalty card or a digital form of payment such as a credit or debit card), information is collected about who purchased the item and when it was purchased. Digital devices like mobile phones can record and communicate the history of their use. Automated data are also generated through clickstream data that record how you navigate through a website or an app. Even in your university courses, when you participate in an online learning management system, data are collected on how often you log in and stay on the site. Automated data are widespread and constantly being collected.

The third category is volunteered data. This refers to data generated and provided by users, though users may be unaware that their data are being collected and analyzed. The most obvious example is participating in social networking sites like Facebook, Instagram, Snapchat, or Twitter, in which users post comments and observations, and upload photos. This practice results in a huge amount of information about the users, their preferences and interests, and their whereabouts. Users may not realize that the data they share about their location, lifestyles, purchases, and opinions are collected and used by social networking sites for marketing and other purposes.

Directed and volunteered data, Kitchin notes (2014: 4–5), can provide insights into urban systems and city lives; however, it is really the *automated* forms that have captured the imagination of those concerned with understanding, building, and managing cities:

> Here the city is envisaged as "constellations of instruments across many scales that are connected through multiple networks which provide continuous data regarding the movement of people and materials" (Batty et al., 2012, p. 482). . . . As such the instrumented city offers the promise of an objectively measured real-time analysis of urban life and infrastructure.

Automated data—which are also generated on digital platforms—can be mapped. For example, by analyzing automated and volunteered data on the social media platform Twitter, geographers have been able to conduct intra-urban geographic analysis of Twitter users in London, Paris, and New York (Adnan and Longley, 2013); the location and distribution of artists and bankers in New York City (Poorthuis and Zook, 2014); and the geographic distribution of French- and English-language content across the provinces of Ontario and Quebec, with particular emphasis on distinctions between the cities of Toronto, Ottawa, Montreal, and Quebec City (Graham and Zook, 2013).

As this discussion has shown, there is a diverse range of data being generated in real-time by firms, residents, and researchers in cities. To understand how digitization affects how cities operate, we can look at data collection and the analysis and utilization of that data at various scales: from city-wide, to the sectoral level, to the individual level. One could argue that the digital city encompasses all three scales, whereas the smart city is really more focused on the city-wide scale.

Digital Technologies and City-Wide Applications

Many city governments have embraced big data to assist in decision-making and service delivery in cities, often referring to these as digital or smart city strategies. Specifically, utility companies have embraced big data to monitor energy systems, water, and lighting. Transportation infrastructure also embraces big data to monitor traffic flows, location, and transportation routes. For example, cities and their transit authorities might use big data to identify where all the buses are in a city at a given time. Mobile phone operators can track customer data including users' location and app use. Big data can be used to assess travel and accommodation patterns to determine where people are staying at any one time, including vacancy rates. Financial institutions and retail chains have embraced big data, as have private surveillance and security firms. Emergency services, policing, and fire response use and collect data. And, of course, social media sites generate huge amounts of big data in terms of opinions, photos, and personal information.

Kingston, Ontario, offers an example of how city governments engage with smart city initiatives in an effort to improve the efficiency of its public services. In February 2018, the City of Kingston announced a public–private partnership with Bell Canada to collect, analyze, and use data to inform and improve municipal decision-making. For Bell—a Canadian telecommunications company that invests billions of dollars each year to improve its infrastructure networks across the country—this is an example of how digitization has opened up new business pathways and partnerships between private firms and public sector entities. Through this initiative, the City of Kingston will access data collected through Internet-enabled sensors. A first priority is to monitor energy use to enable the city to make more informed policy and investment decisions, improve and coordinate maintenance, and reduce its carbon emissions (see Holden and Chang, Chapter 22; Senbel, Liem, and Lesnikowski, Chapter 24). Furthermore, there are plans to install wifi kiosks in public spaces that also provide charging stations for digital devices, as well as direct connections to 911 and emergency services, which presumably will also collect user data (Canadian Press, 2018).

However, these "smart city/digital city" initiatives have drawbacks and introduce other challenges, and thus require critical reflection and observation. Shelton, Zook, and Wiig (2015) remind us to consider smart city initiatives within their historical context and with regard to their spatial implications. They caution that simply applying a smart cities model in an urban setting is not necessarily a path to municipal success (Shelton, Zook, and Wiig, 2015). Kitchin (2014) raises five key concerns that have been echoed by other scholars as well. The first issue is about the politics of big urban data. Data within smart cities are not independent from ideas, techniques, technologies, people, or context. The point here is that there is a *politics* of data and we have to think about the methods used to produce and analyze data. Geographers have always been mindful of how data have been used in the past to create certain types of knowledge and understandings of the city; this is happening now in the era of big data.

The second concern is the privileging of technocratic governance and city development over other forms of analysis. What this means is that big data assume that all aspects of a city can be measured, monitored, and treated as technical problems that can be addressed through technical solutions. Clearly, not all aspects of a city can be measured in this way. We have to be careful that we do not see technological solutions for everything. Indeed, there is a delicate balance to maintain between the need for monitoring and data collection and the need to protect privacy and ensure equitable access to space. We need to be increasingly vigilant that surveillance technology is not used to prevent marginalized people from accessing and enjoying public spaces in Canadian cities.

The third concern is the corporatization of city governance. A lot of big data systems have been produced and sold to cities by large corporations. The concept of the smart city, in fact, was developed by IBM in an effort to expand on consulting opportunities with municipal governments (Spicer and Goodman, 2018). Private firms may see the city system as a growth market and it is imperative to acknowledge that corporate interests are distinct from public sector needs. If smart city solutions are driven more by vendor-push than by city-government-pull, a danger is that we may start to see the marketization of public services, where a city's functions are administered on a for-profit basis. A related concern relates to technological lock-in, where a city is dependent on one corporation to support its smart city initiatives. This could result in a city's development being dictated by the decisions of a single corporation.

The fourth concern relates to security and hackable cities. If ubiquitous and pervasive computing is embedded into city environments, city services and spaces are then dependent on software to function. This can make cities vulnerable to attack. Examples include the Israeli Electric Corporation, which reports that it receives 6000 attempted hacks every second. In October 2012, a major artery in Haifa, Israel, was actually hacked, causing massive traffic chaos for hours (Kitchin, 2014). In 2018, two Ontario towns were the victims of computer hackers. Wasaga Beach paid a ransom of $35,000 in April 2018 in order to receive de-encryption codes following a loss of access to its computer system for seven weeks (Martin, 2018). In September 2018, the town of Midland also paid a ransom to hackers after its computer systems were taken over (Martin, 2018). These examples highlight the very real concern that cities are more vulnerable to cyber-attacks when they embrace big data as part of urban infrastructure management and operations.

The fifth and most provocative point concerns creating a panoptic city. Increasing levels of surveillance are everywhere. Sociologists call this the **surveillance city** (Graham and Wood, 2003). It is easy to track and trace individuals and all their actions, interactions, and transactions. Many of us now accept the practice that if you want to find out about someone quickly, then Instagram, Facebook, and Google are key sources of information. Within minutes, it is possible to learn a tremendous amount about individuals in terms of where they live, their interests, how they interact with people, and their social networks. Through a combination of surveillance, the ubiquitous presence of cameras in public spaces, and other digital technologies, we have created something akin to a "Big Brother" society.

Disrupting Sectors of the Urban Economy

We now turn to our second scale of analysis, which addresses disruptive technologies in certain sectors of the urban economy. Media coverage of disruptive technologies in urban services are hard to avoid as firms such as Uber and Airbnb are transforming the way consumers connect with service providers in commercial transactions (Davies et al., 2017).

Another key point is that activities of digital platform economy firms such as Uber are conflated with the **sharing economy**. The sharing economy has many meanings but is commonly defined in the media as an economic system in which assets or services are shared between private individuals, either free or for a fee, typically mediated through the Internet. Juliet Schor argues that sharing economy initiatives like home-sharing, car-sharing, and ride-sharing are outside a more market-driven approach to the economy because sharing economy activities focus more on the social and environmental opportunities to increase efficiency and community capacity building (Schor, 2017). In recent years, however, people have been moving away from using the term "sharing economy" to describe profit-oriented platform economy activities such as Uber and Airbnb because they are not the same thing (Belk, 2017). As these private sector firms are primarily focused on profit-seeking behaviour and the monetization of goods or services, it is inaccurate to refer to their focus as sharing. For-profit models are increasingly referred to under the terms "digital platform economy, for-profit sharing economy and platform capitalism" (Brail, 2018: 54). Rather than conflate digital, commercial service activities with sharing, others refer to the practices of firms such as Lyft, Airbnb, Facebook, and Etsy as more clearly part of the **platform economy** (Kenney and Zysman, 2016). The platform economy is squarely within the contemporary capitalist economy. Innovation in platform economy firms is characterized by the role that mobile and ICT technologies (smart phones, GPS, rating and credit card platforms) play in the delivery of established services. Platform companies are transforming the way consumers connect with service providers, disrupting old business models in the process.

Implicit in the definition of the platform economy is an emphasis on innovation and the potential to provide new jobs and opportunities for people and places in the digital economy. Others, however, are more skeptical, referring to the platform economy as "neoliberalism on steroids [creating] . . . markets everywhere while also producing a new subjectivity in its participants" (Morozov, 2013). Scholars are concerned about how the rise of platform firms compounds the growing economic and geographic uncertainty of an increasingly urban base of users who "share" their personal items of greatest value in the interest of profit. For instance, Wachsmuth and Weisler (2018: 9) suggest that "[b]y creating higher potential returns to property through the possibility of short-term rentals, Airbnb produces rent gaps, and thereby should be expected to drive gentrification and displacement." Platform economy transactions have the potential to produce significant urban challenges, including gentrification (see Bain and Mark, Chapter 15) and declining employment in traditional sectors such as the taxi industry. Thus, the operating models of platform economy firms can contribute to economic vulnerability and inequality.

Policy-makers also reflect on the implications of disruptive technologies and platform economy firms. One key theme is about equity and access. Who has access to these technologies? Who does not? Who is served? Who is not? For those who cannot afford a smart phone or who are not media savvy, participation in platforms is limited, creating a class division between those that have access and those that do not. A second issue is how these new platforms lead to negative impacts on neighbourhoods. For example, in terms of companies like Airbnb, there is concern, especially in big cities like Vancouver and Montreal, about affordable housing and how home-sharing programs are disrupting communities and neighbourhoods. People have expressed concerns about the commodification of home and domestic space. The third concern has to do with safety and insurance. How do we make sure these services are safe? Who regulates them? The fourth issue relates to innovation. These companies are often identified as symbols of innovation in our economy. But, who benefits from these innovations? Who is profiting from these disruptive technologies? How do platform economy firms and their business models contribute to making the local economy more innovative or lead to job gains or losses? Deep concern and uncertainty for an urban future characterized by platform economy firms leads Kenney and Zysman (2016: 64) to ask, "Will the platform economy, and the reorganization it portends, catalyze economic growth and a surge in productivity driven by a new generation of entrepreneurs? Or will the algorithmically driven reorganization concentrate substantially all of the gains in the hands of those who build the platforms?" These are important questions to think about when considering the future of Canadian cities.

Digital Technologies, Individuals, and Public Space

Now we turn to our third scale of analysis, where we examine how digital platforms and technologies are transforming how individuals interact with urban public space. Public space has been an important topic in urban geography for many decades. Urban geographer Peter Goheen (1998) writes about how public culture was central to an appreciation of the historical rise of the modern city, where public space was an important place to debate the great issues of the day. The public sphere was important because it was where the public was organized and represented, and public space served to mediate the relations between state and society. For Goheen, public space is central to the making of our democratic societies. His research was on the making of the modern city, which saw the emergence of the women's rights movement, the civil rights movement, and human rights movements, among others. Many of these movements and their ideas came about because people could engage in public space, public protest, and have a public conversation.

Today, many people are concerned about the nature of public space in the contemporary city. Marshall Berman, who in 1983 published *All That Is Solid Melts into Air*, said urban public space was important because our modern urban sensibility arises from the ubiquitous and uncontrolled encounters with people and groups in urban public space, especially in the streets (Jacobs, 1961). These uncontrolled encounters with strangers were something that Jane Jacobs also wrote about. In her view, having daily interactions with people you don't know creates a more tolerant and communal society. If, however, we are always on our cellphones or off in our own private worlds, we're not really engaging in this way. Whether an individual is on the bus or the train or in a park, everywhere you look, people are using cellphones and other digital devices. We now must ask ourselves: Is our public space even public anymore when we have so many surveillance cameras recording everything we do all the time? Is it public space if your mobile phone company, wireless data provider, and social media apps know where you are and can track your activities? Should we be concerned about increasing surveillance of public and private space in our digital city? These are important societal questions and exploring them will be important in the future as technology continues to challenge our traditional understandings of how we interact with the public spaces in our cities (see Walker and Blakeley, Chapter 13).

Interpreting the Digital City

Geographers, and particularly economic geographers, are interested in how technological change can influence local economies. There are two strands of significance here: the first is the role that technology-based industries play in terms of attracting and retaining investments, talent, and firms to a particular region and the resulting opportunities and challenges that arise (see also Vinodrai, Chapter 17); the second element is associated with conversations about digital or smart city strategies, including an evaluation of whether or not these produce the desired outcomes, and an assessment of the unintended consequences that operate alongside technology-driven planning principles.

It is widely held that the presence of export-oriented industries create additional employment opportunities in locally oriented industries. Sometimes referred to as a multiplier effect, or a spillover, **cluster**-based strategies often underlie investment attraction efforts to build and support industries whose economic contributions support local opportunities and economic prosperity (see also Vinodrai, Chapter 17; Shearmur, Chapter 18). The technology sector is often used as an example of an industry that benefits jobs and wages in places where it is concentrated. Kemeny and Osman (2018), however, find that while there is a measurable benefit for regions that host technology workers and firms, the benefit is in fact small. As cities and regions look to develop strategies that will benefit their long-term prospects for prosperity and growth, it is important to be aware that technology-driven growth is not necessarily a panacea to urban prosperity.

Another area of debate concerns suggestions of government openness to innovation through economic development strategies. Some suggest that uncertain outcomes surround local government openness to digital economy firms and question whether innovation-friendly government policy has positive or negative impacts on economic development. Donald and Moroz (2017) indicate that one strategy used by proponents of ride-hailing, for instance, is to criticize cities as being outdated or lacking innovativeness if ride-hailing is not permitted to operate. Brail (2017) and Davies et al. (2017) highlight the uneven impacts of platform economy firms on employment and income, which concentrates profit in a small number of firms. They note this as a challenge that needs to be carefully considered in any evaluation of the full economic costs and benefits of the contribution of platform firm to the local economy.

Furthermore, there are growing concerns about the role and influence of the largest technology firms, especially the big four—Amazon, Apple, Facebook, and Google. When compared against all countries, only the United States and China have more financial power than the collective power of these four firms (Galloway, 2017). In terms of their market reach, 55 per cent of consumer product searches begin on Amazon and for every dollar spent in the United States on advertising, 59 cents goes to Facebook or Google. These firms have become increasingly powerful multinational players in technology, retail sales, advertising, hardware, software, services, and more. This reflects both the move towards digitization across a widening number of fields and towards increased prominence of digital influence across the urban landscape.

In fall 2017, Amazon announced it was searching for a second headquarters location, indicating that cities and city-regions across North America were eligible to submit proposals. Canadian cities responded with enthusiasm. According to the request for proposals, the new headquarters location, dubbed HQ2, would ultimately bring 50,000 jobs and US$5 billion in investments to the successful bidder. A total of 238 submissions were received, including at least 13 Canadian submissions spanning the country from coast to coast, although only the Toronto city-region bid was selected as one of 20 finalist locations. While the first round of submissions took place in a highly public forum, with many municipalities sharing their visions (e.g., Halifax, Toronto, Winnipeg), Amazon's subsequent process of selecting one finalist from the top 20 took place under heightened secrecy with little shared information, accompanied by growing public concerns about the impact of this investment.

For example, in Toronto, the opportunity to attract a firm willing to make an investment of this magnitude was viewed as alluring. Yet, once Toronto was selected as a short-listed city, the discussion turned to the challenges paired with the opportunity to be home to Amazon's HQ2. These concerns included the following:

- the ability for the region to build enough housing and transportation infrastructure to accommodate 50,000 new workers plus their partners and families;
- the ability to provide government and other services to support such needs as schools, health care, and daycare;
- the capacity of the regional labour market to provide sufficient high-skilled labour and what impact this might have on talent access for firms already located and invested in the region;
- ethical questions and concerns about the societal costs of encouraging Amazon's growth in terms of loss of traditional retail employment, a poor record of labour practices in Amazon's warehouses, and Mechanical Turk program;[1] and
- the expansion of low-paying, precarious work in non-headquarters activities.

In November 2018, Amazon announced that it had selected two new headquarters locations instead of just one: New York City and Arlington,

Virginia. By splitting their HQ2, Amazon was planning to leverage incentives from both locations worth an estimated US$4.6 billion (Leroy, 2018), while at the same time exploiting their distinct advantages, especially with respect to specialized executive, media, tech, and government talent (Florida, 2018). The Amazon HQ2 search, in retrospect, accomplished a number of things. First, because of the competition-style search, Amazon gained access to a tremendous amount of data and market intelligence on cities across North America. Second, it sparked debate about the role and effectiveness of publicly funded incentives as economic development tools. Third, it reinforced the significance and interconnectedness of talent and place. And in a related vein, it helped to solidify the allure and strength of "superstar" cities such as New York City and Washington, DC, highlighting the growing prominence of a select number of cities and the increasing economic divides between these select cities and those seemingly left behind (Badger, 2018). The Amazon HQ2 search ultimately also demonstrates the power of collective community action and resistance. In January 2019, following protests and refusal by local politicians to support the deal struck by New York's governor and the mayor of New York City, Amazon announced that it would no longer continue to pursue a New York headquarters (Santus, 2019). Instead, the firm suggested that it would instead spread investments and activities across 17 existing North American office and tech hub locations (Amazon, 2019).

Of note, Amazon has continued to invest in new warehousing facilities across Canada. A July 2018 announcement about a new Amazon facility to be located in Ottawa, Ontario, highlighted both the benefits and costs. One benefit to hosting this new warehouse is the creation of 600 full-time warehouse jobs, along with 1500 temporary construction jobs. However, the developer requested that the City of Ottawa waive approximately $800,000 in interest fees on project development charges of $8 million. When the ultimate beneficiary of an $800,000 waiver is a company that has a market valuation in excess of $900 billion, a request for public sector support is a source of controversy.

The rise of Internet-based firms and the digitization of cities go hand in hand. It is increasingly clear that our actions in one realm influence the other. Cities have an important role to play in evaluating economic development opportunities, while at the same time considering and determining how best to guide the broader implications for society. In a report addressing ways for cities to thoughtfully develop smart city strategies that are respectful of personal data and privacy, Bass, Sutherland, and Symons (2018) highlight a pathway for cities to take responsibility for their digital futures. They summarize the roles that cities play in ensuring responsible use of collected data as follows:

1. Leader: Setting the high-level direction for change;
2. Guardian: Making the rules to protect people from harm;
3. Catalyst: Using procurement and funding to create new incentives for responsible data collection and use;
4. Provider: Developing new tools and services; and
5. Connector: Providing opportunities to participate and building local capacity around a cause (Bass, Sutherland, and Symons, 2018: 17).

These examples and their controversies highlight the importance of why decisions about the location and activities of smart city technology firms needs to remain in the hands of the people, that is, the citizens and residents living and working in the cities in question. As mentioned above, technology is not politically benign—it is not about technology for the sake of technology, but it is about technology for people's sake. Cities have different needs, and the roles of government and policy-makers need to be intentional and directive to determine what is best for society, rather than what might be best for profit-seeking firms.

Case Studies: Digital Life in Canadian Cities

Below, we examine four contemporary examples of digital city activities in Canada: (1) the federal government's Smart Cities Challenge; (2) the roll-out of municipal regulations to govern ride-hailing and other private transportation companies; (3) the urban impacts of Airbnb; and (4) the proposal by Sidewalk Labs (an Alphabet Inc. subsidiary) to build a city from "the Internet up" on Toronto's waterfront.

Federal Government Smart Cities Challenge

In November 2017, Canada's federal government announced details for the first round of the country's Smart Cities Challenge (see Taylor and Bradford, Chapter 3). The Smart Cities Challenge is a program intended to be funded with $300 million over 11 years. It calls on "communities of all sizes, large and small, from across Canada to come forward with their best ideas for improving the quality of life of their residents through innovation, data and connected technology" (Government of Canada, n.d.). The challenge is open to all municipalities, regional governments, and Indigenous communities across the country. The goals of the challenge are to

- realize outcomes for residents;
- empower community to innovate;
- forge new partnerships and networks; and
- spread the benefit to all Canadians.

Nearly 200 communities participated by submitting proposals in the first round. Proposals defined smart cities in broad terms, encouraging emphasis on the following focus areas: economic opportunity, empowerment and inclusion, environmental quality, healthy living and recreation, mobility, and safety and security.

In spring 2018, a juried process led to a short-list of 20 proposals that each received $250,000 to further develop their ideas. For example, the cities of Surrey and Vancouver in British Columbia submitted a joint proposal that was short-listed for further development. The proposal addressed two focus areas: (1) mobility, and (2) safety and security. It highlighted the development of two multi-modal, collision-free transportation corridors as a means of demonstrating how Canadian cities can reduce carbon emissions, improve transportation and mobility options to reduce congestion, and contribute to enhanced livability even in the context of continued population growth.

In spring 2019, the first group of four winners was announced: the City of Montreal ($50 million), the City of Guelph and Wellington County ($10 million), Nunavut communities ($10 million) and the Town of Bridgewater, Nova Scotia ($5 million). Each project attempts to leverage technological solutions to improve quality of life in their respective cities and communities.

To date, one positive outcome of the challenge is the number of new partnerships created both among municipalities as well as between municipalities and public, private, education, and non-profit organizations. Overall, the federal government's smart cities challenge appears to encourage municipalities across the country to think about how the deployment of technology and use of data can lead to improved futures for Canadian residents. In its current form, however, it does not appear to engage with questions around privacy, data governance, or surveillance.

Ride-Hailing and Municipal Regulations

Municipal ride-hailing and regulations represent another important example of increased digitization of urban life and the associated role of government in providing oversight and protection. Ride-hailing makes use of the widespread adoption of smart phone technology and GPS and wireless Internet service to enable a digital match between a non-professional driver operating a private vehicle and a passenger. The process is arranged at the tap of a smart phone, the full cost to the rider is confirmed in advance, the payment is digital, and the transaction is facilitated by a digital intermediary. Ride-hailing firms also leverage artificial intelligence through algorithms that direct a wide variety of features, including pricing, driver enticements, and location of available vehicles (Rosenblat, 2018).

Ride-hailing first launched in Canada in 2012, when Uber launched in Toronto and entered other Canadian cities in relatively quick succession. Uber classified itself as a technology company rather than a transportation company, thus making it exempt from expensive taxi laws and regulations. By claiming that its drivers were independent contractors rather than employees, the firm evaded costly worker protections and benefits (Isaac, 2014). Placing itself in a "legal void" (Isaac 2014), Uber shielded itself from industry regulations and employer responsibilities by insisting that existing ground transportation regulations did not apply to Uber. This approach has led to problematic relationships between the company and a host of cities.

Uber's steadfast insistence that it was a technology firm, and not a taxi company, resulted in confrontational legal challenges in several cities, including Toronto. The City of Toronto filed a lawsuit against Uber arguing that the city's ground transportation regulations governing the operation of taxi services applied to Uber. However, in 2015 the Ontario Superior Court of Justice deemed that neither the city's existing taxi regulations nor its regulations governing the operation of limousine services applied to Uber. This decision precipitated the development of regulations specific to ride-hailing in Toronto and across much of the country. As of June 2018, ride-hailing firms operated in 22 of Canada's 30 largest municipalities (Figure 5.1). By January 2019, 20 out of 22 of those municipalities had passed ride-hailing regulations and another six were in the process of developing regulations (Figure 5.2).

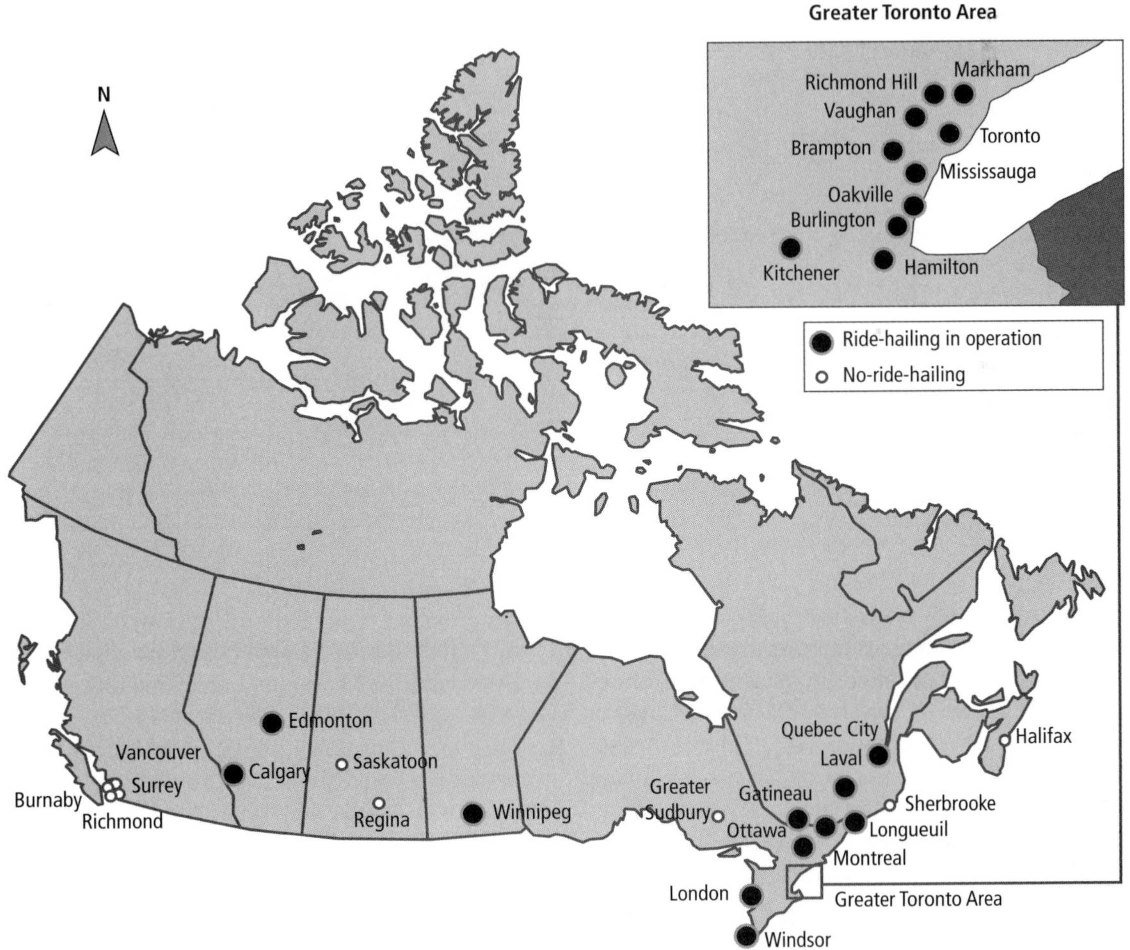

Figure 5.1 Ride-Hailing in Canada

Figure 5.2 Ride-Hailing Regulations in Canada

In early 2017, Statistics Canada released a first series of reports attempting to measure the impact of what it termed the "sharing economy" on the Canadian economy (Statistics Canada, 2017a; Statistics Canada, 2017b).[2] Based on data from macroeconomic accounts, Statistics Canada demonstrated that 7 per cent of Canadians reported using ride-hailing services, spending $241 million. The greatest use of any peer-to-peer ride services or private accommodation services was among 18- to 34-year-olds. On the production side, only 0.3 per cent of the population offered ride-hailing services. Due to limitations in capturing data from unincorporated businesses and non-residents using these services locally, these reports represent only a partial snapshot rather than a full picture. Minimal city-level data were reported. However, the report indicated that 17.6 per cent of Ottawa residents had used ride-hailing services in the past year, followed by 14.8 per cent of Toronto residents (Statistics Canada 2017b).

Acceptance of ride-hailing and the manner in which it is regulated are subject to controversy. First, as ride-hailing has increased in popularity, the taxi sector has seen declines in profits and the value of taxi permits or medallions.[3] Second, labour practices in ride-hailing are heavily criticized for contributing to increases in precarious labour, low wages, and a lack of employment benefits. Third, in markets where public transit exists, ride-hailing may be partly responsible for declining use (and therefore revenues) of public services. And fourth, studies in US cities such as New York and San Francisco have demonstrated that increasing numbers of vehicles operating through ride-hailing networks are contributing to added congestion in already-congested street networks. Vancouver is the largest North American city that does not permit ride-hailing—in British Columbia ride-hailing is provincially regulated and the province is currently conducting consultations to enable ride-hailing operations in the near future. In the meantime, however, investment in active transportation infrastructure such as bike lanes, and improvements to public transit including the opening of the Canada Line in 2009, have translated into an increase in the number of commuters in Vancouver getting to work by walking, cycling, or transit from 57 per cent in 2013 to 59 per cent in 2017 (Zipper, 2019).

It has been suggested that ride-hailing provides an opportunity to reduce private vehicle ownership and potentially carbon emissions and drunk driving, as well as being a means of increasing accessibility to transportation services for residents of low income or racialized communities, groups that taxis have traditionally discriminated against. Additionally, the data collected by ride-hailing firms are of significant value to municipalities when planning transportation routes, understanding key challenges faced by individuals as they move around a city, and assessing opportunities for infrastructure investment. Conversations regarding ways in which ride-hailing firms are required to share data with municipal and other levels of government are increasingly a subject of negotiation. Ride-hailing firms hold a unique dataset that can form an important source in understanding and improving urban transportation.

As the sector evolves, ride-hailing firms are beginning to develop multi-modal networks by expanding into bikes, scooters, and connections with transit to serve first- and last-mile transportation needs. Furthermore, there is also experimentation with new service offerings, including monthly subscription packages and the option to pool rides with other passengers. Sperling, Pike, and Chase (2018) contend that for ride-hailing to contribute positive benefits, it must emphasize the pooling of rides instead of single-passenger trips. It is clear that municipal regulations governing ride-hailing are in a period of transition, and that as the sector matures, so too should policy levers. For example, policies and revenue approaches that incent pooling are a potential next step for regulators. Finally, Sperling, Pike, and Chase (2018) suggest that revising road pricing for all vehicles, and not just ride-hailing vehicles, is an important next step in addressing urban congestion.

Short-Term Rentals

Another recent impact on cities—connected to the platform economy and propelled in part by rising real estate values in central city locations—is the growth of the short-term rental market. Short-term rentals are promoted as a means for homeowners to earn additional income through sharing space in their homes, for a fee. Intermediary firms that play the role of aggregator for rental properties are not a new phenomenon. However, like ride-hailing, the scale and scope of activity based on the rise of the platform economy and the emergence of a single dominant, global firm—Airbnb, launched in 2008—has led to controversy over the impacts of short-term rentals on local real estate markets and calls for short-term rental housing to be regulated.

By 2017, Airbnb listed more than 4 million units and had a market valuation of US$31 billion (Wachsmuth and Weisler, 2018). The platform provides a listing service for hosts (homeowners) and a matching service for guests (home-seekers). It charges hosts a commission for listing their properties, collects rental fees from guests, and can also facilitate the collection and distribution of taxes. Similar to ride-hailing, room-sharing has created a variety of challenges in cities, raising issues about housing affordability and displacement, tax collection, and gentrification (Wachsmuth and Weisler, 2018). Evidence shows that the rapid expansion of short-term room rentals has led to the commodification of housing, with nearly 40 per cent of revenues in 2015 earned by only 9 per cent of firms offering housing on the platform. Wachsmuth and Weisler (2018) indicate that this is evidence that leasing companies rather than individual homeowners are profiting most from this new category of housing.

Short-term rentals are promoted as an opportunity for homeowners to leverage empty bedrooms in their home, or their entire home, as a means of income. Statistics Canada's (2017a) survey of the sharing economy found that only a small number of Canadians offer short-term room rental services, with 69,000 Canadians reporting engagement as hosts. Notably, more than 82 per cent of hosts resided in Ontario, Quebec, or British Columbia (Statistics Canada, 2017b). A study of this sector in Canada's three largest cities found that short-term rentals grew by 50 per cent between 2016 and 2017 (Wachsmuth et al., 2017). For hosts, short-term room rentals provide an opportunity for them to rent properties at a higher nightly rate than they could otherwise secure if rental property leases were longer term. In Montreal, Toronto, and Vancouver, 13,700 full units were listed that are available for short-term rentals for more than 60 nights per year; Wachsmuth et al. (2017) suggest this amounts to the removal of 2 per cent of housing stock in some neighbourhoods.

Given the urban nature of this activity, it is not surprising that Montreal, Toronto, and Vancouver have been most active in developing local regulations to manage short-term room rentals and to tax hosts. In Montreal, the downtown borough of Ville Marie passed a bylaw limiting new permits for tourist residences—defined as residences rented for 31 days or less. Furthermore, the bylaw stipulates that there must be a distance of at least 150 metres separating new rentals from one another. Municipal-level regulation in Ville Marie governs land use and reflects an effort to address the desire for Montreal to be a place of residence for families and communities.

In 2015, the Province of Quebec passed a bylaw requiring people who rent out accommodations for less than a month to apply for a permit and to remit tax revenues to the province. Reports that compliance was weak precipitated a new agreement between the province and Airbnb, and the firm is now collects and remits the accommodation tax to the province on behalf of hosts. Toronto and Vancouver also enacted municipal-level regulations in 2018 to govern short-term room rentals, emphasizing housing affordability, livability, and community over the use of housing as an investment. The case of short-term rentals is instructive as an example of the ways in which digitization can create financial incentives that result in negative repercussions for neighbourhoods and communities.

Sidewalk Labs—the City as Platform

The fourth and final case study of digital life in the city is the Sidewalk Labs proposal in the City of Toronto. In fall 2017, Waterfront Toronto (a tri-government agency with responsibility to revitalize Toronto's waterfront) selected Sidewalk Labs as its initial innovation and funding partner. Sidewalk Labs, a subsidiary of Alphabet Inc. (the parent company of Google), is a firm focused on addressing urban challenges through the application of new technologies and innovation to cities. Sidewalk Labs and Waterfront Toronto are collaborating on a preliminary planning process for the redevelopment of Quayside, a 12-acre site on Toronto's waterfront. Toronto has about 800 acres of waterfront property awaiting redevelopment, a huge and prime stretch of land representing one of the best opportunities in North America to rethink how housing, streets, and infrastructure are built. The 12-acre site is proposed to be a test bed of smart city technology alongside best practices in urbanism. Preliminary plans propose the application of technology-driven solutions applied to the problems of urban life. According to Sidewalk Labs, technology, data, and urban design principles can be used to address challenges such as affordable housing, cold weather mobility, and waste management. While development plans are in progress, some of the potential positive outcomes of the development include environmentally friendly, carbon-neutral development; sensors that separate waste from recycling; modular buildings and flexible land use; self-driving shuttles, shared-ride taxibots, adaptive traffic lights, delivery robots, and heated bike paths and sidewalks; and affordable housing. However, concerns are equally abundant, particularly with respect to data collection and use, privacy and surveillance, genuine public participation, and inclusivity. Among other things, critics suggest that the proposed development promotes the concept of surveillance capitalism, defined by Zuboff (2015: 75) as a "form of information capitalism [that] aims to predict and modify human behavior as a means to produce revenue and market control."

The Sidewalk Labs proposal has become a focal point for highlighting Canadian regulatory weaknesses in governing the collection, use, and monetization of data. Two high-profile resignations from Waterfront Toronto's Digital Advisory Panel, one resignation from Waterfront Toronto's board, and one resignation from Sidewalk Labs have all served to raise the profile and distrust of privately led, data-driven initiatives. Furthermore, there is skepticism about whether a technology firm can contribute to **place-making** and city-building, and how Sidewalk Labs, and Alphabet Inc. more broadly, will profit from the partnership.

Sidewalk Labs submitted a Master Innovation and Development Plan to Waterfront Toronto in June 2019 for approval. If approved, the formal process of application to the City of Toronto would follow. If Sidewalk Labs' innovation and development proposal is ultimately approved, this will be an important development to monitor in the years ahead as it introduces the idea of the city as a platform (Bollier, 2016), where digital networks are profoundly life altering, changing urban life and **governance**. The partnership between Sidewalk Labs and Waterfront Toronto also demonstrates a level of public scrutiny regarding the combination of technology and urbanism that will likely shape future debates about smart cities. Digital networks are clearly changing our relationship to the city.

Discussion and Conclusion

There is no question that cities are in a profound period of digital transformation. As we have shown in this chapter, ubiquitous Internet connections, smart phones, big data, and networked technologies are changing urban life. This chapter introduces concepts associated with the digital city including the smart city, platform economy, and surveillance city. We reviewed three categories of big data: (1) directed, (2) automated, and (3) volunteered. We also reviewed three scales of analysis that are useful when thinking about data and the city—the city scale, the sector scale, and the individual level. We showed how new technologies have had profound impacts on local economies and provided the example of Amazon to demonstrate how global technology firms are occupying a powerful new place in our urban imagination. We then explored in more detail four case studies of digital life in Canadian cities: (1) the federal government's new smart cities challenge, (2) ride-hailing and what it means for municipal regulations, (3) short-term housing rentals, and (4) the Sidewalk Lab's smart city proposal for in Toronto. Collectively, these examples demonstrate the significant implications of digitization for cities, economic development, privacy, regulation, urban policy, and planning.

For decades we have thought of Canadian cities as bounded geophysical spaces governed by a mayor and city councillors with conventional political structures and municipal administrations (see Taylor and Bradford, Chapter 3). Cities have managed what we thought of as basic and unchanging infrastructure in matters such as transit, taxi licensing, and housing. These classic municipal responsibilities, however, are being disrupted and challenged in multiple ways as urban citizens live their lives in hyper-connected virtual spaces and now access traditional goods and services in new and innovative ways. In some instances, multinational technology firms are taking the place of local actors, challenging and disrupting previously localized processes and traditional systems of governance. The rapid pace of change raises more questions than it answers. On the positive side, there may be new opportunities for some Canadian cities in terms of improved mobility and housing options, new jobs and firm formation, and urban design innovation. On the other hand, the changes are amplifying concerns—for instance, over privacy and surveillance—which will undoubtedly have uneven impacts on groups within the city.

Furthermore, questions abound regarding whether the digital **platform economy** can only thrive in an already unequal society, contributing to the persistence of urban inequality. How do we ensure that digitization does not create financial incentives that result in negative repercussions for neighbourhoods and communities? How do we approach policy and revenue schemes that incentivize more environmentally friendly transit and other mobility options? How do we ensure digitization is inclusive and the wealth and advantage it may generate is shared? And related to this, how do we prevent **winner-takes-all urbanism**? What are the distinctive needs of Canada's global centres and mid-size and smaller Canadian cities? How has our relationship with public space changed? These are fundamental and pressing questions that students and scholars of urban geography, planning, and other urban fields will grapple with in the years ahead. One thing is certain: learning about Canadian cities in this period of transition is a constant.

Review Questions

1. What is the platform economy? How has the emergence of a platform economy affected Canadian cities?
2. Thinking back on your day, describe the big data sources that you contributed to, and identify if they were directed, automated, or volunteered? What are some ways that these data might be used by municipal government?
3. How can citizens play a part in ensuring that digitization leads to positive benefits for Canadian cities?

Notes

1. Amazon's Mechanical Turk program is an online platform that pays workers at remote (often home-based) locations on a piecemeal basis to do menial tasks such as categorizing photographs, and at very low levels of pay.
2. At the time of the report's release, news stories suggested that Statistics Canada ought to have used an alternate term rather than the disputed "sharing economy" term to describe these activities (Israel, 2017).
3. A taxi medallion is a permit that allows the owner to operate a taxi in certain markets and can be sold to the highest bidder. The total number of taxi medallions is limited, and taxi medallions historically have appreciated in value in markets where competition for taxis was strong.

References

Adnan, M. and P. Longley. 2013. *Analysis of Twitter Usage in London, Paris and New York City*. At: https://agile-online.org/conference_paper/cds/agile_2013/short_papers/sp_s3.2_adnan.pdf

Amazon. 2019, February 14. "Update on plans for New York City headquarters." At: https://blog.aboutamazon.com/company-news/update-on-plans-for-new-york-city-headquarters

Badger, E. 2018. "In superstar cities, the rich get richer and they get Amazon," 7 November, *New York Times*. At: https://www.nytimes.com/2018/11/07/upshot/in-superstar-cities-the-rich-get-richer-and-they-get-amazon.html

Bass, T., E. Sutherland, and T. Symons. 2018, July. *Reclaiming the Smart City: Personal data, trust and the new commons*. European Union. At: https://media.nesta.org.uk/documents/DECODE-2018_report-smart-cities.pdf

Batty, M., K.W. Axhausen, F. Giannotti, A. Pozdnoukhov, A. Bazzani, M. Wachowicz, et al. 2012. "Smart cities of the future," *European Physical Journal Special Topics*, 214(1): 481–518.

Belk, R. 2017. "Sharing without caring," *Cambridge Journal of Regions, Economy and Society*, 10(2): 249–61.

Berman, M. 1983. *All That Is Solid Melts into Air: The Experience of Modernity*. New York: Verso.

Boiller, D. 2016. "The city as platform: How digital networks are changing urban life and governance," *Washington Institute: Aspen Institute*. At: http://csreports.aspeninstitute.org/documents/CityAsPlatform.pdf

Brail, S. 2017. "Promoting innovation locally: Municipal regulation as barrier or boost?" *Geography Compass*, 11(12). At: https://doi.org/10.1111/gec3.12349

———. 2018. "From renegade to regulated: The digital platform economy, ride-hailing and the case of Toronto," *Canadian Journal of Urban Research* 27(2): 51–63.

Canadian Internet Registration Authority (CIRA). 2018. "Canada's Internet Fact Book." At: https://cira.ca/factbook/canada's-internet-factbook-20

Canadian Press. 2018. "Eastern Ontario city strikes 'Smart City' agreement with Bell." *CBC News*. At: https://www.cbc.ca/news/canada/ottawa/kingston-ontario-bell-smart-city-1.4525892

Davies, A.R., B. Donald, M. Gray, and J. Knox-Hayes. 2017. "Sharing economies: Moving beyond binaries in a digital age," *Cambridge Journal of Regions, Economy and Society*, 10(2): 209–30. At: https://doi.org/10.1093/cjres/rsx005

Donald, B., and A. Moroz. 2017. "The uberization of the city and what it means for local communities," Presented at *Creating Digital Opportunity Annual Conference*, May 3, Montreal, Quebec.

Florida, R. 2018, November 29. "Amazon's HQ2 search was about specialized talent." *Citylab*. At: https://www.citylab.com/life/2018/11/amazon-hq2-specialized-talent-search-nyc-dc/576691/

Galloway, S. 2017. *The Four*. New York: Portfolio.

Goheen, P.G. 1998. "Public space and the geography of the modern city," *Progress in Human Geography*, *22*(4): 479–96.

Government of Canada. n.d. At: https://www.canada.ca/en/privy-council/corporate/clerk/publications/twenty-fifth-annual-report-prime-minister-public-service/delivering-governments-agenda.html

Graham, M., and M. Zook. 2013. "Augmented realities and uneven geographies: Exploring the geolinguistic contours of the Web," *Environment and Planning A: Economy and Space*, 45(1): 77–99. At: https://doi.org/10.1068/a44674

Graham, S., and D. Wood. 2003. "Digitizing surveillance: Categorization, space, inequality," *Critical Social Policy* 23, 227–48.

Isaac, E. 2014. "Disruptive innovation: Risk-shifting and precarity in the age of Uber," *Berkley Roundtable on the International Economy Working Paper*, 2014-7. At: www.brie.berkeley.edu/wp-content/uploads/2015/01/Disruptive-Innovation.pdf

Jacobs, J. 1961. *The Death and Life of Great American Cities*. New York: Random House.

Kemeny, T., and T. Osman. 2018. "The wider impacts of high-technology employment: Evidence from U.S. cities," *Research Policy*, 47(9). At: https://doi.org/10.1016/j.respol.2018.06.005

Kenney, M., and J. Zysman. 2016. "The rise of the platform economy," *Issues in Science and Technology*, 32(3): 61–69.

Kitchin, R. 2014. "The real-time city? Big data and smart urbanism," *GeoJournal* 79(1): 1–14.

Leroy, G. 2018, November 14. "Amazon HQ2, HQ3 subsidy awards costly, not yet fully accounted for," *Good Jobs First*. At: https://www.goodjobsfirst.org/sites/default/files/docs/pdfs/Statement%20day%20after%20HQ2%20finalized%2011-14-18%20V5.pdf

Martin, N. 2018, September 17. "Small Ontario towns pay ransom after hackers hold computer systems hostage." *CBC News*. At: https://www.cbc.ca/news/canada/toronto/small-ontario-towns-pay-ransom-after-hackers-hold-computer-systems-hostage-1.4826545

Morozov, E. 2013. "The 'sharing economy' undermines workers' rights," *Financial Times*, October 14. At: https://www.ft.com/content/92c3021c-34c2-11e3-8148-00144feab7de

Poorthuis, A., and M. Zook. 2014. "Artists and bankers and hipsters, oh my! Mapping tweets in the New York Metropolitan Region," *Cityscape* 16(2): 169–72. At: www.jstor.org/stable/26326893

Rosenblat, A. 2018. *Uberland: How Algorithms Are Rewriting the Rules of Work*. Oakland: University of California Press.

Santus, R. 2019, February 14. "Amazon abandons its plan to build New York City headquarters," *Vice News*. At: https://news.vice.com/en_us/article/j57jgp/amazon-abandons-its-plans-to-build-new-york-city-headquarters

Schor, J.B. 2017. "Does the sharing economy increase inequality within the eighty percent?: findings from a qualitative study of platform providers," *Cambridge Journal of Regions, Economy and Society* 10(2): 263–79.

Shelton, T., M. Zook, and A. Wiig. 2015. "The 'actually existing smart city,'" *Cambridge Journal of Regions, Economy and Society* 8(1): 13–25. At: https://doi.org/10.1093/cjres/rsu026

Sperling, D., S. Pike, and R. Chase. 2018. "Will the transportation revolutions improve our lives—or make them worse?" *Three Revolutions: Steering Automated, Shared, and Electric Vehicles to a Better Future*. Washington, DC: Island Press/Center for Resource Economics. At: https://doi.org/10.5822/978-1-61091-906-7_1

Spicer, Z., and N. Goodman. 2018. "Winning the smart cities challenge with equity, inclusion," *Policy Options*. At: http://policyoptions.irpp.org/magazines/march-2018/winning-smart-cities-challenge-equity-inclusion/

Statistics Canada. 2016. "Life in the fast lane: How are Canadians managing?, 2016." At: https://www150.statcan.gc.ca/n1/daily-quotidien/171114/dq171114a-eng.htm?HPA=1

———. 2017a. "Measuring the sharing economy in the Canadian Macroeconomic accounts." At: www.statcan.gc.ca/pub/13-605-x/2017001/article/14771-eng.htm

———. 2017b. "The sharing economy in Canada." At: www.statcan.gc.ca/daily-quotidien/170228/dq170228b-eng.pdf

Wachsmuth, D., D. Kerrigan, D. Chaney, and A. Shillolo. 2017. "Short-term cities: Airbnb's impact on Canadian housing market," *Urban Politics and Governance Research Group*. School of Urban Planning, McGill University. At: http://upgo.lab.mcgill.ca/airbnb/Shortterm%20Cities%202017-08-10.pdf

———, and A. Weisler. 2018. "Airbnb and the rent gap: Gentrification through the sharing economy," *Environment and Planning A: Economy and Space* 50(6): 1147–70. At: https://doi.org/10.1177/0308518X18778038

Zipper, D. 2019, February 4. "What's it like living in a city without Uber or Lyft?" *Slate*. At: https://slate.com/business/2019/02/uber-lyft-vancouver-no-ride-hail.html

Zuboff, S. 2015. "Big other: Surveillance Capitalism and the Prospects of an Information Civilization," *Journal of Information Technology* 30(1): 75–89.

II Demography, Identity, and Home

The second part of the book deals more specifically with the social geographies of Canadian cities. The chapters focus on where different kinds of people live and work, the reasons for these patterns, and the implications for different kinds of social groups. The first two chapters in this section examine demographic changes. Chapter 6 considers how cities are shaped by age and generational changes, the challenges facing current generations of young adults, such as the millennials, and the implications of age geographies for the future of our cities. Chapter 7 takes a broader perspective, outlining the dominant demographic changes shaping Canadian cities, and how life course and lifestyle changes impact cities, especially through the housing market.

A common theme running through the next three chapters in this section is the question of inclusion and what it would mean to feel at home in the city. Chapter 8 examines immigration. Readers learn how international migration has shaped the geography of urban Canada, the different types of immigrants, and changes in immigration policy. Importantly, the chapter also challenges commonly held assumptions about multiculturalism as a policy, examining its subjective basis and the persistence of prejudice, racism, and exclusion based on ethnicity and race.

Chapter 9 adds a gendered lens to understanding urban Canada. Readers learn to appreciate urban issues through a feminist geography perspective. The chapter outlines changing norms and their impacts on social and economic life. It describes exclusions based on gender and sexuality in Canadian cities, and how changing norms and roles are reflected in labour markets, transportation patterns, housing, neighbourhood change, and identity.

Chapter 10 deals specifically with exclusion from an economic perspective. It tracks growing income inequality and polarization within and between Canadian cities, and the consequences of inequality in cities. Readers learn about the factors contributing to these growing divisions, particularly the role of globalization, deindustrialization, neo-liberalization, and the racialization of poverty.

Although housing issues come up repeatedly throughout the book, the last chapter in this section focuses specifically on housing. Readers learn about how housing is produced, financed, and sold in Canada, and who lives in what kind of housing and why. The chapter explains differences in housing across Canada, declining government involvement in the sector, and three specific housing issues (Indigenous housing, affordability, and environment/health).

After reading the chapters in the second part of the book, readers will have understanding of

- the demographic factors shaping the social geography of Canadian cities;
- the role of immigration and immigration policy in shaping urbanization;
- how changing gender and sexual norms influence urban life;
- the factors shaping inequality and the implications thereof; and
- the character of Canada's housing market and the factors shaping it.

6 Age and Generational Change in the City[1]

Markus Moos

Introduction

"How old are you?" We have all been asked this question, perhaps most often when we were children or as young adults when buying beverages that have age restrictions. As we get older, some see it as impolite to ask somebody's age, and in job interviews doing so is even against the law. Sometimes the question is also used sarcastically when we see someone act in a manner that we think is "immature." In her article, "Act Your Age," sociologist Cheryl Laz (1998) describes how we hold specific expectations of the responsibilities and roles people should fill at different ages. What it means to be "old" or "young" differs across cultures and over time. In other words, Laz distinguishes between the biological and the social meanings of age (also see Pain and Hopkins, 2010). The biological experience of aging is important to consider as our needs for housing, infrastructure, amenities, and services change throughout our lives. But age also defines us socially—for instance, by placing us into specific generations that are believed to have different values, preferences, and lifestyles. The social roles we tend to fill at particular ages shape where and how we reside and spend our time in the city. This chapter considers the importance and relevance of age, and the related concept of generation, in studying and understanding urban landscapes.

The chapter begins with an overview of how the **cohort approach** can be utilized as a method to study generational differences. It compares age groups in different time periods. The discussion considers the kinds of urban environments in which different generations grew up in Canada, followed by an analysis of the geographies of age and age segregation in the largest metropolitan areas. Ageism is discussed as a form of inequality. The chapter then focuses on young adults growing up in different time periods. It tracks changes in employment opportunities and housing decisions of different generations of young adults over time. The chapter ends with a discussion of planning for an aging society and considerations of **age-friendly community** design.

Generational Differences and the Cohort Approach

There is much discussion in the media today about generational differences. Writers often compare the baby boomers (born between the mid-1940s and 1960s) to Generation Y, or millennials (born between the early 1980s and mid-1990s), on work ethic, spending patterns, environmental values, and other characteristics. Such generational comparisons assume that each generation is distinguished from those preceding it because people in the same generation are confronted with "similar opportunities and constraints" (Carr, 2004: 453; Myers, 1999; Twenge,

2006). This assumption is the basis for the **cohort approach**, a common research method used in the social sciences, particularly in housing studies, economics, sociology, and demography (Ryder, 1965). Comparing people of the same age in different time periods is also a useful measure of changing standards of living (Osberg, 2003). The terms **cohort** and *generation* are often used interchangeably to refer to people born in the same time period (Riley, 1987).

Riley's (1987) description perhaps best elucidates the basis of the cohort approach. She explains two "dynamisms" that make aging a process of social change: First, since everyone ages there are necessarily "successive cohorts" aging together. Second, at any given time, societies, and their institutions, are composed of numerous "age strata." Riley describes how these two processes can be visualized as a series of diagonal (cohorts) and horizontal (strata) lines in a two-dimensional space of time (x-axis) and age (y-axis) (Figure 6.1). The implication is that as time passes, society and our cities change simply due to the "dynamisms of aging" (Riley, 1987: 4). As Riley (1987: 4) notes, "the people in a particular age strata are no longer the same people: they have been replaced by younger entrants from more recent cohorts, with more recent life experiences."

Younger cohorts replace older ones in the age strata and selectively transfer emerging behaviours into existing settings, such as the workplace, housing markets, or political institutions. Behaviours and trends that may have been seen as "deviant" or unconventional by one generation become socially acceptable by the next (Mills, 2004). Some of these behaviours will alter the **urban form** as new generations have different ideas about where to live and how to get around the city than previous ones. But it also means that different generations are presented with unique challenges. For

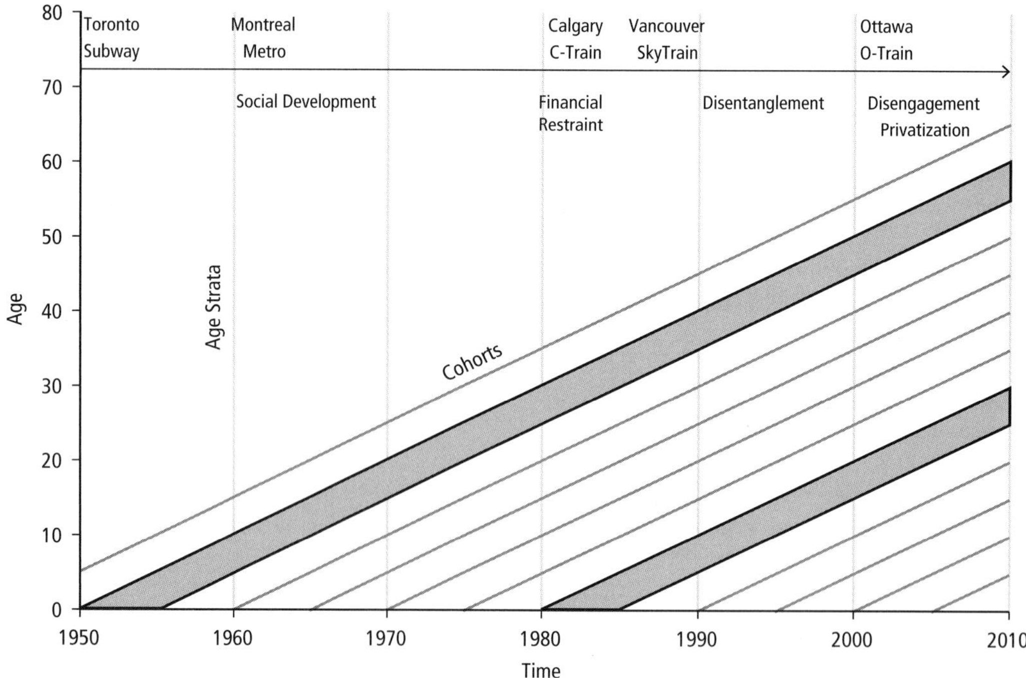

Figure 6.1 Society as Composed of Age Strata and Cohorts

Note: Society is composed of age strata, which consist of different cohorts that age together over time (e.g., shaded areas indicate two generations aging over time). Each cohort experiences unique challenges—for instance, different housing policy contexts (Carroll and Jones, 2000)—and opportunities—for instance, the availability of new public transportation systems.

Source: Based on Riley, 1987.

instance, Figure 6.1 shows four characterizations of the housing policy context in Canada during different time periods, ranging from heavy investment in housing as part of "social development" in the 1960s to subsequently decreasing levels of support and eventual "disengagement and privatization" during the 1990s and 2000s (Carroll and Jones, 2000: 279; see Harris, Chapter 11). As a result, someone looking for housing in the 2000s would encounter higher prices, more condominium apartments, and less government support for assisted housing than someone did in the 1960s. Another example, also shown in Figure 6.1, is transit investment that happens over time so that different generations have different transportation options available to them.

Generational differences are also highly context specific because they are contingent upon the pace of societal change itself (Constable, 1996). In other words, if society changes more slowly, succeeding generations will be more similar than if society changes very quickly. Generational differences are not new. But some have argued that because of rapid changes in technology and social and economic organization in recent years, a period that Giddens (1984) calls "high modernity," generational differences are more amplified than in the past (Furlong and Cartmel, 2007). Two people, one born in 1880 and the other in 1900, experienced cities differently, but the difference is arguably even larger between one born in 1980 and the other in 2000. As Ryder (1965: 855) argued, "In an epoch of change, each person is dominated by his birth date. He derives his philosophy from his historical world, the subculture of his cohort."

However, a growing diversity of **life courses** may also render age a somewhat less useful criterion in understanding socio-spatial differentiation (see Townshend and Walker, Chapter 7). The social construction of the life cycle leads us to associate young adulthood with getting married and having children, which sociologists refer to as "transition markers" (Calvert, 2010). But the decline of traditional norms regarding family formation and marriage, growing educational attainment, and young adults' own changing and varying perception of adulthood question any universality the markers may have held. This "destandardization argument" does remain subject to debate since some scholars suggest that it is a delay in the attainment of markers rather than an increase in the diversity of life courses that is occurring (see Calvert, 2010: 9; Elchardus and Smits, 2006; Shanahan, 2000). Even if our social roles are less connected to age today, not fitting societal expectations remains a source of exclusion as preconceptions remain regarding what someone "should be doing" at a given age (Laz, 1998).

One issue with comparing generations is that doing so can have the unintended effect of overlooking other changes that differentiate society over time, for instance the growth in global migration patterns that has increased the share of immigrants in Canada (see Kobayashi and Preston, Chapter 8). Aging and generational differences help to define the city, but we need to acknowledge that they are not the only factors that determine the characteristics of urban space. Even though not everyone within a generation necessarily has the same history (Pred, 1984)—for instance, having grown up in different places or having been brought up with different values—comparing generations over time is useful because it helps us understand how people's decisions are shaped and constrained by a particular urban and societal context (McDaniel, 2004). For example, unlike an individual born in the 1980s, someone born in the 1950s in Vancouver would not have been able to choose a residential location based on the presence of rapid transit (Figure 6.1). Changing housing policies and transportation contexts shape location and commuting decisions of different cohorts in unique ways.

Growing Up in the City

The theory of generational change suggests that our preferences and values are shaped—at least in part—by the context within which we grow up. Studies have shown links between our residential experiences as children and our residential preferences as adults (Blaauboer, 2011). For instance, Feijten, Hooimeijer, and Mulder (2008: 156), studying households in the Netherlands, found that moves away from the city are triggered by "life events," such as having children, but that "the place of birth turns out to play a decisive part in shaping residential environment choices later in life." Social ties bring people back to the cities or the suburbs where they grew up. In this same vein, growing up in a rural area also made households in the study more likely to move to any rural area. This makes it important to consider the changes in the structure and size of cities experienced by different generations over time. However, residential decisions are influenced by several factors, including income, household composition, ethnicity, and the location of jobs (see Townshend and Walker, Chapter 7). Younger generations also have different values and ideals that create new preferences and location patterns, as was the case in the countercultural movement of the 1960s that brought young artists to the inner city (Ley, 1996). Therefore, where someone grows up is only one variable among others predicting future residential preferences.

Successive generations of residents living in Canada since the beginning of European settlement experienced the country in quite different ways. Until the early 1920s, Canada was primarily a rural nation. Indeed, in 1851, before Confederation, 87 per cent of Canada's population lived in places with fewer than 1000 residents (Statistics Canada, 2011). Most of the population growth since then has occurred in larger towns, cities, and metropolitan areas. The rural population dropped to 51 per cent in 1921, 30 per cent in 1961, 23 per cent in 1991, and 19 per cent in 2011. Those born in the 1950s and 1960s are now in their fifties and sixties—a larger share of this population may hold affinity for rural and small town living than those growing up today in a more urbanized context. Although not everyone in the non-rural population resides in large urban centres, these changes mean that a much larger share of Canada's population today is growing up in proximity to large cities. Some argue that because presently a larger share of the population lives in urban areas near restaurants, shopping, and schools, their experience could continue to fuel demands for municipal services and urban amenities.

To further help us understand the different residential experiences of successive generations, we can compare the location of people from different age groups in two periods (Figure 6.2). The first thing to note here is that there is a growing share of the population residing in larger metropolitan areas (Bourne, 2007). Forty-seven per cent of those 17 and younger in 1976, the oldest of which are in their early sixties today, grew up in places with populations of less than 100,000. By 2016, this share had shrunk to 12 per cent. On the other hand, the percentage of those 17 years or younger residing in places with a population of 1 million or more increased from 27 to 57 per cent; in other words, more young people and children are growing up in large metropolitan areas. Increasing immigration played an important role in these changes since most immigrants and their families have settled in the largest metropolitan areas (see Kobayashi and Preston, Chapter 8).

The second thing to note is that young adults and the working-age population in general are more likely to live in larger urban centres (Figure 6.2). They are attracted to larger urban centres by job prospects, lifestyle, and entertainment amenities (see Skaburskis and Moos, Chapter 19; Bain and Mark, Chapter 15). It is noteworthy, however, that the young and working-age population has become more concentrated in large cities than the population 65 and older. That is to say that the percentage residing in the largest metropolitan areas increased less among those 65 and older than among the rest of the population. We might attribute this to seniors' own past experiences, since they were more likely to have resided in smaller cities and towns when they were younger than is the case for more recent generations. For instance, some have argued that concerns over rapid growth are contributing to the move of some Canadian-born residents away from large metropolitan areas (Ley, 2007).

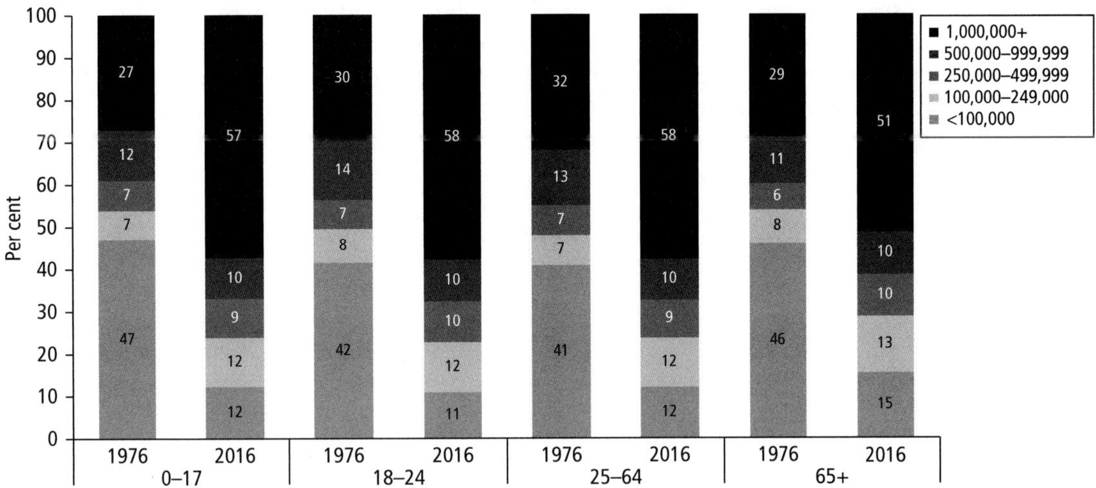

Figure 6.2 The Percentage of Different Age Groups by Community Population Size, 1976 and 2016

Note: The figure shows the percentage of an age group that lives in an urban area (census agglomeration or census metropolitan area) of a particular size. For instance, in 1976, 47 per cent of 0- to 17-year-olds were living in places where the population is less than 100,000, whereas by 2016 this percentage had decreased to 12 per cent.

Source: Statistics Canada, 1976, 2016; Margaret Ellis-Young.

There have also been widespread changes in the internal structure of cities since the 1960s. Downtowns have transitioned from primarily being places of employment, experiencing decline in manufacturing and warehousing employment, to include more residential and recreational amenities (see Bain and Mark, Chapter 15; Ley, 1996). Young adults in particular have been associated with the growth of core area population (see Townshend and Walker, Chapter 7; Moos, 2016). A decline in number of young people holding a driver's licence has led to speculations as to whether this is due to the change in their residential location (i.e., more young adults living in walkable, urban neighbourhoods), to emerging environmental concerns, to declining employment prospects, or to a combination of these factors (Badger, 2013). For instance, a survey by Vancouver's transit agency, TransLink, showed that in 2011, 50 per cent of those 16–19 years of age and 80 per cent of those 20–29 years of age had a driver's licence, a decrease of 10 per cent since 1999 in each age group (Sinoski, 2013). TransLink points to the increasing availability of transit and the high cost of car ownership as explanations.

Since transit use, walking, and cycling are associated with living in higher-density, urban environments, whether or not young adults will continue to register lower car use than previous generations have will depend in part on where they decide to move as they have children (Badger, 2013). Historically at least, urban lifestyles were generally associated with young and non-family households (Van Diepen and Musterd, 2009), although the presence of households with children is increasing somewhat in some downtowns, such as those of Vancouver and Toronto (Willcocks, 2011). A shortage of daycare, schools, playgrounds and other family-oriented amenities is sometimes seen as a barrier to families wanting to raise children downtown. Another factor accounting for a low proportion of families in core areas is the persistence of traditional cultural norms that continue to associate the suburbs, and the availability of private yards, with greater "family friendliness." Political discourse in the City of Toronto highlighted this polarizing issue. One side argued for greater availability of larger sized apartments in the downtown to allow more families to live there, and the other side deemed an urban environment to be unsuitable to raising children (Zerbisias, 2012). Notwithstanding cultural norms, families with children remain more likely to reside in suburbs because

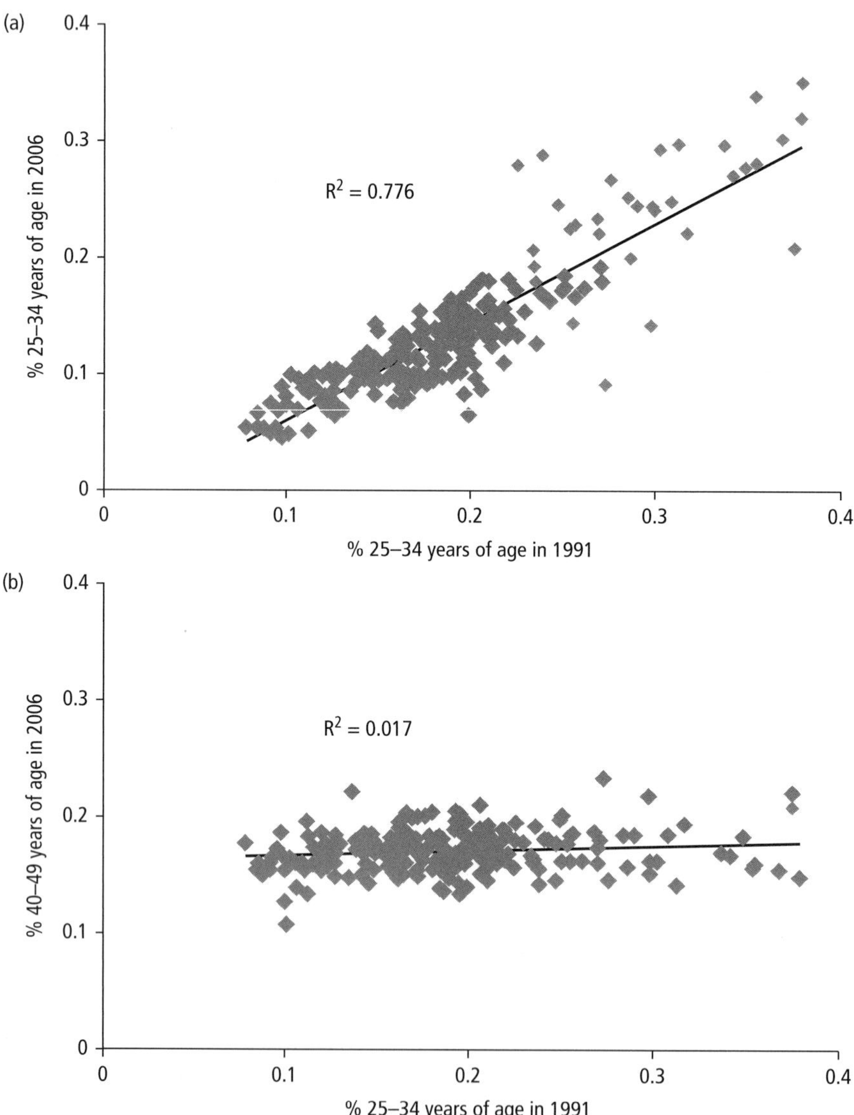

Figure 6.3 Correlation between Age Groups in Vancouver Census Tracts in 1991 and 2006

Source: Statistics Canada, 1991, 2006.

the cost of housing and land (per area) decreases with distance from urban amenities and transit, making it more affordable for larger households to locate in lower-density areas (see Skaburskis and Moos, Chapter 19; Moos, 2016).

There is reason to question whether the growing presence of young adults in higher-density neighbourhoods actually signals a lasting change in preference for urban living, implying that young populations will remain downtown as they age. Figure 6.3 shows the correlation between the percentages of young adults, aged 25–34, residing in Vancouver census tracts (used to define neighbourhoods by Statistics Canada) in 1991 versus 2006 (Figure 6.3a). The figures show a strong association, as evidenced by the high R^2 value, indicating that the percentage of young adults tends to remain high in the same neighbourhoods over time. Figure 6.3 also shows the correlation between the percentage of young adults in 1991 and those 40 to 49 years old in 2006 (Figure 6.3b). If the young adults had remained in the same neighbourhoods between 1991 and 2006, the correlation between these two age groups would be positive. But the correlation is weak, as evidenced by the R^2 value near zero, indicating that there is no relationship between the residential location of 25- to 34-year-olds in 1991 and those 40 to 49 years old in 2006. From this we can at least tentatively infer that new young adults are moving into neighbourhoods where there are already young people and that they change location as they age (Moos, 2014; Moos, 2016). The growing presence of young adults in the downtowns of Canadian metropolitan areas is therefore at least in part a reflection of the delay of child-bearing and an extended young adult life stage (Champion, 2001; Chatterton and Hollands, 2002). Some people may even have fewer or no children to facilitate an urban lifestyle choice (Lauster, 2010; Skaburskis, 1994). As Van Diepen and Musterd (2009: 344) note, "Household structure and urbanity are indisputably related to each other." And, despite important changes, age remains closely connected to household structure.

Geographies of Age and Age Segregation

The linkages between residential location and life stages imply a particular geography of age. There are fewer studies of age than other societal dimensions, such as gender, ethnicity, or social class, and as others have noted the existing literature focuses largely on either the young or the old (Chatterton, 1999; Pain and Hopkins, 2010; Vanderbeck, 2007). Rosenberg and Wilson (2010), in a previous edition of this book, distinguish between "younger and older cities" and between places for youth and older populations within Canadian cities. Some cities, such as Victoria and Quebec City, have a higher share of older residents, while other cities have more children and people of working age, such as Kitchener-Waterloo and Ottawa-Gatineau. At the intra-urban scale, public spaces are also defined by age. Within cities, Rosenberg and Wilson describe how few public places are fully exclusive to any one age group. But there are some spaces that are more frequented by certain age groups—for instance, the presence of young parents with children in playgrounds or teenagers in skateboard parks. Rosenberg and Wilson also describe the challenges of accommodating youth populations in public spaces in cities—that is, those who are too old for playgrounds but too young to spend time in establishments where age of majority is required. They note that young people spending time in public parks or shopping malls can encounter exclusion and stereotyping, being depicted as loiterers or thugs. In response, several Canadian cities have specific programs to help foster youth spaces and activities. One example is a new skateboarding park in Waterloo, Ontario. In the United States, New York City enacted specific measures to counter "discrimination against young people in public places such as restaurants and movie theatres" (Phillips, 2010: 381).

Age segregation is multifaceted because it is both a spatial and temporal phenomenon. The separation of home, work, and schooling means that during the day, many of us interact more with people our own age, while we may interact with all different kinds of age groups within our households (Hagestad and Uhlenberg, 2006; Vanderbeck, 2007), particularly in multigenerational households that are today more common among recent immigrants than among those Canadian born (see Kobayashi and Preston, Chapter 8). Integration of daycares into the workplace and the location of schools, universities, and senior homes in central areas can also increase intergenerational interactions in public places or restaurants. However, the co-location of older and younger residents does not necessarily result in increased social interaction. In fact, such co-location can even lead to tensions, for instance due to the noise from night-time activities in cities where the downtown can be home to both young adults and seniors (Bromley, Tallon, and Thomas, 2005; Chatterton, 1999).

The residential geography of young adults (25 to 29 years of age) and seniors (65 years of age and over) in six of Canada's largest metropolitan areas shows that young adults are most concentrated in the inner cities (Figure 6.4) while seniors are more likely to reside in suburban areas (Figure 6.5). Figure 6.4 and 6.5 use a location quotient to analyze concentration—it measures the percentage of an age group in a census tract divided by the percentage of the same age group in the metropolitan area. In

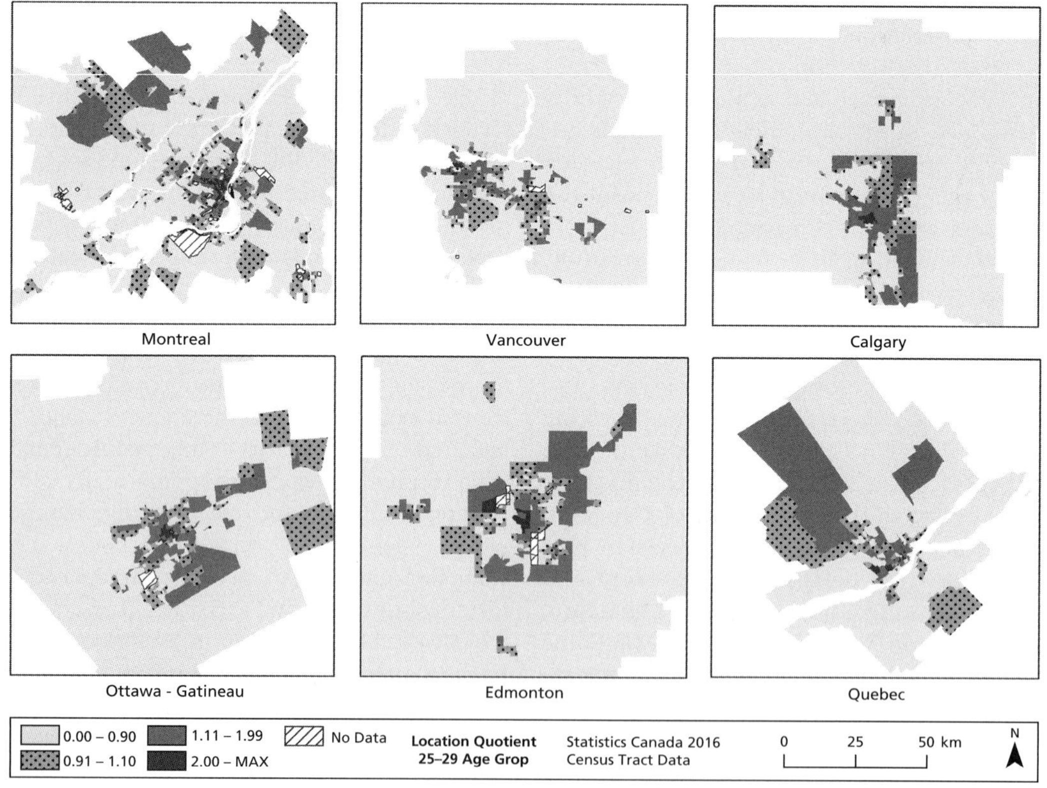

Figure 6.4 The Residential Location of Young Adults in Six Canadian Cities Measured Using Location Quotients, 2016

Source: Margaret Ellis-Young and Robert Walter-Joseph using Statistics Canada census tract data.

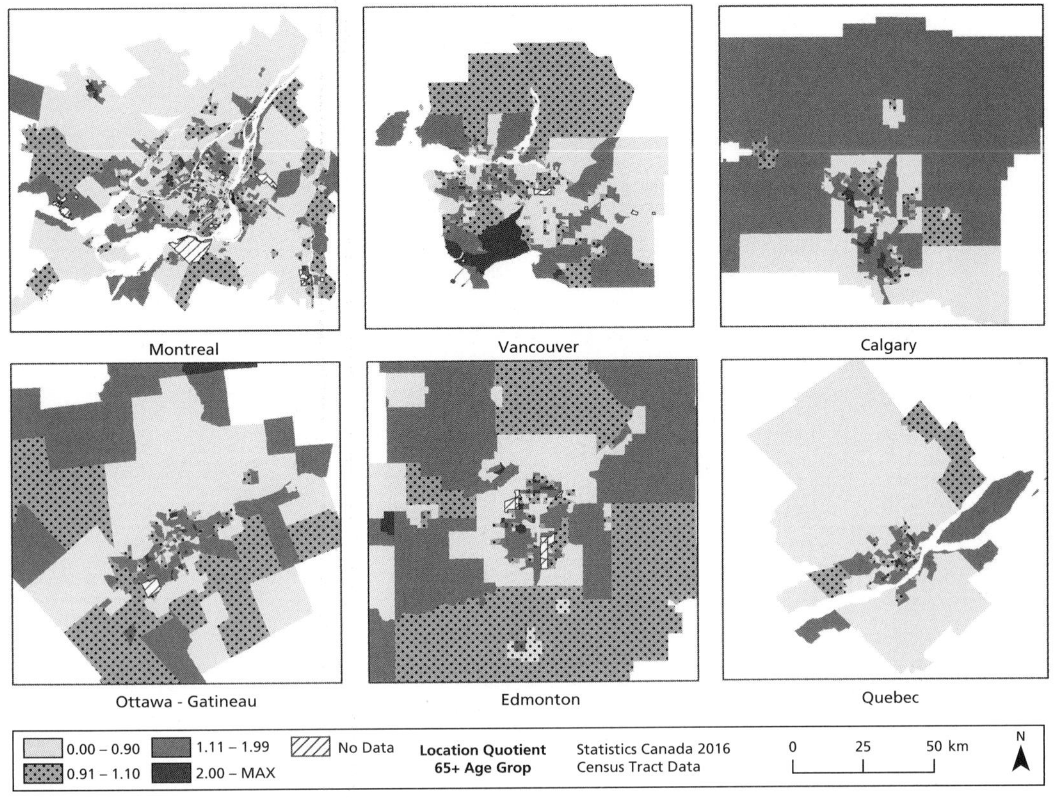

Figure 6.5 The Residential Location of Seniors Measured in Six Canadian Cities Using Location Quotients, 2016

Source: Margaret Ellis-Young and Robert Walter-Joseph using Statistics Canada census tract data.

the largest metropolitan areas, the geography of young adults aligns with higher-density housing, the availability of public transit, and walkable urban areas (see Figure 6.6). In Vancouver, the pattern of young adults extends from the **central business district** (**CBD**) into the suburbs along the SkyTrain transit corridor that connects several urban growth centres (Moos, 2014). While young adults have concentrated in the inner city in the past, this phenomenon has intensified considerably over recent decades, as shown in Toronto (Figure 6.7).

Intensification of land uses also means that the housing stock in the inner city is most suitable for younger households who are delaying marriage and child-bearing or choosing alternate lifestyles altogether. Seniors are most concentrated in old suburbs built between the 1950s and 1970s, in more recent outer suburbs, and in the smaller towns at the outskirts of metropolitan areas (Figures 6.5 and 6.6). But note that young adults are also concentrated in some lower-density outer suburban areas, particularly in the smaller metropolitan areas, such as in Quebec City, where seniors are more present in the inner city. Among young adults, suburban areas are correlated with married households and the presence of children. In some cases the increased presence of young adults in suburbs, as visible in some areas of Toronto, may also relate to the growing share of such individuals still living at home or reporting their parents' home

Figure 6.6 Contrasting Residential Landscapes of Young Adults and Seniors

Left: Toronto neighbourhoods with high concentrations of young adults. Right: Ottawa neighbourhoods with high concentrations of seniors.

address while attending post-secondary education (Figure 6.7).

We can measure the degree of, and changes in, age segregation using a coefficient of localization, which gauges the over- or underrepresentation of a particular age group in different parts of an urban area (Table 6.1). The higher the magnitude of the coefficient, the larger the degree of residential segregation. The results show that residential space is segregated by age but the degree of segregation is certainly not as strong as found with other variables of social differentiation, such as income, immigration, or occupation (Walks, 2001). Across the seven largest metropolitan areas, age segregation is largest for young adults and seniors. There are several changes over time; the largest increases are visible among young adults, which is an outcome of planning policies and demographic changes that have produced high-density residential neighbourhoods in inner cities that are largely targeted at the young. The older populations show the largest decrease in the coefficient of localization, which may be in part a product of the extended time seniors now remain in their own homes, which would increase the variety of age groups found in residential neighbourhoods, and of the dispersal of older populations from the inner cities (Séguin, Apparicio, and Negron, 2008).

For the most part, the segregation of different age groups in the city is mainly a reflection of

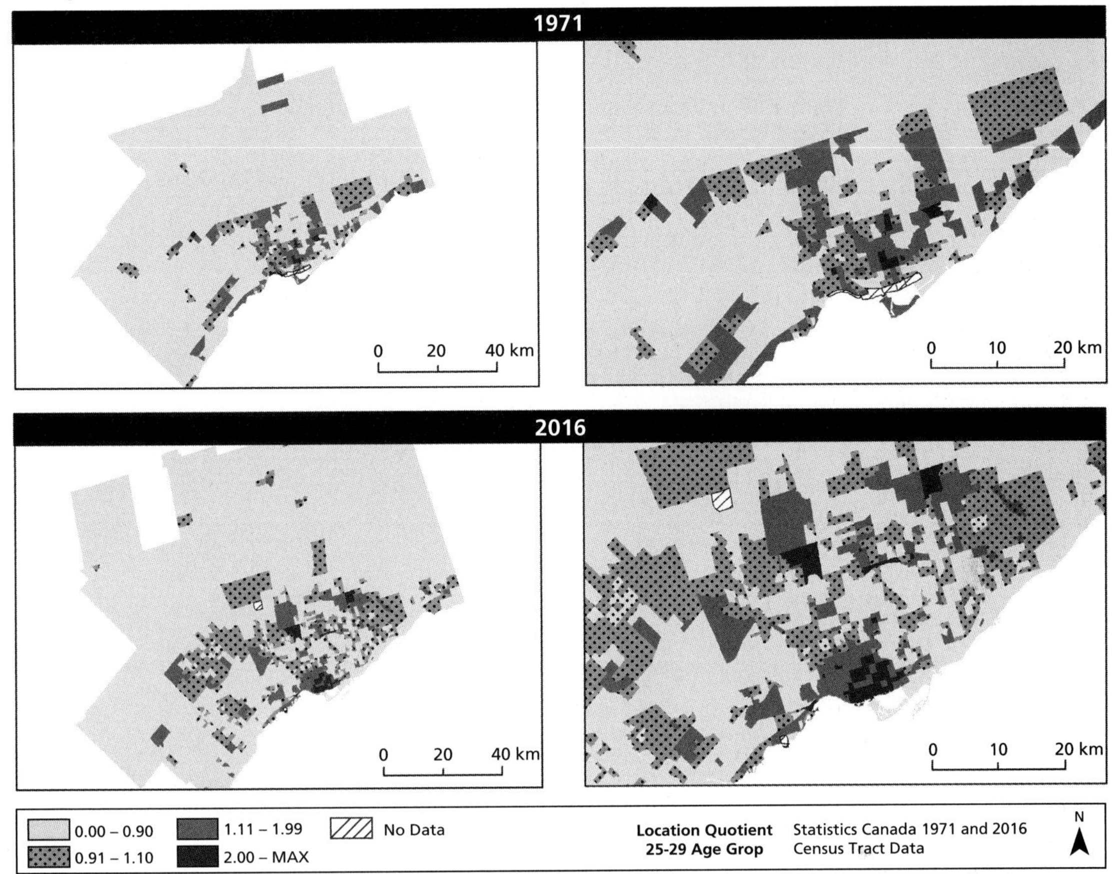

Figure 6.7 The Changing Residential Location of Young Adults in Toronto, 1971 and 2016
Source: Margaret Ellis-Young and Robert Walter-Joseph using Statistics Canada census tract data.

different lifestyles and not so much an indicator of inequality. Ageism is one form of discrimination based on age that produces inequalities (Bisom-Rapp and Sargeant, 2013), and ageism can have a spatial dimension. As Blaikie (2013) notes, the Charter of Rights and Freedoms protects against "discrimination based on age." Any age group can experience age discrimination but often it is the young and the old that experience it. Phillips (2010) indicates that age discrimination is more prevalent in societies placing greater value on economic productivity, as this marginalizes those who are not participating in the workforce.

As others have noted, ageism is a somewhat perplexing type of discrimination because it is targeted at one's past or future self (Joenson, 2013). North and Fiske (2012: 982) point out that "only age encompasses categories that every living person potentially joins" but that we still know little about "prejudice" on the basis of age. Ageism is commonly studied in the context of the workplace. For instance, younger workers are sometimes viewed as less able to handle managerial responsibilities whereas older workers are portrayed as less competent in technological domains despite their actual qualifications (Sargeant, 2013). One survey claims that ageism is "widespread" in Canada and manifests itself in seniors being "seen as less important" or "more ignored than younger generations" (CTV, 2012). Ageism will likely become a more important topic of conversation as the size of the older population grows—that is, as baby boomers age (e.g., Canadian Broadcasting Corporation [CBC], 2012).

Table 6.1 Age Segregation in Canadian Metropolitan Areas Measured Using a Coefficient of Localization, 1971 and 2016

	Toronto		Montreal		Vancouver		Ottawa		Calgary		Edmonton		Quebec City	
Age groups	**Δ1971**	**2016**	**Δ1971**	**2016**	**Δ1971**	**2016**	**Δ1971**	**2016**	**Δ1971**	**2016**	**Δ1971**	**2016**	**Δ1971**	**2016**
<18	0	11	−1	10	−1	11	1	11	−1	12	−1	9	3	13
18–24	0	9	0	8	−3	9	1	11	−9	8	−9	8	2	10
25–29	1	15	5	17	1	16	5	16	3	16	2	15	7	16
30–34	5	15	4	14	7	16	2	12	2	15	1	13	1	11
35–39	3	11	2	10	2	11	0	10	−3	11	−3	9	2	10
40–44	1	8	1	7	−1	6	0	9	−3	9	−3	7	5	10
45–49	−2	6	0	7	−2	5	−3	7	−3	7	−3	6	0	8
50–54	−3	9	−4	8	−4	6	−4	9	−1	13	−2	11	−6	7
55–59	−9	7	−9	7	−5	7	−12	7	−11	10	−10	9	−11	7
60–64	−9	11	−11	9	−8	8	−13	11	−12	15	−10	12	−13	9
65+	−10	16	−8	17	−7	14	−14	18	−14	20	−8	18	−11	18

The table shows the change in coefficient of localization between 1971 and 2016 (Δ1971) and the coefficient of localization in 2016. The coefficient of localization is a measure of segregation of an age group from the rest of the population (see Walks, 2001). It was calculated as follows:

$$CL_g = \sum_{i=1}^{n} |T_i - B_i|/2$$

Where *CL* is the coefficient of localization
g is the age group under observation
i is the census tract
n the number of census tracts
T is the percentage share of the age group under consideration
B is the percentage share of the total population
Source: Statistics Canada, 1971, 2016. Margaret Ellis-Young.

Generation F*CK*D: Debt, Expensive Housing, and Fewer Jobs

The Occupy Wall Street protests, starting in New York and spreading around the globe in 2011, were first and foremost concerned with growing income inequalities (Figure 6.8). But the protests also revealed that some emerging inequalities were in essence generational (Kershaw, 2011). Young adults today are facing more expensive housing markets, lower wages, higher unemployment rates, greater employment uncertainty, higher student and consumer debt, and fewer social services than young adults did in past decades (Moos, 2012, 2013). In Canada, the unemployment rate for youth 15 to 24 was 2.4 times higher in 2012 than for those 25 to 54 years of age, which is the largest generational difference since 1977 (Bernard, 2013). These trends have led some to come up with creative names like "generation squeeze" or "generation F*CK*D" to describe the difficult context faced by young generations today (Bridge, 2011; Kershaw, 2013). Millennials are also labelled by some as "lazy," "whiners," "entitled," and "self-absorbed," but this seems misguided, as others have noted, when in fact many more than in previous generations are university educated yet are working in minimum-wage retail and sales jobs (Fawcett, 2007; Kolm, 2013; Mason, 2011). It may be the case that young peoples' attitudes and expectations are not in line with actual outcomes. As the phrase "oh, young people today . . ." suggests, generational differences have of course long been a source of discussion and, at times, even contempt. One can argue that the many conveniences and opportunities life offers today perhaps lead us to forget the difficulties faced by previous generations. At the same time, economic and social indicators do point to severe challenges confronting the millennials.

The generational wage gap, the difference in earnings between younger and older cohorts in a given time period, is one indicator of the changing economic conditions. Figure 6.9 shows the wage gap and young adults' changing individual income from the mid-1970s to 2016 in Canada (adjusted for inflation). Income has increased over this time period, notwithstanding the decrease during the 1990s due to recessionary times. Significantly, several societal changes took place over this time period that influenced earnings. For example, there has been an increase in dual-earner households as more women entered the labour force (Osberg, 2003). More young adults today are also completing post-secondary education, which raises their incomes, as education is associated with higher earnings; however, higher educational attainment also delays entry into the labour force (Boudarbat, Lemieux, and Riddell, 2010). This means that those 25 to 34 years of age would have lower incomes today than people at the same age did 20 or 30 years ago in part because of increasing educational attainment, which tends to increase earnings later in life (Mincer, 1974).

Elvin Wyly

Figure 6.8 Occupy Wall Street Protestor in Vancouver

Use of profanity by protestors to voice their concern that the world left behind for future generations is "doomed" due to bleak employment outlooks, rising income inequalities, and environmental deterioration.

Figure 6.9 shows the wage gap as a ratio by measuring young adults' incomes as a share of the income of the population 35 years of age and older. The figure indicates that young adults' incomes were about the same and even slightly higher than those of the older population in the late 1970s. While it is not unexpected that young adults' incomes would be lower than for older workers with more experience, the sustained increase in the wage gap over time points to systematic changes. By 2016, young adults' incomes were down to 88 per cent of the incomes of the older population. Statistical analysis of individual earnings shows that this generational wage gap remains even when other changes, such as increasing educational attainment, are taken into account (Boudarbat, Lemieux, and Riddell, 2010). In the Vancouver metropolitan area, for example, young adults are earning about $12,600 less than older workers with otherwise similar characteristics, an increase in the wage gap of $4800 since 1981 (Moos, 2013). The changes have been attributed to economic restructuring and the

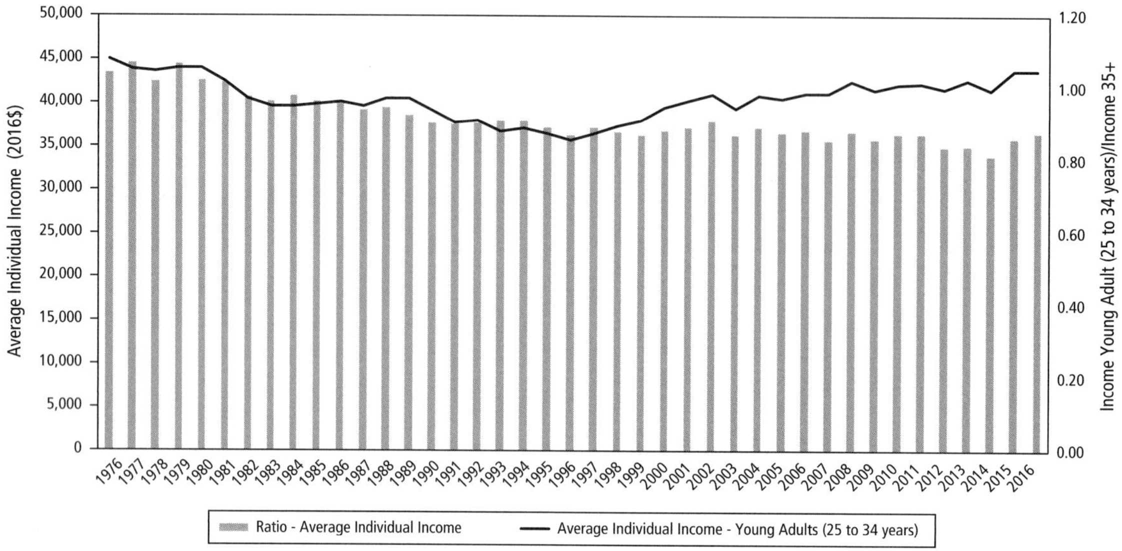

Figure 6.9 Young Adults' Changing Household Income: Total and Ratio Compared to the Older Population in Canada, 1976 to 2016
Source: Statistics Canada, 2018; Margaret Ellis-Young.

neo-liberalization (see **neo-liberalism**) of the **welfare state** that increased employment insecurity and the share of so-called "precarious" and "flexible" employment, where jobs are often less secure and wages generally lower (Myles, Picot, and Wannell, 1993; Vosko, 2006). Educational attainment has become ever more important to gain access to the labour market in an information- and **knowledge-based economy** (Wesson, 2007). But those left out of educational opportunities also face greater obstacles and exclusion than in the past (Hall, 1996).

The increasing wage gap translates into greater difficulties as young adults enter housing markets, competing with the older generations that have greater spending power—although they may not be entering the same local housing markets as we have seen in the geographies of age. One ramification has been the increase of young adults returning to their parental home or remaining there for longer periods than in the past. From 1981 to 2001, the percentage living with parents increased from 12 to 24 per cent for 25- to 29-year-olds and from 5 to 11 per cent for those aged 30 to 34 (Beaupré, Turcotte, and Milan, 2006). Deteriorating economic conditions, rising housing costs, increasing educational attainment, and an increase in the length of the young adult life stage are contributing to this trend (Boyd and Norris, 1999; Chatterton and Hollands, 2002). Moreover, adult children remain at home longer in some immigrant households from Asia and South America in part due to cultural differences (Turcotte, 2006). Also, the growing housing affordability difficulty for young adults is a common issue, widely discussed in the media in Canada's largest cities, where housing prices are highest (Beers, 2007; Rotberg, 2008).

Table 6.2 shows that the percentage of young adults spending more than 30 per cent of their income on shelter costs, a common measure of unaffordability (Lefebvre, 2003), has increased in Toronto and Vancouver between 1986 and 2016. The share of young adults spending more than 30 per cent on housing only increased slightly in Montreal, likely due to lesser price and rent increases and a greater retention of government support to housing affordability programs in Quebec (Germain and Rose, 2000; Walks and Maaranen, 2008).

Table 6.2 Changing Housing Affordability, Home Ownership, and Housing Types among Young Adults, 1986 and 2016

	Toronto		Montreal		Vancouver	
	%	%	%	%	%	%
25-to 34-year-olds	1986	2016	1986	2016	1986	2016
Shelter cost exceeding 30% of income	21.0	42.3	21.8	25.9	31.5	36.9
Owners	40.5	42.0	33.3	36.5	38.5	39.0
Housing types (both rented and owner-occupied):						
Single detached house	32.2	17.7	22.8	18.5	42.0	12.1
Apartment, five or more storeys	32.9	46.6	6.7	9.6	11.0	27.4
Apartment, fewer than five storeys	17.2	16.5	58.3	57.4	31.4	35.2
Semi-detached house	8.3	5.6	4.6	3.3	2.8	1.3
Apartment, duplex	1.7	4.4	3.5	8.0	6.4	15.0
Row house	7.1	9.1	2.7	2.6	5.8	8.8
Other dwellings	0.6	0.1	1.3	0.5	0.7	0.3

Source: Statistics Canada, 1986, 2016; Margaret Ellis-Young.

Interestingly, despite the growing affordability concerns, home ownership among young adults has actually increased or remained almost constant in all three major cities. This is in part an outcome of the growth of the condominium apartment stock now available in most major cities, which was not available to past generations when home ownership was much more directly associated with single-family homes (Harris, 2011). In Montreal, where there are fewer condos than in Vancouver and Toronto, increases in ownership rates among young adults continue to be more directly related to suburban expansion.

But in Vancouver and Toronto, there have been large shifts in the housing types of young adults, from single-family housing to apartment dwellings—not an unexpected phenomenon given the shifts in residential location patterns toward higher-density areas. The high share of apartment dwellings is not too surprising in and of itself given young adults' early stage in the housing and labour markets and their smaller household size (Bailey, 2009; McLeod and Ellis, 1983; Thomas, 2005). Remarkable, however, is the dramatic decline from 42 per cent in 1986 to 12 per cent in 2016 of the proportion of young adults residing in single detached houses in Vancouver. The percentage living in single-family homes was almost 50 per cent in 1981. A similar trend is visible in Toronto, although less pronounced, whereas in Montreal the share of young adults in single-family homes remained almost constant between 1986 and 2006 but decreased slightly by 2016.

The trends reflect important differences among metropolitan areas due to unique housing markets and planning contexts. Prices increased more rapidly in Vancouver and Toronto, while Montreal has had a longer tradition of renting in the central city and fewer planning constraints on suburban expansion (Filion et al., 2010). In Vancouver, some observers have noted that the condominium apartment has become the new bungalow that used to be the stereotypical starter home for baby boomers. Young adults are opting to purchase smaller dwellings in a context of rising costs (Skaburskis, 2001). However, in 2016 about 58 per cent of young adults in Toronto, 63.5 per cent in Montreal, and 61 per cent in Vancouver were renters. Some have argued that increases in owner-occupied higher-density housing forms are an extension of homeownership ideals to the central city (Kern, 2010). However, the data also indicate that not only is a large share of young adults not, or not yet, able to participate in the homeownership market, but also that rental markets may provide one way to access an urban lifestyle in the context of rising inner-city housing costs.

Planning for an Aging Society

The large size of the baby boom generation has steered interest among policy-makers regarding the implications of an aging population for cities. For instance, the Ontario Professional Planners Institute (OPPI) published a report called *Age Friendly Communities: A Call to Action* that outlines anticipated challenges for local community planning (OPPI, 2009). In the document, the OPPI discusses the growing need for more diverse housing options as baby boomers begin to downsize their households. There is also an increasing need for accessibility improvements to accommodate seniors who may be less mobile. The report anticipates growing public transit reliance as people age and lose their ability to keep a driver's licence due to health conditions. Accessibility improvements to sidewalks, streetscapes, parks, and public buildings are called for to ensure that seniors with mobility impairments can still go about their day-to-day activities. The OPPI notes that baby boomers differ from previous generations of seniors in that they have more discretionary income, longer life expectancy, and higher demand for active recreational activities. Florida (2013) argues that seniors today are increasingly moving to large cities to be able to access urban amenities and recreational opportunities. Planning, therefore, can accommodate seniors by locating age-targeted communities and care facilities in central locations near parks, public transit, shopping, and health-care facilities (Figure 6.10). Indeed, particularly

Michael Seasons

Figure 6.10 A retirement community in Waterloo near a park, shopping, recreational amenities, and a light-rail transit stop, all contributing to age-friendly design

in medium-sized cities, baby boomers are a growing target demographic for new downtown condominium apartments. But as we have seen, the overall residential geography of seniors remains associated with suburban landscapes. This raises particular challenges for delivering health-care services and meeting accessibility challenges of seniors aging in place, which is a growing phenomenon (Garvin, 2012). We should also not forget that income is distributed unevenly within older generations and that governments have cut social service provisions, which will make it more difficult for some seniors to afford housing and to maintain consumption-oriented lifestyles.

The research on **age-friendly community** design discusses the need to consider the ways people are able to participate in their communities and engage with other people as they age (Lui et al., 2009; Menec et al., 2011). This literature commonly notes that good health is an important consideration of successful aging but that it is not sufficient. Lui et al. (2009) review various frameworks for age-friendly community design. While details of the frameworks vary, all include physical as well as social dimensions of the urban environment. One framework they review is from the World Health Organization (WHO), which has launched the "Global Age Friendly Cities Project." The WHO framework focuses on "Outdoor spaces and buildings," "Transportation," "Housing," "Communication and information," "Social participation," "Respect and social inclusion," and "Civic participation and employment" (Lui et al., 2009: 118). According to the WHO (2013),

> [i]n an age-friendly community, policies, services and structures related to the physical and social environment are designed to support and enable older people to "age actively"—that is, to live in security, enjoy good health and continue to participate fully in society.

Age-friendly community design thus requires considerations of how people in different age groups can access the physical urban environment as well as of their ability to participate in their communities, employment, and decision-making. The municipality of Saanich, British Columbia, which participated in the WHO project, identified the importance to "provide opportunities for meaningful participation," "recognize the diverse needs," and identify "barriers" to seniors in the community (District of Saanich, 2013). Professor John Lewis, who studies age-friendly community design in the City of Waterloo, which recently received the WHO's age-friendly city designation, argues that barriers are also often attitudinal—people have to become more receptive to having seniors, and people of all ages, participate regularly in all aspects of daily life (Csanday, 2012; Lewis, 2013). As Lui et al. (2009: 119) note, a growing "focus on active participation and engagement of older people is an antidote to the conception of old age as an inevitable period of withdrawal from social roles and relationships."

Conclusion

This chapter dealt with the age and generational dimensions of the urban landscape. Increased spatial segregation of young adults is an indication of the city's growing fragmentation and may be a sign of growing "social distance" between generations that can lead to challenges in social interaction (Hagestad and Uhlenberg, 2006). An example is the workplace, where there may be tensions between younger and older workers, which can even lead to age-specific discrimination or systemic ageism. The growing discourse on generational change deals in part with the challenges of finding common ground between generations (Kolm, 2013; Mason, 2011; Twenge, 2006). It speaks to larger societal challenges of "how to live together" and foster mutual understanding in a metropolitan context, where differences—whether age, class, gender, or ethnicity—are producing stronger socio-spatial divisions.

The study of age cohorts reveals the dramatic structural changes in the inner cities of Canada's major metropolitan areas, which have attracted large shares of young adults to new housing and recreational opportunities. The younger generations in Canada today are better educated than any prior generation and live in a society where standards of living are high.

Housing choices are more abundant, particularly in inner cities. At the same time, young adults are facing greater difficulties in the job market in a context of growing employment uncertainties. Meanwhile, the cost of housing escalates. Young adults live close to public transit and in more walkable neighbourhoods than in the past and compared to other age groups. On average, the aging baby boomers have higher incomes and more housing wealth than any previous generation. In part this wealth is still tied up in their suburban homes, but in some cities baby boomers are beginning to downsize and move into central areas. The context makes it more difficult for younger generations (Beer et al., 2011), and also for those baby boomers with lower incomes, to compete in the housing market. Moreover, the increasing share of seniors produces new challenges in ensuring that our communities remain accessible for all generations. These generational changes reveal how age is socially constructed, as new generations take on different roles, responsibilities, and lifestyles than those in the past. The study of the city through the lens of age helps reveal the unique challenges we face as we inevitably get older.

Review Questions

1. How have young adults' residential locations and housing decisions changed over the past several decades? Do the changes suggest there is increasing age segregation in Canadian cities? What are the implications?
2. Have you ever experienced or witnessed a case of ageism? What is it about the incident that suggests there was discrimination based on age?
3. What is the generational wage gap? And how can it help us understand young adults' changing economic fortunes?

Note

1. Parts of this chapter are based on research and writing from the author's doctoral thesis (Moos, 2012). The author thanks Margaret Ellis-Young, Robert Walter-Joseph, and Michael Seasons for research assistance.

References

Badger, E. 2013. "Millennials lead the trend to less driving, but what happens as they get older?" *The Atlantic Cities* (online magazine), May. At: www.theatlanticcities.com/commute/2013/05/planning-our-transportation-future-millennials-mind/5575/

Bailey, A. 2009. "Population geography: Lifecourse matters," *Progress in Human Geography* 33: 407–18.

Beaupré, P., T. Turcotte, and A. Milan. 2006. "When is junior moving out? Transitions from the parental home to independence," *Canadian Social Trends* 82: 9–16.

Beer, A., T. Clower, D. Faulkner, and C. Paris. 2011. *Consuming Housing? Transitions through the Housing Market in the 21st Century*. Bristol: The Policy Press.

Beers, D. 2007. "Vancouver eats its young," *The Tyee* (online magazine), October. At: http://thetyee.ca/Views/2007/10/23/Vancouver/

Bernard, A. 2013. "Unemployment dynamics among Canada's youth," *Economic Insights* 24: 1–7.

Bisom-Rapp, S. and M. Sargeant. 2013. "Diverging doctrine, converging outcomes: Evaluating age discrimination law in the United Kingdom and the United States," *Loyola University Chicago Law Journal* 44: 717–70.

Blaauboer, M. 2011. "The impact of childhood experiences and family members outside the household on residential environment choices," *Urban Studies* 48: 1635–50.

Blaikie, H. 2013. "Age discrimination internationally: Canada," *Agediscriminaiton.info*. At: www.agediscrimination.info

Boudarbat, B., T. Lemieux, and C. Riddell. 2010. "The evolution of the returns to human capital in Canada, 1980–2005," *Canadian Public Policy* 36: 63–89.

Bourne, L. 2007. *New Urban Divides: How Economic, Social and Demographic Trends Are Creating New Sources of*

Urban Difference in Canada (Research Bulletin No. 33). Toronto: Centre for Urban and Community Studies, University of Toronto.

Boyd, M., and D. Norris. 1999. "The crowded nest: Young adults at home," *Canadian Social Trends* 52: 2–5.

Bridge, T. 2011. "Going gone," *Vancouver Magazine*, November. At: www.vanmag.com/News_and_Features/Gone?page=0%2C0

Bromley, R., A. Tallon, and C. Thomas. 2005. "City centre regeneration through residential development: Contributing to sustainability," *Urban Studies* 42: 2407–29.

Calvert, E. 2010. *Young People's Housing Transitions in Context* (working paper). Southampton: Centre for Population Change, University of Southampton.

Canadian Broadcasting Corporation (CBC). 2012. "Baby boomers push back against age discrimination," *CBC News*. At: www.cbc.ca/news/canada/story/2012/05/29/census-age-discrimination.html

Carr, D.S. 2004. "Psychological well-being across three cohorts: A response to shifting work-family opportunities and expectations? Work and well-being across three cohorts," in O.G. Brim, C.D. Ryff, and R.C. Kessler, eds., *How Healthy Are We? A National Study of Well-being at Midlife*. Chicago, IL: University of Chicago Press.

Carroll, B., and R. Jones. 2000. "The road to innovation, convergence or inertia: Devolution in housing policy in Canada," *Canadian Public Policy* 26: 277–93.

Champion, A.G. 2001. "A changing demographic regime and evolving polycentric urban regions: Consequences for the size, composition and distribution of city populations," *Urban Studies* 38: 657–77.

Chatterton, P. 1999. "University students and city centres: The formation of exclusive geographies," *Geoforum* 30: 117–33.

———, and R. Hollands. 2002. "Theorizing urban playscapes: Producing, regulating and consuming youthful nightlife city spaces," *Urban Studies* 39: 95–116.

Constable, G. 1996. *The Reformation of the Twelfth Century*. New York: Cambridge University Press.

Csanady, A. 2012. "World Health Organization declares Waterloo an 'age-friendly' city," *The Record*, January. At: www.therecord.com/news-story/2597007-world-health-organization-declares-waterloo-an-age-friendly-city/

CTV. 2012. "Ageism widespread in Canada, survey finds," *CTV News*, November. At: www.ctvnews.ca/canada/ageism-widespread-in-canada-survey-finds-1.1021641

District of Saanich. 2013. "Age friendly cities and the municipality of Saanich," *District of Saanich*. At: www.saanich.ca/parkrec/community/agefriendly.html

Elchardus, M., and W. Smits. 2006. "The persistence of the standardized life cycle," *Time and Society* 15: 303.

Fawcett, M. 2007. "Hey boomers, stop banging on us: Young people are tired of your baseless lectures," *The Tyee*, September. At: http://thetyee.ca/Views/2007/09/26/Boomers/

Feijten, P., P. Hooimeijer, and C. Mulder. 2008. "Residential experience and residential environment choice over the life-course," *Urban Studies* 45: 141–62.

Filion, P., T. Bunting, D. Pavlic, and P. Langlois. 2010. "Intensification and sprawl: Residential density trajectories in Canada's largest metropolitan regions," *Urban Geography* 31: 541–69.

Florida, R. 2013. "Why no city can afford to forget about seniors," *The Atlantic Cities*, January. At: www.theatlanticcities.com/neighborhoods/2013/06/why-no-city-can-afford-forget-about-seniors/5789/

Furlong, A., and F. Cartmel. 2007. *Young People and Social Change: New Perspectives*, 2nd edn. Berkshire: McGraw-Hill.

Garvin, T. 2012. "Can we get old here? Senior's perceptions of seasonal constraints of neighbourhood built environments in a northern, winter city," *Geografiska Annaler: Series B, Human Geography* 94: 369–89.

Germain, A., and D. Rose. 2000. *Montreal: The Quest for a Metropolis*. New York: John Wiley & Sons Inc.

Giddens, A. 1984. *The Constitution of Society*. Cambridge: Polity.

Hagestad, G., and P. Uhlenberg. 2006. "Should we be concerned about age segregation? Some theoretical and empirical explorations," *Research on Aging* 28: 638–53.

Hall, P. 1996. *Cities of Tomorrow: An Intellectual History of Urban Planning and Design in the Twentieth Century*, Updated edn. Cambridge: Blackwell.

Harris, D. 2011. "Condominium and the city: The rise of property in Vancouver," *Law & Social Inquiry* 36: 694–726.

Joenson, H. 2013. "We will be different! Ageism and the temporal construction of old age," *The Gerontologist* 53: 198–204.

Kern, L. 2010. "Gendering re-urbanisation: Women and new-build gentrification in Toronto," *Population, Space and Place* 16: 363–79.

Kershaw, P. 2011. "Movement should change focus: Occupy Wall Street zeros in on 'fat cats,' but this thinking overlooks important generational realities," *The Vancouver Sun*, 18 October: A11.

———. 2013. *Generation squeeze*. At: http://gensqueeze.ca/

Kolm, K. 2013. "Why are so many of Canada's young people out of work?" *CBC News*, June. At: www.cbc.ca/news/canada/story/2013/06/12/f-millennial-youth-unemployment.html

Lauster, N. 2010. "A room to grow: The residential density-dependence of childbearing in Europe and the United States," *Canadian Studies in Population* 37: 475–96.

Laz, C. 1998. "Act your age," *Sociological Forum* 13: 85–113.

Lefebvre, S. 2003. "Housing: An income issue," *Canadian Social Trends* 68: 15–18.

Lewis, J. 2013. "Creating age-friendly communities. Diversity and inclusion events: Dialogues on diversity," *Region of Waterloo*. At: www.regionofwaterloo.ca/en/regionalGovernment/resources/DOCS_ADMIN-1374206-v1-john_Lewis_presentation_-_creating_age_friendly_communities.pdf

Ley, D. 2007. "Countervailing immigration and domestic migration in gateway cities: Australian and Canadian variations on an American theme," *Economic Geography* 83: 231–54.

———. 1996. *The New Middle Class and the Remaking of the Central City*. Oxford: Oxford University Press.

Lui, C-W., J.A. Everingham, J. Wrburton, M. Cuthill, and H. Bartlett. 2009. "What makes a community age-friendly: A review of international literature," *Australasian Journal of Ageing* 28, 3: 116–21.

Mason, G. 2011. "Gen Y's pampered losers are just a myth," *The Globe and Mail*, September. At: www.theglobeandmail.com/news/opinions/opinion/gen-ys-pampered-losers-are-just-a-myth/article2175084/

McDaniel, S. 2004. "Generationing gender: Justice and the division of welfare," *Journal of Aging Studies* 18: 27–44.

McLeod, P.B., and J.R. Ellis. 1983. "Alternative approaches to the family life cycle in the analysis of housing consumption," *Journal of Marriage and Family* 45: 699–708.

Menec, V., R. Means, N. Keating, G. Parkhurst, and J. Eales. 2011. "Conceptualizing age-friendly communities," *Canadian Journal on Aging* 30: 479–93.

Mills, M. 2004. "Stability and change: The structuration of partnership histories in Canada, the Netherlands, and the Russian Federation," *European Journal of Population* 20: 141–75.

Mincer, J. 1974. *Schooling, Experience and Earnings*. New York: Columbia University Press.

Moos, M. 2012. *Housing and Location of Young Adults, Then and Now: Consequences of Urban Restructuring in Montreal and Vancouver*. Vancouver, BC: University of British Columbia, Ph.D. thesis.

———. 2013. "Generational dimensions of neo-liberal and post-Fordist restructuring: The changing characteristics of young adults and growing income inequality in Vancouver and Montreal," *International Journal of Urban and Regional Research*. Early view: http://onlinelibrary.wiley.com/doi/10.1111/1468-2427.12088/abstract

———. 2014. "'Generationed' space: Societal restructuring and young adults' changing residential location patterns," *The Canadian Geographer* 58: 11–33.

———. 2016. "From gentrification to youthification? The increasing importance of young age in delineating high-density living," *Urban Studies* 53: 2903–20.

Myers, D. 1999. "Cohort longitudinal estimation of housing careers," *Housing Studies* 14: 473–90.

Myles, J., G. Picot, and T. Wannell. 1993. "Does post-industrialism matter? The Canadian experience," in Esping-Andersen, G., ed., *Changing Classes: Stratification and Mobility in Post-industrial Societies*. London: Sage.

North, M., and S. Fiske. 2012. "An inconvenienced youth? Ageism and its potential intergenerational roots," *Psychological Bulletin* 138: 982–97.

Ontario Professional Planners Institute (OPPI). 2009. "Age friendly communities: A call to action," *Ontario Professional Planners Institute*. At: http://ontarioplanners.ca/PDF/Healthy-Communities/2009/Call-to-Action-Age-Friendly-Communities-June-18-20.aspx

Osberg, L. 2003. "Long run trends in income inequality in the United States, UK, Sweden, Germany and Canada: A birth cohort view," *Easter Economic Journal* 29, 1: 2003.

Pain, R., and P. Hopkins. 2010. "Social geographies of age and ageism," in S. Smith, R. Pain, S. Marston, and J. Jones III, eds., *The Sage Handbook of Social Geographies*. Los Angeles: Sage.

Phillips, B. 2010. *City Lights: Urban-suburban Life in the Global Society*. New York: Oxford University Press.

Pred, A. 1984. "Place as historically contingent process: Structuration and the time-geography of becoming," *Annals of the Association of American Geographers* 74: 279–97.

Riley, M. 1987. "On the significance of age in sociology," *American Sociological Review* 52: 1–14.

Rosenberg, M., and D. Wilson. 2010. "Younger cities, older cities, and cities in the balance: Spaces and places of the younger and older populations," in T. Bunting, P. Filion, and R. Walker, eds., *Canadian Cities in Transition: New Directions in the Twenty-First Century*, 4th edn. New York: Oxford University Press.

Rotberg, H. 2008. *Exploring Vancouverism: The Political Culture of Canada's Lotus Land*. Toronto: Mantua Books.

Ryder, N. 1965. "The cohort as a concept in the study of social change," *American Sociological Review* 30: 843–61.

Sargeant, M. 2013. "Young people and age discrimination," *E-Journal of International and Comparative Labour Studies* 2: 2–16.

Séguin, A-M., P. Apparicio, and P. Negron. 2008. *The Changing Spatial Distribution of the Elderly Population in Eight Canadian Metropolises: A Decreasing Segregation* (Working paper No. 2008-10). Montreal: Centre Urbanisation Culture Société (INRS).

Shanahan, M. 2000. "Pathways to adulthood in changing societies: Variability and mechanisms in life course perspective," *Annual Review of Sociology* 26: 667–92.

Sinoski, K. 2013. "Fewer drivers among urban youth in Metro Vancouver," *Vancouver Sun*, March. At: www.vancouversun.com/Fewer+drivers+among+urban+youth+Metro+Vancouver/8128997/story.html

Skaburskis, A. 1994. "Determinants of Canadian headship rates," *Urban Studies* 31: 1377–89.

———. 2001. "Housing prices and housing density: Do higher prices make more compact cities," *Canadian Journal of Regional Science* 23: 455–80.

Statistics Canada. 1971. Census of Canada—Profiles of census tracts. Accessed through the Canadian Census Analyzer at the University of Toronto. At: http://dc1.chass.utoronto.ca/census/

———. 1976. Census of Canada, 1976: Public use sample tapes on individuals. Accessed through the Canadian Census Analyzer at the University of Toronto. At: http://dc1.chass.utoronto.ca/census/

———. 1986. Census of Canada, 1986: Public use microdata on households and housing. Accessed through the Canadian Census Analyzer at the University of Toronto. At: http://dc1.chass.utoronto.ca/census/

———. 1991. Census of Canada—Profiles of census tracts. Accessed through the Canadian Census Analyzer at the University of Toronto. At: http://dc1.chass.utoronto.ca/census/

———. 2006. Census of Canada—Profiles of census tracts. Accessed through the Canadian Census Analyzer at the University of Toronto. At: http://dc1.chass.utoronto.ca/census/

———. 2011. "Population, urban and rural, by province and territory," *Statistics Canada: Summary Tables*. At: www.statcan.gc.ca/tables-tableaux/sum-som/l01/cst01/demo62a-eng.htm

———. 2016. "2016 census of population—Age (in single years) and average age," *Statistics Canada*. At: https://www12.statcan.gc.ca/datasets/Index-eng.cfm?Temporal=2016&Theme=-1&VNAMEE=Age%20%28in%20single%20years%29%20and%20average%20age%20%28127%29&GA=-1&S=0

———. 2018. "Table 11-10-0239-01. Income of individuals by age group, sex and income source, Canada, provinces and selected census metropolitan areas," *Statistics Canada*. At: https://www150.statcan.gc.ca/t1/tbl1/en/tv.action?pid=1110023901

Thomas, D. 2005. "Socio-demographic factors in the current housing market," *Canadian Economic Observer* 18(10): 3.1–3.9.

Turcotte, M. 2006. "Parents with adult children living at home," *Canadian Social Trends* 80: 2–10.

Twenge, J. 2006. Why Today's Young Americans Are More Confident, Assertive, Entitled—and More Miserable than ever before. New York: Free Press.

Van Diepen, A., and Musterd, S. 2009. "Lifestyles and the city: Connecting daily life to urbanity," *Journal of Housing and the Built Environment* 24: 331–45.

Vanderbeck, R.M. 2007. "Intergenerational geographies: Age relations, segregation and re-engagements," *Geography Compass* 1: 200–21.

Vosko, L., ed. 2006. Precarious Employment: Understanding Labour Market Insecurity in Canada. Montreal: McGill-Queens University Press.

Walks, A. 2001. "The social ecology of the post-Fordist/Global city? Economic restructuring and socio-spatial polarisation in the Toronto urban region," *Urban Studies* 38: 407–47.

———, and R. Maaranen. 2008. The Timing, Patterning, and Forms of Gentrification and Neighbourhood Upgrading in Montreal, Toronto, and Vancouver, 1961 to 2001. Toronto: Centre for Urban and Community Studies, University of Toronto.

Wesson, T. 2007. "The sectoral structure of Canada's economy," in T. Wesson, ed., *Canada and the New World Economic Order: Strategic Briefings for Canadian Enterprise*, 3rd edn. Concord, ON: Captus Press Inc.

Willcocks, C. 2011. Encouraging Family-friendly Condominium Development and Creating Complete Communities in Downtown Toronto. Waterloo, ON: University of Waterloo, M.A. Thesis.

World Health Organization (WHO). 2013. "Global age-friendly cities project," *World Health Organization*. At: www.who.int/ageing/projects/age_friendly_cities/en/index.html

Zerbisias, A. 2012. "Downtown Toronto kids lead enviable lives, their parents tell Holyday," *Toronto Star*, July. At: www.thestar.com/news/gta/2012/07/16/downtown_toronto_kids_lead_enviable_lives_their_parents_tell_holyday.html

7 Urban Change through the Lens of Demography, Life Course, and Lifestyles

Ivan Townshend and Ryan Walker

Introduction

Demographic change is an important driver of social, economic, political, and cultural change. It affects the evolution or emergence of lifestyles, consumption patterns, housing markets, intergenerational relations, types of social inclusion and exclusion, and diversity. In the urban environment we see manifestations of demographic change through time, and we see and experience the spatial outcomes of demography in our everyday lives.

Even a brief introduction suggests that urbanists are working in the dark if they do not have some appreciation for demographic, **life course**, and lifestyle changes affecting Canadian cities. Accordingly, in this chapter we outline a selection of demographic trends that are important structuring parameters of twenty-first-century urbanism in Canada. The first section examines some of the key features of Canadian demographic change. The second follows with a discussion of the life course transitions people typically experience and how these relate to lifestyle changes. The third section brings demographic, life course, and lifestyle changes together through a look at a selected series of effects on the built environment.

Key Forces of Social and Demographic Change in Canadian Cities

In 2016 Canada was home to 35.2 million people (Statistics Canada, 2017a). Canada is a medium-sized country, ranked as only the thirty-eighth largest in population out of 237 countries (*World Factbook*, 2018). Nevertheless, Canada has experienced tremendous population growth in recent decades and in recent years has been the fastest growing of all G7 countries (Statistics Canada, 2016; *Maclean's*, 2017). Indeed, the country's population has more than doubled since 1951. It is important to understand the sources of this growth. Demographers typically allocate population growth or change to three distinctive components: fertility (births), mortality (deaths), and migration (immigration).

Patterns of fertility in Canada have changed dramatically over the past few decades (Figure 7.1). If you are in your twenties or thirties now, think back to your great-grandparents' generations—those most likely born in the 1920s or 1930s. It was common for these people to be born into relatively large families, and, by the time the women born after World War I were in their peak child-bearing

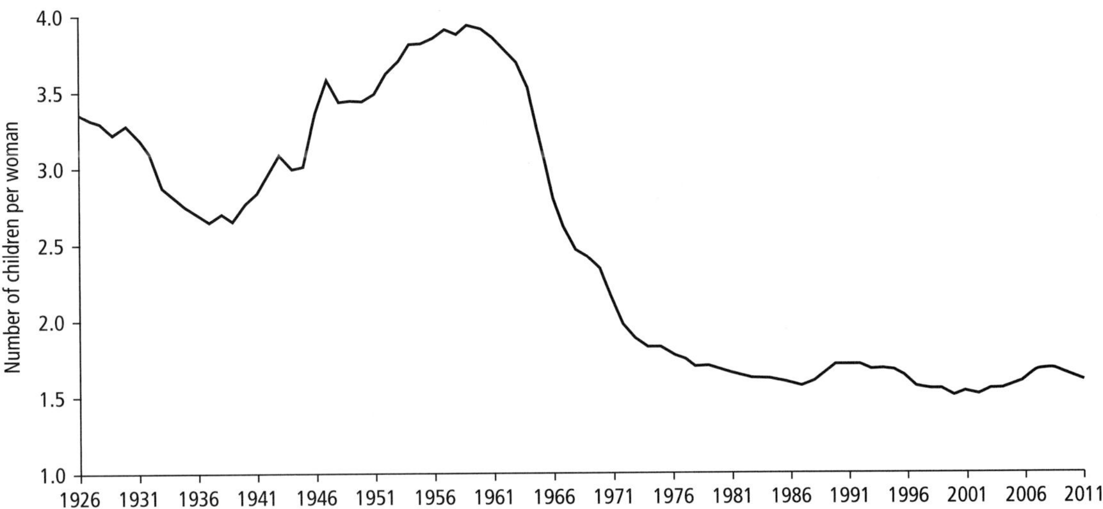

Figure 7.1 Total Fertility Rate, Canada, 1926–2011
Source: Statistics Canada, 2016: 17.

years in the late 1940s to early 1960s, as a group they were producing a greater number of children than at any other time in Canada's history. In the late 1950s, fertility rates peaked at almost four offspring per woman of child-bearing age (Statistics Canada, 2008b). Children of the post-World War II period are part of the biggest **cohort** of children ever born in this country, the "baby boomers." The baby boomers are a notable feature of Canada's age structure and have had an impact on Canadian society at every stage of their progression through the life course, impacting everything from their need for schools, jobs, and housing, as well as their consumer preferences (Foot with Stoffman, 1996). Population pyramids, which depict the age-sex structure of a population, clearly show this large "bulge" of baby boomers, and as they have aged, this bulge has gradually moved through the age structure (Figure 7.2). One analogy used to describe this process is the "pig in the python," a symbolic statement of how, through time, the baby boomers are detectable within the age structure of the population (Dytchwald, 1990).

Figure 7.1 shows that following the introduction of new birth control methods such as the pill in 1961 (legalized for birth control in 1968), and the increased choices of women with respect to the termination of pregnancies, fertility rates began to plummet in the early 1960s and continued a systematic decline until the late 1980s. Despite a very modest increase in fertility during the late 1980s (the baby boom echo), fertility rates continued to decline, reaching the lowest levels in 2002 at just over 1.5 children per woman (Statistics Canada, 2015). In 2016, Canada's total fertility rate was still at 1.5 (Statistics Canada, 2018a), but there are important regional variations (Statistics Canada, 2015), with rates as low as 1.4 in British Columbia, and as high as 3.0 in Nunavut. This pattern of change in fertility through time is common to most of the developed world and is a significant feature of what has been called the second demographic transition (Bourne and Rose, 2001; Lesthaeghe, 1995). There are a number of implications of such fertility trends. Quite simply, fewer babies being born contributes to population aging, increasing the average (median) age, which has risen from 23.9 years in 1921 to 41.2 years in 2016 (Statistics Canada, 2017a). A second implication of low levels of fertility concerns population replacement. Generally, after accounting for the chances of accidental death or infant mortality, the replacement rate (i.e., to maintain a stable population size) is 2.1 children born to

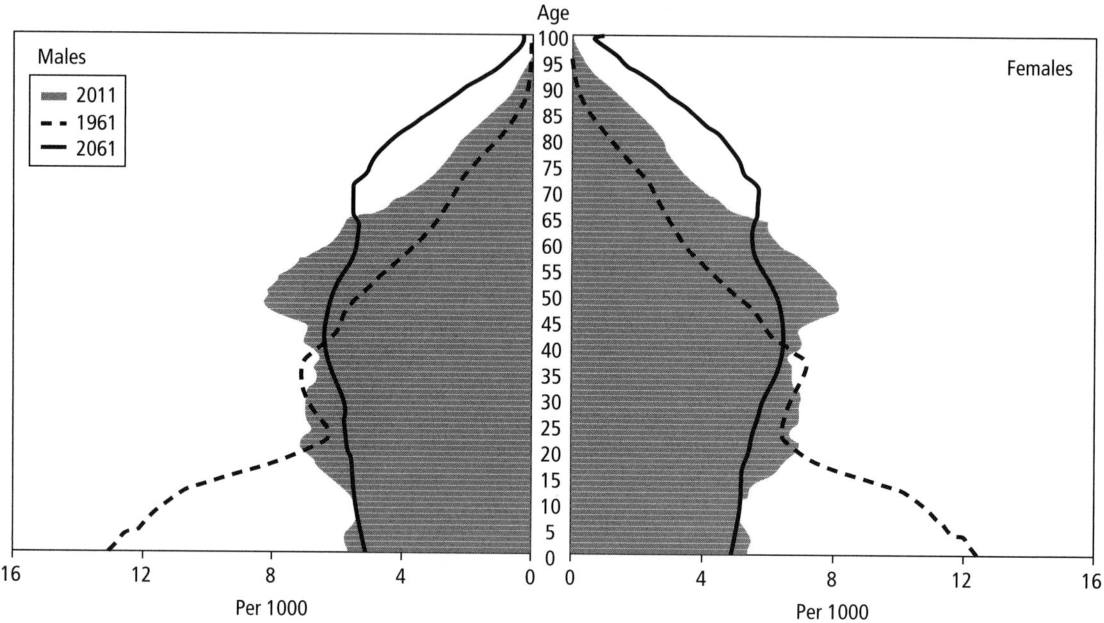

Figure 7.2 Population Pyramids, 1961, 2011, and Projection for 2061
Source: Statistics Canada, 2016: 8.

every woman. Given that fertility rates have been below replacement levels since the late 1960s and are significantly below replacement at present, in the absence of immigration the Canadian population would decline as baby boomers pass away over this and subsequent decades. Immigration is a critical feature of stabilizing population levels as well as population growth in Canada.

The low aggregate figures for fertility rates across the Canadian population mask a demographic issue of great transformative potential in some cities, particularly in western Canada. Canada's Aboriginal population grew by 42.5 per cent between the 2006 and 2016 censuses, which is more than four times the growth rate in the non-Aboriginal population (Statistics Canada, 2017b). High fertility rates and increased life expectancy make up one factor leading to the high Indigenous population growth rate. The second factor is the continued rise in self-reported Aboriginal identity on the census. In cities like Saskatoon, Regina, and Prince Albert, Saskatchewan, children and youth (i.e., aged 24 years and younger) made up just over half of the Aboriginal urban population in 2016 (Statistics Canada, 2017c). The Aboriginal population accounts for a significant part of the total urban population in each of these cities (approximately 11 per cent in Saskatoon, 10 per cent in Regina, 42 per cent in Prince Albert). According to population projections by Statistics Canada, by 2031 Indigenous people could make up between 21 and 24 per cent of the population of Saskatchewan. The corresponding figure in Manitoba is between 18 and 21 per cent (Statistics Canada, 2011).

The second component of population growth is death. Like most industrialized Western societies, Canada has seen profound increases in life expectancy at birth during the nineteenth and twentieth centuries (Decady and Greenberg, 2014). The earliest estimates in Canada (for 1831) placed male life expectancy at only 38.3 years and female life expectancy at 39.8 years (Beaujot, 1991). A century later (1931), life expectancies at birth had increased to 60 years for men and 62.1 years for women. For men, it has risen from 60 in 1931 to 79.8 years in 2015, and for women from 62.1 in 1931 to 83.9 years in 2015 (Nagnur, 1986;

Statistics Canada, 2008a, 2017j). Patterns of survival have also changed as a result of advances in health care as well as improved nutrition and lifestyles. Two key features of survival have contributed to changes in the demographic structure of Canadians: (1) the decline and near elimination of infant mortality, and (2) the increased life expectancy at almost all ages, and especially in old age. Together, these two features have led to an increasing "rectangularization" of the **survival curve** (Fries, 1980; Kraus, 1988; Simmons-Tropea and Osborn, 1993).

In the early part of the twentieth century (1931), infant mortality accounted for a significant portion of death in Canada, with approximately 1 in 10 children not surviving to their first birthday (Statistics Canada, 2018b). By 2015, infant mortality had been virtually eradicated, with fewer than 4.5 children per 1000 dying before age one (Statistics Canada, 2019). The second aspect of survival, the increase in longevity, can be seen at almost every age. In Canada, a man aged 65 in 2013 can expect to live another 19.0 years, while a woman aged 65 can expect to live another 21.9 years (Statistics Canada, 2018b). The aggregate figures for the Canadian population do, however, mask a disparity between Indigenous and non-Indigenous peoples in life expectancy. In 2001, the life expectancy of Indigenous men and women was about five years less than for non-Indigenous Canadians (Statistics Canada, 2008e). Results from the 2011 census, moreover, indicate that the life expectancy of the Indigenous population remains shorter than the non-Indigenous population (Statistics Canada, 2013a).

One notable issue is the disparity in life expectancies and survival patterns between men and women, with women surviving longer than men, although the gap has been closing in recent years. This situation is sometimes referred to as the "feminization of survival," meaning that old age in Canada is increasingly characterized by widowhood. This demographic has implications for the size and composition of elderly households and living arrangements. A second notable issue, when coupled with declining fertility rates, is that greater life expectancy contributes to population aging, reflecting the fact that death is being compressed into a relatively short period at the end of the scale of life expectancy. Canada is indeed an aging society, and in 2016, for the first time ever, the Canadian census reported more seniors (aged 65 and over) in the population (5.9 million or 16.9 per cent) than children aged 0 to 14 years (5.8 million or 16.6 per cent) (*Globe and Mail*, 2017a, 2017b). Changes in survival patterns have had important impacts on life course transitions and lifestyles, aspects of society that will be discussed later in the chapter.

Migratory increase is the third component of population growth. Canada has long been recognized as a country of immigrants, and distinctive waves of immigration to this country have been documented (Boyd and Vickers, 2000; Hiller, 2006; see Kobayashi and Preston, Chapter 8). Figure 7.3 shows the trends in the share of population growth attributable to natural increase and to migration. Prior to 2001, it can be seen that the majority of population growth was the result of natural increase. Between the early 1980s and 2010, the share of migratory increase rose dramatically (Chagnon and Milan, 2011). But an important transition occurred in 2001, and in the 2001–2011 period migratory increase (immigration) became the dominant engine of population growth. Figure 7.1 projects that this trend is likely to strengthen, as population growth from natural increase will continue to decline as the number of deaths approaches the number of births (Statistics Canada 2016: 6). Demographic projections show that immigration is essential to the sustainability and growth of the Canadian population because, unless fertility rates begin to rise again, deaths in Canada will outnumber births (to be expected in an aging population) by about 2030, and the Canadian population could become entirely dependent on immigration for its growth. Immigration has helped reduce the rate of population aging that otherwise would have occurred. More than ever before, immigration has contributed to this country's rich social diversity. In 2016, 7.5 million people, or one in five persons (21.9 per cent) in Canada were immigrants. Some 3.2 million people (9.3 per cent of the population)

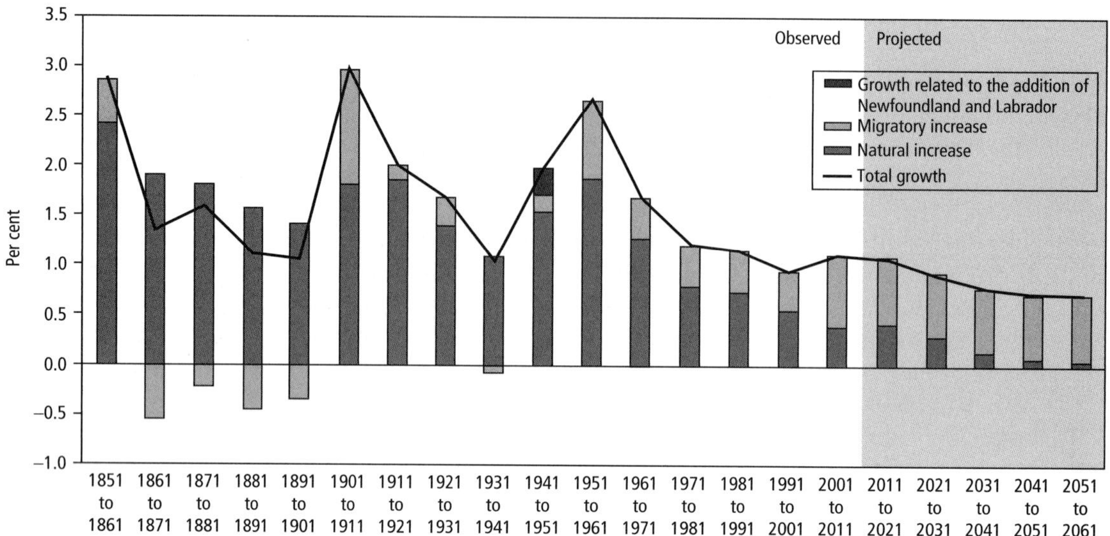

Figure 7.3 Annual Average Growth Rate, Natural Increase, and Migratory Increase per Intercensal Period, Canada, 1851 to 2061

Source: Statistics Canada, 2016: 6.

were immigrants who had arrived since 2001, while 3.5 per cent (1.2 million) of the population were recent immigrants, having arrived between 2011 and 2016 (Statistics Canada, 2017k).

Earlier immigrant streams to Canada had been mainly European in origin so social distance within different waves of immigrants or between immigrants and the host Canadian society were minimized. However, since the 1970s other originating regions, such as Africa, the Caribbean, Latin America, and especially Asia, have been favoured in Canadian immigration policy. Nearly 80 per cent of Canada's new immigrants were from these regions by the 1990s (Bourne and Rose, 2001; Boyd and Vickers, 2000; Halli and Driedger, 1999) (see Kobayashi and Preston, Chapter 8). Increasingly, immigrant streams have been visible minorities with higher levels of social distance from earlier European immigrants, or from charter groups (British and French). As a result, between 1981 (the first time visible minorities were enumerated) and 2016, the share of visible minorities in the population increased from 3.2 million (11.2 per cent) to 7.7 million (22.3 per cent). In short, social diversity on the basis of visible minority status and ethnicity is increasing the social mix in Canada with important implications for urban social life in this country. However, in 2016 three in ten visible minority persons were Canadian born (Statistics Canada, 2008c; Statistics Canada, 2013b, 2017l). In this regard, too, we should point out that the 1.7 million Indigenous Canadians are not considered "visible minorities" in the accounting of Statistics Canada (see Fawcett and Walker, Chapter 4).

While Canada's population is aging overall, the migratory decisions made by both younger and older Canadians regarding their ultimate choice of destination creates interesting city-specific differences (Murdie and Teixeira, 2006). Examples include the movement to retirement centres like Kelowna and Victoria or to relatively younger communities with greater employment opportunities and lower cost of living, such as Calgary and Saskatoon (see Moos, Chapter 6). Domestic migration trends are also an important source of population redistribution and a structuring parameter of Canadian cities as well. Studies have shown that jobs in the mining, oil, and gas sectors and in public administration are key drivers of interprovincial migration (Statistics Canada, 2008a). Some provinces consistently gain from interprovincial migration, others consistently lose, and others such as Saskatchewan and Newfoundland may experience dramatic changes in gains or losses. In some provinces such as Alberta, which has traditionally been susceptible to boom and bust cycles in the oil sector, net in-migration has been evident during boom cycles and net out-migration during bust cycles.

Changes in Life Course and Lifestyles

Fertility, mortality, and migration have introduced important changes in the ways the Canadian population has grown and will continue to grow. These factors point to a society that registers age or generational imbalances in **cohort** sizes, is aging rapidly, and—by necessity, if not by design—is becoming more ethno-culturally diverse. Structural changes in our economy and society have also laid a foundation, however, for shifts in how individuals negotiate life course changes, how they form households and families, and the rich diversity of lifestyles comprising our contemporary society.

A useful starting point for this discussion is the idea of the life course. Individuals typically go through a number of life course transitions and life course stages (Clark, 2007; Hareven and Adams, 1982; Murphy, 1987). Life course perspectives emphasize the importance and uniqueness of individual biography in relation to social and historical time, but at the same time they recognize that at any given time there will be a certain degree of synchronicity of individual and collective biographies. For example, the transition from child or adolescent to adult is typically marked by a series of transition events, such as leaving school, leaving the parental home, gaining full-time employment, and entering a conjugal union (living with a partner as a spouse or common-law couple). Over the past century or so, the normative expectations around the timing and characteristics of such events has changed, as young people today generally spend a longer period of time in childhood and education prior to initiating the independent adulthood stage of life. Likewise, the institution of retirement, coupled with life expectancies in later life, has radically transformed the lives of elderly people and afforded new opportunities for lifestyles of leisure outside of years in the labour force (Atchley, 1992; Markides and Cooper, 1987). A useful schema is to consider a broad life course conceptualization, originally devised as part of the Theory of the Third Age (Laslett, 1987; 1991). This schema recognizes four major stages or "ages" in life. Figure 7.4 is a simplification of some of the key features of these four ages.

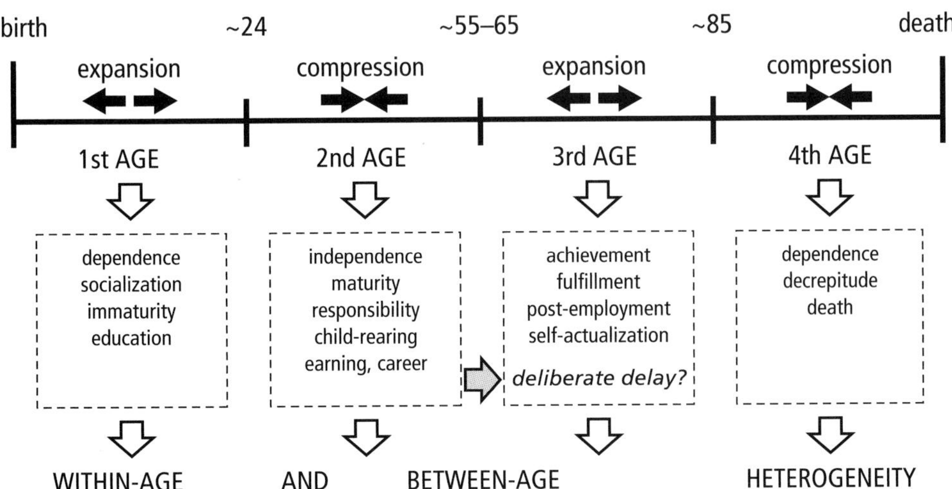

Figure 7.4 Summary Conceptualization of the Third Age Divisions of the Life Course
Source: Townshend, 1997: 21.

The First Age is a time of dependence, socialization, immaturity, and education. Clearly, the First Age begins at birth and, at first, is characterized by babyhood, childhood, and initial instruction. It is a period of socialization and, for the most part, dependence on others. In general, this phase of the life course has been expanded through time as young people delay and elongate the timing of major transitions between the First Age and Second Age. Young people are staying in school longer, pursuing graduate education to a greater extent, delaying entry into the labour force, living in the parental home for a longer period, and postponing or forgoing conjugal unions and child-bearing. In the early 1970s, about three-quarters of Canadian young adults had left school by age 22, whereas by 2006 less than half had left by this age (Clark, 2007). While women often make these transitions at younger ages than men, their delays in transition are often linked to post-secondary education. Over the past few decades, significant increases have occurred in Canadians', especially women's, educational attainment. There have been significant declines in the share of young people who do not complete high school, and negative labour force implications for not completing high school have worsened (Uppal, 2017). Post-secondary education has risen sharply, and since the early 1990s women have outnumbered men in Canadian universities (Statistics Canada, 2008d). Canada is the most highly educated of the Organisation for Economic Co-operation and Development (OECD) countries, and in 2016 more than one in four (28.5 per cent) people aged 25 to 64 had attained a bachelor's degree or higher level of education. There are some important gender dimensions to educational transitions in Canada, with higher rates of post-secondary education among females. In 2016 some 20.9 per cent of females had obtained at least a bachelor's degree, compared to 26.1 per cent for males (Statistics Canada, 2017a). And for the first time in Canadian history, the 2016 census reported that the majority of earned doctorate degrees among the 25- to 34-year-old population were awarded to females. In addition, the decade from 2006 to 2016 saw a significant rise in the share of young men entering the skilled trades sectors (Statistics Canada, 2017e).

The living arrangements of young people have also changed. The share of young adults aged 20 to 24 living in the parental home has increased sharply, rising from 41.5 per cent in 1981 to 62.6 per cent in 2016 (Statistics Canada, 2017d). While there have been humorous portrayals of this phenomenon, such as in the film *Failure to Launch* (Dey, 2006), a recent study of the living arrangements of young people has shown that the propensity to live with a parent has been rising not only for 20- to 24-year-olds, but also for 25- to 29- and 30- to 34-year-olds (leading-edge millennials). For these groups combined (those aged 20 to 34), more than one-third were living with a parent in 2016. While this is an important nationwide trend, its provincial and city-specific impact can vary. In census metropolitans areas (CMAs) such as Toronto, almost half (47.4 per cent) of all young adults aged 20 to 34 were residing with a parent in 2016 (Statistics Canada, 2017d). The propensity for young adults to live as couples has also decreased. For those aged 20 to 24, it has declined from 36.4 per cent in 1981 to 13.4 per cent in 2016, and in 2016 only 39.0 per cent of young people aged 20 to 34 were living as couples and not with a parent (Statistics Canada, 2013c, 2017h).

The Second Age, beginning around 24 years of age for many, is a period of independence, maturity, responsibility, and labour force activity. This typically is a period when most people start a career and a family, attain and maintain some degree of power and authority over others, and enjoy all the social and legal privileges of adulthood. For the population as a whole, this life course phase has become truncated (in relative terms) over the past century. Increases in survival, coupled with the individual ability to cease work at earlier ages—primarily as a result of the institutionalization of personal, corporate, and public retirement savings schemes, and the social sanction of early retirement—have accentuated the tendency toward earlier retirement (McDonald and Wanner, 1984, 1990; Roadburg, 1985). In recent years, however, we may be seeing a reversal of the trend to early retirement or, at least, episodic retirement, partial retirement, or various forms of bridge employment as people exit and re-enter the labour force (Carrière and Galarneau 2011; Hebert and Luong, 2008; Marshall and Ferraco, 2007; Singh and Verna, 2001; Stone, 2006). In general, however, the Second Age has become relatively compressed in time, both because of delayed transitions to labour force entry and independence among young people, but also because of early retirement. Typically, it occupies a smaller fraction of an individual's life course than it did in a historical context.

The same processes that have resulted in the compression of the Second Age of the life course have caused the relative elongation of the Third Age. This phase generally describes active and healthy seniors in society but is marked by a movement out of the labour force, or at least partial retirement. For some, and particularly early retirees, nearly as much of their life will be spent in retirement as was spent in the Second Age. The Third Age is, theoretically and demographically speaking, an unprecedented opportunity for a large share of society (and increasingly large in an aging society) to engage in many years of fulfillment outside of the labour force (Laslett, 1991). Considering advances in life expectancy at all ages, it is quite possible that someone who enters the labour force at 25 and retires at 55 will spend as many years in retirement as in the labour force. This concept was marketed fervently by financial institutions (e.g., London Life's "Freedom 55") and developers in the 1980s and 1990s, and was embraced and planned for by many, although less attention seems to have been given to the Freedom 55 concept in the last decade. The outlook on seniors' lifestyles can be seen in many cities in North America, most notably in new forms of exclusive retirement villages and age-restricted developments.

Although the vast majority of Canadians do not work past the age of 65, and the majority of retirees are early retirees, the 1980s and 1990s may have been overly optimistic about the long-term potential for universal early retirement. While there was a systematic decline in the average age of retirement between the 1970s and 1997, retirement age rose again in the early 2000s but seems to have levelled off since about 2004 (Figure 7.5). However, more recent indicators point to a lengthening of the work life, with a reversal in the pattern of declining share of people over 55 in the labour force. Indeed, the labour force is now considerably older than it was, and there is a larger share of people aged 55 and over in the labour force now than there was in the early 1970s. In 2016 more than one-third of the Canadian labour force was 55 or older, and labour force participation among the 55-plus population was 38 per cent—the highest value ever recorded for this group (Fields et al., 2017). With the legal abolition of mandatory retirement at age 65 in Canada in

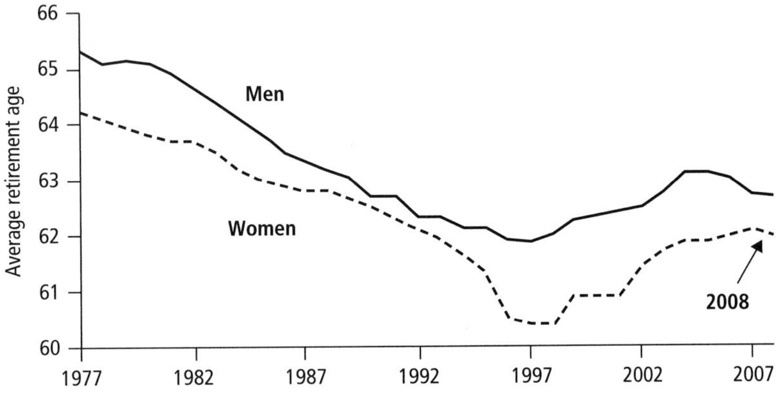

Figure 7.5 Average Retirement Age, Canada, 1977 to 2008

Source: Carrière and Galarneau, 2011: 10.

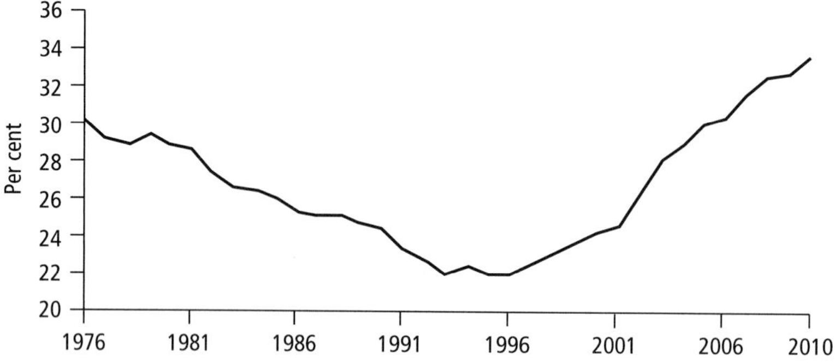

Figure 7.6 Employment Rate Trend for People Aged 55+, 1976 to 2010
Source: Carrière and Galarneau, 2011: 4.

2012, retirement is becoming more complex, with structural and demographic changes in the labour force, multiple entries and exits from retirement, and partial and episodic retirement. It seems that the idealized scenario of declining retirement age (Gower, 1997) and near universal entry into the Third Age at an early retirement age of 55 is a fading ideal, or it is a scenario that will be attainable for only a small minority of Canadians. By 2015 the average retirement age had risen to 63, and news articles with titles such as "Goodbye Freedom 55" proclaimed the end of that dream for many (MacNaughton, 2016). Such changes in the dynamics of an aging labour force and in patterns of retirement will continue to have implications for the social and built form of Canadian cities.

The Fourth Age, marked for many people as the late seventies or early eighties, comprises the later years in life or, more generally, the time when people have been called the "old-old" (Neugarten, 1974). As of 1996, the disability-free life expectancy (for both sexes) of someone aged 65 in Canada was 11.7 years, and it continues to rise (Mandich and Margolis 2014). Generally, it is not until the elderly are in their mid-eighties that we see a majority of seniors experiencing severe disabilities, a period akin to Shakespeare's "seventh age of man," with increasing loss of mobility and independence, health concerns, senility, and decrepitude (Laslett, 1991; Norland, 1994). Either way, for most, this last phase of the life course has been pushed well past the traditional age of retirement (65), and compressed, from an historical point of view, largely as a function of advances in health care and nutrition.

Considerable heterogeneity prevails in the timing of transitions within each of the four ages, as in the social composition of people within each of these phases of the life course, when people may be differentiated by age, socio-economic status, family structures, ethnicity, race, gender, and lifestyles. These differences no doubt create highly variable pathways through the life course, different opportunities and constraints, various inequalities, and different life experiences and biographies. As these generalized life course phases intersect with micro and macro differentiation in our society, they define the potential for myriad life paths and expressions of lifestyles.

One of the most important trends in Canadian society over the past quarter-century has been the revolutionary changes in the structure and composition of households and families. We have already seen how basic demographic changes, such as those of fertility and survival, have been influential in reshaping the age structure of the population, have led to fewer younger people, and have resulted in an aging population. These trends do not occur in isolation from broader forces of societal change, nor from changes in normative behaviours and life choices throughout the life course. The shifting choices that people make with respect to household and lifestyle fundamentally affect the urban experience.

Smaller Households

The size of Canadian households has been declining for more than a century (Rose and Villeneuve, 2006). In 1901, the average household in Canada had five people. In 2016 the average was less than half this size at 2.4 people (Statistics Canada 2017a). This transformation in the size of households is largely attributable to the growth (both absolute and relative) in the number of small households and the decline of large households of four or more people. Between 1981 and 2016 the number of households in Canada grew by 70 per cent, but there are important differences in the growth of households by household size. Figure 7.7 shows how dramatic these changes have been, with the number of single-person households rising from 1.68 million in 1981 to 4.00 million in 2016, an increase of 138 per cent over this time period. One-person households have outnumbered couple households with children since 2011, and the trend is continuing. The number of two-person households has also risen sharply from 2.40 million in 1981 to 4.83 million in 2016, an increase of 101 per cent over 35 years. The number of three- or four-person households has also increased, but only marginally, while the number of households with five or more people has declined. By 2016, the majority (62.6 per cent) of Canadian households were small households of one or two people (Figure 7.8).

Within the smaller household group, the fastest-growing segment has been single-person households. Since the household is generally considered the basic unit of consumption as well as a means of achieving economies of scale in living expenses, there are some important implications for the rise in single-person households, especially in terms of housing affordability (as a result of increased consumption of/demand for smaller units) and income distributions (Miron, 1993). The rise of the single-person household is linked to what some have described as a major shift in the propensity to live alone (Clark, 2007; Rose and Villeneuve, 2006). In 1951 only 2.6 per cent of households were one person. By 1981 this had risen to 20.3 per cent, and by 2016 to 28.2 per cent. Between 1981 and 2016, the share of one-person

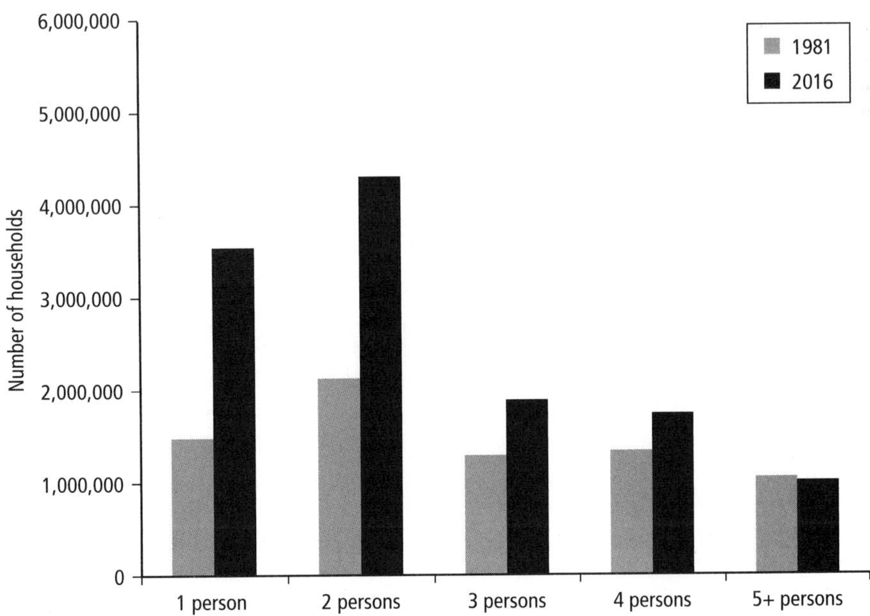

Figure 7.7 Number of Households by Household Size, 1981 and 2016

Source: Compiled from data in Statistics Canada, 1982, 2017a.

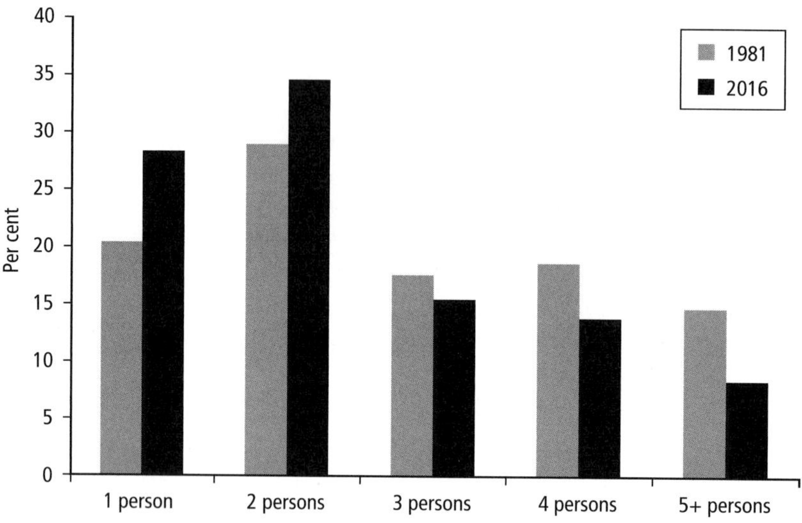

Figure 7.8 Percentage of Households by Size, 1981 and 2016
Source: Compiled from data in Statistics Canada, 1982, 2017a.

households grew by 136 per cent. While living alone may have been a significant phase for people in the early stages of the Second Age (i.e., having just left the "nest") or for some in the later stages of the Third Age, it has become increasingly prevalent for people to live alone in mid-life as well as in later life (Laslett, 1991).

A further reason for the rise in single-person households has been changes in the values and meanings surrounding marriage and family formation and delays in conjugal unions, both of which are linked to the later transition from the First to the Second Age. The age of first marriage for both men and women has been rising steadily since the 1960s, reaching approximately 31.0 years for men and 29.6 years for women by 2008 (Clark, 2007; Milan, 2013). Considerably fewer young adults are entering marriage or other conjugal relationships at an early age. In 1971, for example, 65 per cent of men and 80 per cent of women were in or had been in a conjugal relationship by the age of 25. By 2001 these rates had dropped to 34 and 49 per cent, respectively. Young adults more often are choosing to live alone, or with a roommate, but also, as we noted earlier in this chapter, a substantial share continue to live with at least one parent (Figure 7.9). In essence, they are establishing their own homes and their own families at older ages than previous cohorts of young adults (Statistics Canada 2017d). Interestingly, among young people there is also an increasing tendency to live alone even when coupled—a phenomenon sometimes called "living apart together (LAT)" or "non-cohabiting couples" (Turcotte, 2013). The phenomenon is also seen among other age groups, but at much lower rates. There may be many reasons for this, including work circumstances, choice, or the desire to maintain independence even though coupled. In 2011, 31 per cent of people aged 20 to 24 and 17 per cent of those aged 25 to 29 were in a LAT relationship (Turcotte, 2013).

New Family Configurations and Fluid Conjugal Relationships

There have also been transformations in the meaning and composition of families in Canada (Statistics Canada, 2012a). The historical comparability of family structures has become more complex as revisions to definitions of census families have occurred. For example, in years past the distinction between married and common-law couples (with or without children) seemed important and was presented separately in many census documents. The most recent census, however, collapses these categories, and presents data for "couple families"

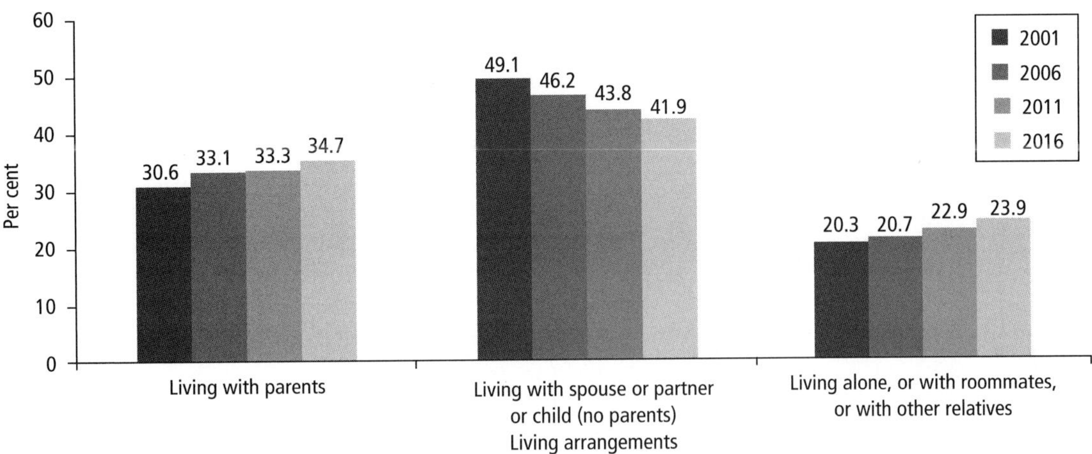

Figure 7.9 Living Arrangements of Young Adults Ages 20 to 34, Canada, 2001 to 2016

Source: Statistics Canada, 2017d: 3.

as one group. which includes married couples, common-law couples, opposite-sex, and same-sex couples. This is a good example of how changing societal norms and values are represented in the different types of data that are presented. The current (2016) designation of census families is shown in Figure 7.10. In 2016 married couple families were still the most prevalent family type, although Figure 7.10 shows a substantial decline in share from 83.1 per cent in 1981 to 65.8 per cent in 2016. This decline has been offset by the rise in share of both common-law couple families (increase of 12.1 per cent) as well as lone-parent families (increase of 5.1 per cent). The presence of children in couple families has also been on the decline. Of all couple families (i.e., both married and common law), the share with children present declined from 64.1 to 51.1 per cent between 1981 and 2016. This may be a function of an aging society and "empty nester" status, as well as declining birth rates. Nevertheless, the rising share of childless couple families (almost half of all couple families in 2016) is also linked to the rising share of small (two-person) households discussed elsewhere in this chapter.

The fluidity with which people move in and out of various types of families also has increased. In part due to separation and divorce but also because common-law unions are less stable than marriages, people more frequently move in and out of relationships throughout their life course, most notably during the Second and Third Age (Milan, 2000, 2013). Less than 10 per cent of marriages end in divorce within the first five years of marriage, whereas about half of common-law unions dissolve within that time frame (Milan, 2000; Statistics Canada, 2008a). Despite the fact that about four in ten marriages end in divorce, which also contributes to the formation of single-person households, approximately 75 per cent will remarry or recouple, creating an episodic pattern of marriage and remarriage. This pattern gives rise to a variety of configurations of blended and stepfamilies, either married or common-law, either of which can also be defined as opposite-sex or same-sex couples.

Recent years have seen a growing social acceptance and institutionalization of the rights of same-sex couples, including the legalization of same-sex marriages in 2005. Canadian same-sex couples were first enumerated with voluntary identification in 2001, and while undoubtedly under-reported, it was estimated that there were just over 34,000 same-sex couples in Canada, about 0.5 per cent of all couple families. By 2016 this count had risen to almost 73,000, or about 0.9 per cent of all couples. These couples are not necessarily all small households or couples without children at home, although this is the case for the vast majority. Twelve per cent of same-sex

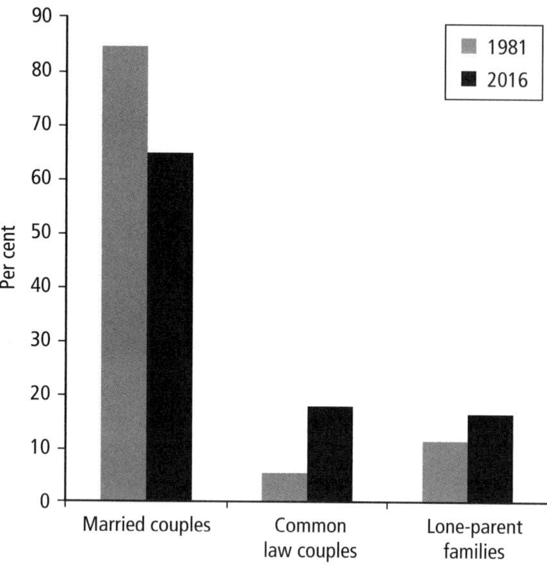

Figure 7.10 Percentage of Census Families by Structure, 1981 and 2016

Source: Compiled from data in Statistics Canada, 1982, 2017a.

couples reported having children living at home in 2016 (Statistics Canada 2017i).

Cities are impacted by these increasingly complex forms of families with different configurations. The living arrangements of intact families, simple stepfamilies, complex stepfamilies, as well as skip-generation families and multigenerational households play an important part in shaping changing family and housing needs (Statistics Canada, 2017g, 2017k). Interestingly, perhaps in part because of the changing ethno-cultural mix in Canada, we are seeing a rising share of multigenerational households, and indeed this is now the fastest-growing form of household (Statistics Canada, 2017f). Figure 7.11 schematically portrays some of the various family configurations that urbanists need to consider in understanding the evolving social characteristics of Canadian cities and neighbourhoods.

The Complexity of Lifestyles

Geographers have long been interested in the ways that people use their time, how time is allocated to different functions in space, how space is segmented with different types of people or consumers, and how lifestyles are manifest in time–space interactions (Chapin, 1965). In recent years there has been a resurgence of interest in spatial lifestyle studies (Schnell and Benjamini, 2001, 2002; Townshend and Davies, 1999). Schnell and Benjamini (2002) remind us that lifestyles are an intersection of time, activity, and space patterns, and so lifestyles can differ on such things as the radius of daily activity; the importance of telecommunications; types and intimacy of social relations; extent of orientation to a home base; and differences in the value placed on family, social life, work, and leisure.

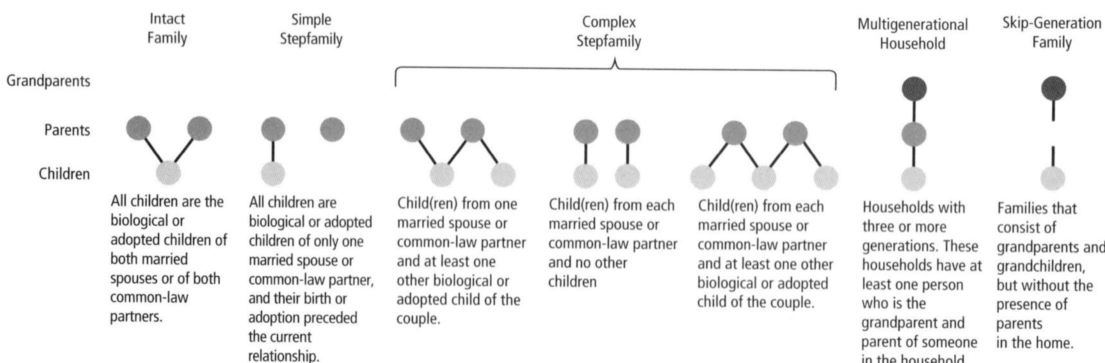

Figure 7.11 Selected Family and Household Arrangements

Source: Adapted by authors from separate figures and definitions in Statistics Canada, 2017m.

Lifestyles are often associated with or labelled in regard to population sub-groups or in terms of market-based geo-demographic segments, so it makes little sense to try to define First Age, Second Age, and Third Age lifestyles, as each of these ages can include a wide array of lifestyles. Market researchers often develop complex segmentation models and typologies to define groups of people within distinctive generations (Statistics Canada, 2012b; Cahill, 2006). Many of these typologies attempt to link key lifestyle traits with behaviours, outlook on life, types of social relationships, and especially consumption preferences. Foot (with Stoffman, 1996), for example, has shown how some of these lifestyle traits have changed through Canadian generations, impacting everything from people's preferences for different types of social relationships to demands for housing. Indeed the creation of consumer profiles through residential postal code geographies has been a booming industry since the 1980s (Burrows 2008; Singleton and Speilman 2014; Webber 2007; Weiss 2000).

Some segmentation analysts have focused on specific generations, such as the "millennials," which may also be referred to as the Net Generation, Screenagers, Digital Natives, Generation Y, or the Echo Boom generation (Alsop, 2008; Howe and Strauss, 2007; see canadianmillennials.ca; see Moos, Chapter 6). Millennials are loosely considered to be those born between the early 1980s and mid-1990s, and so leading-edge millennials may be in the early stages of the Second Age, while the youngest (tailing-edge) millennials were in the late stages of the First Age at the time of the 2016 census. Millennials are the children of the later baby boomers and early baby-bust generation (Statistics Canada, 2012b), and sometimes are called the "most wanted" generation in history. This generation of people is said to exhibit very different lifestyle orientations from previous generations (Howe and Strauss, 2007; Bibby 2009). They have few siblings, have always been considered special or important to their parents, are electronically savvy, value tight social networks, appreciate and value social and ethnic diversity, are group- or team-oriented rather than individualistic, and eschew preferential treatment. They also are high achievers and career-oriented. Some have claimed that these are "trophy kids," both as trophies to their parents (almost all are planned pregnancies) and because they have grown up in school and sports activities where everyone received a trophy for participating. From a demographic and especially geographical perspective, these individuals, as a subset of those in their First Age, have a unique set of preferences, behaviours, form of social networking, consumption potential, and propensity to be involved in civic affairs. As they move through the various transitions and stages of the life course, they will leave their own generation and lifestyle impact, and as Barton and Collins (2012) note, generational shifts in geographic behaviour will have significant impacts on urban morphology. However, while such generalizations may have some validity, there is in fact considerable heterogeneity within the millennials, leading market researchers to "segment" millennials into a variety of sub-groups (see canadianmillennials.ca).

From an urban point of view, some have speculated that the millennials will have an important impact on reshaping inner cities. They are said to have a preference for downtown and inner-city living over suburban living; eschew automobile use and embrace active travel, walkability, and public transit use; engage in urban sustainability initiatives; and appreciate the amenity-rich and socially heterogeneous characteristics of the central city (Belden Russonello Strategists LLC, 2013; Myers 2015; Walter-Joseph, 2015). However, this may be an oversimplification that does not take into account the diversity of lifestyles and residential preferences within the millennial population. For example, a recent study in Calgary mapped the relative concentration of 25- to 29-year-olds in 2011 by census tract and identified three distinctive regions with high concentrations of leading-edge millennials (Burgess, 2017). These areas included high concentrations in central city neighbourhoods with a high density of apartments and condominiums; an area surrounding the central city in predominantly gentrifying neighbourhoods; and a large concentration in a cluster of suburban neighbourhoods—the latter surrounding a new suburban hospital and associated employment centres of the region. This study compared the residential preferences, choices, and constraints of millennials within the three regions across a wide range of social, housing, neighbourhood, environmental, transportation, and personal characteristics. Both similarities and differences among the three regional groups were found, but the study also showed that for some millennials, a suburban lifestyle similar to that of previous generations was preferred. Millennials, therefore, should not be seen as a homogeneous demographic cohort (Counselling Foundation of Canada, 2017).

There are other examples where researchers have studied specific lifestyles and values, such as those of the hippie generation or the yuppies (young urban professionals), and have explored the activities and social relations of "swingles" (swinging singles) in large metropolitan areas. Some have also focused on the lifestyles of the non-cohabiting LAT couples (Partridge, 1973; Smith, 1987; Starr and Carns, 1972; Turcotte, 2013). The urban lifestyle factor of selected demographic groups has also been portrayed in film and television. One example is the film *Bridget Jones's Diary* (*Economist*, 2001; Fielding, 1996). While the film is a light-hearted romantic comedy of the 1990s, it also portrays a very serious side of the urban lifestyle (Maguire, 2001). Bridget Jones, a 30-something Generation-Xer (born in the late 1960s), is a single, educated professional working in the creative economy in a major metropolitan area (London, England). She lives alone in a small flat—a single-person household—and apart from juggling time at the gym and work, has plenty of time, money, and a passion for spending on things fashionable, frivolous, and fun. Her lifestyle is consumption-oriented—dining out, frequenting upscale wine bars, going on weekend holiday getaways, and so forth. She strives to enjoy her career, to find romance, and perhaps at some point to marry. Relatives and coupled friends constantly remind her that "the clock is ticking" and that her delay in marriage and child-bearing may become problematic. Bridget longs for acceptance outside her tightly knit group of friends, all of whom live similar lifestyles and are similarly fun- and consumption-oriented. True as the Bridget character is to life, it has been argued that many singles and other single-person households adopt their friendship network (which for Bridget is a heterogeneous group of professionals and artists, of diverse ethnic backgrounds and sexual orientation) as a surrogate family, in what some have suggested is a new form of "urban tribe" or "fictive kin" (*Economist*, 2001; Martin-Mathews, 2001; Rose and Villeneuve, 2006).

A similar example that underscores single female lifestyles and consumption potential is illustrated by the characters in the American TV series *Sex and the City*. In this show, the single young female (SYF) life, or the "new girl order" is celebrated as the quintessential postmodern lifestyle and is indeed a global trend. As Hymowitz (2007) explains, the new girl order is a combination of trends of delayed marriage, expanded higher education and labour force participation, and increasing urbanization. Together with global media it has produced an international lifestyle such as that portrayed by Carrie Bradshaw in *Sex and the City*.

A more recent portrayal of the demographic lifestyle effect can be seen in the television series *Portlandia*, which debuted in 2011. The series is a parody of the so-called "hipster" lifestyle (Lanham 2003) in Portland, Oregon. As Sappington (2016) notes, the show embraces progressivism, challenges gender constructs, celebrates different types of relationships (heterosexual, homosexual, platonic), and parodies the hipsters who call for humanely raised meat, protest the police, ostracize those who do not bring reusable bags to the grocery store, fight for unpasteurized milk, and embrace left-wing clichés (also see Bain and Mark, Chapter 15).

Beyond the imagery portrayed in these series, however, there is a real or tangible effect of these demographic and lifestyle factors. For instance, it has been noted that the "Bridget Jones Economy" or the Carrie Bradshaw lifestyle is an urban reality in many large cities. This demographic and the associated lifestyle have both a spatial and economic footprint (*Economist*, 2001; Hymowitz, 2007). They are clustered predominantly in central city apartments, often in close proximity to urban nightlife and in areas with high densities of singles. They are key consumers of emerging goods and services in these areas, including functions such as dry cleaners, health food stores, specialty delicatessens, supermarkets specializing in small portions and prepared foods, Internet kiosks, dating services, travel agencies, wine bars, specialty clothing boutiques, and health professionals. Hymowitz (2007) suggests that the SYF phenomenon has even driven diamond merchants to rethink their markets, with one company introducing a "right-hand ring"—a diamond for women with no marital prospects but longing for a rock. In some places, like Tokyo, distinctive central city residential geographies of SYF are emerging, such as entire neighbourhoods and districts defined by condominiums for the SYF lifestyle (Kubo and Yui, 2012). In the case of Portland, Lotking (2018) has suggested a real link between the show *Portlandia* and the in-migration of hipsters to the city and associated gentrification of selected neighbourhoods. Parmett (2018) further argues that *Portlandia*, as a site-specific form of television, has had a significant effect as a new urban regeneration scheme and has been complicit in new waves of gentrification. A key idea in these examples is that the lifestyles of selected demographic segments represent a particular constellation of identity, taste, consumption, style, and leisure. Most importantly, they are spatial lifestyles. Such lifestyles and living arrangements are also triggers of change in the social complexity and social ecology of cities (Davies and Murdie, 1991; Kubo and Yui, 2012; see Walks, Chapter 10; Bain and Mark, Chapter 15; Moos, Chapter 6).

The Third Age also is marked by social and lifestyle diversity, and the concept of "seniors" as a homogeneous group with similar traits fails to capture this diversity. One lifestyle phenomenon that has garnered recent attention from younger hipsters and real estate professionals is the movement of a growing number of aging baby boomers into the central city, a cohort playfully referred to as the "Broken Hip-sters" (*Wall Street Journal*, 2013). The impact of the typically wealthy aging baby boomers on the real estate market of the city centre, and on the younger (and often less affluent) hipsters that have pioneered the urban *lifestyle* there, has yet to be fully understood economically and culturally. As the large baby boomer cohort progresses through the stages of transition to retirement, this group's diversity and consumption potential will be even more noticeable (Foot with Stoffman, 1996), and market researchers and housing developers will continue to refine segments and niche groups within the Third Age (Dytchwald, 1990; Karpel, 1995; Seiler, 1986).

Demography, Lifestyle, and the Built Environment

There is little doubt that age, life cycle stage, and lifestyle have become increasingly commodified, not just in geo-demographic marketing but in the built environment. Developers have responded to a variety of age, preference, and lifestyle niche markets, the result being increasing diversity in the residential landscape. Two of the biggest trends in this area are the growth in condominium ownership and the rise of residential niche communities.

Given the increase in small households without children at home, more housing options today offer a smaller, more convenient, and often maintenance-free lifestyle. Condominium ownership, a type of **common interest development** first adopted in Canada in the late 1960s, has seen a dramatic increase over the past few decades, and in many CMAs the "condoization" of the central city since the early 2000s has reshaped inner-city areas surrounding the central business district. The share of households owning condominiums rose from 3.5 per cent in 1981 to 10.9 per cent in 2006 (Lo, 1996; McKenzie, 1994; 2003; Rea, McKay, and LeVaseur, 2008; Walker and Carter, 2010; see Harris, Chapter 11). Condominium living is considerably more prevalent in large metropolitan areas, with rates as high as 31 per cent in Vancouver, 19 per cent in Toronto, and 13 per cent in Montreal, but smaller CMAs such as St. Catharines-Niagara have much lower rates (5 per cent) (Rosen and Walks, 2013). In some cities, such as Toronto, condominium developments now account for the majority of new housing starts. The condo boom is not just being driven by young people seeking amenity-rich central cities (Burgess 2017), but also by empty nesters downsizing from their suburban homes. Indeed, Rosen and Walks (2013) show that the majority (57 per cent) of condo owners were aged 50 or older.

A second trend in response to the rise in differentiated lifestyles and preferences among those in their Second or Third Age has been innovation in the way residential communities are designed and developed. Many of these changes are linked to new expressions of the privatization of space and communities. Whereas the typical planned community of the 1950s or 1960s may have followed neighbourhood unit design principles, entire subdivisions today are being developed in response to changes in demography and lifestyle preferences, taking on a variety of expressions of product differentiation, gating, and exclusivity (Hodge and Gordon, 2014; Perry, 1929). The rise of gated communities (a type of common interest development) in various countries, including Canada, has been studied in some detail, with many authors expressing concern about the consequences of such developments, especially the loss of the public realm to private space (see Grant and Filion, Chapter 12; Blakely and Snyder, 1997; Blandy et al., 2003; McKenzie, 1994; Punter, 1990; Townshend, 2006; Webster, Glasze, and Frantz, 2002). Blakely and Snyder's (1997) pioneering study of gated communities in the United States identified three main types: (1) lifestyle communities (including retirement communities, golf course communities, and new towns), (2) prestige communities, and (3) security zone communities. Others have attempted to define typologies of gated communities in Canada and internationally (Coy and Pohler, 2002; Glasze, 2002; Grant, 2005; Grant and Mittelsteadt, 2004; Landman, 2006; Raposo, 2006). Despite difficulties in defining such typologies, the function of these designs clearly is to create landscapes of exclusivity and exclusion; as one author noted, these proprietary neighbourhoods are consumed as club goods within a kind of "shareholder democracy" (Glasze, 2002). The supply of and demand for private and gated residential communities will likely continue. Likely, also, is increased product differentiation as developers refine these commodities to appeal to more tightly defined lifestyles and market segments (Lang and Danielsen, 1997). This means that the geographical manifestation of private neighbourhoods in the city will become more complex and more differentiated.

Other forms of residential community privatization are evident in the built environment. Townshend (2006), for example, has discussed the trend toward the thematic development of subdivisions since the 1960s. Initial expressions of these, such as the development of golf course communities, were perhaps in direct response to lifestyle demands and the pursuit of leisure. Along similar lines, a number of developers experimented with "lake communities" in which a private lake, with controlled access, forms the community's recreational nexus. By the 1980s other types of niche communities on a smaller scale began to appear in Canadian cities in the form of retirement villages specifically designed for and marketed to the 55-and-over population (i.e., appealing to the Third Age). Many of these developments, such as "Horizon Villages" with age-restrictive covenants, actively advertised the concept of "Freedom 55" and offered a commodified form of community in which residents could enjoy condominium ownership, maintenance-free living, and a socially homogeneous set of neighbours. The popularity of these types of retirement communities, with particular appeal to elderly single persons, has grown rapidly. In Calgary, for example, almost all residential subdivisions contain at least one of these developments and almost all new subdivisions incorporate plans for at least one retirement community (Townshend, 2006).

A range of different thematic foci for new residential areas has become common. While developers have continued to build lake and golf course communities, other types can be seen. One example is the development of New Urbanist or neotraditional communities (Davies and Townshend, 2015). In response to problems of conventional suburban development patterns, these communities attempt to foster a more traditional urban lifestyle, despite being constructed in new suburban regions (see Grant and Filion, Chapter 12). Following the ideas of leading proponents of **New Urbanism**, such as Andrés Duany and Elizabeth Plater-Zyberk, features of these developments include the return to a smaller or more human scale of mixed housing development and mixed land uses, the return to a narrow grid and more pedestrian-friendly street network, and a reduction in vehicular traffic (Davies and Townshend, 2015; Katz, 1994). Neo-traditional communities usually include provision of a town centre with commercial and local administrative facilities, and aim to promote increased social interaction through design features such as front porches on houses and the placement of garages at the rear of lots. McKenzie Towne in Calgary and Cornell in Markham, just outside the Toronto CMA, are excellent examples of early efforts to adopt New Urbanist principles in Canadian cities.

A second example of new community development has explicitly focused on environment in its design. In these the developer retains some natural environmental features within the design scheme (e.g., wetlands, natural prairie grassland, forest, bird sanctuary). The share of open recreational space in these communities is considerably higher than in other neighbourhood districts. It also is typically more than what is legally required for subdivision approval, with some communities boasting as much as 40–50 per cent of the land area of the district retained as natural environment. Developers seem to have capitalized on the recent trend toward environmental stewardship, so that residents of these areas pay compulsory homeowner association fees to ensure appropriate maintenance and use of the natural features, and to ensure they remain intact in perpetuity. In some cases, a private residents' clubhouse with meeting or sports facilities has been constructed as part of the overall plan.

In the past decade a newer form of specialized community has begun to appear. These are new subdivision-scale areas that are live-work communities on a much larger scale than the kind of live-work arrangements conceptualized, and often criticized, in New Urbanist ventures. Incorporating many of the principles of Smart Growth, and perhaps in response to the growing lifestyle preferences for walkability and live-work proximity, these communities aim to provide a real mix of work and living, and in some ways are more like an office park with residential space included (Davies and Townshend, 2015). One such example is Quarry Park in Calgary, a community designed to contain 1.7 million square feet of office space (in high-rise as well as low-rise buildings) as well as 2200 residential units in a variety of housing forms. The development includes its own lake and a system of pedestrian walkways throughout the area, together with a mini downtown with retail, entertainment, and other services (Figure 7.12). The area is also a form of common interest development, and in reality, however, much of the housing will be unaffordable to some of the employees in the area. Nevertheless, it is an innovative development that aims to address some of the issues of sprawl, as well as respond to lifestyle factors.

Ivan Townshend

Ivan Townshend

Figure 7.12 Quarry Park Development in Calgary

Most of the newer forms of community development discussed above cater to middle- or higher-income residents. Yet Canadian cities have seen a dramatic increase in income inequality and income polarization over the last few decades (see Walks, Chapter 10), and there is a significant housing affordability crisis in Canada (Bunting et al., 2004; Marasco, 2018). Little attention seems to be given to the lifestyles and housing developments of those who are poor or who are becoming increasingly economically marginalized. Certainly the proliferation of secondary suites (basement suites, "granny flats," etc.) represent an important and contested form of housing provision (van der Poorten, 2018), but there are other initiatives that are gaining in importance. For instance "tiny home" communities are beginning to emerge (Craggs, 2017a,b) in some cities, although there seems to be some reluctance on the part of municipalities to allow such developments. Nevertheless, the trend may continue, and a number of organizations are planning to build tiny home villages to provide housing options for those who are poor or for war veterans (https://homesforheroesfoundation.ca/).

Municipalities are now also responding to the housing affordability crisis in novel ways. For instance, the City of Calgary has created Attainable Homes, a non-profit entity that uses innovative financing and assistance with down payments to allow low-income households to enter the

Ivan Townshend

Figure 7.13 "Arrive": An Attainable Homes Development in Bowness, Calgary, June 2018

housing market with a $2000 down payment, and the remainder of the 5 per cent down payment is provided through the program and later recouped through equity gains (Lewsen, 2017; https://attainyourhome.com/). Entire housing developments (almost 1000 homes) have been constructed in nine different neighbourhoods within Calgary. These, such as the Arrive development in the Bowness neighbourhood (Figure 7.13), are high-quality townhome developments that are alleviating some of the housing stress in the city.

Conclusion

This chapter has presented some of the key demographic features of change in Canadian cities and linked them to the malleable boundaries of significant life course transitions. The conceptualization of life course as four "ages" is a useful rubric for tying demographic and lifestyle trends together and for proposing how these might relate to the differentiation and marketing of the built environment. The future of Canadian cities is tied to a greatly diversifying population. Much of that diversity is now quite visible in the form of culture, ethnicity, and Indigeneity, with the importance of each varying regionally. In the Prairie provinces, for example, those cities that do not valorize the rich contemporary cultures of their growing young First Nations and Métis populations will be left behind culturally, socially, and economically in the twenty-first century (Nejad and Walker, 2018; Walker and Belanger, 2013).

Other forms of diversity relate to household sizes, conjugal relationships, and the greater propensity to live alone than has been the case in past decades. How is the Bridget Jones economy likely to continue affecting our inner cities and suburban areas? Will the millennials, the next largest cohort to the baby boomers, really transform the residential geography of Canadian cities? How have other lifestyles, like hipsters, changed the social, economic, and activity spaces of selected Canadian neighbourhoods? How will communities and housing markets respond to an increasing share of childless households? Finally, an aging society means that no form of lifestyle diversity, whether based on culture, ethnicity, Indigeneity, or household preferences, can be understood without inserting the centrality of the age dimension and the rising opportunities being taken by those in their Third Age of life. How will our built environment continue to change to serve the "pig in the python" as it moves through its elongated Third Age? Will those in their Third Age continue to advance suburban development that they have become accustomed to throughout their lives, even if modified to suit new lifestyle preferences (see Grant and Filion, Chapter 12)? Will central cities become increasingly age-contested spaces as seniors and millennials compete for housing in increasingly condoized, amenity-rich, and high-income neighbourhoods. Will racialized visible minorities so essential to population stability and growth in this country experience new forms of segregation in income-declining "ethnoburbs"? Will some in their middle or late Second Age begin to rethink the amenity and locational value of the inner suburbs, and begin to modify these environments through knockdown-rebuild development or a type of "gen-X-trification" that has been identified in Australian cities (Wiesel et al., 2011)? Will the millennial generation opt for newer **urban forms** that turn away from conventional development patterns and instead promote better environmental stewardship (see Rees, Chapter 21; Blay-Palmer and Landman, Chapter 23) and qualitatively "urban" places (see Bain and Mark, Chapter 15; Walker and Blakley, Chapter 13)? Most of the other chapters in this book contemplate the impacts of diversity on Canadian cities in more specific ways, such as immigration (Kobayashi and Preston, Chapter 8), social **polarization** (Walks, Chapter 10), age (Moos, Chapter 6), and Indigeneity (Fawcett and Walker, Chapter 4). Consider this chapter a primer on the demographic and lifestyle changes that run through all phenomena that constitute urban Canada.

Review Questions

1. Describe at least three of the key forces of social and demographic change affecting contemporary Canadian cities.
2. What are some of the impacts that changing lifestyles may have on how we plan and develop our built environment?

References

Alsop, R. 2008. *The Trophy Kids Grow Up: How the Millennial Generation Is Shaking Up the Workplace*. Toronto: Jossey-Bass.

Atchley, R.C. 1982. "Retirement as a social institution," *Annual Review of Sociology* 8: 263–87.

Barton, K.S., and C.O. Collins. 2012. "Generational geographies," in J.P. Stoltman, ed., *21st Century Geography: A Reference Handbook* (Vol. 1, pp. 255–65). Thousand Oaks, CA: Sage.

Beaujot, R. 1991. Population Change in Canada: The Challenges of Policy Adaptation. Toronto: McClelland & Stewart.

Belden Russonello Strategists LLC. 2013. "Americans' Views on their Communities, Housing, and Transportation." At: www.brspoll.com/uploads/files/r-ULI%20housing%202012%20for%20website.pdf

Bibby, R. 2009. *The Emerging Millennials*. Lethbridge, AB: Project Canada Books.

Blakely, E., and M. Snyder. 1997. *Fortress America: Gated Communities in the United States*. Washington, DC: Brookings Institution Press.

Blandy, S., D. Lister, R. Atkinson, and J. Flint. 2003. *Gated Communities: A Systematic Review of the Research Evidence*. Bristol: Economic and Social Research Council, Centre for Neighbourhood Research, CNR Paper 12.

Bourne, L.S., and D. Rose. 2001. "The changing face of Canada: The uneven geographies of population and social change," *Canadian Geographer* 45: 105–19.

Boyd, M., and M. Vickers. 2000. "One hundred years of immigration to Canada," *Canadian Social Trends* (Fall); reprinted in *100 Years of Canadian Society*, Teachers Kit.

Bunting, T., A. Walks and P. Filion. 2004. "The uneven geography of housing affordability stress in Canadian metropolitan areas," *Housing Studies* 19(3): 361–93.

Burgess, T. 2017. Residential Location of Millennials: A Calgary Case Study. M.A. Thesis, University of Lethbridge.

Burrows, R., 2008. "Geodemographics and the construction of differentiated neighbourhood," *Cohesion in Crisis*, 219–37.

Cahill, D. 2006. *Lifestyle Market Segmentation*. Toronto: Routledge.

Carrière, Y., and D. Galarneau. 2011. *Delayed retirement: A new trend?* Component of Statistics Canada Catalogue no. 75-001-X.

Chagnon, J., and M. Milan. 2011. *Population Growth: Canada, Provinces and Territories, 2010*. Ottawa, ON: Statistics Canada, Component of Statistics Canada Catalogue no. 91-209-X.

Chapin, F.S. 1965. *Urban Land Use Planning*, 2nd edn. Urbana, IL: University of Illinois Press.

Clark, W. 2000. "One hundred years of education," *Canadian Social Trends* (Winter); reprinted in *100 Years of Canadian Society*, Teachers Kit.

———. 2007. "Delayed transitions of young adults," *Canadian Social Trends* (Winter).

Counselling Foundation of Canada. 2017. "New survey reveals the diversity of Canada's Millennial generation through their social values." Feb 14, 2017. At: https://www.newswire.ca/news-releases/new-survey-reveals-the-diversity-of-canadas-millennial-generation-through-their-social-values-613786623.html

Coy, M., and M. Pohler. 2002. "Gated communities in Latin American megacities: Case studies in Brazil and Argentina," *Environment and Planning B: Planning and Design* 29: 355–70.

Craggs, S. 2017a. "Tiny house wave comes to Hamilton with new affordable housing project." *CBC News*. 27 February 2017. At:http://www.cbc.ca/news/canada/hamilton/laneway-tiny-units-1.3998639

———. 2017b "Hamilton moves ahead with plan to offer tiny homes for the homeless." *CBC News*. 3 October 2017. At: http://www.cbc.ca/news/canada/hamilton/tiny-houses-1.4320161

Davies, W.K.D., and R.A. Murdie. 1991. "Measuring the social ecology of Canadian cities," in D. Ley and L. Bourne, eds., *The Changing Social Geography of Canadian Cities*. Montreal and Kingston: McGill-Queen's University Press.

———, and I.J. Townshend. 2015. "New urbanisms: From neo-traditional neighbourhoods to New Regionalism," in W.K.D. Davies, ed., *Theme Cities: Solutions for Urban Problems*. London: Springer.

Decady, Y., and L. Greenberg. 2014. Ninety years of change in life expectancy. Statistics Canada, Health Statistics Division, Catalogue no. 82-624-X.

Dey, T., dir. 2006. *Failure to Launch*. Paramount Pictures.

Dytchwald, K. 1990. *Age Wave*. Toronto: Bantam Books.

Economist, The. 2001. "The Bridget Jones economy," 361, 8253: 68–70.

Fielding, H. 1996. *Bridget Jones's Diary*. London: Picador.

Fields, A., S. Uppal, and S. LaRochelle-Côté. 2017. *The impact of aging on labour market participation rates*. Statistics Canada, Insights on Canadian Society. Catalogue no. 75-006-X.

Foot, D., with D. Stoffman. 1996. *Boom, Bust and Echo: How to Profit from the Coming Demographic Shift*. Toronto: Macfarlane Walter and Ross.

Fries, F. 1980. "Aging, natural death, and the compression of morbidity," *New England Journal of Medicine* 303: 130–5.

Glasze, G. 2002. "Gated housing estates in the Arab world: Case studies in Lebanon and Riyadh, Saudi Arabia," *Environment and Planning B: Planning and Design* 29: 321–36.

Globe and Mail. 2017a. "Census 2016: The growing age gap, gender ratios and other key takeaways." 3 May 2017. At: https://www.theglobeandmail.com/news/national/census-2016-statscan/article34882462/

———. 2017b. "Census 2016: For the 1st time, more seniors than children living in Canada." At: https://globalnews.ca/news/3423047/more-seniors-than-children-in-canada-2016-census/

Gower, D. 1997. "Measuring the Age of Retirement." Statistics Canada, *Perspectives*, Summer 1997. Catalogue no. 75-001-XPE. At: https://www150.statcan.gc.ca/n1/pub/75-001-x/1997002/3071-eng.pdf

Grant, J. 2005. "Planning responses to gated communities in Canada," *Housing Studies* 20: 277–89.

———, and L. Mittelsteadt. 2004. "Types of gated communities," *Environment and Planning B: Planning and Design* 31: 913–30.

Halli, S., and L. Driedger, eds. 1999. *Immigrant Canada: Demographic, Economic, and Social Challenges*. Toronto: University of Toronto Press.

Hareven, T.K., and K.J. Adams, eds. 1982. *Aging and Life Course Transitions: An Interdisciplinary Perspective*. New York: Guilford Press.

Hebert, B.P., and M. Luong. 2008. "Bridge employment," *Perspectives on Labour and Income* 9, 11. Statistics Canada Catalogue no. 75-001-tX.

Hiller, H. 2006. *Canadian Society: A Macro Analysis*, 5th edn. Toronto: Pearson Prentice-Hall.

Hodge, G., and D. Gordon. 2014. *Planning Canadian Communities*, 6th edn. Scarborough, ON: Nelson Education.

Howe, N., and W. Strauss. 2007. *Millennials Go to College*, 2nd edn. Ithaca, NY: Paramount Books.

Hymowitz, K.S. 2007. "The New Girl Order." *City Journal* Autumn 2007. At: www.city-journal.org/html/17_4_new_girl_order.html

Karpel, C. 1995. *The Retirement Myth*. New York: Harper Collins.

Katz, P. 1994. *The New Urbanism*. New York: McGraw Hill.

Kraus, A.S. 1988. "Is a compression of morbidity in late life occurring? Examination of death certificate evidence," *Canadian Journal on Aging* 7: 58–70.

Kubo, T., and Y. Yui. 2012. "Changes in the condominium market in Japan after the 1990s," in M. Mantinan and N. Moore, eds., *Changes in the City* (Young Scholars Book 2). Santiago de Compostela: IDEGA-University of Santiago de Compostela and IGU Urban Commission.

Landman, K. 2006. "Privatising public space in post-apartheid South African cities through neighbourhood enclosures," *GeoJournal* 66: 133–46.

Lang, R., and K. Danielsen. 1997. "Gated communities in America: Walling out the world?" *Housing Policy Debate* 8: 867–99.

Lanham, R. 2003. *The Hipster Handbook*. Anchor.

Laslett, P. 1987. "The emergence of the third age," *Ageing and Society* 7: 133–60.

———. 1991. *A Fresh Map of Life: The Emergence of the Third Age*. London: Weidenfeld and Nicolson.

Lesthaeghe, R. 1995. "The second demographic transition in Western societies: An interpretation," in K.O. Mason and A.M. Jensen, eds., *Gender and Family Change in Industrialized Countries*. Toronto: Oxford University Press.

Lewsen, S. 2017. "Calgary housing initiative combines smart architecture with a clever loan program to combat Canada's affordability crisis," *The Globe and Mail*, 28 November 2017. At: https://www.theglobeandmail.com/real-estate/calgary-and-edmonton/in-calgary-an-innovative-housing-program-tackles-affordabilitycrisis/article37109995/

Lo, O. 1996. "Condominium living," *Canadian Social Trends* (Summer): 41.

Lotking, B. 2018. "The *Portlandia* effect: How did the show change the city it satirized?" *Vulture*. Posted 19 March 2018. At: www.vulture.com/2018/03/portlandia-effect-how-did-the-show-change-portland.html

McDonald, P.L., and R.A. Wanner. 1984. "Socioeconomic determinants of early retirement in Canada," *Canadian Journal on Aging* 3:105–16.

———, and ———. 1990. *Retirement in Canada*. Toronto: Butterworths.

McKenzie, E. 1994. *Privatopia: Homeowner Associations and the Rise of Residential Private Government*. New Haven, CT: Yale University Press.

———. 2003. "Common-interest housing in the com-munities of tomorrow," *Housing Policy Debate* 14(1): 203–34.

Maclean's. 2017. "Canada is the fastest growing G7 country." 8 Feb 2017. At: https://www.macleans.ca/news/canada/canada-is-the-fastest-growing-g7-country/

MacNaughton, L. 2016. "Goodbye Freedom 55: More Canadians spend their retirement years working." *CBC News*, 19 Jun 2016. At: www.cbc.ca/news/business/canada-retirement-age-rising-1.3638733

Maguire, S., dir. 2001. *Bridget Jones's Diary*. Miramax Films.

Mandich, S., and R. Margolis. 2014. "Changes in disability-free life expectancy in Canada between 1994 and 2007," *Canadian Studies in Population* 41, no. 1–2 (spring/summer 2014): 192–208.

Markides, K.S., and C.L. Cooper, eds. 1987. *Retirement in Industrialized Societies*. Toronto: John Wiley and Sons.

Marshall, K., and V. Ferraco. 2007. "Participation of older workers," *Perspectives on Labour and Income*, Statistics Canada Catalogue no. 75-001-XIE.

Milan, A. 2000. "One hundred years of families," *Canadian Social Trends* (Spring 2000), reprinted in *100 Years of Canadian Society*, Teachers Kit.

———. 2013. "Marital Status: Overview, 2011." Statistics Canada, Component of Statistics Canada Catalogue no. 91-209-X. At: https://www150.statcan.gc.ca/n1/en/pub/91-209-x/2013001/article/11788-eng.pdf?st=5dpO2Pv1

Miron, J.R. 1993. "Demography, living arrangement, and residential geography," in L.S. Bourne and D. Ley, eds., *The Changing Social Geography of Canadian Cities*.

Montreal and Kingston: McGill-Queen's University Press.

Murdie, R., and C. Teixeira. 2006. "Urban social space," in T. Bunting and P. Filion, eds., *Canadian Cities in Transition: Local through Global Perspectives*, 3rd edn. Toronto: Oxford University Press.

Murphy, M. 1987. "Measuring the family life cycle: Concepts, data and methods," in A. Bryman, B. Bytheway, P. Allatt, and T. Keil, eds., *Rethinking the Life Cycle*. London: Macmillan.

Myers, D. 2015. "Three reinforcing cycles that amplify the rise (and fall?) of Millennial generation impacts in cities." Paper presented at the 55th Annual ACSP Conference, Houston.

Nagnur, D. 1986. "Longevity and historical life tables: 1921–81 (abridged)," *Canada and the Provinces*. Ottawa, ON: Statistics Canada.

Nejad, S., and R. Walker. 2018. "Contemporary urban Indigenous placemaking in Canada," in E. Grant, K. Greenop, A. Refiti, D. Glenn, eds., *The Handbook of Contemporary Indigenous Architecture*, Singapore: Springer.

Neugarten, B. 1974. "Age groups in American society and the rise of the young-old," *Annals of the American Academy of Political and Social Science* 415(1): 187–98.

Norland, J.A. 1994. *Profile of Canada's Seniors*. Ottawa, ON: Statistics Canada, Focus on Canada Series, Catalogue no. 96–312E.

Parmett, H.M. 2018. "Site-specific television as urban renewal: Or, how Portland became *Portlandia*," *International Journal of Cultural Studies* 21(1) 42–56.

Partridge, W. 1973. *The Hippie Ghetto: The Natural History of a Subculture*. New York: Holt, Rinehart and Winston.

Perry, C. 1929. "The neighborhood unit," in *Regional Survey of New York and Its Environs, Regional Survey*, vol. 7. New York: Neighborhood and Community Planning.

Punter, J.V. 1990. "The privatisation of the public realm," *Planning Practice and Research* 5: 9–18.

Raposo, R. 2006. "Gated communities, commodification and aestheticization: The case of the Lisbon metropolitan area," *GeoJournal* 66: 43–56.

Rea, W., D. McKay, and S. LeVaseur. 2008. *Changing Patterns in Canadian Homeownership and Shelter Costs*. Ottawa, ON: Statistics Canada Catalogue no. 97-554-X.

Roadburg, A. 1985. *Aging: Retirement, Leisure and Work in Canada*. Toronto: Methuen.

Rose, D., and P. Villeneuve. 2006. "Life stages, living arrangements, and lifestyles," in T. Bunting and P. Filion, eds., *Canadian Cities in Transition: Local through Global Perspectives*, 3rd edn. Toronto: Oxford University Press.

Rosen, G., and Walks, A. 2013. "Rising cities: Condominium development and the private transformation of the metropolis," *Geoforum*, 49: 160–72.

Sappington, S. 2016. "The far left, liberal utopias and *Portlandia*," *The California Aggie*. At:https://theaggie.org/2016/11/07/the-far-left-liberal-utopias-and-portlandia/

Schnell, I., and Y. Benjamini. 2001. "The socio-spatial isolation of agents in everyday life spaces as an aspect of segregation," *Annals, Association of American Geographers* 91: 622–36.

———, and ———. 2002. "Measuring spatial lifestyles in Israel: A Tel-Aviv–Jaffa case study," in W.K.D. Davies and I. Townshend, eds., *Monitoring Cities: International Perspectives*. Calgary and Berlin: International Geographical Union, Urban Commission.

Seiler, S.R. 1986. "How to develop retirement communities for profit," *Real Estate Review* 16: 70–5.

Simmons-Tropea, D., and R. Osborn. 1993. "Disease, survival and death: The health status of Canada's elderly," in V.W. Marshall, ed., *Aging in Canada: Social Perspectives*. Markham, ON: Fitzhenry & Whiteside.

Singh, G., and A. Verma. 2001. "Is there life after career employment? Labour-market experience of early 'retirees,'" in V.W. Marshall, W.R. Heinz, H. Kruger, and A. Verma, eds., *Restructuring Work and the Life Course*. Toronto: University of Toronto Press.

Singleton, A.D. and Spielman, S.E. 2014. "The past, present, and future of geodemographic research in the United States and United Kingdom," *The Professional Geographer*, 66(4): 558–67.

Smith, N. 1987. "Of yuppies and housing: Gentrification, social restructuring, and the urban dream," *Environment and Planning D: Society and Space* 5: 151–72.

Starr, J., and D. Carns. 1972. "Singles in the city," *Society* 9: 43–8.

Statistics Canada. 1982. *1981 Census of Population, Private Households by Number of Persons, for Canada and Provinces, Urban Size Groups, Rural Non Farm and Rural Farm, 1981*. Catalogue 92-904, Table 3. At: https://archive.org/stream/1981929041982engfra#page/n19/mode/2up

———. 2008a. *Report on the Demographic Situation in Canada 2005 and 2006*. Catalogue no. 91-209-X.

———. 2008b. *Canadian Demographics at a Glance*. Catalogue no. 91-003-XIE.

———. 2008c. *Canada's Ethnocultural Mosaic*. Catalogue no. 97-562-X.

———. 2008d. *Educational Portrait of Canada*. Catalogue no. 97-560-X.

———. 2008e. *Aboriginal Peoples in Canada in 2006: Inuit, Métis and First Nations*. Catalogue no. 97-558-XIE.

———. 2011. *Population Projections by Aboriginal Identity in Canada, 2006 to 2031*. Catalogue no. 91-552-X

———. 2012a. *Fifty Years of Families in Canada: 1961 to 2011*. Catalogue no. 98-312-X2011003.

———. 2012b. *Generations in Canada*. Catalogue no. 98-311-X2011003

———. 2013a. Table 102-0504. Deaths and mortality rates, by age group and sex, Canada, provinces and territories, annual, CANSIM (database).

———. 2013b. *Immigration and Ethnocultural Diversity in Canada*. Catalogue no. 99-010-X2011001.

———. 2013c. *Living Arrangements of Young Adults Aged 20–29*. Catalogue no. 98-312-X2011003.

———. 2015. Total fertility rate (number of children per woman), Canada, provinces and territories, 1981 to 2011. At: https://www150.statcan.gc.ca/n1/pub/91-209-x/2013001/article/11784/tbl/tbl02-eng.htm

———. 2016. *Canadian Demographics at a Glance—Second Edition*. Catalogue no. 91-003-X.

———. 2017a. Canada [Country] and Canada [Country] (table). *Census Profile*. 2016 Census. Statistics Canada Catalogue no. 98-316-X2016001. Ottawa. Released November 29, 2017. At: https://www12.statcan.gc.ca/census-recensement/2016/dp-pd/prof/index.cfm?Lang=E

———. 2017b. *Aboriginal Peoples in Canada: Key Results from the 2016 Census*. Ottawa, ON.

———. 2017c. *Focus on Geography Series, 2016 Census*. Statistics Canada Catalogue no. 98-404-X2016001. Ottawa, ON. Data products, 2016 Census.

———. 2017d. Young adults living with their parents in Canada in 2016. Statistics Canada, Catalogue no. 98-200-X2016008.

———. 2017e. "Education in Canada: Key results from the 2016 Census." Statistics Canada, *The Daily*, 29 November 2017.

———. 2017f. *Families, households and marital status: Key results from the 2016 Census*. Statistics Canada, *The Daily*, August 2, 2017. At: https://www150.statcan.gc.ca/n1/daily-quotidien/170802/dq170802a-eng.pdf

———. 2017g. *Portrait of children's family life in Canada in 2016*. Census in Brief, Catalogue no. 98-200-X2016006. At: http://www12.statcan.gc.ca/census-recensement/2016/as-sa/98-200-x/2016006/98-200-x2016006-eng.pdf

———. 2017h. *2016 Census*. Families, Households and Marital Status Highlight Tables. At: https://www12.statcan.gc.ca/census-recensement/2016/dp-pd/hlt-fst/fam/Table.cfm?Lang=E&T=11&Geo=00&SP=1&view=1&sex=1&presence=3

———. 2017i. *Same-sex couples in Canada in 2016*. Census in Brief, Catalogue no. 98-200-X2016007. At: http://www12.statcan.gc.ca/census-recensement/2016/as-sa/98-200-x/2016007/98-200-x2016007-eng.pdf

———. 2017j. Statistics Canada. Table 39-10-0007-01. Life expectancy and other elements of the life table, Canada and provinces. At: https://www150.statcan.gc.ca/t1/tbl1/en/tv.action?pid=3910000701

———. 2017k. *Immigration and Ethnocultural Diversity Highlight Tables*. Statistics Canada 2016 Census. At: www12.statcan.gc.ca/census-recensement/2016/dp-pd/hlt-fst/imm/Table.cfm?Lang=E&T=11&Geo=00&SP=1&view=2&age=1&sex=1

———. 2017l. "Immigration and ethnocultural diversity: Key results from the 2016 Census." Statistics Canada, *The Daily*, October 25, 2017. At: https://www150.statcan.gc.ca/n1/en/daily-quotidien/171025/dq171025b-eng.pdf?st=smYysG9b

———. 2017m. 2016 Census of Population: Families, households and marital status. Release and Concepts Overview series. Catalogue no. 98-501-X2016004. At: http://www12.statcan.gc.ca/census-recensement/2016/ref/98-501/98-501-x2016004-eng.pdf

———. 2018a. Table 13-10-0418-01. Crude birth rate, age-specific fertility rates and total fertility rate (live births). At: https://www150.statcan.gc.ca/t1/tbl1/en/tv.action?pid=1310041801

———. 2018b. *Report on the Demographic Situation in Canada*, Revised June 14, 2018. Statistics Canada Catalogue no. 91-209X. At: https://www150.statcan.gc.ca/n1/pub/91-209-x/91-209-x2018001-eng.htm

———. 2019. Table 13-10-0713-01. Infant deaths and mortality rates, by age group. At:https://doi.org/10.25318/1310071301-eng

Stone, L. 2006. *New Frontiers of Research on Retirement*. Ottawa, ON: Statistics Canada Catalogue no. 75-511-XIE.

Townshend, I.J. 1997. *An Urban Geography of the Third Age*. Calgary, AB: University of Calgary, Ph.D. Thesis.

———. 2006. "From public neighbourhoods to multi-tier private neighbourhoods: The evolving ecology of neighbourhood privatization in Calgary," *GeoJournal* 66: 103–20.

———, and W.K.D. Davies. 1999. "The derivation of shopper typologies in business revitalization zones," *Papers and Proceedings of the Applied Geography Conferences* 22: 132–45.

Turcotte, M. 2013. *Living apart together*. Insights on Canadian Society, Statistics Canada Catalogue no. 75-006-X.

Uppal, S. 2017. *Young men and women without a high school diploma*. Statistics Canada, Insights on Canadian Society, Catalogue no. 75-006-X.

van der Poorten, K. 2018. "The exclusionary politics of secondary suites in Calgary: Homeowners seeking class monopoly rents." M.A. Thesis, University of Calgary, Department of Geography.

Walker, R., and Y. Belanger. 2013. "Aboriginality and planning in Canada's large Prairie cities," in R. Walker, T. Jojola, and D. Natcher, eds., *Reclaiming Indigenous Planning*. Montreal and Kingston: McGill-Queen's University Press.

———, and T. Carter. 2010. "At home in the city: Housing and neighbourhood transformation," in T. Bunting, P. Filion, and R. Walker, eds., *Canadian Cities in Transition: New Directions in the Twenty-first Century*, 4th edn. Toronto: Oxford University Press.

Wall Street Journal (Online). 2013. "Hip, urban, middle-aged; baby-boomers are moving into trendy urban neighbourhoods, but young residents aren't always thrilled," 8 August: n.p.

Walter-Joseph, R. 2015. "Finding the Generationed City: The Form and Structure of Young Adult Settlement Patterns." (Master of Arts), University of Waterloo. At: https://uwspace.uwaterloo.ca/handle/10012/9275

Webber, R., 2007. The metropolitan habitus: its manifestations, locations, and consumption profiles. *Environment and Planning A*, 39(1): 182–207.

Webster, C., G. Glasze, and K. Frantz. 2002. "Guest editorial: The global spread of gated communities," *Environment and Planning B: Planning and Design* 29: 315–20.

Weiss, M. J. 2000. The Clustered World: How We Live, What We Buy, and What It All Means to Who We Are. Boston, MA: Little, Brown.

Wiesel, I., R. Freestone, S. Pinnegar, and B. Randolph. 2011. "Gen-X-Trification? Generation shifts and the renewal of low-density housing in Sydney's suburbs," in *State of Australian Cities Conference* (pp. 26–29).

World Factbook, The. 2018. Washington, DC: Central Intelligence Agency. At: https://www.cia.gov/library/publications/the-world-factbook/index.html

8 International Migration and Immigration

Remaking the Multicultural Canadian City

Audrey Kobayashi and Valerie Preston

Introduction

Canadian cities are profoundly influenced by international migration, one of the most dynamic processes of social change historically and at the present. Migrants,[1] temporary and permanent, from all over the world join the labour force, choose places to live, send their children to schools, participate in cultural activities, alter consumer choices, and become a part of civic life. They remake the urban landscape, often dramatically, and their stories are part of the dynamic set of social relations that constitute Canadian urban life.

This chapter addresses some of the ways in which international migration—the movement of people across national borders for economic opportunity, refuge, and family unification (Castles, De Haas, and Miller, 2014)—has shaped the geography of Canadian cities, including the varied social, political, and economic contexts that have defined the settlement of immigrants, the roles of family and communities in initiating and supporting newcomers, and the spatial patterns of settlement across cities and within cities. We also discuss the development and ongoing transformation of immigration policy, which regulates the flow of immigrants and affects their ability to adjust to and integrate within Canadian society.

Understanding contemporary immigration is impossible without addressing the historical trends that preceded it. Between 2011 and 2016, 1,212,075 new immigrants settled permanently in Canada. During that period, Canada also received almost 3,000,000 temporary migrants (Immigration, Refugees and Citizenship Canada, 2017). That number of temporary migrants, people entitled to live in Canada for fixed time periods specified by their visas, many of whom had left the country by 2016 after completing their studies or work assignments, is still one of the largest in the postwar period. Immigrants and temporary migrants have increased in cities and towns outside the traditional destinations of Montreal, Toronto, and Vancouver, shifting the settlement map. As in the past, public policies concerning who is admitted to Canada, how long they can stay, and their legal, civic, and social rights, have shaped and responded to the composition of the twenty-first-century migrant population. During the past decade, the pace of policy change has accelerated as international migration flows have grown more unpredictable and policy-makers have sought self-sufficient immigrants who possess skills in demand by Canadian employers and need little help to settle successfully.

Immigration as a Demographic Process

Since their inception in a colonial nation, Canadian cities have been populated by immigrants and generations of their descendants. Montreal was built by French and British settlers, including fur traders and merchants, military and administrative officers, and workers who laboured in the factories, on the docks, and in the transportation system that linked Eastern cities to the rest of Canada (see Filion, Chapter 2). In Toronto, early immigrants included British merchants and administrators, American Loyalists, and workers from many countries, their numbers growing and shifting over time. By 1881, about 88 per cent of Canadians were of British or French ancestry; about 6 per cent, German; and about 3 per cent, Indigenous. No other group exceeded 1 per cent of the total (Applied History Research Group, 1997). The major source of urban growth during the nineteenth and early twentieth centuries was not immigration, however, but a combination of the natural increase of the long-settled population and rural-to-urban migration of the Canadian-born (Buckley, 1960). By 1891, immigrants comprised just 13 per cent of Canada's total population. Nearly half of the total population lived in Ontario, mainly Toronto, and nearly a third lived in Quebec, mainly Montreal. The balance of the country was much more sparsely populated, and cities such as Vancouver and Winnipeg were small. About 80 per cent of the urban population at that time was Canadian born (Applied History Research Group, 1997).

During the late nineteenth and early twentieth centuries, successive groups, mainly Europeans, arrived in Toronto and Montreal. Both cities received large numbers of Irish immigrants during the famine that started in the 1840s. Later came Italian, German, and Jewish immigrants, many of whom joined the construction and garment industries. In Vancouver, the western terminus for the railway that arrived in 1885, sawmills shipped lumber produced with immigrant labour to growing Eastern Canadian cities.

Early in the twentieth century, immigration added significantly to the Canadian population, altering its distribution permanently and unevenly. Many of the nearly two million British immigrants who arrived between 1901 and 1921 found their way to the cities, actively recruited by the Canadian government. But the government also set its priorities on peopling the west (Hawkins, 1988). Between 1896 and 1905, Minister of the Interior Clifford Sifton was responsible for recruiting as many farmers as possible, targeting thousands of Eastern Europeans, whom he deemed inappropriate for urban life but well suited to the harsh conditions of Prairie farming (Knowles, 2006; Simmons, 2010). Winnipeg became the country's most ethnically diverse city, as trainloads of Eastern European newcomers were encouraged to settle the West.

Racialized minorities, however, were largely excluded. Chinese immigrants, who had settled in Vancouver after being recruited to build the railway, were required to pay a fee to enter the country. The so-called "head tax" severely curtailed the entry of Chinese immigrants after 1885. Thousands of Japanese workers hired by Vancouver sawmills established a significant community until their entry, too, was curtailed by a "Gentlemen's Agreement" in 1908. In that year, the government also passed the "Continuous Passage" regulation that prohibited passengers on any ship that had made a stop before arriving in Canada from settling in Canada. The regulation was invoked in 1914 to prevent passengers from India who had arrived on the *Komagata Maru* from disembarking in Vancouver (Knowles, 2006).

By the beginning of World War I, following the influx of over 500,000 newcomers in 1913 alone, people born outside Canada represented about 22 per cent of the national population (Kalbach and McVey, 1971), but their ethnic composition had not altered much. The largest groups were still English (30 per cent), French (28 per cent), Scottish (13 per cent), and Irish (13 per cent), followed distantly by German (3 per cent). Notwithstanding the Prairie influx, the largest immigrant populations were established in the largest cities: Montreal, Toronto, and later Vancouver (MTV), although the distribution, timing, and numbers of immigrants vary considerably among these cities (Simmons, 2010).

The MTV phenomenon continued even as immigration levels dropped during the Great Depression of the 1930s. After World War II, the numbers of migrants fleeing economic and political turmoil in Europe, especially Jewish populations and displaced persons from the Baltic states and Central Europe, rose particularly in Toronto and Montreal. With the 1956 Soviet invasion, large numbers of Hungarians came, particularly to Toronto. A few years later, many refugees from war-torn Vietnam, Laos, and Cambodia settled in French-speaking Montreal (Germain and Rose, 2000). The Vietnam War also saw a less visible demographic shift as tens of thousands of American war resisters headed for Canada and settled in cities and small towns (Hardwick and Mansfield, 2009). During the 1960s, changes to the Immigration Act and its associated regulations removed privileged status for European immigrants. Earthquakes in Portugal encouraged thousands of emigrants to leave, joining compatriots already settled in Montreal and Toronto. When Idi Amin rose to power in Uganda in 1971, many fleeing that country, especially those of South Asian background, took up residence in Canadian cities. When drought struck the Horn of Africa in the 1980s, Canadian planes were sent to collect famine-ravished refugees, and our cities began to receive large numbers of newcomers from Ethiopia, Somalia, and Eritrea. Thousands of Iranians left in the wake of the Ayatollah Khomeini's ascendency in 1979, many settling in the Toronto suburbs of North York and Richmond Hill and in North Vancouver. A decade later, the largest refugee group was from Poland, which was undergoing political and economic upheaval. They were followed by the largest movement of immigrants to date, from Hong Kong to Vancouver and Toronto. By 1996, the number of Hong Kong–born Canadians had reached 241,000.

By the 1990s, another demographic shift had taken place as the majority of immigrants had become racialized minorities, particularly from East Asia but increasingly also from South Asia, Africa, and Latin America. Immigration from the Philippines increased dramatically, notably attracting large numbers of domestic and health-care workers, who settled in smaller cities such as Winnipeg and Calgary as well as in MTV. In 2016, Filipinos were the largest group of recent immigrants, followed closely by those from India and the People's Republic of China (Table 8.1).

Table 8.1 Top 10 Countries of Birth for Recent Immigrants, 1981–2016

2016 Census	2011 NHS*	2006 Census	2001 Census	1991 Census	1981 Census
Philippines	Philippines	PRC	PRC	Hong Kong	United Kingdom
India	PRC	India	India	Poland	Vietnam
People's Republic of China (PRC)	India	Philippines	Philippines	PRC	United States
Iran	United States	Pakistan	Pakistan	India	India
Pakistan	Pakistan	United States	Hong Kong	Philippines	Philippines
United States	United Kingdom	South Korea	Iran	United Kingdom	Jamaica
Syria	Iran	Romania	Taiwan	Vietnam	Hong Kong
United Kingdom	South Korea	Iran	United States	United States	Portugal
France	Colombia	United Kingdom	South Korea	Lebanon	Taiwan
South Korea	Mexico	Colombia	Sri Lanka	Portugal	PRC

Source: Statistics Canada, 2017b: Table 2; 2013: 8 and Chiu, Tran, and Maheux, 2007: 10.

* The National Household Survey was a voluntary survey sent to slightly less than one-third of Canadian households, with an unweighted global response rate of 68.6%. The 2016 long-form census, which was obligatory like earlier censuses, was completed by almost 25% of Canadian households.

Table 8.2 Distribution of Immigrants Between MTV and Other Canadian Locations, 1986–2016

	Per cent of Immigrant Population					
	1986	**1996**	**2001**	**2006**	**2011**	**2016**
Toronto CMA	31.6	35.7	37.3	37.5	37.4	35.9
Montreal CMA	11.8	11.8	11.4	12	12.5	12.4
Vancouver CMA	10	12.7	13.6	13.4	13.5	13.1
MTV	53.4	60.2	62.3	62.9	63.4	61.4
Outside MTV	46.6	39.8	37.7	37.1	36.6	38.6

Sources: Statistics Canada, 2017a, 2013: 12; 2008: 20-1, and Citizenship and Immigration Canada, 2005.

There has also been a spatial shift (Table 8.2). The majority of newcomers, more than 90 per cent in 2016, still settle in metropolitan areas, and Toronto is the destination for almost three in ten, or, 29.4 per cent, of recent immigrants; however, increasing numbers are locating outside MTV. Calgary, Edmonton, and Winnipeg attracted larger shares of recent immigrants in 2016 than they did in 2006 or 2011 (Statistics Canada, 2017a). Smaller metropolitan areas such as Windsor and London have also received more newcomers.

Growing numbers of temporary migrants—international students, **temporary foreign workers**, caregivers, and refugee claimants[2] also live in Canada's cities. They are a vulnerable population with limited rights to residency, political participation, and public services. Since 1987, their numbers have more than doubled from 186,252 to more than 550,000 people admitted in 2016 (Immigration, Refugees and Citizenship Canada, 2017). Of those who arrived in 2011, almost two-thirds planned to live in metropolitan areas; however, they were more dispersed than were immigrants who settled permanently. Sponsored by individual employers in large and small cities and towns and studying in post-secondary institutions and high schools across the country, fewer than half of temporary residents now live in MTV.

Immigration as a Dynamic Social Process

International migration is best understood as an event, or series of events, in the **life course.** A life course approach shows that migration decisions are not random but socially structured (Kobayashi and Preston, 2007). Notwithstanding the complex economic and political contexts that certainly influence migration, decisions to leave one's country take into account numerous factors (Castles, De Haas, and Miller, 2014). Hopes and dreams are tied to people's aspirations for and the availability of work, children's education, lifestyle, and the well-being of an entire—sometimes extended—family. People make decisions around the needs of families, not just individuals, and decisions occur at certain times in the life course: upon entering the labour force, when adopting a life partner or having children, at key points in children's education, and sometimes on retirement. Most people plan their journeys, often waiting for protracted periods to receive authorization from Canadian embassies and consulates abroad. In choosing a destination, migrants also consider their social contacts at the destination, particularly the presence of relatives or an established ethno-cultural community.

Many practices of everyday life, from finding a job to participating in cultural events and worship, are organized through social networks of immigrants and their Canadian-born descendants. A comparative study of Ethiopian and Somali immigrants demonstrated that immigrants who are well established and have dense social networks tend to fare better economically (Danso, 2002). Scholars increasingly interpret migration as the transfer of human capital, including education, job skills, and enabling social networks, from one part of the world to another. Accumulated capital eases the transition to new homes and enriches the social fabric of the destination city as a whole (Castles, De Haas, and Miller, 2014; Massey, 2012).

A century ago international migration was usually a one-way trip. Today it is increasingly **transnational**, meaning that migrants maintain a "transnational field" of social, economic, and political linkages with their countries of origin. They may move back and forth so that migration is circular rather than one-way, and they often use modern modes of communication, such as Skype and instant messaging, to stay in close touch with family, friends, and popular culture. Some become "astronauts," working abroad to support families in Canada. Others send young children to be cared for by grandparents. Still others encourage grown children to move to their parents' country of origin for economic opportunities. Some spend their working lives in Canada and retire to countries of origin. While many immigrants maintain transnational ties to assuage longing for familiar people and culture, others are motivated by discriminatory experiences in Canada that restrict their employment prospects, or by desire to take advantage of Canadian credentials and experience in places enjoying rapid economic growth (Ley, 2013; Preston, Kobayashi, and Man, 2006).

Relations between ethno-cultural groups and the larger society are a crucial aspect of the urban social context for immigrants. The process of immigration has always been socially normative. Newcomers are expected to adapt to, and even assimilate, the norms of the dominant groups. This expectation makes two problematic assumptions, however: (1) that there *is* a norm, based mainly on British and to some degree on French cultural heritages, that provides the social and cultural terms to which newcomers *should* adapt and assimilate; and (2) that members of the dominant groups will indeed welcome newcomers on an equal basis (Guo and Wong, 2015; Hall, 2000). Both assumptions belie the concept of **multiculturalism**, an official policy that eschews normativity but that struggles to overcome prejudices, especially against the most recent arrivals.

Immigration is thus a dynamic social process that can be understood along a number of social dimensions, including relationships within the family, with friends and cultural or faith-based communities, and with the larger society, which is diverse in its attitudes to immigration and in its willingness to change as Canadian social dynamics shift. The social conditions faced by immigrants in different Canadian cities are therefore mixed and often contradictory. Many newcomers enjoy the economic and social benefits of immigration. Destination cities welcome their human capital, including their contributions to a rich and varied cultural landscape. At the same time, newcomers face challenging barriers, not only in overcoming the difficulties of transition to a new place but also due to discriminatory practices and hostile attitudes.

Immigration and Canadian Urban Landscapes

Immigration has its most visible imprint on the residential and commercial landscapes of Canadian cities. Early immigrant communities that grew up around clusters of boarding houses provided both shelter and the basic commercial services that immigrants required, including employment services, translation, and banking. Today, residential and commercial landscapes are layered and varied across and between cities. They can be richly textured landscapes of cultural expression, or they can be segregated and marginalized landscapes of exclusion and poverty. For some early immigrant groups, such as German immigrants in Winnipeg, the visible landscape now contains almost no clues to their arrival and passage. Others, such as Chinese immigrants in the suburbs of Toronto and Vancouver, have transformed the landscape.

The Commercial Landscape

Commercial streetscapes cater to the specific needs of migrant groups, concentrating their participation in the urban economy. Initially located in marginal locations adjacent to the downtown, such areas served cultural and practical needs and provided services not available in the larger community. For example, Pender Street in Vancouver developed as a place populated by Chinese immigrants but shunned by the white citizens (Anderson, 1995). Today it is a colourful landscape dominated by small stores and cafés known outside the neighbourhood as "Chinatown." Similar neighbourhoods, some comprising a few buildings, others several blocks, developed in many Canadian cities. By the 1970s, they symbolized the multicultural city and were often redeveloped as tourist destinations, marketed as much to the dominant population as to immigrants (Lai, 2011).

A few blocks from Pender Street, Vancouver's Powell Street became a commercial strip for Japanese immigrants in the early twentieth century as white residents moved to newer parts of the city, away from the mills and other noxious industries. The transformation of the landscape reflected the needs of recent arrivals: small, single family homes were adapted for businesses providing basic necessities, and the buildings were also extended backwards from the street as boarding houses for the thousands of single Japanese men who moved to Vancouver to find work (Jackson and Kobayashi, 1994) (Figure 8.1). This once thriving area, however, was completely dismantled when Japanese Canadians were forcibly removed during the 1940s.

In Montreal, "The Main" was a strip of businesses on Boulevard St-Laurent that catered to many European immigrants (Germain and Rose, 2000). Dominated by Jewish immigrants in the 1920s and 1930s, this landscape later captured the popular imagination through novels such as *St. Urbain's Horseman* by Mordecai Richler, in which the main character can never entirely leave behind the Montreal neighbourhood where he grew up. Now transformed, "The Main" is home to upscale boutiques, restaurants, and bars.

In Toronto, the Ward was the commercial and residential landscape where immigrants from Irish, Jewish, Chinese, and Italian backgrounds endured unsafe and unhealthy living and working conditions (Lorinc et al., 2015). It has now completely disappeared—expropriated to make way first for City Hall and then for several medical facilities—amid much controversy over how to maintain a "clean" city both morally and physically. It represents one of the first examples of the displacement of minority groups to make way for new versions of public space. The Jewish population relocated along Spadina Avenue, where factories offered employment for men and women and Jewish merchants could avoid restrictive covenants by buying property in Kensington Market (Hiebert, 1993). They were replaced by Portuguese and Chinese immigrants who took over the

Figure 8.1 Powell Street in Vancouver in 1928, showing commercial development by Japanese immigrants

Source: Dominion Photo Company, Vancouver Public Library 21773.

premises of many Jewish merchants, creating a thriving commercial and tourist attraction, often identified as Toronto's "Chinatown."

Wei Li coined the term **ethnoburbs** (Li, 2009) to capture emerging landscapes in suburban areas where many recent newcomers are settling. An ethnoburb is a cluster of residential areas and business districts, home to a heterogeneous population of multiple ethno-cultural groups that includes a concentration of at least one ethnic minority. Unlike earlier concentrations that often developed to serve the consumer needs of a single immigrant group, ethnoburbs house interactions among global, national, and local forces. Over the past three decades, as the Canadian government promoted immigration from Chinese-speaking countries—first from Hong Kong, then China and Taiwan—suburban districts in many cities have been transformed. Still the destination for the dominant population seeking popular "ethnic" food, their main purpose is to serve the consumer needs of a range of growing Chinese ethno-cultural communities (Lo, 2006). Malls covered in Chinese characters where shopkeepers speak first in Cantonese or Mandarin are increasingly common in the suburbs of MTV. Some malls have met with opposition on the grounds that they exclude long-settled residents who do not speak either Chinese language (Preston and Lo, 2000; Zhuang, 2013). In one notorious example, the deputy mayor of Markham commented that the growing Asian population was driving out "the backbone of Markham" (Verma, 2005). The municipal government responded by reiterating its support for diversity and establishing a race relations committee charged with promoting tolerance for diversity and equal treatment for all residents. The incident remains as testimony to the discrimination and racism that still mark contemporary Canadian cities.

The Residential Landscape

According to Carter and Vitiello (2012: 92), the "diversity of newcomers and of housing markets across North America results in such varied patterns that there is no such thing as a 'typical' immigrant housing experience." Yet adequate and affordable housing is a growing concern for virtually all newcomers, because low-cost rental housing is in such short supply (Hulchanski and Shapcott, 2004; see Harris, Chapter 11). In the 1990s, rental units in Montreal were more affordable than in either Toronto or Vancouver (Hiebert et al., 2006); however, average rents in all three metropolitan areas are converging as they increase inexorably. Rising housing costs, shortages of rental housing, low incomes, discrimination by landlords and real estate agents, and cultural preferences for specific housing attributes combine to create different scenarios between cities and between newcomer groups within cities.

Recent scholarship has expanded our understanding of newcomer housing issues, showing that immigration status, income, ethnic and racialized background, and household size and composition affect the security and quality of their housing.[3] At the outset, immigrants sponsored by family members already established in Canada do best in the housing market, often living in multigenerational households that are sometimes crowded but usually safe and affordable (Hiebert, 2011). While immigrants selected to contribute to the Canadian economy on the basis of their qualifications and work experience often struggle initially with high housing costs, refugees, even those sponsored by the Canadian government, encounter enormous barriers to obtaining suitable and affordable housing (Rose and Charette, 2014). Refugee claimants live in the least desirable housing, often paying more than half their income for a dark room in a dank basement with shared kitchen and bathroom (Francis and Hiebert, 2014).

Over time, however, many immigrants find better-paid jobs and move into affordable, suitable, and well-maintained accommodation (Hiebert, 2011). Indeed, some immigrants, particularly those from Italy, Portugal, and Hong Kong, attain higher home ownership rates than those of the population as a whole; however, pathways to home ownership differ. Many Hong Kong immigrants had substantial financial assets that allowed them to purchase spacious suburban housing upon arrival (Ley, 2010; Moos and Skaburskis, 2010). Anxious to send their children to schools with excellent academic reputations, Korean immigrants often struggle to buy properties in Toronto's northern suburbs (Preston et al., 2011). Italian and Portuguese immigrants, on the other hand, often scrimped and saved to own housing (Teixeira, 2007), capitalizing upon connections in the construction industry to ease the strain. Discrimination also places immigrants at a disadvantage (Murdie, 2008a). Although many feel that skin colour, ethno-cultural background, and religion are major reasons for discrimination in Vancouver and Montreal, in Toronto immigrants are more likely to mention that landlords discourage tenants who have large families and many children, or those who are single mothers and dependent on welfare (Preston et al., 2011). It may be that in the Toronto metropolitan area, where almost half the population belongs to a racialized minority (Statistics Canada, 2017b), many immigrants rent from landlords of the same background, reducing the salience of skin colour as a marker of discrimination. Such is certainly not the case in Vancouver or Montreal (Francis and Hiebert, 2014; Hiebert et al., 2006; Rose and Charette, 2014).

Geographers who study the evolution of the early modern city have observed that many immigrants settled initially near the city centre, where low-cost housing was adjacent to industrial plants that hired many newcomers. They moved outward to better-quality and more spacious housing as their financial circumstances permitted (Park and Burgess, 1925; Wirth, 1928). While such patterns still fit some immigrant groups in Canadian cities, settlement patterns are now diverse (Figures 8.2, 8.3, and 8.4). Concentrations of racialized minorities, such as Chinese and South Asian immigrants, have increased in size and number at the same time that the **suburbanization** of immigrants has continued apace (Murdie, 2008b; Teixeira, 2014). In Toronto, some immigrants follow the paths of

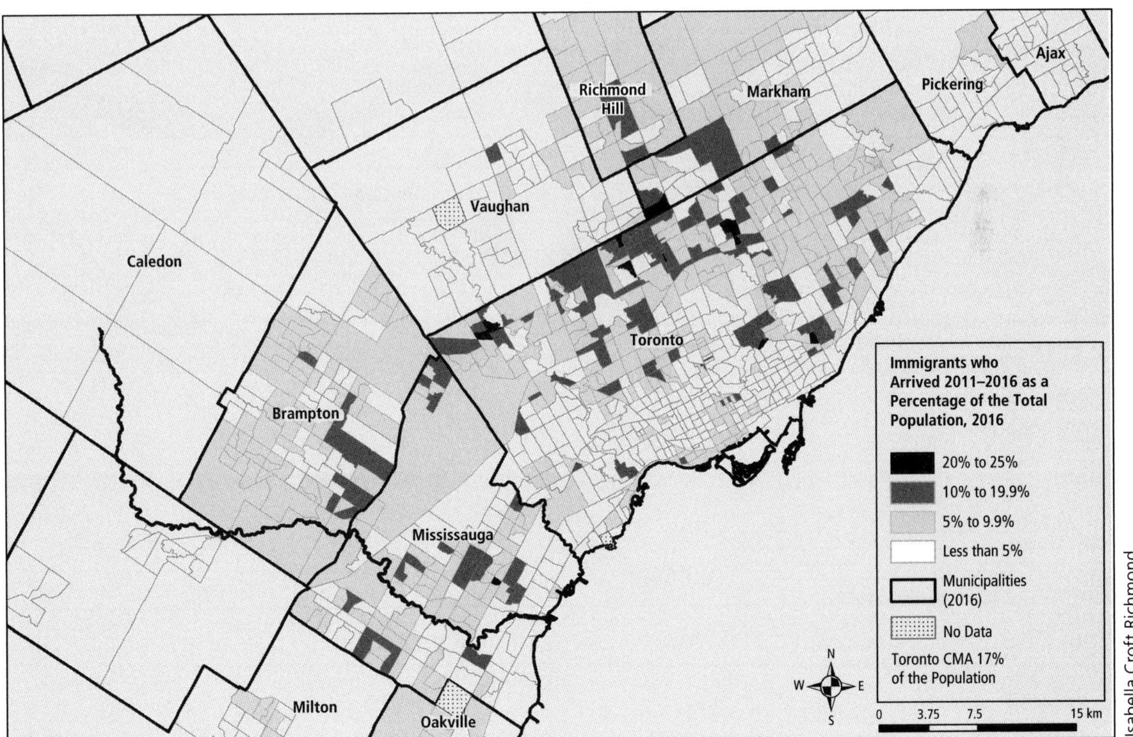

Figure 8.2 Recent Immigrant (2011–2016) Percentage of the Population by Census Tracts, Toronto CMA, 2016

Source: Statistics Canada, Census 2016.

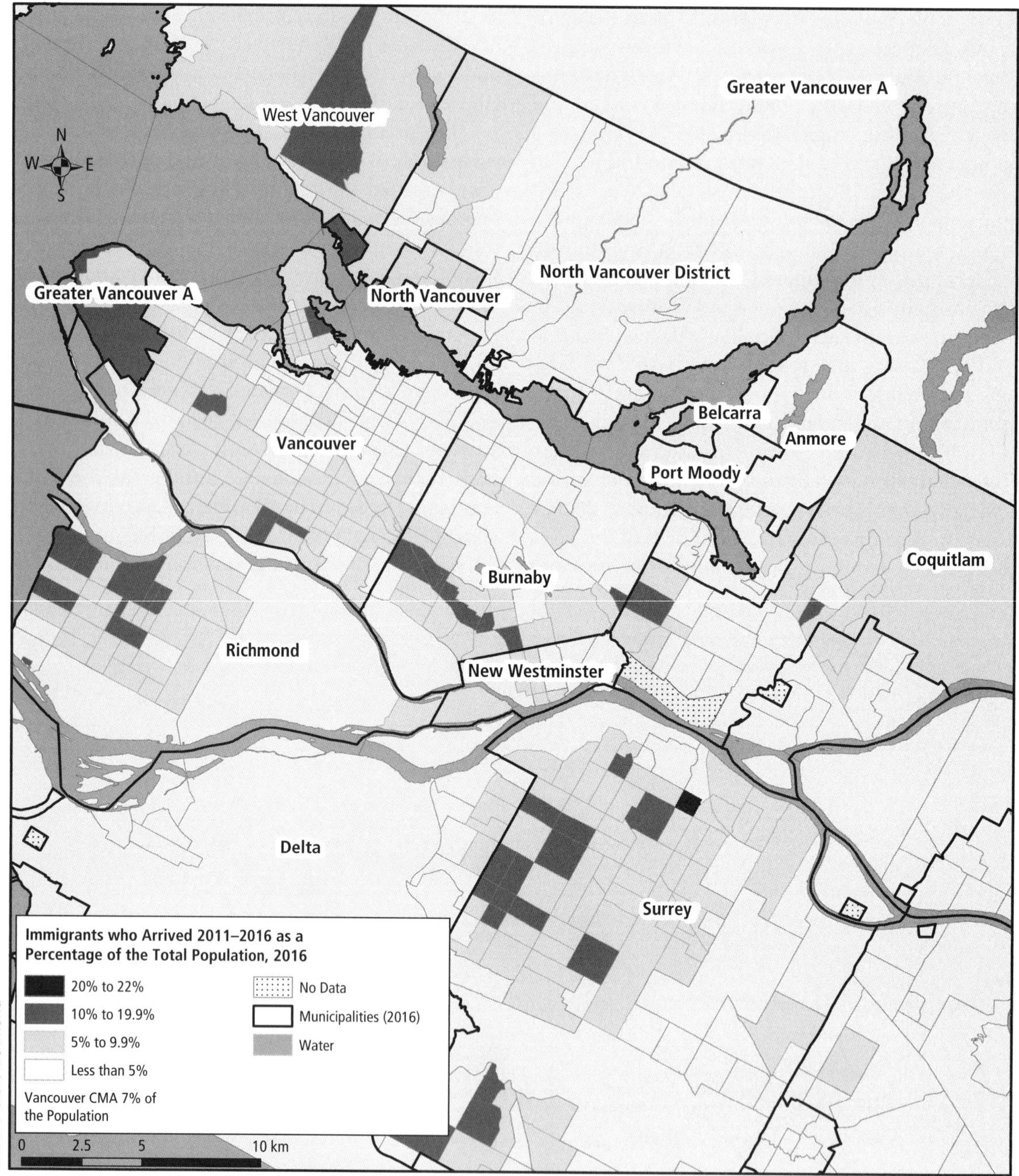

Isabella Croft Richmond

Figure 8.3 Recent Immigrant (2011–2016) Percentage of the Population by Census Tracts, Vancouver CMA, 2016

Source: Statistics Canada, Census 2016.

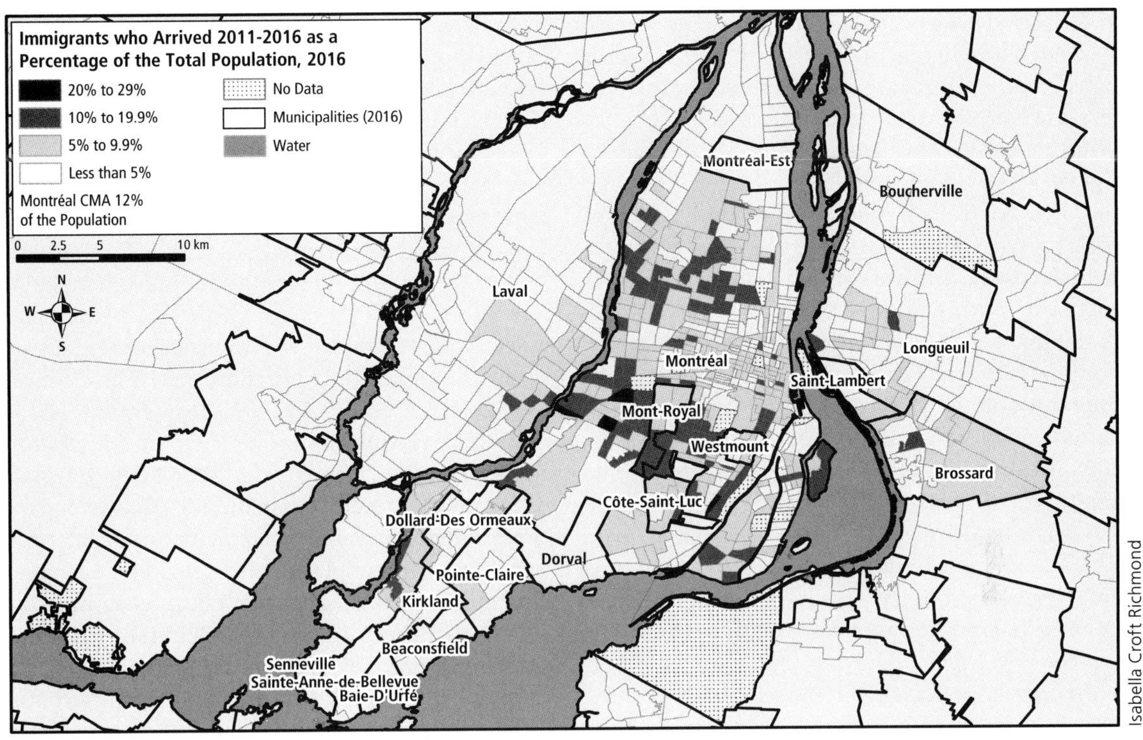

Figure 8.4 Recent Immigrant (2011–2016) Percentage of the Population by Census Tracts, Montreal CMA, 2016

Source: Statistics Canada, Census 2016.

earlier Jewish, Italian, and Portuguese immigrants who moved to newly developed suburbs as their economic circumstances improved, while others are locating directly in the postwar inner suburbs, where rental housing is available (Mensah and Williams, 2013). Still others, particularly Chinese, South Asian, Russian, and Persian immigrants, are typical suburbanites attracted by spacious housing and large lots (Murdie, 2008b; Qadeer, Agrawal, and Lovell, 2010). Suburbanization is well established in Vancouver, where Chinese immigrants from China, Hong Kong, and Taiwan, and South Asians from India and Pakistan are transforming suburban municipalities, such as Richmond and Surrey (Ray, Halseth, and Johnson, 1997; Teixeira, 2014). In Montreal, newcomers are moving beyond the boundaries of the City of Montreal but they remain concentrated on Montreal Island, rather than in the newly developed suburbs to the north (Murdie, 2008b; Rose and Charette, 2014).

The persistent concentrations of immigrants and their children in Canadian cities are often appraised pejoratively as potential **ghettos**, with connotations of poverty, social disorder, and exclusion (Wacquant, 2004; Wirth, 1928).[4] Despite claims that immigrants who live in concentrations fail to integrate into Canadian society, there is no evidence of ghettos in Canadian cities (Walks and Bourne, 2006). Chinese and South Asian concentrations have increased in size and number, especially in the suburbs, but the most recent analyses continue to show that few minorities live in neighbourhoods where they are the majority, and residential concentration is not associated with poverty. There is an expert consensus, however, that many immigrants and their descendants in Canadian cities live in **ethnic enclaves**, spatial concentrations formed by residents' preference to live near others from the same ethno-cultural group, rather than by the processes of exclusion as is the case in so-called ghettos (Murdie, 2008b; Murdie and Ghosh, 2010; Qadeer, Agrawal, and Lovell, 2010).

This relatively benign view of minority concentrations reflects the low levels of segregation of racial and ethnic minorities in Canadian cities compared with those in American cities as well as the low correlations between levels of segregation and poverty. For example, many Italian immigrants and their descendants in Toronto live in Vaughan, an affluent northern suburb (Murdie, 2008b). Concentrations of South Asian immigrants and their children are also readily apparent in the suburban landscapes of Toronto and Vancouver, where developers build large houses to accommodate multigenerational families (Agrawal, 2006). Other scholars, however, have found conflicting evidence (Darden and Fong, 2012; Mensah and Williams, 2013; Teixeira, 2008), emphasizing how racism has shaped residential concentrations for some minorities, especially those of African and Caribbean origin.

Although many newcomers rely on help from friends and family to settle, integration is often enhanced by services that are distributed unevenly across metropolitan areas. Historically, there are often fewer services in the suburbs, and infrequent and slow public transit. Ironically, newcomers who settle in private rental housing are often at a disadvantage compared to those who live near subsidized housing. Many of the services that locate to assist the low-income populations in subsidized housing also serve immigrants (Smith and Ley, 2009). Those who move to the outer suburbs, where the population has grown faster than all types of social infrastructure, often travel long distances to find help or they rely on their own resources and social contacts (Lo et al., 2015).

Recently, migrants' transnational activities have begun to alter the residential landscapes that immigrants construct in Canadian cities. While most immigrants move near friends and family, others are recruited by transnational networks of compatriots, and still others settle with an eye to maintaining economic and social ties to their homelands. Growing numbers of Filipinas, for example, recruit friends and families to work as live-in caregivers in the cities where the recruiters are settled (D'Addario, 2012). For Hong Kongers who wanted to operate businesses in their homeland, proximity to Asia was one of the main attractions of Vancouver, as well as a desire for suburban living (Kobayashi, Preston, and Murnaghan, 2011; Ley, 2003, 2010). The ease with which financial assets can be transferred internationally allows some immigrants to purchase housing as soon as they arrive in Canada, which was the case for many Hong Kong migrants (Ley, 2010; Preston, Kobayashi, and Siemiatycki, 2006). The housing decisions of other transnationals, however, are constrained by their social ties and social relationships at their places of origin. For example, Ghanaian women often rent cheap accommodation so that they can send as much money as possible to support children left in Ghana, a strategy shared by Filipinas (Pratt, 2004; Wong, 2000). Others forgo home ownership in Canada in order to build houses in their homelands (Mensah and Williams, 2013; Wong, 2000).

Transnational ties shape many immigrants' sense of belonging in Canada. Tibetan refugees living near downtown Toronto are typical. For them, home encompasses the altars where daily offerings remind them of religious observances in India and Tibet, the kitchens where Tibetan foods are prepared, and where computers link them to friends and family (Logan and Murdie, 2016).

Landscapes of Inequality

Despite the common perception of being universally accepting, Canadians have been intolerant of "Others."[5] Race, ethnicity, and class have always formed the basis of inequality in Canadian cities, and from the earliest days inequality was associated with immigration. By the 1890s, European immigrants lived in appalling conditions in the poorest areas of Montreal and Toronto. Cities were publicly viewed as places of privilege; and "foreigners"—of Jewish, Italian, Ruthenian, or Macedonian background—as unwanted, unsanitary, and immoral. Around the turn of the twentieth century, urban reform movements arose to "to purge urban centres of some of their worst ills, including prostitution, crime, and disease. In time, the philosophy that urban life was 'bad' and resulted in immoral, unhealthy, and un-British lifestyles affected the focus of future immigration campaigns" (Applied History Research Group, 1997). Eastern Europeans were shunted to the Prairies and, during World War I, many were interned as "enemy aliens."

Anti-"foreign" campaigns have also sparked violence in Canadian cities. In February 1887, the night before Vancouver's first civic election, hundreds of white men attacked a camp of Chinese immigrant workers in an attempt to run them out of town (Donaldson, 2013). Among many instances of violence against Asian immigrants, perhaps the most infamous is the September 1907 race riot that destroyed property on Pender Street and in the neighbouring Japanese immigrant district along Powell Street. When Asian workers, who provided many of the housekeeping and laundry services for Vancouver's white population, staged a work stoppage to encourage an investigation (Gilmore, 2012; Sugimoto, 1973), the subsequent Royal Commission (Canada, 1908) led to the ironic "Gentlemen's Agreement" by which immigration from Japan was curbed. This incident illustrates the collaboration between government and dominant-group citizens to contain what they viewed as "foreign" populations, and one of the first attempts by a racialized minority to protest their treatment and to advocate for their place in the city. In addition, under the slogan "None is too many," European Jews were denied entry to Canada and returned to Europe, where they faced death camps during World War II (Abella and Troper, 1983). And most racialized minorities—whether immigrants or Canadian-born—did not receive the right to vote until the Citizenship Act was passed in 1947.

Study after study (Darden, 2004; Darden and Fong, 2012; Darden, 2015; Francis, 2009; Francis and Hiebert, 2014; Isin and Siemiatycki, 2002; Lai and Huffey, 2009; Li, 2003; Murdie, 2002; Novac et al., 2004; Ray and Preston, 2009; Rose and Charette, 2014; Simmons, 2010; Teixeira, 2008) shows that newcomers have faced prejudice, discrimination, unequal or unfair treatment, and racism throughout Canadian history. Islamophobia—discrimination against Muslim groups who are among the newest arrivals in Canada—is one of the most recent examples. Although the issue came to the fore after the events of 11 September 2001, Arab and Muslim populations have been subject to discrimination for much longer (Abu-Laban and Alladin, 1991; Arat-Koc, 2006; Jamil and Rousseau 2012). Islamophobia is highly visible in controversies over the siting of mosques and the right to wear religious dress while using urban public services and in public places (Abu-Laban and Abu-Laban, 2007; Leroux, 2010).

Racial profiling is not always visible, but it is a powerful means of othering. In Canada, some of the most serious instances concern Muslim-Canadian men who are incarcerated and sometimes tortured with the knowledge and complicity of the Canadian government. For example, Maher Arar, a telecommunications engineer with dual Canadian and Syrian citizenship, was detained by US authorities while changing planes at Kennedy International Airport, en route home to Canada. After being deported to Syria where he was held and tortured for a year, an inquiry found that he had no links to any terrorist organization, and he received an official apology and a compensatory settlement (Mazigh, 2008; Pither, 2008). Young black men also experience racial profiling, particularly in low-income neighbourhoods of major cities where the police regularly stop them. Resistance to the intimidating and questionable police requests for identification (known as "carding") is met with suspicion and the potential threat of arrest (Hayle, Wortley, and Tanner, 2016; Wortley and Owusu-Bempah, 2011). These cases highlight the uneven and unsafe landscapes of justice in Canadian cities, especially for Arabs, Muslims, and blacks, including but not limited to immigrants.

Municipal governments are instrumental in reinforcing or mitigating the insecurity that marks the everyday lives of many immigrants and other racialized minorities. For example, the unwillingness of Toronto police to end carding and the frequency with which the Toronto police share information about people without documents with the Canada Border Security Agency heighten the insecurity of newcomers and minorities. Yet the municipality was one of the first in Canada to declare itself a sanctuary city in which municipal services will be accessible to all residents regardless of their immigration status. The contradictions abound (Bauder, 2017) even in places where there are few newcomers. Many small towns and cities that welcome immigrants as workers, consumers, and investors also struggle to regulate the behaviour of newcomers who are different from the dominant group. For example, the small town of Hérouxville, Quebec, created a code of conduct that banned punishment by stoning and other so-called foreign practices, even though its newcomer population was small, and no such incidents had occurred. In Quebec, municipalities are now debating whether and how their employees, including bus drivers, will enforce provincial legislation that denies publicly funded services to people wearing religious symbols such as the hijab and niqab. These issues extend well beyond traditional municipal responsibilities for land-use planning and delivery of garbage, sewer, water, and other services, but their impact on the integration of newcomers cannot be underestimated, especially in a period when anti-immigrant sentiment and hate crimes against minorities are increasing. As we saw in January 2017 when six men were killed during prayers at a mosque in Quebec City, the insecurity felt by many newcomers is sometimes well founded.

Although there have been many other instances of violence, most of the inequality that is etched upon the urban landscape is neither so violent nor so fleeting. As we have shown above, residential segregation represents the most ubiquitous expression of inequality and inequity in the urban landscape. Nativist opposition to the cultural expression of both residential and commercial landscapes represents a form of violence that strengthens anti-immigrant racism. Places of worship constitute one of the most visible examples. Freedom of religion is a fundamental right guaranteed in the *Charter of Rights and Freedoms.* Religion is also one of the most significant markers of identity, for many surpassing any other form of **citizenship** belonging. In the days following 9/11, however, there were numerous acts of violence, including racist graffiti, personal attacks, and firebombing of mosques and temples, across Canada. The increasing number of non-Christian places of worship—including Hindu, Sikh, Islamic, or Buddhist temples, or mosques—mark the growth and coming of age of immigrant communities (Isin and Siemiatycki, 2002; Mensah, 2008), yet the locations of their places of worship are often contested (Germain and Gagnon, 2003; Hoernig, 2009; Qadeer and Chaudhry, 2000). Religious communities often adapt to place-based opposition by locating in industrial areas rather than in residential areas, and by reaching out to their neighbours in an effort to increase understanding (Hoernig and Zhuang, 2010).

Inequality for immigrants in the urban landscape is most acutely felt in the economic realm. It is taking longer for immigrants to achieve the same earnings as their equally qualified and experienced Canadian-born counterparts (Pendakur and Pendakur, 2015). Seeking affordable housing, many recent immigrants are concentrating in neighbourhoods where low-cost rental housing is available. In Toronto, priority neighbourhoods, where the population is disproportionately low-income, are home to large, racialized immigrant populations (Hulchanski, 2007). With poor transit connections, residents are often isolated from a growing number of suburban jobs and the services that might facilitate their economic integration (Smith and Ley, 2009).

Immigration Policy as a Systemic Process

The Canadian government has always played a dominant role in directing immigration trends, from outright attempts to exclude racialized minorities while encouraging European immigrants to settle on the Prairies, to decisions over which groups to exclude during World War II and how to define refugees during the 1960s and 1970s. The acceptance of American war resisters during the Vietnam War was a contentious policy decision on the part of

Prime Minister Pierre Elliott Trudeau. Over time, immigration policies and procedures have had far-reaching consequences for Canadian cities.

A major shift occurred during the 1960s, when Canada introduced a "points" system, which evaluated each applicant's educational and professional qualifications, work experience, and official language abilities. This system, which limited the power of officials to make arbitrary decisions and removed racial background as an implicit criterion for admission, was used to recruit skilled workers and their families. To accelerate economic development, business-class immigration was introduced to attract experienced entrepreneurs and investors (Ley, 2003). Canadian policies also encouraged family reunification by allowing permanent residents and citizens with sufficient financial resources to sponsor relatives. As a signatory to the 1951 United Nations (UN) Convention on Refugees,[6] Canada also admitted immigrants who met the UN definition of a refugee, including government-assisted (GAR) and privately sponsored (PSR) refugees who were selected by consular officials abroad, and refugee claimants who were recognized as bona fide refugees after their arrival in Canada (UN, 1967).

The effect of the new post-World War II legislation was swift. The number of source countries increased dramatically, and Asian countries quickly replaced European countries as the largest sources of immigrants (see Table 8.1 earlier). People who self-identified as non-Aboriginal or non-white visible minorities accounted for 81.7 per cent of the immigrants who arrived between 2011 and 2016 (Statistics Canada, 2017a). Attracted to Canada's metropolitan areas, the newcomers now constitute large racialized minorities in Montreal, Toronto, and Vancouver. In the Toronto metropolitan area, home to the largest racialized minority population in Canada, one in two residents, or 51.5 per cent, identifies as a visible minority. The numbers are smaller in Vancouver and Montreal—48.9 per cent and 22.6 per cent, respectively. Five other metropolitan areas—Calgary, Edmonton, Ottawa-Gatineau, Winnipeg, and Hamilton—are home to more than 100,000 people who identify as racialized minorities (Statistics Canada, 2017a). This change in the backgrounds of urban residents has drawn attention to discrimination in urban housing and labour markets, the challenge of living with difference for all Canadians, and the success with which some Canadian municipalities are meeting the challenges of multicultural planning, while others are not (Qadeer and Agrawal, 2011).

During the 1990s, the government also reversed longstanding policies that reduced immigrant admissions during periods of economic recession. In response to growing concerns about Canada's declining birth rate and aging population, the government set an annual target for the total number of immigrants, approximately 1 per cent of the population, to be maintained regardless of the state of the Canadian economy. Each year since, the government has admitted between 200,000 and 300,000 immigrants; however, the proportion from each immigrant class—skilled worker and business class, family class, and refugee—has varied. Annual admissions of economic immigrants increased to approximately two-thirds of all permanent residents in 2011, and the percentages for refugees and family-class immigrants fell to approximately 11 per cent and less than 30 per cent of annual admissions, respectively (Citizenship and Immigration Canada, 2012). Since 2015, the total number of immigrants admitted annually has increased slightly, and the mix of immigration classes has changed once again. In 2016, 62,348 convention refugees were admitted, approximately 23.1 per cent of all immigrants, and the number of family-class immigrants also increased. As a result, the number of economic immigrants fell slightly to 155,994 or 52.6 per cent of all those admitted permanently in 2016 (Immigration, Refugees and Citizenship, 2017).

Even at a distance, the extreme and abrupt events that create many refugees—natural disasters, wars, and other violence—have a profound impact on Canadian metropolitan areas. In the summer of 2015, Canada's willingness to admit people who had fled Syria, Afghanistan, and Iraq became a prominent election issue. After a change in government, Canada admitted more than 40,000 Syrian refugees between 4 November 2015 and 29 January 2017. To resettle the newcomers, a public campaign that portrayed

Canada as a welcoming and diverse society and refugee resettlement as part of Canadian identity was launched, and help was solicited from corporations, ordinary Canadians, provincial and municipal governments, and **non-governmental organizations (NGOs)**. Private groups that committed to providing financial and other support to refugee households during their first year in Canada were encouraged to sponsor refugees. The Syrian refugees were settled in urban areas across Canada. Only 39.4 per cent of refugees resettled between January 2015 and April 2018 planned to live in MTV.

Since the election of the Trump administration in the United States in 2016, Canada has faced one of its greatest challenges to successful newcomer settlement. Following the events of 11 September 2001, the two countries signed a "Safe Third Country" agreement that states that neither country will accept refugees who have come through the other country's land border crossings or by air (Canada, 2016). Recently, thousands of people, deeming their chances of asylum in the United States to be limited, began crossing the border on foot, especially in the provinces of Manitoba and Quebec, where there are long stretches of unguarded borders. During 2017, the RCMP intercepted over 20,000 such border crossers, nearly half the total number of refugee claimants from all sources. In the United States, many face the sudden loss of legal temporary residency, and others are taking advantage of their ability to travel to the United States as visitors before moving on to Canada, where they claim asylum.

NGOs and municipalities are struggling to provide basic necessities such as housing even in their main destinations of Montreal, Toronto, and Winnipeg. Small border cities such as Windsor, Niagara Falls, and Sherbrooke that are the initial stops for many claimants often lack services and facilities appropriate for their growing numbers. With limited rights and resources, refugee claimants must often rely on social assistance and other publicly funded supports. Recent policy changes, such as reducing the qualifying period of residency for a work permit from three months to three weeks, help reduce the cost of hosting claimants and respond to demand for workers to do dirty, often dangerous and undesirable jobs, such as staffing the evisceration lines in meat-packing plants (Ross, 2018). By encouraging refugee claimants to work, the government also demonstrates how they may contribute to Canadian society, short-circuiting the anti-immigrant sentiment that is fuelled by a steady stream of media reports about people crossing illegally into Canada. Making it easier for refugee claimants to work is consistent with the turn to neo-liberalism that now underpins many immigration and settlement policies and programs. The recent and welcome renewal of a commitment to refugee resettlement stands in stark contrast to other policy changes that have emphasized employers' needs and encouraged the selection of newcomers who are self-sufficient and can settle without much public support.

For those entering on a permanent status, policies favour well-educated and highly skilled workers who already speak at least one official language or workers already experienced in occupations that are in demand. The expanding role of employers in immigrant selection is readily apparent in the Express Entry system that allows employers to review the skills, qualifications, and experience of potential applicants under the economic class even before they are allowed to apply.

Recognized internationally as an immigrant-receiving country that actively recruits permanent residents, Canada has also become a major destination for temporary residents. Responding to employers' assertions of labour shortages, the number of people admitted on a temporary basis for specific jobs has increased from 294,525 in 1991 to more than 550,000 in 2016 (Citizenship and Immigration Canada, 2012; Immigration, Refugees and Citizenship Canada, 2017).

The growth of temporary migration occurred as some responsibility for immigrant selection and settlement devolved to the provinces. Immigration is a joint responsibility of the federal and provincial governments, but provincial interest in immigration was minimal until late in the twentieth century (Alboim and Cohl, 2012; Leitner and Preston, 2012). The initial and most substantial devolution occurred in 1991 when the Quebec–Canada Accord gave the provincial government of Quebec responsibility for selecting and settling immigrants in that province. In other provinces and territories, the federal government introduced the Provincial Nominee Program, allowing each provincial government to select a specified number of immigrants according to its own criteria as long as the immigrants were also admissible under federal health and security criteria. Provincial Nominee Programs often privilege the needs and views of employers. For example, job offers play a crucial role in decisions to grant permanent residence to people who enter on temporary visas, whether they be international students with a degree from an accredited post-secondary institution in Canada or low-skilled temporary workers (Leitner and Preston, 2012).

The growing emphasis on employers' needs and views and the accompanying shift to a two-stage migration process has dispersed immigrants across Canada, with more newcomers settling in small metropolitan areas such as Winnipeg, Fredericton, and Charlottetown. Since 2000, Edmonton and Calgary also cemented their positions as major immigrant destinations through a combination of Alberta's Provincial Nominee Program and jobs that attracted immigrants from around the world (Statistics Canada, 2017b).

The regionalization of immigration has been accompanied by growing diversity in the settlement supports available in Canadian cities. Permanent residents are eligible to use government-funded services providing job and language training and information about Canadian society. Settlement services are available mainly through NGOs that work on contracts with the federal and provincial governments. The decentralized settlement service system is designed to respond to local needs in a cost-effective manner while ensuring universal access to services; however, the uneven geography of settlement services threatens access. Services remain concentrated in major metropolitan areas. As immigrants settle in growing numbers outside traditional ports of entry in MTV, many find services are in short supply, particularly in the suburbs and small urban areas. There are local exceptions, however. For example, settlement agencies, some of them religiously organized, that developed in Winnipeg in the first two decades of the twentieth century have been revitalized. In Calgary, immigrant-serving agencies have evolved from volunteer organizations that sprang up initially to sponsor refugees. In Toronto's outer suburbs, churches that offer material and moral support are more accessible to many newcomers than traditional service providers (Rose, 2016). As provinces gained more power over settlement services in the first decade of the millennium, eligibility rules and service delivery also began to vary. In Quebec, eligibility for language training extends beyond the acquisition of **citizenship**, while newcomers who become citizens in other provinces and territories no longer qualify for these courses. Across the county, temporary residents are a concern for many municipalities since they are rarely eligible for settlement services and many other crucial public services such as health care.

Settlement services play an immediate role in immigrant integration (Schellenberg and Maheux, 2007), but a multitude of government policies affect the welfare of newcomers. For example, the federal decision to provide medical services only to refugee claimants who posed a risk to public health or a public safety concern was widely criticized because it jeopardized claimants' chances to integrate (Alboim and Cohl, 2012). The housing challenges facing many newcomers reflect the failure to develop a national strategy for affordable housing (Rose and Charette, 2017; Preston et al., 2011). Human rights and employment equity policies also play a crucial role in combating the discrimination that marginalizes and excludes many immigrants, particularly racialized minorities (Lai and Huffey, 2009; Novac et al., 2004; Preston and Ray, 2009).

Conclusion

International migration is an ongoing geographic process with historical, demographic, social, and public policy dimensions. The urban landscape exhibits the cumulative effects of this dynamic process in a variety of ways, including commercial and residential clustering and exclusions, cultural artifacts, and, in some cases, marginalization and invisibility. Immigration is a normative process, regulated by state policies but also subject to the efforts of migrants to find places in their new society. Although Canadian public policy is built around a norm of multiculturalism, according to which newcomers are welcomed and treated as equal Canadians, that norm is often challenged, especially with respect to the newest groups. In recent years, controversies have arisen over the "limits" to multiculturalism—that is, over whether it is possible to go "too far" in accommodating others who are considered not to fit the Canadian norm. Some even claim that multiculturalism actually prevents newcomers from becoming full Canadians because it encourages them to stick with their own and to maintain cultural practices that are "non-Canadian" (Guo and Wong, 2015). Those who oppose or want to limit multiculturalism adopt a range of attitudes from outright hatred and xenophobia, to "multiculturalism within limits," to "multiculturalism for some but not for others," all of which are normative positions that place Others in opposition to a vision of the "ordinary"—or assimilated—Canadian.

These controversies are revealed in Canadian cities: Landscaping changes are contested in Vancouver's most prestigious neighbourhoods (Ley, 1995); requests from Muslims and Orthodox Jews for zoning changes to accommodate their religious establishments are disputed in Montreal (Germain and Gagnon, 2003); and everywhere immigrants complain that landlords object to their cooking and other cultural practices (Francis and Hiebert, 2014; Murdie, 2008a; Rose and Charette, 2014). These examples illustrate the need to understand multiculturalism within a larger context that includes practices of discrimination and racialization, as well as the range of cultural expressions through which contemporary Canadian cities come to be (Guo and Wong, 2015).

Many of the stories that define immigrant experiences have long-established roots, but many also represent recent changes. Migration occurs in a transnational field that involves the source and the destination countries, the immigrants themselves, and the entire population, including the long settled, who live in cities that are the more vibrant, complex, and fascinating for the changes that international migration continues to bring. As labour market conditions shift, as international politics, including armed conflict, create more and more refugees, and as international relations, including the Safe Third Country agreement, alter the balance of destinations, the effects of changing migration patterns are lived out on Canadian streets.

Review Questions

1. How have recent changes in Canadian immigration policies influenced the social geography of Canadian cities?
2. Some people claim that Canadian cities have largely avoided the racial exclusion that occurs in many American cities. Do you think this is true? Be sure to explain the reasons for your opinion.
3. How have various ethno-cultural groups transformed urban landscapes in Canadian cities?

Notes

1. *Migrants* refer to all foreign-born newcomers living in Canada including immigrants who have the right to permanent residence and temporary foreign workers, international students, and refugee claimants whose right to legal residence is circumscribed.
2. In compliance with international law, Canada admits refugee claimants, people who ask to be recognized as *bona fide* refugees according to the United Nations Convention. They are allowed to stay in Canada on a temporary basis until their claims to refugee status are accepted and they become permanent residents or their claims are denied, signalling the end of their right to live in Canada.
3. See Murdie and Logan (2011) for a comprehensive bibliography about recent immigration and housing research in Canada.

4. *Ghetto* referred originally to the forced concentration of Jews in European cities, but it now describes any forced residential concentration of a minority (Wacquant, 2004).
5. *Othering* is a process of differentiation and demarcation by which the dominant group establishes and maintains social distance between itself and groups viewed as subordinate and inferior, henceforth labelled "Other" (Lister, 2004).
6. A *convention refugee* is a person outside his or her country of nationality who is unable or unwilling to return to his or her country of nationality because of a well-founded fear of persecution on the grounds of religion, nationality, membership of a particular social group, or political opinion (United Nations, 1967).

References

Abella, I.M., and H. Troper. 1983. *None Is Too Many: Canada and the Jews of Europe, 1933–1948*. Toronto: University of Toronto Press.

Abu-Laban, B., and M. I. Alladin. 1991. *Beyond the Gulf War: Muslims, Arabs and the West*. Edmonton: MRF Publishers.

Abu-Laban, Y., and B. Abu-Laban. 2007. "Reasonable accommodation in a global village," *Policy Options-Montreal*, 28(8): 28.

Agrawal, S.K. 2006. "Housing adaptations: A study of Asian Indian immigrant homes in Toronto," *Canadian Ethnic Studies*, 38: 117–30.

Alboim, N., and K. Cohl. 2012. *Shaping the Future: Canada's Rapidly Changing Immigration Policies*. Toronto: Maytree Foundation.

Anderson, K. 1995. *Vancouver's Chinatown, Racial Discourse in Canada, 1875–1980*, 2nd edn. Kingston and Montreal: McGill-Queen's University Press.

Applied History Research Group, University of Calgary. 1997. *The Peopling of Canada: 1891–1921*. At: www.ucalgary.ca/applied_history/tutor/canada1891/index.html

Arat-Koc, S. 2006. "Whose transnationalism? Canada, 'clash of civilizations' discourse, and Arab and Muslim Canadians," in V. Satzewich and L. Wong, eds., *Transnational Identities and Practices in Canada*. Vancouver, BC: University of British Columbia Press.

Bauder, H. 2017. "Sanctuary Cities: Policies and Practices in International Perspective," *International Migration*, 55: 174–87.

Buckley, K. 1960. "Historical estimates of internal migration in Canada," in E.F. Beach and J.C. Weldon, eds., *Canadian Political Science Association, Conference on Statistics 1960, Papers*. Toronto: University of Toronto Press.

Canada (Government of). 1908. *Report of W.L. Mackenzie King, C.M.G., Commissioner Appointed to Enquire into the Methods by Which Oriental Labourers Have Been Induced to Come to Canada*. Ottawa: King's Printer.

———. *2016. Canada-US Safe Third Country Agreement. At:* https://www.canada.ca/en/immigration-refugees-citizenship/corporate/mandate/policies-operational-instructions-agreements/agreements/safe-third-country-agreement.html

Carter, T., and D. Vitiello. 2012. "Immigrants, refugees, and housing," in C. Teixeira, W. Li, and A. Kobayashi, eds., *Immigrant Geographies of North American Cities*. Toronto: Oxford University Press.

Castles, S., H. De Haas, and Miller, M.J. 2014. *The Age of Migration: International Population Movements in the Modern World*. New York: Macmillan International Higher Education.

Chui, T., Tran, K., and Maheux, H. 2007. *Immigration in Canada: A Portrait of the Foreign-Born Population, 2006 Census*. Ottawa: Statistics Canada. At: www12.statcan.ca/census-recensement/2006/as-sa/97-557/pdf/97-557-XIE2006001.pdf

Citizenship and Immigration Canada. 2005. *Recent Immigrants in Metropolitan Areas*. Ottawa: Statistics Canada. At: www.cic.gc.ca/english/pdf/research-stats/2001-canada.pdf

———. 2012. *Facts and Figures, Immigration Overview Permanent and Temporary Residents, 2011*. At: www.cic.gc.ca/english/resources/statistics/menu-fact.asp

D'Addario, S. 2012. *Finding Home: Geographical Links between Paid and Unpaid Work for Transnational Care Workers in Toronto's Suburbs*. Toronto: York University, PhD thesis.

Danso, R. 2002. *Ethnic Community Networks, Public Policy, and the Resettlement of Ethiopian and Somali Refugees in Toronto*. Kingston, ON: Queen's University, PhD thesis.

Darden, J.T. 2004. *The Significance of White Supremacy in the Canadian Metropolis of Toronto*. Lewiston: Edwin Mellen Press.

———. 2015. "Homeownership among immigrants in Canada and the United States: Similarities and differences," in C. Teixeira and W. Li, eds., *The Housing and Economic Experiences of Immigrants in US and Canadian Cities*. Toronto: University of Toronto Press.

———, and E. Fong. 2012. "The spatial segregation and socio-economic inequality of immigrant groups," in C. Teixeira, W. Li, and A. Kobayashi, eds., *Immigrant Geographies of North American Cities*. Don Mills, ON: Oxford University Press.

Donaldson, J. 2013. "Remembering Vancouver's first race riot," *The Tyee*, 1 March. At: http://thetyee.ca/Life/2013/03/01/Vancouver-Race-Riot/

Francis, J. 2009. *"You Cannot Settle Like This": The Housing Situation of African Refugees in Metro Vancouver*. Vancouver, BC: Metropolis British Columbia (Working Paper 09-02).

———, and D. Hiebert. 2014. "Shaky foundations: Refugees in Vancouver's housing market," *The Canadian Geographer/Le Géographe canadien*, 58(1): 63–78.

Germain, A., and J.E. Gagnon. 2003. "Minority places of worship and zoning dilemmas in Montreal," *Planning Theory and Practice*, 4: 295–318.

———, and D. Rose. 2000. *Montreal: The Quest for a Metropolis*. New York: Wiley.

Gilmore, J. 2012. "Interpreting social disorder: The case of the 1907 Vancouver riots," *International Journal*, Spring: 485–95.

Guo, S. and L. Wong, eds., 2015. *Revisiting Multiculturalism in Canada: Theories, Policies and Debates*. Rotterdam, Boston, Taipei: Sense Publishers.

Hall, S. 2000. "The multi-cultural question," in B. Hesse, ed., *Unsettled Multiculturalisms, Diasporas, Entanglements, Disruptions*. London: Zed Books.

Hardwick, S.W., and G. Mansfield. 2009. "Discourse, identity, and 'Homeland as Other' at the Borderlands," *Annals of the Association of American Geographers*, 99: 383–405.

Hawkins, F. 1988. *Canada and Immigration*. Kingston and Montreal: McGill-Queen's University Press.

Hayle, S., S. Wortley, and J. Tanner. 2016. "Race, street life, and policing: Implications for racial profiling," *Canadian Journal of Criminology and Criminal Justice*, 58(3): 322–53.

Hiebert, D. 1993. "Jewish immigrants and the garment industry of Toronto: A study of ethnic and class relations," *Annals of the Association of American Geographers*, 83: 243–71.

———. 2011. *Precarious Housing and Hidden Homelessness among Refugees, Asylum Seekers, and Immigrants in Montréal, Toronto, and Vancouver: Introduction and Synthetic Executive Summary*. June. At: http://mbc.metropolis.net/assets/uploads/files/precarious_housing_NATL%20SUMMARY.pdf

———, A. Germain, R. Murdie, V. Preston, J. Renaud, D. Rose, E. Wyly, V. Ferreira, P. Mendez, and A.M. Murnaghan. 2006. *The Housing Situation and Needs of Recent Immigrants in the Montreal, Toronto, and Vancouver CMAs: An Overview*. Ottawa: Canada Mortgage and Housing Corporation.

Hoernig, H.J. 2009. "Planning amidst cultural diversity: Lessons from religious development," *Plan Canada*, 48: 55–59.

———, and Z. Zhuang. 2010. "New diversity: Social change as immigration," in T. Bunting, P. Filion, and R. Walker, eds., *Canadian Cities in Transition: New Directions in the Twenty-first Century*. Toronto: Oxford University Press.

Hulchanski, D. 2007. "The three cities within Toronto: Income polarization among Toronto's neighbourhoods, 1970–2000." Toronto: Centre for Urban and Community Research, Research Bulletin 41. At: www.urbancentre.utoronto.ca/pdfs/researchbulletins/CUCSRB41_Hulchanski_Three_Cities_Toronto.pdf

———, and M. Shapcott. 2004. "Introduction: Finding room in Canada's housing system for all Canadians," in D. Hulchanski and M. Shapcott, eds., *Finding Room: Policy Options for a Canadian Rental Housing Strategy*. Toronto: CUCS Press, Centre for Urban and Community Studies.

Immigration, Refugees and Citizenship Canada. 2017. *2017 Annual Report to Parliament on Immigration*. Ottawa: Immigration, Refugees and Citizenship Canada. At: https://www.canada.ca/en/immigration-refugees-citizenship/corporate/publications-manuals/annual-report-parliament-immigration-2017.html

Isin, E., and M. Siemiatycki. 2002. "Making space for mosques: Struggles for urban citizenship in diasporic Toronto," in S.H. Razack, ed., *Race, Space, and the Law: Unmapping a White Settler Society*. Toronto: Between the Lines.

Jackson, P., and A. Kobayashi. 1994. "Japanese Canadians and the racialization of labour in the British Columbia sawmill industry," *BC Studies: The British Columbian Quarterly*, 103: 33–58.

Jamil, U., and C. Rousseau. 2012. "Subject positioning, fear, and insecurity in South Asian Muslim communities in the war on terror context," *Canadian Review of Sociology/Revue canadienne de sociologie*, 49(4): 370–88.

Kalbach, W.E., and W.W. McVey. 1971. *The Demographic Basis of Canadian Society*, 2nd edn. Toronto: McGraw-Hill Ryerson.

Knowles, V. 2006. *Strangers at Our Gates: Canadian Immigration and Immigration Policy*. Hamilton, ON: Dundurn Press.

Kobayashi, A., and V. Preston. 2007. "Transnationalism through the life course: Hong Kong immigrants in Canada," *Asia Pacific Viewpoint*, 48: 151–67.

———, V. Preston, and A.M. Murnaghan. 2011. "Place, affect, and transnationalism through the voices of Hong Kong immigrants to Canada," *Social & Cultural Geography*, 12: 871–88.

Lai, D.C. 2011. *Chinatowns: Towns within Cities in Canada*. Vancouver, BC: UBC Press.

———, and N. Huffey. 2009. "Experience of discrimination by visible minorities in small communities," *Our Diverse Cities* 6, Spring: 124–9.

Leitner, H., and V. Preston. 2012. "Going local: Canadian and American immigration policy in the new century," in C. Teixeira, W. Li, and A. Kobayashi, eds., *Immigrant Geographies of North American Cities*. Toronto: Oxford University Press.

Leroux, D. 2010. "Québec nationalism and the production of difference: The Bouchard-Taylor Commission, the Hérouxville code of conduct, and Québec's immigrant integration policy," *Quebec Studies*, 49: 107–26.

Ley, D. 1995. "Between Europe and Asia: The case of the missing sequoias," *Cultural Geographies*, 2: 185–210.

———. 2003. "Seeking *homo economicus*: The strange story of Canada's Business Immigration Program," *Annals, Association of American Geographers*, 93: 426–41.

———. 2010. *Millionaire Migrants: Trans-Pacific Life Lines*. West Sussex, UK: Wiley-Blackwell.

———. 2013. "Does transnationalism trump immigrant integration? Evidence from Canada's links with East Asia," *Journal of Ethnic and Migration Studies*, 39: 921–938.

Li, P.S. 2003. *Destination Canada: Immigration Debates and Issues*. Toronto: Oxford University Press.

Li, W. 2009. *Ethnoburb: The New Ethnic Community in Urban America*. Honolulu, HI: University of Hawaii Press.

Lister, R. 2004. *Poverty*. London: Polity Press.

Lo, L. 2006. "Suburban housing and indoor shopping: The production of the contemporary Chinese landscape in Toronto," in W. Li, ed., *From Urban Enclave to Ethnic Suburb: New Asian Communities in Pacific Rim Countries*. Honolulu, HI: University of Hawaii Press.

———, V. Preston, P. Anisef, R. Basu, and S. Wang. 2015. *Social Infrastructure and Vulnerability in the Suburbs*. Toronto: University of Toronto Press.

Logan, J., and R. Murdie. 2016. "Home in Canada? The settlement experiences of Tibetans in Parkdale, Toronto," *Journal of International Migration and Integration*, 17: 95–113.

Lorinc, J., M. McClelland, F. Scheinberg, and T. Taylor, eds., 2015. *The Ward: The Life and Loss of Toronto's First Immigrant Neighbourhood*. Toronto: Coach House Books.

Massey, D.S. 2012. *Worlds in Motion: Understanding International Migration at the End of the Millennium*. New York: Oxford University Press.

Mazigh, M. 2008. *Hope and Despair: My Struggle to Free My Husband, Maher Arar*. Toronto: McClelland & Stewart.

Mensah, J. 2008. "Religious transnationalism among Ghanaian immigrants in Toronto: A binary logistic regression analysis," *The Canadian Geographer*, 52: 309–30.

———, and C.J. Williams, C.J., 2013. "Ghanaian and Somali immigrants in Toronto's rental market: a comparative cultural perspective of housing issues and coping strategies," *Canadian Ethnic Studies*, 45: 115–41.

Moos, M., and A. Skaburskis. 2010. "The globalization of urban housing markets: Immigration and changing housing demand in Vancouver," *Urban Geography*, 31: 724–49.

Murdie, R.A. 2002. "The housing careers of Polish and Somali newcomers in Toronto's rental market," *Housing Studies*, 17: 423–43.

———. 2008a. "Pathways to housing: The experiences of sponsored refugees and refugee claimants in accessing permanent housing in Toronto," *International Migration & Integration*, 9: 81–101.

———. 2008b. *Diversity and Concentration in Canadian Immigration: Trends in Toronto, Montreal and Vancouver, 1971–2006*. Toronto: Centre for Urban and Community Studies, University of Toronto, Research Bulletin 42.

———, and S. Ghosh. 2010. "Does spatial concentration always mean a lack of integration? Exploring ethnic concentration and integration in Toronto," *Journal of Ethnic and Migration Studies*, 36: 293–311.

———, and J. Logan. 2011. *Precarious Housing & Hidden Homelessness among Refugees, Asylum Seekers, and Immigrants: Bibliography and Review of Canadian Literature from 2005 to 2010*. CERIS Working Paper No. 84. Toronto, ON: CERIS—The Ontario Metropolis Centre. At: www.ceris.metropolis.net/wp-content/uploads/pdf/research_publication/working_papers/wp84.pdf

Novac, S., J. Darden, D. Hulchanski, and A. Seguin. 2004. "Housing discrimination in Canada: Stakeholder views and research gaps," in D. Hulchanski and M. Shapcott, eds., *Finding Room: Policy Options for a Canadian Rental Housing Strategy*. Toronto: CUCS Press, Centre for Urban and Community Studies.

Park, R.E., and E.W. Burgess. 1925. *The City: Suggestions for Investigation of Human Behavior in the Urban Environment*. Chicago: University of Chicago Press.

Pendakur, K., and R. Pendakur. 2015. "The colour of money redux: Immigrant/ ethnic earnings disparity in Canada, 1991–2006," in C. Teixeira and W. Li, eds. *The Housing and Economic Experiences of Immigrants in U.S. and Canadian Cities*. Toronto: University of Toronto Press, 227–60.

Pither, K. 2008. *Dark Days: The Story of Four Canadians Tortured in the Name of Fighting Terror*. Toronto: Penguin.

Pratt, G. 2004. *Working Feminism*. Philadelphia: Temple University Press.

Preston, V., and L. Lo. 2000. "Asian theme malls in suburban Toronto: Land use conflict in Richmond Hill," *Canadian Geographer*, 44: 86–94.

———, A. Kobayashi, and G. Man. 2006. "Transnationalism, gender, and civic participation: Canadian case studies of Hong Kong immigrants," *Environment and Planning A*, 38: 1633–51.

———, A. Kobayashi, and M. Siemiatycki. 2006. "Transnational urbanism: Toronto at a crossroads," in V. Satzewich and L. Wong, eds., *Transnational Identities and Practices in Canada*. Vancouver, BC: UBC Press.

———, R.A. Murdie, S. D'Addario, P. Sibanda, and A.M. Murnaghan. 2011. *Precarious Housing and Hidden Homelessness among Refugees, Asylum Seekers, and Immigrants in the Toronto CMA*. CERIS Working Paper No. 87. Toronto: CERIS—The Ontario Metropolis Centre. At: http://mbc.metropolis.net/assets/uploads/files/Precarious_Housing_Toronto_study.pdf

———, and B. Ray. 2009. "Geographies of discrimination: Variations in perceived discomfort and discrimination in Canada's gateway cities," *Journal of Immigrant and Refugee Studies* 7: 228–49.

Qadeer, M.A., and M. Chaudhry. 2000. "The planning system and the development of mosques in the Greater Toronto area," *Plan Canada* 40: 17–21.

———, and S.K. Agrawal. 2011. "The practice of multicultural planning in American and Canadian cities," *Canadian Journal of Urban Research*, 20: 132–56.

———, S.K. Agrawal, and A. Lovell. 2010. "Evolution of ethnic enclaves in the Toronto Metropolitan Area, 2001–2006," *Journal of International Migration and Integration*, 11: 315–39.

Ray, B.K., G. Halseth, and B. Johnson. 1997. "The changing 'face' of the suburbs: Issues of ethnicity and residential change in suburban Vancouver," *International Journal of Urban and Regional Research*, 21: 75–99.

———, and V. Preston. 2009. "Geographies of discrimination: Inter-urban variations in Canada," *Journal of Immigrant & Refugee Studies*, 7: 228–49.

Rose, D., and Charette, A. 2014. "Housing experiences of users of settlement services for newcomers in Montréal: Does immigration status matter," in K. Kilbride, ed., *Immigrant Integration: Research implications for Public Policy.* Toronto: Canadian Scholar's Press.

———, and ———. 2017. Finding housing for the Syrian refugee newcomers in Canadian cities: challenges, initiatives and policy implications. Synthesis report. Montreal: INRS-CUS. At: http://espace.inrs.ca/6458/1/FindinghousingforSyrianrefugees-report.pdf

Rose, J.J. 2016. Immigrant Identities and Geographies of Belonging: Jamaican Immigrant Organizations in Toronto. Ph.D. dissertation, Department of Geography, York University.

Ross, S. 2018. "Canada puts asylum seekers to work as system feels pressure," *Toronto Star,* May 21.

Schellenberg, G., and H. Maheux. 2007. "Immigrants' perspectives on their first four years in Canada: Highlights from three waves of the Longitudinal Survey of Immigrants to Canada," *Canadian Social Trends,* April: 2–17.

Simmons, A. 2010. *Immigration and Canada: Global and Transnational Perspectives.* Toronto: Canadian Scholars Press.

Smith, H., and D. Ley. 2009. "Even in Canada? The multiscalar construction and experience of concentrated immigrant poverty in gateway cities," *Annals of the Association of American Geographers,* 98: 686–713.

Statistics Canada. 2008. *Canada's Ethnocultural Mosaic, 2006 Census.* Ottawa, ON: Statistics Canada. At: https://www12.statcan.gc.ca/census-recensement/2006/as-sa/97-562/index-eng.cfm

———. 2013. *National Household Survey.* Ottawa: Statistics Canada. At: www12.statcan.gc.ca/nhs-enm/ 2011/as-sa/index-eng.cfm

———. 2017a. *Focus on Geography Series, 2016 Census.* Statistics Canada Catalogue no. 98-404-X2016001. Ottawa, ON. Data products, 2016 Census.

———. 2017b. "Immigration and ethnocultural diversity: Key results from the 2016 Census," *The Daily,* October 25. At: https://www150.statcan.gc.ca/n1/en/daily-quotidien/171025/dq171025b-eng.pdf?st=VHltJft-

Sugimoto, H. 1973. "The Vancouver Riot and its international significance," *The Pacific Northwest Quarterly,* 64: 163–74.

Teixeira, C. 2007. "Residential experiences and the culture of suburbanization: A case study of Portuguese homebuyers in Mississauga," *Housing Studies,* 22: 495–521.

———. 2008. "Barriers and outcomes in the housing searches of new immigrants and refugees: A case study of 'Black' Africans in Toronto's rental market," *Journal of Housing and the Built Environment,* 23: 253–176.

———. 2014. "Living on the "edge of the suburbs" of Vancouver: A case study of the housing experiences and coping strategies of recent immigrants in Surrey and Richmond," *The Canadian Geographer/Le Géographe canadien,* 58: 168–87.

United Nations. 1967. *Protocol Relating to the Status of Refugees,* 31 January. United Nations, Treaty Series, vol. 606, p. 267. At: www.refworld.org/docid/3ae6b3ae4.html

Verma, S. 2005. "Multicultural Markham, 10 years on, immigrants' role divided town in 1995, now, Chinese-Canadians right at home," *Toronto Star,* 2 July.

Wacquant, L. 2004. "What is a ghetto? Constructing a sociological concept," in N.J. Smelser and P.B. Baltes, eds., *International Encyclopedia of the Social and Behavioral Sciences.* London: Pergamon.

Walks, R., and L.S. Bourne. 2006. "Ghettos in Canada's cities? Racial segregation, ethnic enclaves and poverty concentration in Canadian urban areas," *The Canadian Geographer,* 50: 273–97.

Wirth, L. 1928. *The Ghetto.* Chicago: University of Chicago Press.

Wong, M. 2000. "Ghanaian women in Toronto's labour market: Negotiating gendered roles and transnational household strategies," *Canadian Ethnic Studies,* 32: 45–74.

Wortley, S., and A. Owusu-Bempah. 2011. "Crime and justice: The experiences of black Canadians," in B. Perry, ed., *Diversity, Crime, and Justice in Canada.* Toronto: Oxford University Press.

Zhuang, Z.C., 2013. "Rethinking multicultural planning: an empirical study of ethnic retailing," *Canadian Journal of Urban Research,* 22(2): 90-116.

9 Gender, Sexuality, and the City

Damaris Rose

Introduction

For over a century, the urban spaces of cities, transformed by mass urbanization and economic diversification, have been key arenas where society's dominant visions about *gender* and *sexuality* have been constructed, destabilized, reinforced and renegotiated. The concept of *gender* refers to the roles, behaviour, activities, and responsibilities that societies and cultures deem appropriate for people biologically assigned to the female or male sex to fulfill in their key arenas and institutions—such as the family, the workplace, the community, civic and political bodies, and public space. Individuals identify with and outwardly express gender based on their internalization, or their contestation, of these norms. The concept of *sexuality* refers not only to biologically driven sexual desire, but also to the sets of sexual behaviours that societies and cultures deem to be "normal" or "abnormal," in private and public domains, based on biological sex assignment, differences, and to the identities that people take on in relation to these expectations. In fact, issues of gender and sexuality underpin many of the topics and urban processes that form the subjects of chapters in this book.

Beginning in the 1970s, a diverse body of urban research has adopted a feminist perspective, seeking to understand the roles of the urban in shaping gendered social inequalities and gendered dimensions of identity (for reviews, see McDowell, 1999; Parker, 2011). Themes addressed include how urban spatial arrangements have often reflected and reinforced stereotyped gender roles, even when these no longer correspond to the realities of today's urban households; gendered inequities in access to housing and other crucial resources; gendered aspects of urban fear in public space; and how gendered constructions of space and place intersect with those based on social class and racialization (see Kobayashi and Preston, Chapter 8). Research on sexualities and urban space emerged in the 1980s and has flourished in recent years (for reviews, see Johnston, 2015; Peake, 2010), much of it aligned with the goal of "queer-ing" urban analysis, which means seeing to uncover and critique the ways that heteronormativity (Peake, 2016) undergirds many aspects of the organization of urban life. This concept refers to the entrenchment of heterosexual forms of attachment, sex, and reproduction in institutions, policies, and expectations of behaviour in public space. This translates into heterosexism by implicitly "coding" urban spaces as heterosexual, such that gay, lesbian, transgender (trans), and gender non-conforming individuals are made to feel, at best, out of place. This has led researchers in this field to study how lesbians, gays, bisexuals, and transgender people adapt to, make use of, and contest their "outsider" status by creating their own urban "scenes" (Valentine and Skelton, 2003).

Feminist urban studies and the sexualities and space literatures at first tended to move along parallel tracks (Domosh, 1999; Knopp, 2007). Recent work has begun to build connections, a key one lying in the idea that heteronormativity corresponds to a "moral heartland" at the intersection of what society deems to be "good" or "normal" gender relations and "good" or "normal" sexualities (Hubbard, 2008; Oswin, 2010; Wright, 2010). In Western, highly urbanized societies, such as Canada's, for most of the twentieth century the landscape of this "moral centre" was dominated by the heterosexual nuclear family with a strongly gender-typed role division between male patriarch and breadwinner and female nurturer and homemaker. Moreover, the idea of heterosexuality and homosexuality as binary opposites became increasingly entrenched. People who did not marry, and who lived autonomously from family members (parents, relatives, siblings) were placed outside this intersection of gender and sexual norms. All faced various forms of "othering," marginalization, or social ostracism (e.g., being referred to as "spinsters," "unwed mothers," "male drifters," etc.) depending on the perceived threat they posed to prevailing norms.

However, those found to be engaging in gay or lesbian practices faced outright repression and punitive measures lasting until the gay rights movements, which emerged, like **second-wave feminism**, in the context of the social and cultural ferment of the 1960s. It is important to bear in mind that Canadian provinces and territories did not outlaw discrimination based on sexual orientation until the 1980s, 1990s, or even later in some cases. The federal government did so in 1996, but it took another 10 years of activism and court challenges before it legalized same-sex marriage. Moreover, transgender people and others whose gender identity and/or gender expression did not correspond with their biological sex assignment at birth did not gain full federal human rights protection until 2017.

Two dominant narratives, in creative tension with each other, traverse both the feminist geography and sexualities and space literatures: the liberating potential associated with life in a large metropolitan centre compared to rural and small town living, versus the constraining and controlling qualities of the heteronormative organization of space within the metropolis (Bondi and Rose, 2003; Cattan and Leroy, 2010; Garber, 2000; Muller Myrdahl, 2013; Wilson, 2001). In this chapter, I explore this interplay of "freedom" and "constraint" within topic areas that have received considerable attention in feminist or queer geographies. First, I discuss the linkages between urbanization and women's opportunities for economic autonomy via labour market participation, and examine the supposition that large cities are magnets for gays and lesbians. Second, I review and update research on a key dimension of metropolitan, especially suburban, life in which gender-based spatial constraints have been extensively documented—urban mobility, in particular travel-to-work. Third, I discuss the challenges facing lone-parent families in metropolitan space—a family form that has been widely subjected to economic marginality, in large part because of its outsider status in a heteronormative society geared toward the two-parent nuclear family. Finally, I turn to gendered dimensions of **gentrification**, and discuss central city neighbourhoods as resources for sexual minorities.

Urbanization, Gender and Sexuality in Historical Perspective: Challenges to and Reframings of Heteronormativity

The social, cultural, and political upheavals of the 1960s (Campbell, Clément, and Kealey, 2012) are often seen as a turning point for women in Canadian society because they accelerated their "individualization" (Jones, Marsden, and Tepperman, 1990)—their ability to implement their own choices about their sexuality, fertility, marital arrangements, and, of course, their opportunities for paid work and career employment. Recent historical research, however, shows that this process was set in motion half a century earlier and was closely linked to urban development. The first half of the twentieth century was an era of major socio-economic and demographic upheavals punctuated by the two world wars and the Great Depression. Across Canada, mass urbanization was fuelled by out-migration from struggling agricultural communities, successive waves of international immigration, and new labour needs. In large numbers, young single women flocked to central Canada's major urban centres to work in light industry and domestic service, typically boarding with relatives or employers while sending remittances to their parental families (Gauvreau, Olson, and Thornton, 2008) while their male counterparts gravitated to western Canada's centres of resource industries.

During and after World War I, the rise of corporate capitalism, technological changes in white-collar work, and the rise of the consumer society among the expanding urban middle classes all fuelled the growth of the service economy. Better education opened these clerical and sales jobs to increasing numbers of young women (England and Boyer, 2009; Srigley, 2010). In the first half of the twentieth century, the feminization of clerical work grew in line with the growth of the urban population and more rapidly than the feminization of the labour force as a whole. Young, single working women gained new freedoms to circulate in public space, and increasingly asserted their rights to join in urban consumer culture and recreational activities with their peers (Boyer, 1998; Comptois, 2007; Srigley, 2010). These new gendered mobilities generated "moral panics" among bourgeois social reformers, clergy, and public-opinion shapers about young women deemed at risk of drifting into areas of the city associated with prostitution, as well as fears that some might become "too independent" and lose interest in marriage (Myers, 2006; Strange, 1995).

Religious and philanthropic organizations set up outreach programs to encourage young women migrants to live in "approved" boarding homes or institutional residences (Kirkland, 2006; Piché, 1999). By the 1920s, however, housing developers in big cities were responding to the rising demand for small self-contained rental apartments. These rental units offered an alternative to boarding and lodging for female office workers (who were not well paid) sharing with peers to make them affordable, thereby creating living spaces that gave them more autonomy, as well as the option of companionship (Dennis, 2006/7; Rose, Dillon, and Caron, 2016). Both residential and public spaces were thus key sites where young women, at least those from Anglo-Protestant backgrounds, negotiated individualized "modern" identities, including greater heterosexual freedoms (Comacchio, 2006). Nevertheless, they faced very strong social pressures to conform to a heteronormative vision of their future life course, involving marriage, departure from the workforce, and full-time domesticity (Kirkland, 2006; Strange, 1995: 179–85). "Spinsters" who remained unmarried over the longer term faced economic precariousness, since most occupations open to women paid poorly. Moreover, to maintain their employment they had to demonstrate "moral" (i.e., sexual) "respectability" in their personal life—living in a boarding house run by a widow, for example (O'Hanlon, 2002; Tallentire, 2006).

Compared to single women, married women faced a more constrained relationship to the urban spaces of the early- to mid-twentieth-century city. Prior to World War II, women were shut out of almost all Canadian workplaces upon marriage. Big labour unions, corporate employers, and governments supported mass production with labour relations based on the concept of a "family wage," earned by a single male breadwinner. The application of this concept would generate enough disposable income to fuel the spread of a middle-class consumer society based on home ownership and mass consumption of standardized goods. Integral to this economic and cultural paradigm (referred to in political-economic theory as **Fordism**) was an ethos of conformity, epitomized by the spread of tract suburbs, reinforcing private/public space divisions, where women were to devote themselves exclusively to raising heterosexual families. Mass media, advertisers, government agencies, social science experts, housing reformers, and urban planners strongly encouraged this confined, home-centred way of life for both middle-class women and those in upwardly mobile working-class households; for the latter, often raised in poor housing conditions, it represented a considerable improvement in living standards and domestic working conditions (Séguin, 1989).

While women's domestic expertise was respected, and marital relationships were now supposed to be companionate, women's economic dependency perpetuated old, established patriarchal relationships (Mackenzie and Rose, 1983; Mendes, 2010; Strong-Boag, 2002 [1991]). And, of course, the automobile was central to this paradigm. In poorer working-class families, women often continued older traditions of engaging in various forms of home-based employment to make ends meet (Baillargeon, 1999), or, especially in the case of immigrant women, went out to work in female-typed manufacturing jobs in the garment industry.

From the 1940s onwards, steadily increasing numbers of married women did join the paid workforce—especially in clerical work and consumer services—in rapidly increasing numbers (Dominion Bureau of Statistics, 1954: 702–3). While the rising cost of living associated with the consumer society was one key motive, middle-class women's growing access to higher education and their mounting frustration with what Betty Friedan, the American pioneer of second-wave feminism, famously called "the problem with no name," led growing numbers to seek out more fulfilling employment opportunities in a diversity of professional and administrative fields. Women's access to higher-status and better-paying employment opened up greatly in the 1970s and 1980s, especially in large metropolitan cities and provincial capitals experiencing rapid expansion of advanced tertiary employment (Rose and Villeneuve, 1993). Access to well-paid managerial positions, in the financial sector for instance, has taken longer to achieve but has been growing steadily since the 1990s (Statistics Canada, 2018). These changes brought a large increase in dual-earner married couples who had to consider both partners' needs when deciding on a residential location, as well as a large share of women, young or in mid-life, with the economic resources to opt to live without a domestic partner. In sum, urbanization and related processes entrenched the dominance of the heterosexual nuclear family, yet also created the conditions for some women to spend at least some years of their adult lives not conforming to this norm.

Over the same period, metropolitan cities also became crucibles for the building of gay, lesbian, and queer identities and support networks in semi-clandestine spaces such as bars and boarding houses (Chauncey, 2004; Meyerowitz, 1990). Research supports the view of the metropolis as magnet for people seeking refuge from rural communities and small towns where one could not avoid heteronormative cultures, heterosexist workplaces, and patriarchal extended family structures that limited the freedoms of adult children who did not marry (Christie, 2004; Lewis and Mills, 2016; Muller Myrdahl, 2013; Podmore, 2006). This is in line with a more general argument about the role of the metropolis in individual self-affirmation first made by early-twentieth-century sociologist Georg Simmel (1950 [1903]), who proposed that, due to its size, high density, and social diversity, the metropolis affords anonymity, but also the opportunity to make oneself visible to others with whom one has affinities. With a critical mass of gays, lesbians, and people with transgender or non-conforming gender identities, they could organize themselves into subcultures invisible to heterosexuals, the police, and other institutions involved in punishing transgressive sexualities.

Gay male migration to large metropolitan centres in the United States in the post–World War II period was linked to growing service sector employment opportunities, which were less associated with traditional expressions of masculinity than was work in heavy industries (D'Emilio, 1998 [1983]; Lauria and Knopp, 1985). Some of these cities were also cultural and artistic hubs, sectors long associated with gay communities. Government services, whose highest concentrations are in national and state or provincial capitals, are also a key sector of professional and white-collar employment. The Cold War period of the 1950s and 1960s was a time of severe surveillance and exclusion of presumed gay men from government jobs, but subsequently, capital cities evolved into hubs of gay activism for legal rights. Ottawa's government services sector acquired a reputation for relative tolerance of "responsible" middle-class gay men, who developed strong professional networks to increase their inclusion (Lewis, 2012; Lewis and Mills, 2016). In short, gay men's migration to such cities embodied both a quest for stable and satisfying employment in workplaces where they could feel accepted and their search for community and affective attachments within a homophobic society.

Since 2001, the Canadian census has collected information about people living in same-sex couples. Although this indicator gives only a partial measure of Canada's LGBTQ+ (lesbian, gay, bisexual, transgender, queer, and others) populations, it is useful in that if people are comfortable making their same-sex partnering visible by setting up house together, this is a measure of the social acceptance of (certain) LGBTQ+ people in the cities where they live. A Statistics Canada analysis compares how same-sex couples are distributed between urban centres of different sizes compared with the distribution of the population as a whole (Statistics Canada, 2017a). It shows that although same-sex couples are still statistically overrepresented in the five largest census metropolitan areas (CMAs; Toronto, Montreal, Vancouver, Calgary, and Ottawa-Gatineau), this relative concentration decreased between 2001and 2016 while increasing in a number of other large and mid-size CMAs. Toronto, not surprisingly, has more gay and lesbian couples than any other CMA in absolute numbers, but it falls some way short of the top five in terms of the percentage of same-sex couples; its cultural diversity may play a role in reducing the reporting of same-sex domestic partnerships. Gay and lesbian couples are still relatively under-represented in Southern Ontario CMAs whose economy was traditionally manufacturing-based, as well as in Prairie CMAs, but the degree of under-representation is shrinking.

If we break out the census data used in the Statistics Canada study by gender, some important nuances emerge. Living in a very large metropolitan area seems to matter more to male than to female same-sex couples (in 2016, 61 per cent of the former but only 46 per cent of the latter were living in one of the five largest CMAs). As well, since 2011, these cities' share of the total has remained unchanged for male couples while declining by 2.5 percentage points for female couples (Statistics Canada, 2017d). Moreover, when we look at CMAs with a critical mass (500+) of same-sex couples, we find marked gender differences in the ranking of CMAs in terms of the prevalence of same-sex couples compared to couples overall. For males, four of the top five are very large CMAs: Montreal (1), where (1.60 per cent of male spouses are part of a same-sex couple); Quebec City (2); Vancouver (3); and the Ontario part of the Ottawa-Gatineau CMA (4). The mid-sized CMAs of Halifax and Victoria tie for fifth place. In contrast, for females, mid-sized CMAs occupy the top spots: Halifax (1), where (1.61 per cent of female spouses are part of a same-sex couple); Victoria (2); the Quebec part of the Ottawa-Gatineau CMA (3); Kingston (4); and Quebec City (5). Across these differences, however, is a striking similarity: most of the cities on both lists are national/provincial capitals and/or have a large university presence, suggesting that the question of "gay-friendly" occupations (Lewis and Mills, 2016) remains relevant today.

Over the past few decades, queer communities have strengthened and thrived in a variety of smaller cities, subject to a critical mass to ensure community supports and to avoid social isolation (Hulko and Hovanes, 2018; Muller Myrdahl, 2013). As well, small towns' share of Canada's same-sex couples has grown since 2006, even though small towns are home to a decreasing share of Canada's population overall (Statistics Canada, 2017a). The LGBTQ+ couples identified by census data are likely to be in mid-life, economically established, and "out" (Lewis and Mills, 2016: 2486). Not enough is known about the migration patterns of LGBTQ+ individuals not living as a couple. Organizing for services and rights was historically very challenging for lesbians and gays in small towns and rural areas (Millward, 2012). While smaller places' socially conservative reputations endure in queer circles (Hulko and Hovanes, 2018), it is noteworthy that the first two municipalities to install a Trans Pride flag across a city crosswalk were small and mid-sized cities (Whitehorse and Lethbridge; Stevenson, 2017). Moreover, Canada's first openly transgender mayor was voted in by the electors of a small village in western Quebec (Cole, 2017).

Suburban Development and Gendered Mobilities

Rapid metropolitan growth, the increasing demographic, socio-economic, and ethno-cultural diversification of suburbs, and the rising interest in **urban sustainability** have all contributed in recent years to renewed interest in the relationships between transportation and mobility, individual opportunity, and systemic social exclusion. This makes it opportune to revisit one of the first issues in urban studies to be extensively studied from a feminist perspective: how well do transportation networks mesh with the spatial arrangements—planned or otherwise—of housing, places of employment, commercial and community services, public facilities, and other land uses in metropolitan cities? And who is affected by inadequacies in these linkages?

The primary responsibility for reproductive work and caring tasks still falls mainly to women. This is why access to efficient, affordable, and safe transportation, not only for daily travel from home to employment, but also for travel to childcare facilities, shopping and personal services, is a gendered issue. And, in large urban contexts, it is made more burdensome, especially for low-income women, by the spatial dispersion of employment and the places where these services are carried out (Lee et al., 2017). Strong spatial separation between different functional land uses has characterized outer suburban morphology since the 1950s. Transportation planners typically designed mass rapid transit systems for radial commuting between middle-income suburbs and downtown cores, with cross-town suburban bus service (if provided) as an afterthought. According to time-budget and travel diary research, this spatial "mobility of care" means that the daily mobility of women with young children—including those with employment—is more likely than that of men or women without young children to involve multi-purpose trips and "trip chaining": stops to pick up and drop off of children, shop, check on an elderly relative, and so on (Bernard et al. 1997; Gendered Innovations in Science, Medicine, and Engineering Team, n.d.). Middle-class suburban women in nuclear families have shorter work trips than their male counterparts and seek to reduce their commute times in order to make time for domestic responsibilities (Preston and McLafferty, 1993; Preston and McLafferty, 2016). Sharing of daily reproductive tasks between parents is more common among millennials than among previous generations (Dyck, 2005; Statistics Canada, 2012), but is often impractical because jobs held by men tend to involve longer and less flexible hours.

This high mobility of care work has different impacts for women depending on income and racialized immigrant or minority status. In middle-class households, women's growing access to professional occupations has been accompanied by an increase in their **automobility** (Featherstone, 2004). As personal aspirations combined with rising house prices lead more suburban-based middle-class women to seek a "breadwinner's" salary, their commuting trips tend to lengthen. Greater Montreal, for instance, has seen a sharp rise in women driving to work, especially from the region's very low-density outer suburbs (see Filion et al., 2010), whereas men are beginning to shift toward greater use of public transportation (St-Pierre, 2007). For women, acquiring their own car became a key to better access to community and commercial services beyond the immediate neighbourhood, and, as these women joined the labour force, car ownership facilitated their time management and coordination of paid work and care work. For example, in Quebec City, where the configuration of highway corridors has led to scattered linear patterns of employment growth, which are poorly served by public transportation, increased car ownership was an enabling factor in women's increased labour force participation between the 1970s and 1990s (Vandersmissen, Villeneuve, and Thériault, 2001). Compared with the "spatial captivity" of mid-twentieth-century suburban "housewives," this rising automobility symbolizes a type of freedom associated with women's individualization. However, it is tempered by ongoing gender-typed imbalances in time spent on reproductive work (Statistics Canada, 2012).

Some employed women now belong to a more affluent fraction of the middle class that can afford to live in elite or gentrified neighbourhoods close to downtown that are well served by public transit, so both men and women benefit from short commute times. However, many women in "working poor" families face lengthy and inconvenient trips to work because of a more distant residential location and much greater dependency on public transportation (Preston and McLafferty, 1993). The displacement and rental inflation caused by gentrification of convenient central neighbourhoods worsens this situation. Censuses and surveys show that those who use public transportation to get to work face longer trip times than car commuters. Moreover, the quality of transit users' commuting experiences varies depending on the mode, quality, and state of maintenance of these forms of public infrastructure—we can think of the contrast between using a high-speed commuter line versus making two or three bus connections. Recent immigrant women in Toronto, Montreal, and Vancouver, who are concentrated in low-paid service sector jobs often involving non-standard working hours, are one of the most transit-dependent groups in Canadian society, even more so than recent immigrant men (Heisz and Schellenberg, 2004). In 2016, in Toronto, for instance, among non-immigrant car commuters, 46 per cent are female, but this drops to 31 per cent among recent immigrants (Statistics Canada, 2017c). Racialized immigrant and minority women often live in poor and underserved suburban apartment districts from which they have far to travel to get a job paying a living wage (Fiedler, Schuurman, and Hyndman, 2006; see Kobayashi and Preston, Chapter 8), which is a case of what labour market geographers refer to as "spatial mismatch." Improving public transportation access in such neighbourhoods is an important focus of racialized suburban women's activism (Paradis, 2018) and an urgent urban social infrastructure policy issue (see Walks, Chapter 10). As well, fear about personal safety while using public transportation, a major theme of feminist organizing in Canada and internationally for more inclusive cities (Whitzman, Andrew, and Viswanath, 2014) is both gendered and racialized. Moreover, in outer suburbs, small cities, and rural areas, access to employment, adult education, and community services and activities, including those that help newcomers settle, is very challenging without a car, but owning one is a major drain on low-income families' expenses (Farber et al., 2018).

Finally, age intersects with gender as a determinant of car use for commuting to work (Figure 9.1). Younger, presumably single, people, who are more likely to live in a central neighbourhood, are less likely to go to work by car, but car use rates increase greatly in the mid-life years when most have families and live in the suburbs. In Canada's most car-dependent CMAs, Quebec City and Calgary, both men and women have very high rates of car commuting in all age groups, but both in these cities and in those with lower rates of car commuting, the rates drop off faster for older women workers than for males. Are older women opting, more than their male peers, to work close to home? Do older women feel less comfortable driving? Are those who are now living alone more likely to move from the suburbs to a condo in the city? As people are remaining in the workforce beyond age 65, these questions are becoming important.

Placing a gendered lens on urban transportation issues thus highlights how greater personal spatial mobility can generate greater accessibility to diverse social resources and opportunities, including better jobs and a wider range of community and commercial services. It also generates the less tangible benefits of a greater sense of autonomy and empowerment, which are important for newcomer women, for example (Ray and Rose, 2000). However, lengthy multi-stop trips to get a day's paid work and reproductive labour done can be stressful and burdensome. Conversely, "mobility deprivation" is a significant dimension of social exclusion of some low-income groups, such as sole-support mothers living in underserved low-income suburbs (Morency et al., 2011). We examine lone-parent families in the next section. Finally, critiques of twentieth-century urbanism developed by the 1980s generation of feminist researchers envisioned equitable and socially inclusive planning models whereby people of different income levels and age groups would be able to access essential services and social activities without being dependent on the automobile

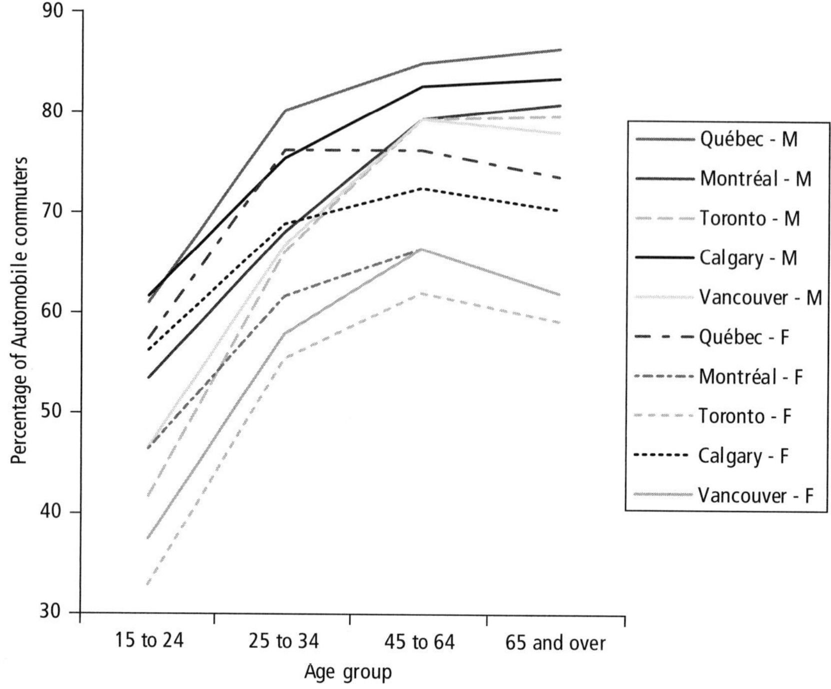

Figure 9.1 Percentage of Employed Females and Males Who Commute to a Fixed Place of Work by Car, Truck, or Van, by Age Group, Selected CMAS, 2016
Source: Statistics Canada, 2017c. Data Tables, 98-400-X2016330.

(Wekerle, 1985). In the twenty-first century, as parts of the outer suburban zone of many Canadian metropolitan areas are being densified (see Filion, Chapter 2; Addie, Fiedler, and Keil, Chapter 16) and transit-oriented development (TOD) hubs are being created around commuter rail stations in the name of sustainable development, it is important to re-engage with these ideas. Do the mixed land-use complexes around TOD hubs include daycare centres and social and affordable housing? Are lower-income outer suburbs being connected to the new transit nodes by feeder bus routes? Are municipalities requiring the developers of new market housing in the TODs to cater to various household and family types and different generations, or merely allowing them to target higher-income groups or to bring the "urban lifestyle" condo, associated with gentrification to the suburbs (see Bain and Mark, Chapter 15; Addie, Fiedler, and Keil, Chapter 16)?

The Gendered Challenges of Lone Parenthood in Urban Settings

In Canadian society, as elsewhere, conjugal relationships have become more fluid in recent decades, and family types more varied. The liberalization of divorce laws in 1968 sparked a marked growth in the number of lone-parent families compared to two-parent families. Previously, solo parenting had mainly been a consequence of widowhood, which declined with rising life expectancy after World War II (Bradbury, 2000; Statistics Canada, 2017b). As of 2016, among families with at least one child under 25 and living in a major urban centre, almost one-quarter are headed by a lone parent (Statistics Canada, 2017d). Just over half of solo parents are divorced or legally separated; almost one-third have never married, many of whom were in common-law unions (Milan and Bohnert, 2012a). What has changed little, however, is the highly gendered nature of lone parenthood. In 2016, in the major urban centres, female-headed lone-parent families were almost four times as numerous as their male-headed counterparts (Statistics Canada, 2017d), a ratio that has changed very little in the past half-century. Even though shared custody arrangements have been on the rise over the last two decades (Milan and Bohnert, 2012b), the decision as to who will rear the children after relationship breakdown is still shaped by powerful and enduring socio-cultural norms of femininity and masculinity, in which child-care in the home and other "routine" aspects of parenting are still deemed to be primarily women's responsibility. As well, some lone-parent families are formed as a result of women and children fleeing domestic violence.

While lone mothers are less likely to be stigmatized as sexually "irresponsible" than in the mid-twentieth century, beliefs persist that this family type is inherently deprived and less adequate for child-raising than the two-parent nuclear family. As recently as 2010, a socially conservative think-tank report ranking Canadian cities' "family friendliness" gave a lower score to those with a high percentage of lone-parent families (Walberg and Mrozek, 2010)! Groups organizing for better living conditions and services for lone parents in low-income neighbourhoods contest this viewpoint because it presents them as passive victims of circumstance (Paradis, 2018). Moreover, it has heteronormative and ethnocentric undertones: in urban Canada, lone parenting is more prevalent than average among those with Caribbean origins and urban Indigenous Canadians—these cultural groups have strong traditions of supporting child-raising through extended kinship networks (Mata, 2011; Peters, 2000). From a feminist perspective, gendered and racialized poverty rather than family type are the root causes of the materially and socially deprived conditions that many lone-parent families have to deal with, such as having to live in deteriorating postwar rental housing in the inner suburbs of metropolitan areas, which are deficient in essential community and commercial services and transportation infrastructure (Paradis, 2018). While middle-income lone parents still face a major income disadvantage relative to two-parent families, they have far greater housing and neighbourhood choices, and their children do not experience a higher rate of problems in their schooling and social integration than those growing up in two-parent families (Ambert, 2006).

The economic situation of female lone parents varies greatly depending on education and age. Those under 25 typically have not been able to complete their education and face further barriers to employment due to the young age of their children and lack of affordable child-care; because they rely on social assistance, they are predominantly in a very low-income situation. While poverty can be life-long for some, low-income rates decrease in each successive age group over 25. Those who become lone parents later in life are more likely to have access to their share of significant family assets built up over a number of years of domestic partnership (Ambert, 2006). Earnings of employed female lone parents have increased considerably in recent decades due to higher education levels and greater access to full-time rather than part-time work. Nevertheless, and reflective of the persistent overall gender gap in earnings in Canadian society (Moyser, 2017), they earn much less than their male counterparts (LaRochelle-Côté, Gougeon, and Pinard, 2009), and this is also true for their total income from all sources, regardless of age group and education (Canada Mortgage and Housing Corporation, 2014).

Figure 9.2 displays estimates of median total income from all sources on an annual basis over the past 40 years for two types of families with all children under 18: lone-parent families (female and male are combined in this data set) and couples with children, for Canada, Toronto, and Montreal. It underscores the scale of the income disadvantage faced by lone parents, which has actually widened slightly over this time frame. This is probably because in Canada since the 1970s, female lone parents' employment rates have not increased as fast as those of mothers in two-parent families, and having a very young child reduces the employment rate more for lone mothers than

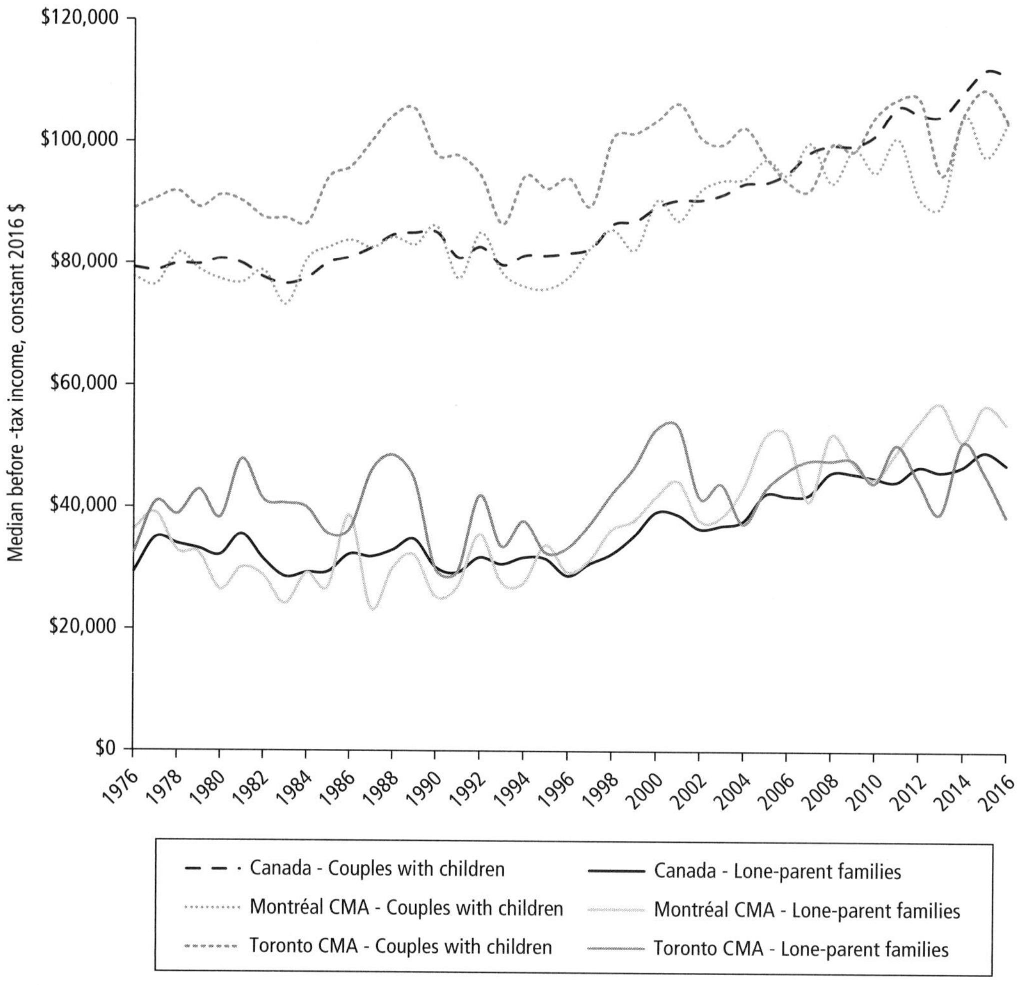

Figure 9.2 Median Income, from All Sources, before Tax, in Constant 2016 Dollars, for Couples with Children and Lone-Parent Families, Canada, Montreal CMA, and Toronto CMA, 1976–2016

Notes:

1. The families included are those where the children (by birth, adopted, step, or foster) of the major income earner are under age 18. The family may also include other relatives.
2. The data are estimates. Statistics Canada rates the quality of all data points in this chart as "acceptable" or higher, except for the values for lone-parent-families in Montreal for 1998, 2002, and 2006, and in Toronto for 1998, 2007, and 2012, which are to be used with caution.

Source: Statistics Canada, Table: 11-10-0191-01 (formerly CANSIM 206-0021) Table 206-0021—Income statistics by economic family type and income source, Canada, provinces and selected census metropolitan areas (CMAs), annual, https://www150.statcan.gc.ca/t1/tbl1/en/tv.action?pid=1110019101 (accessed: Nov. 6, 2018).

for those in two-parent families (inferred from Williams, 2010). Difficulties in access to suitable and affordable child-care arrangements are no doubt in play here. Since the Quebec government implemented a highly subsidized daycare program in 1997, the labour force participation of women in that province with very young children has overtaken the rates in the other three largest provinces; consequently, poverty rates of lone-mother families have fallen considerably (Friendly and Prentice, 2009; Proulx et al., 2011). Perhaps for the same reason, Figure 9.2 also reveals that median incomes of Montreal's lone parents overtook those in Toronto in the early 2000s—a CMA with much higher housing costs.

The economic and emotional precariousness often associated with becoming a lone parent have important repercussions for their housing situations and needs. If the family has to move to a different neighbourhood in order to find a cheaper, easier-to-maintain home, this can be disruptive for children's school attendance, daily routines, and social networks. In the case of flight from domestic violence, non-profit organizations (supported by government and charitable funding) play a crucial role in providing safe and supportive temporary shelter and transitional accommodation. Overall, in Canada's CMAs and census agglomerations (CAs), female single-parent households aged under 65 were fairly evenly split between homeowners and renters (52 per cent vs. 48 per cent), according to 2011 estimates (Canada Mortgage and Housing Corporation, unpublished 2011 National Household Survey data, Core Table D3). **Core housing need** (see Harris, Chapter 11) is almost twice as common among female- as among male-led lone-parent families (Canada Mortgage and Housing Corporation, 2014). A shelter cost-to-income ratio of 50 per cent or more is not uncommon, making families vulnerable to losing their housing if there is an unexpected drop in income from one month to the next (see Walks, Chapter 10). Female-led lone-parent families who are recent immigrant and refugee newcomers to Canada may have to tolerate substandard housing conditions or "double-up" with relatives and friends in crowded conditions as a coping strategy to avoid homelessness or pool resources to pay for housing (Paradis, 2018; Walsh et al., 2016). Lone parents' housing choices may be limited by discrimination by private landlords, especially if they are also newcomers to Canada and living on social assistance, although this is illegal under provincial human rights codes (Lauster and Easterbrook, 2011; Rose and Charette, 2014). Female lone-parent families are statistically overrepresented in subsidized rental housing; nevertheless, in 2011 an estimated 75 per cent of female lone-parent renters did not receive any kind of subsidy for their housing (Canada Mortgage and Housing Corporation unpublished 2011 National Household Survey data, Core Table D3). **Social housing** can provide security of tenure, rent stability—either rent-geared-to-income or a low-end-of-market rent that increases only slowly—and, sometimes, companionship and mutual support (see Harris, Chapter 11). Over the years, lone parents have been at the forefront of activism to develop housing co-operatives (Wekerle, 1993) and defend rent-geared-to-income public housing against stigmatization, neglect and, more recently, the loss of housing units when social housing is converted into mixed-tenure developments.

Gendered Dimensions of Gentrification

Scholars such as David Ley (1996) have convincingly shown how the settlement of "new bohemians"—artists, students, and others rooted in the "countercultural" social movements of the 1960s—in a good number of inner-city neighbourhoods in Canadian cities was an important factor in paving the way for those neighbourhoods' subsequent gentrification (see Bain and Mark, Chapter 15). In the context of debates around gentrification in the 1980s and 1990s, feminist geographers argued that the changes in gender relations and identities that also began to foment in the 1960s and accelerate in the next two decades also contributed to early stages of gentrification (Bondi, 1999; Rose, 2010 [1984]). For instance, in the 1980s, during the early stages of gentrification in Montreal's Plateau Mont-Royal district, women were overrepresented among professionals living there, and those women professionals were more likely to be single or lone parents compared to women professionals overall in the Montreal CMA (Rose, 1989; Rose and Le Bourdais, 1986). In a modern variant of the early-twentieth-century pattern of single women establishing their independence in metropolitan urban space, discussed earlier in this chapter, this research pointed to a sub-group of economically self-supporting women, on their own or with children, attracted to diverse and affordable inner-city neighbourhoods. These neighbourhoods were located close to downtown professional employment in advanced tertiary sectors, such as government services, specialized education, and health institutions, where career opportunities for female professionals expanded from the 1970s to 1990s (Rose and Villeneuve, 1993). Many of these professional positions were, however, precarious (e.g., short-term contracts).

Both functionally and culturally, such neighbourhoods offered more comfortable spaces for these "non-traditional" women compared to classic postwar low-density suburbia. They facilitated the participation of mothers of young children, without access to a car, in informal networks, thereby reducing isolation and fostering mutual aid. Such women were among the key players in collective urban movements to develop services and infrastructure to build up the supportive qualities of these neighbourhoods for modest-income households—such as parent-run daycare centres, food and housing co-operatives, or, in Quebec, "informal" forms of co-ownership tenure (Rose, 2010 [1984]; Wekerle, 1985, 1993). A more contemporary version of this phenomenon has been identified in a "pre-gentrifying" Toronto neighbourhood where small, female-led commercial enterprises are embedded in gender-typed mutual aid networks that also outreach to the area's most precarious residents severely hit by **neo-liberal** social services cuts (Kern, 2012). However, in many cases, and especially in cities that have successfully converted their economic base to a post-industrial one with a strong downtown employment core, the gentrification of these types of neighbourhoods has accelerated and become much more upscale, decreasing their socio-economic diversity. Influxes of affluent dual-earner professional/managerial couples have helped fuel inflation in local real estate markets, contributing to displacement of low-income households and increasing the **polarization** of household incomes within the inner city (see Walks, Chapter 10). For instance, 2016 census data show that the proportion of lone-parent families in Montreal's Plateau Mont-Royal borough has now fallen below the city average; to find affordable rents, such families now have to live in suburban apartment districts. Parents in these affluent couples have less of a stake in non-profit service provision compared to the early-stage gentrifiers, since they can afford to purchase child-care and other resources at market rates. Gendered understandings of gentrification thus always need to be placed in relation to dynamics associated with social class.

The ways that gentrification takes place have diversified since the mid-1990s with the increasingly widespread phenomenon of **new-build gentrification**, wherein the local state primes the redevelopment of former industrial sites, including large tracts adjacent to downtown cores, by inviting and offering incentives to private developers to create new residential zones comprised essentially of condominium buildings, sometimes at very high densities. This is part of a broader urban economic revitalization strategy of land-use intensification aimed at supporting the emergence of post-industrial "consumption landscapes" that are supposed to reboot and sustain urban economies on the basis of leisure, tourism, cultural, and other **new economy** industries (see Vinodrai, Chapter 17). Such residential densification has reawakened interest in gendered aspects of gentrification because it not only targets the affluent but also builds small, relatively "affordable" condos for the growing number of middle-income men and women living alone (however, with rapidly inflating condo prices in some central cities, more affordable condos may now be found in TODs in suburban locations, as discussed earlier in the chapter). Including people living alone in the homeownership market is an important indicator that solo households are now, unlike in most of the twentieth century, viewed as part of the social mainstream, across age groups. The "starter condo" phenomenon has some noteworthy gendered dimensions. According to the 2011 National Household Survey, in every age group, when women homeowners living alone are compared to their male counterparts, the solo female owners are much more likely to own a condo than a non-condo dwelling (for instance, 62 per cent of female owners versus 48 per cent of male owners aged 30 to 34 are condo owners). In absolute numbers, however, below the age of 50, there are still more male than female solo condo owners (Canada Mortgage and Housing Corporation, unpublished 2011 National Household Survey data, Core Table A7).

The downtown condo not only offers low maintenance and the promise of an "urban lifestyle," it has become an investment instrument for young and older adults on their own, counting on continued real estate inflation to give them future economic security that can no longer be counted on through the labour market and retirement pensions. Making this purchase is thus a marker of "acting responsibly" as well as reinforcing personal self-esteem (for Toronto, see Kern, 2010; for Montreal, see Rose 2009), set within the broader context of the **neo-liberal** political economy, which favours marketized and individualized solutions over collectivized forms of social security. Moreover, large-scale condo development of land bordering urban cores is an important dimension of neo-liberal urbanism whereby local states increasingly count on private for-profit development to deliver urban services and amenities that are key to social reproduction and quality of life but are only accessible to the few that can afford such locations (see Skaburskis and Moos, Chapter 19). Kern (2010) identifies a gendered aspect of this in a study of Toronto condo developers' marketing pitches, which played on the century-old dual narratives of freedom and constraint associated with downtown heteronormative lifestyles of young single, and predominantly white, working women. While advertisements stressed their freedom to frequent safe after-hours meeting places, such as wine bars, sometimes representing women in a sexually objectified way, they also played on gendered experiences of urban fear by emphasizing the buildings' security features (see also Wright et al., 2015). A gendered rationale is thus offered for securitized and highly exclusionary forms of "revitalization" led by private funding, but it is an individualistic rationale rather than a social vision for developing *public* policy to make downtown neighbourhoods safe and inclusive for all. The displacement and exclusion of low-income residents resulting from gentrification affects both men and women. However, when neighbourhoods with a high percentage of older adults are gentrified, it is important to note that, since women are overrepresented among the low-income elderly, they are particularly affected by evictions, the loss of familiar shops, and local social institutions (Burns et al., 2012). As a corollary, we find female senior citizens taking lead roles in anti-gentrification activism (for an example from Vancouver's Chinatown, see Carnegie Community Action Project, 2017).

Gay Villages and Queer Neighbourhoods: Growth and Challenges

As discussed earlier in this chapter, queer subcultures were found in large cities decades before the legalization of private gay and lesbian practices and the winning of civil rights for LGBTQ+ individuals. These individuals gathered, had personal encounters, and built networks in private and clandestine spaces; as well, various "mainstream" public spaces were coded as "furtive" and anonymous meeting places for casual gay encounters (Cattan and Leroy, 2013; Podmore, 2006). These transgressive spaces were in districts adjacent to the downtowns of major metropolitan cities that were characterized by transient populations, and mixes of household types, socio-economic status, and land uses. Neighbourhoods of furnished rooms and boarding houses were home to diverse subcultures, including artists and other bohemians, but also to discreet lesbian residential communities (Meyerowitz, 1990).

"Gay villages," however, are a much more recent phenomenon. While their histories and geographies vary from city to city, the gay liberation movement of the 1960s empowered gay and lesbian establishments to make their presence more visible both at the scale of the city and in society at large. Often, police repression on the basis of local laws about "public morality" forced gay nightlife establishments out of downtown entertainment districts that were supposed to be for heterosexual meeting, dating, and coupling (Hubbard, 2008), and into more economically marginal neighbourhoods somewhat removed from the downtown core (Nash, 2006; Podmore, 2006). Creating strong spatial concentrations of gay residents and critical mass of social venues and businesses was also a way to gain representation of their interests in local politics and economic development (Castells, 1983). Thus, gay villages would foster pride and empowerment; they would facilitate "coming out" in a non-threatening context; and, through "safety in numbers," they would afford some protection against homophobic oppression and the public violence frequently inflicted on gays and other queer people until well into the 1990s (Burnett, 2014), and which has not entirely disappeared. In Ottawa, for example, some gay men feel that the absence of a gay village—which means that they have no public places to socialize in couples without standing out as "different"—increases their vulnerability to "gay-bashing" by vigilantes from outlying socially conservative communities (Lewis, 2012). Even today, restaurants and bars still tend to be implicitly coded as heterosexual spaces unless they are located in identifiably gay enclaves, such that lesbian and gay couples expressing ordinary gestures of affection in such locales may be made to feel out of place (Browne, 2007).

Nevertheless, while LGBTQ+ communities share experiences of heteronormative exclusion, they are internally diverse. In some cities, and in certain time periods, lesbian, bisexual, transgender, or people with non-conforming sexualities, as well as LGBTQ+ racialized minorities, have not felt fully comfortable, accepted, or adequately protected against violence in spatial enclaves that are predominantly gay and male, especially as these enclaves have become more "mainstream" and affluent (see below). In some cases, transgender and other queer-identifying groups have begun creating subtle and discreet queer territorialities elsewhere in the city (Nash, 2013a). A key issue for LGBTQ+ organizations is how to share material and cultural space in gay villages (Podmore, 2013b), so that they are, or remain, inclusive spaces for all sexual minorities, regardless of gender, class, and ethno-racial background, and whether the wider society now considers their sexual identities and bodily expressions to be socially acceptable—in the image of the urbane gay "metrosexual" couple—or dissident and transgressive. With the legalization of gay marriage in 2005, some argue that society has incorporated "homonormative" practices (gay or lesbian relationships between people who identify with their biological sex assignment) into a heteronormative world that is still not inclusive of other queer minorities (Podmore, 2013a). A related question concerns the type of neighbourhood where lesbian and gay couples with children can feel accepted (Nash and Gorman-Murray, 2014).

Gay villages were initially characterized by socio-economic mix. That remains the case for some, such as the one in Montreal, which is located in an old working-class neighbourhood with strong community organizations and a high proportion of public housing. Although it is today a regional, national, and international destination, many of its commercial and cultural establishments still cater to LGBTQ+ people with limited financial resources (Giraud, 2013). Nevertheless, both residential and commercial gentrification are powerful forces shaping gay villages. One scenario is that of heterosexual gentrification displacing spatially concentrated gay and lesbian communities that do not manage to establish a strong economic and symbolic power base (Doan and Higgins, 2011; Podmore, 2006). Another scenario is where affluent gay male couples accelerate residential and/or commercial gentrification of gay neighbourhoods, on account of their very high disposable incomes. Personal preference and lifecycle stage affect whether such couples opt to locate in gay villages or in gentrifying neighbourhoods not strongly associated with gay communities (Giraud, 2013; Ray, 2004; Ray and Rose, 2000).

The challenges to gay villages' capacity to be inclusive of diverse queer sexualities go beyond gentrification, however. Since the mid-1990s, policies for the **revitalization** of the economies and central city landscapes of old metropolitan cities have increasingly leaned toward developing specialized consumer items and services and cultural production to appeal to residents and visitors with high disposable incomes (see Bain and Mark, Chapter 15). Municipal governments have thus come to value and embrace gay villages, gay pride events, and other cultural manifestations of gay and lesbian identities, as much for their revenue-generating potential as for reasons of equity and inclusion. As cities become national and international gay tourism destinations, this often draws in heterosexual visitors as well, which raises the question as to whether these places then remain as LGBTQ+ "safe spaces" (Doan, 2011). Cities are also harnessing the "gayness" of people and neighbourhoods as a neo-liberal economic development tool for the promotion of their city as a "creative city" able to compete in the international "knowledge economy" (Nash, 2013b; see Vinodrai, Chapter 17). Richard Florida's (2002) extraordinarily influential "creative cities" thesis has brought some segments of gay communities into the mainstream in ways that they could never have envisaged a few decades ago. This thesis posits that in every economic and social epoch, certain sub-groups of the population, trained or with inherent talents in particular kinds of occupations and industrial sectors, were the "carriers of creativity." For urban economies to flourish, then, cities needed to attract this **creative class** by offering them an excellent quality of life: good environmental amenities plus a social climate embracing "diversity" and a cosmopolitan ethos. Florida attracted particular attention for his "diversity index" of major North American and world cities in which a major element weighting the score favourably was the presence of a large gay population; he argued that this increased the city's attractiveness to creative class members in general. This thesis has been subject to many critiques (Peck, 2005) (see also Vinodrai, Chapter 17), but sexualities and urban space researchers express serious concerns with two aspects in particular (Muller Myrdahl, 2011; Nash, 2013b). First, it appropriates gay and lesbian identity struggles for economic gain. Second, it implicitly only includes white, middle-class, and homonormative individuals, who can now be full and privileged citizens of the entrepreneurial city. Nevertheless, urban-based activism to support diverse queer identities continues within and beyond gay-identified districts (Lewis 2013).

Finally, organizations supporting transgender rights are increasingly lobbying and working for public spaces and institutions, wherever they are located in the city, to be respectful and inclusive of bodies that do not conform to sexual binaries and binaries of gender expression. This particularly concerns gender-neutral or inclusive public washrooms and swimming pool change areas (see for example, City of Vancouver, 2014), seemingly small and mundane issues that, as municipal governments are coming to realize, make a significant difference to some daily lives and to the inclusionary or exclusionary image of their city.

Conclusion

We have seen in this chapter that issues of gender and sexuality crosscut many of the urban issues addressed in other chapters in the present volume. By this, I mean three things. First, we see that new insights can be gained on numerous aspects of urban development and socio-spatial dynamics by analyzing them with the aid of concepts developed by researchers in the feminist geography and sexualities and space traditions. Second, we see that urban contexts shape experiences of inequality and injustice related to gender and sexuality in important ways. Third, over recent decades, the accumulation of individual and collective endeavours to challenge and overcome gendered and sexed barriers to equity and inclusion in the public and private spheres has significantly changed the social, cultural, and economic dynamics of Canadian cities and neighbourhoods. For the future, research in these fields has important implications for how we think about urban sustainability.

Review Questions

1. How are changes in gender roles and especially in the participation of women in the labour force reflected in the social geography of cities and in women's daily mobility patterns?
2. What are the causes and consequences of higher poverty levels among women?
3. What are the reasons for the appearance of gay villages in some large Canadian cities? What are the impacts of these villages on LGBTQ+ communities and on the economy of cities?

References

Ambert, A-M. 2006. One Parent Families: Characteristics, Causes, Consequences, and Issues, Contemporary Family Trends. Ottawa: Vanier Institute of the Family.

Baillargeon, D. 1999. *Making Do: Women, Family and Home in Montreal during the Great Depression*. Waterloo, ON: Wilfrid Laurier University Press.

Bernard, A., A-M. Séguin, Y. Bussière, and A. Polacchini. 1997. "Household structure and mobility patterns of women in O-D surveys: Methods and results based on the case studies of Montreal and Paris," in Federal Highway Administration United States Department of Transportation, Office of Highway Policy Information, ed., *Women's Travel Issues: Proceedings from the Second National Conference* (held in Baltimore, Oct. 1996). Washington, DC: Federal Highway Administration.

Bondi, L. 1999. "Gender, class, and gentrification: Enriching the debate," *Environment and Planning D: Society and Space* 17: 261–82.

———, and D. Rose. 2003. "Constructing gender, constructing the urban: A review of Anglo-American feminist urban geography," *Gender, Place and Culture* 10: 229–45.

Boyer, K. 1998. "Place and the politics of virtue: Clerical work, corporate anxiety, and changing meanings of public womanhood in early twentieth-century Montreal," *Gender, Place & Culture* 5: 261–75.

Bradbury, B. 2000. "Single parenthood in the past: Canadian census categories, 1891–1951, and the 'normal' family," *Historical Methods: A Journal of Quantitative and Interdisciplinary History* 33: 211–17.

Browne, K. 2007. "(Re)making the other, heterosexualising everyday space," *Environment and Planning A* 39: 996–1014.

Burnett, R. 2014. "Prejudice to pride: The forgotten murder of Joe Rose," *Montreal Gazette,* 16 March.

Burns, V.F., J.-P. Lavoie, and D. Rose. 2012. "Revisiting the role of neighbourhood change in social exclusion and inclusion of older people," *Journal of Aging Research* 2012 (Article ID 148287): 12 pages. doi: 10.1155/2012/148287.

Campbell, L., D. Clément, and G.S. Kealey, eds. 2012. *Debating Dissent: Canada and the 1960s*. Toronto: University of Toronto Press.

Canada Mortgage and Housing Corporation. 2014. *Canadian Housing Observer 2014*. Ottawa: CMHC (catalogue no. NH2-1/2014E). At: https://www.cmhc-schl.gc.ca/odpub/pdf/68189.pdf.

Carnegie Community Action Project, 2017. *We Are Too Poor to Afford Anything. Retail Gentrification Mapping Report*. Vancouver: Carnegie Community Centre Association. At: www.carnegieaction.org/wp-content/uploads/2017/02/CED-REPORT-PRINT.pdf.

Castells, M. 1983. *The City and the Grassroots*. Berkeley, CA: University of California Press.

Cattan, N., and S. Leroy. 2010. "La ville négociée : Les homosexuel(le)s dans l'espace public parisien," *Cahiers de géographie du Québec* 54 (151): 9–24.

———, and———. 2013. *Atlas mondiale des sexualités. Libertés, plaisirs et interdits*. Paris: Éditions Autrement.

Chauncey, G. 1994. *Gay New York: Gender, Urban Culture, and the Makings of the Gay Male World, 1890–1940*. New York: Basic Books.

Christie, N. 2004. "Bachelors and spinsters: Introduction," in N. Christie and M. Gauvreau, eds., *Mapping the Margins: The Family and Social Discipline in Canada, 1700–1975*. Montreal and Kingston: McGill-Queen's University Press.

City of Vancouver. 2014. *Trans* and Gender Variant Inclusion Working Group—Final Report*. Vancouver: City of Vancouver Board of Parks and Recreation. Presented to Park Board Meeting, April 28, 2014. At: http://vancouver.ca/files/cov/REPORT-TGVIWorking-GroupReport-2014-04-28.pdf

Cole, Y. 2017. "Calgary intersection features Transgender Pride flag crosswalk." *Calgary Herald*, August 24. At: http://calgaryherald.com/news/local-news/calgary-intersection-features-transgender-pride-flag-crosswalk

Comacchio, C. 2006. *Dominion of Youth: Adolescence and the Making of Modern Canada, 1920-1950*. Waterloo: Wilfrid Laurier University Press.

Comtois, P. 2007. *Jeunes femmes et loisirs commerciaux durant les années folles (1919-1929): étude des discours ecclésiastiques et journalistiques*, Master's thesis, Histoire, Université du Québec à Montréal. At: www.archipel.uqam.ca/817/

D'Emilio, J. 1998 [1983]. *Sexual Politics, Sexual Communities: The Making of a Homosexual Minority in the United States, 1940–1970*, 2nd edn. Chicago, IL: University of Chicago Press.

Dennis, R. 2006/7. "Working women downtown: Single women in Toronto 1900–1930," *London Journal of Canadian Studies* 22: 35–58.

Doan, P.L., ed. 2011. Queerying Planning: Challenging Heteronormative Assumptions and Reframing Planning Practice. Farnham: Ashgate.

———, and H. Higgins. 2011. "The demise of queer space? Resurgent gentrification and the assimilation of LGBT neighborhoods," *Journal of Planning Education and Research* 3: 6–25.

Dominion Bureau of Statistics. 1954. *Canada Year Book, 1954*. Ottawa: DBS.

Domosh, M. 1999. "Sexing feminist geography," *Progress in Human Geography* 23: 429–36.

Dyck, I. 2005. "Feminist geography, the 'everyday,' and local-global relations: Hidden spaces of place-making," *The Canadian Geographer* 49: 233–43.

England, K., and K. Boyer. 2009. "Women's work: The feminization and shifting meanings of clerical work," *Journal of Social History* 43: 307–40.

Farber, S., A. Mifsud, J. Allen, M.J. Widener, K. B. Newbold, and M. Moniruzzaman. 2018. "Transportation barriers to Syrian newcomer participation and settlement in Durham Region," *Journal of Transport Geography* 68: 181–92.

Featherstone, M. 2004. "Automobilities: An introduction," *Theory, Culture & Society* 21: 1–24.

Fiedler, R., N. Schuurman, and J. Hyndman. 2006. "Hidden homelessness: An indicator-based approach for examining the geographies of recent immigrants at-risk of homelessness in Greater Vancouver," *Cities* 23: 205–16.

Filion, P., T. Bunting, D. Pavlic, and P. Langlois. 2010. "Intensification and sprawl: Residential density trajectories in Canada's largest metropolitan regions," *Urban Geography* 31: 541–69.

Florida, R. 2002. *The Rise of the Creative Class: And How It's Transforming Work, Leisure, Community and Everyday Life*. New York: Basic Books.

Friendly, M., and S. Prentice. 2009. *About Canada: Childcare*. Winnipeg, MB: Fernwood.

Garber, J.A. 2000. "'Not named or identified': Politics and the search for anonymity in the city," in K.B. Miranne and A.H. Young, eds., *Gendering the City: Women, Boundaries, and Visions or Urban Life*. Lanham, MD: Rowman and Littlefield.

Gauvreau, D., S. Olson, and P. Thornton. 2008. "The harsh welcome of an industrial city: Immigrant women in Montreal, 1880–1902," *Social History/Histoire sociale* 40(80): 345–80.

Gendered Innovations in Science, Medicine, and Engineering Team. No date. *Public Transportation: Rethinking Concepts and Theories*. European Commission, Stanford University, and National Science Foundation. At http://genderedinnovations.stanford.edu/case-studies/transportation.html

Giraud, C. 2013. "Le 'Village Gai' de Montréal. Une aventure urbaine minoritaire," *Espaces et sociétés* 154: 33-48.

Heisz, A., and G. Schellenberg. 2004. *Public Transit Use among Immigrants*. Ottawa, ON: Statistics Canada, Analytical Studies Branch research paper series, Catalogue no. 11F0019MIE—No. 224.

Hubbard, P. 2008. "Here, there, everywhere: The ubiquitous geographies of heteronormativity," *Geography Compass* 2: 640–58.

Hulko, W., and J. Hovanes. 2018. "Intersectionality in the lives of LGBTQ Youth: Identifying as LGBTQ and finding community in small cities and rural towns," *Journal of Homosexuality* 65(4): 427-55.

Johnston, L. 2015. "Gender and sexuality I: Genderqueer geographies?" *Progress in Human Geography* 40(5): 668-78.

Jones, C.L., L.R. Marsden, and L. Tepperman. 1990. *Lives of Their Own: The Individualization of Women's Lives, Studies in Canadian Sociology*. Toronto: Oxford University Press.

Kern, L. 2010. *Sex and the Revitalized City: Gender, Condominium Development, and Urban Citizenship*. Vancouver, BC: University of British Columbia Press.

. 2012. "All aboard? Women working the spaces of gentrification in Toronto's Junction," *Gender, Place & Culture* 20: 510–27.

Kirkland, E. 2006. "A home away from home: Defining, regulating, and challenging femininity at the Julia Drummond residence in Montreal, 1920–1971," *Urban History Review* 34: 3–16.

Knopp, L. 2007. "On the relationship between queer and feminist geographies," *The Professional Geographer* 59: 47–55.

LaRochelle-Côté, S., P. Gougeon, and D. Pinard. 2009. "Changes in parental work time and earnings," *Perspectives on Labour and Income* (Statistics Canada, Cat. no. 75-001-X) (October): 5–16.

Lauria, M., and L. Knopp. 1985. "Toward an analysis of the role of gay communities in the urban renaissance," *Urban Geography* 6: 152–69.

Lauster, N., and A. Easterbrook. 2011. "No room for new families? A field experiment measuring rental discrimination against same-sex couples and single parents," *Social Problems* 58: 389–409.

Lee, J., I. Vojnovic, and S.C. Grady. 2017. "The 'transportation disadvantaged': Urban form, gender and automobile versus non-automobile travel in the Detroit region," *Urban Studies*, online pre-publication.

Lewis, N.M. 2012. "Gay in a 'government town': The settlement and regulation of gay-identified men in Ottawa, Canada," *Gender, Place & Culture* 19: 291–312.

———. 2013. "Ottawa's Le/The Village: Creating a gaybourhood amidst the 'death of the village,'" *Geoforum* 49: 233-42.

———, and S. Mills. 2016. "Seeking security: Gay labour migration and uneven landscapes of work," *Environment and Planning A: Economy and Space* 48(12): 2484–503.

Ley, D. 1996. *The New Middle Class and the Remaking of the Central City*. London: Oxford University Press.

McDowell, L. 1999. *Gender, Identity and Place: Understanding Feminist Geographies*. Minneapolis, MN: University of Minnesota Press.

Mackenzie, S., and D. Rose. 1983. "Industrial change, the domestic economy and home life," in J. Anderson, S. Duncan, and R. Hudson, eds., *Redundant Spaces in Cities and Regions*. London: Academic Press.

Mata, F. 2011. Lone-Parent Status among Ethnic Groups in Canada: Data Explorations on its Prevalence, Composition and Generational Persistence Aspects, Working Paper Series. Vancouver, BC: Metropolis British Columbia (Working Paper Series, 11-17).

Mendes, K. 2010. "Reading *Chatelaine*: Dr. Marion Hilliard and 1950s women's health advice," *Canadian Journal of Communication* 35: 515–31.

Meyerowitz, J. 1990. "Sexual geography and gender economy: The furnished room districts of Chicago, 1890–1930," *Gender & History* 2: 174–296.

Milan, A., and N. Bohnert. 2012a. *Fifty years of families in Canada: 1961 to 2011. Families, households and marital status, 2011 Census of Population, Census in Brief*. Ottawa, ON: Statistics Canada (Catalogue no. 98-312-X2011003).

———,———. 2012b. *Portrait of Families and Living Arrangements in Canada. Families, Households and Marital Status, 2011 Census of Population, Analytical Document*. Ottawa, ON: Statistics Canada (Catalogue no. 98-312-X2011001).

Millward, L. 2012. "Making a scene: Struggles over lesbian place-making in anglophone Canada, 1964–1984," *Women's History Review* 21: 553–69.

Morency, C., A. Paez, M.J. Roorda, R. Mercado, and S. Farber. 2011. "Distance traveled in three Canadian cities: Spatial analysis from the perspective of vulnerable population segments," *Journal of Transport Geography* 19: 39–50.

Moyser, M.. 2017. Women and Paid Work (Women in Canada: A Gender-based Statistical Report), Women in Canada: A Gender-based Statistical Report. Ottawa: Statistics Canada (cat. no. 89-503-X).

Muller Myrdahl, T. 2011. "Queerying creative cities," in P.L. Doan, ed., *Queerying Planning: Challenging Heteronormative Assumptions and Reframing Planning Practice*. Farnham: Ashgate.

———. 2013. "Ordinary (small) cities and LGBQ lives," *ACME: An International E-Journal for Critical Geographies* 12: 279–304.

Myers, T. 2006. *Caught: Montreal's Modern Girls and the Law, 1869–1945*. Toronto: University of Toronto Press.

Nash, C.J. 2006. "Toronto's gay village (1969–1982): Plotting the politics of gay identity," *The Canadian Geographer* 50:1–16.

———. 2013a. "Queering neighbourhoods: Politics and practice in Toronto," *ACME: An International E-Journal for Critical Geographies* 12: 193–219.

———. 2013b. "The age of the "post-mo"? Toronto's gay Village and a new generation," *Geoforum* 49: 243–52.

———, and A. Gorman-Murray. 2014. "LGBT neighbourhoods and 'new mobilities': Towards understanding transformations in sexual and gendered urban landscapes," *International Journal of Urban and Regional Research* 38(3): 756–72.

O'Hanlon, S. 2002. "'All found they used to call it': Genteel boarding houses in early twentieth-century Melbourne," *Urban History* 29: 239–53.

Oswin, N. 2010. "The modern model family at home in Singapore: A queer geography," *Transactions of the Institute of British Geographers* 35: 256–68.

Parker, B. 2011. "Material matters: Gender and the city," *Geography Compass* 5(6): 433–47.

Peake, L. 2010. "Gender, race, sexuality," in S.J. Smith, R. Pain, S.A. Marston, and J.P. Jones III, eds., *The Sage Handbook of Social Geographies*. London: Sage.

———. 2016. "Heteronormativity," in International Encyclopedia of Geography: People, the Earth, Environment and Technology. London and New York: Wiley.

Peck, J. 2005. "Struggling with the creative class," *International Journal of Urban and Regional Research* 29: 740–70.

Peters, E. 2000. "'The two major living realities': Urban service needs of First Nations women in Canadian cities," in K.B. Miranne and A.H. Young, eds., *Gendering the City: Women, Boundaries, and Visions or Urban Life*. Lanham, MD: Rowman and Littlefield.

Piché, L. 1999. "La jeunesse ouvrière catholique féminine un lieu de formation sociale et d'action communautaire, 1931–1966," *Revue d'histoire de l'Amérique française* 52: 481–506.

Podmore, J. 2006. "Gone 'underground'? Lesbian visibility and the consolidation of queer space in Montreal," *Social and Cultural Geography* 7: 595–625.

———. 2013a. "Critical commentary: Sexualities landscapes beyond homonormativity," *Geoforum* 49: 263–67.

———. 2013b. "Lesbians as village 'queers': The transformation of Montréal's lesbian nightlife in the 1990s," *ACME: An International E-Journal for Critical Geographies* 12 (2): 220–49.

Preston, V., and S. McLafferty. 1993. "Gender differences in commuting at suburban and central locations," *Canadian Journal of Regional Science* 16: 237–59.

———and———. 2016. "Revisiting gender, race, and commuting in New York," *Annals of the Association of American Geographers* 106(2): 300–10.

Proulx, C., S. Faustmann, H. Raiq, and A. van den Berg. 2011. "Internal diversity in social policy regimes: The case of Canada's four major provinces," in G. Fréchet, D. Gauvreau, and J. Poirier, eds., *Statistiques sociales, pauvreté et exclusion sociale: Hommage à Paul Bernard*. Montreal: Presses de l'Université de Montréal.

Ray, B. 2004. "A diversity paradox: Montréal's gay village," *Our Diverse Cities* 2004(1): 72–5.

———, and D. Rose. 2000. "Cities of the everyday: Socio-spatial perspectives on gender, difference and diversity," in T. Bunting and P. Filion, eds., *Canadian Cities in Transition: The Twenty-First Century*, 3rd edn. Toronto: Oxford University Press Canada.

Rose, D. 1989. "A feminist perspective of employment restructuring and gentrification: The case of Montréal," in J. Wolch and M. Dear, eds., *The Power of Geography: How Territory Shapes Social Life*. Boston: Unwin Hyman.

———. 2009. "L'accès à la propriété et ses significations chez les ménages d'une seule personne: Une étude de cas montréalaise," in J. Charbonneau, A. Germain, and M. Molgat, eds., *Habiter seul: Un nouveau mode de vie?* Quebec: Presses de l'Université Laval.

———. 2010 [1984]. "Rethinking gentrification: Beyond the uneven development of Marxist urban theory," in L. Lees, T. Slater, and E. Wyly, eds., *The Gentrification Reader*. London & New York: Routledge.

———, and A. Charette. 2014. "Housing experiences of users of settlement services for newcomers in Montréal: Does immigration status matter?" In K. Kilbride, ed., *Immigrant Integration: Research Implications for Public Policy*. Toronto: Canadian Scholars' Press.

———, L. Dillon, and M. Caron. 2016. "Lives of their own, a place of their own? The living arrangements of 'business girls' in early twentieth-century Canadian cities," *British Journal of Canadian Studies* 29(2): 225–48.

———, and C. Le Bourdais. 1986. "The changing conditions of female single parenthood in Montréal's inner city and suburban neighborhood," *Urban Resources* 3(2): 45–52.

———, and P. Villeneuve. 1993. "Work, labour markets and households in transition," in L.S. Bourne and D. Ley, eds., *The Changing Social Geography of Canadian Cities*. Montreal and Kingston: McGill-Queen's University Press.

Séguin, A.-M. 1989. "Madame Ford et l'espace: lecture féministe de la suburbanisation," *Recherches féministes* 2(1): 51–68.

Simmel, G. 1950 [1903]. "The Metropolis and mental life," in K.H. Wolff, ed., *The Sociology of Georg Simmel*. New York: Free Press.

Srigley, K. 2010. *Breadwinning Daughters: Young Working Women in a Depression-era City, 1929–1939, Studies in Gender & History*. Toronto: University of Toronto Press.

St-Pierre, B. 2007. *Déplacements des personnes dans la grande région de Montréal. Scénario prévisionnel 2026 tendanciel* (Mtl03-26T_DGMO2006n). Québec, QC: Ministère des transports du Québec, Service de la modélisation des systèmes de transport. At: www.mtq.gouv.qc.ca/portal/page/portal/Librairie/Publications/fr/ministere/recherche/enquetes/montreal/previsions2007.pdf

Statistics Canada. 2012. *General Social Survey: Table 7: Time spent on household domestic work, by working arrangement, Canada, 2010*. Ottawa, ON: Statistics Canada. At: www.statcan.gc.ca/pub/89-503-x/2010001/article/11546/tbl/tbl007-eng.htm

———. 2017a. *Census in Brief: Same-sex Couples in Canada in 2016, Census in Brief*. Ottawa: Statistics Canada. At : www12.statcan.gc.ca/census-recensement/2016/as-sa/98-200-x/2016007/98-200-x2016007-eng.pdf

———. 2017b *Lone-parent families: The new face of an old phenomenon*. At: www.statcan.gc.ca/pub/11-630-x/index-eng.htm

———. 2017c. *2016 Census. Data Tables: Main Mode of Commuting (10), Immigrant Status and Period of Immigration (9), Sex (3) and Age (5) for the Employed Labour Force Aged 15 Years and Over Having a Usual Place of Work or No Fixed Workplace Address, in Private Households of Canada, Provinces and Territories, Census Metropolitan Areas and Census Agglomerations, 2016 Census—25% Sample Data. Table no. 98-400-X2016330.* Ottawa: Statistics Canada. At: www12.statcan.gc.ca/census-recensement/2016/dp-pd/dt-td/Rp-eng.cfm?LANG=E&APATH=3&DETAIL=0&DIM=0&FL=A&FREE=0&GC=0&GID=0&GK=0&GRP=1&PID=110717&PRID=10&PTYPE=109445&S=0&SHOWALL=0&SUB=0&Temporal=2017&THEME=125&VID=0&VNAMEE=&VNAMEF=

———. 2017d. *2016 Census. Data Tables: Marital Status and Opposite-/Same-sex Status (9), Age (16) and Sex (3) for the Population 15 Years and Over in Private Households of Canada, Provinces and Territories, Census Metropolitan Areas and Census Agglomerations, 2016 and 2011 Censuses—100% Data*. Table no. 98-400-X2016027. Ottawa: Statistics Canada. At: http://www12.statcan.gc.ca/census-recensement/2016/dp-pd/dt-td/Rp-eng.cfm?LANG=E&APATH=3&DETAIL=0&DIM=0&FL=A&FREE=0&GC=0&GID=0&GK=0&GRP=1&PID=109646&PRID=10&PTYPE=109445&S=0&SHOWALL=0&SUB=0&Temporal=2016&THEME=117&VID=0&VNAMEE=&VNAMEF=

———. 2018. CANSIM database, Table 282-0142: Labour Force Survey estimates (LFS), by National Occupational Classification (NOC) and sex, annual. At: www5.statcan.gc.ca/cansim/a26?lang=eng&id=2820142

Stevenson, V. 2017. "'We're writing history': Canada's 1st trans mayor elected in village west of Montreal. Julie Lemieux says voters in Très-Saint-Redempteur wanted change and put 'skills before status.'" *CBC News*, 10 November. At: www.cbc.ca/news/canada/montreal/we-re-writing-history-canada-s-1st-trans-mayor-elected-in-village-west-of-montreal-1.4397903.

Strange, C. 1995. *Toronto's Girl Problem: The Perils and Pleasures of the City, 1880–1930, Studies in Gender and History*. Toronto: University of Toronto Press.

Strong-Boag, V. 2002 [1991]. "Home dreams: Women and the suburban experiment in Canada, 1945–60," in V. Strong-Boag, M. Gleason, and A. Perry, eds., *Rethinking Canada: The Promise of Women's History*, 4th edn. Oxford, UK: Oxford University Press.

Tallentire, J. 2006. Everyday Athenas: Strategies of Survival and Identity for Ever-Single Women in British Columbia, 1880–1930. Vancouver, BC: University of British Columbia, PhD thesis.

Valentine, G., and T. Skelton. 2003. "Finding oneself, losing oneself: The lesbian and gay 'scene' as a paradoxical space," *International Journal of Urban and Regional Research* 27: 849–66.

Vandersmissen, M-H., P. Villeneuve, and M. Thériault. 2001. "Mobilité et accessibilité: Leurs effets sur l'insertion professionnelle des femmes," *L'Espace géographique* 30: 289–305.

Walberg, R., and A. Mrozek. 2010. Canada's Top Family-friendly Cities. Canada Is a Great Place to Live. But What Are the Best Cities for Families? A New Report Card Finds Out. Ottawa, ON: Institute of Marriage and Family Canada.

Walsh, C.A., J. Hanley, N. Ives, and S.R. Hordyk. 2016. "Exploring the experiences of newcomer women with insecure housing in Montréal, Canada," *Journal of International Migration and Integration* 17(3): 887-904.

Wekerle, G.R. 1985. "From refuge to service center: Neighborhoods that support women," *Sociological Focus* 18(2): 79–85.

———. 1993. "Responding to diversity: Housing developed by and for women," *Canadian Journal of Urban Research* 2: 95–111.

Whitzman, C., C. Andrew, and K. Viswanath. 2014. "Partnerships for women's safety in the city: 'four legs for a good table,'" *Environment and Urbanization* 26(2): 443-56.

Williams, C. 2010. *Economic Well-being, Women in Canada: A Gender-based Statistical Report*. Ottawa, ON: Statistics Canada (Catalogue no. 89-503-X).

Wilson, E. 2001. "The invisible flâneur (revised version)," in E. Wilson, ed., *The Contradictions of Culture: Cities, Culture, Women*. London: Sage.

Wright, M.W. 2010. "Gender and geography II: Bridging the gap—feminist, queer, and the geographical imaginary," *Progress in Human Geography* 34: 56–66.

Wright, J., A. Glasbeek, and E. van der Meulen. 2015. "Securing the home: Gender, CCTV and the hybridized space of apartment buildings," *Theoretical Criminology* 19(1): 95-111.

10 Urban Divisions: Inequality, Neighbourhood Poverty, and Homelessness in the Canadian City

Alan Walks

Introduction

How well a society ensures equality and equity among its citizens is a key measure of its health. Unfortunately, for over three decades Canada and its cities have been growing more unequal, and becoming increasingly segregated, over time. Inequality grew in Canada as a whole particularly rapidly during the 1990s and early 2000s (Organisation for Economic Co-operation and Development [OECD], 2008), after which inequality has grown only slowly (OECD, 2016). However, inequality in Canada remains above the average for countries of the OECD, and the ability of the tax and transfer system to reduce inequality has diminished since 2010 (OECD, 2016). Income inequality has been growing at multiple spatial scales from the national level, to the provincial and metropolitan scales, on down to the level of neighbourhoods. It is foremost in Canada's cities where the factors driving social inequality and polarization are most salient.

This chapter interrogates the forms that inequality and polarization take in Canadian urban areas. It begins by detailing the extent of shifts in income inequality and poverty between and within Canadian metropolitan areas. It examines the processes that have been producing and reproducing inequality, both those articulated structurally among households and individuals, and spatially between neighbourhoods. The relationship between the spatial concentration of poverty, racialization processes, and immigrant settlement are explored. The chapter then examines problems and forms of homelessness, and the factors driving housing affordability stress, homelessness, and shelter poverty. The chapter ends with a discussion of the implications of the trends outlined, and the potential policy responses that might address growing social inequality.

Trajectories of Inequality

Comparing trends from the mid-1970s onward, Canada's overall level of income inequality has been rising since the early 1980s, but really began rising in earnest in the early 1990s (Heisz, 2007; see also Figure 1 in Walks, 2013a). Virtually all of the national growth in inequality is due to rising inequality among adults of working age, as the level of inequality among those aged 65 and over has continued to decline. Over time, the Canadian political economy has done a fairly decent (if slow) job at reducing inequality among seniors, but a poor job at preventing rising inequality among everyone else.

National shifts in income inequality are reflected in rising inequality among households, and individuals, within Canada's census metropolitan areas, or CMAs (Figure 10.1). Trends reveal a continued incline toward increasing social inequality, particularly in Canada's most globally connected cities, although the data required to calculate the 2016 levels is not yet available. Inequality has

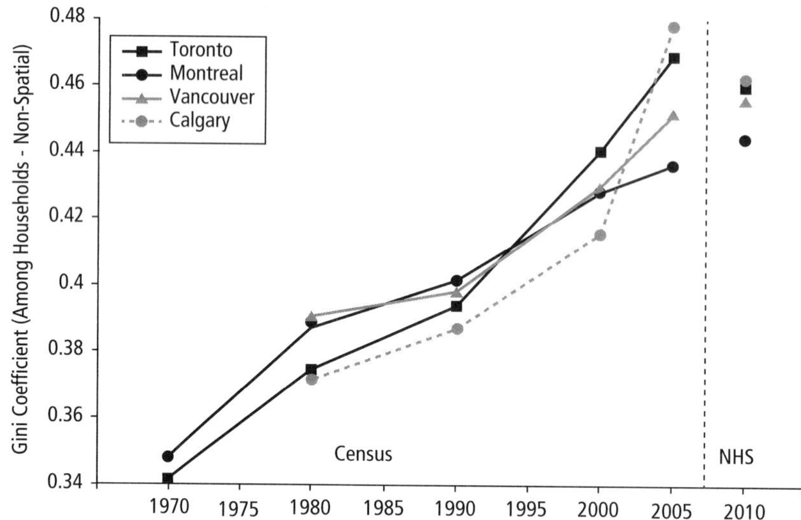

Figure 10.1 Level of Income Inequality (Gini) among Households, Four of the Largest CMAs, 1970–2010

Notes: Units of analysis are households. Coefficients are Gini concentration ratios. The only CMAs in the 1970 PUMF are Toronto and Montreal. Income data in the census are always for the year previous (thus, the 1981 census contains income data for 1980). The 2016 raw census files had not yet been released by Statistics Canada at the time of completion of this chapter.

Source: Calculated by the author with the help of Dylan Simone using the raw Census of Canada files (1981, 1991, 2001, and the 2011 NHS) and of Richard Maaranan from the 1971 Census of Canada Public Use Microsample Files (PUMF).

grown due to both increasing poverty, and to higher incomes for the wealthy. While not strictly a poverty line, Statistics Canada's **low income cut-offs** (LICOs) are often used as such. While the rate of low income has waxed and waned according to the economic fortunes of each metropolitan area, the general trend for many CMAs is slow growth upwards since the mid-1980s (see Figure 6.2a in Walks, 2011). Between 2005 and 2015 most CMAs reveal stagnant or slowly declining proportions of lower-income households in general (Table 10.1), which has reduced some of the tendencies toward income polarization, although there has been continued growth of the numbers of households with the very lowest incomes, as discussed below.

It is the growth of high incomes that has been most responsible for recent increases in household inequality. Between 1970 and 2005, the proportion of Canadian households that reported annual incomes surpassing $100,000 (in constant 2000 dollars) more than quadrupled, from only 3 per cent to 16 per cent, with the vast majority of these high-income earners living in Canada's largest metropolitan areas. Since 2005, as shown in Table 10.1, every single CMA in Canada witnessed an increase in the proportion with high incomes of over $125,000 (in 2015 constant dollars), and this occurred despite declining household sizes. Metropolitan areas in Western Canada, particularly those in Alberta like Calgary and Edmonton, but also Vancouver and Winnipeg, saw the most rapid increases in high-income households. Such trends reflect a more general phenomenon in which the wealthy have seen their incomes continually rise since the early 1990s (Saez and Veall, 2005). Across Canada, between 1980 and 2010 the income share of the top 1 per cent of earners grew from 8.9 per cent to 13.6 per cent, while the wealthiest 0.1 per cent of earners saw their incomes more than double (World Inequality Database, 2018). In turn, the proportion of households with middling incomes has slowly but consistently declined among CMAs in virtually every decade (see also Table 6 in Walks, 2013a, for a more historical perspective).

Table 10.1 Percentage of Households with Low, Median and High Incomes, 2005 and 2015

	Low < $30,000/year (%)		Medium $30,000 to $124,999/year (%)		High > $125,000/year (%)	
	2005	**2015**	**2005**	**2015**	**2005**	**2015**
Halifax	20.6	18.1	63.1	61.8	16.3	20.1
Quebec	21.7	17.3	65.7	65.6	12.6	17.1
Montreal	23.9	20.7	62.2	62.1	13.9	17.2
Ottawa	15.9	14.6	58.9	57.6	25.2	27.8
Toronto	17.3	16.6	57.6	56.1	25.0	27.3
Hamilton	17.8	16.2	60.6	58.9	21.6	24.9
Winnipeg	21.3	16.9	64.0	62.6	14.7	20.6
Calgary	13.2	10.4	59.8	51.9	27.0	37.7
Edmonton	16.2	11.3	61.2	54.2	22.6	34.5
Vancouver	21.2	19.0	59.5	56.8	19.3	24.2
Canada	20.7	17.8	62.2	60.2	17.1	22.0

Source: Calculated by the author from Statistics Canada, 2016 Census of Population, Statistics Canada Catalogue no. 98-400-X2016097.

Factors Propelling Urban Social Inequality

Perhaps the most important factors that have been fuelling income inequality in Canada's cities are the effects of **globalization** and **deindustrialization** of the workforce. Some large Canadian metropolitan regions, including Toronto, Vancouver, and Montreal, represent second-tier "global" or "world" cities specializing in **producer services** with strong connections to networks of global financial, information and migration flows, while others such as Calgary have become global centres controlling the production of distribution of scarce resources (in this case, oil). These metropolitan areas have generally become richer over time. But many smaller cities that specialized in what are now declining industrial sectors, or as transportation depots for agricultural products, have seen their economies grow weaker under the onslaught of globalization and deindustrialization. Deindustrialization is the process whereby the proportion of total economic activity, and of the employed labour force, is declining. Absolute forms of deindustrialization also involve absolute reductions in the number of those employed in manufacturing, whereas relative forms involve declining proportions but still-growing numbers of workers. While driving new flows of investment and providing local employment booms, continental integration introduces a level of vulnerability for local workers who face the threat of plant relocation and restructuring of the production complexes and the shift of production to the United States. The loss of manufacturing jobs has been one of the most important factors explaining the rise in earnings inequality among workers at the CMA level (Bolton and Breau, 2012).

Globalization is expected to lead to increasing labour force bifurcation and polarization (Sassen, 2001). Jobs linked to international finance and producer services experience rising earnings, while those working in most personal services, retail, and manufacturing experience falling wages. The efforts of those in what is termed the

"creative class" to attain an increasing proportion of the total pie for themselves within the context of a globalizing economy is one key trend driving labour market polarization (see Vinodrai, Chapter 17; Skaburskis and Moos, Chapter 19). A culture of compensation paid in yearly bonuses within the financial services industry and some other sectors has produced a sense of entitlement among a new managerial elite increasingly detached from the realities of most workers, and which increasingly resembles the parasitic leisure class examined by Veblen a century ago (2007 [1899]). The practice of paying even mid-level management with shares, bonuses, and stock options turns them into quasi-entrepreneurs in whose economic interest it is to create greater divisions between themselves and lower-paid workers (Sennett, 1998). When the data for different occupational sectors is disaggregated at the CMA level, the only jobs that have seen their relative incomes increase in any significant way since 1980 are managerial, administrative, and business occupations, particularly in Calgary and Toronto. Meanwhile, the incomes of most professionals (lawyers, engineers, doctors, teachers) have largely remained flat, and those employed in manufacturing, sales, and services have all tended to suffer from declining relative earnings (see Figure 6.4 in Walks, 2011).

Immigration

A number of Canadian cities have become quite racially and ethnically diverse, due to persistently high rates of immigration from East and South Asia, the Middle East, Africa, and Latin America (see Kobayashi and Preston, Chapter 8). The geography of immigration flows overwhelmingly favour the largest "global" cities, particularly Toronto and Vancouver, while other cities, including Quebec City and many smaller cities, receive far fewer immigrants. There is a pattern by which the incomes of recent immigrants have declined fairly consistently since 1980, relative to native-born Canadians, a pattern that has only recently slowed in a few CMAs (Figure 10.2a). In 1980, recent immigrants (arriving in the previous 10 years) in the eight largest cities on average earned 85 per cent of native-born Canadians' incomes. By 2015, this had declined to 61 per cent, a 24 per cent drop, with the steepest drop occurring in Toronto (32 per cent). It takes more than 20 years for many immigrants, including Chinese immigrants, to close the earnings gap with the rest of the population (Wang and Lo, 2005). This in turn fuels a similar divergence in income between members of visible minorities and the white population, the gap for which has grown by 26 per cent on average, and by 28 per cent in Toronto (Figure 10.2b). The racialization of poverty, and the correspondence between race and economic success, is producing new forms of social exclusion in Canada's cities (Galabuzzi, 2006), not the least of which includes growing racial spatial segregation (Walks and Bourne, 2006).

Neo-Liberalism

Neo-liberalism refers to an ideological commitment to the concept of individual liberty and the institution of private property, market-based solutions to public policy, and an antipathy to any form of state intervention in the economy, and hence privatization of state resources (see Harvey, 2005; Peck and Tickell, 2002). *Neo-liberalization* involves the transformation of state functions in line with the ideology of neo-liberalism. Peck and Tickell (2002) distinguish two forms of neo-liberalization: (1) "roll-back" of the **welfare state**, often involving the elimination or privatization of state assets and service delivery functions (public transportation, public housing), and (2) a "roll-out" form in which public policies and service delivery models are restructured according to market-based logics. Keil (2009) posits the existence of a third variant—"roll with it" neo-liberalization, in which political realities dictate that even left-leaning subjects critical of neo-liberalism must work with it in negotiating public policy reforms and pursuing public goals: in doing so they co-construct various hybrid forms of actually existing neo-liberalism.

In Canada, neo-liberalism has been tied to the integration of Canada's economy with that of the United States, largely accomplished through

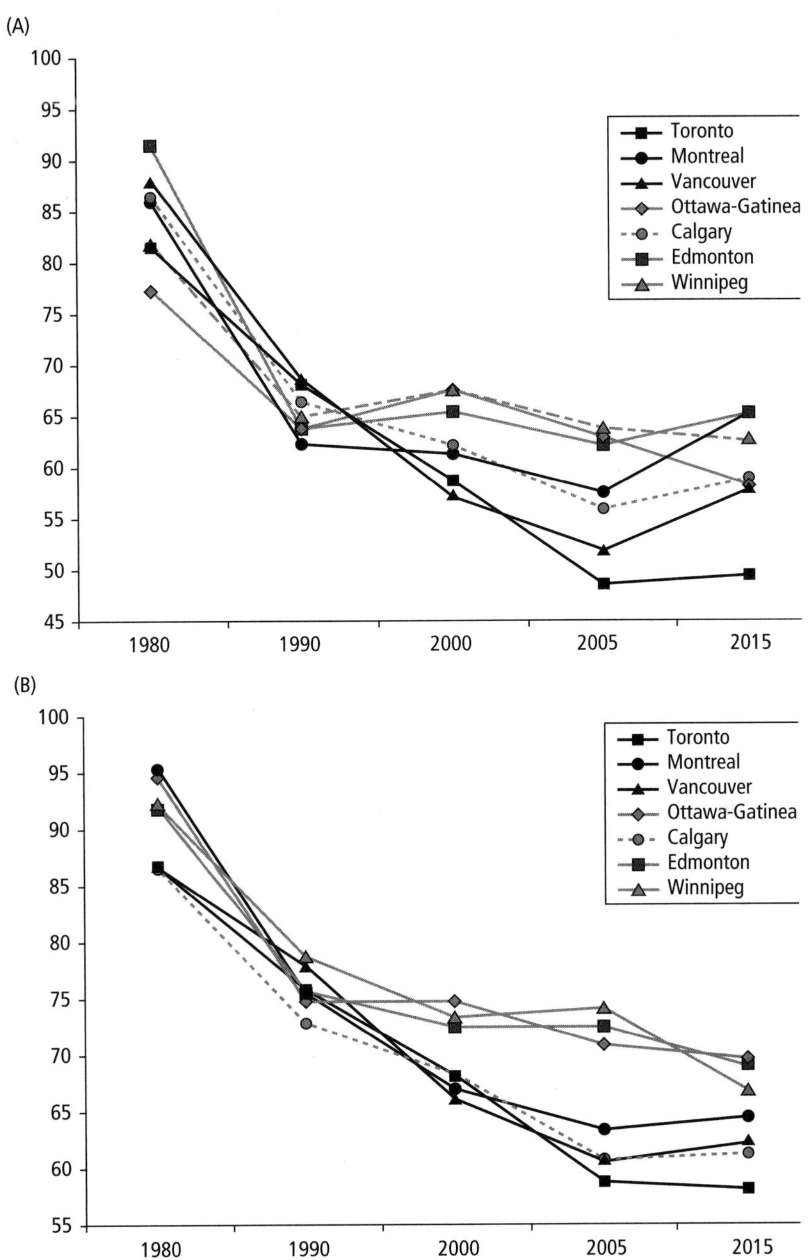

Figure 10.2 Average Relative Incomes of (a) Recent Immigrants and (b) Visible Minorities, Seven of the Largest CMAs, 1980–2015

Source: Calculated by the author using Census of Canada Special Tabulations files, the 2016 Census of Canada Data Tables (http://www12.statcan.gc.ca/census-recensement/2016) and with the help of Richard Maaranen, the Census of Canada Public Use Microsample Files (PUMF), various years.

the reduction of barriers to trade, and flows of labour and capital, between the two countries. Once the North American Free Trade Agreement (NAFTA) was implemented, authors argued that Canada would be compelled to bring many social policies in line with those of the United States in order to maintain the competitiveness of its industries (Courchene and Telmer, 1998; Hurtig, 2002). Indeed, since the 1990s federal and provincial governments have sought to limit or reduce the range of social protections inherent in Canada's welfare state, to cut taxes, and to restructure the system of benefits. While the federal government made its first small moves to "roll back" the federal state in the late 1980s, it more aggressively moved to restructure the entire welfare state system in the 1990s with the restructuring of transfers to the provinces for health and education. Reductions in employment insurance benefits, the downloading of responsibility for certain federal programs, and income tax reductions, all followed. These were then matched by neo-liberal policy reforms enacted by provincial governments in the mid-1990s, starting with Alberta (1993) and Ontario (1995). Such changes led to drastic reductions in (and the restructuring of) welfare benefits and housing subsidies, public transit funding, public education funding, and transfers to municipalities, while outside of Quebec and British Columbia new **social housing** construction came to a virtual halt.

One result of neo-liberalization is that Canada's welfare state and tax structure became less redistributive and equitable. Whereas the redistributive effect of government taxes and transfers grew over the period between 1976 and 1994, largely offsetting the rise in market-based income inequality, from 1995 onwards a persistent reduction in the redistributive effects of the welfare state and taxation is evident (Heisz, 2007: 46). This is also precisely the time when the richest 20 per cent of Canada's population saw their relative income, as well as their proportionate share of income, start to grow, while the relative incomes of the other 80 per cent remained stagnant or declined (Heisz, 2007: 33, 43).

Neo-liberalism is also implicated in another salient phenomenon: financialization of the economy and the growth of household indebtedness. **Financialization** refers to the situation in which the financial sector increases in importance in relation to other economic sectors, and the "pattern of accumulation in which profits accrue primarily through financial channels rather than through trade and commodity production" (Krippner, 2005: 174). Neo-liberalization across the developed world has involved the deregulation of banks and of finance, the increasing importance of the financial sector within developed-nation economies, and a commitment to a low-inflation and low-interest-rate policy (Krippner, 2011). In addition to funding the carry-trade (borrowing in a low-interest-rate country like the United States or Japan, and using the money to invest in growing sectors in higher-interest rate economies) and in turn the shift of industrial investment abroad, the easy credit facilitated by financial deregulation encourages profligate and predatory lending (the pushing of high-interest-rate debt onto those with weak ability to pay), as well as local speculation in housing and other asset markets (Walks, 2010, 2014a). In Canada such processes have accompanied changes to mortgage finance and insurance rules that encourage investment in residential housing (Walks, 2013b, 2014a, 2024b). This has led to record levels of household debt across the country, which by 2017 had reached over 172 per cent of annual household disposable income, almost triple the level in the early 1980s (Figure 10.3). Unfortunately, Canada has a regressive distribution of debt: poorer households have higher levels of debt (Meh et al., 2009), and debt is associated with increasing income segregation (Walks, 2013b).

Furthermore, although neo-liberalism harkens back to the tenets of classical liberalism—in which those who take risks should suffer any losses—Harvey (2005) notes that in practice neo-liberalism has meant that the large lenders and investors are bailed out in the name of saving the system, while borrowers and workers have to make up for losses. The coordinated bailouts of the finance and real estate sectors that were implemented across the developed world in response to the global financial crisis, including those that took place in Canada (Walks, 2010, 2014a), effectively distributed the losses of the

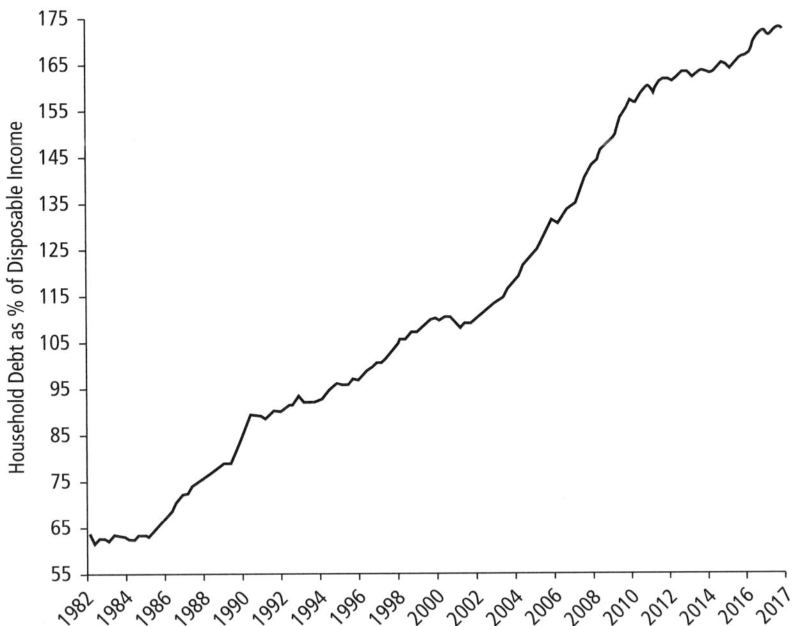

Figure 10.3 Household Debt as a Proportion of Household Disposable Income, Canada, 1982–2017

Source: Compiled and calculated by the author from Statistics Canada CANSIM II database, Table 3780051 (1982–1990) and Table 3780123 (1990–2017).

financial system across the general public, while privatizing the gains in the hands of a few corporations and wealthy individuals (Alessandri and Haldane, 2009). Future research is necessary to determine the real and lasting effects of such shifts on the articulation of poverty and inequality. The 2016 census data suggests it has made the rich all that much richer, while saddling a significant proportion of the population with record amounts of debt.

Shifting Household Structures

The composition of households, as well as living arrangements, aging, and other demographics, also plays a role in the articulation of inequality (see Bain and Mark, Chapter 15; Moos, Chapter 6; Rose, Chapter 9). The proliferation of single-person households, and the growth of female lone-parent families (from 12 per cent in 1981 to over 16 per cent of all families), coupled with the decline of families with children (from 56 to 38.7 per cent of all families between 1981 and 2006), has introduced greater vulnerability and inequality into the urban household sector. From the perspective of inequality, a key development has been class-based marriage pairing based on employment status, with the standard one-earner household of the early postwar period—invoking the stereotypical image of a male breadwinner supporting a nuclear family with a stay-at-home housewife—now overshadowed by the growth of both two-earner and no-earner households (Figure 10.4). The coupling of highly educated and well-paid men and women of educated and wealthy parents is feeding class polarization (Myles, 2010). While the wealthy search out each other for mates, those less educated and less wealthy find it more difficult to meet prospective partners from different socio-economic groups, and increasingly partner with those with similar class and educational backgrounds. As family composition becomes more polarized, it is increasingly those who can draw on two or more incomes who are able to

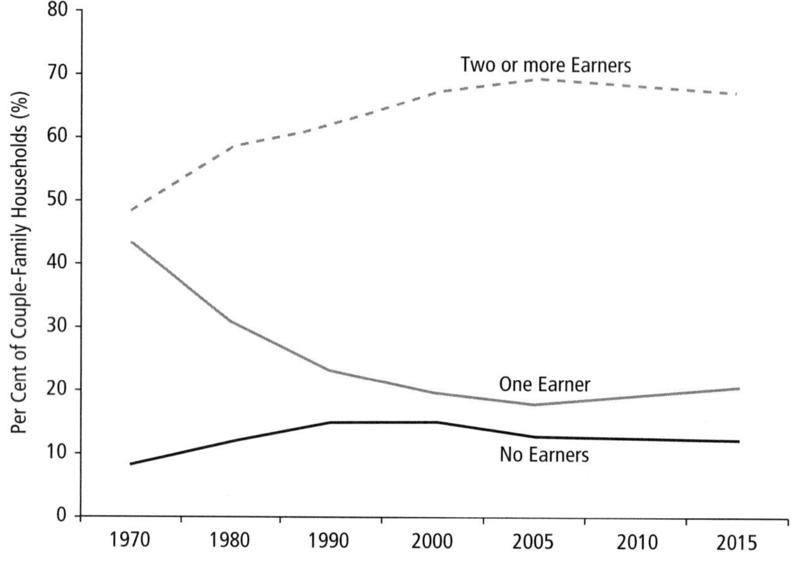

Figure 10.4 Family Households by Employment Status, Canada, 1970–2015

Source: Bourne, 1993; Census of Canada, 1991, 1996, 2001, 2006, 2016; Statistics Canada CANSIM II Tables 111-0016 through 111-0032.

compete in housing markets, inadvertently driving up living costs (and mortgage debt levels) for everyone else.

Socio-Spatial Polarization of the City

Growing social inequality in Canadian society is reflected in the socio-spatial polarization of its cities. Residential segregation among neighbourhoods by class and housing tenure provides one of the more salient articulations of socio-spatial polarization. When the real estate market determines access to space, high-income households with rising incomes are able to outbid others for housing and location (see Skaburskis and Moos, Chapter 19), displacing low-income households from desirable neighbourhoods, driving up housing values, and leading to greater neighbourhood segregation. The census tracts defined by Statistics Canada are typically used as proxies for neighbourhoods, given the lack of a consistent definition or theoretical delineation of the latter.[1] There is a clear and consistent trend toward socio-spatial income segregation among neighbourhoods in Canada's metropolitan areas (Figure 10.5), particularly starting in the 1990s. After years of stability (and even slight declines) in income segregation over the 1970s—a period in which it has been said that looking for inequality was about as exciting as watching the grass grow (MacLachlan and Sawada, 1997)—inequality began ascending. This is particularly evident in faster-growing and more globally connected cities. The level of inequality as measured by the Gini coefficient—the gold standard in inequality measures—rose in Toronto and Calgary by over 60 per cent, and in Vancouver by over 30 per cent.

Inequality among neighbourhoods is spatially articulated in different ways in each city. The early postwar period was characterized by the spread of the middle class into the suburbs, while lower-income households remained in the inner cities where most of the services for the poor or for new migrants to the city are traditionally located. In some Canadian cities, including Winnipeg, neighbourhoods close to the **central business district** have continued to house many

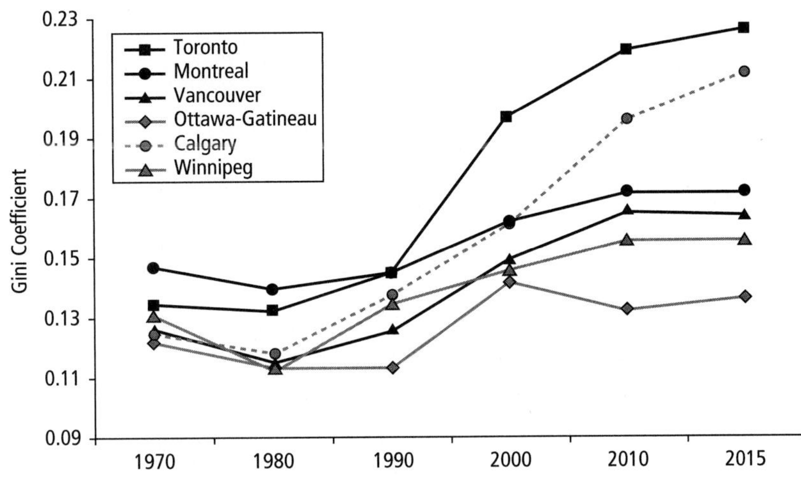

Figure 10.5 Level of Income Inequality (Gini) among Neighbourhoods, Six of the Largest CMAs, 1970–2015

Source: Calculated by the author with the help of Richard Maaranen, from Census of Canada (1970 through 2000, and 2015), and CRA Taxfiler data for 2010.

of the poorest residents. However, since the 1970s the inner cities in most of the larger metropolitan areas have been experiencing **gentrification**, in which housing that once contained a mix of family types and incomes, and which once housed tenants, is converted into space for ever-wealthier households (see Bain and Mark, Chapter 15; Skaburskis and Moos, Chapter 19). As the rental stock is converted to owner-occupied housing, and affordable rental housing disappears from the inner city, low-income households are displaced to the suburbs, where there are fewer services and accessibility to public transit is lower.

Research demonstrates that gentrification is by now well established in a number of Canadian cities (Ley, 1996; Walks and Maaranen, 2008a). Economic restructuring and deindustrialization, as well as the neo-liberal policy reforms and the deregulation of finance, which stimulated speculation in housing, have contributed to gentrification processes in cities (Keil, 2009; Lees, Slater, and Wyly, 2007). And as manufacturing work disappears from the inner cities and is replaced in part by high-end jobs in the FIRE sector (finance, insurance, real estate) and other "quaternary" sector industries, it is the established and gentrifying neighbourhoods in the inner cities that become the target of high-income households, as well as the site of new condominium development (Ley, 1996; Rosen and Walks, 2013; Walks and Maaranen, 2008a). With the gentrification of the inner-city housing stock, and the lack of significant new investments in social housing or rental apartments, low-income households and new in-migrants become concentrated in the older suburbs, where a sizeable proportion of the social housing and of the higher-density market rental housing stock is located in some Canadian cities. These areas now house the greatest proportion of low-income households, and are now functioning as the new immigrant-reception neighbourhoods in the face of gentrification (Walks and Maaranen, 2008b).

Such processes are producing new spatial patterns of urban disadvantage and neighbourhood change. This is clearly evident within the (amalgamated) City of Toronto (Figure 10.6). Traditionally wealthy areas in the very centre of the city north of Bloor Street have continued to see their incomes rise, while many neighbourhoods south of the Bloor-Danforth subway have witnessed gentrification and are now joining their counterparts to the north in the high-income club. At the

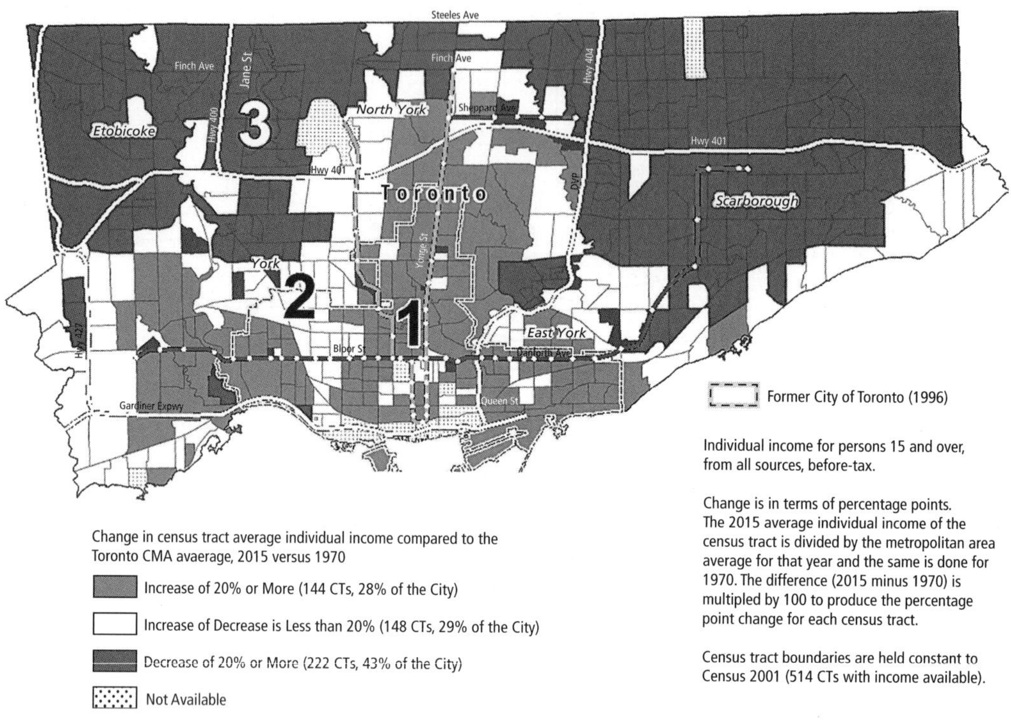

Figure 10.6 The "Three Cities" in Toronto: Neighbourhood Income Change, 1970–2015
Source: Adapted and updated from Hulchanski, 2007, by Richard Maaranen using the 2016 Census.

same time, many neighbourhoods in the older ring of postwar suburbs of Etobicoke, North York, and Scarborough have seen incomes persistently decline over time, and increasingly concentrate ever-greater proportions of low-income households, single parents, poor seniors, and recent immigrants (Hulchanski, 2007). As a result of such processes, poverty has been slowly shifting outward into the postwar suburbs in Toronto and a number of other Canadian cities (Ades, Apparicio, and Seguin, 2012; Walks, 2001).

From Concentrated Neighbourhood Poverty to Ghettos?

A question that often arises when discussing neighbourhood change and socio-spatial inequalities in Canadian cities is whether increasing income segregation is accompanied by rising levels of racial segregation, and whether this might be producing **ghettos** along the lines of the black or Latino ghettos in the United States. Strictly speaking, a ghetto is a neighbourhood in which a subjected racialized minority is compelled to live due to discrimination (see Johnston et al., 2002; Massey and Denton, 1993; Philpott, 1978). The ghetto is thus a concept distinct from the **slum**—a neighbourhood in which poverty is concentrated, often due to the spatial concentration of poor-quality and under-maintained rental housing (Johnston et al., 2002; Massey and Denton, 1993; Philpott, 1978). In the United States, the degree of spatial overlap between the actual location and functioning of ghettos and slums means that high-poverty neighbourhoods and ghettos are often conflated, or fused via the concepts of the "ghetto-poor" or the "outcast ghetto" (Jargowsky, 1997; Marcuse, 1997; Wilson, 1987). Kazemipur and Halli (2000) draw on this US-based discourse in suggesting that the increasing concentration of poverty—the growth of census tracts in which 40 per cent or more of the residents have incomes below the poverty line (LICO in Canada)—could be leading to a process of "ghettoization" in Canada.

The socio-spatial polarization of the city also raises important questions about "neighbourhood effects": the idea that the local neighbourhood affects the life chances of those who reside there. Neighbourhoods provide a basis for the formation of social relationships and networks and accessibility to jobs and information, are a source of amenities and services, and provide a major part of the context against which urban residents live. When poverty becomes concentrated in place, increased social needs may put pressure on local schools; and, if neighbourhoods in an urban municipality disproportionately concentrate residents who depend on public services, this puts pressure on the fiscal health of municipalities. Higher-poverty neighbourhoods cannot support as many retail establishments, potentially impacting local residents' access to quality food and other resources. When geographically isolated, higher-poverty neighbourhoods may limit access to employment or social contact with other urban residents, thus producing a "spatial mismatch" between the locations of those in need and the locations of low-skill jobs and key amenities on which they depend (Ihlanfeldt and Sjoquist 1998; Smith and Ley, 2008). Some of the literature, however, suggests that benefits may flow to ethnic groups who are concentrated in space, and that there might be both "good" and "bad" forms of segregation (Peach, 1996). Spatial concentration that is voluntary (rather than compelled) may provide greater access to job opportunities, friends and family, preferred retail amenities and religious institutions, and services in a language one understands: in such cases, ethnic concentration might on the whole provide significant benefits.

While visible minorities have generally remained segregated and new immigrants continue to face significant challenges, the evidence does not indicate that any process of ghettoization is occurring in Canada's cities (Walks, 2014b; Walks and Bourne, 2006). Instead of being formed predominantly through discrimination, as is the case for the US ghetto, many neighbourhoods with high concentrations of single ethnic groups in Canada's cities have middle incomes, and are formed through voluntary action and positive choice, a process which Myles and Hou (2004) call "locational attainment" (Myles and Hou, 2004). Indeed, there are many newer suburban areas surrounding Canada's cities that fit this criteria. Li (1998) has coined the term **ethnoburb** to refer to the formation of such ethnic communities in the suburbs of North American cities. Of course, this is not to say that there is no overlap between concentrations of racialized communities and poverty. However, the latter occur in places where no single ethnic group dominates, but instead where the poorest members of each different minority group are compelled to live because of low incomes, often in poor-quality private rental housing, and in older social housing estates dating from the 1960s and 1970s (Walks, 2014b; Walks and Bourne, 2006). It is in these neighbourhoods, many located within the inner suburbs, that spatial accessibility and isolation increasingly play important additive roles in shaping the experience of concentrated neighbourhood poverty among low-income minorities, Indigenous people, and immigrants (Smith and Ley, 2008).

Homelessness and the Canadian City

The social distribution of housing wealth and affordability is an increasingly important issue given rising socio-spatial inequality and neoliberal approaches to finance and urban **governance**. The proportion of Canadian households that own their own home increased significantly and consistently over the postwar period, in large part due to access to subsidized mortgage credit (see Harris, Chapter 11). This has allowed those who can afford home ownership, which between 2008 and 2015 has included many lower-income households due to federal mortgage insurance policies (Walks, 2014a), the ability to amass more wealth and retain more income than they otherwise would. But the flipside is that those who

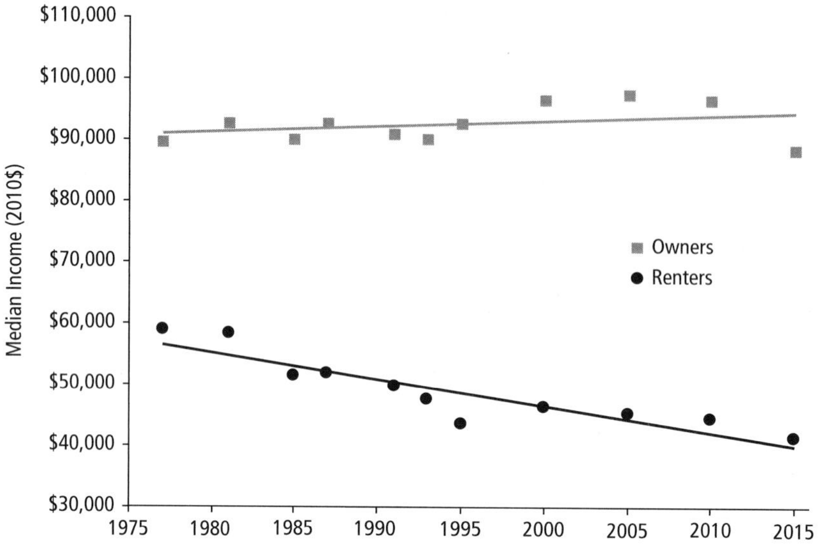

Figure 10.7 Trajectories of Median Income, by Housing Tenure, Canada, 1976–2015
Source: CMHC, 2001; Census of Canada, 1996, 2001, 2006, 2015; Survey of Household Spending, 2011.

cannot afford to be homeowners have, as a whole, become progressively poorer, and have had to find shelter in an increasingly residualized rental housing market (Figure 10.7).

One result of the divisions growing in urban society is homelessness, which is a form of extreme and often invisible poverty. The United Nations defines homelessness as either (1) having no place to call home and being forced to sleep either outside or in a temporary shelter, or (2) having access to housing that is lacking in one or more of the following: sanitation, protection from the elements, safe water, security of tenure, affordability, personal safety, and accessibility to daily needs. Homelessness thus embodies not only those who literally have no home, but also those who do have some form of shelter but whose present housing situation is precarious and insufficient. The difference between these situations is often conceptualized as one between *absolute (or "literal") homelessness* (not having any home) and *relative homelessness* (precarious or insufficient housing). Discrepancies in precise definitions are important because alternate definitions can yield diverging estimates of the extent of homelessness and they support different courses of remedial action. Those whose housing is sufficient but who are paying so much on housing that it cuts into their ability to meet other daily needs are experiencing *housing affordability stress*, and thus are considered *at risk of homelessness*. Stone (1993) calls the latter situation **shelter poverty**, because in such cases housing costs are a cause of households' current poverty.

Absolute homelessness can be distinguished by its duration, with three archetypical forms (Kuhn and Culhane, 1998):

1. *Chronic homelessness* involves those who remain homeless over extended periods of time; it consumes the greatest proportion of shelter resources.
2. *Episodic homelessness* concerns repeated forays into and out of homelessness; it is the most difficult form of homelessness to treat.
3. *Temporary homelessness* typically entails brief periods of homelessness; it is the easiest form to target with preventive or emergency services.

A recent report on the state of homelessness in Canada estimates that of the roughly 200,000 people who are estimated to experience homelessness in any given year, roughly 2–3.5 per cent involve chronic homelessness, and between 3 and 10 per cent involve episodic forms of homelessness, with the remainder (between 86 and 95 per cent) falling into the category of temporary homelessness (Gaetz et al., 2013: 7).

Most experts acknowledge the existence of a *continuum of residential security* between homefulness at one extreme and absolute homelessness at the other (Figure 10.8). Home ownership is the housing situation that best characterizes "homefulness" in Canada, and indeed, as discussed by Harris in Chapter 11, this form of tenure is what federal housing policy has mainly been geared to promote (see also Walks, 2014a). At the other extreme is the situation most stereotypical of the homeless—"sleeping rough." This is the housing option of last resort, typically adopted when attempts to double-up with as many family and friends as possible are exhausted. Those who find themselves having to sleep rough often take refuge on park benches, under bridges, or in tents, bus shelters, parking garages, and cars. Between these extremes lie varying levels of relative homelessness, each of which can be further distinguished by the level of state or community involvement in their provision (including that provided by **non-governmental organizations**). State and non-market forms of housing (emergency, transitional, supportive, and social/co-operative housing) provide significant options for households to help prevent them from falling into shelter poverty and/or absolute homelessness, particularly those people with special needs and those facing problems in the labour market. Gaetz et al. (2013) estimate that in 2009 147,000 unique individuals, or 1 in 230 Canadians, stayed in an emergency shelter at least once.

The contemporary literature conceptualizes the causes of homelessness as deriving from pathways that conceptually link macro structural factors to individual circumstance and life worlds (Clapham, 2002). Such approaches bridge old debates between agency approaches that explain homelessness as the result of individual failings (substance abuse, job loss, disability, depression, mental health problems, etc.) and structural explanations that pin homelessness on employment restructuring and inequitable housing markets. As O'Flaherty noted more than 25 years ago (1993), in any single city it is those individuals who suffer from one or more of the individual-level problems listed above who typically first fall into homelessness when the economy worsens, while the total number of homeless among different cities is strongly predicted by their rate and length of unemployment. Studies of the characteristics of the homeless in Canada's cities contradict stereotypes of the lazy male "hobo" with substance-abuse issues. In fact, virtually all homeless adults receive some income, and upwards of two-fifths are employed (City of Calgary, 2012; City of Toronto, 2009: 37). A common trigger inducing homelessness is the reduction or elimination of social assistance (SPCW, 2011), as well as rising rents/eviction, and loss of other forms of income (GVRSCH, 2010). While it is true that a disproportionate proportion of the homeless suffer

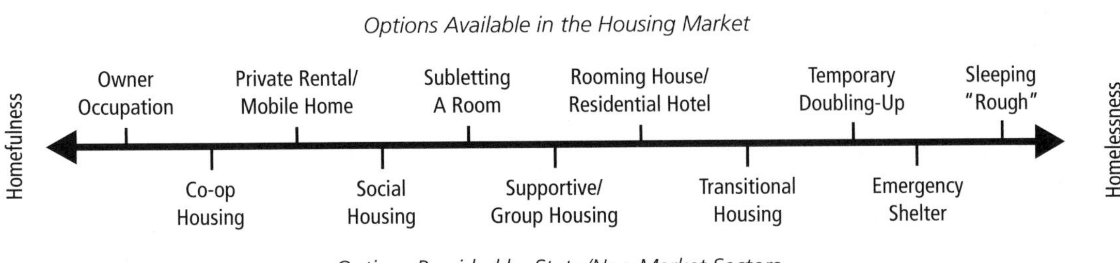

Figure 10.8 Continuum of Homefulness to Homelessness

from mental health problems, it is unclear how much of this predates, or alternatively is caused by, the homeless situation. One survey found that 42 per cent of the homeless suffered abuse as a child (City of Calgary, 2012: 13). A large proportion—upwards of one-fifth of the homeless, depending on the year—tend to be children and youth (City of Calgary, 2012: 13; Gaetz et al., 2013), and homelessness has risen more quickly among families with children than any other group (City of Calgary, 2012; City of Toronto 2013b; Gaetz et al., 2013), requiring resources to be put into new specialized family shelters (Gallimore et al., 2007). Indigenous people are overrepresented among the homeless in virtually every Canadian city (Gaetz et al., 2013).

The growth of housing affordability stress and homelessness is the result of the many factors and changes occurring within Canadian cities that have already been discussed above (Figure 10.9).

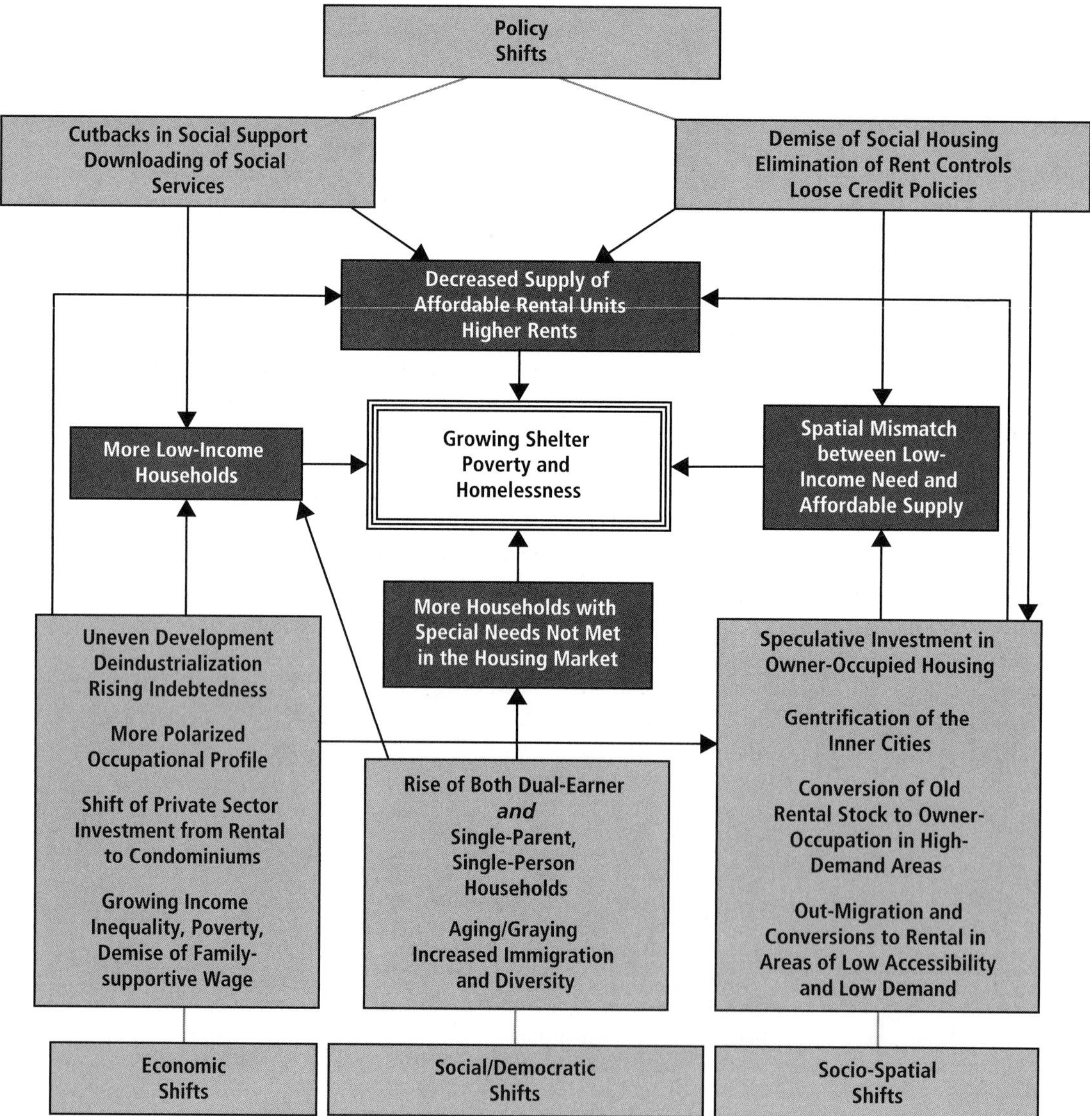

Figure 10.9 Factors Inducing Homelessness and Housing Affordability Stress

First of all are broadly based economic changes, including deindustrialization and globalization, which lead to more low-income households in certain cities, yet also upward pressure on housing markets in those metropolitan areas that have been successful in adapting to globalization, which then also typically experience gentrification and the de-conversion of rental units into owner-occupation. Second are those shifts occurring within the household sector, in which the increasing diversity of household types and sizes—the outcome of rising divorce and family dissolution rates over the postwar period—results in greater dispersion in ability to pay, with many specialized needs not easily met in the regular housing market. Third, gentrification and the spatial restructuring of the city has meant the erosion of rental units in the most accessible locations, producing a spatial mismatch between the demand and supply of rental units. With the rise of home ownership, and the shift away from building new rental housing (and toward the condominium as the preferred tenure for new high-rise buildings; see Rosen and Walks, 2013), renters have either had to pay very high rents for premium new units, or must settle for residualized units in areas of mostly low accessibility. Spatial mismatches are exacerbated by immigrant flows that strongly favour certain cities and neighbourhoods over others.

Finally, each of these factors is exacerbated by the effects of political decisions and public policies enacted since the 1980s associated with neo-liberal governance. These include the elimination of rent controls; declining rental subsidies; the drastic drop in new social rental construction; cutbacks in welfare support; the downloading of social housing and other social services to municipalities (which are least able to afford to maintain them); and loose credit policies that have encouraged speculation in the housing market, funded gentrification, and led to unprecedented house price inflation. This puts a squeeze on Canada's rental market, reducing the number of affordable units. In Canada, 310,000 low-rent units were lost alone during the early 1990s (Pomeroy, 2004), while Ontario lost another 45,000 during the late 1990s (Shapcott, 2004). The mismatch between the demand and supply of affordable units is so great in Toronto that the city reported over 86,000 households on the wait list for social housing in 2012—more than the total number of such units actually existing in the city—and a 22-year wait at the going rate of absorption (City of Toronto, 2013b).

Counting and Alleviating Homelessness

The numbers of homeless present on any given night, including those in temporary shelters and those sleeping rough/doubling-up, varies across Canadian cities (Table 10.2). While lower than comparable figures for US cities, which range between 21 and 77 per 10,000, the evidence suggests that homelessness has been growing. Some cities, including Ottawa, report that homelessness has been growing consistently, with increased pressure on the city's shelter system making it difficult to even find a bed (Figure 10.10). Calgary's experience has been one shared by many Canadian cities (Figure 10.11). After witnessing rapid growth in the numbers of homeless over the 2000s, Calgary has seen fluctuations but no consistent reduction in the numbers of homeless.

A key principle in addressing the growth of homelessness and housing unaffordability includes the application of "housing-first" policies, in which the provision of permanent housing is made a priority, along with better integration of public services for those who are homeless. In Toronto, the implementation of the Streets-to-Homes program since the mid-2000s is largely considered a success, with 3881 people permanently housed through this program between 2005 and 2012, reducing the stress on the shelter system (City of Toronto, 2013b). Likewise, housing-first policies are at the centre of Calgary's 10-year plans to end homelessness, finding permanent housing for 92 per cent of those in the program within 12 months, and providing them with significantly improved access to health care and other public services (City of Calgary, 2012). Other cities, including Edmonton and Vancouver, have adopted similar approaches and are also seeing improvements in their state of homelessness and poverty

Table 10.2 Counts of the Homeless, Canadian Cities, 2010–2016

City	Study year	Population (#)	Total annual homeless (#)	Number per 10,000	Unsheltered (#)	Sheltered (#)
On a Single Night:						
Halifax (Metro)	2015	403,131	284	7	48	183
Montreal	2015	1,704,694	3,016	18	429	2,587
Toronto (City)	2013	2,731,571	5,253	19	447	4,806
Winnipeg	2011	663,617	765	12	350	415
Saskatoon	2013	222,189	379	12	72	307
Red Deer	2012	90,564	279	31	184	93
Calgary	2016	1,235,171	3,430	28	194	3,236
Edmonton	2016	812,201	1,752	27	388	1,370
Vancouver (City)	2016	603,502	1,847	31	539	1,308
For the Whole Year:						
Halifax	2011	390,096	1,973	51	na	1,973
Saint John	2012	70,063	341	49	na	341
Ottawa	2012	883,391	7,308	83	na	7308
Sudbury	2010	160,274	1,014	63	na	1,014
Saskatoon	2010	222,189	2,700	122	na	2,700

Source: ATEH, 2012; Campbell et al. 2016; City of Toronto, 2009, 2013a, 2013b; CUISR, 2010; Gaetz et al., 2013; Greater Sudbury, 2011; Latimer et al., 2015; NSHHN, 2012; SJHDC, 2013; SPCW, 2011; Thompson, 2016; UWS, 2013.

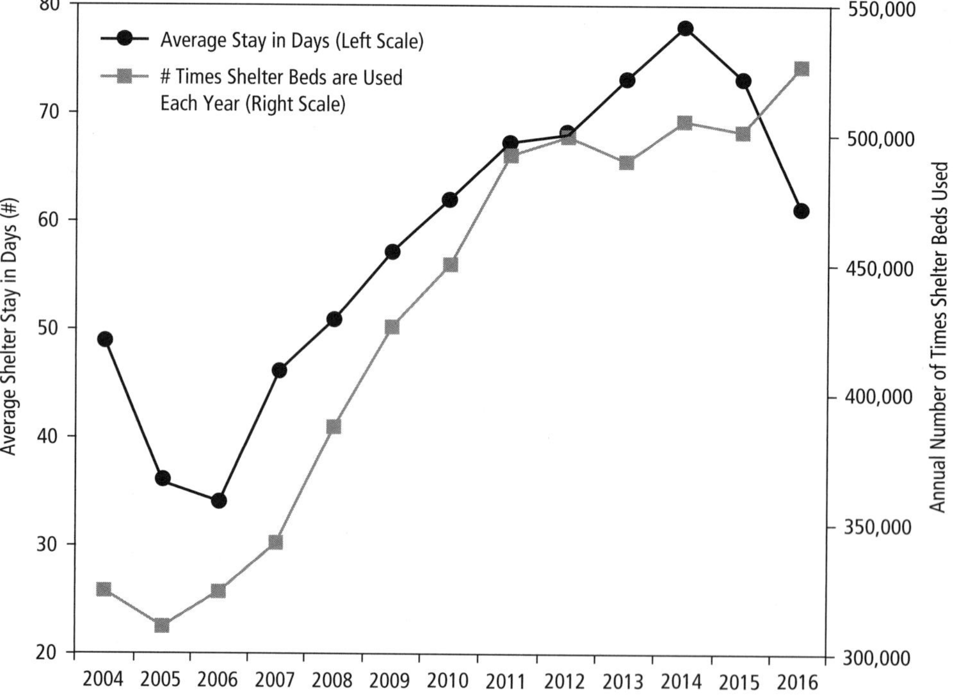

Figure 10.10 Increasing Use of Shelter Beds, and Longer Stays, Ottawa, 2004–2016
Source: Alliance to End Homelessness, various years.

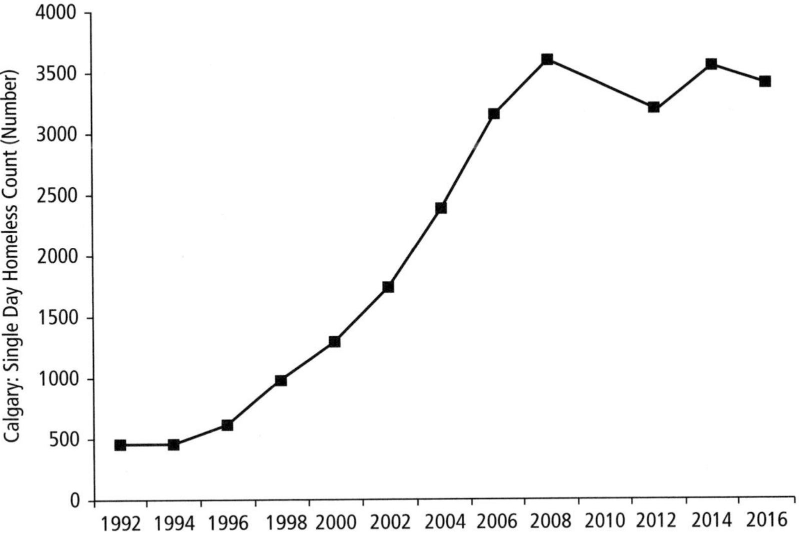

Figure 10.11 Counts of the Homeless, Calgary, 1992–2016

Source: City of Calgary, 2012, 2016

Notes: Data from Calgary represent both those sleeping rough and those in emergency shelters. No single-day snapshot was taken in Calgary in 2010.

(Gaetz et al., 2013). These efforts have been helped by federal and provincial initiatives; however, neither the federal government, nor any provincial government, has yet adopted the "1 per cent solution" advocated by the Federation of Canadian Municipalities (FCM), pushing for an additional 1 per cent of the annual budget to be spent on affordable housing. This would provide for stable funding to build upwards of 10,000 new affordable rental units per year. If these units were built in the inner cities, they would help meet demand in exactly those places suffering from overheated housing markets, unaffordable rents, and homelessness. This would also help limit gentrification, maintain local social mix, and promote a more equitable and just city.

Conclusion

The effect of globalization, economic restructuring, deindustrialization, the racialization of poverty, demographic shifts, and neo-liberal policies are all factors that, in combination, continue to drive increases in social inequality throughout urban Canada. As the middle class declines, the feeling that everyone is in the same boat wanes, potentially driving fears of those in other classes and leading to a politics of blame and scapegoating. Social inequality can spur a series of self-reinforcing feedback loops, in which the responses of individuals feed into the system to further aggravate the problems caused by polarization. As they become wealthier, in part from the debt payments being made by the poor, rich households can, for instance, attain more power to influence the political process to push for policy reforms in their favour, such as tax and public spending cuts. Such a process further drives the processes facilitating the transfer of power to the wealthy, while impacting the quality of urban services that the majority depend on. The residualization of the older rental housing stock makes renting less desirable, limiting rent levels and the incentives to build new rental units, while at the same time fuelling the filtering of housing and concentrating the poor in increasingly stigmatized rental districts. These developments feed on each other, driving complex positive feedback loops between income inequality and neighbourhood polarization.

To reverse the growth of social and spatial inequality, and ameliorate the problems caused by unaffordable housing, a multi-pronged and multi-scaled response is needed. In particular, governments need to refocus on efforts that reduce wage inequality, shift the balance of profits away from financial sectors, and build affordable housing. Governments need to increase the redistributive function of the tax system and public services, and items of collective consumption and public infrastructure, upon which the majority of urban residents depend, should be maintained so that they are universally accessible, and the privatizations of strategic public infrastructure reversed. Revenue generation, particularly for social services and other "soft" services, should be shifted from property taxation and other regressive forms of taxation, onto the progressive income tax. Special efforts are required to counter the declining incomes of new immigrants and Indigenous people, and the racialization of poverty in the larger cities, and to enhance gender and racial equity. Property rights must be democratized, and the right to the city, and to full participation in society, will need to be implemented in order to prevent Canada's cities from becoming more unequal and more polarized.

Review Questions

1. What are the factors bringing about heightened income inequality in Canadian cities?
2. Is homelessness an individual or a societal problem? Explain why.
3. Is it the government's role to address income inequality? Why or why not?

Note

1. Census tracts contain between 2000 and 8000 people, and have been constructed in order to contain a similar social demographic with boundaries based on recognizable physical features and transportation routes. Because census tract boundaries have largely been maintained over time, with the outer boundaries reflected in any divisions, census tracts are the most appropriate units for examining neighbourhood changes.

References

Ades, J., P. Apparicio, and A-M. Seguin. 2012. "Are new patterns low-income distribution emerging in Canadian metropolitan areas?" *Canadian Geographer* 56(3): 339–61.

Alessandri, P., and A.G. Haldane. 2009. *Banking on the State*. London: Bank of England. Working Paper. At: www.bankofengland.co.uk/publications/speeches/2009/speech409.pdf

ATEH (Alliance to End Homelessness). Various years. *Report Card/Progress Report on Ending Homelessness in Ottawa*. At: https://static1.squarespace.com/static/5a4d46cdb1ffb6b826e6d6aa/t/5a7333568165f596c6f77b85/1517499224005/Final+2015+Alliance+Progress+Report+%28EN%29.pdf

Bolton, K., and S. Breau. 2012. "Growing unequal? Changes in the distribution of earnings across Canadian cities," *Urban Studies* 49(6): 1377–96.

Bourne, L.S. 1993. "Close together and worlds apart: An analysis of changes in the ecology of income in Canadian cities," *Urban Studies* 30(8): 1293–1317.

Campbell, R., N. Falvo, and M. Smith. 2016. *Point-in-time Count Report 2016*. Calgary: Calgary Homeless Foundation. At: https://docs.wixstatic.com/ugd/ff2744_0d7addfe44964389914cc396884c9fdb.pdf

City of Calgary. 2012. *The State of Homelessness in Calgary, 2012*. At: http://calgaryhomeless.com/assets/research/The-State-of-Homelessnessonlineversion.pdf

City of Toronto. 2009. *Street Needs Assessment Results, 2009*. Toronto: City of Toronto, Shelter, Support and Housing Administration. At: www.toronto.ca/legdocs/mmis/2010/cd/bgrd/backgroundfile-29123.pdf

———. 2013a. *Total Numbers of Unique Individuals using the Shelter System 2000–2008*. Toronto: City of Toronto, official data. At: www.toronto.ca/legdocs/mmis/2009/cd/bgrd/backgroundfile-24924.pdf

———. 2013b. *Quick Facts about Homelessness and Housing in Toronto*. Toronto: City of Toronto Shelter, Support and Housing Administration. At: www.toronto.ca/housing/pdf/quickfactsinfographic.pdf

Clapham, D. 2002. "Housing pathways: A postmodern analytical framework," *Housing, Theory and Society* 19(1): 57–68.

CMHC (Canada Mortgage and Housing Corporation). 2001. *Residualization of Rental Tenure: Attitudes of Private Landlords towards Low-Income Households*. Ottawa: CMHC Research Highlights, Socio-Economic Series, issue 93.

Courchene, T.J. and C. Telmer. 1998. *From Heartland to North American Region State: The Social, Fiscal and Federal Evolution of Ontario*. Toronto: Centre for Public Management, University of Toronto.

CUISR (Community-University Institute for Social Research). 2010. *Saskatoon Homeless Individuals and Families Information System (HIFIS) Report Card 2010*. Saskatoon: Community-University Institute for Social Research. At: www.homelesshub.ca/Library/Saskatoon-Homeless-Individuals-and-Families-Information-System-Report-Card-2010-53660.aspx

Gaetz, S., J. Donaldson, T. Richter, T., and T. Gulliver. 2013. *The State of Homelessness in Canada*. Ottawa: Canadian Homelessness Research Network Press. At: www.homelesshub.ca/Library/The-State-of-Homelessness-in-Canada-2013-55941.aspx

Galabuzi, G-E. 2006. *Canada's Economic Apartheid: The Social Exclusion of Racialized Groups in the New Century*. Toronto: Canadian Scholars Press.

Gallimore, C., B. Kreps, and T. Morasse. 2007. *On Any Given Night: Measuring Homelessness in Hamilton, 2007 Edition*. Hamilton: Social Planning and Research Council of Hamilton. At: www.homelesshub.ca/Library/On-Any-Given-Night-Measuring-Homelessnessin-Hamilton-2007-Edition-36447.aspx

Greater Sudbury. 2011. *Sudbury Report on Homelessness 2010*. Sudbury: Greater Sudbury. At: www.homelesshub.ca/Search.aspx?search=Sudbury+homelessness

GVRSCH (Greater Vancouver Regional Steering Committee on Homelessness). 2010. *Homelessness in Metro Vancouver: A Comparative Community Profile*. Vancouver: GVRSCH. At: www.homelesshub.ca/Library/Homelessness-in-Metro-Vancouver-A-Comparative-Community-Profile-53968.aspx

Harvey, D. 2005. *A Brief History of Neo-liberalism*. Oxford: Oxford University Press.

Heisz, A. 2007. *Income Inequality and Redistribution in Canada: 1976 to 2004*. Ottawa: Statistics Canada. Cat. No. 11F0019MIE Research Paper No. 298

Hulchanski, D. 2007. *The Three Cities in Toronto*. Toronto: University of Toronto Centre for Urban and Community Studies Research Bulletin 41.

Hurtig, M. 2002. *The Vanishing Country*. Toronto: McClelland and Stewart Ltd.

Ihlanfeldt, K., and D. Sjoquist. 1998. "The spatial mismatch hypothesis: A review of recent studies and their implications for welfare reform," *Housing Policy Debate* 9(4): 849–92.

Jargowsky, P. 1997. *Poverty and Place: Ghettos, Barrios, and the American City*. New York: Russell Sage.

Johnston, R., Forrest, J. and Poulsen, M. 2002. "Are there ethnic enclaves/ghettos in English cities?" *Urban Studies* 39(4): 591–618.

Kazemipur, A, and S.S. Halli. 2000. *The New Poverty in Canada: Ethnic Groups and Ghetto Neighbourhoods*. Toronto: Thompson Educational Publishing.

Keil, R. 2009. "The urban politics of roll-with-it neo-liberalization," *City* 13(2–3): 230–45.

Krippner, G. 2005. "The financialization of the American economy," *Socio-Economic Review* 3: 173–208.

———. 2011. *Capitalizing on Crisis: The Political Origins of the Rise of Finance*. Cambridge, MA: Harvard University Press.

Kuhn, R., and D.P. Culhane, 1998. "Applying cluster analysis to test a typology of homelessness by pattern of shelter utilization: Results from the analysis of administrative data," *American Journal of Community Psychology* 26: 207–32.

Latimer, E., J. McGregor, C. Methot, and A. Smith. 2015. *Je compte MTL 2015: Dénombrement des personnes en situation d'itinérance à Montréal le 24 mars 2015*. City of Montreal. At: http://ville.montreal.qc.ca/pls/portal/docs/PAGE/D_SOCIAL_FR/MEDIA/DOCUMENTS/RAPPORT_DENOMBREMENT_PERSONNES_SITUATION_ITINERANCE.PDF

Lees, L., T. Slater, and E. Wyly. 2007. *Gentrification*. London: Routledge.

Ley, D. 1996. *The New Middle Class and the Re-Making of the Central City*. New York: Oxford University Press.

Li, W. 1998. "Anatomy of a new ethnic settlement: The Chinese ethnoburb of Los Angeles," *Urban Studies* 35(3): 479–501.

MacLachlan, I., and R. Sawada. 1997. "Measures of income inequality and social polarization in Canadian metropolitan areas," *Canadian Geographer* 41(4): 377–97.

Marcuse, P. 1997. "The enclave, the citadel, and the ghetto: What has changed in the post-Fordist U.S. city?" *Urban Affairs Review* 33(2): 228–64.

Massey, D., and N. Denton. 1993. *American Apartheid: Segregation and the Making of the Underclass*. Cambridge, MA: Harvard University Press.

Meh, C.A., Y. Terajima, D.Z. Chen, and T. Carter. 2009. *Household Debt, Assets, and Income in Canada: A Microdata Study*. Ottawa: Bank of Canada. Discussion paper 2009-7.

Myles, J. 2010. "The inequality surge: Changes in the family life course are the main cause," *Inroads* 26 (Winter/Spring): 66–73.

———, and F. Hou. 2004. "Changing colours: Spatial assimilation and new racial minority immigrants," *Canadian Journal of Sociology* 29(1): 29–58.

NSHHN (Nova Scotia Housing and Homelessness Network). 2012. *Halifax Report Card on Homelessness 2012*. Halifax: Nova Scotia Housing and Homelessness Network. At: www.homelesshub.ca/Library/Halifax-Report-Card-On-Homelessness-2012-55308.aspx

OECD. 2008. Growing Unequal? Income Distribution and Poverty in OECD Countries. Zurich: OECD. Report 8108051E.

———. 2016. *Income Inequality Remains High in the Face of Weak Recovery*. OECD. November 2016. At: www.oecd.org/social/OECD2016-Income-Inequality-Update.pdf

O'Flaherty, B. 1993. "Wrong person and wrong place: For homelessness, the conjunction is what matters," *Journal of Housing Economics* 13(1): 1–15.

Peach, C. 1996. "Good segregation, bad segregation," *Planning Perspectives* 11(4): 379–98.

Peck, J., and A. Tickell. 2002. "Neo-liberalizing space," *Antipode* 34(3): 380–404.

Philpott, T.L. 1978. The Slum and the Ghetto: Neighbourhood Deterioration and Middle Class Reform, Chicago 1880–1930. New York: Oxford University Press.

Pomeroy, S. 2004. "Toward a comprehensive affordable housing strategy for Canada," in D. Hulchanski and M. Shapcott, eds. *Finding Room: Policy Options for a Canadian Rental Housing Strategy*. Toronto: University of Toronto Centre for Urban and Community Studies.

Rosen, G., and R.A. Walks. 2013. "Rising cities: Condominium development and the private transformation of the metropolis," *Geoforum* 49: 160–72.

Saez, E., and M.R. Veall. 2005. "The evolution of high incomes in Northern America: Lessons from Canadian evidence," *The American Economic Review* 95(3): 831–49.

Sassen, S. 2001. *Global City: New York, London, Tokyo*, 2nd edn. Princeton, NJ: Princeton University Press.

Sennett, R.1998. *The Corrosion of Character: The Personal Consequences of Work in the New Capitalism*. New York: W.W. Norton.

Shapcott, M. 2004. "Where are we going? Recent federal and provincial housing policy," in D. Hulchanski and M. Shapcott, eds., *Finding Room: Policy Options for a Canadian Rental Housing Strategy*. Toronto: University of Toronto Centre for Urban and Community Studies.

SJHDC (Saint John Human Development Council). 2013. *Experiencing Homelessness in Saint John, New Brunswick*. Saint John: Human Development Council. At: www.homelesshub.ca/Library/2013-Saint-John-Homelessness-Report-Card-55680.aspx

Smith, H., and D. Ley, 2008. Even in Canada? The multiscalar construction and experience of concentrated immigrant poverty in gateway cities. *Annals of the Association of American Geographers* 98(3): 686–713.

SPCW (Social Planning Council of Winnipeg). 2011. *A Place to Call Home: Homelessness in Winnipeg, 2011*. Winnipeg: Social Planning Council of Winnipeg. At: www.homelesshub.ca/ResourceFiles/A%20Place%20to%20Call%20Home_FINAL.pdf

Stone, M.E. 1993. *Shelter Poverty: New Ideas on Housing Affordability*. Philadelphia: Temple University Press.

Thompson, M. 2016. *Vancouver Homeless Count, 2016*. Vancouver: M. Thompson Consulting. At: http://vancouver.ca/files/cov/homeless-count-2016-report.pdf

United Way of Saskatoon (UWS). 2013. *Saskatoon Plan to End Homelessness*. Saskatoon: United Way of Saskatoon. At: www.unitedwaysaskatoon.ca/me/uploads/2013/04/P2EH-Report-Final.pdf

Veblen, T. 2007 [1899]. *Theory of the Leisure Class*. Oxford: Oxford University Press.

Walks, A. 2001. "The social ecology of the post-fordist/global city? Economic restructuring and socio-spatial polarization in the Toronto urban region," *Urban Studies* 38(3): 407–47.

———. 2010. "Bailing out the wealthy: Responses to the financial crisis, Ponzi neoliberalism, and the city," *Human Geography* 3(3): 54–84.

———. 2011. "Economic restructuring and trajectories of socio-spatial polarization in the twenty-first century Canadian city," in L.S. Bourne, T. Hutton, R. Shearmur, and J. Simmons, eds., *Canadian Urban Regions: Trajectories of Growth and Change*. Don Mills, ON: Oxford University Press.

———. 2013a. *Income Inequality and Polarization in Canada's Cities: An Examination and New Form of Measurement*. Toronto: University of Toronto Cities Centre, Research Paper #227.

———. 2013b. "Mapping the urban debtscape: The geography of household debt in Canadian cities," *Urban Geography* 34(2): 153–87.

———. 2014a. "Canada's housing bubble story: Mortgage securitization, the state, and the global financial crisis," *International Journal of Urban and Regional Research* 38(1): 256–84.

———. 2014b. "Gentrification, social mix, and the immigrant-reception function of inner-city neighbourhoods: Evidence from Canadian globalizing cities," in K.R. Good, L. Turgeon, and T. Triadafilopoulos, eds., *Segmented Cities? How Urban Contexts Shape Ethnic and Nationalist Politics*. Vancouver: UBC Press.

———, and Bourne, L.S. 2006. "Ghettos in Canada's cities? Racial segregation, ethnic enclaves and poverty concentration in Canadian urban areas," *The Canadian Geographer* 50(3): 273–297.

———, and Maaranen, 2008a. *The Timing, Patterning and Forms of Gentrification and Neighbourhood Upgrading in Montreal, Toronto, and Vancouver 1961 to 2001*. Toronto: University of Toronto Cities Centre Research Report 211.

———, and Maaranen, 2008b. "Gentrification, social mix, and social polarization: Testing the linkages in large Canadian cities," *Urban Geography* 29(4): 293–26.

Wang, S.G., and L. Lo. 2005. Chinese immigrants in Canada: Their changing composition and economic performance. *International Migration* 43(3): 35–71.

Wilson, W.J. 1987. The Truly Disadvantaged: The Inner City, the Underclass, and Public Policy. Chicago: University of Chicago Press.

World Inequality Database. 2019. At: https://wid.world

11 Housing: Dreams and Nightmares

Richard Harris

Introduction

Housing occupies 30 per cent of the land area of Canadian towns and cities, and so dwellings help frame the character of each place: the urbane **row plexes** of Montreal, the graceful limestone homes of old Kingston, the interwar bungalows of Vancouver, the modern condominiums of Toronto, the garage-fronted homes that characterize postwar suburbs from coast to coast. These dwellings have other kinds of significance. Housing is the largest item in most people's budget. Owning a home is the Canadian dream but, including furnishings, appliances, utilities, and property taxes, it absorbs a quarter of household income; it is the owners' largest asset. Most Canadians value the comfort, privacy, and autonomy that housing provides, but four walls can hide loneliness and abuse. For better or worse, we spend most of our lives at home and we care a lot about how we are housed.

This chapter aims to show and explain six key aspects of the subject: (1) how housing has been produced, financed, and sold; (2) how it is occupied; (3) how it is used by different types of people; (4) geographical variations at the regional and local scales; (5) the evolution of public policy, especially at the federal and local levels; and finally, (6) the ongoing issues of Indigenous housing, prices and affordability, energy use, and public health.

For Canadians, two aspects of housing are of fundamental importance: price, which determines its *affordability*, and *tenure*, whether it is owned or rented. The two are linked because most Canadians aspire to own their own home, but lately rising prices have made this a challenge. In the past, affordable housing became available to lower-income households through the process of *filtering*: homes built for the middle class slowly deteriorated, becoming more affordable to others. Increasingly, however, this process has faltered, making government interventions more important, whether directly through *social housing* or via indirect mechanisms.

The Production, Financing, and Sale of Housing

Critics have complained that house-building resists innovation (Harris and Buzzelli, 2005). True, most dwellings are still erected on site (McKellar, 1993), and the pace of technological change has been slow. But it happens: a century ago, electrical wiring became the norm; later, iron pipes were replaced by plastic while wallboard supplanted lath and plaster; insulation has improved; roof trusses are now assembled off-site; and, after 1945, power tools became ubiquitous. It is also true that many builders remain small and apparently inefficient. In Ontario in 1998, more than 98 per cent of them erected fewer than 100 units/year and were responsible for 74 per cent of all housing starts (Buzzelli, 2001). Builders rely on subcontractors, both for physical tasks such as digging basements and for professional services, like accounting (Harris and Buzzelli, 2005). For this reason they operate with little capital equipment: a truck, some power tools, and a cell phone. It is still easy to become—and to fail as—a builder: during the 1990s in Ontario, on average one-third of the builders active in any year had entered the industry in the previous 12 months; another third would be gone in a year's time (Harris and Buzzelli, 2003). Construction has always been favoured by immigrants with limited capital, and it still is (Walton-Roberts and Hiebert, 1997). But the industry is surprisingly efficient. Subcontracting enables entrepreneurs and tradespeople to specialize. Suppliers deliver to sites just-in-time and, with builders and contractors, form dense and flexible production networks, akin to those in other industries.

More dramatic shifts have affected building organization, finance, and land development (Figure 11.1). Three changes are noteworthy. First,

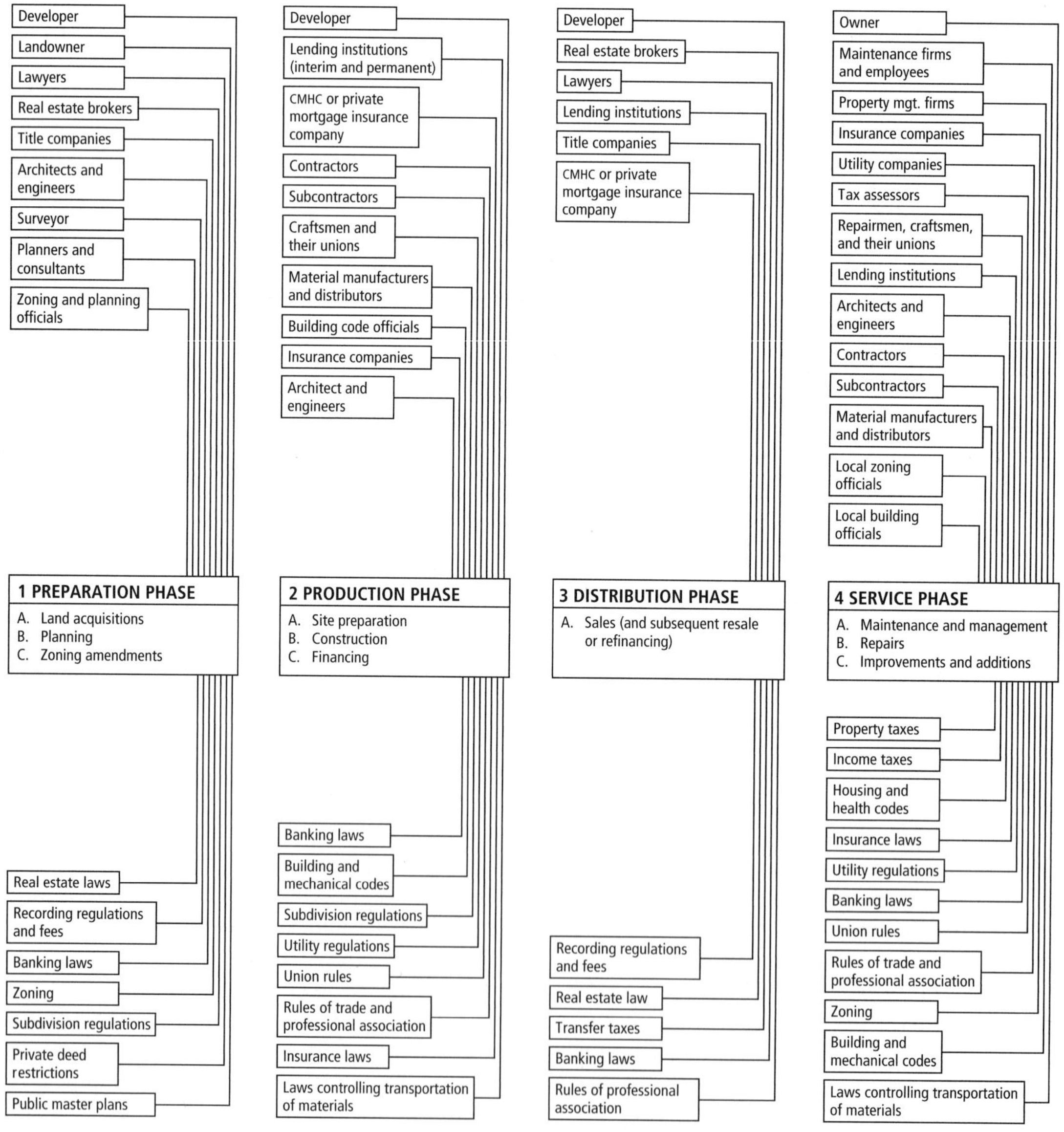

Figure 11.1 The Structures of Residential Building Provision

Source: Adapted from Report of the President's Committee on Urban Housing (the Kaiser Committee), *A Decent Home* (Washington, 1968), 115.

speculative builders (who build first and then seek buyers) have grown in importance. Once, many households commissioned homes from architects and/or custom builders. Low-income households built their own, with help from family or neighbours. In the building boom of the 1900s, a third of new homes in Toronto were self-built, and in western cities the proportion was probably higher (Harris, 2004). Today, custom work makes up a fraction of new construction and owner-building is important only in the Atlantic region (Rowe, 1989).

A second factor that has reshaped housing is finance. As late as the 1940s, many households relied on savings. Few borrowed more than half the value of their property: mortgage ratios were below 50 per cent. Borrowers obtained credit from other individuals, including family, friends, or people contacted through local lawyers or, in Quebec, *notaires publiques* (Harris, 2004). In the 1930s, to revive construction and the wider economy, the federal government passed the first in a series of Housing Acts that revolutionized home finance. In 1946 the Central (now Canada) Mortgage and Housing Corporation (CMHC) was set up, mainly to grant and insure mortgages. Buyers were encouraged to borrow; mortgage terms were relaxed to allow 25-year loans on high ratio (80–90 per cent) mortgages. This helped to restore and then exceed the level of urban home ownership achieved before 1945 (Figure 11.2). Institutional lenders—first insurance companies and after 1954 the banks—soon dominated the mortgage market, except among some immigrant groups (Murdie, 1986). By 2016, banks alone held 77 per cent of residential mortgage debt.

A third change has been the rise of the developer. A century ago, land was subdivided, sold in parcels to speculators, sold again to small builders, built upon, and eventually sold to the first occupants. The result was a varied landscape where small and large dwellings might be juxtaposed. This still happens in some **exurban** areas. But, especially within large metropolitan areas, most house-building is supervised by developers who control the process from land subdivision to final sale. Developers emerged in the 1920s, shaping areas like Kingsway Park in Etobicoke (Toronto) (Paterson, 1984) and Westdale in Hamilton (Weaver, 1978). After World War II, they became dominant in building offices, shopping centres, apartments and condominiums, as well as suburban homes. Everywhere, the result was the planned subdivision.

Balancing the uniformity of large-scale development has been the growth of home renovation. In 1966, renovations and repairs to the existing homes accounted for 18 per cent of housing investment (the remainder was for new construction). By 1976, this proportion had increased to 23 per cent; and by 2011, to 48 per cent (Fallis, 1993; Statistics Canada, 2012, Table 6.4). If the land component of new housing is excluded, renovation expenditures exceeded new construction by 1981 (Clayton Research Associates, 1988: 6). Renovations are more significant than published data suggest since much occurs through Do-It-Yourself (DIY). It is supported by building suppliers who target consumers, offering advice as well as materials. Home Depot and Lowe's are US-based chains that have been resisted by Canadian Tire

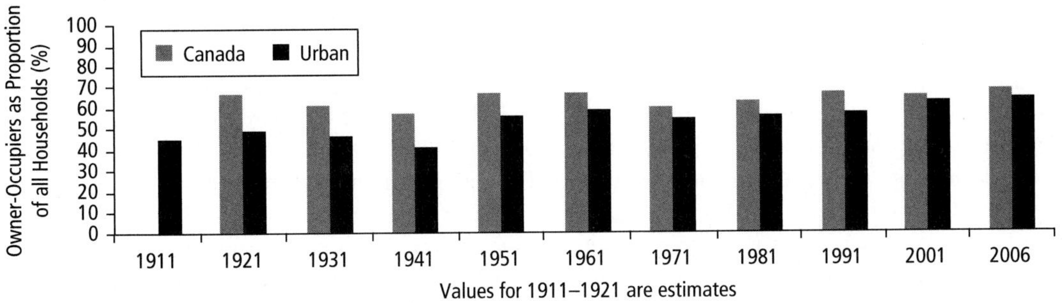

Figure 11.2 The Historical Trend in Home Ownership in Canada, 1911–2006

Source: Canadian Census, various years.

and Rona, a Quebec company (Yakabuski, 2004). Other renovations are done by contractors whose receipts are unrecorded and not declared for tax purposes: the underground economy accounts for about 15 per cent of new construction, 56 per cent of renovations, and 67 per cent of repairs (Canadian Home Builders' Association, 2003). Hard to measure, demand for the services of small contractors and subcontractors remains strong.

The growth of speculative builders has made marketing more important. In the early twentieth century, builders and homeowners used local newspapers to sell houses. From the 1920s, speculative builders built model homes. Today, they blur the lines with custom building: offering variations on a few basic models, they build only when a client has signed a contract. Some new homes and most existing homes are sold, while many apartments are rented, through agents. In major cities, "multiple listing services" (MLS) are owned by local real estate boards. They record information on the price and characteristics of dwellings that are on the market. Member agents use these listings to identify properties that might interest their clients. Using the MLS guarantees that a property will be seen by many agents and potential buyers. This advantage costs the seller a commission of 5–6 per cent of selling price, which is one reason why homeowners move less often than tenants. In 2011, the federal Competition Bureau forced boards to share some of this information, enabling agents to offer varying, and cheaper, mixes of services. After appeal, the decision has stood, but in 2018, the impact has still been limited.

The Occupation of Housing

By definition, housing units are occupied by one or more households (Miron, 1988, 1993). These may be a social entity—an individual, a couple, or a nuclear family—but not always. In the past, many families took in lodgers, though this became uncommon after the 1940s (Harris, 1994: 35) Then, too, "the family" might include grandparents or siblings of the parents, especially among immigrants. The household is malleable, changing with economic circumstances, social mores, and market conditions.

Household formation depends on incomes and the age composition of the population. High rates of household growth after 1945 reflected rising incomes: one-third of the increase in 1951–1981 was due to new living arrangements enabled by affluence (Miron, 1988). In time, it also reflected the maturation of the baby boom generation. This age **cohort** sought rental accommodation, fuelling the apartment boom of the 1960s. As they married, saved, and had children (though fewer than their parents) they acquired homes, pushing prices up, but ensuring that ownership rates remained high (Foot and Stoffman, 1996: Chapter 2). In the 1990s, as the "echo" generation entered the housing market, high prices compelled many to stay or return home (Mitchell and Gee, 1996). After 2001, low interest rates pushed prices higher. Between 2001 and 2016, the proportion of 20- to 34-year-olds living with at least one parent increased from 30.6 per cent to 34.7 per cent. Those able to buy property fuelled the market for condominiums, many as singles (Moos, 2014). The latter included young women, a growing force in professional occupations (Kern, 2010). The result was a major condo boom (Rosen and Walks, 2013).

Filtering and Neighbourhood Change

Condominiums have changed how part of the housing market works. Traditionally, new households rarely occupied new housing. Young people with modest savings bought starter homes that were older and deteriorated, lacking some conveniences, and therefore cheaper. So did moderate-income households, of whatever age. For decades, experts and policy-makers assumed that this "filtering down" process would deliver housing to those unable to afford new homes (Bourne, 1981: 149–60) (Figure 11.3). In the 1920s, the idea of filtering influenced Chicago sociologists, who assumed that immigrants must occupy inner-city housing that had filtered far down market.

Until the 1970s, filtering was common. In Vancouver the Chinatown and Eastside districts,

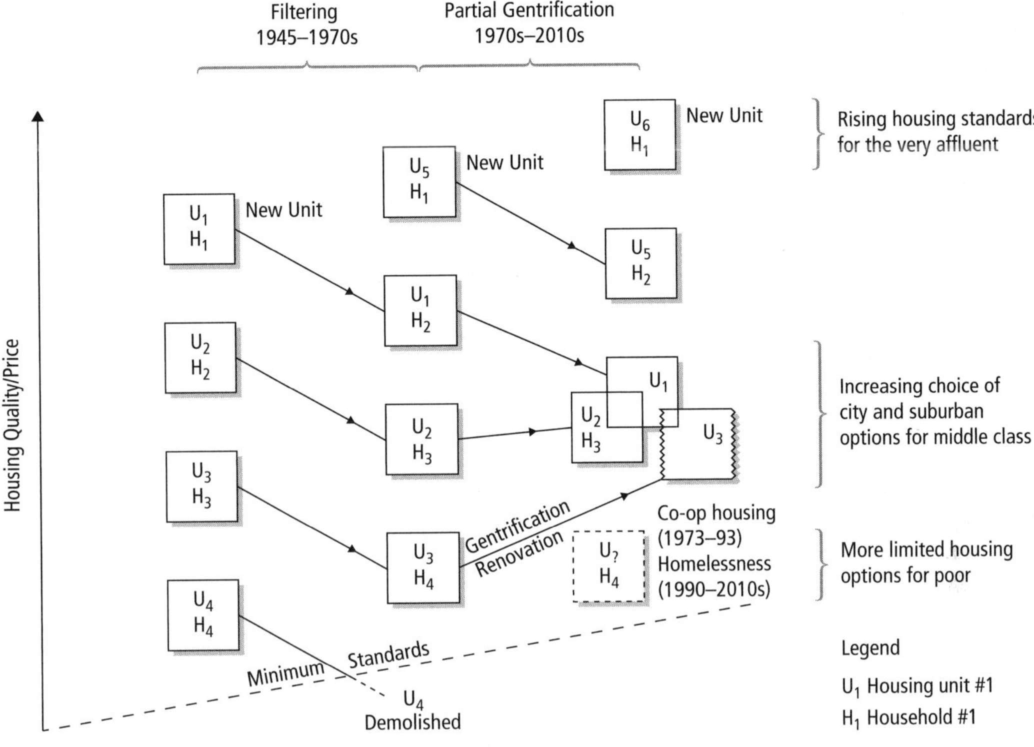

Figure 11.3 Patterns of Urban Filtering in Canada since 1945

in Toronto the Kensington-Spadina neighbourhoods, and in Montreal the St. Urbain corridor, exemplify this process. In cities like Hamilton and Winnipeg, this pattern continued longer. Elsewhere, the **gentrification** of inner neighbourhoods, including those just mentioned, has reversed the process (Ley, 1997; Rosen and Walks, 2013). Classic gentrification involved the upgrading of existing dwellings; notably in Vancouver and Toronto, it has segued into a condo boom that has seen redevelopment, sometimes of industrial and commercial land (Harris, 2011; Lehrer and Wieditz, 2008). One consequence has been the loss of cheap rental housing, together with the displacement of immigrants, single parents, and others with low incomes. Increasingly, the demand for inexpensive housing is being shifted to inner suburbs, where filtering still occurs (Grant, Walks, and Ramos, 2018). In many central neighbourhoods, however, reverse-filtering has disrupted the fragile market mechanism by which the poor were housed.

Residential Mobility and Neighbourhood Change

People move when they perceive their existing dwelling to be less desirable than an affordable alternative. Many factors can contribute to dissatisfaction (Brown and Moore, 1970). These include job relocation, changes in income or family size, or a deteriorating neighbourhood. In relocating to the suburbs after 1945, Canadians often moved closer to suburbanizing jobs, acquiring the larger homes in more spacious surroundings that growing families needed and that rising incomes could support (Clark, 1966; Michelson, 1977). The decline of some city neighbourhoods added a push. Immigrants, endorsing home ownership and family-focused values, have helped perpetuate the suburban trend (Teixeira, 2006). This preference often carries into the second generation (Kataure and Walton-Roberts, 2013).

Gentrification, too, may be explained by changes in job location and household structure (Ley, 1997). Its visible aspects—singles and gay bars, upscale specialty stores and restaurants—reflect a cosmopolitan lifestyle. These patterns have been shaped by office growth in central cities, and by the desire of employees to live close by. Between 2006 and 2011, employment in Toronto's downtown core grew by 14.2 per cent, with a net increase of 3600 businesses (TD Economics, 2013). The resident population increased by about the same amount. Then, too, changes in household structure have caused many to re-evaluate inner-city living. More households consist of singles, or two earners with no children. Schools and open space—except perhaps somewhere to walk a dog—are low priorities; **walkability**, street life, and meeting places are what matter. A central location also makes sense for single parents, since inner areas are better served by transit and the community services (notably daycare) that such households need (Rose, 1989; Rose and LeBourdais, 1986).

The growing labour force participation of women has had complicated effects on these considerations (Hanson and Pratt, 1995; Jarvis, Pratt and Wu, 2001). With two earners and two workplaces, adults must compromise and negotiate about where to live. Women still compromise more than men. Many are employed part-time and/or in jobs—such as clerical work, nursing, and teaching—that are available in scattered locations (Armstrong and Armstrong, 1994). Most households still decide where to live mainly on the basis of where the man works. The woman then seeks employment nearby to avoid interfering with her greater responsibilities for housework and child-rearing (Michelson, 1984). As more women build careers and/or live alone, however, their preferences and options have changed, complicating the household's location decision (Mok, 2007). Some minimize commuting to leave time for household chores and are attracted to easily maintained condominiums, whether in the city or the suburbs (Rose, 1989). Alternatively, the rise of assortative mating has encouraged "power couples" to locate in larger metropolitan centres, where they have fuelled gentrification (Skaburskis, 2012). Even so, women usually do not commute as far as men.

Social Differences in Housing Occupancy

Social groups occupy different types of housing. Some "social" housing is provided to the needy, but most is allocated via the market. Income differences determine what type of housing—and in what location—households can afford.

Owners, managers, and middle-class professionals live in relatively large and well-equipped homes, and tend to own. In Canada there are various incentives for home ownership. Unlike other assets, homes are exempt from capital gains tax. For this and other reasons, in 2009 the average homeowner received seven times the government subsidy of the average tenant; in Ontario, total subsidies to homeowners were four times greater than expenditures on **social housing** (Clayton, 2010). Not surprisingly, then, except for those who move often or who find property maintenance burdensome, households prefer to buy. Since they can afford to, most managers do in fact own their own homes. Home ownership rates among blue-collar workers are lower, although higher than their income might suggest: workers have especially strong aspirations to own their own home. Vulnerable to layoff, with little control over their work environment, and with manual skills, they view the home as a source of financial security and a place of personal autonomy (Harris, 2003). Those with the lowest incomes, including the unskilled and welfare poor, occupy the worst, typically rental, accommodation, and some districts of such housing are badly deteriorated. Housing abandonment, however, is rare in Canada.

Differences in the housing situation of men and women began to receive serious attention in the 1980s (McClain and Doyle, 1983). In the 1950s, most adults lived in husband–wife households, so men and women occupied the same dwellings. The home meant something different to each: women experienced it as a domestic workplace and perhaps as a showplace for homemaking talents, while men viewed it as a haven from work and sometimes as a DIY project (Strong-Boag, 1991; Strong-Boag et al., 1999). But both lived in the same structure. With rising divorce rates and more single-parent households, mostly headed by women, this is no longer true. An increasing proportion of households living below the poverty line, typically in poor-quality rental housing, are headed by women (Rose and Wexler, 1993; Watson, 1986). In 2015, the incidence of child poverty ranged from 11.2 per cent in two-parent households, through 25.5 per cent in lone-father households, to 42.0 per cent in lone-mother households (Statistics Canada, 2017).

Immigrants and ethnic minorities also differ in their housing situation. In the past—although not always today—immigrants had low incomes and occupied modest housing. They strove to establish themselves in their adopted country, and tolerated crowding to save capital to acquire property. Indeed, until the 1990s, immigrants were more likely than native-born Canadians to own their own home. Since then, however, immigrants have fared worse in the job markets, and by the 2000s their homeownership edge had disappeared (Haan, 2005). Even so, saving money by pooling resources in multigenerational households, and by incurring debt, they are able to improve their ownership position rapidly, year-by-year (Hiebert, 2017; Simone and Walks, 2019). Even refugees are able to substantially improve their housing conditions and ownership position soon after arrival (Hiebert, 2017). But it can be misleading to generalize. Some recent immigrants, notably those from Hong Kong, are affluent (Ley, 2010). Some groups, including Italians and Portuguese, have attached special value to home ownership. Observers have suggested the existence of a cultural difference between Canada's two "charter" groups, and that Montreal once had a lower home ownership rate than Toronto because French-Canadians did not value home ownership. Research has challenged this view (Choko and Harris, 1990), but in general cultural as well as economic factors influence the types of housing people aspire to and occupy.

The Meaning and Uses of the Home

Once built and occupied, houses acquire meaning. Consumers use material goods to construct and express their identities, and to signal membership in socially distinct groups (Bourdieu, 1984; Jackson, 1999; Miller, 1987). Dwellings are important in this regard because they are valuable and visible. Frontages present the social self; interiors embody personal traits, especially for women (Belk, 1988; Leslie and Reimer, 2003; Sadella, Verschure and Burroughs, 1987). The balance has shifted as homes became more important for recreation. Gas barbecues and swimming pools testify to the recreational uses of the back yard. Radio, television, computers, and the Internet have made home leisure more attractive. Homes have grown and developed new rooms, in part to house this technology (Simon and Holdsworth, 1993). As feminists have emphasized, however, the home is also important as a place of work.

Unpaid Work

The home has always been a workplace, especially for women (Bradbury, 1993; MacKenzie and Rose, 1983). People, clothes, floors, and dishes have to be kept (fairly) clean. Food has to be bought, and sometimes cooked. Children create tasks: changing diapers, teaching, ferrying, and consoling. Some technologies have saved domestic labour: doing the wash by hand, or fetching water from a well, took time. But the differences between past and present can be overstated (Cowan, 1983). There is more to keep clean: single detached homes built in the 2000s were over 50 per cent larger than those built in 1946–60, even though households have declined in size, from an average of 3.9 persons in 1961 to 2.4 in 2016. Appliances have raised standards of cleanliness and culinary expertise, creating work. Dishwashers must be bought and maintained. Automobiles have encouraged us to build at lower densities, and sprawl compels us to buy and maintain them. To consume more than earlier generations we have to work hard.

Unpaid work is also performed on the home, chiefly by men. As late as the 1950s, many families built homes with their own hands (Harris, 2004). Today, more modest DIY projects are common. Some maintenance is carried out by almost everyone, including tenants when landlords offer rent discounts (Krohn, Fleming, and Manzer, 1977), but owner-occupiers have the greatest opportunities and incentives. Many buyers hope that DIY will save them money, build capital, and allow self-expression. Then, too, ownership can change people's lifestyle, encouraging DIY as part of a more home-centred life (Michelson, 1977: 268–69).

Paid Work

The home is also a place of paid employment. Historically, many women earned income by taking in lodgers. Recently, other types of home employment have increased. Licensed daycare has lagged behind the number of young mothers in the labour force, accounting for only 33 per cent of the arrangements working parents make (Sinha, 2014). The remainder rely on home daycare or arrangements with grandparents and nannies, much of which falls within the underground economy. The Internet and digital technology has made possible new types of home employment. The recent growth of small businesses, many run out of the home, has been supplemented by telework. In 2000, 10.2 per cent of employees did at least some of their work from home (averaging eight hours a week), as did 50 per cent of the self-employed. By 2008 these percentages had increased to 11.2 per cent and 60 per cent, respectively (Turcotte, 2011). Combined, by then, the number of teleworkers had risen to 3.5 million. Lately, however, this trend has levelled off.

Homeworkers place new demands on homes and neighbourhoods. Many alter their homes to accommodate work needs, typically by improved lighting and wiring (Gurstein, 1995: 34). Almost a third of respondents to a survey in 1994 had renovated existing space, while 12 per cent had added a new room for home-based work purposes. Use of neighbourhood facilities also changed. Home-based workers make more use of post offices, delivery services, and copy centres, while many contravene zoning regulations. In terms of public health and safety, paid work in the home is a growing issue.

Work at home is part of a gender division of labour. Since the nineteenth century, men have commuted, and worked on the home, while women have laboured in it, whether for love or money. This pattern is weakening, but persists. Women spend a good deal of time working outside the home, but they still do most of the housework, even in households where both adults work full-time (Armstrong and Armstrong, 1994; Michelson, 1984). In 1986, in dual-earner households women did 75 per cent of the unpaid housework and earned 29 per cent of the household income (Houle, Turcotte, and Wendt, 2017). By 2015 the proportions were, respectively, 61 per cent and 38 per cent. This was a significant shift but, more than men, women juggle work to accommodate domestic responsibilities (Hanson and Pratt, 1995). Some are attracted to teleworking for this reason: in the mid-1990s women made up about 45 per cent of the labour force (rising to 48 per cent in 2016) but 55 per cent of those who worked for pay at home (Gurstein, 1995: 12). The persistence of geographical differences between men and women is nowhere more apparent than in the use of the home.

A Geographical Perspective

Especially in Canada, it can be misleading to generalize about housing. True, with the deplorable exception of First Nations reserves, Canadians occupy some of the best, and best-equipped, housing in the world: almost all urban dwellings are structurally sound, and can boast piped water, electricity, and central heating. But the form these houses take, and how much they cost, varies greatly from place to place.

The Uniqueness of Each Place

The size of Canada, and the immobility of housing, guarantees that housing market conditions vary enormously from place to place (Bourne and Bunting, 1993). A local shortage cannot be met from a surplus elsewhere, especially since most major urban areas are beyond commuting distance of one another. This affects vacancies, rents, and house prices. Across metropolitan areas, rental vacancy rates averaged 3.0 per cent in late 2017, but ranged from a low of 0.2 per cent in Kelowna and Abbotsford to a high of 9.6 per cent in Saskatoon (CMHC, 2017a). House prices are also variable, and often out of sync. Prices in Calgary rose rapidly during the OPEC-induced oil boom of the 1970s and slumped in the early 1980s; in Toronto and southern Ontario, prices boomed in the prosperous 1980s, but dropped sharply in the early 1990s; in Montreal, they were depressed for a generation after the Parti Québécois's first victory in 1976. Recently, economic growth and low interest rates have pushed prices up everywhere (Figure 11.4). Five-year mortgage rates fell from 9 per cent in 1995 to 7 per cent in 2002 and by 2017 had reached an all-time low of just over 2 per cent, creating a sellers' market. Canada-wide, the average price of homes sold through the Multiple Listings Service increased 114 per cent between January 2005 and January 2017. This nationwide boom has been unusual, and average prices still vary enormously: in 2017 they were almost three times as high in Vancouver ($906,000) as in Montreal ($314,000). Single-industry towns have the most volatile housing markets, but even the largest centres are susceptible to booms and busts.

The progression of European settlement in Canada has helped distinguish urban housing submarkets. Eastern cities have the highest proportion of older housing, developed before the automobile, including more dense, multi-occupancy housing with higher levels of tenancy. In western cities, neighbourhoods of single detached homes may be found close to the **central business district (CBD)**. As each city was settled by different immigrants, varying architectural styles predominate. Nineteenth-century Montrealers favoured "plexes," superimposed dwellings with separate entrances; Torontonians favoured gabled row houses; while Vancouverites drew on California bungalow and British "Tudorbethan" models (Holdsworth, 1977). Even postwar ranch styles found regional expressions, notably in Quebec (Morisset and Noppen, 2004). The historical geography of the country helped create a unique housing stock in each centre.

Generic Differences between Places

If each place is unique, some geographical variations are generic, common to certain types of places. The most basic difference is that between

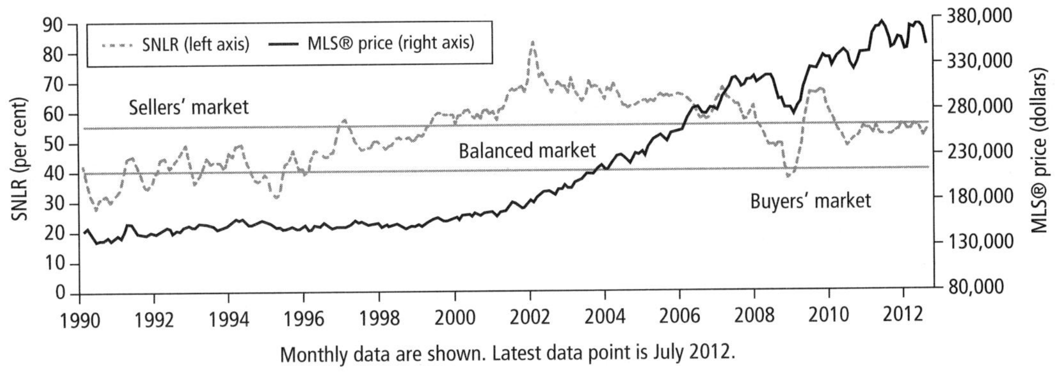

Figure 11.4 The Rise of a Seller's Market in Urban Housing since 1990

Source: Canada Mortgage and Housing Corporation (CMHC). Data from the Canadian Real Estate Association.

rural and urban areas. Urban land is more expensive than rural, and so urban housing requires a high proportion of residents' incomes, is less likely to be owner-occupied, and includes more multi-family dwellings. In 2001 the home ownership rate was lower in **census metropolitan areas (CMAs)** (61 per cent) than in smaller urban centres (68 per cent), and much lower than in rural areas and places with a population of less than 10,000 (76 per cent). Contrasts in the proportion of multi-unit dwellings—respectively 52, 36, and 18 per cent—were even more striking.

The difference between urban and rural housing is one of degree, and suburbs lie in between. During the 1960s, proportionately more apartments were built in Canadian than in US suburbs, but even in this country suburbs contain a high proportion of detached, owner-occupied homes (Harris, 2004: Chapter 2). This is due to lower land prices and the fact that postwar suburbs developed when incomes were relatively high. City-suburban differences in home ownership are associated with variations in the uses to which the home is put. Owners do more unpaid work around the home than tenants, partly because they are more likely to have children, and are also more likely to become active in the community (Cox, 1982). Suburbanites depend on their cars, and automobility is in turn linked to political orientation (Walks, 2015). Generic differences between city and suburban housing, then, go with different ways of life.

Because of urban-rural differences, urbanization affected the types of housing that Canadians occupied until the trend tapered off in the 1970s.[1] It raised the proportion of multi-unit structures during the 1920s and 1960s. It also depressed the level of owner-occupation, though this was counteracted by rising incomes and the suburban trend. Across Canada, the home ownership rate fell from 66 per cent to 62 per cent between 1951 and 1981, rebounding to 68.4 per cent in 2006 (Figure 11.2). Because of soaring prices, it had slipped to 67.8 by 2016. This occurred despite the popularity of condominiums, which enable households to own real estate in row and multi-unit structures. Across all urban areas, 46 per cent of dwellings completed in 2016 were condo units, mostly in multi-unit buildings. The share was highest in the largest centres, notably Toronto (72 per cent), and lower in mid-sized (e.g., Calgary at 46 per cent) and smaller (e.g., Halifax at 5 per cent) places (CMHC, 2017a: Table 21). In the major centres, at least a fifth of condo units are rented out, but most are owner-occupied and are transforming the geography of housing tenure in Canadian cities.

After 1945, suburban development eroded the character of each place. Everywhere, suburbs looked similar: the first appearance of the generic term "suburbia" in the *Globe and Mail* was in 1946 (Harris, 2004). A new National Building Code, the growth of municipal planning and of large land developers, and the popularity of ranch and split-level styles, produced a standard environment (Relph, 1987). Today, much of Mississauga and Kingston Township in Ontario and Richmond, BC, look alike. Since the 1970s, however, neo-historical styles have become increasingly popular. These draw on local and regional architectural traditions: in Toronto, more "Victorian" homes date from the late twentieth than from the late nineteenth century. Beneath the surface variety, however, methods of land development have become more streamlined than ever.

Housing and the State

In Canada, housing is produced by the private sector and mostly owned by private corporations or individuals, but its character and use are profoundly influenced by the state. The rights associated with owner-occupation and tenancy, for example, are defined by government and enforced in the courts. And, through political lobbying, they can be changed, as is illustrated by the continuing evolution of tenant law, and the creation of condominium and co-operative tenures.

The Growth, and Faltering, of Government Activity

All three levels of government have played a growing role in the housing market over the past century. Municipalities were the first to become involved because they were most directly affected by poor conditions. Industrial urbanization, coupled with low wages, led to overcrowding and poor sanitation. Infectious diseases threatened everyone, including the middle class. Frame construction led to serious fires. Municipalities began to control the way houses were built, maintained, and occupied (Hodge and Gordon, 2013). Suburbs, especially those of immigrants and workers, adopted controls slowly, but most had done so by the 1950s.

After 1900, municipalities also took control of land use. Once, land was developed piecemeal, with homes, stores, and small factories sometimes jumbled together. The value of a fine home, as a result, might be undermined by a neighbour's shack or warehouse. Subdividers perceived a demand for districts where undesirable users were prohibited. They used legal covenants to prohibit certain uses or building materials and, sometimes, specific residents. On the west coast the Chinese were targeted; down east, African-Canadians; frequently, Jews (Harris, 2004). In 1948, the Supreme Court declared ethnic covenants illegal, and soon the provinces were requiring municipalities to develop systematic land-use plans. Recently, these have been supplemented again by private covenants, although not as commonly as in the United States (Filion and Alexander, 1994). Some are associated with **common interest developments (CID)**, where, rather like condominiums, buyers acquire a stake in (and responsibility for) shared facilities. Especially on the west coast, some are age-restricted for retirees; a few are gated (Grant, 2005). The CID trend may continue as the proportion of households headed by people aged 65-plus rises from 21 per cent in 2006 to a projected level of 31 per cent in 2026.

Everywhere, municipal zoning controls how land parcels may be used (Hodge and Gordon, 2013). Rules are not always followed. In older districts, "non-conforming" uses persist, including secondary suites in basements or attics (Harris and Kinsella, 2017), while the ubiquity of the Internet has enabled the rapid rise of short-term rentals, notably through Airbnb (Wachsmuth and Weisler, 2018). Anywhere, zoning may be changed if developers or local residents make a strong case. Vancouver has been a leader in promoting laneway and second units in areas once zoned as single-family dwellings (Lauster, 2016). Ad hoc changes require public hearings, an everyday staple of grassroots politics. Just as significant as overt controls are implicit biases in property taxes and development charges that favour single-family as opposed to multi-family dwellings and suburban sprawl over urban redevelopment (Blais, 2010). Some of the most important effects of policy are unintended.

Permanent federal involvement in housing dates from the Great Depression (Bacher, 1993; Harris, 2004). In 1935, the government passed a Dominion Housing Act (DHA) that provided amortized mortgage loans to boost demand for the building industry; most buyers were affluent (Belec, 1997). In 1946, the CMHC was created to administer federal policy. Its mortgage loan activity has shaped suburban development ever since (Poapst, 1993; Pomeroy and Falvo, 2013; Walks, 2013).

Unlike the United States, Canada did not build public housing during the Depression, and only in 1964 did it make the financing attractive enough for provinces to participate. In the next five years there was a boom in public housing construction, that is, of subsidized rental accommodation owned by public agencies (Patterson, 1993). Units for seniors were well-received but those for families were opposed by neighbours. Large schemes exacerbated social problems and attracted stigma, especially those housing visible minorities. In Toronto, black people at first occupied only 4 per cent of units managed by the Metro Toronto Housing Authority (Murdie, 1994). By 1986 the figure had soared to 27 per cent, five times this group's proportion of the Toronto population. By the early 1970s, opposition, and high unit costs, led to the federal program's demise (Patterson, 1993).

In the early 1970s, innovative programs encouraged the rehabilitation of existing housing and new types of social housing, chiefly cooperatives and municipal non-profits (Patterson, 1993). (Social housing refers to units subsidized for lower-income households.) In co-ops, ownership is shared and projects are often socially mixed. Since most were built in older, mixed neighbourhoods, they avoided social stigma (Skelton, 1994). However, project mix reduced the number of low-income families housed. In the late 1980s, the federal government reduced support for social housing and in 1993 froze funding for new projects, except on Indigenous reserves (Pomeroy and Falvo, 2013). Few provincial governments picked up the slack, and Ontario devolved responsibility for project management to municipalities. The construction of social housing virtually ceased (Carter, 1997; Shapcott, 2004; Wolfe, 1998). In Hamilton, Ontario, for example, no units were built between 1996 and 2003 (Hamilton, 2004: 37–39).

One consequence of funding cuts was the growth of homelessness, broadly defined (Gaetz et al., 2016). Many who are not on the street are, nevertheless, housed precariously. Without a place of their own, they surf the couches of friends and kin (Peters, 2012). Rates of homelessness are highest among the mentally disabled and among Indigenous Canadians. Deinstitutionalized in the 1970s, many with mental disorders have had to fend for themselves, comprising a large proportion of people on the street or in prison (Schiff, Schiff, and Schneider, 2010). Indigenous people are not much better off. In 2008, they comprised 2.5 per cent of Calgary's population but 36 per cent of its visible homeless (Belanger, Head, and Awosoga, 2012: 30). They made up 15 per cent of the homeless population in Toronto, 30 per cent in Ottawa, 80 per cent in Regina, and 95 per cent in Yellowknife. In part for this reason, in 1999 the federal government developed a National Homelessness Initiative. In 2000, social housing programs were reframed as an Affordable Housing Initiative (AHI) that avoided long-term subsidies or financial commitments (Pomeroy and Falvo, 2013). Following the financial crisis in 2008, additional funds were earmarked as part of Canada's Economic Action Plan. By 2012, 51,843 affordable units had been funded under the AHI, accounting for about 3 per cent of housing starts during the 2000s (CMHC, 2012: 5-3). A cost-effective program has been Housing First, which houses the homeless and thereby reduces related costs, such as health care (MacNaughton et al., 2017).

To access federal funds, provinces entered cost-sharing agreements. Quebec and BC acted promptly but other provinces lagged (Hackworth, 2008). Overall, although Canadians take pride in their social safety net, their governments have built and subsidized less social housing than those of most other industrialized nations—including the United States (Harris, 2000; Shapcott, 2004). It is unclear whether a new National Housing Strategy, announced in November 2017, will change this situation (Government of Canada, 2017). It promises substantial long-term funds, cost-shared with the provinces. These are for rent supplements, vouchers that recipients can use for apartments in any location, thereby avoiding the stigma of segregation, and also for inclusionary housing, where developers include affordable units within market-priced developments. Some argue that the National Housing Strategy is incomplete, doing little to encourage the production of rental accommodation, which is currently unprofitable (CMHC, 2017b). It also neglects the affordability problems of young home buyers. However, even its critics see it as a valuable initiative.

The Purposes and Consequences of Housing Policy

In Canada, the main purposes of housing policy have been economic, a fact implied in the naming of CMHC: it is a "corporation" concerned with "mortgages." This is not surprising. Building 10,000 new homes creates 19,337 person-years of work, including 10,533 in construction and 3087 in manufacturing (CMHC, 2010:19). Federal policy has consistently sought to strengthen the private sector, including corporate land development (Bacher, 1993; Hackworth, 2008). Actions during the 1930s were designed to revive the building industry and facilitate mortgage lending. The first president of the CMHC had been an executive at Sun Life. After 1945, policies favoured large developers over small builders, and financial institutions over individual lenders, although the latter held half of all mortgage debt into the early 1950s (Harris, 2004). The CMHC has consistently made a profit from its mortgage operations (Shapcott, 2004). However, it arguably received a bailout in 2008 and may be troubled if the current housing bubble bursts (Pomeroy and Falvo, 2013; Walks, 2013).

The market bias of policy is a mixed blessing. It has helped produce safe, energy-efficient, and roomy housing for most Canadians, but at a price: the promotion of mortgages has created a nation of indebted consumers living in low-density suburban homes, with unfortunate environmental and health consequences. Policy has also been socially regressive, notably through the subsidies to owner-occupation (Clayton, 2010; Steele, 1993). Housing policy has not been an important part of Canada's social safety net. It has neglected many who cannot help themselves. Past biases have become self-perpetuating. Most Canadian households own their own home and would resist any attack on the subsidies they enjoy. Periodically, conservative politicians promise them new subsidies, although economists deplore the idea. The constituency for the status quo is both numerous and powerful.

Current Issues

Currently, three issues dominate the public debate about housing in Canada: (1) conditions for the mentally disabled and for Indigenous people; (2) the long-running real estate boom, which threatens affordability and financial stability, turning dreams to nightmares; and (3) effects on environment and health.

Until the 1950s, the housing problem was one of poor housing conditions and overcrowding. Since then, municipal regulations have raised standards of construction and maintenance. Today, disabled and Indigenous peoples are the main groups who commonly experience substandard or crowded housing, a fact dramatized by media coverage of conditions in Attawapiskat, and by the "Idle No More" movement in 2012–13. In 2001, only 2.5 per cent of the units occupied by non-Indigenous households was inadequate (in need of major repairs), and 1.4 per cent failed the National Occupancy Standard (NOS) (CMHC, 2004).[2] The equivalent ratios for Indigenous households off reserves were 6.8 and 4.9 per cent, respectively; for those on reserves the proportions were 22.5 and 10.3 per cent.

Mostly, however, the key issue is affordability. For individual households, the greatest challenge is felt by tenants and those on low incomes. **Core housing need** is said to exist when a household must spend at least 30 per cent of before-tax income for an adequate, uncrowded dwelling. On this criterion, 12.7 per cent of all urban Canadian households were in need in 2015, unchanged in a decade. At 19 per cent in 2011, however, the incidence of need among Indigenous people was almost double that of other Canadians (CMHC, 2014). In general, the problem of affordability is felt most urgently by immigrants and by tenants. In 2011, the incidence of core need among immigrants (17 per cent) was significantly higher than among the native born (11 per cent) (CMHC, 2014). The 2016 census revealed that while 18 per cent of homeowners spent more than 30 per cent of their income on housing, the proportion among tenants was 41 per cent.

In the longer run, affordability issues and the indebtedness of homeowners may pose a larger economic problem. The level of household debt has risen steadily since the 1980s, exceeding an unprecedented 160 per cent of **gross domestic product (GDP)** by 2010 (Walks, 2013). Exceptionally low interest rates have kept manageable the level of personal debt as a proportion of disposable income, but by November 2017 even this had risen to the unprecedented level of 171 per cent. If the economy stalls, if house prices dip (as many believe they will), and if interest rates rise (as they will eventually), many borrowers, lenders, and the CMHC, as the country's largest insurer of mortgage debt, may be in trouble. As the world learned in 2008, the fallout from a burst house price bubble affects everyone.

Canadians have viewed larger houses as a sign of progress, but researchers now challenge this view. Oil price increases in the early 1970s made energy costs an issue. The federal government funded a short-lived program to promote energy conservation in existing dwellings; in 1973, this criterion was added to the National Building Code; in 1981 the National Home Builders' Association and National Resources Canada initiated the "R2000" building program for new homes and this has helped make house-builders unusually effective in limiting **greenhouse gas emissions** (Canadian Home Builders' Association, 2012, 2013). International awareness of the links between energy consumption and global warming has grown. Oil and gas are the main fuels used for home heating, and home energy consumption is an environmental issue. A federal program helps property owners to improve energy efficiency. But most Canadian dwellings are detached and, per cubic metre, these use 80 per cent more energy than those in multi-unit structures. High-rise condominiums, and **New Urbanist** suburbs that include row dwellings, are more environmentally friendly, but carbon emissions will remain a housing-related issue in Canadian cities (Gordon and Tamminga, 2002; Grant, 2006).

Large homes and reliance on the automobile are associated with suburban sprawl that brings health and environmental problems (Frumkin, Frank, and Jackson, 2004). Homes and subdivisions save effort: we drive to the store for a carton of milk; order fast food from the driver's seat; and change channels from the couch. On average, Canadians do less, eat more, and grow more overweight and obese by the year, raising the incidence of heart disease and other health problems. Such issues can be mitigated through urban design that promotes walking, cycling, and public transit, but these require higher densities and smaller dwellings. Apart from building more affordable and energy-efficient housing, Canadians will have to change their expectations. Less is sometimes more.

Review Questions

1. What are "affordable" housing and "social" housing?
2. How can "filtering" help low-income households, and where is it happening?
3. How do federal and local governments affect the housing that Canadians occupy?

Notes

1. The number of people living in cities continues to grow, but the proportion has changed little since the 1970s.
2. The definition of the NOS is complex. See CMHC (2012: 5–16).

Recommended Websites

Canadian Housing and Renewal Association
www.chra-achru.ca/en/
Canada Mortgage and Housing Corporation
www.cmhc-schl.gc.ca/
Canadian Real Estate Association
www.crea.ca/
The Homeless Hub
http://homelesshub.ca/

References

Armstrong, P., and H. Armstrong. 1994. *The Double Ghetto*, 3rd edn. Toronto: McClelland & Stewart.

Bacher, J. 1993. *Keeping to the Marketplace: The Evolution of Canadian Housing Policy*. Montreal and Kingston: McGill-Queen's University Press.

Belanger, Y., G.W. Head, and O. Awosoga. 2012. *Assessing urban aboriginal housing and homelessness in Canada*. Ottawa: National Association of Friendship Centres and the Office of the Federal Interlocutor for Métis and Non-Status Indians. At: www.homelesshub.ca/Library/Assessing-Urban-Aboriginal-Housing-and-Homelessness-in-Canada-54375.aspx

Belec, J. 1997. "The Dominion Housing Act," *Urban History Review* 25(2): 53–62.

Belk, R. 1988. "Possessions and the extended self," *Journal of Consumer Research* 15: 139–68.

Blais, P. 2010. *Perverse Cities: Hidden Subsidies, Wonky Policy, and Urban Sprawl*. Vancouver: UBC Press.

Bourdieu, P. 1984. *Distinction. A Social Critique of the Judgment of Taste*. London: Routledge Kegan Paul.

Bourne, L.S. 1981. *The Geography of Housing*. London: Edward Arnold.

———, and T. Bunting. 1993. "Housing provision, residential development and neighbourhood dynamics," in L.S. Bourne and D. Ley, eds., *The Social Geography of Canadian Cities*. Montreal and Kingston: McGill-Queen's University Press.

Bradbury, B. 1993. Working Families: Age, Gender, and Daily Survival in Industrializing Montreal. Toronto: McClelland & Stewart.

Brown, L.A., and E.G. Moore. 1970. "The intra-urban migration process: A perspective," *Geografiska Annaler* 28: 1–13.

Buzzelli, M. 2001. "Firm size structure in North American housebuilding. Persistent deconcentration, 1945–98," *Environment and Planning A* 33: 533–50.

Canada Mortgage and Housing Corporation (CMHC). 2004. *2001 Census Housing Series Issue 6: Aboriginal Households. Research Highlight. Socio-Economic Series 04-036*. Ottawa: CMHC. At: www.cmhc-schl.gc.ca:50104/b2c/b2c/init.do?language=en

———. 2010. "Housing and the Economy," in *Canadian Housing Observer*. Ottawa: CMHC. At: www.cmhc.ca/en/corp/about/cahoob/upload/chapter2_housing_economy_2010.pdf

———. 2012. *Canadian Housing Observer*. Ottawa: CMHC. At: www.cmhc.ca/en/corp/about/cahoob/cahoob_001.cfm

———. 2014. *Characteristics of Households in Core Housing Need, Canada, Provinces, Territories and Metropolitan Areas, 2001, 2006, 2011*. Ottawa: CMHC. At: https://www.cmhc-schl.gc.ca/en/hoficlincl/homain/stda/data/data_024.cfm

———. 2017a. *Rental Market Report. Canada Highlights*. Ottawa: CMHC. At: https://www.cmhc-schl.gc.ca/odpub/esub/64667/64667_2017_A01.pdf?fr=1513882201923

———. 2017b. "The economics of new purpose-built rental housing development in selected Canadian markets," *Research Insight*. October. Ottawa: CMHC. At: https://www.cmhc-schl.gc.ca/odpub/pdf/69184.pdf?fr=1514927917322

Canadian Home Builders' Association (CHBA). 2003. *The Need for Action. Pre-Budget Submission*. Ottawa: CHBA.

———. 2012. *Recognizing the Residential Sector's Contribution to Greenhouse Gas Reduction*. Ottawa: CHBA. At: http://chba.ca/uploads/policy%20archive/2012/Housing%20and%20Canada%27s%20GHG%20Emission%20Targets.pdf

———. 2013. *Welcome Home to R2000*. Ottawa: CHBA. At: www.chba.ca/r-2000.aspx

Carter, T. 1997. "Current practices for procuring affordable housing. The Canadian context," *Housing Policy Debate* 8: 593–631.

Carter, T.S., C. Polevychok, and J. Osborne. 2009. "The role of housing and neighborhood in the re-settlement process. A case study of refugee households in Winnipeg," *The Canadian Geographer* 53: 305–22.

Choko, M., and R. Harris. 1990. "The local culture of property: A comparative history of housing tenure in Montreal and Toronto," *Annals of the Association of American Geographers* 80: 73–95.

Clark, S.D. 1966. *The Suburban Society*. Toronto: University of Toronto Press.

Clayton, F.A. 2010. *Government Subsidies to Homeowners versus Renters in Ontario and Canada*. At: www.cfaafcapi.org/pd2/Submission.Supplementary_Eng.pdf

Clayton Research Associates. 1988. *The Changing Housing Industry in Canada, 1946–2001*. Ottawa: CMHC. At: www.chba.ca/uploads/Policy%20Archive/1988/1988-ChangingIndustry.pdf

Cowan, R. 1983. *More Work for Mother: The Ironies of Household Technology from the Open Hearth to the Microwave*. New York: Basic Books.

Cox, K. 1982. "Housing tenure and neighborhood activism," *Urban Affairs Quarterly* 18: 107–29.

Fallis, G. 1993. "Postwar Changes in the Supply-Side of Housing," in J. Miron, ed., *House, Home and Community. Housing Progress in Canada since 1945*. Montreal and Kingston: McGill-Queen's University Press.

Filion, P., and M. Alexander. 1994. "Restrictive covenants: Hidden obstacles," *Plan Canada* 35(1): 33–7.

Foot, D., and D. Stoffman. 1996. *Boom, Bust, and Echo: How to Profit from the Coming Demographic Shift*. Toronto: Macfarlane, Walter and Ross.

Frumkin, H., L. Frank, and R. Jackson. 2004. *Urban Sprawl and Public Health. Designing, Planning and Building for Healthy Communities*. Washington, DC: Island Press.

Gaetz, S., E. Dej., T. Richter, and M. Redman. 2016. *The State of Homelessness in Canada 2016*. Toronto: Canadian Observatory on Homelessness. At: http://homelesshub.ca/sites/default/files/SOHC16_final_20Oct2016.pdf

Gordon, D., and K. Tamminga. 2002. "Large-scale traditional neighborhood development and pre-emptive ecosystem planning. The Markham experience, 1989–2001," *Journal of Urban Design* 7: 321–40.

Government of Canada. 2017. *Canada's National Housing Strategy. A Place to Call Home*. At: https://www.placetocallhome.ca/pdfs/Canada-National-Housing-Strategy.pdf

Grant, J. 2005. "Planning responses to gated communities in Canada," *Housing Studies* 20: 273–85.

———. 2006. *Planning the Good Community: New Urbanism in Theory and Practice*. New York: Routledge.

———, J., R.A. Walks, and H. Ramos, eds. 2018. *Changing Neighbourhoods: Social and Spatial Polarization in Canada's Cities*. Vancouver: UBC Press.

Gurstein, P. 1995. *Planning for Telework and Home-based Employment: A Canadian Survey on Integrating Work into Residential Environments*. Ottawa and Vancouver: CMHC and Centre for Human Settlements, University of British Columbia.

Haan, M. 2005. *The Decline of the Immigrant Homeownership Advantage. Life Cycle, Declining Fortunes and Changing Housing Careers in Montreal, Toronto and Vancouver, 1981–2001*. Ottawa: Statistics Canada. At: http://dspace.cigilibrary.org/jspui/bitstream/123456789/301/1/The%20Decline%20of%20the%20Immigrant%20Homeownership%20Advantage.pdf?

Hackworth, J. 2008. "The durability of roll-out neoliberalism under centre-left governance. The case of Ontario's social housing sector," *Studies in Political Economy* 81: 7–26.

Hamilton (City of). 2004. *Keys to the Home. A Housing Strategy for Hamilton*. Hamilton: The City. At: www.hamilton.ca/NR/rdonlyres/0ED608D0-7D8F-4CBF-83F5-B6B88102A689/0/KeystotheHome.pdf

Hanson, S., and G. Pratt. 1995. *Gender, Work, and Space*. New York: Routledge.

Harris, D.C. 2011. "Condominium and the city. The rise of property in Vancouver," *Law and Social Inquiry* 36: 694–726.

Harris, R. 1994. "The flexible house: The housing backlog and the persistence of lodging, 1891–1951," *Social Science History* 18: 31–53.

———. 2000. "More American than the United States: Housing in urban Canada in the twentieth century," *Journal of Urban History* 26: 456–78.

———. 2003. "The suburban worker in the history of labor," *International Labor and Working Class History* 64: 8–24.

———. 2004. *Creeping Conformity. How Canada Became Suburban, 1900–1960*. Toronto: University of Toronto Press.

———, and M. Buzzelli. 2003. "Small is transient. House-building firms in Ontario, Canada 1978–1998," *Housing Studies* 18: 369–86.

———, and M. Buzzelli. 2005. "House building in the machine age, 1920s–1970s. Realities and perceptions of modernisation in North America and Australia," *Business History* 47(1): 59–85.

———, and K. Kinsella. 2017. "Secondary suites. A survey of evidence and municipal policy," *The Canadian Geographer* 61(4): 493–509.

Hiebert, D. 2017. "Immigrants and refugees in the housing markets of Montreal, Toronto and Vancouver, 2011," *Canadian Journal of Urban Research* 26(2): 52–78.

Hodge, G., and D. Gordon. 2013. *Planning Canadian Communities*. Toronto: Nelson.

Holdsworth, D. 1977. "House and home in Vancouver: Images of West Coast urbanism, 1881–1929," in A.F.J. Artibise and G. Stelter, eds., *The Canadian City: Essays in Urban History*. Toronto: McClelland & Stewart.

Houle, P., M. Turcotte, and M. Wendt. 2017. *Changes in Parents' Participation in Domestic Tasks and Care for Children from 1986 to 2015*. Ottawa: Statistics Canada. At: www.statcan.gc.ca/pub/89-652-x/89-652-x2017001-eng.pdf

Jackson, P. 1999. "Commodity cultures: The traffic in things," *Transactions, Institute of British Geographers* 24: 95–108.

Jarvis, H., A.C. Pratt, and P. Wu. 2001. *The Secret Life of Cities: The Social Reproduction of Everyday Life*. Harlow: Prentice-Hall.

Kataure, V., and M. Walton-Roberts. 2013. "The housing preferences and location choices of second-generation South Asians living in ethnic enclaves," *South Asian Diaspora* 5 (1): 57–76.

Kern, L. 2010. *Sex and the Revitalized City. Gender, Condominium Development, and Urban Citizenship*. Vancouver: UBC Press.

Krohn, R.G., B. Fleming, and M. Manzer. 1977. *The Other Economy: The Internal Logic of Local Rental Housing*. Toronto: Peter Martin.

Lauster, N. 2016. *Death and Life of the Single Family House*. Philadelphia: Temple University Press.

Lehrer, U., and T. Wieditz. 2009. "Condominium development and gentrification. The relationship between policies, building activities and socio-economic development in Toronto," *Canadian Journal of Urban Research* 18: 82–13.

Leslie, D., and S. Reimer. 2003. "Gender, modern design, and home consumption," *Environment and Planning D Society and Space* 21: 293–316.

Ley, D. 1997. *The New Middle Class and the Remaking of the Central City*. New York: Oxford University Press.

———. 2010. *Millionaire Migrants. Trans-Pacific Life Lines*. Chichester: Wiley-Blackwell.

McClain, J., and C. Doyle. 1983. *Women as Housing Consumers*. Ottawa: Canada Mortgage and Housing Corporation.

McKellar, J. 1993. "Building technology and the production process," in J. Miron, ed., *House, Home and Community: Housing Progress in Canada since 1945*. Montreal and Kingston: McGill-Queen's University Press.

MacKenzie, S., and D. Rose. 1983. "Industrial change, the domestic economy and home life," in J. Anderson et al., eds., *Redundant Spaces in Cities and Regions*. London: Academic Press.

MacNaughton, E., et al. 2017. *Final Cross-Site Report on Housing First Implementation in Six Canadian Communities*. Waterloo, ON: Wilfrid Laurier University. At: http://homelesshub.ca/sites/default/files/PHSI%20Cross-Site%20Case%20Study%20Final%20July%2025%202017.pdf

Michelson, W. 1977. *Environmental Choice, Human Behavior and Residential Satisfaction*. New York: Oxford University Press.

———. 1984. *From Sun to Sun: Daily Obligations and Community Structure in the Lives of Employed Mothers and Their Families*. Totowa, NJ: Rowman and Allanheld.

Miller, D. 1987. *Material Culture and Mass Consumption*. Oxford: Blackwell.

Miron, J. 1988. *Demographic Change, Household Formation and Housing Demand: Canada's Postwar Experience*. Montreal and Kingston: McGill-Queen's University Press.

———. 1993. "Demographic change, household formation and housing demand: Canada's postwar experience," in J. Miron, ed., *House, Home and Community. Housing Progress in Canada since 1945*. Montreal and Kingston: McGill-Queen's University Press.

Mitchell, B., and E.M. Gee. 1996. "Young adults returning home: Implications for social policy," in B. Galaway and J. Hudson, eds., *Youth in Transition: Perspectives on Research and Policy*. Toronto: Thompson.

Mok, D. 2007. 'Do two-earner households base their choice of residential location on both incomes?' *Urban Studies* 44(4): 723–50.

Moos, M. 2014. "'Generationed' space: Societal restructuring and young adults' changing location patterns," *The Canadian Geographer* 58(1): 11–33.

Morisset, L.K., and Noppen, L. 2004. "Le bungalow Québécois, monument vernaculaire. De l'espace urbain a l'identité domestique," *Cahiers de géographie du Québec* 48: 127–54.

Murdie, R.A. 1986. "Local strategies in resale home financing in the Toronto housing market," *Urban Studies* 28: 465–83.

———. 1994. "Blacks in near-ghettoes? Black visible minority population in Metropolitan Toronto Housing Authority public housing units," *Housing Studies* 9: 435–58.

Paterson, R. 1984. "The development of an interwar suburb: Kingsway Park, Etobicoke," *Urban History Review* 13: 225–35.

Patterson, J. 1993. "Housing and community development policies," in J. Miron, ed., *House, Home and Community. Housing Progress in Canada since 1945*. Montreal and Kingston: McGill-Queen's University Press.

Peters, E. 2012. "'I like to let them have their time.' Hidden homeless First Nations people in the city and their management of household relationships," *Social and Cultural Geography* 13: 321–38.

Poapst, J.V. 1993. "Financing of postwar housing," in J. Miron, ed., *House, Home and Community. Housing Progress in Canada since 1945*. Montreal and Kingston: McGill-Queen's University Press.

Pomeroy, S., and N. Falvo. 2013. "Pragmatism and political expediency: Housing policy under the Harper regime," in G.B. Doern and C. Stoney, eds., *How Ottawa Spends*. Kingston and Montreal: McGill-Queen's University Press.

Relph, E.C. 1987. *The Modern Urban Landscape*. Baltimore, MD: Johns Hopkins University Press.

Report of the President's Committee on Urban Housing (the Kaiser Committee). 1968. *A Decent Home*. Washington: President's Committee on Urban Housing.

Rose, D. 1989. "A feminist perspective on employment restructuring and gentrification: The case of Montreal," in J. Wolch and M. Dear, eds., *The Power of Geography: How Territory Shapes Social Life*. Boston, MA: Unwin Hyman.

———, and C. Le Bourdais. 1986. "Changing conditions of female single parenthood in Montreal's inner city and suburban neighborhoods," *Urban Resources* 3(2): 45–52.

———, and M. Wexler. 1993. "Post-war social and economic changes and housing adequacy," in J. Miron, ed., *House, Home and Community. Housing Progress in Canada since 1945*. Montreal and Kingston: McGill-Queen's University Press.

Rosen, G., and A. Walks. 2013. "Rising cities. Condominium development and the private transformation of the metropolis," *Geoforum* 49: 160–72.

Rowe, A. 1989. "Self-help housing provision: Production, consumption, accumulation and policy in Atlantic Canada," *Housing Studies* 4: 75–91.

Sadella, E.K., B. Verschure, and J. Burroughs. 1987. "Identity symbolism in housing," *Environment and Behavior* 19: 569–87.

Schiff, R., J.W. Schiff, and B. Schneider. 2010. "Housing for the disabled mentally ill: Moving beyond homogeneity," *Canadian Journal of Urban Research* 19: 108–28.

Shapcott, M. 2004. "Where are we going? Recent federal and provincial housing policy," in J.D. Hulchanski and M. Shapcott, eds., *Finding Room: Policy Options for a Canadian Rental Housing Strategy*. Toronto: University of Toronto, Centre for Urban and Community Studies.

Simon, J., and D. Holdsworth. 1993. "Housing form and use of domestic space," in J. Miron, ed., *House, Home and Community. Housing Progress in Canada since 1945*. Montreal and Kingston: McGill-Queen's University Press.

Simone, D., and A. Walks. 2019. "Immigration, race, mortgage lending, the geography of debt in Canada's global cities," *Geoforum* 98: 286–99.

Sinha, M. 2014. *Child Care in Canada*. Ottawa: Statistics Canada. At: www.statcan.gc.ca/pub/89-652-x/89-652-x2014005-eng.pdf

Skaburskis, A. 2012. "Gentrification and Toronto's changing household characteristics and income distribution," *Journal of Planning Education and Research* 32(2): 191–203.

Skelton, I. 1994. "The geographic distribution of social housing in Ontario, Canada: Comparing public housing and locally sponsored, third sector housing," *Housing Studies* 11: 189–206.

Steele, M. 1993. "Incomes, prices and tenure choice," in J. Miron, ed., *House, Home and Community. Housing Progress in Canada since 1945*. Montreal and Kingston: McGill-Queen's University Press.

Statistics Canada. 2012. *Canada Year Book*. Ottawa: Statistics Canada. At: https://www150.statcan.gc.ca/n1/pub/11-402-x/2012000/chap/construction/construction-eng.htm

———. 2017. *Census in Brief. Children living in low-income households*. Ottawa: Statistics Canada. At: https://www12.statcan.gc.ca/census-recensement/2016/as-sa/98-200-x/2016012/98-200-x2016012-eng.cfm

Strong-Boag, V. 1991. "Home dreams: Women and the suburban experiment in Canada, 1945–60," *Canadian Historical Review* 72: 471–504.

———, K. England, and L. Johnson. 1999. "What women's spaces? Women in Australian, British, Canadian and US suburbs," in R. Harris and P. Larkham, eds., *Changing Suburbs: Foundation, Form, and Function*. London: E and FN Spon.

TD Economics. 2013. Toronto—a return to the core. Toronto: TD Bank. At: www.td.com/document/PDF/economics/special/ff0113_toronto.pdf

Teixeira, C. 2006. "Residential experiences and the culture of suburbanization: A case study of Portuguese homebuyers in Mississauga," *Housing Studies* 22: 495–521.

Turcotte, M. 2011. *Working at home: An update*. Canadian Social Trends No. 91. Ottawa: Statistics Canada. At: www.statcan.gc.ca/pub/11-008-x/2011001/article/11366-eng.pdf

Wachsmuth, D. and A. Weisler. 2018. "Airbnb and the rent gap. Gentrification through the sharing economy," *Environment and Planning*. doi: 10.1177/0308518X18778038.

Walks, A. 2013. "Canada's housing bubble story: Mortgage securitization, the state and the global financial crisis," *International Journal of Urban and Regional Research* 38(1): 256–84.

———. 2015. *The Urban Political Economy and Ecology of Automobility. Driving Cities, Driving Inequality, Driving Politics*. New York: Routledge.

Walton-Roberts, M., and D. Hiebert. 1997. "Immigration, entrepreneurship, and the family. Indo-Canadian entrepreneurship in the construction industry of Greater Vancouver," *Canadian Journal of Regional Science* 20: 119–40.

Watson, S. 1986. *Housing and Homelessness: A Feminist Perspective*. London: Routledge and Kegan Paul.

Weaver, J. 1978. "From land assembly to social maturity: The suburban life of Westdale (Hamilton), Ontario, 1911–1951," *Histoire sociale/Social History* 11: 411–40.

Wolfe, J. 1998. "Canadian housing policy in the 1990s," *Housing Studies* 13: 121–34.

Yakabuski, K. 2004. "Bob the builder," *Report on Business* October: 84–92, 94, 96.

III Urban Form, Structure, and Design

Why do cities look the way they do? Why do they take a particular physical form? The five chapters in the third part of this book all consider the changing physical characteristics of cities in one way or another. Moving beyond changes in the physical appearance of cities, the chapters consider the reasons for these changes and their implications for society at large. The first chapter in this part of the book, Chapter 12, examines emerging urban forms. The chapter emphasizes the processes that are shaping new urban forms and the regulatory context within which cities change. They highlight the growing focus in urban policy on creating mixed-use centres, residential growth in the downtowns of most major metropolitan areas, as well as continued low-density, auto-oriented suburban expansion, such as power centres and office parks.

Chapter 13 considers the characteristics and diverse roles of public spaces. The chapter highlights the role of squares, streets, and sidewalks for public celebrations, protest, and commerce, urging readers to think about public space as more than simply spaces we move through on our way somewhere else. Readers learn about how public space functions, the ways in which we can measure these functions, and the physical design and programming of public space in select Canadian cities.

Chapter 14 deals with transport. Readers learn to think about cities as places that facilitate accessibility to diverse opportunities. Accessibility is defined as the ease of travel and the destinations that can be reached from a particular location. Accessibility is therefore a function of both the transport system, as well as land-use patterns. Chapter 14 details the history and current state of Canada's urban transport systems, highlighting modal shifts and increasing attention by policy-makers to the sustainability of transport patterns. Basics of travel demand modelling used in transport planning and evaluation are introduced.

Chapter 15 zeroes in on the inner city, which has changed substantially over recent decades due to cultural and economic transformations. The chapter details the growing focus on providing arts and cultural amenities in inner cities to attract highly skilled workers, and the resulting growth in condominium apartments. Readers are urged to think about who can live in the revitalized inner city and who has access to amenities. Readers will learn about gentrification and how it unfolds in specific cases, and its implications for vulnerable populations.

Finally, Chapter 16 focuses on the outer parts of the Canadian cities, the suburbs. Readers learn that the majority of Canada's urban population actually lives in suburban places. Suburbanization is defined as a process of spatial expansion at the city's edge. The authors differentiate *suburbanization* from *suburbanisms*—the latter defined as particular ways of living that take distinct forms in different times and places. Chapter 16 details the factors that have contributed to the historic and ongoing suburbanization in different regions of Canada. Readers learn about the political changes in suburbs, the growing diversity of suburban landscapes (for instance, through immigration), and the emergence of "post-suburbia," which is neither like the historic inner city nor the stereotypical, cookie-cutter suburbs.

After reading the chapters in Part III of this book, readers will have understanding of

- the changing physical form of cities and the reasons for these changes;
- the role and uses of public spaces;
- the characteristics of Canada's urban transport system and the factors shaping accessibility;
- the factors shaping inner-city change and how they contribute to growing inequality and exclusion; and
- the changing characteristics of suburbs, the nature of suburbanization, and the growing diversity of Canadian suburban life.

12 The Market, Planning, and Emerging Urban Forms

Jill L. Grant and Pierre Filion

Introduction

The present chapter is about urban innovations, which begs the question: innovations relative to what? Many of the emerging **urban forms** we document are reactions to the domination of twentieth-century urban development by modernist planning. The last century was indeed marked by an all-out effort to adapt cities to the car and to take advantage of automobile-induced accessibility to reduce urban density and separate land uses—often in large monofunctional zones—so as to minimize negative externalities. The most advanced form of automobile-focused planning is found in suburbs developed according to the super block model, structured by high-capacity arterials. It is difficult in such areas to depart from low-density, spatially segregated, automobile-dependent land uses because of the deeply embedded dynamic between land use and car reliance.

Since the 1960s, planning has reacted with growing intensity against the modernist automobile-oriented planning model. Over recent decades, in responding to concerns about the environment, quality of life (e.g., long automobile commutes), the health effects of sedentary lifestyles, and development costs, city planners have proposed forms of urban development that reflect **smart growth** and sustainable urbanization principles. Proposals voiced in recent plans are largely antithetical to development patterns promoted by modernist automobile-oriented planning. They indeed mark a shift to the promotion of higher density and multi-functionality, and of public transit, cycling, and walking. In reaction to a trend towards decentralized urbanization, recent plans recommend a recentralization of urban areas. Some of the emerging urban forms described in the chapter relate to these planning objectives.

But planning attempts to achieve smart growth and sustainable urbanization are not the sole source of new urban forms. Other such forms consist of refinements to car-oriented patterns of development, which accentuate rather than attenuate automobile dependency. In this chapter we connect urban form innovations to society-wide technical, value, and demographic trends as filtered through planning processes and the market. We attribute diverging trajectories taken by the emerging urban forms we describe in this chapter to contradictions in these trends and tensions between planning and market dynamics.

Factors Contributing to Urban Change

We might expect components of the city that are sold or rented to behave like other consumer products, which are the object of constant efforts at innovation in response to changing circumstances or simply to cravings for novelty (Baudrillard,

1994; McDonald and McMillen, 2007). Cities are inherently stable because innovations must jibe with infrastructure networks and dominant technologies while conforming to regulations, such as building codes, zoning bylaws, and infrastructure standards. Moreover, the home-purchasing process itself induces conservatism (Lipson, 2006; Steacy, 1987); a home is not just a commodity but also an investment. Accordingly, the cautious homebuyer considers the potential preferences of the subsequent purchaser who will determine the house's resale value. Lending institutions, moreover, shy away from unusual house designs, preferring proven styles. Buzzelli and Harris (2006) suggest that local builders tend to avoid innovations that may increase their risks, and Harris (2004) documented the role of the Canada Mortgage and Housing Corporation (CMHC) in standardizing suburban development after the 1940s.

These circumstances account for a rate of urban change that lags behind most consumer products. While consumers typically replace computers and cars relatively frequently, houses last many decades, even centuries. In cities, circumstances favour urban stability, not innovation. Yet major forces of change can challenge the inherent inertia of cities.

Over the past decades, economic transformations have had profound impacts on cities. For example, the decline in manufacturing changed demand for land in urban areas while altering employment structure. Rising consumption of services represents a related economic change with powerful urban reverberations (Gallouj, 2002; Industry Canada, 1996). The mushrooming of restaurants, gyms, health centres, and countless other categories of service establishments reveals the effects of the transition. In many cities, rising levels of service consumption contribute to the growing appeal of "revitalized" residential areas in the core and the inner city (Meligrana and Skaburskis, 2005; Skaburskis and Moos, 2008). Deepening income **polarization** resulting from the loss of middle-class jobs and reduced transfer payments is having a dramatic effect on the social geography of Canadian cities, as the number of middle-class neighbourhoods dwindles (Hulchanski, 2007; Walks and Bourne, 2006; Walks and Maaranen, 2008).

In the competitive context of urban development in high-growth areas, some developers and builders responded by targeting projects at niche markets. New urban forms and practices have emerged as developers have looked for ways to reduce development costs, achieve economies of scale, control the chain of production, and improve the marketability of projects. To lower the costs of urban infrastructure, some developers turn to building private streets and services sold to occupants through condominium corporations. Such economic factors contribute to new patterns of social and spatial sorting in urban landscapes.

Technological innovations have been at the core of twentieth-century urban transformations, which consisted mostly of the adaptation of the city to the car. Things are not as clear in the present century, as attempts to depart from car-oriented urban development are not accompanied by the presence on the horizon of an alternative to the car. In fact, the main technological innovation presently under development is a variant of the car—the autonomous vehicle. The revolution in information technology that has taken place over recent decades has had surprisingly little transformative effect on urban patterns. This may be due to the fact that to reach their full impact, technological innovations must be associated with attendant organizational innovations. In this sense, we may be on the cusp of major urban transformations as organizational formulas making use of information technologies are refined. Consider the impact that Uber and similar platforms are having on urban transportation, the effect of online shopping on brick-and-mortar retailing, and the rapid rise of the sharing economy (Moulaert, 2009; Sundararajan, 2016).

Changing cultural values and practices have had significant effects on urban form over the past two or three decades. Demographic shifts induced some forms of urban innovation. Over the period under consideration, several demographic trends had a major impact on housing (Bourne and Rose, 2001; Foot with Stoffman, 1996; Isaacs et al., 2007). First, the demographic structure changed as birth rates fell and longevity increased, leading to aging of the population (see Moos, Chapter 6; Townshend and Walker, Chapter 7). Second, household size dropped steadily as more people chose to live alone and divorce rates increased, although household size has plateaued and even increased in high-cost residential markets (Statistics Canada, 2017). Third, immigration levels increased with more immigrants coming from Asia and other parts of the world; these immigrants were accustomed to high-density living although many may have come to this country to partake in the Canadian dream perceived as consisting of access to a single-family home and one or more cars (see Kobayashi and Preston, Chapter 8). These trends influenced options for recent forms of residential development. For instance, households with no children or with older members constitute a significant proportion of the target market for gated communities (Maxwell, 2004). As well, the popularity of condominium developments during the last few decades owes a debt to the growth in small and immigrant households.

Consumer preferences reflect cultural values. While many Canadians continue to look for homes in car-oriented conventional suburbs, recent decades have shown that a segment of the population shifted its priorities. Urbanism—a commitment to participating in urban life—inspired a resurgence of interest in downtown and inner-city living (Meligrana and Skaburskis, 2005; Skaburskis, 2006; see Bain and Mark, Chapter 15). These consumers want specialty services like coffee shops; they seek access to public transportation; they willingly give up residential space for urban amenities (the city is their living room); they drive the growth in the development of mixed nodes. Other households appreciate elements of traditional towns and provide a market for **New Urbanism** communities. Households seeking privacy and security may purchase homes in private or gated enclaves. The greater diversity in values associated with residential choices supports the trend toward niche markets and specialization in urban forms.

In the immediate postwar period, consumers and governments seemed to believe that energy and materials were endless, technology could solve any problem, and the future promised unlimited opportunities. For many Canadians, that optimism has succumbed to concerns about pollution, climate change, dwindling fossil fuels, economic recession, urban gridlock, and health risks from urban living (see Rees, Chapter 21). Cultural values associated with environmentalism—commitment to environmental protection and appreciation—have increasingly influenced public policy and local planning (see Holden and Chang, Chapter 22; Senbel, Liem, and Lesnikowski, Chapter 24). Environmentalism supports efforts to limit land consumption and to reduce reliance on the automobile. Recent trends to protect waterways, wetlands, and forested areas reflect, in part, the shift to these values. Urban planning gives increasing attention to climate change mitigation and adaptation, especially in urban areas that are increasingly susceptible to flooding. At the same time, however, Canadians have proven remarkably resistant to the idea of giving up their cars and trucks.

Public Policy Responses

Governments respond to prevailing economic and cultural trends. As early as the 1970s, cities like Toronto, Vancouver, and Ottawa were adopting policies promoting urban **intensification** and environmental protection. Those policies took some time—and changing economic conditions—to influence urban outcomes. Over the past 40 years, governments have strengthened policies and improved implementation mechanisms. Dominant philosophies of what it takes to make cities and economies thrive shifted: whereas economic development strategies once recommended land-use segregation, decongestion, and garden suburbs, contemporary development strategies advocate density and mixed uses. Spatial planning at a large scale—including amalgamated city-regions—is making a comeback as urban regions seek to improve their competitiveness in a global market.

Laws, policies, and regulations have influenced urban form outcomes in various ways. For instance, one legal innovation had a profound effect on urban form outcomes over the past decades. In the late 1960s and 1970s, provincial governments across Canada adopted legislation to permit condominium tenure—a form of property ownership that allows people to own units within shared property managed by a corporation of which they are members. By the 1980s, condominium tenure had become the predominant form of high-density residential development (Hulchanski, 1988). High housing costs and declining household sizes confirmed the trend during the 1990s and 2000s (see Harris, Chapter 11). And by 2016, 13.3 per cent of Canadian households lived in condominiums, with a peak of 30 per cent in the Vancouver census metropolitan area (CMA) (Canadian Broadcasting Corporation [CBC], 2017). With 31.5 per cent of housing starts in 2016 in condominium tenure, the potential for private residential enclaves—whether gated communities in the suburbs, private courts in redeveloped urban **brownfield sites,** or high-rise towers near transit stations—was growing rapidly (CMHC, 2016).

The rise of **neo-liberal** philosophies in the 1980s and 1990s encouraged governments to change their ways of managing urban development (see Taylor and Bradford, Chapter 3). Many municipalities transferred responsibility for building urban infrastructure to developers, who then passed on the costs to purchasers. To reduce the cost of roads in large development sites zoned for multi-family housing, developers often created bare-land condominium projects with attached or free-standing homes sharing private access streets. The condominium projects offered the additional attraction to municipalities that residents would remain responsible for road maintenance, snow clearing, and garbage collection. In an era when municipalities were struggling to meet their financial obligations while keeping property taxes as low as possible, private communities proved attractive.

By the 1990s plans and policies in most Canadian cities promoted greater densities, compact form, multi-functionality, mixed housing, environmental protection, quality urban design, and transportation options to promote the values of efficiency, competitiveness, and diversity. Demonstration projects and master-planned developments provided opportunities to introduce innovative planning and design practices, like complete communities. Responding to economic and demographic pressures and to changing values on the part of the public, governments adopted plans that called for urban intensification and reduced dependence on the car (e.g., Metro Vancouver, 2007; Ontario, 2006). At the same time, however, market forces continued to fight for space for auto-oriented development forms that met the needs of business. The objectives of plans were only partly achieved. Heavy dependence on the automobile persists, and so-called complete communities generally fail to attain their social-mix goals.

The Emergence of New Urban Forms

In response to the factors driving urban change, we see several important emerging urban forms in Canada. The forms differ according to three criteria:

1. the extent to which they mix a range of uses or tend toward reproducing a segregated pattern of uses;
2. the degree to which commercial or residential uses dominate the mix of uses; and
3. the location within the city (urban core or suburban fringe).

Figure 12.1 illustrates the new forms that are emerging. New forms cluster in two general types: predominantly mixed use (left side of graphic), and predominantly segregated use (right side). Public policy and planning agendas advocate mixing uses; public investment in infrastructure in urban areas experiencing rapid growth promotes mixed-use urban forms. The new, segregated-use patterns reflect the persistent resistance of economic and market forces, even in the face of policy efforts, to shifts away from auto-dependency.

Within the general types are further categories of new forms that differ according to the types of uses that dominate within them, and depending on where they are in the city. Mixed-use projects include mixed nodes and complete communities. Segregated-use forms include auto-oriented pods and private communities.

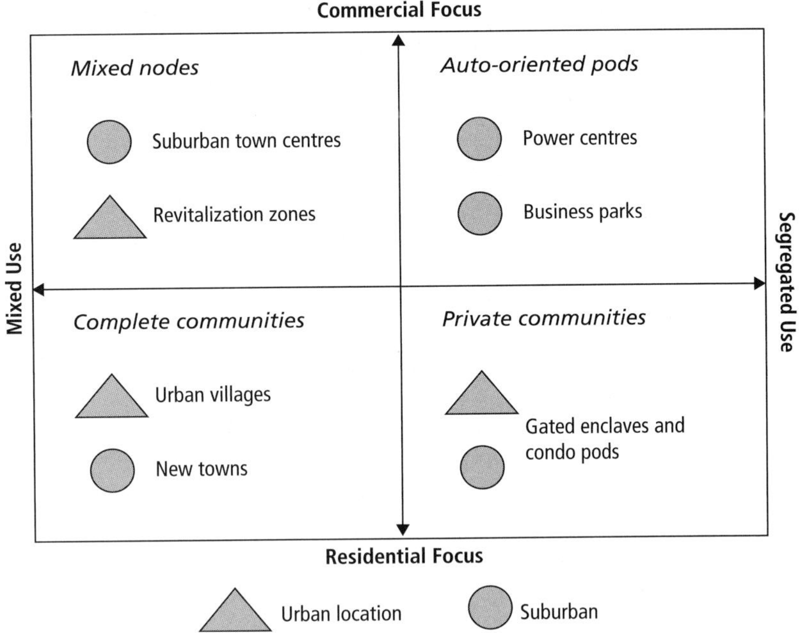

Figure 12.1 Emerging New Urban Forms

New urban forms vary by the degree of integration of uses, by the type of use that dominates the mix, and by location in the city.

Mixed-Use Forms

Governments seek to integrate uses by enabling **mixed-use development.** Mixed-use forms reflect the philosophy of Jane Jacobs (1961) and the growing influence of sustainable development, New Urbanism, and smart-growth principles in urban planning. Mixed-use projects are most common in high-growth areas seeking to facilitate higher-density land use and transportation options. In some cases, the projects reuse parts of cities hollowed out by the decline in manufacturing. Most projects combine residential and commercial uses, and sometimes add civic uses. Uses may be vertically integrated (within the building) and/or horizontally mixed (within the block). In mixed nodes, commercial uses represent the more common use, with residential being ancillary. In complete communities, residential uses tend to occupy more territory, but commercial and other uses are planned to provide services needed for daily living.

Mixed Nodes

In this chapter, "nodes" are understood as medium- to high-density multi-functional developments, usually with planned access to public transportation networks. Plans see nodes as transit-oriented developments linked to each other by high-quality public transportation networks, an objective that is, however, generally unmet. The planning and implementation of nodes stem from the adherence of governments since the 1970s to urban intensification in an effort to manage sprawl and accommodate more residents and uses within developed areas. The public sector thus played a key role in launching nodes.

We consider two types of nodes: suburban and downtown. Suburban nodes (variously labelled suburban town centre, sub-centre, suburban downtown, or regional city centre) purport to reproduce the dynamics found in successful traditional downtowns, characterized by intense pedestrian-based interaction between diverse land uses. The second type of nodal development involves the premier node of major metropolitan regions, the downtown area. We demonstrate how the large-scale redevelopment of abandoned or underutilized industrial, commercial, or institutional sites recently enhanced the density and diversity of downtowns. These efforts enriched downtowns with new residential, recreational, and cultural uses.

Suburban Town Centres

Suburban town centres are planned intensification nodes located in large suburbs on the periphery of major city-regions. Local authorities attempt to mimic the dynamics of traditional downtowns, typically by including retail, office space, public sector institutions and services, and housing uses. To encourage public transit patronage, nodes may include rail transit stations and/or assume the role of bus service hubs. Their multi-functionality and density distinguish these nodes from the suburban low-density and segregated land-use norm (see Addie, Fiedler, and Keil, Chapter 16).

The nodal approach is a keystone of metropolitan region planning across North America, often linked to the ideal of transit-oriented development (Filion, Kramer, and Sands, 2016). Calgary and Edmonton, for example, promote transit-oriented development around certain stations of their light-rail systems, and Ottawa does the same for stations on its Transitway, a system of roads for the exclusive use of buses, much of which is currently being upgraded to light-rail status (Calgary, 2008; Edmonton, 2009; Ottawa, 2007). Transit-oriented development remains in its initial stages in Canada, although suburban town centres can be perceived as large transit-oriented developments when they are spurs on high-capacity public transit networks.

Virtually all North American urban regions adopt some version of the nodal model in their metropolitan-scale planning. Toronto and Vancouver, where regional administrations have pursued a nodal strategy for over 25 years, are among those North American metropolitan regions where nodal development is most advanced (Metro Vancouver, 2008; Ontario, 2006; Raad, 2006).

Toronto features three major suburban nodes—North York City Centre, Scarborough Town Centre, and Mississauga City Centre—as well as many smaller ones (Figure 12.2). Additional major nodes, Markham City Centre and Vaughn Metropolitan Centre, are growing quickly. In the Vancouver metropolitan region, eight major nodes or regional town centres are at different stages of development: Lonsdale, Richmond City Centre, Metrotown, New Westminster, Coquitlam Town Centre, Surrey Centre, Maple Ridge Town Centre, and Langley Town Centre. The approach is different in Montreal, where the current metropolitan plan proposes intensified development in 155 identified transit-oriented development sites (Communauté Métropolitaine de Montréal [CMM], 2012). In contrast with the Toronto and Vancouver metropolitan strategies, there is no attempt at concentrating high-density and mixed-use development in a few major suburban centres so as to achieve a critical mass.

Factors accounting for the development of suburban nodes include the presence of rail public transit, a large regional shopping mall, planning support, and the economic performance of the metropolitan region.

How well do the suburban centres function? Toronto's major nodes meet some of their objectives but not to the extent anticipated in planning documents (Filion, 2007). Over the past 30 years, a near-stalling of high-density office development has threatened their multi-functionality. Meanwhile, residential development, in the form of high-rise condominiums, has boomed since the early 2000s, causing an imbalance between an expanding residential function and a stagnating employment base. Moreover, while suburban nodes offer more transit options than other suburban areas, workers and residents remain dependent on automobiles. Nodes suffer from Canadians' enduring reliance on driving for shopping, and from poor access to public transit in the suburban environments that constitute local catchment areas. Indeed, much of the space in nodes serves the automobile, with deleterious effects on walking: for example, office workers within Scarborough Town Centre and Mississauga City Centre do not patronize local restaurants as much as they could

Figure 12.2 Mississauga City Centre

Like many suburban nodes, Mississauga City Centre developed around a regional mall and its parking area.

if large expanses of surface parking did not separate offices from restaurants (Filion, 2001). The best-laid plans to induce mixed use and to reduce auto-oriented development patterns continue to face significant challenges in practice. For example, such plans often assume that the achievement of a walkable and multi-functional environment will generate pedestrian movement within suburban centres. In reality, despite favourable walking conditions, many will still rely on the automobile in order to access a wider range of activities than those present in the centre.

Downtown Revitalization Zones

The traditional postwar image of the downtown entailed a **central business district (CBD)** occupied exclusively by offices, retailing, and ancillary activities. While zoning contributed to land-use specialization, market forces drove the process. Until recently, office buildings, retail uses, and hotels outbid other activities and thus collectively took over downtowns. Middle- and upper-income households migrated to uniformly residential suburban districts (Murphy, 1972). Downtown Calgary comes close to retaining the conventional CBD model, although residential development has recently dotted its edges. The financial districts of Toronto and Montreal (where the highest office buildings are found) remain highly specialized (Gad, 1991). Meanwhile, other areas have moved increasingly to a nodal model of mixed uses downtown.

Planning attempts to depart from the CBD model began in Toronto in the 1970s as the city responded to the infrastructure overload caused by a massive wave of office development. Authorities attributed worsening traffic and transit congestion to the growing concentration of downtown employment. Planners argued that housing more residents downtown would reduce in-bound commuting. Downtown residents could walk to work and other core area activities. The presence of housing would ensure that the downtown would remain lively in the evenings rather than emptying after office hours (Toronto, 1974). Developers received bonuses allowing them to build higher and larger buildings if they included housing components in their projects. As a result, the mid- to late-1980s witnessed a surge in downtown residential development, followed by massive condominium construction in the 2000s. Railway and industrial lands left fallow for decades now feature condominium developments.

Vancouver also actively encouraged high-density housing development in its core area. It first allowed high-rise residential redevelopment in the West End, an amenity-rich neighbourhood between Stanley Park and downtown. High-rise apartment construction proceeded apace in the 1960s and 1970s. A second high-rise residential boom occurred in the 1990s with redevelopment of the Expo 86 site and of other waterfront land adjacent to downtown (Berelowitz, 2005; Punter, 2003). New high-rise residential buildings spread through the downtown. Vancouver, like Toronto, registers accelerated downtown residential growth. Its 2016 downtown population approximates 110,000 (Table 12.1). Downtown Vancouver developed its own style of high-density residential development, referred to as "Vancouverism": narrow glass towers of condominium units sit on podiums that provide three- or four-storey facades at street level. Ground-level uses include townhouses, stores, cafés, and restaurants. Public policy requires that developers provide affordable housing and family units in downtown condominium structures. In some cases, new downtown schools have been integrated in the developments. Downtown Vancouver arguably offers one of the most livable high-density residential formulas (Punter, 2003) (Figure 12.3).

Downtown residential intensification strategies have largely achieved their planning objectives in the largest cities. Downtown nodes have become livelier, with a wide range of activities (including supermarkets) that cater to the growing resident population. Also, transportation objectives have been met. In downtown Toronto, for example, 35.6 per cent of home-based journeys are made on foot, more than combined auto-driver and auto-passenger modal shares (29.2 per cent) (Joint Program in Transportation, 2011). Fostering downtown residential concentrations thus offers a clear environmental benefit in reduced vehicle use. Lively downtowns attract young professionals and the employers seeking their services: some indicators suggest that the kinds of businesses and workers occupying downtown offices differ from those selecting suburban highway-related locations (Gad, 1985; Ley, 1996).

Table 12.1 Downtown Population in Large Canadian Census Metropolitan Areas (CMAs) (with a 2016 Population Exceeding One Million), 1971–2016*

CMA	1971	1991	2001	2011	2016	1971–1991	1991–2011	2001–2016	1971–2016
Toronto	63,110	81,646	111,122	150,967	185,929	29.4%	84.9%	67.3%	194.62%
Montreal	95,619	54,747	80,378	91,002	99,800	−42.7%	66.2%	24.2%	4.4%
Vancouver	42,499	34,442	51,340	100,187	110,107	−19%	190.9%	114.5%	159.1%
Ottawa	92,737	73,972	79,506	78,917	82,425	−20.2%	6.7%	3.7%	−11.1%
Calgary	68,742	66,604	70,263	77,211	83,915	−3.1%	15.9%	19.4%	22.1%
Edmonton	39,041	27,594	36,005	44,236	45,019	−29.3%	60.3%	25%	15.3%

*Includes census tracts mostly located within 2 km of the peak value intersection.

Source: Compiled by authors from 1971, 1991, 2001, 2011, and 2016 Statistics Canada census data.

Pierre Filion

Figure 12.3 High-rise condominiums, Vancouver.
Vancouver has experienced massive downtown residential growth, much of it in the form of high-rise condominium buildings.

However, urban nodes are not without problems. Despite Vancouver's attempts to promote social diversity, its downtown has become mainly an upper- and middle-income enclave due to the high cost of most condo units and the replacement of low-income housing with new condominium towers. Threats to social diversity are even more serious in Toronto, where little effort was made to provide units suited to families and/or that are affordable in new condominium developments. In addition, the growing presence of condominium developments may limit downtown functional diversity as housing crowds out office buildings (Boddy, 2005; McCullough, 2005). Unfortunately, nodal development has not alleviated social polarization, nor can it guarantee continuing diversity of uses in the context of market processes that may favour specialization. With the increasing appeal of walkability and public transit, especially among young adults, sectors offering favourable walking and transit conditions, such as downtowns, are the object of gentrification.

As shown in Table 12.1, other large Canadian metropolitan regions have not experienced anything close to the downtown residential intensification that took place in Toronto and Vancouver. These findings suggest that large-scale, condominium-based downtown intensification requires particular conditions. These include rich downtown amenities (entertainment and culture in Toronto and natural features in Vancouver), sustained demographic and economic growth, high real estate values due to land scarcity, and accommodating planning interventions. Mid-size cities have similarly struggled to try to reverse processes hollowing their cores. While cities like London, Ontario, and Winnipeg struggle with the **revitalization** of their centres, others, such as Halifax and Kingston, have had some success in luring new residents downtown (Burayidi, 2001; Filion et al., 2004; Grant and Gregory, 2016).

Complete Communities

While nodes reflect attempts at creating or enhancing poles of attraction within the urban environment, complete communities are new areas envisioned as complete small towns or urban villages. Complete communities bring to local neighbourhoods the recentralization principles attempting to achieve multi-functionality discussed above in the context of major suburban nodes and downtowns. In contrast to conventional development trends that generated homogeneously residential suburbs, complete communities plan for a mix of uses, housing, and densities. Most involve suburban locations, but some entail comprehensive redevelopment of brownfield sites. The new complete communities reflect largely interchangeable principles associated with the contemporary development paradigms of New Urbanism, sustainable development, and smart growth (Grant and Scott, 2012). The complete community creates a miniature urban structure with a town centre of commercial and office uses, sets high design standards, implements a mix of housing types, creates a connected street network (often supplemented with rear service lanes), and includes a system of open and public spaces. These are large master-planned projects intended to host hundreds to thousands of homes and the services residents require to meet their daily needs. Despite the claim to be "complete," however, most plan few employment opportunities within the development.

In some cases, like Cornell in Markham (in the Toronto metropolitan region), provincial and municipal governments played a significant role in assisting with planning and developing projects. In other cases, like Calgary and Surrey (to the east of Vancouver), municipal policy adopted the principles promoting complete communities advocated by project developers. In 2006, Ontario made smart growth principles provincial policy for the Greater Golden Horseshoe area. While most of the development and building of new communities remains in the hands of the private sector, various levels of government set a policy context that promotes complete communities as a desired form of urban development.

Since the mid-1990s, New Urbanism principles have influenced residential development in many parts of Canada (Grant, 2003). New Urbanism originated in the United States with the development of projects such as Seaside, Florida, and Kentlands, Maryland. Its proponents advocate building complete and compact communities with a mix of uses and housing types, pedestrian-oriented streets, and an attractive public realm (Duany, Plater-Zyberk and Speck, 2000). Many of the projects feature traditional architectural and urban design qualities. A 2008 study identified 42 New Urbanist communities in Canada, mostly in rapidly growing suburban areas in Ontario, Alberta, and British Columbia (Grant and Bohdanow, 2008).

In Ontario the provincial government set the policy context for promoting New Urbanism in the early 1990s by arranging for design competitions and providing a large parcel in Markham for a demonstration project. Cornell, designed by Duany Plater-Zyberk Associates, revealed the potential of New Urbanism to support environmental policies and objectives to protect sensitive areas (Gordon and Tamminga, 2002).

The East Clayton community in Surrey established an ambitious environmental agenda in its construction. Designers employed sustainable development principles to guide the site and landscape design. Green infrastructure manages storm water on the site, and park designations protect waterways from development (Condon, 2003). New Urbanism principles influenced the urban design: small lots, traditional architectural designs, and rear service lanes are common features throughout the project.

Developers began work on McKenzie Towne in southeast Calgary in the mid-1990s. Creating a New Urbanism community was a private sector marketing strategy to give the project a strong identity in a region where many developments constructed artificial lakes to anchor subdivisions. Despite a promising start, however, McKenzie Towne struggled in the marketplace. The distance of the development from Calgary's city centre and the continuing lure of large suburban houses with attached garages made small homes on small lots a challenge to sell. The developer returned to a conventional development concept after building three phases. Another Calgary new community, Garrison Woods, proved more successful. Its development was launched in the late 1990s, when Canada Lands began planning to transform a former military base in Calgary into New Urbanism communities. The commercial centre, built adjacent to a successful commercial district and existing transportation networks, attracted a range of uses. The project sold out two years ahead of schedule. Property values increased rapidly, though, pricing homes out of the range of the average Calgary household.

Complete communities ideally contain a robust mix of uses and housing types and enjoy good access to transportation options. In addition, an attractive and well-connected street network encourages walking. Each community includes a town centre with civic, commercial, office, and residential uses. Businesses in the town centres in developments like McKenzie Towne and Cornell, however, struggled to survive on the limited trade available within the developments. Early applications of the New Urbanism model located the mixed-use area centrally in the development, but later developments moved the commercial district close to major arterials to take advantage of wider markets (Figure 12.4). While many of the projects

Pierre Filion

Figure 12.4 The commercial area of Cornell (Markham, Ontario).

The amount and range of retailing is very limited in New Urbanism developments, as seen here in Cornell's commercial area.

have some access to public transportation, few are well sited for rapid transit to major employment centres. The mix of housing types provided has not resulted in the degree of socio-economic diversity that planners hoped to achieve in designing "complete" communities (Grant and Perrott, 2009). Although the projects attract a range of household types (from single persons to multigenerational families), they produce relatively homogeneous income levels. Indeed, McKenzie Towne and Cornell offered reasonably affordable housing in their early years of development, but escalation of prices in the Toronto and Calgary markets soon priced homes out of range of many households. Apart from a small number of well-known projects, few of the Canadian New Urbanism communities have lived up to the potential of the concept. None of the projects is fully "complete" in terms of the original design.

Segregated-Use Forms

While government policy over the past decades has generally encouraged mixing uses, market pressures continue to promote conventional forms of segregated use—as in suburban residential pods—and to develop new forms of separation. New segregated-use forms reflect developers' continuing efforts to find niche markets to appeal to consumers and to raise returns on development. The forms presume that most users will have access to private automobiles.

Although we classify these forms as segregated, we recognize that some projects may involve a small admixture of uses other than the dominant ones. The projects are occurring in cities of varying sizes across Canada. While the forms reflect market dynamics at work, they occur because government policies have made space for them. New practices in retail and commercial developments may take advantage of land that local governments previously designated for future industrial use.

Auto-Oriented Pods

In the 1960s through the 1980s, many cities designated, zoned, or developed land for industrial use. The decline in manufacturing in Canada over the last few decades, however, slowed demand for industrial land; many communities were oversupplied in land they originally hoped would house industrial uses. Looking for alternative uses for the surplus land, many accepted proposals for big-box retail outlets or for business parks.

Power Centres

Most mid- and large-sized cities now feature one or more **power centres**: concentrations of big-box and factory outlet retail stores, entertainment and recreational facilities, and hotels. The new power centres generally locate on the urban periphery near major highway transportation corridors and junctions. They feature large parking areas, sometimes shared by several retailers. By contrast with the suburban nodes promoted by public policy and investment, power centres continue the tradition of auto-oriented land-use segregation. They become significant shopping destinations, catering to consumers looking for bargains and willing to drive some distance to find what they want (Lorch, 2005).

The power centre reflects a new form in its scale and mix of commercial uses. The pattern and location of retail uses in cities is constantly changing in response to economic conditions and consumer behaviour (Zukin, 2005). A century ago, residential neighbourhoods enjoyed a wide mix of uses, including local grocers and barbers. By the early twenty-first century, the neighbourhood mom-and-pop corner store struggled to survive in the face of competition from chain outlets on arterial streets. The shopping malls of the postwar economic boom looked frayed around the edges: some closed or converted to discount outlets or call centres; others renovated in the hopes of staving off redundancy. The pattern of commercial uses transformed over the decades not only in the largest urban centres, but even in smaller settlements. Consumers looking for bargains and variety drove the success of the new power centres.

Most of the power centres have appeared in suburban or urban fringe locations. For instance, in Halifax, the Bayer's Lake big-box centre was developed in the 1980s on land prepared for industrial uses. With industry and warehouses in the region preferring to buy land in Burnside

Industrial Park in Dartmouth (across the harbour), Halifax council decided to permit a big-box retailer to open in Bayer's Lake. Within a decade the area had become a big-box power centre subject to considerable traffic congestion. In the early 2000s, now amalgamated as a regional municipality with Dartmouth and the county, Halifax approved a new power centre at Dartmouth Crossing: the new site dwarfs other regional shopping facilities and serves as a tourism destination for the Maritimes region. In 2009, the project developers applied to Halifax Regional Municipality for a zoning change to permit them to add a residential component to the project. Most Canadian power centres have limited residential uses, but many "leisure centres" in American cities have added condominium units.

While the power centres in Canada are generally in the suburbs, developers have shown some interest in proposing them in central places, even in Toronto. The City of Toronto opposed an effort to transform a former film studio in Leslieville into a power centre, citing policy aimed at protecting the area for employment uses (Vincent, 2009). Even in contexts where municipal policy clearly discourages such auto-oriented commercial concentrations, developers continue to promote segregated development models.

A variation on the power centre theme has arisen over the last decade. This new retailing formula attempts to reproduce traditional walking-oriented commercial street settings as well as public squares. It includes offices and some high-density housing. These centres appear to be an amalgam of power centres and mixed-use suburban centres. But despite their apparent walking orientation, they share with conventional power centres a nearly exclusive dependence on automobile accessibility. Examples of this new retail formula are Shops on Don Mills in Toronto and DIX30 in suburban Montreal.

Although a relatively new development formula, power centres may be approaching the end of their life cycle as a form of retailing. All forms of retailing are presently confronting the competition from online retailing, which is slowing new retailing development in Canadian cities.

Business Parks

The growth in business parks—sometimes called office parks or research parks—similarly involves new uses for lands previously held for industrial activities. Searching for less expensive land with ample parking for employees, office developers have looked to the periphery of urban regions to create upscale business parks since the 1980s (Lang, 2003). These projects do not include warehousing or heavy manufacturing facilities. Many feature natural or environmental amenities; the business parks house growing concentrations of office uses, accompanied by business-oriented retail and service uses. Landscaping and design standards in business parks may screen parking lots from view, but these developments generally are low density and auto-oriented. The campus configuration of business parks and their easy car access by a wide suburban labour pool have proved especially popular with businesses. The stalling of office development in many downtowns and suburban nodes is in part a consequence of the success of business parks.

Private Communities

While government policy in many larger cities advocates complete communities, market-driven factors are creating new types of developments in cities of varying sizes. The popularity of condominium tenure underpins the building of private communities. Private communities involve clusters of buildings in areas zoned for medium- to high-density housing with land held in common ownership. Sometimes units share recreational amenities, but the communities are not complete because they rarely contain commercial or civic uses. Many of these residential districts have private streets, and some are gated. Private enclaves are appearing in urban cores and suburban areas.

For most of the twentieth century, open-concept community designs were popular in Canada. Postwar subdivisions often avoided fences or hedges, preferring unbounded lawns that facilitated views and access. By the 1980s, though, gated communities—developments restricting public access to internal streets—began to appear in parts of British Columbia. The popularity of gated communities in the United States at this time arguably inspired their development in Canada. The trend accelerated by the late 1990s, with some development companies exploiting a niche market for affluent retirees choosing to move into condominium townhouse projects. By 2003, Canada had thousands of private communities and hundreds of gated enclaves (Grant, Greene, and Maxwell, 2004). Two-thirds of Canadian gated projects are in British Columbia: provincial laws in BC and Alberta make it possible for developers to target projects for seniors, a population that finds private enclaves attractive.

Municipal authorities eager to reduce the costs of providing services such as garbage collection, snow removal, and road maintenance have proven sympathetic to private townhouse condominium developments and have directly or indirectly facilitated gating. In high-growth areas, medium-density townhouse projects have become a common development form providing more affordable housing options. Many such projects employ private streets built by developers and managed by condominium corporations. Those private streets, which some municipalities call "shared access driveways," can easily be gated to restrict access when residents decide the cost of doing so is worthwhile.

Most enclaves occur in suburban areas, but some appear in urban districts where sites are redeveloped for high-end residential use. The smallest private enclaves have only a few homes around a cul-de-sac, but the largest have more than a thousand units. Most enclaves, however, have fewer than 100 units and feature modest clubhouses or pool facilities for the exclusive use of residents. Many involve small households of high socio-economic status seeking privacy, exclusivity, and a degree of security. Large gated projects may house social elites and sometimes include recreational amenities, such as private golf courses, lakes, or marinas. Few Canadian gated communities, in contrast to their US counterparts, employ guards.

Critics suggest that private communities exacerbate social and spatial fragmentation in suburban landscapes and encourage automobile use (Grant and Curran, 2007). Certainly, the growth in gated communities parallels advancing social polarization in Canadian society and exacerbates the spatial fragmentation that already characterizes suburban areas. Gated developments like Swan Lake in Markham, with over 1000 homes planned, force pedestrians, cyclists, and drivers to go around a large enclosed area. As an urban form, the private enclave constitutes a space marked for privacy and social distance.

Planning policies that promote urban connectivity, social integration, and mixed uses have slowed but not prevented the proliferation of private and gated communities. Developers continue to respond to consumer demands for privacy. Having permitted ground-oriented condominium development originally as a way to address the need for more affordable housing for Canadians, governments now find themselves unable to prevent an urban form that generates some significant social and spatial concerns (Figure 12.5).

Conclusion

Our investigation has revealed two types of urban innovations: those that conform to smart growth and urban sustainability planning principles (mixed nodes, suburban town centres, downtown revitalization zones, complete communities) and those that are consistent with automobile-oriented urban patterns (power centres, business parks, private communities). The first type of emerging urban forms feature intensification, mixed use, and a focus on public transit and walkability. These urban forms are clearly a reaction against low density and functionally specialized land uses associated with generalized car use. But while planning has been successful in raising densities, as evidenced in the downtowns and suburban centres of major urban areas, it has not fared as well with respect to walkability and public transit use in suburban centres, a consequence of enduring reliance on the car. The first type of emerging urban forms has also pointed to planning difficulties in achieving social mix. Among these urban forms, downtowns of large cities are most socially segregated due to intense gentrification.

These findings suggest that planning is best apt to achieve its objectives when aligned with market trends, which themselves reflect evolving demography and values. For example, smaller

Figure 12.5 A gated community in Calgary

households and the entry of millennials in the housing market have fuelled the downtown condo boom. The need for compatibility between planning and market trends explains the difficulty for planning to implement urban innovations that fully conform to its principles.

The second type of emerging urban forms accentuate urban features associated with car use, especially low density and functional separation. These urban forms, which are primarily market driven, thus contradict the smart growth and urban sustainability principles advanced by planning. They expose a tension between the principles planning defends and the need for planning to accommodate market trends. Despite the fact they clashed with the principles it promotes, planning did authorize this second type of urban innovations.

Review Questions

1. What emerging trends are influencing public policy responses in cities? What forms are local responses taking?
2. What urban forms are surfacing in the contemporary city? To what extent are these new urban forms likely to alter the course of urban development?
3. Among the new urban forms described in the chapter, which are most likely to influence future urban patterns?

References

Baudrillard, J. 1994. *Simulacra and Simulation*. Ann Arbor, MI: University of Michigan Press.

Berelowitz, L. 2005. *Dream City: Vancouver and the Global Imagination*. Vancouver, BC: Douglas &McIntyre.

Boddy, T. 2005. "'Downtown' a fool's paradise? Council policy has encouraged the construction of new condos but not new office towers," *Vancouver Sun*, 10 Aug., B2.

Bourne, L.S., and D. Rose. 2001. "The changing face of Canada: The uneven geographies of population and social change," *Canadian Geographer* 45: 105–19.

Burayidi, M., ed. 2001. *Downtowns: Revitalizing the Centers of Small Communities*. New York: Routledge.

Buzzelli, M., and R. Harris. 2006. "Cities as the industrial districts of housebuilding," *International Journal of Urban and Regional Research* 30: 894–917.

Calgary, City of. 2008. *Transit Oriented Development (TOD): Development around Rapid Transit Stations*. Calgary: City of Calgary. At: www.calgary.ca/portal/server.pt/gateway/PTARGS_0_2_527219_0_0_18/Transit+Oriented+Development+.htm

Canada Mortgage and Housing Corporation (CMHC). 2016. *Canada Housing Observer: Housing Market Indicators, Canada, 1990-2016*. Ottawa: CMHC. At: www.cmhc-schl.gc.ca/en/hoficcincl/homain/stda/data/data_001.cfm

Canadian Broadcasting Corporation (CBC). 2017. "1.9 million Canadian households live in condos census data show," October 25. At: www.cbc.ca/news/business/census-housing-1.4370757

Communauté Métropolitaine de Montréal (CMM). 2012. *An attractive, competitive and sustainable Greater Montréal*. Montreal: CMM. At: http://cmm.qc.ca/fileadmin/user_upload/pmad2012/documentation/20120813_PMAD_eng.pdf

Condon, P. 2003. "Green municipal engineering for sustainable communities," *Municipal Engineer* 156: 3–10.

Duany, A., E. Plater-Zyberk, and J. Speck. 2000. *Suburban Nation: The Rise of Sprawl and the Decline of the American Dream*. New York: North Point Press.

Edmonton, City of. 2009. *Stadium Station Transit Oriented Development*. Edmonton: City of Edmonton. Mar. At: www.edmonton.ca/city_government/planning_development/stadium-station-transit-oriented-development.aspx

Filion, P. 2001. "Suburban mixed-use centres and urban dispersion: What difference do they make," *Environment and Planning A* 33: 141–60.

———. 2007. *The Urban Growth Centre Strategy in the Greater Golden Horseshoe: Lessons from Downtowns, Nodes and Corridors*. Toronto: Neptis Foundation.

———, H. Hoernig, T. Bunting, and G. Sands. 2004. "The successful few: Healthy downtowns of small metropolitan regions," *Journal of the American Planning Association* 70: 328–43.

———, A. Kramer, and G. Sands. 2016. "Recentralization as an alternative to urban dispersion: Transformative planning in a neoliberal city," International Journal of Urban and Regional Research 40: 658–78.

Foot, D.K., with D. Stoffman. 1996. *Boom, Bust and Echo: Profiting from the Demographic Shift in the New Millennium*. Toronto: Macfarlane Walter and Ross.

Gad, G. 1985. "Office location dynamics in Toronto: Suburbanization and central district specialization," *Urban Geography* 6: 331–51.

———, 1991. "Toronto's financial district," *Canadian Geographer* 35: 203–7.

Gallouj, F. 2002. *Innovation in the Service Economy: The New Wealth of Nations*. Cheltenham, UK: Edward Elgar.

Gordon, D.L.A., and K. Tamminga. 2002. "Large-scale traditional neighbourhood development and pre-emptive ecosystem planning: The Markham experience 1989–2001," *Journal of Urban Design* 7(2): 41–54.

Grant, J. 2003. "Exploring the influence of new urbanism in community planning practice," *Journal of Architectural and Planning Research* 20: 234–53.

———, and S. Bohdanow. 2008. "New urbanism communities in Canada: A survey," *Journal of Urbanism* 1: 111–30.

———, and A. Curran. 2007. "Privatised suburbia: The planning implications of private roads," *Environment and Planning B: Planning and Design* 34: 740–54.

———, K. Greene, and D.K. Maxwell. 2004. "The planning and policy implications of gated communities," *Canadian Journal of Urban Research, Canadian Planning and Policy* 13, 1 (suppl.): 70–88.

———, and W. Gregory. 2016. "Who lives downtown? Neighbourhood change in central Halifax, 1951-2011," *International Planning Studies* 21: 176–90.

———, and K. Perrott. 2009. "Producing diversity in a new urbanism community: Policy and practice," *Town Planning Review* 80: 267–89.

———, and D.E. Scott. 2012. "Complete communities and the Canadian dream: Representations of suburban aspirations," *Canadian Journal of Urban Research* 21, 1 (Supplement), 132–57.

Harris, R. 2004. *Creeping Conformity: How Canada Became Suburban, 1900–1960*. Toronto: University of Toronto Press.

Hulchanski, J.D. 1988. "The evolution of property rights and housing tenure in postwar Canada: Implications for housing policy," *Urban Law and Policy* 9: 1350–6.

———. 2007. *The Three Cities within Toronto: Income Polarization among Toronto's Neighbourhoods, 1970–2000*. Toronto: Centre for Urban and Community Studies, University of Toronto (Research Bulletin 41). At: www.urbancentre.utoronto.ca/pdfs/researchbulletins/CUCSRB41_Hulchanski_Three_Cities_Toronto.pdf

Industry Canada. 1996. *Canada's Service Economy*. Ottawa: Industry Canada.

Isaacs, B., G. Miller, G. Harris, and I. Ferguson. 2007. "Bracing for the demographic tsunami: How will seniors fare in an aging society," *Plan Canada* 47, 4: 20–1.

Jacobs, J. 1961. *The Death and Life of Great American Cities*. New York: Random House.

Joint Program in Transportation. 2011. "Data Management, 2011 Transportation Tomorrow Survey." At: dmg.utoronto.ca

Lang, R.E. 2003. *Edgeless Cities: Exploring the Elusive Metropolis*. Washington: Brookings Institution Press.

Ley, D. 1996. *The New Middle Class and the Remaking of the Central City*. Oxford: Oxford University Press.

Lipson, B.D. 2006. *The Art of the Real Estate Deal*. Toronto: Thomson Carswell.

Lorch, B. 2005. "Auto-dependent induced shopping: Exploring the relationship between power centre morphology and consumer spatial behaviour," *Canadian Journal of Urban Research* 14: 364–83.

Maxwell, D.K. 2004. "Gated communities: Selling the good life," *Plan Canada* 44, 4: 20–2.

McCullough, M. 2005. "Office space suffers lack of demand: Real Estate 1 Cheapest condos now worth more per square foot than the priciest office space," *Vancouver Sun*, 30 Apr., G3.

McDonald, J.F., and P. McMillen. 2007. *Urban Economics and Real Estate: Theory and Policy*. Malden, MA: Blackwell.

Meligrana, J., and A. Skaburskis. 2005. "Extent, location and profiles of continuing gentrification in Canadian metropolitan areas, 1981–2001," *Urban Studies* 42: 1569–92.

Metro Vancouver. 2007. *Choosing a Sustainable Future for Metro Vancouver: Options for Metro Vancouver's Growth Management Strategy*. Burnaby, BC: Metro Vancouver. At: www.metrovancouver.org/about/publications/Publications/RGS_Options.pdf

———. 2008. *Our Livable Region 2040: Metro Vancouver Growth Strategy (Preliminary Draft)*. Burnaby, BC: Metro Vancouver.

Moulaert, F. 2009. "Social innovation: Institutionally embedded, territorially (re)produced," in D. MacCallum and S. Vicari Haddock, eds. *Social Innovation and Territorial Development*. London: Routledge.

Murphy, R.E. 1972. *The Central Business District*. Chicago: Aldine-Atherton.

Ontario, Ministry of Public Infrastructure Renewal. 2006. *Growth Plan for the Greater Golden Horseshoe*. Toronto: Government of Ontario.

Ottawa, City of. 2007. *Transit-oriented Development Guidelines*. Ottawa, ON: City of Ottawa. At: www.ottawa.ca/residents/planning/design_plan_guidelines/completed/transit/index_en.html

Punter, J. 2003. *The Vancouver Achievement: Urban Planning and Design*. Vancouver, BC: University of British Columbia Press.

Raad, T. 2006. "Turning transit stations into transit villages: Examples from Greater Vancouver," *Plan Canada* 46(2): 25–8.

Skaburskis, A. 2006. "Filtering, city change and the supply of low-priced housing in Canada," *Urban Studies* 43: 533–58.

———, and M. Moos. 2008. "The redistribution of residential property values in Montreal, Toronto, and Vancouver: Examining neoclassical and Marxist views on changing investment patterns," *Environment and Planning A* 40: 905–27.

Statistics Canada. 2017. "The shift to smaller households over the past century," *The Daily*, 3 Mar. At: www.statcan.gc.ca/pub/11-630-x2015008-eng.htm

Steacy, R. 1987. *Canadian Real Estate*. Toronto: Stoddart.

Sundararajan, A. 2016. The Sharing Economy: The End of Employment and the Rise of Crowd-based Capitalism. Cambridge, MA: MIT Press.

Toronto, City of, Planning Board. 1974. *Core Area Task Force: Report and Recommendations*. Toronto: City of Toronto, Planning Board.

Vincent, D. 2009. "OMB rejects big box plans in Leslieville," *Toronto Star*, 5 Mar. At: www.thestar.com/news/gta/article/596848

Walks, R.A. and L.S. Bourne. 2006. "Ghettos in Canada's cities? Racial segregation, ethnic enclaves and poverty concentration in Canadian urban areas," *Canadian Geographer* 50: 273–97.

———, and R. Maaranen. 2008. "Gentrification, social mix, and social polarization: Testing the linkages in large Canadian cities," *Urban Geography* 29: 293–326.

Zukin, S. 2005. *Point of Purchase: How Shopping Changed American Culture*. New York: Routledge.

13 Public Space in the City Centre

Design, Activity, and Measurement

Ryan Walker and Jill Blakley

Introduction

Our chapter examines one of the most dynamic aspects of our urban age—public space. Jan Gehl and Lars Gemzoe (2008) characterize the uses of public space by referring to its three Ms: movement, market, and meeting. The most well-known public space in Western history may be the agora of the ancient Greeks, serving as the main public square for meeting and assembly and as the central marketplace (Madanipour, 2003). Canada's oldest public market, Market Square in Kingston, has served in that lineage of function in the socio-cultural, political, and economic lives of citizens for hundreds of years, and before that as a First Nations meeting place (Lynch and Ley, 2010). Today you will find people shopping at its public market and, depending on the season, meeting to ice-skate or watch open-air films. Such programming and management of activities in public spaces is a key feature that we return to throughout the chapter. The same physical space can vary remarkably over time and between seasons based on its programming.

The most prominent of public spaces in any city is its streets. In Regina on 24 November 2013, Albert Street—known in the city centre as the Green Mile in honour of the province's Canadian Football League (CFL) team—came alive with thousands of people to celebrate the Saskatchewan Roughriders' victory in the Grey Cup match against the Hamilton Tiger-Cats. It was the first time the Grey Cup had been won by the Saskatchewan team while playing at their home stadium. Young and old, families, friends, strangers from across Saskatchewan gathered in the streets of the capital city to share with each other the victory of their team and to trumpet their pride of place to the rest of Canada. On 10 February 2018, thousands of people gathered in the streets and squares of city centres across Canada to protest a verdict reached the evening before by an all-white jury at the Battle-ford, Saskatchewan, courthouse, presided over by a white judge, which found Gerald Stanley, a 56-year-old white man, not guilty of any crime. Stanley fatally shot Colten Boushie in the head at close range; Boushie was a 22-year-old Cree man who had driven onto Stanley's property with a group of friends. Stanley and Boushie, their communities, and the location of the trial were all on Treaty Six territory. Spadina Crescent East in Saskatoon's city centre was just one of many streets and squares across Canada that became a site of solidarity against the violence of racism in our society and systemic racism in the criminal justice system. Indeed, practising **citizenship**—the active animation of state–society relations—is a principal purpose of public space. While streets are most commonly associated with movement, it is critical that we understand their role equally as a meeting place for protest, for celebration, and for public life generally, including commerce.

In the next two sections, we discuss the concept of public space and some of the measurement and metrics for better understanding how our public spaces are functioning. We then focus on the physical design of public space and programming of activities (e.g., arts, commerce, recreation, debate) for enhancing the movement, market, and meeting dimensions of public life in Canadian cities. The design and programming of streets is our main focus, followed by a shorter examination of squares. We look at city centres—the heart of a city's identity and image, commerce, and neighbourhoods—although the ideas apply to other areas of the city (also see Senbel, Liem, and Lesnikowski, Chapter 24; and Bain and Mark, Chapter 15). We return to the three *M*s of public space throughout as a useful tool for structuring our discussion. While we use examples from across Canada, most are from Saskatoon and Montreal, the two cities where we have directed our most recent empirical attention. Some important public space topics are not covered in this chapter. One is waterfront redevelopment, which is an important aspect of post-industrial public space (re-)design (e.g., Galland and Hansen, 2012; Gordon, 1997; Greenberg, 2011; see Bain and Mark, Chapter 15 on waterfront re-development and Brail and Donald, Chapter 5, Vinodrai, Chapter 17, and Shearmur, Chapter 18, for discussions of the post-industrial urban economy). Another topic on public space that this chapter does not address is the design of urban parks and their integration with regional natural processes (e.g., Hough, 2004; McHarg, 1969; see Rees, Chapter 21; Holden and Chang, Chapter 22).

The Role of Public Space

> The values attaching to public space are those with which the generality of the citizenry endows it. Citizens create meaningful public space by expressing their attitudes, asserting their claims and using it for their own purposes. It thereby becomes a meaningful public resource. The process is a dynamic one, for meanings and uses are always liable to change. Renegotiation of understandings is ongoing; contention accompanies the process. (Goheen, 1998: 479)

Peter Goheen (1998), a well-respected urban historical geographer, attributes much importance to the role of public space in modern urbanism. Citing Georges-Eugène Haussmann's massive urban redevelopment of historic central Paris in the mid- to late-nineteenth century, Goheen explains how the city became available to ordinary citizens to a much greater and unrestricted extent than before. Streets or boulevards, in particular, became the places of regular and uncontrolled encounters among the citizenry (Goheen, 1998). Today there is concern that the practice of citizenship in streets and squares is diminishing, replaced in prominence by the passive citizen-spectator, spectacle, and a **commodification** of public space, where citizen-consumers engage in individualized pursuits for leisure offered as much by private companies as by public authorities (e.g., Goheen, 1998; Madanipour, 2003; Zukin, 2003). Public space is also a receptacle for the structural failures of our modern economy and social **welfare state**, when seen through the rise in homelessness over the past quarter-century in the cities of Canada, a wealthy country (see Walks, Chapter 10).

The importance of symbolism in constructing social reality looms large in the design of public spaces in our cities. For example, state power and authority are asserted symbolically through selections of who or what to commemorate in the naming and monuments of public spaces (Madanipour, 2003). State power and authority are also asserted overtly, perhaps no more starkly than in 2001, when federal, provincial, and municipal governments erected a three-metre-high concrete and chain-link barricade around the centre of Quebec City and brought in police and military personnel in riot gear, supported by helicopters and armoured vehicles. This state control of public space was done to keep the public away from heads of state and their officials gathered for the third Summit of the Americas. Different social groups, private corporations, and public authorities construct different layers of meaning in public spaces and their uses, symbols, monuments, and heritage in dynamic and continuous ways over time (Madanipour, 2003). Contention does indeed accompany the process (Goheen, 1998).

Public space is best understood in conjunction with the concept of "place," which encompasses a meeting point for social and cultural relations with physical form. As Nicholas Lynch and David Ley (2010) explain, place is connected intimately to identity and belonging (Relph, 1976). They discuss how people's love of place flourishes from their attachment of meaning to relationships between culture, environment, and time in space (Tuan, 1974). Lynch and Ley (2010: 327) argue that a sense of place is created by a combination of many facets of the built and socio-cultural environment and is physically and socially constructed "with continuously circulating meanings among different groups reflecting history and supporting identities."

Place-making and the value of urban design have become a priority of many public authorities, citizens, and the private sector. This is particularly true in an era of city-"concepting," branding, and marketing in an effort to attract and retain workers in a post-industrial international economy, often referred to as the **creative class**, or to augment urban tourism (Carmona, de Magalhaes, and Edwards, 2002; Florida, 2004; see Vinodrai, Chapter 17; Shearmur, Chapter 18). Project for Public Spaces (PPS), a non-profit organization based in New York City, argues that about 80 per cent of the success of most public spaces is attributable to their management and programming by responsible authorities (e.g., municipality, community or other non-profit organization) (PPS, 2005). The drawing power of public spaces—where not only will people move through for necessary activities, but will also stay for optional activities (Gehl, 2010)—requires good design, management, and programming. In the following section, we move from our conceptual discussion to look at how direct observation and empirical measurement can enhance our understanding of how public spaces function.

Measurement and Metrics of How Public Spaces Are Functioning

The urbanists whose conceptual insights into city spaces and public life we treasure in the classroom learned most of what they know by direct observation. Jane Jacobs, Jan Gehl, William Whyte, and Allan Jacobs, for example, did a lot of walking around, sitting, watching, measuring, taking film and photographs, making notes, sketching, and speaking with citizens using the spaces. In our own public space and activity research program (e.g., City of Saskatoon, 2011), we use a set of methods that have much in common with—and were inspired by—those used by Gehl and his team (see Gehl and Svarre, 2013, for an introduction to their toolkit) and by PPS (2005). Our methods are grouped into three categories: site survey, direct observation, and public surveys.

In the first category (i.e., site survey) are inventories of street furniture, window and sidewalk displays, public art placement, landmarks, street trees, soft and hard landscaping, and snow clearance, among other things. Also included are evaluations of ground floor frontage, test-runs through cycling and pedestrian infrastructure, transit stop improvements, and revitalized laneways. Our research team also conducts sensory perception surveys that develop understanding of comfort, safety, and enjoyment in the public realm.

In the second category of methods (i.e., direct observation), we include active transport counts in winter and summer of pedestrians and cyclists, and intersection studies that examine how cyclists, pedestrians, and motorists are interacting at intersections, including the frequency of use at different times of day (Figure 13.1). Stationary activity mapping of public spaces is done to examine how people are using them; where they situate themselves, for how long, and at what times of the day; and changes in the use of public space by season.

Some cities have introduced automated approaches as a substitute for direct observation to understand how the cycling network is being used. Montreal, for example, has permanently

Figure 13.1 Direct observation methods in public space and activity research, Saskatoon city centre

installed Eco-Counters along some of its cycling lanes, providing daily cyclist counts for every day of the year since 2008. In 2013 the city also launched Mon RésoVélo, a free smart phone app that records the routes and speeds of cyclists, sends it to the city, and in turn is used by the city to make data-driven investments in cycling infrastructure (see Brail and Donald, Chapter 5).

For the third category of methods (i.e., public surveys), we use intercept surveys of people moving through or staying in public spaces to learn about their perceptions of comfort, safety, and enjoyment, as well as their mode of travel to the city centre. Finally, participatory mapping is done with people in the city centre to learn which areas they perceive as being safe or dangerous, and which areas they treasure and dislike most.

The composite understanding developed through this series of methods provides an updatable longitudinal database that can be used to target improvements to the public realm (e.g., new cycling infrastructure), measure the impact of new investments in the public (e.g., a parklet) and private realms (e.g., redevelopment of a prominent corner parcel on a block). The results of these methods can also be used to monitor at regular intervals (e.g., every five years) how public space, activity, and **urban form** are interacting in the city centre and whether public space planning goals are being met.

The Design and Programming of Streets

> Streets and their sidewalks, the main public places of a city, are its most vital organs. Think of a city and what comes to mind? Its streets. If a city's streets look interesting, the city looks interesting; if they look dull, the city looks dull. (Jacobs, 1961: 29)

In the previous section, we provided an overview of measurements and metrics used to understand how public spaces are functioning. In this section, we examine what these approaches have taught scholars and practitioners about the movement, market, and meeting functions of streets, the primary public space of any city.

Many cities still plan for the efficient movement of private automobiles but give only secondary consideration to public transit, and a very distant third consideration to bicycle network infrastructure. The pedestrian experience and active transportation options are generally overshadowed by the Canadian quest to move more cars, more quickly. For public health, financial, and environmental reasons, cities need to accelerate the overlay of the already well-designed automobile infrastructure with an equally effective infrastructure for transit, cycling, and pedestrian travel. For example, a study of the safety of different types of cycling infrastructure in Toronto and Vancouver showed that cyclists are nine times more likely to be injured on a major street with parked cars and no bike infrastructure than on a dedicated cycling lane on a major street separated by a physical barrier, such as a curb, median, or bollards (Teschke et al., 2012). Major streets with bike lanes marked with solid or dotted lines and no parked cars presented about half the risk to cyclists compared to those with parked cars and no cycling infrastructure. By comparison, shared lane infrastructure, like "sharrows" or shared bike-bus lanes, showed negligible risk reductions to cyclists (Teschke et al., 2012).[1]

Montreal ranks first in North America (tied with Vancouver), and eighteenth worldwide, in the 2019 Copenhagenize Index of bicycle-friendly cities, based on a review of 118 cities (Copenhagenize Design Co., 2019). Montreal created some of the north–south segments of its cycling network in the mid-1980s, from the Old Port to near Laval; however, its current approach to transportation planning was catalyzed by the launch of its *Réinventer Montréal: Transportation Plan 2008* (City of Montreal, 2008). As a result, active transportation and transit became the focus of significant new investment (see El-Geneidy and Grisé, Chapter 14). The city's network of cycling infrastructure includes lanes separated from vehicle traffic by medians (Figure 13.2); lanes painted on streets between parked and moving vehicles and next to moving vehicles without parked cars; sharrows; and, along the Quartier des Spectacles, a bike lane that is adjacent to pedestrians on the sidewalk. Montreal's cycling infrastructure network has very good connectivity and continuity and is expanding yearly.

One action put forward in *Réinventer Montréal* was to create a self-service bicycle system as part of its public transit system. In 2008 BIXI (a bicycle-sharing system) was implemented by a private non-profit company created by the City of Montreal. By 2018, there were over 500 BIXI stations and over 6000 bikes around the city. Each station has a payment system and a series of bike docks that house the bikes. Bikes are rented for short intervals (e.g., up to 45 minutes) at one dock, then ridden and returned to any dock in the system. There are over 35,000 BIXI members in Montreal, and usage has gone from just over one million single trips in 2009 to nearly five million trips in 2017 (BIXI Montréal, 2018; Damant-Sirois, Grimsrud, and El-Geneidy, 2014). The BIXI model has since been exported to cities around North America (e.g., Boston, Toronto) and overseas (e.g., London, Melbourne).

Another action that came out of *Réinventer Montréal* is keeping a portion of the cycling lane network clear during the winter. This *Réseau Blanc* comprises areas of the city centre where cycling is most prevalent, which has helped increase ridership numbers during winter months. City centres typically have higher cycling trips than other areas of the city. Saskatoon is similar in this regard and has twice as many residents in the city centre who choose cycling as their means of travel (i.e., 5 per cent) as the city overall (City of Saskatoon, 2011). Cities that want to get a toehold for improving cycling infrastructure overall start in the city centre.

Complete Streets

A progressive model for creating a full spectrum of movement infrastructure on our streets is

Ville de Montréal

Figure 13.2 Bicycle lanes separated from vehicle traffic by median, rue University, Montreal

known as **complete street design**. The Centre for Active Transportation (2012a: 4) defines it as follows: "Complete Streets provide for all road users—pedestrians, bicyclists, transit users, and motorists of all ages and abilities." Faced with transforming current city centre streets—from automobile and parking lanes to complete streets with vehicle, parking, cycling, and transit lanes—political will can be difficult to muster. As more cities achieve success, however, other jurisdictions are emboldened into action. In 2010, for example, the City of Winnipeg redesigned Assiniboine Avenue in its city centre as a complete street. Assiniboine Avenue includes the city's first separated bike lanes and improvements to the pedestrian infrastructure, such as corner bulbs (i.e., curb extensions) at intersections to reduce the crossing distance. Count data collected in 2009, 2011, and 2012 showed a 65 per cent increase in cyclists during the afternoon rush hour (The Centre for Active Transportation, 2012b).

Re-apportioning the street right-of-way to provide safe and dedicated passage to multiple modes can involve removing one or more vehicle lanes in order to accommodate new bike or bus lanes, or removing some on-street parking to improve pedestrian infrastructure, such as corner bulbs at intersections and mid-block crossings, bus bulbs, new or improved medians, or widened sidewalks (Figure 13.3). There are many possible reconfigurations, and the design of a particular street requires an empirical understanding of the users, the street's role in the city's transportation system, and spatial parameters of the particular site. Complete streets have the capacity to move more people through the street on the full spectrum of properly appointed transportation infrastructure, each inducing new demand to match the infrastructure supplied for it.

Existing Street Section

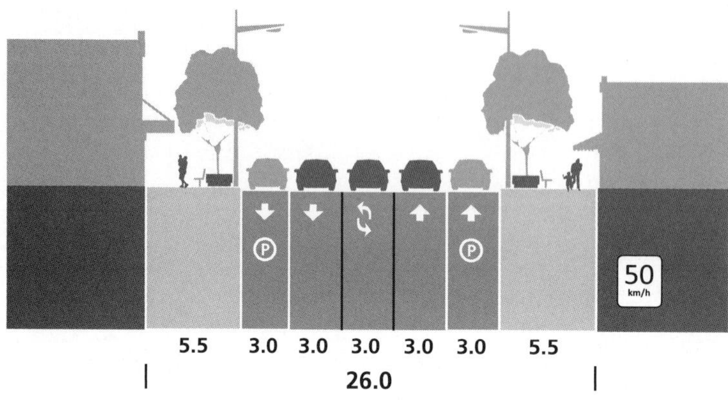

Complete Street Section

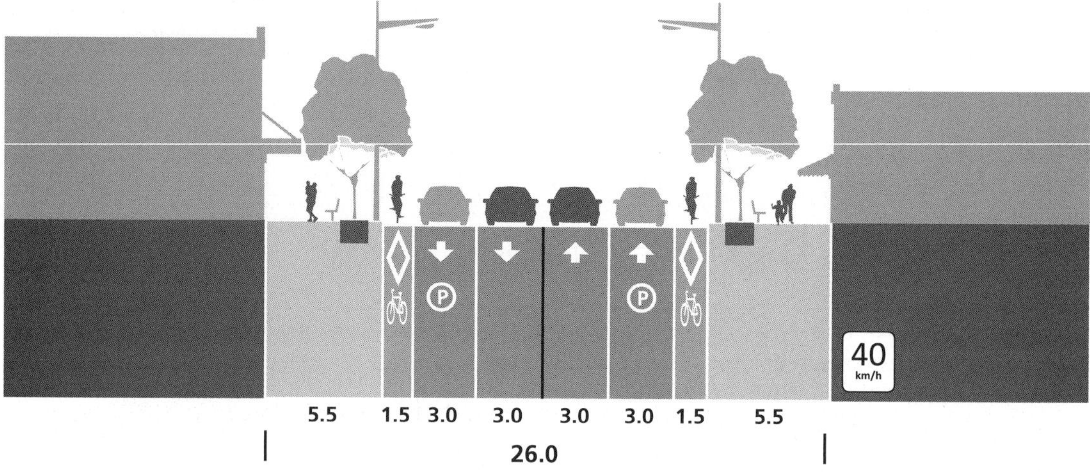

Figure 13.3 Example of Re-Design to "Complete Street" Standards
Source: The Centre for Active Transportation, 2012a: 10–11.

Shared Streets

The concept of complete street design only addresses movement, one of the public space functions of streets. Therefore, it could be said that the complete street design concept is, in fact, incomplete. Streets are public spaces that include meeting and market functions as well. **Shared street design** is a complementary concept to complete streets, and there is room for both models of street recalibration in any city.

The shared street concept designs for its most vulnerable user, the pedestrian, although it is also used by other modes, including bikes and cars. Rather than designating separate lanes for cyclists and cars, as well as sidewalks for pedestrians, the design is open and the right-of-way shared by all users. Curbs and lane markings are either non-existent or very subtle. To demarcate areas where vehicles are *not* permitted, bollards may be used. Paving materials and patterns are selected that are aesthetically pleasing in appearance, texture, and sometimes even in the sound they make as vehicles and bicycles drive over them. The paving is used for the full width of the street right-of-way to evoke urban quality at a human scale and invite pedestrians to make full use of the street.

The same is true of street lights and other street furniture. Places for people to congregate and stay, with public art and other interpretive elements often punctuate the space along the street. All of these design features are geared to the human scale, with lower speed limits and design cues that tame vehicle traffic with the intrigue and uncertainty created by the shared and active space (Engwicht, 1993). Popular city centre examples include the stretch of Sainte-Catherine Street in Montreal that runs through the Quartier des Spectacles and of King Street in Kitchener in front of City Hall. Both are used extensively as event and festival spaces, blending the movement, meeting, and market functions of public space.

Perhaps the most famous type of shared street design is the *woonerf*, which originated about 40 years ago in the Netherlands (Ben-Joseph, 1995). This type of shared street design is most common in areas with high residential use, combining neighbourhood-scale social space that is safe for children to play in with movement of local traffic, designed for pedestrian-priority rather than vehicles. The *woonerf* has started to appear in Canadian cities. Ottawa (e.g., a section of Cambridge Street), Montreal (Saint-Pierre), Toronto (West Don Lands), and Winnipeg (John Hirsch Place) are four examples of cities that have adopted them (Figure 13.4).

Design Qualities of Great Streets

Based on his examination of "great streets" internationally, Allan Jacobs (1993: 270–92) identifies eight categories of design qualities that he argues great streets require: (1) places for people to walk

Ryan Walker

Figure 13.4 John Hirsch Place Woonerf, Winnipeg

with some leisure; (2) physical comfort; (3) definition; (4) qualities that engage the eyes; (5) transparency; (6) complementarity; (7) maintenance; and (8) quality of construction and design. We adapt and elaborate upon some of those qualities next.

1. Places for People to Walk with Comfort

Sidewalks must accommodate people who walk at varying paces: those in a hurry; those at leisure; those who wish to stop mid-course to look more closely through a window, at public art, or at a building; or those who wish to stop to chat with other people. City planners need to carefully consider whether the city centre sidewalks are wide enough to enable the kinds of public space improvements, movement, and stationary activities they aim to foster. If the space is overcrowded, neither leisurely nor hurried pedestrian travel is feasible (Jacobs, 1993). However, sidewalks that are slightly overcrowded are better than sidewalks that are empty. Put another way, it is better to be too lively for the space available than to be lifeless (Gehl, 2010). People are attracted to spaces full of other people.

Sidewalks need separation from vehicles to keep the pedestrian environment secure and pleasant, with curbs, trees that are planted sufficiently close in succession, other landscaping, and perhaps a row for parked cars or a cycling lane to create a buffer between pedestrians and those using the street for rapid movement. The 20th Street West Streetscape Improvement Project in Saskatoon's city centre, for example, began installing bus bulbs and corner bulbs along this commercial street in 2013 (Figure 13.5). These design features improve transit efficiency and pedestrian safety when embarking and alighting from buses and crossing at intersections. These features also improve the pedestrian realm with a variety of associated public art installations and amenity strips. Heated sidewalks and outdoor fireplaces in some locations, and heated transit shelters are proposed in Saskatoon's new City Centre Plan (City of Saskatoon, 2013) to enhance streets in this winter city.

City of Saskatoon

Figure 13.5 Corner bus bulb nearing end of construction, 20th Street West Streetscape Improvement Project, Saskatoon

Safety and trust on a street full of strangers is an important dimension of walking comfortably through city streets. Jane Jacobs (1961) explains how sidewalks with people at all times of the day can create the conditions for safety and trust. Her now famous concept of having many "eyes upon the street" from the windows and doorways of buildings with a variety of uses oriented to the street is fundamental to its safety. Participatory mapping undertaken with people in Saskatoon's city centre showed that the streets with the highest residential density—where windows, doorways, and balconies are oriented to the street—were perceived to be most safe. Areas with surface parking lots lining a significant portion of the block-face and little residential density were perceived as unsafe (City of Saskatoon, 2011).

2. Definition

Streets should create a sense of being in an "urban room." Christopher Alexander, Sara Ishikawa, and Murray Silverstein explain that outdoor spaces—of which the street is the most prominent—can be either positively or negatively defined:

> Outdoor space is negative when it is shapeless, the residue left behind when buildings—which are generally viewed as positive—are placed on the land. An outdoor space is positive when it has a distinct and definite shape, as definite as the shape of a room, and when its shape is as important as the shapes of the buildings which surround it. (1977: 518)

Streets are defined in combination between their vertical (e.g., height of buildings) and horizontal dimensions (e.g., street width from building facades on one side of the street to the buildings on the other side) (Jacobs, 1993). The wider the street, the more building height and mass it takes to define the street and create the human-scale sensory perception of enclosure within an urban room. On wide streets, closely spaced trees on medians and sidewalks can have an intervening influence and enhance definition. From his observation of great streets around the world, Jacobs (1993: 279–80) concludes that "buildings along streets are likely to provide a sense of definition when height to horizontal distance ratios are at least 1:4 with the viewer looking at a 30-degree angle to the right or left of the direction of the street." At a ratio of 1:5 or smaller, the street will not have a sense of definition (Figure 13.6).

Terminal vistas also help create definition. Examples include a sightline to a prominent natural feature or landscape, a roundabout with a sculpture, a building or landmark, or a public square. Serial views that emerge as one moves down a gradually curving street (e.g., due to physical topography or street pattern) are another way that definition is enhanced with visual endpoints along the street wall. Finally, gateways can enhance street definition and improve the legibility of movement into and out of distinctive urban precincts (Alexander, Ishikawa, and Silverstein, 1977). They can be literal gateways, for which many Chinatown precincts in Canadian cities are famous (Figure 13.7). Alternatively, they can be created by using building-design guidelines

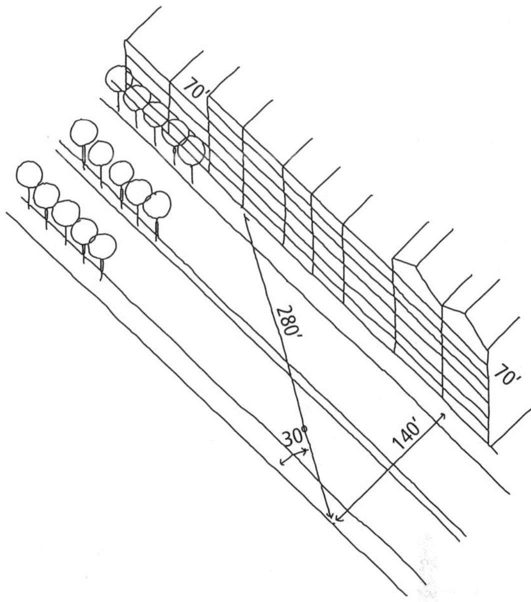

Figure 13.6 Street Definition: Height to Horizontal Distance Ratio

Source: Jacobs, Allan B. *Great Streets, Street Definition: Height to Horizontal Distance Ratio*, © 1993 Massachusetts Institute of Technology, by permission of The MIT Press.

(e.g., directing the prominence of **massing** and facade at the lot line, roof treatment, materials, lighting) to create distinction on corner lots at intersections that mark the entryway to precincts and thus frame the street beyond. As Jacobs (1961: 380) explains, many streets "need visual interruptions, cutting off the indefinite distant view and at the same time visually heightening and celebrating intense street use by giving it a hint of enclosure and entity."

A challenge to street definition in some Canadian cities is the prevalence of surface parking lots in places where buildings should be. These holes in the street wall can undermine the urban quality of an entire block, which would otherwise be well-defined. Saskatoon faces a remarkable challenge in this regard, with surface parking covering roughly 26 per cent of developable land in its city centre (City of Saskatoon, 2011). Unfortunately, its property tax assessment structure does not include a premium on the inherent value of zoned developable land in the city centre. This and the absence of regulations against using

iStock.com/Photo Italia LLC

Figure 13.7 South Paifang, Chinatown, Montreal

building land as surface parking lead to the demolition, rather than up-keeping, of older buildings in the commercial rental market. As well, there is a tendency to hold land indefinitely to use as under-capitalized, income-generating (often gravel) parking lots (see Skaburskis and Moos, Chapter 19).

3. Design Qualities That Stimulate and Create Transparency

The human field of vision has evolved to observe the world on a frontal horizontal plane while moving through it on foot at an average of roughly 5 km/h, ranging from slightly faster in the winter to a little bit slower in the summer (Gehl, 2010). By shifting our eyes, without tilting our head, our vertical field of vision is roughly 50–55 degrees upward from horizontal and 70–80 degrees downward. Our eyes naturally shift their focal point as many as a hundred times over the course of a minute, communicating curiosity and interest psychologically in proportion to changing visual stimuli. Quality physical design promotes activity in the public realm by being a draw for people. In turn, people are the item of most interest and greatest stimulation to other people (Gehl, 2010; Whyte, 1980).

Many units, windows, doorways, and street furniture are important design features, where we can see, hear, and (when close enough) smell people inside, outside, and in liminal spaces throughout our 100-metre horizontal "social field of vision" (Gehl, 2010). Stimulating paving surfaces, vegetation, awnings, and a complexity of lines and texture in building facades all determine the degree to which we experience quality in our streetscapes. And the design of the first three stories of buildings matters most in defining the urban room, given the field of vision we have as human beings. Vertical relief in building facades with narrow and active shop fronts—rather than long, horizontally undifferentiated passive walls—have the effect of creating a good sensory rhythm for the pedestrian as visual stimuli change more often and walking distances seem shorter as well as more interesting (Gehl, 2010). The kinds of cities that are memorable and that retain residents and attract visitors are those that put time into urban quality considerations. Vancouver is a good example of this with its high-quality public space design, framing mountain views, and high-rises constructed to have a narrow tower and street-oriented podium (base) designed to enhance the streetscape at ground level and preserve significant sightlines.

Window displays, signage, doorways with people shuffling in and out between the private and public realms of the street enhance public life on both sides of the threshold (Whyte, 1980). The sight of activity is an incentive for activity; as a result, seeing inside buildings that frame the street has the effect of enlarging the pedestrian realm and making it richer (Alexander, Ishikawa, and Silverstein, 1977). The City of Saskatoon has operationalized transparency as one of the key dimensions of an attractive exchange zone at the ground floor between buildings and the street (City of Saskatoon, 2011; Gehl, 2010). In Saskatoon, attractive ground floor frontages along street blocks in the city centre have 11–20 units, or shop fronts, per 100 metres, with many entrances and few or no closed or passive units (e.g., bank or law-office windows with blinds drawn). Neutral frontages have 6–10 units per 100 metres, with some closed or passive units; dull frontages have 1–5 units per 100 metres with many closed or passive units (Figure 13.8). On the best commercial streets, the edges between private and public space will be soft, and merchandise or services will be displayed or performed in the area of sidewalk or street in front of the shops (e.g., books from the bookshop on display on the sidewalk, café seating on the street or sidewalk, clothing racks brought outside the shop doorway on nice days, sandwich-board signage with a restaurant's daily specials) (Alexander, Ishikawa, and Silverstein, 1977; Gehl, 2010; Jacobs, 1993; Whyte, 1980).

Attractive frontages in the City Centre

Active frontages are predominantly found in the core of the retail and entertainment district along 2nd Avenue South and 21st Street East, as well as parts of Broadway Avenue and 20th Street West. These streets are busy with shops, restaurants and bars, as well as attractive streetscapes. In total there are over 5.3km of attractive street frontages in the City Centre.

Evaluation:
Narrow units
Many entrances
(10–20 units per 100 m)
A diversity of functions
Few closed or passive units
Interesting façades
Quality materials & details

Neutral frontages in the City Centre

The City Centre has a notable amount of neutral ground floor frontages. Many of the these buildings are found in the commercial district, some are office uses. Most of the neutral frontages are medium-sized buildings, with fewer units and entrances, with some fenestration and architectural detail.

Evaluation:
Mixture of small/large units
(6–10 units per 100 m)
Some diversity of functions
Some closed, passive units
Average façades
Decent materials & details

Dull frontages in the City Centre

The City Centre has many inactive and dull ground floor frontages. Most of the these are found along the western edge of the downtown where larger buildings have been developed along the former railway lands. The long blank walls of the Midtown Plaza present a significant challenge as vital and active urban edge. The recently built Galaxy Cinemas, although smaller in size, similarly lacks active frontage along 20th Street and most of 2nd Avenue.

Evaluation:
Larger units, few entrances
(1–5 units per 100 m)
Little diversity of functions
Many closed, passive units
Mostly unattractive façades
Few or no details

City of Saskatoon, 2011: 74

Figure 13.8 Ground floor frontage typology for Saskatoon's city centre

4. Complementarity

To create a streetscape that is positively defined and transparent, buildings must be respectful of the street character overall and re-enforce that character. Having no prescriptive guidelines on building size, massing, lot coverage, materials, or placement pattern of windows and doorways can leave a streetscape at serious risk of long-term damage when individual building projects treat their site as though it existed in isolation. The vast majority of users of the public realm will not set foot within any specific building within this realm; most public realm users only engage with the exterior a building contributes to the streetscape it helps to frame. As a result, meeting the needs of the users of the public realm outside the building—just like meeting the needs of tenants inside the building—is part of the responsibility of developing real estate in a city. Property developers have every right to develop their land parcel, but, we would argue, do not have the right to sidestep the public interest in maintaining a high-quality streetscape when doing so.

Canadian cities have been devising tools to protect complementarity in urban design. The Halifax Regional Municipality's HRMbyDesign downtown design plan and Design Review Committee (DRC), for example, ensure that development applications comply with minimum standards set for built form and public space. The DRC has approval authority for the qualitative aspects of built form; it is rare if not unique in Canada for a DRC to have authority rather than serving in an advisory capacity to the municipality's development officer. Advisory DRCs are more common, and they can work well, especially when they provide formative input in early stages of the development proposal. They can also provide a summative recommendation to the city's development officer when the application and site plan is submitted.

The City of Saskatoon's architectural control overlay zoning district for the Broadway Avenue commercial area in the city centre—a 2011 "Great Street" winner in the Great Places in Canada competition—is an example of design guidelines to safeguard the public interest. The initiative originated out of the community-driven Broadway 360° Development Plan. A DRC advises the city's development officer on building projects according to 14 design guidelines, which include building expression, orientation and placement, street wall, heritage contexts, corner sites, storefronts, residential street access units, roof treatment, above-grade parking, material and architectural quality, sidewalk cafés, building lighting, signage, and sustainable design.

Although complementarity is critically important, Jane Jacobs (1961) has argued convincingly for the role of landmarks—most often buildings—sited sparingly through streetscapes to enhance overall urban quality. Examples might include a historic church, courthouse, library, or heritage building that sits differently on its lot than its neighbours. The overall positive effect of landmarks like these on enhancing streetscapes will be better if they are not clustered together—in a civic complex for example—and are instead spread throughout different streets (Jacobs, 1961). Not only the buildings themselves but also the uses they house within them are best spread widely. A block that contains mostly civic buildings, and little else, neither assists with a diversity of uses at the fine-grained scale that creates urban vitality nor assists with spreading landmark buildings through the city to provide well-conceived punctuation to streetscapes.

We must end the discussion on complementarity with a return to the fundamentals for city life: density and diversity. One cannot expect to have a great street—with vitality, safety, and public life—if there is insufficient density and diversity of residents, commerce, and institutional uses to "give each other close-grained and lively support" (Jacobs, 1961: 377). No design treatment will make up for the absence of these things, although these things can make up for deficits in design.

Programming Streets—Art, Culture, and Commerce

Lively streets, well-designed with the physical attributes discussed above, are best supported by density and diversity along and in adjacent areas, and enhanced with programming in art and design, culture, and commerce. Together, this will augment the street's meeting and market functions.

Street furniture itself can be artfully conceived and built to carry through design elements that create a street identity using tactics of emphasis and suggestion (Jacobs, 1961). The design of garbage bins, benches, bus shelters, street lights, bollards, and maps or historic interpretive panels can all be designed in a style that carries through a theme enhancing the spirit of place. For example, tree grates along the street at River Landing, the signature public space of Saskatoon's city centre, were designed in consultation with Cree and Dakota Elders to reflect the secular activities that would have been undertaken at that place prior to urbanization (see Fawcett and Walker, Chapter 4). In downtown Kitchener, bike racks are in the shape of bicycles and painted in vibrant colours, providing both public art and bike parking. The design of main street in Kitchener's downtown allows the street to be easily closed off to traffic for concerts, markets, and festivals as part of the city's regular programming of events. With thoughtful placement, the advantage of triangulation can occur, where a bench next to a map or interpretive panel or next to a bike rack, garbage bin, or an intriguing tree grate can prompt casual encounters between strangers, whether a conversation or only a smile or nod (Whyte, 1980). Each encounter is a contribution to public activity and the relations of trust so fundamental to vibrant and safe streets.

The streetscape design joining the pavilions of the Musée des Beaux Arts along rue Sherbrooke and Avenue du Musée in Montreal provides an excellent example of place-specific public art (Cartiere, 2010). Undertaken by the City of Montreal, the Place du Musée project was inspired by the cultural presence of this eminent art gallery and the nearby northern backdrop of Mont-Royal, one of the city's most distinctive place-making natural landmarks (Figure 13.9).

Denise Caron/Ville de Montréal

Figure 13.9 Place du Musée project, avenue du Musée, Montreal

City urban design staff worked with an artist to inspire concepts and create models for limestone blocks that would serve as seating and areas for play, and as plinths for sculptures. The limestone blocks mirror the geological composition of Mont-Royal and are situated to emerge out of the ground along the sidewalks as though extending from the mountain itself. Some blocks are cut with a five-degree angle and are clustered in groups of six, interspersed with vegetation and trees that mirror species from the mountain. A terraced sidewalk along Avenue du Musée serves as a sculpture garden that leads toward Mont-Royal, carrying the art gallery into the street. At night, the sculptures are lit by street lights that are directed as spotlights on each sculpture, creating a theatrical effect much as you would see in an exhibit inside the gallery. The segment of Avenue du Musée closest to Sherbrooke is closed to vehicle traffic in summer and itself becomes the set for an additional installation of street art. The project creates a brilliant backdrop for public activity, triangulation prompting casual encounters among strangers, and an artful interpretation and re-enforcement of this specific cultural and natural place. In "winter cities" particularly, the use of light in urban design initiatives is an effective way to bring a feeling of warmth to the public realm when daylight hours are short and the display of light can be very powerful and heartening.

Streets and laneways are becoming some of the most exciting places for art and performance in Canadian city centres. Public art is diversifying beyond the occasional sculpture, mural, fountain, or monument, and it is enhancing public life in interactive and place-specific ways (see Bain and Mark, Chapter 15). Contemporary art and performance of a scale often confined to museums, galleries, and auditoriums are flowing into public space and producing "an in-between space where the divisions between contemporary art, audiences and everyday life are, at least temporarily, re-imagined and considerably altered" (Borsa, 2004, 2006: 132). In Saskatoon, for over ten years, the Free Flow Dance Theatre Company has transformed rear laneways in the city centre into the site for its show, Back Alley Antics. There, the lines between audience and performer blur, and the intimate human scale of the laneway streetscape becomes the setting for the production of space through public art and activity (Figure 13.10). Similarly, rear laneways and fire escapes in the city centres of Edmonton and Calgary have been transformed by the Mercury Opera company into sites for public enjoyment of opera, engaging local business owners as well in a variety of pre- and post-show events, which carry elements of performance to the shops themselves. In 2011, Mercury Opera turned the Bay/Enterprise Square LRT station in downtown Edmonton into the site of one of its opera performances, "104 Underground," where the musicians arrived and departed by transit as part of the performance, playing a roster

Marcia Provenzano

Figure 13.10 Free Flow Dance Theatre Company's "Back Alley Antics," Saskatoon

of popular opera pieces, engaging local businesses along 104th Street as partners (Figure 13.11).

Rear laneways are some of the most underused public spaces of our city centres. Artists have figured this out, as indicated above, but others are catching on. The intimate human-scale of laneways, and their ability to provide better shelter and more direct mid-block routes in the cold of winter makes them attractive year-round prospects for enhancing movement, market, and meeting (Akkerman, 2013; City of Saskatoon, 2011). Behind commercial streets with rising rents like Broadway Avenue in Saskatoon, they offer the possibility of new commercial frontage that hits a price point accessible to small-scaled or start-up businesses that can no longer afford the front-of-house.

Vadim Bulitko

Figure 13.11 Mercury Opera's "104 Underground," Edmonton

Initiatives led by local shop owners and **business improvement areas** have also tapped into new vessels for culture and commerce on the streets. For four weeks spanning August–September 2012, the Downtown Yonge Business Improvement Area in conjunction with Toronto Ward 27 Councillor Kristyn Wong-Tam hosted "Celebrate Yonge." Lanes were reduced from four to two in order to expand the pedestrian space and create 11 themed event areas that contained art installations, street furniture, lounges, and patios for the segment of Yonge Street between Queen Street and Gerrard Street in the city centre (City of Toronto, 2013). Pedestrian traffic increased, vehicle traffic decreased, and there was roughly a 40 per cent increase in sales (Greenberg, 2013), supporting the argument that people—not cars—drive commerce.

In September 2013, Saskatoon businessman Dave Denny celebrated the one-hundredth anniversary of his Drinkle No. 3 building on Third Avenue downtown by closing the block between 21st and 22nd Streets for a day to hold DrinkleFest (Figure 13.12). He worked with local organizations like Saskatoon Cycles and Great Places to install a temporary bike lane and bike box as a demonstration project for **complete street design**, and stage a variety of other place-making initiatives to engage citizens. Street performances and a talk by Better Block co-founder and tactical urbanist Jason Roberts were held. A restaurant in the building moved patio seating onto the street, occupying on-street parking spaces along the shop front. While the street was only closed to vehicle traffic for the one-day DrinkleFest celebration, the restaurant was permitted by the City to keep its on-street dining installed for two weeks as a demonstration project, an initiative that was profitable for both the restaurant and public life. Examples like DrinkleFest and Celebrate Yonge teach us how interconnected the movement, market and meeting functions of the street should be. It is only in our recent past that we separated these functions and privileged movement (of cars) over

Michelle Berg/The StarPhoenix

Figure 13.12 DrinkleFest Better Block party, Third Avenue, Saskatoon

the others. Initiatives like these help to repair the damage done by the disaggregation of the three *M*s.

The creation of parklets provides a final example of a street-side public space intervention—again by reclaiming on-street parking spaces—that is catching on in Canadian cities. San Francisco is often given pioneering credit for formalizing the parklet as a type of permanent (or at least seasonally recurrent) public space starting in 2010. The City of San Francisco (2013: 1) defines it thus:

> A parklet repurposes part of the street into a public space for people. They are intended as aesthetic enhancements to the streetscape, providing an economical solution to the need for increased public open space. Parklets provide amenities like seating, planting, bike parking, and art.

Parklets are typically funded and maintained in some type of partnership between local businesses, residents, community organizations, and civic authorities. An important distinction between parklets and on-street café dining, for example, is that parklets are public spaces. Anyone can enjoy them without buying anything. Montreal's first (seasonally) permanent parklet was created in 2013 along boulevard Saint-Laurent in the Plateau-Mont-Royal neighbourhood. It occupies the space of three on-street parking spots and comprises a painted shipping container with openings throughout, with seating, and with planter boxes.

In the next section we continue our examination of what the observation of the movement, market, and meeting functions of public spaces have taught scholars and practitioners by discussing public squares, one of the signature spaces of cities.

The Design and Programming of Squares

Squares are public spaces at the intersection of two or more streets in an urban area where people can sit and relax, walk, people-watch, and meet others. They may serve variously as a hardscaped marketplace or space for ceremony and civics, or an area of lush and landscaped trees, grass, and gardens. Like streets, good physical design is necessary but insufficient on its own to ensure a vibrant square: "What stands out most is that design is only a small fraction of what goes into making a great square. To really succeed, a square must take into account a host of factors that extend beyond its physical dimensions" (PPS, 2013: n.p.).

Factors present in great squares include a key landmark feature, such as a fountain or other touchable water feature, or a prominent sculpture or civic building (PPS, 2005; Whyte, 1980). A great square will have several smaller places within it that attract a variety of different people at different times of the day, such as cafés, food vendors, play areas for children, enclosed off-leash areas for dogs, or performance spaces. If a square is one dimensional, it will be empty most of the time. Thoughtful placement of amenities like public art, lighting, benches, and garbage bins are all important, some of which can be movable (e.g., some of the seating, tables, games) to accommodate seasonal changes in programming (PPS, 2013). A square may house a skating rink in the winter and open-air movies in the summer (like Kingston's Market Square); it may have a market space that shifts over the course of the year from an open-air food and produce market in the summer to covered (e.g., heated tents) selling local artisanal crafts and gifts during the holiday season in winter.

Easy access for pedestrians is important for a vibrant square, with transit and cycling infrastructure nearby and pedestrian-scale street design at crossings that form the entry points to the square, as well as along the edges. Attention to the movement functions of streets nearby will enhance the meeting and market functions of a square. Specifically, vehicle traffic should be tamed along the edges. A square is not something for vehicles to drive by quickly. It is, rather, a feature of the urban environment that must invite slow, safe, and deliberate attention to movement at a pedestrian's scale.

Management of the space—its upkeep, safety, and programming—is fundamental to success, and often this comes with a partnership between public (e.g., city hall, community association) and private (e.g., local business improvement district) sectors (PPS, 2013). Local community members who use and value the space are a key source of expertise on the features of the square that work best and what improvements might be made to design and programming.

"The life of a public square forms naturally around its edge. If the edge fails, then the space never becomes lively" (Alexander, Ishikawa, and Silverstein, 1977: 600). The edges of a square, then, require much attention, in both design and programming. People naturally gravitate toward the edges of public space to linger. They like to have their back to something and look outward toward the space before them and beyond. Creating pockets of activity (e.g., food and craft vendors, heritage and art installations) around the edges, starting at points of entry to the square, are a technique for building vibrancy and use (Alexander, Ishikawa, and Silverstein, 1977). People will enter, linger, and perhaps then move out into other areas of the square. One common error in the design of squares is that they are too large and undifferentiated. With people most focused upon the edges and entry points, vast expanses of open space can defeat the human scale required for comfortable enjoyment. On the other hand, if a larger overall space is designed in such a way as to create precincts of different uses, attractions, and amenities throughout, then the effect can be positive. Many of the principles of definition that are relevant in great streets are also transferable to squares.

Edges should provide shade when it is warm and sunlight when it is cold, and primary (e.g., benches) and secondary seating (e.g., ledges along planters, niches in walls, steps) where people can stay and engage with the public space in front of them (Gehl, 2010; Whyte, 1980). People tend to want to linger in a square, standing or sitting, in the direct line of movement through the space (Whyte, 1980), to have lunch, read a book. Most of all, people want to see and be seen by other people. Just as in streets, edges of a square can be important exchange zones between adjacent buildings and the square's public space (Gehl, 2010). Café seating can spill out of a building fronting a square in the spring, summer, and fall. In the winter, the seating is all indoors, but the presence of hot chocolate so close by makes skating in the square that much more enjoyable. This is certainly the case at one of the most popular wintertime public spaces in Saskatoon's city centre, the Meewasin Valley Authority's public skating rink in the plaza next to the Bessborough Hotel, named the best outdoor skating rink in Canada by *Reader's Digest* in 2006 (Figure 13.13). With hot drinks available across the street and edges created by an outdoor fireplace, warming hut with wood-burning stove, ornate lighting (including coloured tree lights), and the magnificent hotel defining one side and the river valley another, the invitation to public activity even on the coldest days and evenings of winter is irresistible. It is not uncommon to see brides and grooms skating on it for wedding photos. The raw materials of urban design—including two of the most potent for winter cities, notably, colour and light—are masterfully combined.

Finally, the density and diversity of residents, commerce, and institutional uses that surround the square are fundamental to populating it with people at all times of the day (Jacobs, 1961). William Whyte (1980: 16) estimates that the "market radius" around a square is about three blocks for people working in the area. Within that distance, workers looking to sit outside for lunch

Nola Stein/Meewasin Valley Authority

Figure 13.13 Cameco Meewasin Skating Rink @Nutrien Plaza, Saskatoon

Ryan Walker

Figure 13.14 Square Saint-Louis, Montreal

will populate the square. Adult and children residents of the area will also walk with some frequency within that radius, making a square a regular node in their common living patterns. Square Saint-Louis—a popular and vibrant public space—and its surroundings in Montreal provide a good example of the "close-grained and lively support" discussed by Jacobs (1961: 377) that is so pivotal in thriving city spaces (Figure 13.14). Immediately bordering the square are medium-density housing, food, retail and office shop fronts, and BIXI docks. One block away is a Metro (subway) station. Well within three blocks of the square is high-density housing, a university satellite campus, a military regiment, government offices for health and social services, a tourism institute, two schools, a rehabilitation hospital, and vibrant commercial streets connecting on the north and south (Saint-Denis) and west (Prince-Arthur, a pedestrian street) sides of the square. The design of the square itself includes most of the features described in this section for vibrant squares.

Conclusion

Canada's most popular urbanism magazine, *Spacing*, assembled a panel of urbanists across the country to rank the 100 best public spaces in Canadian cities in its 2011 national issue (e.g., University Avenue in Charlottetown; Place d'Youville in Quebec City; Nathan Phillips Square in Toronto; 13th Avenue in Regina; Fernwood Square in Victoria). The Canadian

Institute of Planners Great Places in Canada contest awards the best public spaces in communities across the country, nominated and voted on each year. Public space ignites our emotions and attachment to place. It continues to be the connective tissue of our cities, our local economy, and our **citizenship**. The balance of proportions may have shifted over time between the movement, market, and meeting functions of public space, with vehicle movement dominating the street in the past several decades and consumption (market) and spectacle overtaking **citizenship** practices (meeting) in the public square. However, all the functions of public space remain, much as in the agora and the streets of the ancient Greeks, or Haussmann's boulevards in Paris. Our knowledge of public space in the context of social and cultural theory is constantly evolving. It is contentious precisely because all of the currents of the urban age take the stage at one point or another in public space, whether it is homelessness, celebration, leisure, or protest. While knowledge of physical design is well developed, the programming and management dimension of public space research has received far less attention, and is one of the areas where further empirical and educational focus would be highly valuable (Nikitin, 2011).

The study of the interaction between public space, activity, and urban form is fascinating, and some of the methods introduced here, and the lessons learned, will hopefully inspire readers to undertake field observation as a necessary (and enjoyable) dimension of understanding how Canadian cities function. Adding new layers of depth and complexity to the conceptualization, design, and programming of public spaces will enrich Canada's comparatively young (by international standards) cities. We hope this chapter motivates further exploration of public space theory and practice in all of its fascinating intricacy.

Acknowledgements

The authors appreciate the assistance given by Nancy Shoiry, Yannick Roy (cycling infrastructure), Wade Eide (Place du Musée), and François Gagné at Ville de Montréal. The authors thank David Gordon at Queen's University for sharing his perspective on light as a winter design treatment and Mike Velonas at Meewasin Valley Authority for his insight on street design. The authors recognize Jan Gehl at Gehl Architects for inspiring a deeper knowledge of how public spaces function on his visit to Saskatoon in 2009 and acknowledge Jeanna South, Rick Howse, Chris Hardwicke, and Mark Sterling for their collaboration on the public spaces, activity, and urban form work in Saskatoon.

Review Questions

1. The concept of "complete streets" focuses predominantly on only one of the three Ms of public space. Which one is it, and how might this influential concept in street design become truly "complete"?
2. Programming is an often overlooked dimension of enhancing public life and activity in public space. What are local artists, citizens, and business people teaching us about successfully programming public space?
3. Think of examples in your city centre of streets that exhibit good "definition," "stimulating design quality," and "complementarity." Think of examples that do not.
4. Describe some of the features that a square in the city centre requires in order to induce public life and activity. Do you have any in your city centre that exhibit these features or that fail to do so?

Note

1. A "sharrow" is a shared lane marking to communicate to both cyclists and motorists the lateral positioning of cyclists in the centre of the shared traffic lane. A "sharrow" does *not* denote a dedicated cycling lane, a common misconception. It simply informs motorists (and cyclists) that cyclists may ride in the full lane of traffic and need not ride pressed up closely against the curb. The word "sharrow" combines *share* and *arrow*, and the name is credited to a cycling-program official in San Francisco.

References

Akkerman, A. 2013. "Reclaiming the back alley," *Public Sector Digest: Intelligence for the Public Sector* Fall: 6–10.

Alexander, C., S. Ishikawa, M. Silverstein, with M. Jacobson, I. Fiksdahl-King, and S. Angel. 1977. *A Pattern Language: Towns, Buildings, Construction*. New York: Oxford University Press.

Ben-Joseph, E. 1995. "Changing the residential street scene: Adapting the shared street (woonerf) concept to the suburban environment," *Journal of the American Planning Association* 61: 504–15.

BIXI Montréal. 2018. Sommaire Financier et d'Activités BIXI Montréal: Année 2017: Regard sur 2018. Montreal: BIXI Montréal.

Borsa, J. 2004. "The ephemeral collection: Public repositories or sites for creative endeavour," in A. Kiendl, ed., *Obsession, Compulsion, Collection: On Objects, Display Culture, and Interpretation*. Banff, AB: The Banff Centre Press.

———. 2006. *Site- and Context-oriented Curating: Staging Encounters between Art and Everyday Realities*. Montreal: Concordia University, unpublished PhD thesis.

Carmona, M., C. de Magalhaes, and M. Edwards. 2002. "Stakeholder views on value and urban design," *Journal of Urban Design* 7: 145–69.

Cartiere, C. 2010. *Re/Placing Public Art: The Role of Place-specificity in New Genre Public Art*. Saarbrucken: VDM Verlag Dr. Muller.

Centre for Active Transportation, The. 2012a. *Complete Streets by Design: Toronto Streets Re-designed for All Ages and Abilities*. Toronto: The Centre for Active Transportation.

———. 2012b. "Assiniboine Avenue, Winnipeg," *Complete Streets for Canada*. At: www.completestreetsforcanada.ca/examples/assiniboine-avenue-winnipeg

City of Montreal. 2008. *Réinventer Montréal: Transportation Plan 2008*. Montreal: City of Montreal.

City of San Francisco. 2013. *San Francisco Parklet Manual*. San Francisco: City of San Francisco.

City of Saskatoon. 2011. *Public Spaces, Activity and Urban Form: Strategic Framework—City Centre Plan Phase 1*. Saskatoon: City of Saskatoon.

———. 2013. *City Centre Plan: Strategic Framework—City Centre Plan Phase 3*. Saskatoon: City of Saskatoon.

City of Toronto. 2013. *Celebrate Yonge—Post-event Review*. Toronto: City of Toronto.

Copenhagenize Design Co. 2019. *The 2019 Copenhagenize Index of Bicycle-Friendly Cities*. At: http://copenhagenizeindex.eu

Damant-Sirois, G., M. Grimsrud, and A. El-Geneidy. 2014. "What's your type: A multidimensional cyclist typology," paper presented at Transportation Research Board annual meeting. Washington, DC: January.

Engwicht, D. 1993. *Reclaiming Our Cities and Towns: Better Living with Less Traffic*. Gabriola Island, BC: New Society Publishers.

Florida, R. 2004. *Cities and the Creative Class*. London: Routledge.

Galland, D., and C. Hansen. 2012. "The roles of planning in waterfront redevelopment: From plan-led and market-driven styles to hybrid planning?" *Planning Practice and Research* 27: 203–25.

Gehl, J. 2010. *Cities for People*. Washington, DC: Island Press.

———, and L. Gemzoe. 2008. *New City Spaces*. Copenhagen: Danish Architectural Press.

———, and B. Svarre. 2013. *How to Study Public Life*. Washington, DC: Island Press.

Goheen, P. 1998. "Public space and the geography of the modern city," *Progress in Human Geography* 22: 479–96.

Gordon, D. 1997. "Managing the changing political environment in urban waterfront redevelopment," *Urban Studies* 34: 61–83.

Greenberg, K. 2011. *Walking Home: The Life and Lessons of a City Builder*. Toronto: Vintage Canada.

———. 2013. "Students and emerging leaders," presentation at Council for Canadian Urbanism National City-building Leadership Summit. Halifax, NS: September.

Hough, M. 2004. *Cities and Natural Process: A Basis for Sustainability*. London: Routledge.

Jacobs, A. 1993. *Great Streets*. Cambridge: MIT Press.

Jacobs, J. 1961. *The Death and Life of Great American Cities*. New York: Vintage Books.

Lynch, N., and D. Ley. 2010. "The changing meanings of urban places," in T. Bunting, P. Filion, and R. Walker, eds., *Canadian Cities in Transition: New Directions in the Twenty-first Century*, 4th edn. Toronto: Oxford University Press.

Madanipour, A. 2003. "Why are the design and development of public spaces significant for cities?" in A. Cuthbert, ed., *Designing Cities: Critical Readings in Urban Design*. Malden: Blackwell Publishing.

McHarg, I. 1969. *Design with Nature*. Garden City, NY: Natural History Press.

Nikitin, C. 2011. *Personal Communication*, March 1, Vice-President for Downtowns, Project for Public Spaces and 2011 Planner in Residence, University of Saskatchewan, Saskatoon.

Project for Public Spaces (PPS). 2005. *How to Turn a Place Around: A Handbook for Creating Successful Public Spaces*. New York: Project for Public Spaces, Inc.

———. 2013. *10 Principles for Successful Squares*. At: www.pps.org/reference/squaresprinciples/

Relph, E. 1976. *Place and Placelessness*. London: Pion.

Teschke, K., M. Harris, C. Reynolds, M. Winters, S. Babul, M. Chipman, M. Cusimano, J. Brubacher, G. Hunte, S. Friedman, M. Monro, H. Shen, L. Vernich, and P. Cripton. 2012. "Route infrastructure and the risk of injuries to bicyclists: A case-crossover study," *American Journal of Public Health* 102: 2336–43.

Tuan, Y. 1974. "Space and place: Humanistic perspective," *Progress in Geography* 6: 211–52.

Whyte, W. 1980. *The Social Life of Small Urban Spaces*. New York: Project for Public Spaces.

Zukin, S. 2003. "The urban landscape," in A. Cuthbert, ed., *Designing Cities: Critical Readings in Urban Design*. Malden, MA: Blackwell Publishing.

14 Integrated Transport and Land-Use Planning

Connecting Individuals to Opportunities

Ahmed El-Geneidy and Emily Grisé

Introduction

There are many reasons why people and firms locate in cities. For some, it may be the bright lights; for others, it may be the necessity to find work—there are perhaps as many reasons to locate in a city as there are people and firms. One of the most critical reasons is that a city enables people, households, and firms to gain access to each other (see Filion, Chapter 2). This chapter looks at transport in Canadian cities through the lens of *accessibility*, a notion that helps in understanding and conceptualizing the complex relation between transport and land use in a city and their impacts on city organization, development, and planning to achieve more sustainable outcomes. The study of transport and land-use interactions more realistically reflects the behaviour and perceptions of individuals and firms, as the transport system itself is not what is important—rather, it is the fact that the transport system connects individuals to desired destinations.

The following section describes and defines the key terms of *transport* and *land use* and how they relate to accessibility. The second section examines the history of transport in Canadian cities, and the role of transport and land use in their creation and development. This is followed by a discussion of the current state of transport in Canadian cities in comparison to other urban regions around the world in terms of congestion levels and mode share. The fourth section concentrates on the activity dimension of the land-use and transport cycle: we will show how travel demand is typically modelled, how this has changed over time, and how planners and engineers seek to improve their modelling techniques today. The fifth section moves the attention away from more mobility-focused issues to examine accessibility measures and how they are calculated. The chapter ends with a discussion of the future of the land-use and transport system in Canadian cities.

Transport, Land Use, and Accessibility

Alexander et al. (1977, p. 62) once expressed the importance of "[putting] the magic of the city within the reach of everyone in a metropolitan area." Alexander and his colleagues were promoting an ideal city in which a person could easily reach desired destinations—or *opportunities*. In the field of transport and land use, this concept is known as *accessibility*. Both people and firms locate in cities in order to gain access to destinations (opportunities) with which they can, and want, to interact. Accessibility, or the ease of reaching valued destinations (Hansen, 1959), is often confused with mobility—defined as the ability to move from one place to another (Handy, 1994).

Since the widespread use of the private automobile, and in particular since World War II, transport planning has been dominated by a car-oriented and mobility-focused paradigm.

Within this paradigm, speed, efficiency, and fluidity of automobile transport have been central concerns, using indicators such as congestion as one of the main measures of mobility. For example, the Texas Transport Institute (TTI) generates an annual urban mobility report for cities in the United States, which receives wide media attention every year. The report includes various measures of congestion, such as delays and total number of hours wasted per year due to congestion. In contrast, the concept of accessibility is concerned not only with how easy it is to get from one place to another, but also with what one can reach. The use of accessibility measures often leads to different conclusions about the same transport and land-use situation compared to mobility measures when they are used alone. While the concept of accessibility has been around for more than 50 years and has been a common element in the goals and objectives of transport plans (Boisjoly and El-Geneidy, 2017; Handy, 2002), its actual adoption in transport planning has been limited until recently. Access Across America is a new initiative that started recently at the University of Minnesota where measures of accessibility across the 49 largest metropolitan regions in the United States are generated every year for public transport, automobile, and walking (Owen, Murphy, and Levinson, 2017).

Since accessibility evaluates not only the ease with which one can go from one place to another but also what can be reached, it is considered one of the most comprehensive planning performance measures and an important contributor to better understanding the complex relationship between land-use and transport systems in a region (Wachs and Kumagai, 1973). This relationship has been simplified by Giuliano and Agarwal (2017) (Figure 14.1).

Transport and land use are mutually dependent, and consequently changes to the transport system influence changes in land use, and similarly changes in land use will affect transport (Figure 14.1). Characteristics of the transport system (e.g., network and reach) influence the level of accessibility in a region or the ease of reaching opportunities. The accessibility of the urban system in turn shapes the land use or the distribution of activities (e.g., households and firms), as it is assumed that firms and households have a preference to locate in places with good accessibility. The location of activities and the transport system connecting individuals to activities affect activity patterns, which in turn shape the travel patterns of individuals in a region. Lastly, travel patterns, which can be expressed as flows on the transport network, influence regional mobility levels (i.e., congestion), and future changes in levels of accessibility will result.

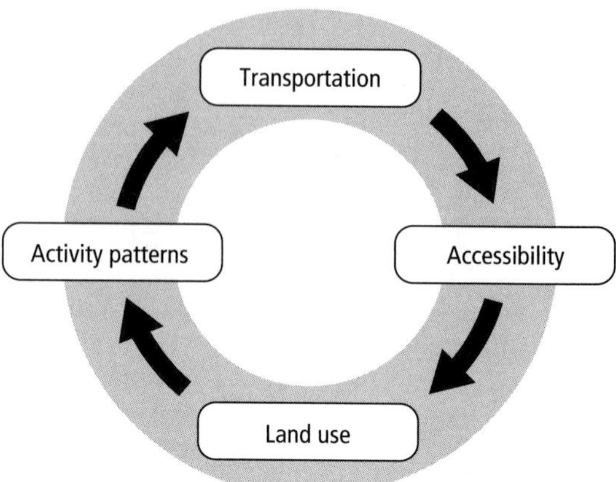

Figure 14.1 The Transport and Land-Use Cycle
Source: Adapted from: Giuliano, G., & Agarwal, A. 2017. "Land use impacts of transport investments: Highway and transit." In G. Genevieve & S. Hanson (Eds), *The Geography of Urban Transport* (Fourth edn). New York: The Guilford Press.

Transport Systems in Canadian Cities

History of Transport Systems and City Structure in Canada

Development and growth of cities in Canada (and indeed everywhere) were largely shaped by their access to different transport modes. For example, Montreal owes its origin to the fact that it was the furthest inland port that trans-Atlantic ships could reach (before the construction of the Lachine Canal, and the St Lawrence Seaway afterwards) because of the rapids surrounding the Island of Montreal (Shearmur and Rantisi, 2011). Winnipeg owes its origin to being at the junction of the Red and Assiniboine Rivers—a critical link along continental fur-trading routes (Artibise, 1977).

Until the 1860s, most internal transport within Canadian cities was done by foot, by horse, or by cart (Nader, 1975). These cities were typically quite small, dense, and characterized by juxtaposing and overlapping residential, commercial, and industrial uses, with the centres of these cities located near the terminals of long-distance transport routes. The first public transport systems began appearing in Canadian cities in the mid-1800s with the development of horse-drawn streetcar routes. By the end of the nineteenth century, electric streetcars replaced their horse-drawn predecessors. The streetcar systems allowed households to locate further away from the urban centres since it was now possible to travel greater distances, allowing households access to cheaper land and larger residences while still maintaining good accessibility to jobs in the urban core. Most economic activity, however, remained concentrated in the urban core because transport of goods did not experience the same improvements as public passenger transport (i.e., freight was not moved on streetcars, so rapid increases in mobility did not occur). This resulted in residential development around streetcar lines, creating "radial" or star-like urban development patterns. Commercial development also concentrated along streetcar routes, where people waiting to embark or upon disembarking took advantage of easy access to shops and services, reinforcing their commercial viability (see Filion, Chapter 2). The link between innovations in transport technology and the resulting changes to land use can be characterized as a positive feedback system (Levinson, 2008), where the construction of rail stations motivated population density and new development, and vice versa.

Despite the significant role of the development of the rail network in shaping present-day city form, no other mode of transport has impacted land-use or travel behaviour as dramatically as the private automobile (Wachs, 1993). The internal combustion engine, already in use at the turn of the twentieth century, did not revolutionize the transport and land-use system until the 1920s, when personal automobiles gained popularity. The rise in popularity of personal automobiles at this time can largely be attributed to the first mass-produced Ford Model T, which largely reduced the cost of owning a private automobile. The internal combustion engine also allowed an increase in freight transport by truck. The main implication of automobiles and trucks was to improve accessibility to large areas around the traditional urban cores, thus allowing for residential development and the decentralization of goods-handling activities in these areas. Since gas-powered vehicles were much less limited in where they could go compared to streetcars, development spread further and filled in areas previously undeveloped between streetcar lines, giving cities a more concentric development as in the pre-streetcar era. Several cities started to abandon streetcars due to lack of demand and increased cost, yet some cities, such as Toronto, did fight the demolition of streetcars as they saw the positive impact they do play in a region. Today Toronto has 11 streetcar lines running across the city and operating some newly purchased vehicles along the same old routes.

After World War II, this tendency toward expanding cities was exacerbated by the construction of controlled access freeways. Even greater stretches of undeveloped land now had much greater accessibility by car and were developed into low-density suburbs (Perl and Kenworthy, 2010). The freeway system not only enticed households to move in great numbers to these suburban locations but also encouraged manufacturing and other industries to suburbanize. Large regional shopping centres developed as well to serve the rapidly suburbanizing regions of urban Canada (Yeates and Garner, 1980). The result of this increasing decentralization of households and industry is commonly referred to as **urban sprawl**. *Sprawl* is a difficult term to define, but easy to recognize. While it has been characterized in many different ways, common elements to its characterization are low-density urban development, where residential, commercial, industrial, and institutional uses tend to be separated, thus developing a dependence on the private automobile for movement between necessary daily functions in the city and making public transport and active transport mode choices inefficient (see Filion, Chapter 2; Addie, Fiedler, and Keil, Chapter 16).

Current Transport System in Canada

Currently, many Canadian cities are suffering from the secondary effects that came with the post-World War II city-building decisions favouring the private automobile and creating sprawl. Congestion along freeways and major arterials is a common element in most Canadian cities. Private car ownership has become the norm among most socio-economic groups, while commuting by car is becoming a real burden to the population for various reasons, including stress (Legrain, Eluru, and El-Geneidy, 2015; Loong, van Lierop, and El-Geneidy, 2017). The proportion of trips made using different transport modes—called mode share—is one common indicator used to describe transport systems. Overall, Canada is an automobile-oriented country, with 79.5 per cent of work trips made by car or truck (Statistics Canada, 2016b). Mode share for public transport has slowly but steadily been on the rise since the first collection of this information in the 1996 census. In 1996 10.1 per cent of Canadians used public transport to commute to work, while this number increased to 12.4 per cent in 2016. The opposite trend is observed for the combined mode share for cycling and walking, falling from 8.1 per cent in 1996 to 6.9 per cent in 2016. Together, public transport, walking, and cycling are referred to as "active" or "sustainable" transport modes. While the term *sustainable* can have many elements, sustainable transport is most often referred to in the context of environmental sustainability. These modes are considered sustainable because, unlike personal forms of motorized transport, they generally produce fewer (as in the case of public transport) or no emissions (walking and cycling). Furthermore, over the last decade, there has been increased research interest in the relationships between the use of active or sustainable transport and individual health. One of the most widely recognized benefits of active modes of travel is the increase in daily physical activity and accordingly the potential of reducing obesity levels (Wasfi, Dasgupta, Orpana, and Ross, 2016). Additionally, there is a growing body of literature related to wider health benefits associated with non-automobile travel, including positive influences on an individuals' subjective well-being and perceived quality of life that have been associated with adopting active modes of travel (Sallis et al., 2015; van Wee and Ettema, 2016).

Figure 14.2 shows the variation in 2016 public transport mode shares by the population of **census metropolitan areas (CMAs)** (Statistics Canada, 2016a). The graph indicates that a larger proportion of work commutes are done using public transport as the size of the CMA increases. Public transport mode share increases up to over 20 per cent in the largest Canadian CMAs, such as Toronto (24.3 per cent), Montreal (22.3 per cent), and Vancouver (20.4 per cent), while it is between 10 and 20 per cent for smaller CMAs, such as Ottawa-Gatineau, Edmonton, Calgary, and Winnipeg. However, the link between population and transit mode share does not hold true for all cities. Consider some of the smaller CMAs, such as Halifax and Victoria, that have public transport mode shares above 10 per cent even though their populations are approximately the same as other CMAs, like London or Regina, that have low shares of transit users (less than 8 per cent). The high mode share by public transport in metropolitan areas can be linked mainly to population size, congestion levels, cultures, and socio-economic disparities. While these city-level, "aggregate" mode share statistics are informative, they hide a great deal of variation in mode share within cities, across different demographics as well as across different kinds of trips. Within cities, public transport, cycling, and walking mode shares tend to be lowest in suburban areas, and highest toward the city core.

International comparisons can highlight areas in which the Canadian transport system could improve, or bring to light aspects that are well advanced but may be further encouraged. Table 14.1 shows the mode share of commute trips for six Canadian cities and for other cities in the United States and Europe, focusing on the percentage of commute trips made by public transport and the two other main active forms of transport (i.e., cycling and walking). Canada's largest CMAs are roughly halfway between US and European cities in terms of public transport mode share (11 to 24 per cent). This share is higher than the typical city in the United States (except for New York), whereas it is typically much higher in Europe. We see similar

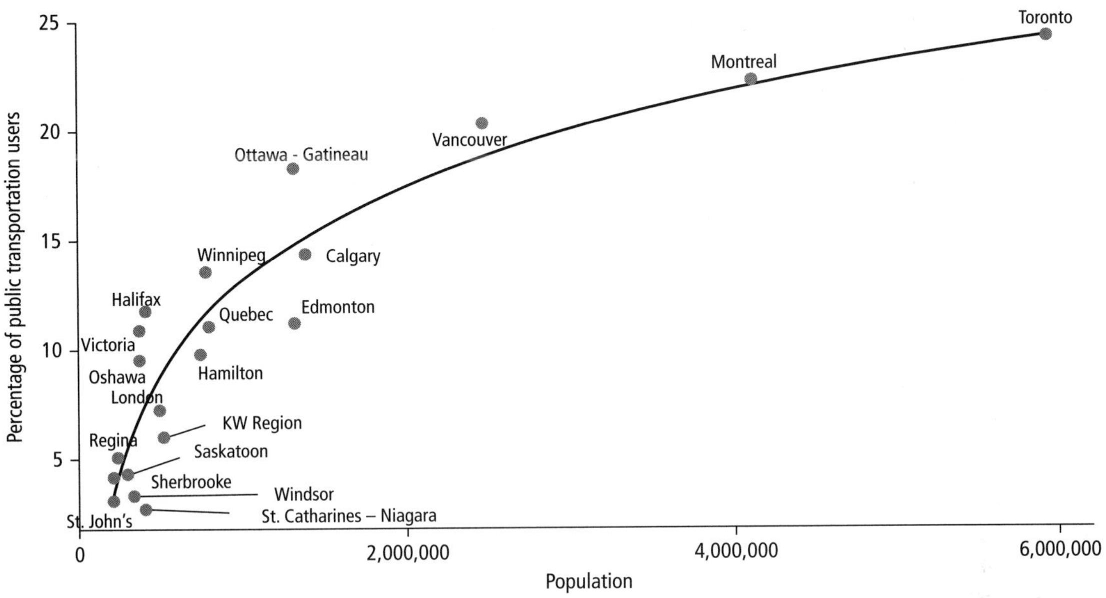

Figure 14.2 Share of Public Transport Users in Canadian CMAs by Population in 2016
Data source: Statistics Canada (2016a).

patterns in terms of cycling and walking in the six selected Canadian CMAs and cities in the United States; however, in Europe, they increase considerably. These differences in active mode shares are likely related to several factors, including the age (and therefore structure) of the city, land-use distribution, the domination of automobile culture, or the harsh winters in some Canadian cities.

It is also worthwhile to see how mode shares have been evolving over time. Figure 14.3 shows public transport mode share for commuters in CMAs with populations over one million. The data were obtained from the 1996, 2001, 2006, and 2016 censuses and from the National Household Survey for the year 2011 (Statistics Canada, 1996, 2001, 2006, 2011, 2016a).[1] For Canada's six largest CMAs, we can observe a pattern of increase in the share of daily trips by public transport, with the exception of Ottawa-Gatineau and Calgary, where these cities experienced a decline in the share of daily trips by public transport between 2011 and 2016. With the exception of these two cities, the trend for increasing public transport mode share is encouraging, since public transport is a more sustainable form of transport for cities than private automobiles; moreover, this shows that Canadian cities and citizens are capable of changing their mode shares. At the same time, based on a comparison with other regions around the world and with Europe in particular (Table 14.1), there is still room for improvement given that higher levels have been achieved in other cities.

Another common indicator used to characterize a transport system is the level of congestion.[2] Issues of congestion have always been prominent in transport planning, especially within the mobility paradigm of car-oriented North American cities. In this context, congestion is considered problematic since it compromises mobility (as well as accessibility) but also because congestion is seen today as detrimental to the environment and to public health. Indeed, high congestion means higher greenhouse gas (GHG) and other emissions from private vehicles. Congestion is, moreover, a cause of stress to road users. Table 14.2 summarizes the congestion levels for Canada's six largest CMAs, and compares the Canadian context to other cities in the US and Europe. Based on these measures, there is quite a bit of variation in how Canadian cities "perform" when compared with

Table 14.1 Percentage of Commute Trips Using Public Transport, and Cycling and Walking in Selected Major Canadian, US, and European Cities

Country/ continent	City	Commute mode share: Public transport	Commute mode share: Cycling and walking	Year
Canada	Toronto	24.3	6.7	2016
	Montreal	22.3	7.2	2016
	Vancouver	20.4	9.1	2016
	Ottawa-Gatineau	18.3	8.7	2016
	Calgary	14.4	6.2	2016
	Edmonton	11.3	4.7	2016
United States	New York	56.6	11.1	2016
	Chicago	27.8	8.3	2016
	Los Angeles	10.1	4.7	2016
Europe	Madrid	63.0	24.0	2015
	Paris	80.0	60.0	2015
	Stockholm	65.0	50.0	2015

Note: Respondents were given the option of selecting more than one means of transport for their commute to work; therefore, the modal shares may rise to over 100 per cent.
Source for American cities: U.S. Census Bureau, 2012-2016 American Community Survey 5-Year Estimates (United States Census Bureau, 2016).
Source for Canadian cities: 2016 Canadian Census (Statistics Canada, 2016a).
Source for European cities: Eurostat (European Commission, 2017).

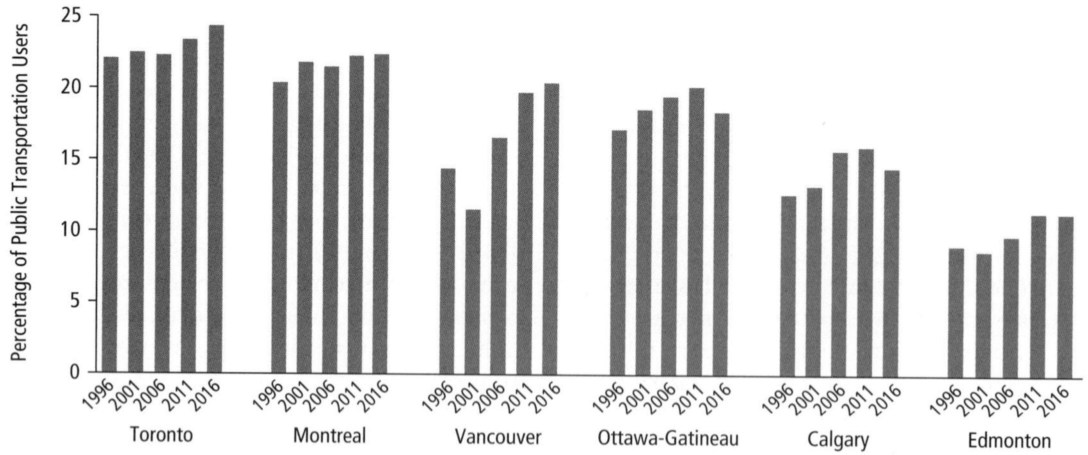

Figure 14.3 Change in Share of Daily Trips by Public Transport in Canadian CMAs with Populations over One Million, 1996–2016

Source: Statistics Canada, 1996, 2001, 2006, 2011, 2016a.

Table 14.2 Congestion Levels and Delay per Hour Driven in Peak Period, 2016

Country/ continent	City	Congestion level (%)	Delay per hour driven in peak period (minutes)
Canada	Toronto	30	34
	Montreal	29	33
	Vancouver	39	38
	Ottawa-Gatineau	28	33
	Calgary	20	21
	Edmonton	20	19
United States	New York	35	34
	Chicago	26	29
	Los Angeles	45	44
Europe	London	40	40
	Madrid	25	27
	Paris	38	40
	Milan	30	35
	Stockholm	28	33
	Moscow	44	49
South Africa	Johannesburg	30	37

Source: Based on TomTom North American Congestion Index: TomTom International B.V. At: http://www.tomtom.com/lib/doc/trafficindex/2012-0704-TomTom-Congestion-index-2012Q1namerica-km.pdf

other cities around the world. Vancouver, the most congested CMA in Canada, approaches the same congestion levels as Los Angeles and Paris.

Travel Demand Modelling

One of the most important roles of transport planning is to forecast travel demand. Travel demand modelling therefore involves a number of statistical and mathematical techniques designed to estimate what future travel demand will look like given different transport policy and investment scenarios. In the past, and since the development of these techniques at the end of the 1950s, the primary indicators of interest for travel demand modelling have been concerned with mobility. As such, different transport scenarios are evaluated based on their ability to reduce congestion and to increase speed and fluidity—particularly during peak periods. As we will show, traditional methods typically take the land-use system as a given and use this to estimate activity patterns. These activity patterns are what determine overall transport demand, and this, in conjunction with the transport system, results in various measures of mobility, such as congestion and travel times between parts of a region. While mobility is a critical part of accessibility, accessibility has not been central to traditional travel demand analysis. In fact, the link between accessibility and the land-use system has often been ignored. The traditional approach to travel demand modelling has been dominated by the four-step model, outlined in Figure 14.4 (Meyer and Miller, 2000). You can find detailed descriptions of the modelling process in many planning textbooks (e.g., Meyer and Miller, 2000; Ortuzar and Willumsen, 2011; Vuchic, 2005).

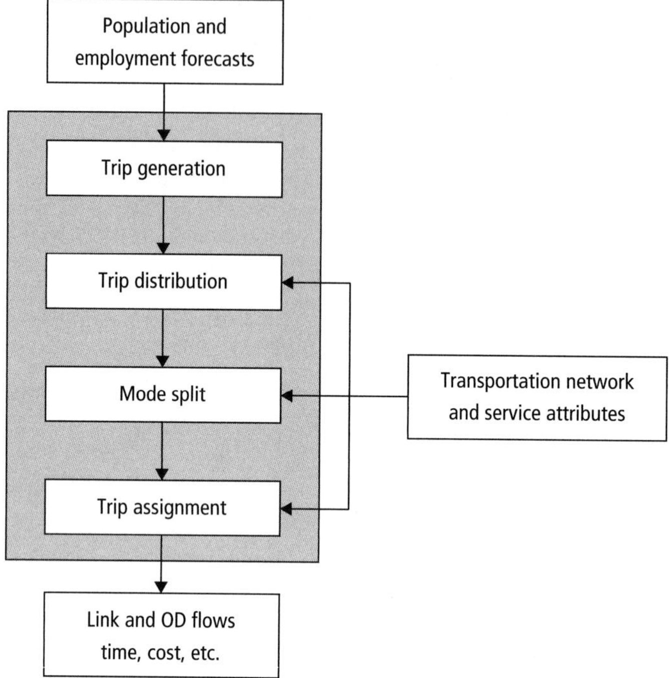

Figure 14.4 Four-Step Travel Demand Model

Source: Meyer, Michael and Eric Miller. *Urban Transport Planning*, 2/e © 2001, McGraw Hill Education.

In order to go through the four-step modelling process, the region under study is first divided into transport analysis zones (TAZs). These zones can be thought of as neighbourhoods where trips both originate and end. Note that the model is trip-based, meaning that transport demand is represented by individual and independent trips. Inputs to the model are generally forecasts of the number of people and employment in the different TAZs, as well as information on the existing transport network, including its service attributes. The steps of the model are as follows:

1. Trip generation determines the number of trips produced by, and coming to, every TAZ. In other words, it predicts total flows coming into and going out of each zone without stating the origin or the destination zone of these flows.
2. Trip distribution matches origin with destination flows. It links the trips generated from one zone with trips ending in another zone.
3. Mode choice calculates the proportion of trips between all origin and destination pairs that uses a given available transport mode (automobile, transit, walking, cycling, etc.).
4. Route assignment assigns trips between an origin and destination by a particular mode to a route using the respective modal network. Route assignment can be static or dynamic depending on the complexity of the model being used. Dynamic traffic assignment uses finer increments of time (minutes instead of hours, e.g., during rush hour) and more explicitly models vehicle behaviour.

The main results of the four-step process are as follows: predicted traffic flows by mode between different zones of the urban system (origin-destination [OD] flows); predicted traffic flows along road segments (link flows); travel times between zones (OD travel times); travel times along links (link travel times) as well as link and OD measures of congestion. As mentioned above, congestion is measured as the difference between travel times during free flow conditions (when there is little traffic, e.g., at night) and congested travel times (during rush hour). The model might be executed several times to evaluate different scenarios.

The four-step modelling process was developed, and is used, primarily for modelling passenger transport. Freight transport, however, is an increasingly important component of urban transport demand. Yet it has been virtually ignored due to its complexity and lack of data for several decades. As a result, freight modelling represents an important direction for future research in transport planning (Ortuzar and Willumsen, 2011).

Although the four-step model has been used extensively since its development, it suffers from a number of weaknesses that new methods are looking to overcome. Its first disadvantage is that it oversimplifies the complex processes involved in the travel decision-making and trip generation process. Since it concentrates on explaining trips between geographically aggregate TAZs, it is difficult (in some cases impossible) to model individual or household behaviour that is the basis of travel demand and that can be very complex. The four-step model is thus referred to as a trip-based model since it concentrates on modelling aggregate trips (aggregated to the level of the TAZ). With this approach, the trips made are abstracted from the people making them.

In response to this limitation, transport planners and engineers have been working to improve this method by developing activity-based models. Activity-based modelling, a new approach currently used by several agencies, is built upon the behaviour and activities of individuals. As such, the activities of each individual as well as his or her trips (their number, their mode, etc.) are simulated as a function of individual characteristics, as well as of the transport system itself. In addition to providing a richer and better understanding of travel behaviour, activity-based models address another weakness of traditional methods—that the traditional modelling framework cannot account for **induced demand** (i.e., extra trips that are made when new infrastructure is built).

Induced demand is composed of two elements. The first element refers to trips that are generated simply due to the presence of new infrastructure. Assuming this new infrastructure makes travel easier, some people will decide to make trips they did not make before. This part of induced demand cannot be captured easily by the regular travel demand model (McNally, 2000), but it can be captured through an activity-based approach. The second aspect of induced demand is related to households or firms attracted to locate near the new infrastructure to benefit from the higher levels of mobility or accessibility that is offered by its presence. This change in land use will lead to an increase in the number of trips as well. Such an increase also cannot be well captured through the regular four-step model; rather, it requires an integrated land-use and transport modelling approach.

Integrated transport and land-use models are models that help forecast the distribution of households and firms as a function of various factors, including the transport system. While land-use forecasting has evolved more slowly than transport modelling, the availability of increasing amounts of geographic data and computing power has resulted in the development of various land-use and integrated transport and land-use models (for examples, see Iacono, Levinson, and El-Geneidy 2008).

Another major drawback of the four-step travel demand model has been the focus on modelling private automobile flows while ignoring other modes of transport. To address this shortfall, engineers and planners in the past few decades have increasingly incorporated public transport in their models and, most recently, cycling and walking (Currans and Clifton, 2015). This shift in focus is related to the rise of various arguments stating that cities cannot, and will not, manage to simply "build their way out" of congestion—a radical and new argument for the field of transport and mobility studies (Downs, 2004).

Finally, some planners and engineers argue that instead of focusing on the construction of new infrastructure and its future impacts, transport planning should focus on **travel demand management** through better planning of land use and concentrating on individual behaviours and the rationale behind them.

Destinations Are the Goal: Accessibility Measures

Although some people travel simply for the joy of it or to use a particular mode, reaching a desired destination is most often the goal of travel in urban areas. Accessibility, which is the ease of reaching these valued destinations (Hansen, 1959), is not easily communicated or measured as a number, which may explain why planners have continued, until recently, to concentrate on mobility issues. In spite of this difficulty, research has shown that accessibility within a region does have noticeable effects on travel behaviour and can have considerable economic impacts on home values (El-Geneidy and Levinson, 2006; Franklin and Waddell, 2003; Levinson, 1998; Manaugh and El-Geneidy, 2012). Accessibility is therefore important to consider in urban and transport planning. Indeed, transport systems are designed to help people and firms participate in activities distributed across space and over time. In order to discern whether this is occurring successfully or equitably, however, some type of measure is required. Accessibility is a useful concept for this, as it indicates the collective performance of the land-use and transport systems and thus determines how well that complex system serves its residents. As a result, a large number of review studies classify and evaluate accessibility measures (in the measurement meaning of the word) according to various criteria (Baradaran and Ramjerdi, 2001; Cerdá and El-Geneidy, 2010; El-Geneidy and Levinson, 2006; Geurs, 2006; Geurs and Ritsema van Eck, 2001; Handy and Niemeier, 1997; Koenig, 1980).

Most accessibility measures are built around the same basic components. On the one hand, the activity component is a measure of the land-use system, which is itself represented by the presence of destinations, such as jobs, restaurants, daycares, health-care facilities, and households. These destinations, or opportunities, can also be weighted to account for their attractiveness or competition effects in relation to each other. On the other hand, the transport component consists of the transport system using a particular mode of transport (e.g., car, walking, cycling). The transport component is then quantified—for example, by calculating the distance separating the origin and destination or by estimating the travel time between these two points, in congested and uncongested contexts. Congested travel times are generally obtained from travel demand models, as presented in the previous section. Many precise and theoretically sound measures of accessibility have been developed, but they are not necessarily based on readily available data; moreover, some are complex to calculate and interpret. Note that accessibility measures do *not* replace measures of mobility. Regions should work on generating both measures since they complement each other in generating a better picture of a region's transport and land-use systems.

One of the most commonly used measures of accessibility is the cumulative opportunity measure, which was among the earliest to be developed and is also among the simplest to calculate and communicate (Vickerman, 1974; Wachs and Kumagai, 1973). Cumulative opportunity reflects the number of opportunities available from a predetermined point within a given travel time or travel distance, using a given mode of transport or combination of modes. For example, cumulative opportunity can be used to generate accessibility to jobs by car. To do so, the number of jobs that can be reached by car in a certain threshold of travel time, such as 30 minutes, is counted for a given area. Cumulative opportunity can also be used to generate, for instance, accessibility to restaurants by bicycle. This is done by counting the number of restaurants a person can reach from a given point by bicycle within a certain threshold of travel time. As an example of the cumulative opportunities measure, Figure 14.5 shows accessibility to jobs by public transport in a 45-minute travel time threshold for the Greater Vancouver Region in 2016. The number of jobs—and, thus, job accessibility—decreases as one moves out from the downtown area and away from the dense transport network provided in the central city (the SkyTrain [rapid transit] network, in particular).

An alternative to the cumulative opportunity measure is the gravity-based opportunity measure (Hansen, 1959), which is also a commonly used measure of accessibility. In this measure, destinations are weighted according to their proximity to the point of origin. As a result,

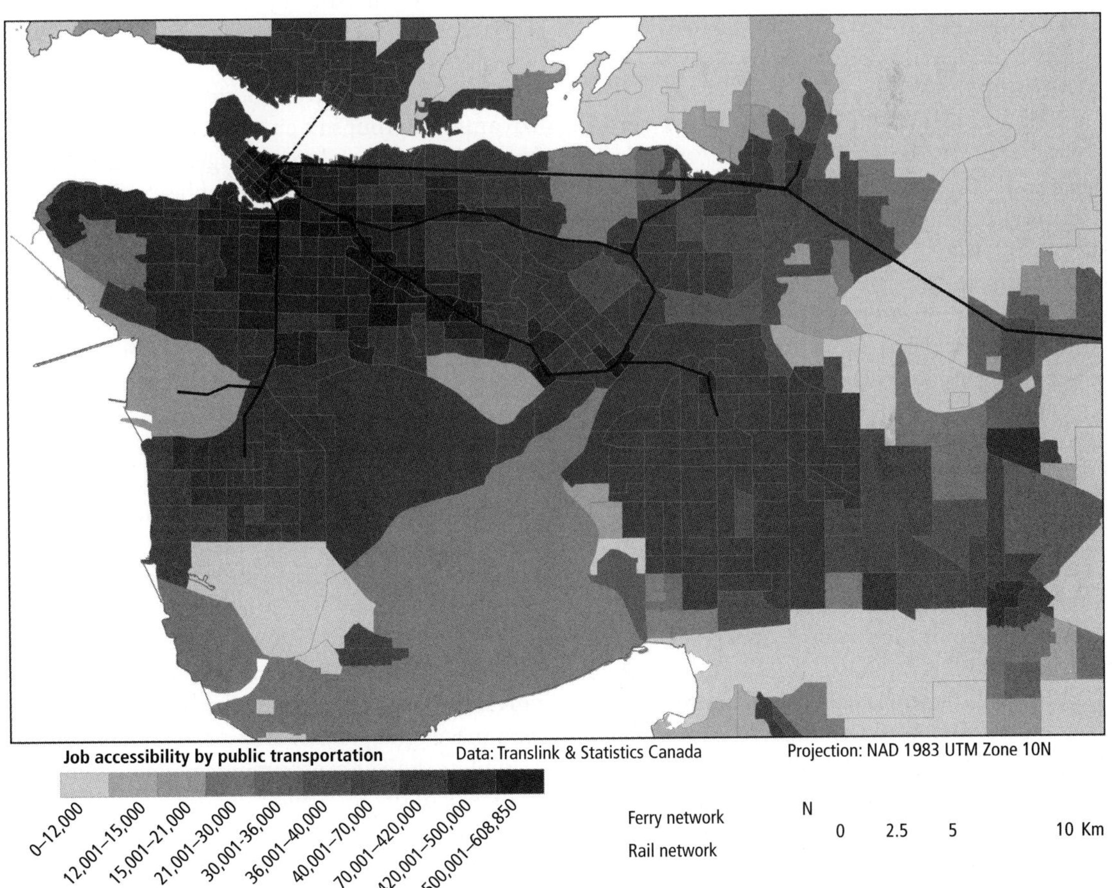

Figure 14.5 Accessibility to Jobs within 45 Minutes Travel Time by Transit, Greater Vancouver Region

Source: Data from Translink and Statistics Canada.

distance or travel time affects the value of a destination. An empirically determined distance or travel-time decay function for a given transport mode can be derived based on travel surveys. The gravity-based measure of accessibility presents several advantages. First, it is more theoretically sound than the cumulative measure because it discounts opportunities according to their distance or travel time from an origin, rather than using an artificially determined threshold (e.g., cumulative measure includes opportunities within 30 minutes of travel time but not those requiring 31 minutes or more). Second, the gravity measure is more representative of how users perceive the transport system because it strikes a balance between the utility of a destination and the cost of travelling to it from a given origin (Miller, 2005).

The gravity measure also presents certain disadvantages compared with the cumulative measure. Although it is widely accepted in the transport planning literature, the gravity measure is relatively complex to calculate and can be difficult to interpret and explain to the general public or to decision-makers. In contrast, cumulative measures allow for easy comparison of accessibility across modes and types of destinations, thereby simplifying the interpretation of results and their discussion in public forums. Finally, cumulative opportunities, as opposed to gravity-based measures, do not rely on assumptions about the *value* of destinations to users but only assume the chosen threshold in travel time or distance. Some argue that this is a more objective approach (Geurs and Ritsema van Eck, 2001).

Accessibility measures are increasingly used to conduct social equity assessments. In such cases, studies have looked at accessibility in a more disaggregated manner—namely, by focusing on a specific transport mode and/or a specific group of users. For example, Grisé, Boisjoly, Maguire, and El-Geneidy (2019) considered accessibility by public transport for people in a wheelchair in order to highlight and quantify the performance of the public transport network in a region, in terms of providing service to people in a wheelchair, and compared that level of service to an individual not in a wheelchair. After constraining the public transport network to only stations and vehicles that are universally accessible (i.e., free of stairs or other physical barriers), the authors measured to what extent these barriers across the public transport network impact the ability of an individual travelling in a wheelchair to access jobs[3] in 2011. The results for the Island of Montreal are shown in Figure 14.6, highlighting priority areas for the construction of elevators in the rail network or the need to introduce accessible vehicles on bus or streetcar routes. On average, wheelchair users in Montreal only have access to 46 per cent of the jobs that are accessible to transit users who are not in a wheelchair. Measuring the level of transit (in) accessibility currently experienced by these disadvantaged groups in cities such as Montreal can benefit transport planning agencies that aim to promote equity within the public transport network based on service supply and provision for wheelchair users.

Another example of the application of accessibility in transport planning is through evaluating the benefits of proposed projects in a transport plan. Figure 14.7 shows two maps visualizing the accessibility to jobs in the Greater Toronto-Hamilton Area (GTHA) before and after the complete implementation of "The Big Move" plan in the year 2031. The map demonstrates areas within the region that will benefit the most from higher levels of job accessibility by transit, if the plan is successfully implemented. The 2031 transport plan for the (GTHA) (Metrolinx, 2008) is a bold plan organized around the belief that transport within the GTHA must be transformed. The plan places the priority first on public transport, second on active transport, and only third on the efficiency of the road and highway network.

Taken together, Canadian cities seem to be embracing the shift away from the mobility-fixated approach to transport planning and toward a comprehensive framework of sustainable accessibility. Although the automobile remains the most common mode of transport in Canadian cities, there is a growing awareness of the need for

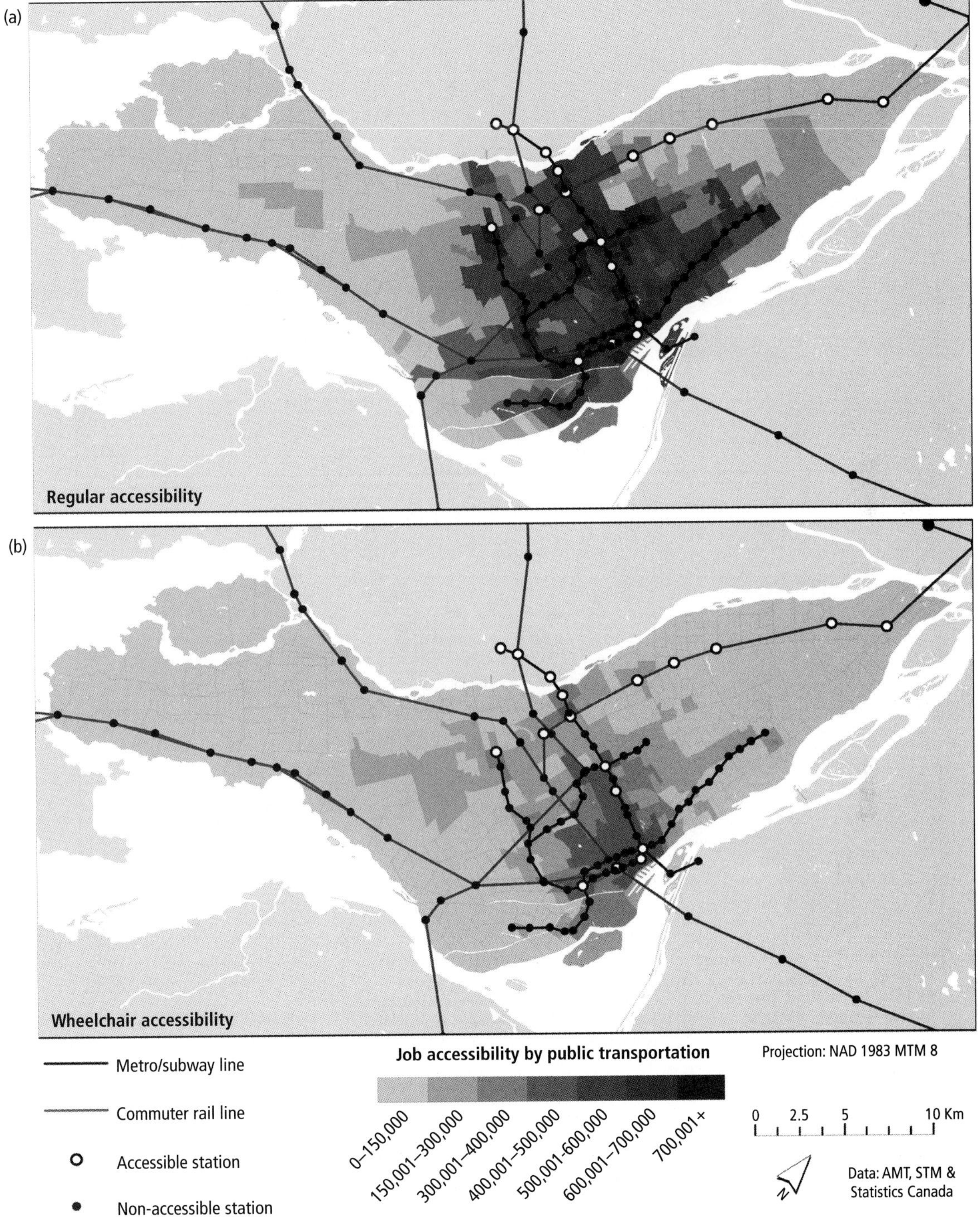

Figure 14.6 Job Accessibility by Public Transport in Montreal

Source: Grisé, E., Boisjoly, G., Maguire, M., and El-Geneidy, A. (2019). *Transportation Research Part A: Policy and Practice.*

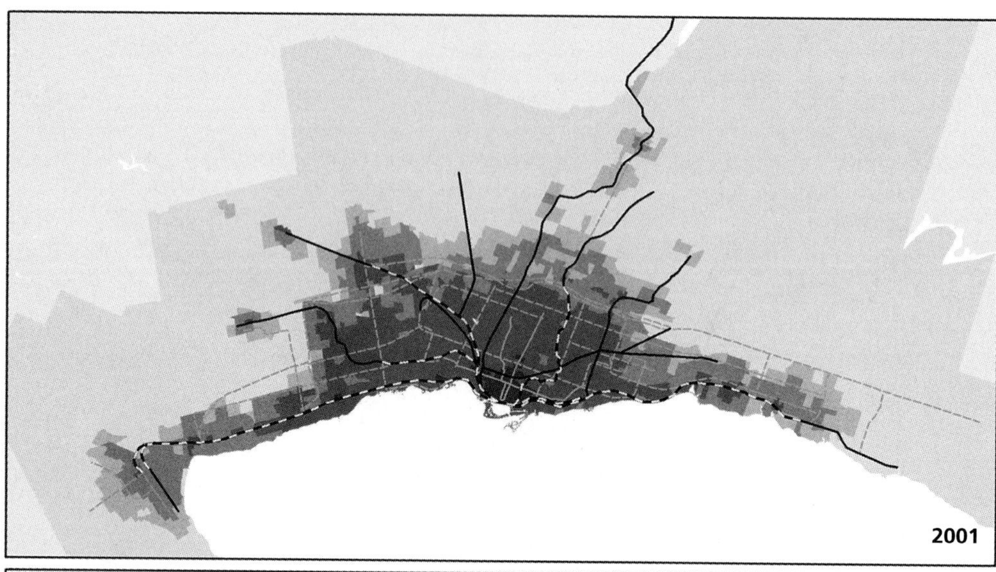

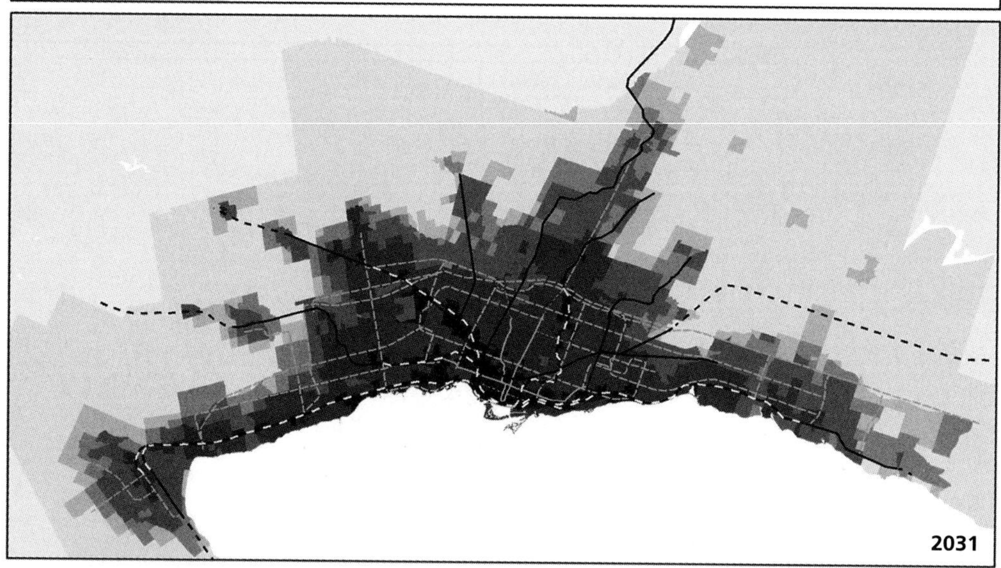

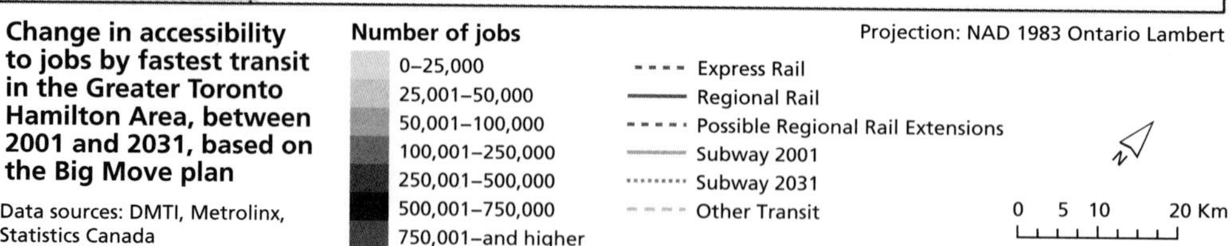

Figure 14.7 Accessibility to Jobs by Fastest Transit—Based on Theoretical Project Implementations of the 2031 Regional Transport Plan

Source: Manaugh, K, N. Foth, D. Willis and A. El-Geneidy. 2012. Greater Toronto Destination Access: Past and Future Land Use and Transport Interaction in the gtha. Toronto: Metrolinx.

public transport and active modes. Planners and city officials are increasingly aware of the powerful impact that transport can have on the environment and well-being of residents, and are taking this into account in the policies put forward and the prioritization given to reducing emissions and encouraging public transport, walking, and cycling as privileged modes of transport in the city. In addition, several of the transport plans mentioned earlier are beginning to make explicit connections between land-use and transport systems. This is essential given that the concept of sustainability combines the promotion of environmental, economic, cultural, and also social equity goals.

Walkability Scores

Accessibility can be measured at different scales (regional, sub-regional, neighbourhood, etc.) depending on the mode examined (Iacono, Krizek, and El-Geneidy, 2010). For example, a local measure of accessibility is the walkability of a given urban area (Manaugh and El-Geneidy, 2012). A walkability score represents the proximity of different amenities as well as the ease to get around as a pedestrian to reach these opportunities. There are several walkability scores generated in the transport literature (Manaugh and El-Geneidy, 2011). A simple and easy-to-obtain measure of local accessibility is the Walk Score, obtained from walkscore.com, which provides walkability scores for any address in Canada, the United States, and Australia. Walk Score has become a popular online source for the general public to gain insight into the walkability of an area. The Walk Score generates a score between 0 and 100 based on the presence of nearby amenities in 13 separate categories (e.g., food, cafés, libraries, parks, cinemas, etc.). There are limitations associated with how Walk Score is measured: for example, straight-line distances (as opposed to distances along sidewalks) are used; street design characteristics are not taken into account (such as quality of pedestrian infrastructure or aesthetics); and it is difficult to distinguish among types of restaurants, food stores, and retail, meaning that a convenience store is "worth" the same as a grocery store (Walk Score, 2018). Walk Score uses a simple gravity-based measure to weigh nearby locations higher than those more distant. The system's major strength is its speed: an online query takes a few seconds to complete for an inputted location. Validation of the accuracy of the Walk Score algorithm has been the subject of extensive research. Carr, Dunsiger, and Marcus (2011) published research that supports the accuracy of the Walk Score algorithm, while Manaugh and El-Geneidy (2011) linked it to actual walking behaviour and found that the Walk Score explains the probability of walking in a neighbourhood just as well as other complex measures—if not better, in some cases.

Using an extensive sample of Walk Score data collected across Canadian CMAs, we measured average Walk Score for the downtown region of Canada's 20 most populated CMAs to compare how walkable city centres are across Canada. Using the location of the city's tallest building as the centre of every city, we calculated the average Walk Score of all postal codes with a centroid that was under five kilometres from the city centre. Figure 14.8 presents the average downtown Walk Score of Canadian CMAs relative to the CMA population.

Toronto, Montreal, and Vancouver show average Walk Scores of above 85 in their downtown core, which awards these cities with the classification of very walkable due to the density of amenities located downtown and the compact street network. On the other hand, the observed Walk Scores in less populated CMAs are highly variable among Canadian CMAs. For example, St. John's, Sherbrooke, and St. Catharines–Niagara have an average Walk Score below 55, whereas Windsor and Kitchener-Cambridge-Waterloo have a Walk Score of around 80 in the downtown core. These observed differences are due in part to the structure of these cities, where some may not possess a highly developed city centre with a concentration of activities and amenities. A less pronounced city centre and more sprawling development patterns would reduce walkability

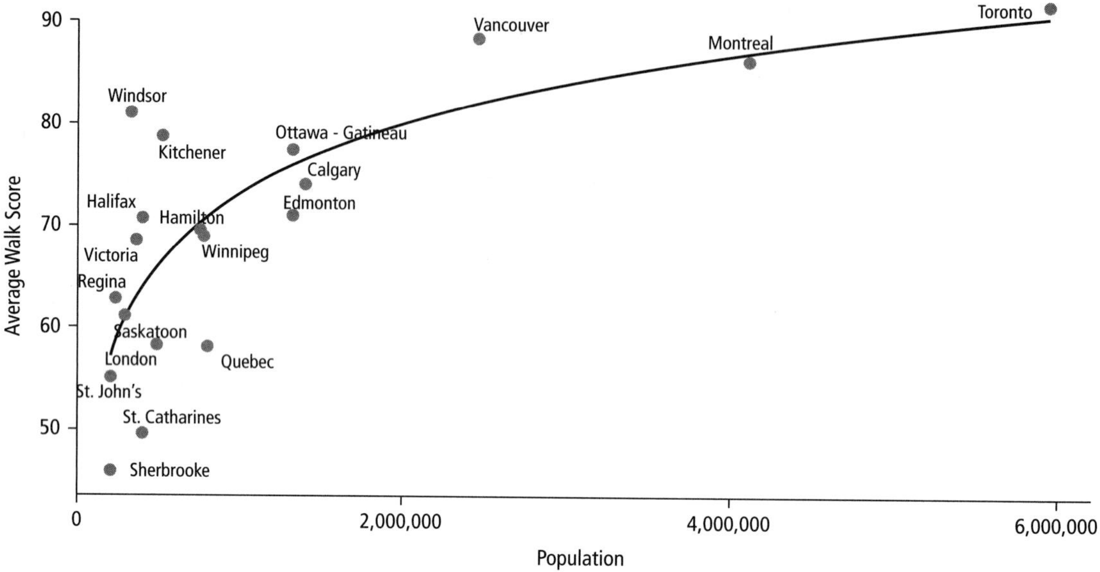

Figure 14.8 Average Walk Score of City Centres in Canadian CMAs by Population in 2016
Data source: Walk Score, 2016.

and likely the actual walking behaviour of residents in that region.

It is insightful not only to compare CMAs to each other in terms of their overall walkability but also to examine the variation of Walk Scores within each CMA. Figure 14.9 shows the Walk Score for Toronto, Montreal, and Vancouver as the distance from the centre of the CMA increases. For example, we measured the average Walk Score for neighbourhoods that are located up to 5 kilometres from the city centre, neighbourhoods between 5 and 10 kilometres from the city centre, and so on. All three cities display high walkability scores, around 90 and more, in the centre of the city as well as a few kilometres around the core. These large CMAs are thus characterized by good accessibility by foot in their central areas. This is not surprising given the dense street network and high concentration of amenities, including parks, businesses, restaurants, schools, and so on, in their urban cores. For Montreal and Vancouver, walkability decreases relatively quickly, down to about 70, between 5 and 10 kilometres from the centre. In comparison, the average Walk Score of Toronto neighbourhoods that are between 5 and 10 kilometres from the city centre is 80. These neighbourhoods can be classified as "inner suburban neighbourhoods" with good public transport access and have a higher population density and access to amenities than Toronto's suburban neighbourhoods. However, at neighbourhoods located between 20 and 25 kilometres away from the centre, walkability decreases to around 50 in each of these cities. This average walkability score is similar to smaller Canadian CMAs such as St. John's and London.

As walkability is a measure of local accessibility to opportunities, it is highly variable within a city. Therefore, we would expect that travel patterns, specifically walking and cycling behaviour, would also be highly variable across a city. Total mode share for active transport in Toronto (walking and cycling) was 6.7 per cent in 2016; however, if we break this number down by different regions within the city and look more broadly at travel behaviour for all trip purposes, mode share is highly

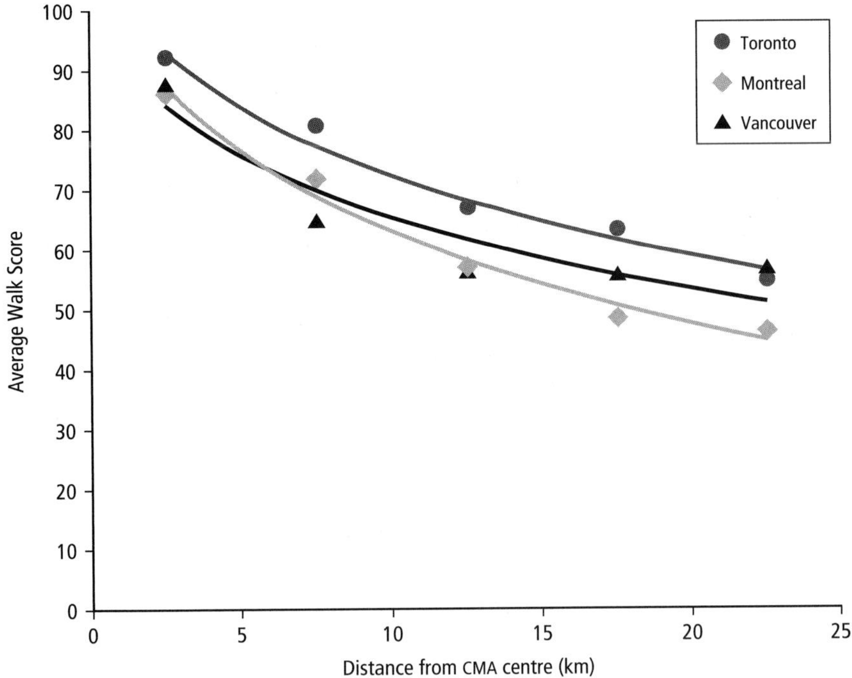

Figure 14.9 Walk Scores for the Largest Three CMAs in Canada

Data source: Walk Score, 2013.

variable. According to the major travel survey of people's daily trip patterns in Toronto (Transport Tomorrow Survey), in the downtown core of Toronto 26 per cent of all trips are achieved by either walking or cycling, whereas in Toronto's suburban neighbourhood Etobicoke 5 per cent of all trips are by walking or cycling (City of Toronto, 2012). It is important also to acknowledge that other elements may be also influencing mode split, such as weather, culture, socio-demographic characteristics, and traveller motivations and values.

The Future of Transport in Canada

The Canadian transport system today faces several challenges, one of the most important being the need to mitigate the environmental impacts of the transport sector. Overall, the transport sector contributes a large share of Canada's total CO_2 emissions. For example, in 2015 total national emissions were 722,000,000 tonnes of CO_2 equivalent, while emissions from the transport sector amounted to 173,000,000 tonnes of CO_2, which represents about 24 per cent of total emissions (Environment and Climate Change Canada, 2017a). The transport sector is the second largest emitter of greenhouse gases (GHG), which is only marginally lower than the oil and gas sector (26 per cent). GHG emissions from the transport sector have grown by 42 per cent between 1990 and 2015, and this growth by transport mode is displayed in Figure 14.10 (Environment and Climate Change Canada, 2017b). Between 1990 and 2015, total passenger emissions grew by 17 per cent; however, there has been a significant shift during this time period away from smaller and more fuel-efficient cars towards heavier and less fuel-efficient vehicles (light trucks), thus impacting GHG emission trends. Emissions from cars declined by 23 per cent, while emissions

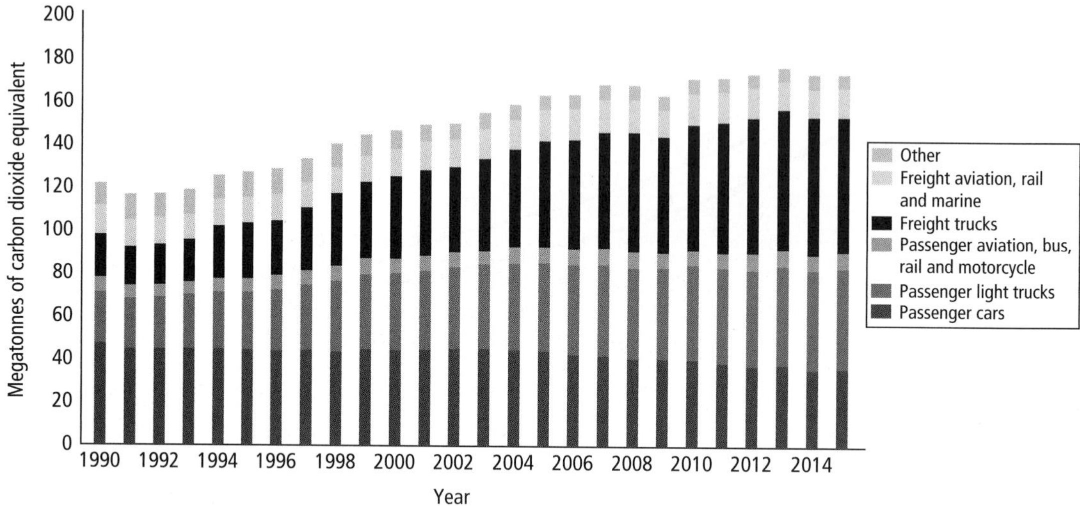

Figure 14.10 Change in GHG Emissions from the Transport Sector in Canada by Mode of Transport between 1990 and 2015

Data source: Environment and Climate Change Canada, 2017b.

from light trucks (trucks, vans, and sport utility vehicles) have doubled. While between 1990 and 2015 newly manufactured vehicles have been produced to be significantly more fuel efficient, the adoption of alternatively fuelled vehicle technologies, particularly electric vehicles (EVs), will likely be paramount in helping Canada achieve climate change targets. In 2012 there were only 2500 EVs on the road in Canada, while in 2017 there were 29,000 EVs on the road, representing a 10-fold increase in EVs on the road in only five years (International Energy Agency, 2017). In 2016, the proportion of electric cars sold relative to internal combustion cars was 0.59 per cent, placing Canada ninth in the world for its share of EVs, with Norway leading the world at 28.76 per cent, and the United States at 0.91 per cent (International Energy Agency, 2017). Canada has set a target for the sale of EVs to be 30 per cent by 2030, with the expectation that the significant penetration of EVs in the automotive market will contribute to meeting reduction targets for GHG emissions in the transport sector.

Canada's outlook on electric vehicles is also linked to the adoption of automated vehicles (AVs), which are mostly being developed on EVs (Standing Senate Committee on Energy, 2017). AVs are predicted to provide a range of benefits to the well-being of individuals, including improved safety through reduced collisions and fatalities and enhanced mobility to individuals who would otherwise not drive (e.g., older adults). However, the impact on cities that will arise with the emergence of new transport technologies and innovations (such as AVs), new economic models of car ownership, and mobility-as-a-service (e.g., Uber and Lyft) is relatively unknown (see Brail and Donald, Chapter 5, for a fuller discussion of this subject). Levinson and Krizek (2017) outline the anticipated impacts of these innovations on the land use of cities—namely that AVs will increase the capacity of roads (allowing for closer following and narrower lanes) and reduce required space for parking (i.e., cars can be more compactly parked or drop-offs/pick-ups can be programmed). Ideally, these land-use changes will reduce land dedicated to roads, allowing for the reallocation of space for bicycle lanes, pedestrian paths, transit right-of-way, and green space. Alternatively, AVs will change how we perceive driving—the stress associated with driving and the loss of productivity that arises from sitting in traffic will be eliminated, thus allowing us to multi-task during commute times. Accordingly, our thresholds for time allocated to commuting may increase, thus triggering an expansion of metropolitan areas, as previous major technological advancements (e.g., the internal combustion engine) encouraged individuals to choose to live further distances away from the downtown core. Planners and policy-makers have an important role in mitigating the detrimental impacts associated with further sprawl surrounding cities—namely, increased sedentary behaviour, increased congestion, and reduced social interaction.

Conclusion

No form of urban mobility has affected land use as dramatically as the invention and widespread usage of the automobile. The auto-dependent culture that presently exists in many cities around the world today has left city planners and policy-makers with significant issues to grapple with including chronic congestion, unsatisfactory public transport service, population health impacts, inadequate transport for certain population groups in certain areas, and negative environmental impacts. While major advances in technology, such as EVs, will mitigate the environmental and health concerns associated with emissions from the transport sector, these new transport technologies, including AVs, are expected to significantly impact our levels of mobility. Advancements to our mobility can provide us with access to more opportunities; alternatively, however, greater levels of mobility can be an instrument to facilitate greater levels of sprawl in metropolitan areas. Canadian cities should embrace these new transport technologies and mobility services (e.g., on-demand ride service, car- and bike-share systems) through careful integrated land use and transport planning and through clear objectives that aim to advance the social well-being of residents.

The shift that has been occurring in transport planning in the last decade—from mobility-focused planning, which consisted chiefly of car-oriented emphasis on mobility issues and concerns over how to travel in the fastest and most efficient way possible, to accessibility-focused planning, which is directed instead at the ease with which people can reach desired opportunities—prioritizes the need of individuals to participate in the full spectrum of urban life and activities (see Walker and Blakley, Chapter 13). Planning for accessibility rather than mobility more realistically encompasses the needs and perceptions of individuals and firms, as the transport system is needed to provide individuals with access to opportunities that are spatially and temporally dispersed. The main advantage of the concept and measurement of accessibility over mobility is that it brings together the location of people, places, and activities, as well as the design of the transport network in one indicator that helps us to understand the impacts of such interactions on location patterns, travel behaviour, and, most importantly, quality of life.

Sustainable modes of transport, including public transport, cycling, and walking, have become more prominent in Canadian cities and cities worldwide. Successfully promoting these modes will require major changes in the built environment (e.g., higher-density and mixed-use neighbourhoods, and a greater proportion of our streets allocated to sustainable transport modes), in people's behaviour, and in the approaches and priorities adopted on a political and economic level. Moreover, actions towards promoting sustainable modes of transport will require collaboration among many different governmental levels and key stakeholders. Inspired by other cities around the world, and based on sound and innovative policies, Canadian cities must carry on toward progressive and sustainable transport, while carefully acknowledging the important role that the automobile is likely to play in urban transport in the future, especially with the imminent introduction of electric cars and autonomous vehicles.

Review Questions

1. What are the different methods used in travel demand modelling?
2. What are the differences between mobility and accessibility, and what are the implications of these differences on transport planning?
3. How can accessibility be used in transport planning?

Notes

1. For the 2011 Canadian Census, the obligatory "long form" of the census, in which the question on transport mode to work has traditionally been asked, was replaced with the National Household Survey (NHS). Since it was not obligatory to answer the NHS, it is unclear to what degree the 2011 figures are comparable with previous censuses.
2. Congestion levels here are defined as the increase in overall travel times compared to a free flow situation, in minutes. For example, a congestion level of 12 per cent corresponds to 12 per cent longer travel times compared to a free flow situation. *Free flow* here refers to times of the day in which there is little or no traffic, such as at night.
3. While this study focused on public transport trips for individuals in a wheelchair, findings from this research are largely applicable for many other individuals including older adults, individuals travelling with a stroller, or travellers with heavy bags or luggage.

References

Alexander, C., S. Ishikawa, M. Silverstein, M. Jacobsen, I. Fisksdahl-King, and A. Angel. 1977. *A Pattern Language: Towns, Buildings, Construction*. New York: Oxford University Press.

Artibise, A. 1977. *Winnipeg: An Illustrated History*. Toronto: James Lorimer & Company.

Baradaran, S., and F. Ramjerdi. 2001. "Performance of accessibility measures in Europe," *Journal of Transportation Statistics* 4(2/3), 31–48.

Boisjoly, G., and A. El-Geneidy. 2017. "How to get there? A critical assessment of accessibility objectives and indicators in metropolitan transportation plans," *Transport Policy* 55, 38–50.

Carr, L., S. Dunsiger, and B. Marcus. 2011. "Validation of walk score for estimating access to walkable amenities," *British Journal of Sports Medicine* 45(14): 1144–48.

Cerdá, A., and A. El-Geneidy. 2010. *Mesures d'accessibilité: Mesures de performance pour la planification de l'utilisation du sol et du transport dans la région métropolitaine de Montréal*. Montréal: Ministère des Transports du Québec.

City of Toronto. 2012. *Road to health: Improving walking and cycling in Toronto*. At: https://www.toronto.ca/legdocs/mmis/2012/hl/bgrd/backgroundfile-46520.pdf

Currans, K., and K. Clifton. 2015. "Using household travel surveys to adjust ITE trip generation rates," *Journal of Transport & Land Use* 8(1): 85–119.

Downs, A. 2004. *Still Stuck in Traffic*. Washington, DC: Brookings Institution.

El-Geneidy, A., and D. Levinson. 2006. *Access to Destinations: Development of Accessibility Measures*. Minnesota, MN: Minnesota Department of Transportation.

Environment and Climate Change Canada. 2017a. *Canadian Environmental Sustainability Indicators Greenhouse Gas Emissions*. At: www.ec.gc.ca/indicateurs-indicators/FBF8455E-66C1-4691-9333-5D304E66918D/GHGEmissions_EN.pdf

———. 2017b. *Canadian Environmental Sustainability Indicators: Greenhouse Gas Emissions*. At: www.ec.gc.ca/indicateurs-indicators/default asp?lang=En&n=FBF8455E-1

European Commission. 2017. *Eurostat Regional Yearbook: 2017 Edition*. At: https://ec.europa.eu/eurostat/documents/3217494/8222062/KS-HA-17-001-EN-N.pdf/eaebe7fa-0c80-45af-ab41-0f806c433763

Franklin, J., and P. Waddell. 2003. "A hedonic regression of home prices in King County, Washington using activity-specific accessibility measures." Paper presented at the Transportation Research Board 82nd Annual Meeting, Washington DC.

Geurs, K. 2006. *Accessibility, Land-Use And Transport: Accessibility Evaluations of Land-Use and Transport Developments and Policy Strategies*. Utrecht: Utrecht University.

———, and J. Ritsema van Eck. 2001. "Accessibility measures: Review and applications. Evaluation of accessibility impacts of land-use transport scenarios, and related social and economic impacts," in RIVM, ed., *Report*. Bilthoven, Holland: National Institute of Public Health and the Environment.

Giuliano, G., and A. Agarwal. 2017. "Land use impacts of transportation investments: Highway and transit," in G. Genevieve and S. Hanson, eds., *The Geography of Urban Transportation*, 4th edn. New York: The Guilford Press.

Grisé, E., G. Boisjoly, M. Maguire, and A. El-Geneidy. 2019. "Elevating access: Comparing accessibility to jobs by public transport for individuals with and without a physical disability," *Transportation Research A: Policy and Practice* 125, 280–93.

Handy, S. 1994. "Highway blues: Nothing a little accessibility can't cure," *Access* 5, 3–7.

———. 2002. "Accessibility- vs. mobility-enhancing strategies for addressing automobile dependence in the U.S." At: www.des.ucdavis.edu/faculty/handy/ECMT_report.pdf

———, and D. Niemeier. 1997. "Measuring accessibility: An exploration of issues and alternatives," *Environment and Planning A*, 29(7): 1175–94.

Hansen, W. 1959. "How accessibility shape land use," *Journal of the American Institute of Planners* 25(2): 73–76.

Iacono, M., K. Krizek, and A. El-Geneidy. 2010. "Measuring non-motorized accessibility: Issues, alternatives, and execution," *Journal of Transport Geography*, 18(1): 133–40.

———, D. Levinson, and A. El-Geneidy. 2008. "Models of transportation and land use change: A guide to the territory," *Journal of Planning Literature* 22(4): 323–40.

International Energy Agency. 2017. *Global EV Outlook 2017: Two Million and Counting*. At: https://www.iea.org/publications/freepublications/publication/GlobalEVOutlook2017.pdf

Koenig, J.G. 1980. "Indicators of urban accessibility: Theory and application," *Transportation* 9, 145–72.

Legrain, A., N. Eluru, and A. El-Geneidy, A. 2015. "Am stressed, must travel: The relationship between mode choice and commuting stress," *Transportation Research Part F: Traffic Psychology and Behaviour* 34, 145–50.

Levinson, D. 1998. "Accessibility and the journey to work," *Journal of Transport Geography* 6(1): 11–21.

———. 2008. "Density and dispersion: The co-development of land use and rail in London," *Journal of Economic Geography* 8(1): 55–77.

———, and K. Krizek. 2017. *The End of Traffic and the Future of Access: A Roadmap to the New Transport Landscape*. Network Design Laboratory.

Loong, C., D. van Lierop, and A. El-Geneidy. 2017. "On time and ready to go: An analysis of commuters' punctuality and energy levels at work or school," *Transportation Research Part F: Traffic Psychology and Behaviour* 45, 1–13.

Manaugh, K., and A. El-Geneidy, 2011. "Validating walkability indices: How do different households respond to the walkability of their neighbourhood?" *Transportation Research Part D: Transport and Environment* 16(4), 309–15.

——— and ———. 2012. "What makes travel 'local': Defining and understanding local travel behavior," *Journal of Transport and Land Use* 5(3), 15–27.

McNally, M.G. 2000. "The activity-based approach," in D.A. Hensher and K.J. Button, eds., *Handbook of Transport Modelling*. New York: Elsevier.

Metrolinx. 2008. *The Big Move*. Toronto: Metrolinx.

Meyer, M., and M. Miller. 2000. *Urban Transportation Planning*. Toronto: McGraw-Hill.

Miller, E.J. 2005. "An integrated framework for modelling short- and long-run household decision-making," in H.J.P. Timmermans, ed., *Progress in Activity-Based Analysis*. Amsterdam: Elsevier.

Nader, G.A. 1975. *Cities of Canada* (Vol. 1). Toronto: McMillan of Canada.

Ortuzar, J., and L. Willumsen. 2011. *Modelling Transport*, 4th edn. Chippenham, Wiltshire, UK: John Wiley and Sons.

Owen, A., B. Murphy, and D. Levinson. 2017. *Access Across America: Transit 2016*. At: http://access.umn.edu/research/america/transit/2016/index.html

Perl, A., and J. Kenworthy. 2010. "The Canadian city at a crossroads—between 'Passage' and 'Place'," in T. Bunting, P. Filion, and R. Walker, eds., *Canadian Cities in Transition: New Directions in the Twenty-First Century*, 4th edn. Don Mills, ON: Oxford University Press.

Sallis, J., J. Spoon, N. Cavill, J. Engelberg, K. Gebel, M. Parker, . . . D. Ding. 2015. "Co-benefits of designing communities for active living: an exploration of literature," *International Journal of Behavioral Nutrition and Physical Activity* 12: 30. doi:10.1186/s12966-015-0188-2

Shearmur, R., and N. Rantisi. 2011. "Montreal: Rising again from the same ashes," in L.S. Bourne, T. Hutton, R. Shearmur, and J. Simmons, eds., *Canadian Urban Regions: Trajectories of Growth and Change*. Toronto: Oxford University Press Canada.

Standing Senate Committee on Energy, the Environment and Natural Resources. 2017. *Decarbonizing Transportation in Canada*. At: https://sencanada.ca/content/sen/committee/421/ENEV/reports/TransportationReport_FINAL_e.pdf

Statistics Canada. 1996. *Profile of Census Tracts*. Ottawa: Statistics Canada.

———. 2001. *Profile of Census Tracts*. Ottawa: Statistics Canada.

———. 2006. *Profile of Census Tracts*. Ottawa: Statistics Canada.

———. 2011. *Commuting to Work: National Household Survey 2011*. Ottawa: Statistics Canada.

———. 2016a. *2016 Profile of Census Metropolitan Area & Census Agglomerations.* Ottawa: Statistics Canada.

———. 2016b. *Journey to work: Key results from the 2016 Census.* At: www.statcan.gc.ca/daily-quotidien/171129/dq171129c-eng.htm

United States Census Bureau. 2016. *Commuting characteristics by sex: 2012-2016 American Community Survey 5-Year Estimates.* At: https://factfinder.census.gov/faces/affhelp/jsf/pages/metadata.xhtml?lang=en&-type=dataset&id=dataset.en.ACS_16_5YR

van Wee, B., and D. Ettema. 2016. "Travel behaviour and health: A conceptual model and research agenda," *Journal of Transport & Health* 3(3): 240–48.

Vickerman, R.W. 1974. "Accessibility, attraction and potential: A review of some concepts and their use in determining mobility," *Environment and Planning A* 6, 675–91.

Vuchic, V. 2005. *Urban Transit—Operations, Planning and Economics.* Hoboken, NJ: John Wiley and Sons.

Wachs, M. 1993. "Learning from Los Angeles: Transport, urban form, and air quality," *Transportation* 20, 329–54.

———, and T. Kumagai. 1973. "Physical accessibility as a social indicator," *Socioeconomic Planning Science* 7, 327–456.

Walk Score. 2013. Data Services. At: https://www.walkscore.com/professional/research.php

———. 2016. Data Services. At: https://www.walkscore.com/professional/research.php

———. 2018. *Walkscore.* At: www.walkscore.com/

Wasfi, R., K. Dasgupta, H. Orpana, and N. Ross. 2016. "Neighborhood walkability and body mass index trajectories: Longitudinal study of Canadians," *American Journal of Public Health* 106(5): 934–40.

Yeates, M., and B. Garner. 1980. *The North American City.* New York: Harper & Row.

15 Re-Imaging, Re-Elevating, and Re-Placing the Urban

Gentrification and the Transformation of Canadian Inner Cities

Alison Bain and Bryan Mark

Introduction

This chapter explores the new structural dynamics of metropolitan change at work in the inner cities of large Canadian cities, with particular attention to the transformative power of gentrification. As the first two chapters in this volume illustrate, it is inner cities that have experienced dramatic changes over the past quarter-century. And, in the past decade, with the apparent dominance of the new neo-liberal urban agenda, this transformation process has accelerated.

In this chapter, the prefix *re-* (meaning "once more," "again," or "afresh") is used to emphasize that cities, and neighbourhoods within them, are works in progress shaped by, and reflective of, prevailing political, economic, and social forces. In the case of inner cities in many large Canadian metropolitan areas, this chapter demonstrates how gentrification has transformed formerly industrialized and centralized portions of cityscapes into upscale and fashionable post-industrial places. This chapter asks what the contemporary built landscapes of core parts of Canada's largest urban centres reveal about who and what are privileged there. It argues that the increasing property values, widening income inequalities, and socio-spatial changes that accompany gentrification have limited the capacities of different, often lower-income, marginalized groups to both access and live in city centres.

Urban Development in Transition

A Neighbourhood Typology

In an earlier edition of this book, Ley and Frost (2006) document the historical development of the Canadian inner city as concept, region, and object of study. They present a fourfold typology of different kinds of districts: (1) declining; (2) stable; (3) revitalized; and (4) redeveloped.

Neighbourhoods experiencing decline usually have the following characteristics: housing stock that is physically deteriorated; declining property values; a greater proportion of rental rather than home ownership; population loss from the out-migration of higher-income groups; high deprivation indicators (e.g., high unemployment, single-parent households, low levels of high school completion, and elevated levels of social assistance); and an increasing presence of marginalized social groups (e.g., immigrants, refugees, mentally ill, substance abusers, sex-trade workers, homeless) and the services to support them (e.g., dollar stores, charity shops, food banks, and drop-in centres). In contrast, more stable neighbourhoods tend to have lower population turnover, higher levels of home ownership, well-maintained properties, dynamic voluntary organizations and community institutions that support durable social networks, and a strong sense of place.

All neighbourhoods sit somewhere on a continuum of decline to stability to revitalization. The extent of change that a neighbourhood experiences depends upon the age of the physical environment, the age of the residents, and the movement of households (Knox and McCarthy, 2012). Differences usually exist between neighbourhoods in terms of the rate and type of change, and the investment opportunities available. For example, a neighbourhood with a good physical stock of properties that is socially and demographically stable is unlikely to attract significant reinvestment. If, however, a profit can be made by changing the character of a neighbourhood or land use within it to better meet market demand, then developers and urban planners may consider it ripe for redevelopment and reinvestment.

Although the inner city is at no time homogeneous, Ley and Frost's typology represents a useful transitional view of inner-city neighbourhoods. They move from a discussion of the 1960s late-modern period, when deterioration and decline seemed all-pervasive across inner-city zones, to the early twenty-first century when inner-city redevelopment and **revitalization/regeneration** have become celebrated characteristics of creative and competitive cities.

The Optics of the Twenty-First-Century Competitive City

Contemporary urban development practices are self-consciously entrepreneurial, characterized by flagship strategies centred on profit, spectacle, and consumption (Hall, 2003; Logan and Molotch, 1987). Civic leaders take great care to use architectural form and urban design to promote positive and high-quality images of places (Harvey, 1989; Zukin, 1995). This formal attention to **place-making** creates a distinctive development style that is notable in Canada's largest and fastest-growing metropolitan areas, where branding is applied to whole cities in an effort to develop a distinctive civic image that appeals to tourists, investors, and mobile members of the knowledge economy—what Florida (2002) has referred to as the **creative class** (see also Vinodrai, Chapter 17; Shearmur, Chapter 18). The "boosterist" goal of many municipalities is to replace perceptions of the city as a place of disinvestment, decay, crime, and poverty left behind by the post–World War II suburban transitions with images of growth, vitality, and prosperity (Avraham, 2004; Short, 1999). In a global economy, city and neighbourhood images are of concern, particularly in the best-known and most central parts of the metropolitan region, because they influence where new businesses and residents locate. This, in turn, has an impact upon inward investment and the infrastructure of daily life. But it is not just the built and physical environments that are central to imaging a city or neighbourhood; the lifestyles and symbolic economies that exist in these spaces through recreation, leisure, and cultural activities also contribute to a city and neighbourhood's image and perceived quality of life (Zukin, 1995, 2004).

"Urban spectacularisation, which serves both for real-estate speculation and for political propaganda" has become an instrument of economic development (Vaz and Jacques, 2007: 249). In addition to a bias toward the better-off in society, spectacle can have an inversely proportional relationship to popular participation: the more spectacular the interventions in urban revitalization, the less the participation of the population (Vaz and Jacques, 2007). Thus, who is included and who is excluded, who wins and who loses in the urban revitalization process are important considerations. Is this so-called circus serving those in need of bread and other necessities of life (Kipfer and Keil, 2002)? The assumed trickle-down mechanisms may not work: poorer local residents may become further marginalized, their visions of less exclusive, corporate, or authoritarian urban futures sidelined or ignored by urban elites (see Walks, Chapter 10).

Re-Imaging Canadian Cities through Spectacle: From Culture to Condominiums

Over the past two decades, urban planners and municipal politicians across North America and Europe have favoured a flexible creative capital model of urban development that privileges knowledge, creativity, and commodified difference as a means of civic renewal (Gertler et al., 2002; Markusen and King, 2003). Turning away from the more traditional primary and secondary sectors of the economy for support, this new model seeks instead to use urban amenities to attract mobile and educated professionals (Florida, 2002, 2005; Pratt 2009; Vinodrai, Chapter 17; Shearmur, Chapter 18). The iconic "creative" citizens of the new **knowledge-based economy** are presented as the "storm troopers" of city branding; they are said to be attracted to cities that offer "the 3Ts": technology, talent, and tolerance. But as Wilson and Keil (2008) question, tolerance of whom and by whom? They argue that the "real" creative class is the poor—who use everyday resourcefulness to navigate torn social safety nets, make homes in neglected neighbourhoods, and work in low-wage service sector jobs. For Wilson and Keil (2008: 846), the concept of the creative class is a perverse fantasy in which "the white affluent are the historic and current bearers of civility, tradition, and good culture"; and the creative capital model of urban development is just another fast-track public policy option that privileges "the desires and aspirations of capital and the affluent." In a quest to lure knowledge workers, Canadian cities are remaking themselves as trendy centres of arts and culture, consumption, and play. They are marketing themselves as places that provide stimulation, diversity, and a richness of experience that inspire creativity and innovation.

In large Canadian cities where significant inner-city redevelopment has occurred, the nineteenth-century, middle-class stigma once associated with living among the supposed deprivation, pollution, and pathologies of the working classes has lost some of its potency. Large cities in Canada's urban system have not lost population nor become hollowed out because of racial tensions, violence, or poor schools. In fact, some of urban Canada's most expensive real estate can be found in core area neighbourhoods (see Skaburskis and Moos, Chapter 19). Recognizing that land is a finite resource, many sizeable Canadian cities have focused development on brownfield and infill sites, a process often referred to as **intensification**. For example, in the Province of Ontario's Growth Plan for the Greater Golden Horseshoe prepared under the Places to Grow Act (2005), intensification is defined as "the development of property, site or area at a higher density than currently exists through redevelopment, including the reuse of **brownfield sites**; the development of vacant and/or underutilized lots within previously developed areas; infill development, or the expansion or conversion of existing buildings" (www.placestogrow.ca). The Growth Plan establishes intensification and density targets for 25 urban growth centres (400, 200, and 150 people and jobs per hectare) as part of an integrated package of policies to foster the creation of more compact, **mixed-use**, and transit-supportive communities. The highest density target has been applied to the City of Toronto (Downtown Toronto, Etobicoke Centre, North York Centre, Scarborough Centre, and Yonge-Eglinton Centre), which has a large population and high-frequency public transit subway service. Much of this intensification, particularly in Toronto, has taken the form of privately developed condominiums.

"Condo-mania" has a firm grip on large cities across the country. Many urban planning departments are overrun with proposals for high-density residential living and must contend with condominium development as a major component of municipal comprehensive plans (Kern, 2007; Lehrer, 2008). Inevitably, such intensive residential development involves substantial physical and social change to the fabric of the city.

There are many potential benefits associated with increased residential densities: improved energy and land conservation; maximization of existing infrastructure; reduced reliance on the private automobile; improved economic capacity through the increased circulation of money in the local economy; an increased tax base; and the social and safety dimensions of a more vibrant street life. But drawbacks exist, also. It has been argued, for example, that intensification in the form of high-rise condominium towers that commonly cater to childless singles or couples from the middle and upper classes who seek maintenance-free living has created "vertical gated communities" (see Townshend and Walker, Chapter 7; Grant and Filion, Chapter 12). Condominium fees cover the costs of private, residents-only facilities (e.g., gym, swimming pool, spa, games room, movie theatre) and security systems, creating islands of wealth and privilege where residents may seldom interact with each other or with the people and spaces of the surrounding neighbourhood. People insulated from one another lose a sense of community and of commitment to a common social project with shared social justice and social sustainability goals (e.g., equity, inclusion, participatory processes, and access to services, housing, and employment) (Quastel, Moos, and Lynch, 2012). As a result, planners must consider ways of fostering social interaction at the neighbourhood scale over the longer term, perhaps by providing free, accessible, inclusive, well-maintained public spaces (e.g., parks and parkettes) that support a variety of uses and involve a range of different social groups. Intensified residential development can be a positive contribution to city life, if cities have the financial means and the political will to adequately invest in providing and maintaining public infrastructure (e.g., developing dedicated bicycle routes and public transit). Indeed, Danyluk and Ley (2007) have shown that higher-density central city neighbourhoods in Canada tend to be home to pedestrians and cyclists who rely on non-motorized modes of commuting.

In his book *Cities for People*, Danish architect and urban designer Jan Gehl (2010) advocates that urban planners attend to the human scale by creating cities that are pedestrian- and bicycle-friendly. Built environments where attention has been paid to experiential details encourage people to walk slowly, stop frequently, and linger. Widening sidewalks, laying textured surfaces, planting shade trees, improving street crossings, and providing comfortable street furniture and good lighting are all practical recommendations for improving the public realm and increasing the amount of time that people spend in public space without necessarily having to purchase anything (see Walker and Blakley, Chapter 13).

The quality of the public realm is a key component of livability (see **livable cities**), which, in conjunction more recently with sustainability goals (e.g., densification, social mixing, and transit accessibility) has driven twenty-first-century urban planning policy agendas in North America. Livability, in contrast to sustainability, does not refer to the future but, rather, to the current state of the environment and the quality of life it offers to local residents. Cities like Vancouver, which often appear high on world rankings of livable cities, have paid attention to retaining a distinctive local character, ensuring connectivity and ease of movement, promoting higher-density living, supporting a diversity of land uses, and creating a high-quality public realm. In privileging a sustainability agenda, Vancouver has established the Greenest Capital Action Team (GCAT) with the intent of becoming the greenest city in the world by 2020.

But for whom are cities livable and sustainable? While densification may lay "at the heart of a process of cultural and economic restructuring," it can also lead "to the gentrification and revalorization of walkable and transit-accessible spaces" that privilege middle-class priorities (Quastel, Moos, and Lynch, 2012: 1077). Thus, the potential exists to exacerbate spatial injustice and offer few environmental benefits for more marginal social groups.

Place-Making at the Water's Edge

Much condominium construction is concentrated along river-, lake-, or oceanfront property. The reclaiming of waterfront sites began in the 1980s, and today these sites are considered valuable urban amenities with a central role to play in contemporary city-making (Marshall, 2001). Waterfronts are complex places that have hosted many functions over time. Historically, the working part of the city was on the waterfront—industry located here and ship and rail yards interfaced. Urban waterfronts consisted of dirty, messy, and contaminated sites and so were largely neglected. As a result, for much of the twentieth century, Canada's urban waterfronts were underutilized, separated from the rest of the city by transportation corridors. Today, however, waterfronts have been revalued as city assets, and efforts are underway to recapture these areas by reconfiguring connections between the older, original city centre and the water's edge (Bunce, 2009; Desfor and Ledley, 2011).

A noteworthy example of post-industrial city-making through waterfront redevelopment is Vancouver. Since the 1990s, Vancouver has gained an international reputation for its high standard of urban planning and design practice, so much so that when North American architects and planners promote the idea of a high residential density, high-public amenity central city, they call it "Vancouverism" (Boddy, 2005). In Vancouver, the ocean is brought into the heart of the city and together with the mountains provides the raw material for the "cult of the view," which, in conjunction with a series of overlapping official "view corridors," has driven high-rise residential development in the inner city for the past 50 years (Berelowitz, 2005). Historic neighbourhoods in the inner city (e.g., Strathcona, Gastown, and Chinatown) remain largely intact as pedestrian-friendly, bicycle-friendly places because Vancouver has managed to avoid the worst of traditional North American urban renewal: highways, elevated and underground pedestrian systems, large shopping malls, and big-box retail. Within Vancouver's municipal boundaries, no highways disrupt the regular street grid or cut the city off from its waterfront. Since the 1970s and accelerating after Expo 86 and continuing beyond the 2010 Winter Olympics, Vancouver's waterfront has been strategically transformed from industrial and railway use to residential and recreational use for the middle- and upper-classes who can afford to live downtown (Hutton, 2004). The construction of public pathways along the seawall has repositioned recreational activity in a continuous corridor along the inner-city waterfront. The downtown street grid and urban fabric have gradually been extended to the water's edge and filled in with tall, glass condominium towers known locally as "see-throughs" (Coupland, 2000). But the city has also faced criticism for privileging residential development over office development, thus jeopardizing its commercial land base, elevating land values, and creating more of an expensive and exclusionary recreational resort than a working city.

A centrepiece of Vancouver's waterfront renewal is Granville Island, a 14-hectare site managed by the Canada Mortgage and Housing Corporation (CMHC). This former industrial site beneath the southern viaduct of the Granville Bridge was transformed in the 1980s into a commercial, arts, entertainment, and recreation destination, with the Emily Carr University of Art + Design as a key anchor. While the university moved its campus off of Granville Island in 2017, the *Granville Island 2040: Bridging Past & Future* (Stevenson, 2017) comprehensive planning vision commissioned by the CMHC designates the repurposing of the North Building as an arts and innovation hub and a performance and installation space and the South Building for a single institutional arts user (https://granvilleisland2040.ca). Despite the loss of Emily Carr, the public farmers' market remains an important focal point in combination with an eclectic mix of studios, galleries, theatres, restaurants, and specialty shops in converted warehouses. Granville Island has been characterized as a "creative cultural resource" (Griffin, 2017) but also as a privatized public place of conspicuous consumption and play (Ley, 1996a). This island of wealth, investment, and pleasure stands in contrast to the poverty, homelessness, and health crises of drug use and HIV transmission experienced in Vancouver's Downtown Eastside (DTES), perhaps Canada's poorest inner-city neighbourhood (Burnett, 2014). Stigmatized as a "skid row" district of concentrated poverty and single-room occupancy hotel housing, the DTES gained international notoriety for the disappearance and murder of 69 sex-trade workers (Robertson, 2007). The DTES is a diverse neighbourhood that has housed a variety of populations over time, each attaching different social meanings to the neighbourhood (Burke, 2006). Yet gentrification and property development processes unfolding here work through policymakers, service providers, and police to discipline places and racialized and sexualized bodies into a more uniform, middle- and upper-class urban landscape (Robertson, 2007).

While Vancouver may arguably be North America's most livable city, this positive affinity for the city is likely only shared by those who can afford the high real estate costs. As in most other fast-growing, large Canadian cities, the process of urban redevelopment and modernist planning practice has physically reorganized poverty and affluence in the city, often concentrating poor residents in neighbourhoods that are poorly serviced, far from employment opportunities, and physically separated from the rest of the city (Walks, 2001). The image of a new, affluent city of leisure seeks to hide such marginal spaces from view.

What Lies Behind the Shimmering Facades?

In a **neo-liberal** era of municipal financial constraint, where discourses of urban entrepreneurialism and competitiveness dominate and business elites hold the reins of power in public–private partnerships, market forces are readily accommodated while social justice objectives that could help to meet the basic needs of the most marginal populations are all too easily abandoned (Miles, 2005a, 2005b). The large-scale cultural and residential redevelopment projects discussed in this chapter can create huge commercial gains for developers and landlords and increase the city's tax base. However, "[w] hat this branding of commodified space also shares over time is the fact that little or none of the increased values accrue to the residents who have been decanted and displaced" (Evans, 2007: 198–200). Social displacements and dramatic inequities have been especially apparent in western Canada, where the oil boom accelerated urban growth.

The inflow of workers and money to cities such as Saskatoon, Calgary, and Vancouver has caused local housing markets to climb to record highs, thus minimizing the availability of affordable accommodation (see Walks, Chapter 10). In Calgary, rental apartments have been converted into luxury condominiums and blocks of lower-end residences have been bulldozed to make room for condominium towers. From 1994 to 2008, the Calgary Homeless Foundation documented that Calgary had the fastest-growing homeless population in Canada. In an effort to combat homelessness, Calgary produced an ambitious *10-Year Plan to End Homelessness*, which it sought to implement from 2008 to 2018 (Southwick, 2018). Over the decade, the city has prioritized a housing-first strategy that focused on the most vulnerable subpopulations (e.g., youth, women, families, and Indigenous people) with long-term needs being able to access affordable housing. A key component of the plan involved the development of a Homeless Management Information System, which makes data on homelessness and affordable housing available in real time to facilitate inter-agency coordination in the design and implementation of programs. While research by public policy scholars at the University of Calgary has shown that the number of overnight stays by adults in emergency shelters has declined by 40 per cent, demand for counselling and employment services and meals remains high, as does the need for affordable housing (Rumboldt, 2017).

The Social Tectonics of Gentrifying Urban Landscapes

The story of inner-city neighbourhood change through gentrification and other kinds of upgrading is not a new one. It is a story that is both international and national in scope, having an impact upon cities diverse in population, age, size, and location (Lynch and Pottie-Sherman, 2017). The following discussion reviews the Anglo-American gentrification literature, documents a shift in scholarly attention from residential to retail gentrification, and attends to the social tensions and exclusions that mark its urban landscapes.

In 1964, the German-born English sociologist Ruth Glass published a report on neighbourhood change in London, England, in which she documented how more affluent and educated people, the gentry, were buying and renovating old mews and cottages, upscaling the residential landscape and replacing the original blue-collar population. She coined the term *gentrification* to refer to this restructuring of place from lower- to higher-socio-economic-status residents, triggering a physical displacement of former working-class residents. Her findings remain relevant today—affluent homeowners continue to extract social and economic capital from real estate in devalued inner-city neighbourhoods and price out less affluent residents.

Subsequent urban scholarship sought to model the gentrification process in coherent stages (Caulfield, 1994). The first stage includes "marginal gentrifiers" (Rose, 1984), a lower-middle class of professionals (including artists) and students who are often non-family groups who rent inexpensive housing in socially diverse inner-city neighbourhoods. The second stage includes "early gentrifiers," a group of highly educated social and cultural workers, who may be first-time homeowners using sweat equity (their own manual labour) to renovate their homes. The third stage, "gentrification proper," includes increasingly affluent middle-class professional households and developers exploiting the "rent gap" to maximize the greatest profit from the land. The "advanced gentrification" stage includes major redevelopment projects funded by "super-gentrifying" financiers who re-gentrify areas previously gentrified (Butler and Lees, 2006). The classic stage model offers significant explanatory power, but it fails to concern itself with the location and welfare of displaced incumbents. The gentrification process is also not so neat and tidy. In practice, gentrification is uneven and fragmented (Ley and Dobson, 2008).

Much of the early literature on gentrification focused on the causes of the process. It positioned the socio-cultural (consumption-side) explanations of neighbourhood change against economic (production-side) explanations. Consumption-side explanations celebrate the emancipatory potential of gentrification to bring different people together and to create opportunities that foster social interaction and cultural diversity. For David Ley (1996a, 1996b), it is the "new middle class"—affluent, young, child-free couples in professional, managerial, and advanced service occupations with cultural capital and non-conformist lifestyles and politics who prefer inner-city neighbourhoods in close proximity to work, trendy shops, and leisure activities—who are the instigators of neighbourhood change.

On the other side of the debate, Neil Smith (1979, 1996) attributes the movement of capital into and out of the built environment as the main agent of neighbourhood change (rather than the gentrifiers themselves). With his rent gap thesis, Smith argued that gentrification occurs when there is a gap between the decreasing actual ground rent (land value) in neighbourhoods experiencing decline and the potential ground rent that the land could generate, if it were redeveloped. When the gap is wide enough, a property is ripe for reinvestment, which can initiate a chain reaction process of gentrification in a neighbourhood. The rent gap model has been critiqued because gentrification often does not occur where the gap is deepest; the property development industry is risk-adverse and gentrification usually occurs based on the relative risk that developers and builders are willing to take. Production-side explanations of neighbourhood change assume that gentrification is largely the result of structural forces at play within local, national, and global political and economic arenas. For Smith, the intent of gentrifiers is revanchist—seeking spatial revenge. Invariably, in any neighbourhood different social groups will have different priorities and values that can come into conflict; however, in production-side explanations gentrification is framed as a class struggle in which the middle-class fight the working-class to claim inner-city neighbourhood space. Smith (2002) went on to argue that gentrification is a "global urban strategy" that is fundamentally connected to new socio-economic policies of urban neo-liberalism and state restructuring that privilege capitalist production over social reproduction. Thus, civic leaders increasingly favour urban revitalization that caters to more affluent residents over the housing and quality of life and subsistence needs of less wealthy citizens.

Yet the capital versus culture divide is not so straightforward. As Sharon Zukin (1982) shows in her classic analysis of the conversion of manufacturing buildings to upscale loft living spaces in the warehouse district *So*uth of *Ho*uston Street in Lower Manhattan (SoHo), capital and culture meet in the production of gentrified space. Indeed, loft conversions are now a global economic and cultural phenomenon—they have contributed to the deindustrialization and gentrification of urban centres around the world. Lofts provide "artistic," high-end, central city living space close to urban amenities (e.g., waterfronts, entertainment, leisure, food, and culture) for a select few, but this commodification of urban space prices out other homeowners and renters and changes the social and political context of local neighbourhoods.

As the gentrification literature has proliferated over the past half-century, urban scholars have privileged the study of housing dynamics to the neglect of commercial and retail change. As Hubbard (2017: 2) notes, "local shopping streets [are] presumed to reflect the socio-economic character of surrounding neighbourhoods rather than in a sense driving the changes in those spaces." Yet the retail fabric of local shopping streets is a key battleground of gentrification. Retail gentrification refers to the replacement of "downscale" shops catering to poorer populations and ethnic minorities—a convenience grocery store, an appliance repair shop, or a working-class bar, which televises the local news—by higher end businesses. Through cappuccinos and microbrewery pints, commercial upscale growth and the tastes of the new urban middle class are envisioning new beginnings on streets, destroying working-class, lower-income origins and reinforcing social distinctions (Zukin, 2010).

While retail gentrification is non-residential, new businesses and shops on a local shopping street are functionally interconnected to the demographic of the surrounding residential neighbourhood, both driving and responding to the settlement of higher-income people and the displacement of lower-income people. Those who live around a local shopping street often are the largest group of shoppers, whose needs and tastes determine the economic success of local retail. As sites of gentrification, the intrusion of trendy restaurants, boutiques, art galleries, and coffee shops into the social fabric of cities form visible spaces of lifestyle and cultural consumption, which emerge out of the highly prescriptive global toolkit of urban revitalization (Zukin, 1991; Zukin et al., 2016). New retail entrepreneurs are attracted to gentrifying neighbourhoods by the profit potential derived from an upward shift in social status and the greater disposable income that accompanies it. As vendors of symbolic goods and services and creators of spaces of sociability for a select few, these "culturepreneurs" (Lange, 2006) create the conditions for consumers to realize their "good taste" in curated products and to reinforce their social distinction (Bourdieu, 1984). By perpetuating class hierarchies through cultural consumption, retail entrepreneurs engage in the process of social reproduction, creating a local environment that defines and defends social privilege.

More critically, the arrival of high-end retail spaces has social justice consequences at the local scale. Retail upscaling and the escalating rental rates that accompany it, displaces the low-end stores that serve poor and newcomer populations and the social networks they support. The quiet closure of a neighbourhood barber who proudly displays hand-written letters of gratitude from loyal customers in his shop window, and its quick replacement by an organic frozen yogurt shop, not only disrupts everyday routines, but also destroys important social worlds and bonds of trust that have been built over time. New conspicuously middle-class businesses send powerful social messages about who appears to own and belong on the shopping street and who is out of place, potentially fostering feelings of discomfort, disconnection, and exclusion for some long-time residents. As Takahashi (2017: 596) found in Toronto's rapidly gentrifying Little Portugal neighbourhood, the social tectonics between commercial gentrifiers and long-time ethnic store owners proved unstable and antagonistic—"socially divided even though they are spatially inseparable."

Although the commercial landscapes of local shopping streets seem to be controlled by the invisible hand of the market in conjunction with private stakeholders (e.g., store owners, shoppers, and suppliers), the visible hand of the state still wields a powerful grip over the character of the street through the professional practices of urban planning and urban design. At the municipal level, zoning laws are legal instruments that shape landscapes of consumption by organizing commercial activity and controlling what types of stores are allowed to operate in a particular neighbourhood. Theoretically, zoning is supposed to safeguard the public good; however, as "rezoning becomes the city government's preferred tool of redevelopment" zoning bylaws also increasingly conform to market forces (Zukin, 2010: 29).

A different socio-spatial strategy for managing and controlling local shopping streets manifests through business improvement associations (BIAs). These are a form of public–private partnership comprised of store owners and building owners, in which members pay a mandatory tax to the municipality, which is then returned to the BIA to "clean up" and brand the commercial street as a shopping destination. Entrepreneurial tactics such as painting over graffiti, removing litter, lobbying for anti-panhandling bylaws, and installing sleep-resistant benches are all used to control perceived "urban disorder." Through such strategic micro-interventions into the urban fabric, BIAs seek to immunize local shopping streets against cultural triggers of urban anxiety, imitating the sanitization strategies of shopping malls and amusement parks. BIAs also attract shoppers by beautifying and branding the streetscape with hanging flower baskets and lampposts decorated with welcome banners. The simultaneous sanitization and aestheticization of commercial streets undergirds the drive of many BIAs to increase property values.

While the local state wields the power to regulate how a local shopping street is spatially ordered and structured, dominant local media narratives publicized in print media, broadcast media, websites, blogs—and increasingly, social media—also create influential urban images and filter definitions of "the urban good life" (Zukin, 2009; Hyde, 2014). Relying on words and images, media representations socially construct a language of cultural power that embodies a rhetoric of upscale growth and middle-class, market-oriented desires. As cultural authorities, the media's hunger for new content translates into favourable representations of neighbourhoods and commercial streets, which drive the seemingly never-ending desire for distinction and authenticity and glorify processes of urban renaissance. There is also a danger that media discourses can trivialize gentrification's social destruction as a mere "aesthetic code of a new urban lifestyle" (Zukin, 2009: 549), initiating a convenient collective amnesia about who and what has been displaced.

Ossington Avenue: Consuming Hipster Culture in Toronto

Like Whyte Avenue in Edmonton, 16th Street in Calgary, 20th Street West in Saskatoon, James Street North in Hamilton, and Portland Street in Dartmouth, Ossington Avenue in downtown Toronto is framed in the media as an epicentre of hipster culture. The hipster is an urban "type"—"flannel-shirted, flat-white coffee consuming bearded" white figure "indelibly associated with urban life, being found in the neighbourhoods taken to embody cosmopolitan and metropolitan values" (Hubbard, 2016: 2)—who adopts fashionable styles in opposition to mainstream consumer culture and is wealthier than the residents of multicultural and working-class inner-city communities that they have colonized (Maly and Varis, 2016). Media debates question whether hipsters are a symptom of gentrification or a causal agent. In her discussion of the pretentious lifestyle known as "hipster urbanism,"

Cowen (2006: 22) describes hipsters as agents of neo-liberalism who claim the inner city as their own through "banal, mimetic creativity." While hipsters' rejection of mainstream commercialism purports to challenge oppressive capitalist norms, their niche consumption does not resist the economic structure of the prevailing neo-liberal moment and can actually reinforce it (Maly and Varis, 2016).

A north-south commercial street, Ossington Avenue between Queen Street West and Dundas Street West is situated near the Centre for Addiction and Mental Health (CAMH). Since the turn of the twenty-first century, and following the redevelopment and opening up of the CAMH site, the commercial street has intensely gentrified. Once a commercial landscape characterized by automobile repair shops, Portuguese businesses, and karaoke bars, this "artery of crime" (O'Kane, 2016) has transformed into, "a downtown west-end mecca for foodies and fashion lovers alike" (Archer, 2015). Culturally catapulting from a non-destination with an unwelcoming reputation to a gritty consumer spectacle, Ossington Avenue has been reshaped by capital investment, becoming an elite cultural playground of conspicuous consumption.

The form of urban commercial change on Ossington Avenue follows "the Williamsburg paradigm" in Brooklyn, New York, wherein the "old authenticity" of the streetscape, including karaoke bars and automobile manufacturing, leads to a new bourgeois bohemian arrangement of Asian fusion cuisine and Parisian-inspired, locally sourced butcher shops (Zukin, 2010). The commercial revolution unleashed by new entrepreneurs' social, cultural, and economic capital has refashioned the visible face of Ossington as an upscale *terroir* for boutique fashion, furniture, cocktail bars, and expensive restaurants. The cultural breath exhaled by such conspicuous taste aromatizes Ossington's new habitus with an air of pretentious cosmopolitanism. Moreover, the street's cultural field of commercial fitness and the uptick in the number of fitness and yoga clubs redefines the cultural physique of the neighbourhood by providing the commercial space for health clubs to sell their fitness lifestyle.

In Ossington's gentrifying retail ecosystem, tattooed baristas serve "Babyccinos" of foamed milk to residents' children, and microbrewery bar staff pour beer underneath an electric canopy of Edison light bulbs. Such a synergy between new retail capital investment and the cultural currency of new upscale shops feeds the elitist appeal of Ossington's symbolic economy. The influx of restaurateurs' "artisanal" commerce on Ossington is worth noting as the critical density of new restaurants and their design contributes to the production of this local shopping street's visual flavour and style. The concentration of high-end restaurants plating food that is exclusively local, seasonal, and organic flaunts the economic and cultural capital of its bourgeois patrons, establishing Ossington as a distinctive place to eat in Toronto. Indeed, with the production of the "Ossington lifestyle," this commercial street got "too hip, too fast" (Wingrove, 2009). Fearing that the street's retail change was out of control, city council issued a one-year moratorium on new bars and restaurants opening on Ossington Avenue. Civic leaders' decision to tame the wild private investment of restaurateurs underscores the frantic pace of gentrification on this local shopping street.

This hipster story of Ossington Avenue and the celebration of its commercial gentrification is driven and glorified by local, national, and corporate media, and blogs and wikis who publish glowing restaurant reviews. In the digital public realm, mobile critics who write positive mini consumer essays on social media platforms such as Yelp.com produce discursive investments in both individual restaurants and the street's gentrified urban image (Zukin et al., 2017). Absent in these romanticized narratives are the stories and lived experiences of displaced marginalized residents (e.g., cultural workers, rooming-house residents, mental health consumers, and refugees) and business owners.

Strategies of Collective Resistance to Gentrification

A particularly disturbing dimension of gentrification is the assumption that this urban transformative process is inevitable. If we want to build more socially just and inclusive cities, it is crucial that the gentrification-as-inevitable narrative be interrogated and challenged. Writing about gentrification resistance in London, England, urban geographer Loretta Lees (2014) in collaboration with community groups has created an anti-gentrification handbook that responds to the demolition and gentrification of public housing. The authors outline a series of tactics to combat the regeneration of affordable housing and the displacement of tenants. In order to successfully oppose the spread and destruction of gentrification, the authors emphasize the power of collaboration, urging tenants to question what is occurring and to share information with each other (e.g., neighbours, local friends, neighbourhood shops, and the tenants' association).

It is also important that local residents be attentive to signals of gentrification, such as an increase in "for sale" and "for rent" signs, media and real estate claims the area is "up and coming," and/or replacement of low-priced shops with expensive stores. The authors encourage community organizing as a tool to cohere a collective voice that challenges demolition projects and developers' profit-driven regeneration schemes. Telling one's story to local newspapers, media, and on social media is another way to publicize community demands and to garner attention from municipal politicians and local, national, and even international media. In this process, the power of storytelling and the cultivation of support and trust between community members becomes a source of collective inspiration that can strengthen grassroots planning by and for local communities.

A strategic community alternative to "staying put" is the legal model of urban Community Land Trusts (CLTs). With the growing unaffordability of Canadian cities, CLTs are an innovative, legally binding way to make real estate affordable for low-income people in perpetuity. A focus on equitable access, ownership, and stewardship of the land is at the core of CLTs' structure. Davis (2010) outlines four key aspects of CLTs as a unique method for land ownership and stewardship. First, land is appreciated for its use value and common socio-cultural and ecological heritage rather than its exchange value and individual rights to land. Second, land is permanently removed from land speculation through restrictions enforced on the resale of land. Third, land is held by non-profit, non-governmental organizations who provide the terms of building ownership and rental tenure. Buildings on the land are either owned or leased by CLT residents. Finally, ground leases act as the legal contracts between the building owner, homeowner, and tenant and the CLT organization.

In the urban context, CLTs protect valuable and highly profitable city land from market pressures such as land inflation and rising housing prices, while protecting community-based interests and prioritizing the land for community benefit (Bunce and Aslam, 2016; Walker and Carter, 2010). CLTs do not just prioritize affordable housing preservation; they also acquire land for local business incubation, urban agriculture, and other community economic development initiatives.

A land trust has emerged recently in Toronto—Parkdale Neighbourhood Land Trust (PNLT). Located on the southwestern edge of Toronto's downtown, just a few kilometres west of Ossington Avenue and CAMH, the neighbourhood of Parkdale has faced persistent territorial stigmatization following the deinstitutionalization of psychiatric patients during the 1980s from the then-Queen Street Centre for Addiction and Mental Health. Parkdale, and its high concentration of privately owned but affordable apartments and Victorian-styled rooming houses, became a crucial rental housing stock that residentially absorbed the exodus of discharged mental health patients (Slater, 2004). Now, however, as many apartment buildings are acquired by corporate landlords (i.e., MetCap Living Management Inc.) and Victorian rooming houses de-converted and re-valorized as opulent, single-family mansions, many low-income residents are at risk of eviction, displacement, and homelessness.

As the state chronically neglects the social housing question (Abbruzzese, 2016), the PNLT has intervened to protect the longevity of Parkdale's affordable housing. In their 2017 report, *No Room for Unkept Promises*, the PNLT outlines a 10-year strategy that calls for the preservation and rehabilitation of at-risk rooming houses through non-profit acquisition. Purchasing rooming houses would remove this housing stock from the competitive real estate market and ensure that units remain safely affordable for low-income residents. Currently, PNLT's only parcel of community-owned land is the Milky Way Garden, a shared urban agricultural space, which cultivates healthy food and nourishes community building and cultural sustenance amongst diverse groups in Parkdale.

While the Milky Way garden represents the desire of social rootedness, delivering affordable housing through the land trust would ensure community members can stay in place over the long term. While PNLT's prioritization of the rooming house rental market recognizes the social value of these increasingly vulnerable units, such a place-based strategy seeks to ameliorate rather than eliminate displacement and does not deal with the structural roots of uncontrolled gentrification. Indeed, it is without the political will of elected leaders, and the power of state laws to protect people's right to stay put, that market forces under the guise of gentrification will continue to displace residents and destroy communities.

Conclusion

To withstand waves of national and international competitive urbanism, Canadian cities have relied heavily on branding and spectacularization to reinvent and redevelop inner-city spaces. Across the country, luxury condominiums, retail districts, and festival waterfronts have been created in redundant industrial areas. Such urban development and policy choices have a consequential geography that creates lasting structures of unevenly distributed advantage and disadvantage (Soja, 2010). In the face of such apparent consensus on how Canada's largest cities should be re-imaged, re-elevated, and re-placed, one sees the danger of creating generic, homogenizing, single-thought cities where differences and conflict are neither tolerated nor valued. For it is the urban citizens "excluded from this spectacularisation process [who] possess perhaps the key to its reversal, which would entail . . . popular participation . . . and deep acquaintance with urban spaces" and the great diversity of people who desire to use them for more than consumptive purposes (Vaz and Jacques, 2007: 250).

Popular participation and a deep, rather than a shallow, engagement with urban spaces can be encouraged through socio-spatial improvisational opportunities—spaces of social interaction across difference that are changeable, malleable, and affordable, that encourage spontaneous and intuitive activities, and that support risk-taking (Bain, 2003). They afford small-scale experiential initiatives and local interventions in the urban fabric that allow for unexpected encounters with differences. They nurture alternative, non-hipster and non-consumptive ways of being and behaving that "disrupt the authoritative structures that govern them" (Jonsson, 2006: 37). As in-between socio-spatial opportunities that hold the potential for plurality and contradiction within them, they can be interpreted as a more socially sustainable alternative to spectacular space. Such localized spaces of possibility (whether they be land trusts, informal community cultural hubs, **community gardens**, or tool libraries) are the sites from which grassroots urban activism, optimism, and creativity that resist dominant ideologies and practices can emerge.

However, the opportunities for finding and securing affordable housing and improvisational socio-space opportunities in Canadian cities are rapidly diminishing. They used to be found in lower-income, working-class, inner-city neighbourhoods. But gentrification, the condominium boom, and waterfront redevelopment projects have all intensified residential development in the downtown core to meet middle- and upper-class consumptive and experiential practices, and in the process marginal social groups have been displaced. Such processes of displacement need to be continually and actively challenged if inner-city neighbourhoods are to remain socially diverse, equitable, and inclusive places that can function as incubators for culture, entrepreneurship, and sociability for more than just the middle class.

Review Questions

1. Why is the prefix *re-* (as in *re-imaging*, *re-elevating*, and *re-placing*) used with regard to inner-city change?
2. Explain the relationship between the competitive city and the inner city as "spectacle."
3. How has gentrification been differently theorized by urban scholars?

References

Abbruzzese, T. 2016. "'Build Toronto' (not social housing): Neglecting the social housing question in a competitive city region," in Keil et al. *Governing Cities Through Regions: Canadian and European Perspectives*. Waterloo: Wilfrid Laurier University Press.

Archer, B. 2015. "Montreal vs. Toronto: Comparing Ice Cream Parlours," *Air Canada: En Route Magazine*, 26 June.

Avraham, E. 2004. "Media strategies for improving an unfavorable city image," *Cities* 21, 6: 471–9.

Bain, A.L. 2003. "Constructing contemporary artistic identities in Toronto neighbourhoods," *Canadian Geographer* 47: 303–17.

Berelowitz, L. 2005. *Dream City: Vancouver and the Global Imagination*. Vancouver: Douglas & McIntyre.

Boddy, T. 2005. "Vancouverism and its discontents," *Vancouver Review*. At: www.vancouverreview.com/past_articles/vancouverism.htm

Bourdieu, P. 1984. *Distinction: A Social Critique of the Judgment of Taste*. R. Nice (trans.). Cambridge MA: Harvard University Press.

Bunce, S. 2009. "Developing sustainability: Sustainability policy and gentrification on Toronto's waterfront," *Local Environments* 17: 651–67.

———, and F.C. Aslam. 2016. "Land trusts and the protection and stewardship of land in Canada: exploring non-governmental land trust practices and the role of urban community land trusts," *Canadian Journal of Urban Research* 25: 23–34.

Burke, A. 2006. "Beneath and before: Continuum of publicness in public art," *Social and Cultural Geography* 7: 949–64.

Burnett, K. 2014. "Commodifying poverty: gentrification and consumption in Vancouver's Downtown Eastside," *Urban Geography* 35: 157–76.

Butler, T., and L. Lees. 2006. "Super-gentrification in Barnsbury, London: Globalisation and gentrifying global elites at the neighbourhood level," *Transactions of the Institute of British Geographers* 31: 467–87.

Caulfield, J. 1994. *City Form and Everyday Life: Toronto's Gentrification and Critical Social Practice*. Toronto: University of Toronto Press.

Coupland, D. 2000. *City of Glass: Douglas Coupland's Vancouver*. Vancouver: Douglas & McIntyre.

Cowen, D. 2006. "Hipster urbanism," *Relay: A Socialist Project Review* 13: 22-23.

Danyluk, M., and D. Ley. 2007. "Modalities of the new middle class: Ideology and behavior in the journey to work from gentrified neighbourhoods in Canada," *Urban Studies* 44: 2195–210.

Davis, J.E. 2010. "Origins and Evolution of the Community Land Trust in the United States," in J. Davis, ed., *The Community Land Trust Reader*. Cambridge: Lincoln Institute of Land Policy.

Desfor, G., and J. Ledley, eds. 2011 *Reshaping Toronto's Waterfront*. Toronto: University of Toronto Press.

Evans, G. 2007. "Branding the city of culture—The death of city planning?" in J. Monclus and M. Guardia, eds., *Culture, Urbanism and Planning*. Aldershot: Ashgate.

Florida, R. 2002. *The Rise of the Creative Class: And How It's Transforming Work, Leisure, Community and Everyday Life*. New York: Basic Books.

———. 2005. *The Flight of the Creative Class*. New York: Harper Business.

Gehl, J. 2010. *Cities for People*. Washington: Island Press.

Gertler, M., R. Florida, G. Gates, and T. Vinodrai. 2002. *Competing on Creativity: Placing Ontario's Cities in a North American Context*. Toronto: Ontario Ministry of Enterprise, Opportunity, and Innovation.

Glass, R. 1964. "Introduction: Aspects of change," in Centre for Urban Studies, ed., *London: aspects of change*. London: MacKibbon and Kee.

Griffin, K. 2017. "Granville Island expects to thrive despite loss of Emily Carr University," *Vancouver Sun*: 6 September. At: https://vancouversun.com/business/local-business/granville-island-expects-to-thrive-despite-loss-of-emily-carr-university

Hall, T. 2003. "Art and urban change: Public art in urban regeneration," in A. Blunt et al., eds., *Cultural Geography in Practice*. New York: Arnold.

Harvey, D. 1989. "From managerialism to entrepreneurialism: The transformation in urban governance in late capitalism," *Geografiska Annaler: Series B, Human Geography* 71: 3–17.

Hubbard, P. 2016. "Hipsters on our high streets: Consuming the gentrification frontier," *Sociological Research Online* 21: 1–6.

———. 2017. *The Battle for the High Street: Retail Gentrification, Class & Disgust*. Basingstoke, Palgrave.

Hutton, T. 2004. "Post-industrialism, post-modernism and the reproduction of Vancouver's central area: Retheorising the 21st-century city," *Urban Studies* 41: 1953–82.

Hyde, Z. 2014. "Omnivorous gentrification: Restaurant reviews and neighborhood change in the downtown eastside of Vancouver," *City & Community* 13: 341–59.

Jonsson, T. 2006. "Space invaders: There goes the neighbourhood," *In/Site* 15: 37–9.

Kern, L. 2007. "Reshaping the boundaries of public and private life: Gender, condominium development, and the neo-liberalization of urban living," *Urban Geography* 28: 657–81.

Kipfer, S., and R. Keil. 2002. "Toronto Inc? Planning the competitive city in the new Toronto," *Antipode* 34: 227–64.

Knox, P., and L. McCarthy. 2012. *Urbanization: An Introduction to Urban Geography*, 3rd edn. Boston, MA: Pearson.

Lange, B. 2006. "Landscapes of scenes: Socio-spatial strategies of culturepreneurs in Berlin," in T.S. Terkenli and A.M. d'Hauteserre, eds., *Landscapes of a New Cultural Economy of Space*. Dordrecht: Springer.

Lees, L. 2014. *Staying Put: An Anti-Gentrification Handbook for Council Estates in London*. London: Calverts Cooperative. At: www.justspace.org.uk

Lehrer, U. 2008. "Urban renaissance and resistance in Toronto," in L. Porter and K. Shaw, eds., *Whose Urban Renaissance? An International Comparison of Urban Regeneration Strategies*. London: Routledge.

Ley, D. 1993. "Past elites and present gentry: Neighbourhoods of privilege in Canadian cities," in L. Bourne and D. Ley, eds., *The Changing Social Geography of Canadian Cities*. Montreal and Kingston: McGill-Queen's University Press.

———. 1996a. *The New Middle Class and the Remaking of the Central City*. Oxford: Oxford University Press.

———. 1996b. "The new middle class in Canadian central cities," in J. Caulfield and L. Peake, eds., *City Lives and City Forms: Critical Research and Canadian Urbanism*. Toronto: University of Toronto Press.

———, and H. Frost. 2006. "The inner city," in T. Bunting and P. Filion, eds. 2006. *Canadian Cities in Transition: Local through Global Perspectives*. Toronto: Oxford University Press.

———, and C. Dobson. 2008. "Are there limits to gentrification? The contexts of impeded gentrification in Vancouver," *Urban Studies* 45: 2471–98.

Logan, J., and H. Molotch. 1987. *Urban Fortunes: The Political Economy of Place*. Berkeley, CA: University of California Press.

Lynch, N., and Y. Pottie-Sherman. 2017. "Gentrification, gated communities, and social mixing," in A.L. Bain and L. Peake, eds., *Urbanization in a Global Context*. Don Mills: Oxford University Press.

Maly, I., and P. Varis. 2016. "The 21st century hipster: On micro-populations in times of super diversity," *European Journal of Cultural Studies* 19: 637–53.

Markusen, A., and D. King. 2003. *The Artistic Dividend: The Hidden Contributions of the Arts to the Regional Economy*. Minneapolis, MN: Project on Regional and Industrial Economics, Humphrey Institute, University of Minnesota.

Marshall, R., ed. 2001. *Waterfronts in Post-Industrial Cities*. London: Spon Press.

Miles, M. 2005a. "Interruptions: Testing the rhetoric of culturally led urban redevelopment," *Urban Studies* 42: 889–911.

Miles, S. 2005b. "Creativity, culture and urban development: Toronto examined," *disP—The Planning Review* 162, 3: 70–87.

O'Kane, J. 2016. "Shinola, the coffee shop, and the price of doing business on Toronto's Ossington" *The Globe and Mail*, July 19.

Pratt, A. 2009. "Urban regeneration: From the arts 'feel good' factor to the cultural economy," *Urban Studies* 46: 1041–61.

Quastel, N., M. Moos, and N. Lynch. 2012. "Sustainability-as-density and the return of the social: The case of Vancouver, British Columbia," *Urban Geography* 33: 1055–84.

Robertson, L. 2007. "Taming space: Drug use, HIV, and homemaking in Downtown Eastside, Vancouver," *Gender, Place and Culture* 14: 527–49.

Rose, D. 1984. "Rethinking gentrification: Beyond the uneven development of Marxist urban theory," *Environment and Planning D: Society and Space* 2: 47–74.

Rumboldt, R. 2017. "Calgary homeless population still 'living on the edge,' despite drop in numbers," *CBC News*, 21 September. At: https://www.cbc.ca/news/canada/calgary/calgary-homelessness-study-emergency-shelter-usage-1.4299089

Short, J.R. 1999. "Urban imagineers: Boosterism and the representation of cities," in A. Jonas and D. Wilson, eds., *The Urban Growth Machine: Critical Perspectives Two Decades Later*. Albany, NY: State University of New York Press.

Slater, T. 2004. "Municipally managed gentrification in South Parkdale," *Canadian Geographer* 48: 303–25.

Smith, N. 1979. "Toward a theory of gentrification: A back to the city movement by capital, not people," *Journal of the American Planners Association* 45: 538–48.

———.1996. The New Urban Frontier: Gentrification and the Revanchist City. New York: Routledge.

———. 2002. "New globalism, new urbanism: gentrification as global urban strategy," *Antipode* 3: 427–450.

Stevenson, M. 2017. Granville Island 2040: Bridging Past & Future: A Final Report on Comprehensive Planning and Vision for the Future of Granville Island. Ottawa: Canada Mortgage and Housing Corporation.

Soja, E. 2010. *Seeking Spatial Justice*. Minneapolis, MN: University of Minnesota Press.

Southwick, R. 2018. "Calgary's 10-year plan to end homelessness has run out of time," *CBC News*, 29 January. At: https://www.cbc.ca/news/canada/calgary/calgary-10-year-plan-end-homelessness-1.4507625

Takahashi, K. 2017. "Toronto's Little Portugal: gentrification and social relations among local entrepreneurs," *Urban Geography* 38: 578-605.

Vaz, L.F., and P.B. Jacques. 2007. "Contemporary urban spectacularisation," in J. Monclus and M. Guardia, eds., *Culture, Urbanism and Planning*. Aldershot: Ashgate.

Walker, R. and Carter, T. 2010. "At home in the city: Housing and neighbourhood transformation," in T. Bunting, P. Filion, and R. Walker, eds., *Canadian Cities in Transition: New Directions in the Twenty-first Century*. Toronto: Oxford University Press.

Walks, A. 2001. "The social ecology of the post-Fordist/global city? Economic restructuring and socio-spatial polarization in the Toronto urban region," *Urban Studies* 38: 407–47.

Wilson, D., and R. Keil. 2008. "The real creative class," *Social and Cultural Geography* 9: 841–7.

Wingrove, J. 2009. "Too hip, too fast: City puts brakes on Ossington development." *The Globe and Mail*, May 27. Zukin, S. 1982. *Loft living: Capital and Culture in Urban Change*. New Brunswick, NJ: Rutgers University Press.

———. 1991. *Landscapes of power: from Detroit to Disney world*. Berkeley, CA: University of California Press.

———. 1995. *The Cultures of Cities*. Cambridge, MA: Blackwell.

———. 2004. *Point of Purchase: How Shopping Changed American Culture*. New York: Routledge.

———. 2009. "Changing Landscapes of Power: Opulence and the Crisis of Authenticity," *International Journal of Urban and Regional Research* 33: 543-53.

———. 2010. *Naked City: The Death and Life of Authentic Urban Places*. Oxford: Oxford University Press.

———, P. Kasinitz, and X. Chen. 2016. eds., *Global Cities, Local Streets*. New York: Routledge.

———, Linderman, S., and Hurson, L. 2017. "The omnivore's neighborhood? Online restaurant reviews, race, and gentrification," *Journal of Consumer Culture* 17: 459-79.

16 Cities on the Edge

Suburban Constellations in Canada

Jean-Paul D. Addie, Robert S. Fiedler, and Roger Keil

Introduction

Statistically speaking, a majority of Canadians now live in urban environments. However, most of our nation's urbanities, in fact, live in places that are profoundly suburban—whether considered in terms of the built environment, their location on the periphery of large urban centres, small town, or even villages, or the lifestyles such constellations support (Gordon and Janzen, 2013). And if we take into account **exurban** sprawl, there is even more dispersion (see Table 16.1). Those who move to major urban areas from elsewhere in Canada (rural and remote areas, as well as small towns) and those who arrive from distant shores (the majority of the approximately 250,000 immigrants who arrive each year), now tend to settle directly in suburban and exurban areas found around the country's largest cities.[1] That latter trend, which started in the final decades of the last century, has changed not only the face of our urban peripheries, but also the demographic and cultural composition of Canada overall. Throughout this volume, we have read stories of **walkability**, downtown living, creative economies, increased transit use, condominiums, and revived cultural scenes in downtowns and adjacent inner-city neighbourhoods. Re-urbanization (i.e., the increase in population and density in traditional inner cities) remains a vital process, at least in some of our urban centres, but in what follows, we demonstrate that this is only one side of the story. Welcome to suburbia: the most diverse and dynamic frontier of the Canadian city.

In this chapter, we are interested in exploring the metropolitan fringe through the intertwined lenses of suburbanization and suburbanisms. In simple terms, **suburbanization** is the non-central growth of populations and economic activity combined with the spatial extensions of urban areas. In contrast to these more quantifiable phenomena, the notion of **suburbanisms** captures the idea that these expanding and evolving landscapes generate a multiplicity of qualitatively distinct "suburban ways of life" (Ekers, Hamel, and Keil, 2012; Keil, 2017). In other words, contemporary suburbia is characterized by the emergence of diverse and non-traditional ways of living that are fundamentally different from those classically understood in relation to both the city and the countryside (see Moos and Walter-Joseph, 2017). To better understand suburbanisms as a plural phenomenon, an *Atlas of Suburbanisms*, created at the University of Waterloo, has mapped its dimensions across the Canadian metropolitan landscape—city, suburb, exurb, and rural fringe. The project distinguished between suburbs as places and suburbanisms as ways of living by examining the 25 largest metropolitan areas in Canada at the dissemination area (da) level via three sets of variables: (1) the built-form/commute-mode dimension; (2) the domesticity dimension; (3) and the social status dimension (Moos and Kramer, 2012; see also Moos and Mendez, 2014).[2] The resulting maps,

Table 16.1 Population Distribution in Different Urban and Suburban Realms in Canada

CMA	2011 population	Active core (%)	Transit suburb (%)	Auto suburb (%)	Exurban (%)
Toronto	5,583,064	11.0	14.0	72.0	3.0
Montreal	3,824,221	10.5	13.5	72.0	4.0
Vancouver	2,313,328	16.0	12.0	67.0	6.0
Ottawa	1,236,324	12.0	10.5	65.0	12.5
Calgary	1,214,839	12.9	2.6	80.0	4.6
Edmonton	1,159,869	11.0	12.0	66.0	11.0
Quebec City	765,706	15.0	6.0	65.0	14.0
Winnipeg	730,018	12.0	5.0	75.0	8.0
Hamilton	721,053	11.0	10.0	71.0	8.0
Kitchener-Waterloo	477,160	10.0	9.0	77.0	4.0
London	474,786	14.0	14.0	58.0	14.0
St Catherines	392,184	15.0	2.0	69.0	14.0
Halifax	390,317	14.0	18.0	44.0	24.0
Oshawa	356,177	3.0	13.0	77.0	7.0
Victoria	344,615	17.0	11.0	67.0	4.0

Source: Statistics Canada, 2011 Census; www.canada.com/news/infrographics/suburbs.suburbs.html (based on analysis by David Gordon, Professor and Director: Queen's University School of Urban and Regional Planning).

shown in Figure 16.1, make it apparent that characteristics associated with suburbs and suburban living are found across the entire metropolitan fabric of Canada, including in central cities or urban core areas (albeit less intensely than elsewhere in metropolitan space). In these terms, we are a suburban nation in an urbanized country.

Uncovering the Historical Geographies of Canadian Suburbs and Suburbanization

How did we become suburban? Being a settler society, the country's history of suburbanization entails a taking of Indigenous land. This means we can approach the suburban from the unsettled resource fringe of the empire; from the world of staples, beaver pelts, wheat, and wood, to the Indigenous lands and frozen northern tundra from which neither cities nor suburbs are usually viewed. Land in modern settler societies like in the Americas has had pre-colonial histories of Indigenous use. Lorenzo Veracini has noted "that suburbia re-enacts settlement" (2012: 340). This is certainly true for "the single-family house, the nuclear family, the separation between work and home, and the separation between gendered spaces" (Veracini, 2012: 340), but the Canadian case entails more aspects of suburbanization such as sprawling resource economies, mixed morphologies, and socio-demographic and economic diversity (Keil, 2018).

Indigenous land becomes part of the operational landscape of the—resource-based—settler capitalism of the Canadian state; Indigenous people become marginalized out of consciousness or even material existence, as Fernandes and Matharu viscerally capture in their work "Otherness" (see Figure 16.2). At the same time, the taking and subdivision of land has led to a

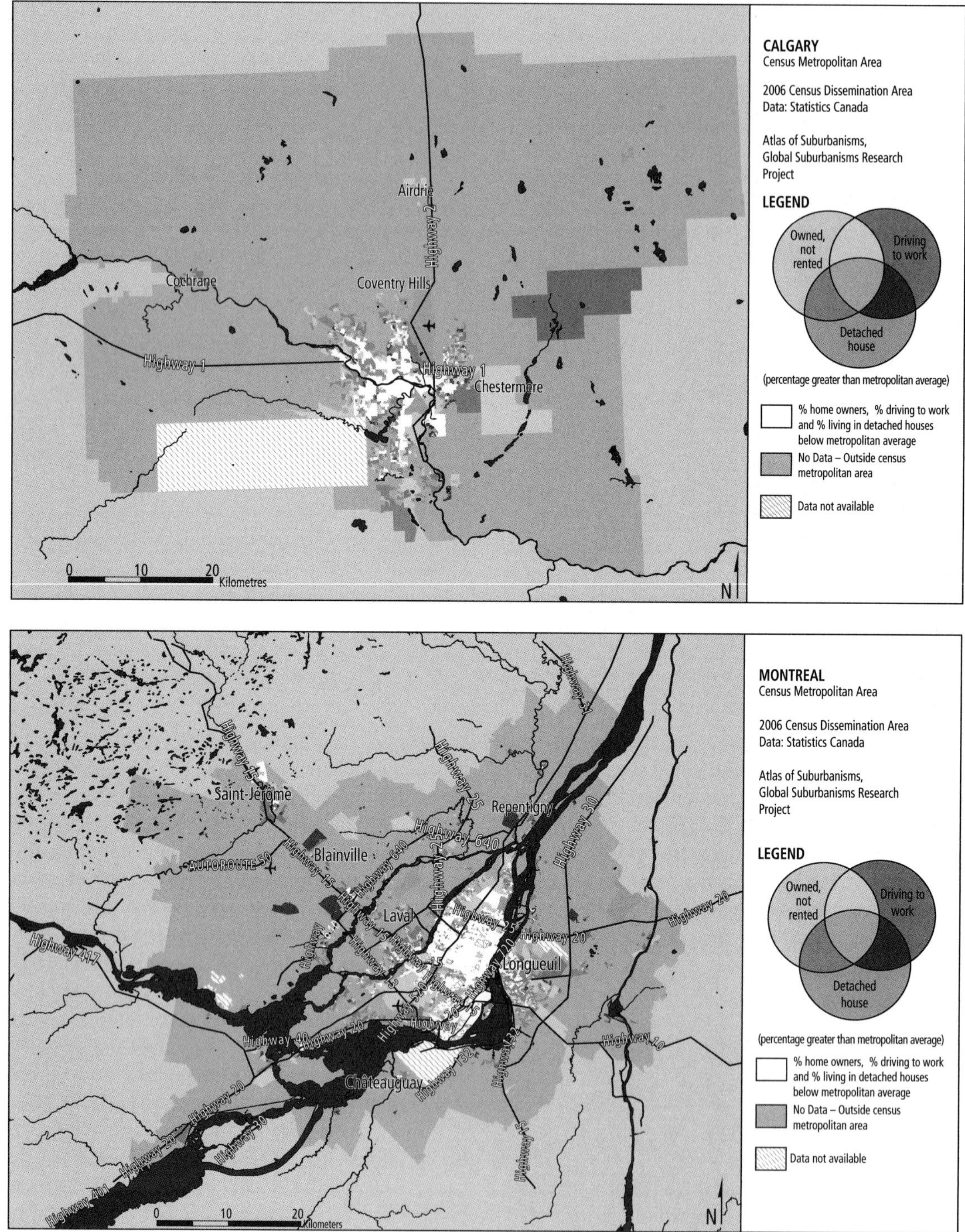

Figure 16.1 Built Form and Commute Type in the Calgary and Montreal CMAs.

Source: Moos and Kramer, 2012.

Note: Colour maps available online https://uwaterloo.ca/atlas-of-suburbanisms

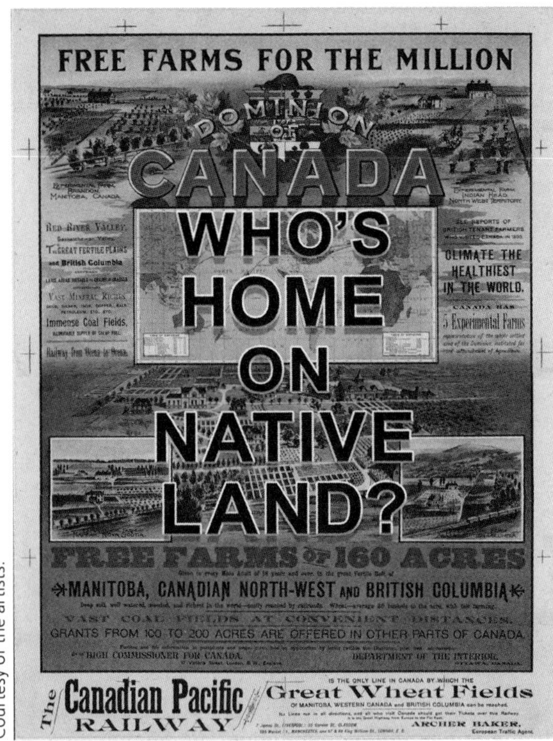

Courtesy of the artists.

Figure 16.2 "Otherness" (Marilyn Fernandes and Pamila Matharu) Taking a Page, 2017. Backlit vinyl, 43" x 89". Commissioned by the Art Gallery of York University.

tremendous gap in long-term wealth in Canada. Political conflict has erupted over the compensation and reconciliation of land ownership denied to Indigenous communities as well as African-Canadians. As generational wealth has largely bypassed these communities, the map of suburban home- and asset-ownership reveals landscapes shaped by, and reproducing, white privilege—a form of less-conscious or intentional racism that Laura Pulido (2000: 15) defines as "the hegemonic structures, practices, and ideologies that reproduce whites' privileged status . . . because they accrue social and economic benefits by maintaining the status quo."

The building of cities and suburbs in the bush, or in previously Indigenous agricultural landscapes, has a long and lasting history in Canada. Colonization of the land, especially where it is tied to resource exploitation in the North, is primarily suburbanization before urbanization. In the resource periphery, suburban subdivisions emerge where no city yet exists (see Figure 16.3). Rob Shields calls the product of such voracious land taking in the bush "feral suburbs" and observes for the tar sands capital of Fort McMurray, Alberta, that "strangeness vies with normality as cultural and economic forms of shift-work and suburban life are remade" at that particular frontier of suburbanism (Shields, 2012: 206).

Beginning our historical account of Canadian suburbs in such "feral" peripheries is significant for two reasons. First, it starts to shine an importation light on the place of Indigenous peoples, communities, and landscapes in Canada's urban history. Second, such a post-colonial reading presents a stark juxtaposition to the environments and ways of living that typically come to mind when we think about suburbia: the "ticky-tacky" houses, white picket fences, and safe, homogeneous, and decidedly middle-class aspirations of the postwar North American suburb. The pervasiveness of such suburban imaginaries is easy to understand. In the far-reaching postwar transformation of Canadian cities, no change has been more impactful than the shift from relatively compact forms of urban development, built when walking or public transit were the principal means for getting around, to the more sprawling, large-scale suburban forms that emerged alongside the mass adoption of private automobile-based travel after World War II. The changes were drastic and abrupt. They altered society, the economy, and **governance** at all scales. As urban historian Robert Fishman (1987: 183) has noted, the postwar suburban building boom "was so powerful that it was like a tide that washed over all precedents. It was as if suburbanization *began* in 1945."

The prominence of postwar suburbanization has a number of implications for how we typically see our suburbs. Periodic calls to "re-invent the suburbs" (see Bourne, 1996) are inevitably directed at the multiple stereotypes and clichés that have attached themselves to postwar suburbia. In the 1950s young families flooded into

Roger Keil

Figure 16.3 "Feral suburbia," Kelowna, British Columbia

new suburban developments filled with modest homes, automobile usage proliferated, and television extended the reach of mass culture and consumerism. Chain stores in shopping plazas and suburban malls began to replace independent shops and street-based retailing in newly built suburban areas. The newness and initial sameness of postwar suburbs led urban-based intellectuals, writers, artists, and social commentators to deride them as sterile, banal, and conformist places. Initially viewed as the solution to urban problems such as poverty, crime, vice, and pollution, the suburb now seemed to urbanists to be a hellish place devoid of culture and home to all manner of new social pathologies (Nicolaides, 2006). Since the 1970s, various attempts have been made to demonstrate that "the suburbs" have evolved into a complex and variegated socio-spatial landscape. But stereotypes and clichés that originated in the 1950s, and have gained almost mythical status in the decades since, continue to powerfully shape how suburbs are represented and understood.

Rather than attempt to disprove "the suburban myth" (see Donaldson, 1969), it is more productive to see cities and their suburbs as both real and imagined places, and consider why certain representations of them endure, and the degree to which they reflect the realities of "actually existing" suburbs (past and present). A vital task, therefore, is to question the often taken-for-granted assumption that cities and suburbs are opposites (Schafran, 2013). If instead we conceive of suburbs as transitional spaces—a middle landscape between city and open countryside (Rowe, 1991)—then it becomes easier to consider "city" and "suburb" as existing within a continuum of urban spaces, rather than set them off against each other as categorically different places. It also encourages us to view suburbs or suburban areas as places that evolve: in response to the lifecycle of residents; as political, economic, or social conditions change; and as their relative location within a metropolitan region is altered by continued urban growth and spatial expansion (McManus and Ethington, 2007). They are sites of creativity, vibrancy, and dynamism—characteristics traditionally associated with the city rather than its peripheral "other" (Bain, 2013).

The challenge is to better understand when and where stereotypes and clichés, often masquerading as conventional wisdom, obscure contemporary suburban realities and distort suburban histories. German planning theorist Tomas Sieverts (2003: 48, 17) argues that the "myth of the old city" or a "one-sided love for the historical city" amongst urbanists distorts how suburbs are perceived. In North America, a preoccupation with the vast, sprawling landscapes built around cities after World War II colours how we understand prior waves of suburbanization. This is particularly true when older suburban landscapes, such as streetcar suburbs, are folded into the urban realm and washed of their suburban-ness. We need to understand that many areas now within the city limits were once peripheral and suburban (see Harris and Lewis, 1998a). Though it is impossible to know the transformations and change that await today's suburbs, the urban past tells us they will transform and change in response to internal needs, metropolitan pressures, and wider social and economic forces.

Revisionist scholarship on North American suburbanization has emphasized that prior to World War II "differences between cities and the suburbs as a whole were quite minor and were dwarfed by variations within the city and among suburbs" (Harris and Lewis, 2001: 284). Further, texts and commentary on suburbanization and metropolitan development from the first half of the twentieth century unsettle the notion that the only (or even main) participants in the suburban trend were affluent commuters (Wunsch, 1995). Multiple factors converge to explain the diversity of Canadian suburbia prior to the Second World War, including the availability of cheap land, industrial decentralization, and working-class desire for homeownership (see Harris, 1996; Lewis, 2000). Street railways also played an important role in allowing urban development to extend further outwards from the core, and enable a wider range of workers to live beyond the limits of the walking city (Warner, 1962). In addition to the development and expansion of street railways in Canadian cities, the episodic nature of urban land development—real estate booms and busts—underwrote the variation in lot-sizes and housing styles found in many pre-World War II suburban districts (McCann, 1999). The reappraisal of (mostly American) suburban histories increasingly demonstrates that an overemphasis on elite or middle-class suburbanization in suburban histories has distorted who and what is to be found in the suburban periphery. The suburb as a landscape of residential privilege and exclusion has long co-existed with "other" forms of suburban life, including lower-income, self-built, industrial, ethnically segregated, and new immigrant suburbs.

That "other" suburbs existed does not mean that individual suburbs were diverse, however. Historical research suggests that the metropolitan periphery during first half of the twentieth century was a segmented social space in which individual suburbs could be relatively homogeneous at the same time as the whole (suburbia) was diverse. Rather than resembling the simple zonal pattern of increasing affluence with distance from the urban core, an assumption built into the concentric model of the Chicago School (see Harris and Lewis, 1998b), the complexity of Canadian metropolis prior to World War II was shaped by the decentralization of industry along rail corridors, and the segmentation of social space along class, ethnic/racial, and religious lines (Harris, 1996, 2004; Lewis, 2000). The conventional image of suburban affluence and conformity, shaped heavily by postwar suburban stereotypes, has made it easier to elide the historical importance of industrial and working-class suburbanization.

At the same time, the historical development of Canadian suburbanization has been characterized by regional diversity (McCann, 2006). Differences in provincial planning practices and the role of the distinctive suburban strategies employed by large regional land syndicates (notably the Hudson's Bay company and the Canadian Pacific Railroad) in the pre-World War I era shaped the overall suburban form we now find everywhere. Postwar suburbanization did not simply happen in the form that it did because of world-historical forces, though broad structural factors, such as social class stratification (emergence of the middle class), patterns of industrialization (large-scale, **Fordist** production facilities), technological change (automobilization), and the governance of land ownership (far-reaching government programs such as mortgage subsidies), did play a significant role. The suburbs were also forged by various interests, values, and beliefs—of suburban residents, land developers, builders, planners, financial organizations and governments—that combined to change the "pluralistic" character of early Canadian suburbs, usually toward a more regulated, planned, and middle-class (though not completely) urban space. In most cases this led to lower-income, self-built forms of suburbanization being squeezed out by the proliferation of corporate, mass-produced landscapes (McCann, 1999).

The Diverse Canadian Suburb: Immigration, Class, and Gender

Although Canadian suburban development adopted a more regulated planned and institutional form in the postwar era, our nation's urban periphery now presents a broad picture of social diversity that has decisively moved suburban Canada from the classical trajectory of Anglo-American suburbanization to present a remarkable new model of urban development (Pitter and Lorinc, 2016; Lo et al., 2015). No longer are the outskirts of cities more homogeneous and straightforwardly wealthier than cities: they are produced, experienced, and understood in differentiated terms relative to the class, gender, and socio-cultural background of the individuals and communities who call them home.

The immigrant experience and the heterogeneity of new suburban populations increasingly define Canadian suburbs. Whereas previously new immigrants settled in the inner city and moved to the suburbs as their economic situation permitted, the contemporary pattern is for (notably non-European) immigrants to move directly to the suburbs. The rise of nicknames like "Brownstown" and "Bramladesh" for Brampton, Ontario, signify places in our suburban world where visible minorities are now the majority (Ahmed-Ullah, 2016). Immigrants are therefore generating qualitative changes to suburban living, which destabilize stereotypes of white middle-class nuclear families residing in single-family homes as the suburban norm. This trend takes two main forms. First, in the old, formerly white and middle-class suburbs of the postwar years, there are now large and remarkably diverse non-white and immigrant populations attracted by low-rent apartments, affordable single-family homes, and the availability of public transit. Here change is registered visually by the appearance of new places of worship and via the now kaleidoscopic diversity of signage for ethnic businesses found in strip malls and modest plazas located along major arterial roads. Such socio-cultural place-making parallels and defies more conventional account of retrofitting suburbia in morphological terms (Dodson, 2014; Dunham-Jones and Williamson, 2009).

Second, in newer, outer suburbs, more affluent immigrant households are shaping new suburban spaces, rather than adapting existing ones to meet new needs. One manifestation of this is a novel form of suburban development: the **ethnoburb** (Li, 2009). Chinese immigration, for example, has altered the suburban landscape in both Toronto and Vancouver in profound ways. In both cities, new Chinese immigrants increasingly bypass traditional reception areas in the inner city and settle directly in peripheral communities. In Richmond (south of Vancouver) and Markham (north of Toronto), significant concentrations of Chinese immigrant populations have formed. Accompanying the development of these places into ethnoburbs has been the emergence of distinctive and novel suburban forms such as "Asian theme malls" (Preston and Lo, 2000; Ley, 2011). This reflects a shift away from traditional geographies of Chinese business and commercial activity, which concentrated in inner-city Chinatowns (Wang, 1999). Now well established as ethnoburbs, the initial transformations in Richmond and Markham revealed tensions between official

multiculturalism (the ideology and policy) and its actual practice as an everyday urban reality on the ground in Canadian communities (Wood and Gilbert, 2005).

Traditional understandings of home and neighbourhood may explain the resistance of established suburban residents to the changes prompted by the influx of newcomers to previously white suburban areas. In two (quite different) residential subdivision developments in Surrey (a mostly lower-middle-class or working-class suburban city about 40 minutes east of Vancouver), white residents avoided the use of explicitly racialized language to explain their residential locational choices and preferences (Dowling, 1998). Feelings about perceived "others"—non-traditional families or Indo-Canadians—were made apparent, instead, through the use of a local "socio-spatial vocabulary" that linked specific suburban places to marginalized or racialized groups.

In north Scarborough, Richmond Hill, and Markham, the transformation of the suburban landscape in the 1980s and 1990s was marked by planning conflicts over proposals for new Chinese-oriented shopping centres (Preston and Lo, 2000). Opposition was usually framed by residents around traffic, parking, and character/scale of the proposed developments, but political debate and newspaper coverage at the time reveals at least an undercurrent of racial intolerance; anxiety about the scale and pace of ethnic change was also a factor. Proposals to establish new places of worship in the suburbs by minority groups can also spur community opposition, though increasingly ethnic communities in the suburbs are locating their churches, temples, gurdwaras, mosques, and other cultural institutions in industrial areas (Hackworth and Stein, 2012).

Given the heavy Canadian reliance on immigration for population growth, the anticipated flow of newcomers into mostly suburban environments has far-reaching implications. This is particularly the case if we consider the social, economic, and political context into which immigrants arrive. These are fragile **neo-liberal** times characterized by widening income inequality, the growth of precarious employment, and rising levels of housing stress and homelessness (Walks, 2009). Among South Asian and Chinese Canadians, there is a clear trend towards suburban residency (Hiebert, 2000), which may be at least partly explained by higher incidences of multigenerational and/or multi-family living arrangements. It is likely that multigenerational and/or multi-family households find larger suburban homes a preferred form of accommodation. But the diversity of immigrants arriving results in highly uneven suburban geographies of poverty and housing stress. For example, low-rent apartments in suburban areas are often highly localized within large tracts of mostly single-family homes, creating micro-geographies of acute housing need and poverty in otherwise comfortably housed areas (Bunting, Walks, and Filion, 2004; Fiedler, Schuurman, and Hyndman, 2006). The picture is further complicated by evidence of housing need and poverty that is thinly distributed across areas where residents are generally well-housed—a phenomenon that may be explained by the growing role of basement and secondary suites as a form of low-rent housing in suburban areas, though this is somewhat speculative (see Tanasescu, 2009).

This is not to suggest that the inner city no longer faces problems. It is more to highlight that social problems such as poverty, homelessness, food insecurity, and so forth, have become suburban problems too—albeit with different implications for policy-makers, service providers, and those who are marginalized, poor, and suburban. It is also to emphasize structural changes to postwar, inner suburban areas, which have experienced considerable **deindustrialization** since the 1980s (Walks, 2001). These changes have been identified, particularly as they relate to income inequality and neighbourhood poverty, in influential reports and studies (Hulchanski, 2010; United Way, 2004).

As the suburbanization of poverty continues to move outward, it is now an emergent reality in newer, outlying suburban areas. The concept of "priority neighbourhoods," introduced as part of Toronto's Strong Neighbourhoods Strategy, has been exported to allegedly wealthy suburbs where social distress has not previously been acknowledged or presumed to exist (Baluja, 2011). This has yet to translate, as it has in relation to older, closer-in suburban areas, into the talk of suburban **ghettos**, as is sometimes evoked in media coverage of certain neighbourhoods, such as Toronto's Jane and Finch, Kingston-Galloway, or Malvern neighbourhoods. These have become objects of perhaps misguided top-down policies transferred from other cases to pump resources into places that are considered breeding grounds of poverty, educational failure, and crime (Black, 2011). Putting this aside, it would seem that poverty and social exclusion are on the radar of policy-makers, which is important given that suburban municipalities have been slow to build the kinds of social infrastructure available in inner-city areas (Lo et al., 2015).

Gender also has a profound impact on the way suburban lives are enacted and understood. Feminist geographers, though, have long argued that suburban studies have largely failed to consider suburbs as "women's spaces" or critically examine the gendered nature of discourses on suburbanization (Strong-Boag et al., 1999). Conventional accounts have focused on the suburbs as a male paradise, a haven from the hustle and bustle of the city and a separation of work life from family life. The suburbs were closely implicated in the Victorian defence of the ideal middle-class family through the spatial separation of "private" homes and "public" cities. Accordingly, the inner city came to represent locales of the deviant, the poor, the recent immigrant and racially marginalized, and, in particular, a place of female sin.

The design of suburbs and suburban housing has enacted (and reinforced) gendered divisions of labour. "Traditional" suburbia was underpinned by women's unpaid domestic work. Over time, these "bourgeois utopias" have become harder to find as women increasingly entered the paid workforce in order to maintain middle-class lifestyles in the face of rising costs and stagnant or declining real wages (see Rose and Villeneuve, 1998). Paradoxically, increased work outside the home by women has resulted in contradictions between the ". . . assumptions underpinning suburban communities and the reality of female residents' lives" (Strong-Boag et al., 1999: 178). These contradictions include the increased isolation of stay-at-home mothers and the difficulties experienced by working mothers trying to manage the multiple demands of family life, wage labour, and long commutes. This contrasts with the active role of women in negotiating and constructing the suburban landscape and highlights the differing experiences of women in the suburbs, with some women finding the "burden of being 'good' wives and mothers worsened by a landscape that has regularly ignored them," while "[o]thers have flourished, able to mobilize community and kin resources in ways they experience as rewarding" (Strong-Boag et al., 1999: 179). In the absence of traditional family structures, social relationships had to be reconfigured in new suburban neighbourhoods. For example, women whose childcare needs were not being met in these emerging communities engaged in reciprocal economies to perform the "invisible work" done at home (Pratt, 2003). **New Urbanist** developments like Cornell in the Toronto suburb of Markham have done little to change the overall patterns of gender relations seen in classical suburban neighbourhoods (Markovich and Hendler, 2006). While such developments post higher-density patterns and different design features, which could suggest different lifestyles, in reality their residents behave just like their counterparts in conventional suburbs.

There is much controversy about the gender aspects of the suburban landscape. In terms of safety, some groups of women have been found to have a preference for inner-city urban design, for abundance of "populated" public spaces and sense of community that makes them feel safer—more at home—in contrast with images of bleak, unpopulated suburban landscapes that make them feel isolated and vulnerable (Kern, 2005, 2010). But race and class privilege may act as a blindspot to urban violence more broadly. Spaces of whiteness or middle-classness are constructed in Toronto through associating urban violence with "other" neighbourhoods like inner suburban Jane-Finch or inner-city Parkdale (Kern, 2005).

Complicating Suburban Form and Function

At the same time as the social geography of Canadian suburbia is being radically reconfigured, contemporary suburbanization is also marked by deep transformations in the structure and function of urban periphery. Suburban constellations such as **edge cities** (Garreau, 1991) or **technoburbs** (Fishman, 1987) and suburban downtowns (Filion and Gad, 2006)—the sprawling mixed-use suburban zones on the urban periphery that are automobile dependent, highway oriented, computer network-enabled, and relatively autonomous from older central cities—have been identified over the past three decades and have received increased attention in relation to contemporary preoccupations such as metropolitan integration and suburban sustainability (Atkinson, 2007; Keil and Addie, 2016; Lang and Knox, 2009). Yet much suburban development, as captured in Figure 16.4, now takes place in a dynamic landscape of mature and maturing suburban realms that neither resembles the old inner city nor the cookie cutter suburbs of pop-culture stereotypes (Harris, 2010). Significantly transformed since their original development, these dynamic yet cacophonous suburban spaces are often neglected in favour of a focus on the densification of the downtown and the continued sprawl of subdivisions across greenfields in the outer reaches of Canadian cities (Macdonald and Keil, 2012).

Most Canadians now live, work, and play in a quite undefined and non-descript middle landscape, where everything seems to happen at once: large-scale infrastructure like highways and airports are next to residential quarters; all manner of service infrastructures including universities and high-tech corridors sit adjacent to low-rent

Ute Lehrer

Figure 16.4 Driving down Highway 400: Toronto's Post-suburbia

apartments; parks and parking are side by side; high-speed highways and food and transit deserts define the same space; religious mega-structures are across the street from ethnic mini-malls. We can think of these spaces through a number of emergent hybrid conceptual lenses that are now redefining the terrains of suburban studies: "post-suburbanization" (Teaford, 1997; Phelps and Wu, 2011); the "in-between city" (Sieverts, 2003; Young, Wood, and Keil, 2011); "metroburbia" (Knox, 2008); or "exopolis" (Soja, 2000). Although some observers have stressed the "endurance" or "obduracy" of the suburban social morphology (Filion, 2018), we suggest here that the notion of something distinctly *post*-suburban captures a sense of a shift from previous suburban processes and the emergence of a new mode of urbanization that breaks from our traditional views of the relationship between the metropolis and its core (Lucy and Phillips, 1997). If suburbanization was once a process associated with social segregation and distance-making between classes, ethnicities, races, economic activities, and so forth (Maher, 2004), a partial reverse is now underway that casts (post-)suburbanization as "a process of adventurous mixing and reshuffling of the urban, while the city (as much as it exists as a recognizable unit) becomes the rarefied monoculture of condominium-dwelling creatives that operate in safe, sterile, predictable environments" (Keil, 2017: 47–8).

Post-suburban landscapes are dynamic and diverse spaces that are difficult to pin down in conceptual or geographic terms (Phelps, Wood, and Valler, 2010). Post-suburbia, as such, may best be read as an attempt to conceptualize an incremental shift from pre-existing suburban forms and processes, just as the postwar suburbs presented an evolution from earlier urban and industrial settlement patterns. It is a space of multiple speeds and scales of movement that offer the potential to retrofit, reconfigure, and re-imagine auto-centric and atomized suburban places. Importantly, while emergent post-suburban landscapes produce new nodes within polycentric urban regions, they do not hold the same functional logics or spatial practices as the historical centre city or even postwar suburbia (Archer, 2011). The interactive patterns of contemporary suburbs "are less like its blocky spatial layout and more like the entwined overlay of paths and nodes in a rainforest, where clearings and connections for different uses are mixed together, connected by twisting links, lacking any easy visible order" (Kolb, 2008: 160). The typical "in-between" landscapes found in the urban periphery are in perpetual transition, structured both by the continuation of existing urban traditions and the implementation of new experiments and innovations (Young et al., 2011).

Acknowledging the capacity of the urban periphery to support experimentation and adaptability is vital when we approach suburbia as a space that both houses and generates the infrastructure needed by the twenty-first-century metropolis. Suburbs are populated by a multitude of infrastructures that not only serve the immediate environment but support multi-scalar urban constellations and higher-order purposes. Indeed, it is useful to consider the differentiated ways suburban infrastructures are constructed in, of, and for, suburb space (Addie, 2016). The development of the inner city as a space of bohemian creativity, urbane living, and corporate capital accumulation is itself premised upon the relocating certain functions and land uses to the urban periphery, from premium networked and recreation spaces like airports, intermodal yards, and golf courses, to noxious and unaesthetic functions including garbage dumps, power stations, and water treatment facilities. On one hand, such complex assemblages plague peripheral areas that are subject to post-suburbanization with "myriad constellations of infrastructure problematiques that link to a host of political, social and environmental issues" (Filion and Keil, 2017: 8). The urban periphery is a site of enhanced risk reflecting both socio-technical vulnerabilities (collapsing bridges, blackouts, propane explosions) and social exclusions for those unable to afford access to premium services and utilities (Graham and Marvin, 2001; Keil and Young, 2009). Yet on the other hand, emergent forms of social and technological innovation are also taking shape, with global implications—for example, the tremendous success of the TransMileno Bus Rapid Transit (brt) project in Bogota, Columbia (Montero, 2017), a vital example for brt systems now in operation across the Greater Toronto Area, and the prospect of light-rail transit (LRT) networks forming a foundation for (post-) suburban mobility in the twenty-first century. Elsewhere, transformations in inner suburban Toronto reveal how landscapes that appear to be placeless (see Relph, 1976) can foster a sense of place as new users adapt what exists to meet changing needs. Postwar factories can be reborn as infill housing (or as big-box **power centres**) while strip malls—their retail equivalent—function as landing pads for immigrant entrepreneurs and "soft targets" for intensified development. In either case, old factories and strip malls are seen by many as obsolete reminders of Toronto's postwar embrace of the car—spaces that might be rendered more "urban" and "productive" via compact, **mixed-use development**.

The road to this more post-suburban future is not unproblematic. In addition to skepticism about the scale of the transformation required and whether it can produce the more complete streets and communities that urbanists and others rightly advocate, there is reason to pause and consider what and who might be displaced, as well as attend to how the process of change itself plays out in specific instances. Perhaps shabby and unloved, the old factories and strip malls of the in-between city perform an important function: they provide low-cost spaces in an otherwise expensive and increasingly exclusionary city. The potential or likely impact of redevelopment on low-income residents and economically marginal businesses or activities is often absent from plans. Likewise, vocal opposition at public meetings often skews strongly toward more privileged and powerful groups whose efforts are principally directed toward preserving or enhancing certain qualities, physical and social, of their local communities and limiting perceived encroachment of the "urban" in the form of increased high-density housing (Poppe and Young, 2015). That can leave the impression that opposition to **intensification** is nimbyism, directed toward exclusionary aims. Reducing it to that, however, also serves as convenient cover for those interests that stand to profit from densification and elides social justice concerns linked to displacement. This highlights a need to be attentive to the politics of post-suburban transformation and change.

Suburban Politics, Political Suburbs

Suburban and post-suburban communities are animated by a diverse collection of political contradictions (Hamel and Keil, 2016). These are emergent at different scales: between economic growth and the provision of collective consumption amenities, and between continued development and environmental conservation and the forces of amalgamation and secession (Phelps and Wood, 2011). Peck, Siemiatycki, and Wyly (2014: 389) pithily surmise that we are now talking about "a complexification and folding in of suburbanizing cultures and rationalities, as opposed to a linear process of centrifugal succession." In this context, we must recognize that the governance of suburbs is a process that combines the interplay of government, market, and private activities (Ekers et al., 2012). In Canada, historically, these activities have been mixed in a particular way:

> There is now a truly Canadian story developing from coast to coast which is both internally differentiated (in a federalist context) and shows some remarkable similarities across the nation's (sub)urban reach. Cities, regions and their suburbs are now recognized as central to the governance of the vast territory of Canada, which is beginning to understand itself as a primarily urban country as the majority of its citizens now live in some form of urban or suburban area (Keil et al., 2015: 102).

The governance of these suburbanizing urban regions is constituted through a new melange of politics that depart from the business of usual, in which central interests prevailed. An "in-between" type of politics has emerged that colonizes political spaces in the emergent post-suburban region:

> The idea of post-suburbia includes the understanding that the traditional dichotomy in urban politics of taking either redeveloping inner cities or newly built suburbs as the natural arenas of urban action has to be challenged. Instead, post-suburbia now validates both the overlooked spaces in-between and the emerging metropolitan spaces of which they are part (Young and Keil, 2014: 1606).

The implications of such transitions are likely to increasingly shape the national political agenda since the voter-rich suburbs are now vital strategic battlegrounds that decide Canadian elections. Reminiscent, at first, of 1990s campaigns, when white middle-class voters in conservative ridings north of Toronto formed the power base of a neo-liberal Tory government under Mike Harris (Keil, 2002), the suburban voters that were wooed in the 2011 federal election were new immigrants, visible minorities, and the elderly. Contentiously termed "ethnic voters" were audiences for stump speeches in suburbs like Surrey, British Columbia, and Brampton or Markham, Ontario, where politicians visited folkloristic and religious events. Issues such as social conservatism, immigration and settlement, and pension concerns took central stage as it was assumed that they would resonate with the local, diverse electorate (Friesen, 2011). Some progressive candidates also made inroads into the electoral base of the ethnoburbs as a younger generation of visible minority New Democrat politicians were elected in provincial and federal politics in Ontario. Most recently, this was the case, for example, during the Ontario provincial election when three suburban Brampton ridings were carried by NDP candidates, all of them people of colour (CBC News, 2011; *The Globe and Mail*, 2018).

Toward the end of the 2015 federal election, which returned the Liberals to power, the received wisdom that Canada's "two solitudes" would be urban and suburban in the twenty-first century was less than certain, though it was still possible to claim "[i]t is undeniable that the suburbs are key to this election" (Heartfield, 2015). Liberal gains across the country in suburban areas were key to their victory and signalled to many that the growing political strength of the suburbs was not ushering in an era of Conservative political dominance, a notion that had taken hold since the previous election. It is too soon, however, to draw strong conclusions. Analysis of 1965, 1984, and 2000 Canadian federal election results show that inner cities have become more likely to vote for left-wing parties, while suburban areas increasingly support right-wing parties (Walks, 2005, 2006).

But official elections and formal politics are not the only arena of new suburban politics in Canada. A new generation of activists has begun to change the style and substance of suburban political discourse. Ethnically diverse and multigenerational organizations, such as the citizens environmental coalition "Sustainable Vaughan," have added to and challenged the traditional development-driven political agenda in the suburbs. A lively debate has started in the suburbs on the character of their culture and politics in which young, often second-generation immigrants are putting traditional notions of inner and outer cities to the test (Alang, 2011). The opening up of new understandings of ethnoracial relationships and the politics that accompany them are now on the agenda in Canadian suburbs. Long gone are the times where politics could safely be located along socio-demographic lines as was often assumed in the past. We will consequently need to pay attention to the "strategies of surveillance, dispersal, and consumption" that contextualize much of politics and governance in the post-suburban landscape of today (Quinby, 2011: 139). In contrast to the perception of the suburbs as a space ruled by rational choice, personal freedom, economic autonomy, and land ownership, politics in the urban periphery now deals with corporate power, lack of collective consumption services, the presence of a strong local state (social programs and police presence in marginalized communities), and poverty. Increasingly, it will also have to grapple with fundamental political ecologies surrounding the future of auto-dependent landscapes (Atkinson, 2007) and climate change (Keil, 2016). In this reorientation, suburban politics may be less defined against the traditional imaginaries of the inner city and the landscape of power that those built. Suburban politics may now establish itself, municipality by municipality, community by community, in its own right, as a self-confident part of the post-suburban region.

If a closer examination of contemporary Canadian suburban political space indicates a disconnect between the predominant imaginaries and realities of the metropolis, this will have distinct ramifications for both political practices engendered in negotiating diverse suburban spaces, and in integrating the political-economic realities between the urban core and its surrounding periphery within particular geographical contexts. Collective action in "the city" has been well studied, particularly with regard to the **gentrification** of inner-city neighbourhoods or the restructuring/rescaling of urban politics. The self-conscious use of inner-city neighbourhood space as a marker of lifestyle and political distinctions, as non-suburban or mass society oriented or more specifically as an "oppositional space" in relation "automobility," "neo-liberalism," and "rationalities of ignorance" (Walks, 2006, 2015), begs the question: are the suburbs as conservative and reactionary as this presupposes, and perhaps more importantly, given the social diversity of many suburban areas, what new socio-political movements might be present (or emerging) within the changing, increasingly post-suburban dynamic? What might a "right to the suburbs" look like (Carpio et al., 2011)?

Emergent struggles around intensification and redevelopment not motivated by an interest in protecting or enhancing property values, offers one possible version (Charmes and Keil, 2015). Another can be glimpsed in diasporic activism that occasionally renders a particular kind of subaltern cosmopolitanism visible in suburban Canada (Basu and Fiedler, 2017). Not all claims over the "right to the suburbs" are necessarily progressive, though. There is good reason to focus on the amorphous neo-liberal populism embodied in many places by a "stop the war on the car" politics, which accompanies and responds to infrastructural investments and land-use planning oriented to smart growth intensification and its vision of denser, more compact suburbs that are most inclusive, more walkable and transit-oriented, and less auto-dependant. We therefore need to be more attentive to how intensification plays out in practice. A recent plan update for the Metrotown area of Burnaby, a Metro Vancouver suburb, calls for significant densification of an area home to a considerable concentration of relatively affordable low-rise rental apartments. An opposition group that emerged, the Alliance Against Displacement, noted that residents "are afraid the Metrotown plan will accelerate what they call 'demovictions'" because market-driven policies (increased height and density permissions in land-use plans and zoning) creates a strong financial incentive for demolition of low-rise apartment blocks, so they can be replaced by much larger condominium towers, a process already underway in the area (Saltman, 2017).

Resistance and collective action of a very different sort can also emerge in suburban areas where particular immigrant groups coalesce in sufficient numbers (Basu and Fiedler, 2017). In 2009, an expatriate protest movement in Toronto blocked the Gardiner Expressway to draw attention to the massacre of Tamil civilians in Sri Lanka as that country's long-running civil war entered its final stages. Prior to that, throughout certain parts of inner suburban Scarborough, messages and imagery were displayed on storefronts and in public spaces. Rallies were also held at major intersections. Similarly, protests by Bangladeshi Hindu, Christian, and Buddhist residents over treatment of fellow religious minorities in Bangladesh started at a local temple in Scarborough before moving to downtown Toronto and then to Ottawa, the national capital. These are cases in which new migrants in the suburbs claim and use public spaces such as grocery stores, libraries, restaurants, community centres, street corners, and places of worship to foster a multifarious public arena in which integrative multiplicity can take hold for themselves and other marginalized groups. They also reveal the alternate globalities of subaltern cosmopolitanism and suggest, along with activism directed at preventing displacement of low- and modest-income renters via "demovictions," how post-suburbia might be complicated, rescaled, and restructured as a political space.

Conclusion

To find the Canadian suburb today, we have to look towards the post-colony. This entails mainly two strident advances.[3] One is to free ourselves from the theoretical baggage of the urban gaze towards the suburbs and free suburbanisms from the fetters of the urban-centred yardstick. And the second advance is to look beyond the "arrival cities" of the liberal mind to the reality of the subaltern in the Canadian periphery. The Canadian suburb has arrived in the **global city**. In the suburban ring around Toronto, where now more than half of the region's people live, the majority of the population will soon be non-white and new immigrants throughout. Those suburban cities, from Mississauga and Brampton in the west to Markham in the east, have become home to large minorities or even majorities of East and South Asians—Chinese, Tamils, Pakistani, Indians—as well as hundreds of thousands of migrants from elsewhere in those parts of the world that mostly feed Canadian immigration. Accordingly, suburbs in Canada today trouble the conventional use of the concepts of multiculturalism and diversity that seem innate to Canadian urbanism. As Jay Pitter tells us in her introduction to a critical new book on the subject, we have entered a period of "hyper-diversity." Politically, this entails an "intersectionality [that] requires us to acknowledge that our cities contain diversities within diversities within diversities. They are deeply complicated places" (Pitter, 2016: 9).

Suburbanization processes today cannot be understood, as they used to be, as outcomes mostly of national policy and **Fordist-Keynesian** economics. Suburban constellations are now in the crosshairs of international migration, flows of global capital and labour, locational decisions by transnational corporations, diasporic networks of businesses and communities, and cultural diversities. Equally, suburbanisms as diverse ways of life are beginning to reshape the urban periphery in Canada. While the *Atlas of Suburbanisms* continues to show a sustained prevalence of the automobile-oriented suburban form in the outer reaches of small and large urban regions in Canada, the container space of cul-de-sacs and shopping malls is filled with new social and cultural practices and even politics in the ever-changing geographies of the "arrival city" (Saunders, 2010). As the conclusion to a recent collection of essays on global suburbanization suggests,

> to arrive in the global suburb is no longer an original experience. It is not terra incognita, empty unmarked space. The moving trucks taking the huddled masses to the air and light of the periphery have long disappeared. Arriving in the suburb is getting home to the metropolitan future that is most likely ours for some time to come (Keil, 2013: 201).

Indeed, Canadian suburbs have ceased to be a derivative of the North American suburban phenomenon. They are also not mere extensions of European-influenced metropolitan landscapes of tower neighbourhoods and bungalow estates (see Phelps, 2017). Under the influence of changing immigration patterns and economic **globalization**, and under the severe stresses of neo-liberal urbanization, the Canadian periphery has entered a phase of suburbanization that is both universal in its appearance, unique in its outcomes, and unprecedented in its myriad suburban ways of life. Therein lies the particular character of Canadian suburbanization and suburbanisms today, and with it, the future of Canadian urban society.

Review Questions

1. Suburbs are sometimes portrayed as being homogeneous entities. Why is this depiction of Canadian suburbs no longer accurate? Was it ever accurate?
2. How has immigration changed Canadian suburbs?
3. In what ways do suburban high-rise apartment developments challenge traditional depictions of suburban ways of living?
4. How do you think the suburbs in your city will change over the coming years? What factors are contributing to these changes?

Notes

1. According to Statistics published by Citizenship and Immigration Canada for the period 2003-2012, the average number of new permanent residents (immigrants) each year was 249,455. Yearly intake during this period ranged from a low of 221,349 in 2003 to a high of 280,689 in 2010. See "Facts and Figures 2012—Immigration Overview: Permanent and Temporary Residents," at www.cic.gc.ca/english/resources/statistics/facts2012/index.asp
2. Dissemination areas (das) are the smallest standard geographical unit for which Statistics Canada makes long-form census data available for mapping and geographical analysis. They vary in geographic size, but typically contain a population of 400 to 700 persons.
3. This borrows from joint work with Sean Hertel.

References

Addie, J.-P.D. 2016. "Theorizing suburban infrastructure: A framework for critical and comparative analysis," *Transactions of the Institute of British Geographers* 41(3): 273–85.

Ahmed-Ullah, N. 2016. "Brampton a.k.a Browntown," in J. Pitter and J. Lorinc, eds., *Subdivided: City-Building in an Age of Hyper-Diversity*. Toronto: Coach House Books.

Alang, N. 2011. "Why white scenesters hate the 'burbs,'" *Toronto Standard*. At: www.torontostandard.com/the-sprawl/why-white-scenesters-hate-the-%E2%80%98burbs

Archer, J. 2011. "Everyday suburbia: Lives and practices," *Public* 43(1): 10.

Atkinson, A. 2007. "Cities after oil: Sustainable development and energy futures," *City* 11(2): 201–13.

Bain, A. (2013). Creative margins: Cultural production in Canadian suburbs. Toronto: University of Toronto Press.

Baluja, B. 2011. "Poverty hides in the suburbs: Will 'priority neighbourhoods' help?" *The Globe and Mail*, April 20: A11.

Basu, R., and R.S. Fiedler. 2017. "Integrative multiplicity through subaltern realities: Exploring diversity through public spaces in Scarborough," *Urban Geography* 38(1): 25–46.

Black, S. 2011. "Life in 'Third City': Nasty brutish and short," *Toronto Star*, June 8: A27.

Bourne, L.S. 1996. "Reinventing the suburbs: Old myths and new realities," *Progress in Planning* 46: 163–84.

Bunting, T., R.A. Walks, and P. Filion. 2004. "The uneven geography of housing affordability stress in Canadian Metropolitan Areas," *Housing Studies* 19(3): 361–94.

Carpio, G., C. Irazabel, and L. Pulido. 2011. "Right to the suburb? Rethinking Lefebvre and immigrant activism," *Journal of Urban Affairs* 33(2): 185–208.

CBC News. 2011. "Canada's first Tamil MP looks forward to challenge," *CBC News: Toronto*. At: www.cbc.ca/news/canada/toronto/story/2011/05/03/sitsabaiesan-ndp-election.html

Charmes, E., and R. Keil. 2015. "The politics of post-suburban densification in Canada and France," *International Journal of Urban and Regional Research* 39(3): 581–602.

Dodson, J. 2014. "Suburbia under an energy transition: A socio-technical perspective," *Urban Studies* 51(7): 1487–505.

Donaldson, S. 1969. *The suburban myth*. New York: Columbia University Press.

Dowling, R. 1998. "Neotraditionalism in the suburban landscape: Cultural geographies of exclusion in Vancouver, Canada," *Urban Geography* 19: 105–22.

Dunham-Jones, E., and J. Williamson. 2009. *Retrofitting Suburbia: Urban Design Solutions for Redesigning Auburbs*. Hoboken, NJ: Wiley & Sons.

Ekers, M., P. Hamel, and R. Keil. 2012. "Governing suburbia: Modalities and mechanisms of suburban governance," *Regional Studies* 46(3): 405–22.

Fiedler, R.S., N. Schuurman, and J. Hyndman. 2006. "Hidden homelessness: An indicator-based approach for examining the geographies of recent immigrants at-risk of homelessness in Greater Vancouver," *Cities* 23(3): 205–16.

Filion, P. 2018. "Enduring features of the North American suburb: Built form, automobile orientation, suburban culture and political mobilization," *Urban Planning* 3(4): 4–14.

———, and G. Gad. 2006. "Urban and suburban downtowns: Trajectories of growth and decline," in T. Bunting and P. Filion, eds., *Canadian Cities in Transition: Local through Global Perspectives*, 3rd edn. Toronto: Oxford University Press.

———, and R. Keil. 2017. "Contested infrastructure: Inequity and innovation in the global suburb," *Urban Policy and Research* 35(1): 7–19.

Fishman, R. 1987. *Bourgeois Utopias: The Rise and Fall of Suburbia*. New York: Basic Books.

Friesen, J. 2011. "Brampton-Springdale: A suburban boom city grows, a political battlefield emerges," *The Globe and Mail*, April 15: A1. At: http://m.theglobeandmail.com/news/politics/in-brampton-a-suburban-boom-city-grows-on-fertile-electoral-ground/article1986393/?service=mobile

Garreau, J. 1991. *Edge City: Life on the New Frontier*. New York: Doubleday.

Globe and Mail, The. 2018. "Ontario election results 2018: A map of the results." June 7. At: www.theglobeandmail.com/canada/article-ontario-election-results-2018-a-map-of-the-live-results/

Graham, S., and S. Marvin. 2001. *Splintering Urbanism: Networked Infrastructures, Technological Mobilities and the Urban Condition*. New York: Routledge.

Gordon, D.L., and M. Janzen. 2013. Suburban nation? Estimating the size of Canada's suburban population. *Journal of Architectural and Planning Research* 30(3): 197–20.

Hackworth, J., and K. Stein. 2012. "The collision of faith and economic development in Toronto's inner suburban industrial districts," *Urban Affairs Review* 48(1): 37–63.

Hamel, P., and Keil, R., eds. 2015. *Suburban Governance: A Global View*. Toronto: University of Toronto Press.

Harris, R. 1996. *Unplanned Suburbs: Toronto's American Tragedy, 1900 to 1950*. Baltimore: John Hopkins University Press.

Harris, R. 2004. *Creeping Conformity: How Canada Became Suburban, 1900–1960*. Toronto: University of Toronto Press.

Harris, R. 2010. "Meaningful types in a world of suburbs," in M. Clapson and R. Hutchinson eds., *Suburbanization in Global Society*. Bingley: Emerald Group Publishing.

———, and R. Lewis. 1998a. "How the past matters: North American cities in the twentieth century," *Journal of Urban Affairs* 20: 159–74.

——— and ———. 1998b. "Constructing a fault(y) zone: Misrepresentations of American cities and suburbs, 1900–1950," *Annals of the Association of American Geographers* 88(4): 622–39.

——— and ———. 2001. "The geography of North American cities and suburbs, 1900–1950: A new synthesis," *Journal of Urban History* 27: 262–92.

Heartfield, K. 2015. "Conservatives' hold on suburbs does not appear unshakeable," *Ottawa Citizen*. October 13. At: http://ottawacitizen.com/opinion/columnists/the-fight-for-ottawas-suburbs

Hiebert, D. 2000. "Immigration and the changing Canadian city," *Canadian Geographer* 44(1): 25–43.

Hulchanski, D. 2010. *The Three Cities within Toronto: Income Polarization among Toronto's Neighborhoods, 1970–2005*. Toronto: University of Toronto Centre for Urban and Community Studies.

Keil, R. 2002. "'Common Sense' neoliberalism: Progressive conservative urbanism in Toronto, Canada," *Antipode* 34: 578–601.

———. 2013. "Escape from the burbs?" in R. Keil (ed.) *Suburban Constellations: Governance, Land and Infrastructure in the 21st Century*. Berlin: Jovis Verlag.

———. 2016. "Twelve theses on the suburb as the chief artifact of the anthropocene," *Suburbanicity*. At: http://suburbs.info.yorku.ca/2016/04/twelve-theses-on-the-suburb-as-the-chief-artifact-of-the-anthropocene/

———. 2017. *Suburban planet*. Cambridge: Polity Press.

———. 2018. "Canadian suburbia: From the periphery of empire to the frontier of the sub/urban Century," *Zeitschrift für Kanadastudien* 38(2): 47–64.

———, and J.-P.D. Addie. 2016. "'It's not going to be suburban, it's going to be all urban': Assembling post-suburbia in the Toronto and Chicago regions," *International Journal of Urban and Regional Research* 39(5): 892–911.

———, P. Hamel, E. Chou, and K. Williams. 2015. "Modalities of suburban governance in Canada," in P. Hamel and R. Keil, eds., *Suburban governance: A global view*. Toronto: University of Toronto Press.

———, and D. Young. 2009. "Fringe explosions: Risk and vulnerability in Canada's new in-between urban landscape," *Canadian Geographer* 53(4): 488–99.

Kern, L. 2005. "In place at home in the city: Connecting privilege, safety and belonging for women in Toronto," *Gender Place and Culture* 12(3): 357–77.

———. 2010. *Sex and the Revitalized City: Gender, Condominium Development, and Urban Citizenship*. Vancouver: UBC Press.

Knox, P.L. 2008. *Metroburbia, USA*. Piscataway: Rutgers University Press.

Kolb, D. 2008. *Sprawling Places*. Athens, GA: University of Georgia Press.

Lang, R., and P.K. Knox. 2009. The new metropolis: Rethinking megalopolis. *Regional Studies* 43(6): 789–802.

Lewis, R. 2000. *Manufacturing Montreal: The Making of an Industrial Landscape 1850 to 1930*. Baltimore: John Hopkins University Press.

Ley, D. 2011. *Millionaire Migrants: Trans-Pacific Life Lines.* Malden: Wiley-Blackwell.

Li, W. 2009. *Ethnoburb: The New Ethnic Community in Urban America.* Honolulu: University of Hawai'i Press.

Lo, L., V. Preston, P. Anisef, R. Basu, and S. Wang. 2015. *Social Infrastructure and Vulnerability in the Suburbs.* Toronto: University of Toronto Press.

Lucy, W.H., and D.L. Phillips. 1997. The postsuburban era comes to Richmond: City decline, suburban transition and exurban growth. *Landscape and Urban Planning* 36: 259–75.

McCann, L. 1999. "Suburbs of desire: The suburban landscape of Canadian cities, c.1900-1950," in R. Harris and P.J. Larkham, eds., *Changing Suburbs: Suburban Foundation, Form and Function.* London: E and FN Spon.

———. 2006. "A regional perspective on Canadian suburbanization: Reflections on Richard Harris's 'Creeping conformity,'" *Urban History Review* 35(1): 32–45.

Macdonald, S., and R. Keil. 2012. "The Ontario Greenbelt: Shifting the scales of the sustainability fix?" *Professional Geographer* 64(1): 125–45.

McManus, R., and P.J. Etherington. 2007. "Suburbs in transition: New approaches to suburban history," *Urban History* 34: 317–37.

Maher, K.H. 2004. "Borders and social distinction in the global suburb," *American Quarterly* 56(3): 781–806.

Markovich, J., and S. Hendler. 2006. "Beyond 'soccer moms,'" *Journal of Planning Education and Research* 25(4): 410–27.

Montero, S. 2017. "Study tours and inter-city policy learning: Mobilizing Bogota's transportation policies in Guadalajara," *Environment and Planning A* 49(2): 332–50.

Moos, M. and A. Kramer. 2012. *Atlas of Suburbanisms.* At: http://env-blogs.uwaterloo.ca/atlas/

———, and P. Mendez. 2015. "Suburban ways of living and the geography of income: How homeownership, single-family dwellings and automobile use define the metropolitan social space," *Urban Studies* 52(10): 1864–82.

———, and R. Walter-Joseph. 2017. *Still Detached and Subdivided? Suburban Ways of Living in 21st Century North America.* Berlin: Jovis.

Nicolaides, B.M. 2006. "How Hell moved from the city to the suburbs: Urban scholars and changing perceptions of authentic community," in K.M. Kruse and T.J. Sugrue, eds., *The New Suburban History.* Chicago: University of Chicago Press.

Peck, J., Siemiatycki, E., and Wyly, E. 2014. "Vancouver's suburban involution," *City: Analysis of urban trends, culture, theory, policy, action* 18(4–5): 386–415.

Phelps, N.A., ed. 2017. *Old Europe, New Suburbanization? Governance, Land, and Infrastructure in European Suburbanization.* Toronto: University of Toronto Press.

———, and A. Wood. 2011. "The new post-suburban politics?" *Urban Studies* 48(12): 2591–610.

———, ———, and D. Valler. 2010. "A post-suburban world? An outline of a research agenda," *Environment & Planning A* 42: 366–83.

———, and F. Wu, eds. 2011. *International Perspectives on Suburbanization: A Post-Suburban World?* Basingstoke: Palgrave Macmillan.

Pitter, J. 2016. "Introduction," in J. Pitter and J. Lorinc, eds., *Subdivided: City-Building in an Age of Hyper-Diversity.* Toronto: Coach House Books.

———, and J. Lorinc, eds. 2016. *Subdivided: City-Building in an Age of Hyper-Diversity.* Toronto: Coach House Books.

Poppe, W., and D. Young. 2015. "The politics of place: Place-making versus densification in Toronto's tower neighbourhoods," *International Journal of Urban and Regional Research* 39(3): 613–21.

Pratt, G. 2003. "Valuing childcare: Troubles in suburbia," *Antipode* 35: 581–602.

Preston, V., and L. Lo. 2000. "Asian theme malls in suburban Toronto: land use conflict in Richmond Hill," *Canadian Geographer* 44(2): 182–90.

Pulido, L. 2000. "Re-thinking environmental racism: White privilege and urban development in Southern California," *Annals of the Association of American Geographers* 90(1): 12–40.

Quinby, R. 2011. *Time and the Suburbs: The Politics of Built Environments and the Future of Dissent.* Winnipeg: Arbeiter Ring Publishing.

Relph, E. 1976. *Place and Placelessness.* London: Pion.

Rose, D., and P. Villeneuve. 1998. "Engendering class in the metropolitan city: Occupational pairings and income disparities among two-earner couples," *Urban Geography* 19(2): 123–59.

Rowe, P.G. 1991. *Making a Middle Landscape.* Cambridge: The MIT Press.

Saltman, J. 2017. "Metrotown Downtown Plan raises fears of affordability, demovictions," *Vancouver Sun.* February 22. At: http://vancouversun.com/news/local-news/metrotown-downtown-plan-raises-fears-of-affordability-demovictions

Saunders, D. 2010. *Arrival City: How the Largest Migration in History is Reshaping Our World.* New York: Vintage Books.

Schafran, A. 2013. "Discourse and dystopia, American style: The rise of 'slumburbia' in a time of crisis," *City: Analysis of urban trends, culture, theory, policy, action* 17(2): 130–48.

Shields, R. 2012. "Feral suburbs: Cultural topologies of social reproduction, Fort Murray, Canada," *International Journal of Cultural Studies* 15(3): 205–15.

Sieverts, T. 2003. Cities without Cities: An Interpretation of the Zwischenstadt. London: Spon Press.

Soja, E.W. 2000. *Post-metropolis.* Cambridge: Blackwell.

Strong-Boag, V., I. Dyck, K. England, and L. Johnson. 1999. "What women's spaces? Women in Australian, British, Canadian and US suburbs," in R. Harris and P.J. Larkham, eds., *Changing Suburbs: Suburban Foundation, Form and Function.* London: E and FN Spon.

Tanasescu, A. 2009. "Informal housing in the heart of the New West: An examination of state toleration of illegality in Calgary," *North American Dialogue* 12(2): 1–11.

Teaford, J. 1997. *Post-Suburbia: Government and Polities in Edge Cities*. Baltimore: Johns Hopkins University Press.

United Way. 2004. *Poverty by Postal Code: The Geography of Neighbourhood Poverty, 1981-2001*. Toronto: United Way of Greater Toronto and Canadian Council on Social Development.

Veracini, L. 2012. "Suburbia, settler colonialism and the world turned inside out," *Housing, Theory and Society* 29(4): 339-57.

Walks, R.A. 2001. "The social ecology of the post-Fordist/ global city? Economic restructuring and socio-spatial polarisation in the Toronto Urban Region," *Urban Studies* 38(3): 407-47.

———. 2005. "The city-suburban cleavage in Canadian federal politics," *Canadian Journal of Political Science* 38(2): 383-414.

———. 2006. "The causes of city-suburban political polarization? A Canadian case study," *Annals of the Association of American Geographers* 96(2): 390-414.

———. 2009. "The urban in fragile, uncertain, neoliberal times: Towards new geographies of social justice," *Canadian Geographer* 53(3): 345-56.

———. 2015. "Stopping the 'war on the car': Neoliberalism, Fordism, and the politics of automobility in Toronto," *Mobilities* 10(3): 402-22.

Wang, S.G. 1999. "Chinese commercial activity in the Toronto CMA: New development patterns and impacts," *Canadian Geographer* 43(1): 19-35.

Warner, S.B. 1962. *Streetcar Suburbs: The Process of Growth in Boston, 1870-1900*. Cambridge: Harvard University Press and The MIT Press.

Wood, P.K., and L. Gilbert. 2005. "Multiculturalism in Canada: Accidental discourse, alternative vision, urban practice," *International Journal of Urban and Regional Research* 29(3): 679-91.

Wunsch, J.L. 1995. "The suburban cliché," *Journal of Social History* 28: 643-58.

Young D., and R. Keil. 2014. "Locating the urban in-between: Tracking the urban politics of infrastructure in Toronto," *International Journal of Urban and Regional Research* 38(5): 1589-608.

———, P. Wood, and R. Keil, eds. 2011. *In-Between Infrastructure: Urban Connectivity in an Age of Vulnerability* Kelowna, BC: Praxis (e)Press.

IV Economies of Cities

The chapters in Part IV of this book consider the city through an economic lens. Readers learn about the changing dimensions of Canadian urban economies and the factors shaping them. The Chapter 17 tracks the sectoral and occupational changes associated with deindustrialization leading to the emergence of what has become known as the new economy or knowledge-based economy. Readers learn how structural changes in the economy, and policy, play out in different Canadian urban contexts, as well as about initiatives used by various institutions to respond to and facilitate economic changes. Highlights are the decline and changing dimensions of manufacturing, the growth of more knowledge-intensive economic activity, and the growth of creative and cultural sectors.

Chapter 18 focuses on one specific dimension of economic activity that is receiving growing attention by policy-makers: innovation. Actors and institutions of all kinds are increasingly emphasizing the role of cities in producing innovation to generate economic value and growth. Readers will learn how to critically evaluate the presumed link between urbanization and innovation. The chapter unpacks the diversity of places and sectors where innovation occurs. Innovation is defined and examples of its measurement are provided. The chapter shows that cities play an important role in innovation, but also reminds readers that innovation is certainly not restricted to urban areas nor is it exclusive to the private sector.

Chapter 19 deals with the factors that have an impact on the economic value of land, and how land value influences urban development and vice versa. Readers learn about the characteristics of urban land markets and the models economists use to better understand the factors shaping the value of land. The broader societal context shaping property and land markets is considered, including the implications for efficiency and equity. Readers learn about property value gradients in cities, particularly the revalorization (and gentrification) of inner-city land, as well as the different economic explanations offered for this trend.

Chapter 20 challenges the commonly held assumption that cities are places of growth. Readers learn how a large share of the Canadian urban system is actually declining or growing quite slowly. The chapter explains what is meant by decline and slow growth, and where it is occurring. The chapter highlights the growing difference among cities within Canada, contributing to disparities among those that grow quickly and those that do not. Readers learn about the growing importance of managing decline and policies that respond to conditions of decline or slow growth through planning and economic development interventions. The chapter outlines the challenges and opportunities associated with slow growth and decline, and the policies being used to address them.

After reading the chapters in Part IV, readers will have understanding of

- the changing characteristics and role of cities as economic spaces;
- the ways in which innovation comes about and the conditions, places, and sectors conducive to innovation;
- the factors shaping the economic value of urban land and the changing characteristics of urban land and property markets; and
- slow-growth and no-growth cities and their urban policy responses.

17 The New Economy of Canadian Cities?

Employment, Creativity, and Industrial Change

Tara Vinodrai

Introduction

The economies of cities are dynamic and ever changing, responding to and being shaped by economic, social, political, technological, and cultural forces. While there is no question that cities are important social, cultural, and political spaces, it is their key role as economic spaces that is the focus of this chapter. Cities have always been important centres for commerce and business; however, the types of work and economic activities taking place in cities have evolved and changed over time. Since the 1970s, cities in Canada and other advanced capitalist societies have undergone substantial deindustrialization and economic restructuring marked by the growth of employment in both the high- and low-order service sectors, the growing importance of highly educated workers, and the decline in "blue-collar" jobs. These economic changes have been accompanied by shifts in the labour market, characterized by declining levels of private sector unionization and the transformation of employment relations, including the increase of flexible, part-time, temporary, and precarious work. It is common for scholars, policy-makers, and the popular and business press to refer to the suite of changes as the **new economy**, the **knowledge-based economy**, or—more controversially—the creative economy; these terms become shorthand to describe the contemporary economy.

In this chapter, I examine how economic change has been experienced in Canadian cities and describe some of the initiative and strategies used by these cities in response. The chapter begins by discussing the broader context in which the economic landscapes of Canadian cities have changed over the past four decades. The chapter proceeds to document some key socio-economic transformations in the Canadian economy since the 1970s. The third section more closely examines three key dynamics that characterize the economic transitions in Canadian cities: the changing landscape of manufacturing; the transformation of urban economies towards more innovation-based and **knowledge-intensive economic activity** (see also Shearmur, Chapter 18); and the ascendance of creative and cultural activities. The chapter emphasizes how these economic changes have not affected all Canadian cities in the same way. While the broader shift from a natural resource-based or manufacturing-based economy to one based on tertiary- and quaternary-order service activities generally holds true across Canadian cities, the contours of this change are **path-dependent** (that is, shaped by past history and events) and vary from city to city based on their unique local histories, institutions, politics, economic development strategies, and regional contexts. Paying attention to how different Canadian cities have developed and adapted over time is therefore important.

The Transition to a New Economy

Before looking at the specific economic changes in Canadian cities, we must understand how advanced capitalist societies have evolved over the past several decades. There is general agreement that the nature of production and its associated labour market structures, employment relations, and forms of **governance** have evolved away from those dominating the post-World War II period, characterized as Fordism (Amin, 1994; Bryson and Henry, 2005). In the narrowest sense, Fordism is associated with a set of production practices pioneered by Henry Ford and applied to the automotive manufacturing industry in the early twentieth century. These practices included a detailed division of labour resulting in simplified, deskilled tasks requiring minimal job training; highly standardized and routinized manufacturing; and dedicated machinery organized in an assembly line model allowing for mass production. The ascendance and widespread adoption of this organizational form rested not only on the production practices themselves but also on an accompanying set of institutional and organizational practices that enabled their success. **Fordism** is therefore understood more broadly by social scientists as an economy-wide set of institutional architectures that sustained production and promoted economic growth in North America and Europe from the end of World War II until the early 1970s.

What the French Regulation School refers to as the "regime of accumulation" (in this case, Fordist mass production) requires a complementary "mode of regulation" (Boyer, 2005; Amin, 1994). In this case, most advanced, capitalist countries found an appropriate regulatory apparatus in the form of the Keynesian (see **Keynesianism**) **welfare state** articulated through strong national state intervention, regulation and control, and social welfare provision (Boyer, 2005; Tickell and Peck, 1992). Jane Jenson (1990) has argued that Canada's particular form of Fordism, what she refers to as "permeable Fordism," was produced through a set of historical processes and negotiated between government, business, and the organized labour movement (e.g., unions) beginning in the interwar period; it was designed to deliver higher wages, full employment, and minimal labour unrest, as well as a set of social programs to support social reproduction. However, in the early 1970s, it became apparent that these particular institutional arrangements were increasingly strained across advanced, capitalist economies as profit levels declined, global competition intensified, and broader economic restructuring resulting from rising oil prices placed the system in crisis (Harvey, 1990).

In seeking to characterize the subsequent developments, scholars have referred to the coming of post-industrial (Bell, 1973) or risk society (Beck, 2000), **post-Fordism** (see Amin, 1994), an era of flexible specialization (Piore and Sabel, 1984) or flexible accumulation (Harvey, 1990), and—more recently—an emerging form of **cognitive-cultural capitalism** (Scott, 2007). Despite using different terms, there is widespread agreement that the economy has changed substantially over the past decades. The most striking shifts have been the tremendous growth of service-based industries, alongside the growing importance of human capital and advanced technologies. Scott (2007: 1466) notes that:

> much of productive activity today involves digital technologies and flexible organization sustaining the expansion of sectors that thrive on innovation, product diversity and the provision of personalized services. . . . Labor processes have come to depend more and more on intellectual and affective human assets (at both high and low levels of remuneration), and are increasingly less focused on bluntly routinized mental or manual forms of work.

Scott (2007) identifies technology-intensive manufacturing; services; fashion-oriented, neo-artisanal production; and cultural-products industries as the key drivers of growth and innovation in the contemporary economy.

This long term trend of **deindustrialization** is attributed to the introduction of new technologies allowing for significant automation, the outsourcing and relocation of production-oriented activities to offshore locations, and changing patterns of consumption including demand for improved and differentiated services, as well as goods that require more creative and service-based inputs in their production and distribution. More recently, the introduction of potentially disruptive technologies, such as artificial intelligence, autonomous vehicles, algorithmic and machine learning, blockchain, and the rise of the **platform economy,** have also been identified as drivers of economic change (see Brail and Donald, Chapter 5).

The New Economy of Cities

What role do cities play in this new economy? Observers identify two competing trends that raise questions about the continued vitality of cities in an age of **globalization** and disruptive technologies. On the one hand, transportation and communication technologies have allowed for shifts in terms of the locational choices of firms, industries, and—increasingly—workers (Moos and Skaburskis, 2007). In this view, economic activities are no longer tied to particular locations; capital and labour are increasingly footloose—able to relocate to any location. On the other hand, scholars have argued that place—and especially cities—has become even more important in the contemporary economy (Glaeser, 2011; Storper, 2013; Florida, 2017). Despite technological advances in transportation and communications that reduce the costs and barriers to interaction over space and time, many of the industries critical to producing the content, tools, and infrastructure of this new economy are agglomerated in particular urban locations. In other words, economic activity—particularly the forms associated with the new economy—remains spatially concentrated in cities. I outline three key sets of ideas that help us understand the features of new economy cities: (1) the clustering or agglomeration of knowledge-intensive activities; (2) the importance of human capital (or the **creative class**); and (3) the growth of the cultural economy of cities.

Clustering and Agglomeration

Spatial agglomeration in industrial districts—what Porter (1998) calls "clustering"—enables firms involved in similar activities to draw on shared local resources and collective infrastructure, reduce transaction costs, and gain other efficiencies (see Shearmur, Chapter 18). Firms benefit from local access to deep pools of specialized, skilled labour; proximity to suppliers, service providers, and other related and supporting industries; and closeness to sophisticated customers who provide demand-side impetus for developing and improving products and services, although this latter characteristic remains contested (Wolfe and Gertler, 2004). In addition to these traditional agglomeration economies or externalities, firms benefit from knowledge spillovers, the ability to monitor their competition, and access to local institutional supports. In other words, agglomeration enables learning, knowledge flows, co-operation and competition. Of course, firms are connected to other places through global production networks, complex divisions of labour, and supply and commodity chains, and markets and firms access knowledge through both local and non-local partners and networks (Bathelt, Malmberg, and Maskell, 2004; Wolfe and Gertler, 2004, 2016). However, there are compelling reasons why firms often remain agglomerated in particular cities. Most fundamental among these—from an economic perspective—is the emphasis placed on knowledge and learning to innovation and the creation of economic value in the contemporary era.

Cities provide many of the necessary conditions to support the innovation, learning, and knowledge generation and circulation processes integral to the development of dynamic clusters or specialized agglomerations of activity. Moreover, scholars increasingly recognize that learning and innovation—critical for firms' competitiveness—are inherently social processes that rely on interaction between different economic actors, such as firms, industry and professional associations, universities, government, private research and development laboratories, technology transfer offices, unions, venture capitalists, and other organizations (Wolfe and Gertler, 2016). While these interactions can be in person (e.g., meetings, conferences, trade shows, other events) or electronically mediated (e.g., email, videoconferencing, social media, virtual environments), there is agreement that the primarily tacit nature of a lot of knowledge demands that some of the most important interactions occur face-to-face in order for understanding, learning, and the sharing of ideas to take place. Cities, because of the density of activity, provide an ideal environment where such interactions are easily made possible and are more likely to occur since norms and values are shared (see Shearmur, Chapter 18, for a counterargument).

Human Capital and the Creative Class Debate

Human capital, particularly highly skilled labour—referred to as knowledge workers, talent, the creative class (Florida, 2002), the new class (Gouldner, 1979), or the new middle class (Ley, 1996)—should not be overlooked as an important embodied input into the innovation process. These workers often provide specialized skills, creativity, ideas, and know-how to knowledge-intensive production and the innovation process. Cities are viewed as the preferred location for these workers. Richard Florida's (2002) bestselling and controversial book *The Rise of the Creative Class* is often associated with this view. Florida suggests that creativity is the key driver of the contemporary economy and that a particular (and growing) group of workers (the creative class) are driving economic growth. Place is central to his argument. Echoing Jane Jacobs and other urban theorists, Florida suggests that cities are ideal environments to foster and support creativity.

The creative class is attracted to cities, but not just any cities; those cities that have high-quality social environments, low barriers to entry into labour markets and networks, and are diverse and open to differences will be particularly attractive. To capture this dynamic, Florida invokes the 3T's (technology, talent, and tolerance) to suggest that cities that achieve high scores on these variables will be the winners in today's economy. And while Florida's broad observations about the transformation of the labour market and its geography are not entirely original (cf. Bell, 1973; Gouldner, 1979; Ley, 1996), the creative class label (or brand) has resonated with policy-makers, practitioners, and the general public. How many urban scholars are ever mentioned in *Vanity Fair*? Moreover, urban policy-makers in Canada and further afield have very quickly (and often uncritically) adopted a "creative class policy script" that shifts focus away from creating a favourable business climate towards one favourable to attracting the urban elite (Peck, 2005).

The creative class thesis has been highly influential and highly controversial. While some critique has focused on Florida's own role in promoting his perspective (e.g., through his consultancy work and speaking tours), others raise issues related to theory, methods, empirical evidence, and policy translation (Peck, 2005; Shearmur, 2007; Storper and Scott, 2009). Issues have been raised concerning the mobility patterns of creative workers (Deslisle and Shearmur, 2010), the absence of differentiation amongst creative workers (Markusen, 2006), the applicability to smaller cities (Lewis and Donald, 2010), and the lack of attention paid to outcomes in different institutional contexts (Vinodrai, 2013). Importantly, even Florida (2017) in his more recent work notes that there is a troubling dark side to catering to the creative class, as it potentially excludes and displaces less educated, lower-skilled workers, contributing to gentrification and growing inequality (Peck, 2005; McCann, 2007; Florida, 2017).

The creative class thesis has spurred empirical inquiry that both confirms and disrupts the creative class thesis, and Canadian scholars have been actively involved in this debate (cf. Grant, 2014; Wolfe and Gertler, 2016). Gertler et al.'s (2002) study was the first to explore whether or not the correlations and relationships observed in Florida's work on the 3T's held outside of the US context. They found similar relationships between these key variables in Canadian cities, although there were important national differences between Canada and the United States related to educational attainment and immigration. Lewis and Donald (2010) argue that smaller cities are marginalized by creative class theory and policy, advocating for a perspective that emphasizes sustainability and livability, qualities of place offered by smaller Canadian cities such as Kingston, Ontario. A collaborative national research project involving Canadian geographers, planners, and other social scientists studied the dynamics of talent attraction and retention across 15 Canadian cities. Overwhelmingly, the evidence suggests that talent attraction and retention dynamics are nuanced and complex, dependent on both individual characteristics (e.g., occupation, ethnic identity) and place-specific context (e.g., location, city size, local institutions and culture) (Grant, 2014).

The Cultural Economy of Cities

Related to these discussions of creativity is the growing importance of intangible inputs (such as symbolic, design, and aesthetic content) rather than tangible assets in the production of goods and services. One group of industries is assuming a more prominent role: cultural-products industries, which include film and television production, music recording, book publishing, video game production, live theatre, and other entertainment (Scott, 2004, 2007). Studies consistently demonstrate that these industries have high levels of creative content, are engaged in a constant search for novelty, and are often susceptible to rapid shifts in consumer demands, necessitating higher rates of innovation and easy access to information about changing tastes and styles (Vinodrai and Keddy, 2015). These creative and cultural activities tend to agglomerate in major urban centres for several reasons. First, firms need constant access to cutting-edge knowledge and pools of highly skilled labour. Second, workers benefit from being able to develop their careers while living in "cool," diverse urban environments and neighbourhoods, although this can lead to polarizing, gentrification dynamics (Ley, 1996; see Walks, Chapter 10; Bain and Marks, Chapter 15). Third, these urban environments provide inspiration and, in turn, elements of place become embedded in outputs (Molotch, 2002; Lloyd, 2006). Fourth, this type of work often involves flexible organizational forms such as project-based work, which expose workers (despite their professional status) to high levels of individual risk through self-employment, contract-based work, and freelancing (Vinodrai and Keddy, 2015). The nature of the industries and of work itself requires well-developed (local) social and knowledge networks to access key information about jobs, projects, styles, and leading-edge developments—all of which are most readily available in large, diverse cities.

The above discussion foregrounds how broader economic changes (re)shape urban economies and labour markets and highlights how cities have become even more important sites of economic activity in the contemporary era. The remainder of this chapter documents how Canada's urban economies have changed since the crisis of Fordism and the rise of a knowledge-based economy, marked by the dominance of new forms of work, industrial activity, and technologies.

Post-Industrial Cities? The Economic Transformation of Canadian Cities

The story of deindustrialization in Canada is not straightforward. This section examines how deindustrialization and the growth of the knowledge-based economy have occurred unevenly across Canada's urban system (resulting in urban decline in some cases; see Hall and Donald, Chapter 20). I focus on three significant changes that epitomize this economic transition. First, given the wrenching deindustrialization experienced by many Canadian cities, I examine the evolving manufacturing landscape, paying attention to Ontario's cities where these changes have been most acute. Second, given the increasing emphasis on innovation and knowledge to the growth of cities, I explore the economic development paths taken by different Canadian cities in responding to these changes. Finally, given the increased attention paid to culture and creativity in cities, I examine the cultural industries, where these activities are prominent and well-developed in Canada's largest urban centres.

Manufacturing in Canada and Its Cities

In line with broader structural changes in advanced economies described above, the Canadian economy has undergone substantial restructuring over the past four decades. Table 17.1 compares employment gains and losses across 16 sectors of the Canadian economy between 1976 and 2018. The overwhelming majority of the employment gains were in the services-producing sector, and the highest annual growth rates were in professional, scientific, and technical services (4.9 per cent) and business, building, and other support services (4.3 per cent); followed by health care and social assistance (3.0 per cent); information, culture, and recreation industries (3.0 per cent); finance, insurance, and real estate (2.0 per cent); and accommodation and food services (2.2 per cent). Agriculture and manufacturing both experienced employment losses in absolute and relative terms. Indeed, manufacturing was once an important source of employment in Canada, providing a large number of stable, well-paying jobs and contributing to national exports and gross domestic product (GDP). However, as Figure 17.1

Table 17.1 Employment (000s) by industry in Canada, 1976–2018

	1976	**2018**	**Change**	**% Change**	**Annual growth rate**
Goods-producing sector	3,371	3,929	558	16.5	0.4
Agriculture	464	277	(187)	(40.2)	(1.4)
Forestry, fishing, mining, quarrying, oil and gas	255	341	86	33.7	0.8
Utilities	110	145	35	31.9	0.8
Construction	682	1,438	756	110.9	2.0
Manufacturing	1,861	1,728	(133)	(7.1)	(0.2)
Services-producing sector	6,377	14,729	8,353	131.0	2.3
Trade	1,572	2,795	1,223	77.8	1.6
Transportation and warehousing	563	991	428	76.0	1.5
Finance, insurance, real estate, and leasing	526	1,174	648	123.2	2.2
Professional, scientific, and technical services	253	1,467	1,214	480.7	4.9
Business, building, and other support services	161	777	616	381.5	4.3
Educational services	677	1,325	649	95.9	1.8
Health care and social assistance	794	2,407	1,613	203.3	3.0
Accommodation and food services	347	787	440	127.0	2.2
Information, culture, and recreation	413	1,235	822	198.8	3.0
Other services	427	803	376	88.1	1.7
Public administration	645	969	324	50.3	1.1
Total, all industries	**9,748**	**18,658**	**8,910**	**91.4**	**1.8**

Source: Statistics Canada, CANSIM Table 14-10-0023-01 Labour force characteristics by industry, annual (× 1000) (author's calculations).

Figure 17.1 Manufacturing Employment in Canada, 1976–2018

Source: Statistics Canada, CANSIM Table 14-10-0023-01 Labour force characteristics by industry, annual (× 1000) (author's calculations).

demonstrates, between 1976 and 2018, manufacturing employment fluctuated, even as it declined in relative terms from 19.1 per cent of Canadian employment in 1976 to 9.3 per cent in 2018. Because of this trend, observers are often quick to decry the death of manufacturing. However, manufacturing still accounts for just under one in ten jobs in Canada, suggesting that we should be careful about dismissing the importance of this sector.

These broad shifts in industrial structure and the decline in the relative importance of manufacturing have not affected all Canadian cities in the same way. Table 17.2 shows that, in 1971, manufacturing accounted for more than one-third of employment in Brantford (40 per cent), Kitchener (39.1 per cent), Oshawa (39.2 per cent), Hamilton (34.5 per cent), Windsor (33 per cent), and St. Catharines–Niagara (33 per cent). By 2011, these percentages had dropped to 17.4 per cent in Brantford, 16.2 per cent in Kitchener, 8.6 per cent in Oshawa, 11.7 per cent in Hamilton, 20.7 per cent in Windsor, and 9.1 per cent in St. Catharines–Niagara. These dramatic changes in the relative importance of manufacturing reflect large employment losses primarily experienced in southern Ontario and Quebec, the historic manufacturing centres of Canada. For example, between 1971 and 2016, manufacturing employment declined by 24,315 in St. Catharines–Niagara and 30,160 in Hamilton. This can be attributed to significant restructuring in the dominant industries in these cities: automotive and steel.

A similar portrait emerges when examining the occupations most closely associated with the production side of manufacturing: processing and machining occupations (considered the core of blue-collar work). Table 17.3 shows that, between 1971 and 2016, processing and machining occupations declined in their relative importance across Canadian cities, yet these shifts were uneven across Canadian cities. In Toronto, the absolute number of processing and machining jobs declined (42,585), and by 2016, workers in these occupations accounted for a smaller proportion of employment, declining from 14.6 per cent to 4.7 per cent. Montreal experienced a sharp absolute decline in this type of employment (87,965), leading to a decline from 15.7 per cent to 4.0 per cent

Table 17.2 Employment in manufacturing industries in Canadian cities, 1971–2016

	1971		2016		1971–2016	
	Employment	**%**	**Employment**	**%**	**Change**	**% Change**
Brantford	14,245	40.0	11,460	17.4	(2,785)	(22.6)
Calgary	19,830	11.2	37,960	5.1	18,130	(6.0)
Edmonton	24,965	11.4	41,700	6.0	16,735	(5.4)
Guelph	8,245	29.6	15,240	18.5	6,995	(11.1)
Halifax	7,885	8.3	9,340	4.5	1,455	(3.8)
Hamilton	73,255	34.5	43,095	11.7	(30,160)	(22.8)
Kingston	4,490	12.0	3,390	4.4	(1,100)	(7.6)
Kitchener	41,360	39.1	44,030	16.2	2,670	(22.9)
London	28,670	22.1	26,670	11.1	(2,000)	(11.0)
Montreal	276,770	25.6	203,200	10.0	(73,570)	(15.6)
Oshawa	19,375	39.2	16,030	8.6	(3,345)	(30.6)
Ottawa–Gatineau	20,670	7.9	21,385	3.2	715	(4.8)
Peterborough	8,285	30.7	4,265	7.6	(4,020)	(23.0)
Quebec City	21,910	12.3	30,875	7.4	8,965	(4.9)
Regina	5,815	9.4	6,210	4.9	395	(4.5)
Saint John	6,935	16.3	3,690	6.2	(3,245)	(10.0)
Sarnia	9,465	29.0	4,990	11.7	(4,475)	(17.3)
Saskatoon	5,230	9.8	8,405	5.4	3,175	(4.4)
Sault Ste Marie	11,235	34.4	3,480	10.3	(7,755)	(24.1)
Sherbrooke	6,890	21.4	13,805	13.7	6,915	(7.7)
St Catharines–Niagara	41,355	33.0	17,040	9.1	(24,315)	(23.9)
St John's	3,335	6.9	4,095	4.0	760	(3.0)
Sudbury	8,030	12.9	3,465	4.4	(4,565)	(8.5)
Thunder Bay	7,430	15.9	3,080	5.4	(4,350)	(10.5)
Toronto	315,570	25.4	272,695	9.1	(42,875)	(16.2)
Trois-Rivières	9,680	28.0	8,410	11.9	(1,270)	(16.1)
Vancouver	78,750	16.6	80,925	6.3	2,175	(10.3)
Victoria	7,135	8.8	6,355	3.4	(780)	(5.4)
Windsor	34,820	33.0	30,950	20.7	(3,870)	(12.3)
Winnipeg	44,425	18.2	34,485	8.7	(9,940)	(9.5)
Canada	*1,593,595*	*19.6*	*1,508,870*	*8.8*	*(84,725)*	*(10.8)*

Source: Statistics Canada, *Census of Population, 1971 and 2016* (author's calculations).

Table 17.3 Employment in Processing and Machining Occupations in Canadian Cities, 1971–2016

	1971		2016		1971–2016	
	Employment	**%**	**Employment**	**%**	**Change**	**% Change**
Toronto	182,255	14.6	139,670	4.7	(42,585)	(10.0)
Montreal	169,310	15.7	81,345	4.0	(87,965)	(11.7)
Vancouver	59,000	12.4	36,660	2.9	(22,340)	(9.6)
Kitchener	27,565	26.0	23,110	8.5	(4,455)	(17.5)
Edmonton	22,150	10.1	19,580	2.8	(2,570)	(7.3)
Hamilton	42,335	19.9	19,245	5.2	(23,090)	(14.7)
Winnipeg	32,670	13.4	17,450	4.4	(15,220)	(9.0)
Calgary	15,505	8.7	16,935	2.3	1,430	(6.4)
London	17,780	13.7	13,985	5.8	(3,795)	(7.9)
Windsor	23,920	22.7	16,105	10.8	(7,815)	(11.9)
Quebec City	17,105	9.6	2,985	2.7	(14,120)	(6.9)
Oshawa	11,870	24.0	8,485	4.6	(3,385)	(19.5)
St Catharines–Niagara	27,620	22.0	8,165	4.3	(19,455)	(17.7)
Ottawa–Gatineau	16,580	6.4	7,450	1.1	(9,130)	(5.3)
Guelph	5,515	19.8	8,395	10.2	2,880	(9.6)
Brantford	8,980	25.2	6,205	9.4	(2,775)	(15.8)
Sherbrooke	4,865	15.1	5,920	5.9	1,055	(9.2)
Saskatoon	4,560	8.6	4,300	2.8	(260)	(5.8)
Trois-Rivières	6,850	19.8	3,805	5.4	(3,045)	(14.4)
Halifax	6,705	7.0	3,455	1.6	(3,250)	(5.4)
Regina	4,655	7.6	3,015	2.4	(1,640)	(5.2)
Peterborough	4,775	17.7	2,415	4.3	(2,360)	(13.4)
Victoria	6,875	8.5	2,310	1.2	(4,565)	(7.3)
St John's	3,585	7.5	1,945	1.9	(1,640)	(5.6)
Sarnia	4,860	14.9	2,465	5.8	(2,395)	(9.1)
Saint John	4,875	11.4	1,750	3.0	(3,125)	(8.5)
Sault Ste Marie	6,540	20.0	1,385	4.1	(5,155)	(15.9)
Sudbury	7,845	12.6	1,380	1.8	(6,465)	(10.9)
Kingston	3,545	9.5	1,550	2.0	(1,995)	(7.5)
Thunder Bay	6,440	13.8	1,400	2.5	(5,040)	(11.4)
Canada	*1,210,025*	*14.0*	753,370	*4.4*	*(456,655)*	*(9.7)*

Source: Statistics Canada, *Census of Population, 1971 and 2016* (author's calculations).

in the proportion of the labour force accounted for by processing and machining occupations. Many of Ontario's small and mid-sized cities saw both the number and share of employment in processing and machining occupations decrease. For example, St. Catharines–Niagara and Hamilton saw employment in processing and machining occupations decrease by 19,455 (a 17.7 per cent decrease in share of employment) and 23,090 (a 14.7 per cent decrease in share of employment) respectively.

Manufacturing remains important in several Canadian cities, and—in some cases—deindustrialization has not been an important dynamic of local economic change. For example, western Canadian cities like Vancouver and Calgary were never major manufacturing sites and their economies are more connected to extractive and natural resource-based industries (Barnes and Hutton, 2009; Hutton, 2010). Even though cities like Toronto and Montreal have witnessed significant changes in their manufacturing base, manufacturing industries still account for a similar proportion of employment in each city (9.1 per cent and 10.0 per cent respectively), including manufacturing in areas such as biotechnology, pharmaceuticals, medical equipment and assistive technologies (MAT), and other life sciences.

The life sciences provide an interesting case of the evolution of manufacturing, including the strong links between advanced manufacturing and related services. Toronto and Montreal both have firms engaged in a range of life science activities. Montreal is more specialized in pharmaceuticals, whereas Toronto has a more diverse mix of activities. In each city, the life sciences industries are embedded in a broader regional and national innovation system and benefit from an array of institutional supports (Gertler and Vinodrai, 2009; Lowe and Gertler, 2005, 2009). National strategies and policies such as the 1983 National Biotechnology Strategy, tax incentives to stimulate R&D expenditures (particularly the Scientific Research and Economic Development tax credits), and—in Montreal—the creation of National Research Council (NRC) laboratories focused on biotechnology were critical to the industry's development. Different provincial policies related to public health care also contribute to each city's development trajectory. Quebec's provincial drug plan provided a 15-year exclusive approval guarantee for brand-name drugs, even if generic alternatives were available. Since the 1970s, however, Ontario has required patients to purchase cheaper, generic versions of drugs when available. Consequently, Montreal has a high proportion of brand-name drug producers, whereas generic producers are more prominent in Toronto.

At the local level, publicly funded research occurring in the universities and teaching hospitals has created opportunities for entrepreneurship and innovation. For instance, Toronto has a long history of medical research and innovation at the University of Toronto and elsewhere in the city, including the founding of Dow Pharmaceuticals in the 1880s, the establishment of the Connaught Laboratories at the University of Toronto in 1914, the discovery of insulin in the early twentieth century, as well as other medical breakthroughs. Moreover, Lowe and Gertler (2005: 26) suggest the diversity of economic activity in Toronto, including "a wide range of sophisticated service industries, including finance and professional/**producer services**, [and] a strong manufacturing base in industries such as automotive, food products, electronics, specialized machinery and aerospace," has provided opportunities for cross-sectoral knowledge exchange and convergence between different types of technologies leading to the diversity of highly innovative life science activities.

Manufacturing is also still important in a number of smaller and mid-sized cities in Ontario, in part because the automotive industry (including automotive parts suppliers and car assembly plants) has historically been one of Ontario's strongest economic sectors. While the automotive industry remains an important employer in cities such as Oshawa, Kitchener, and Windsor, the respective locations of General Motors (GM), Toyota, and Ford plants, the industry continues to face challenges. In November 2018, GM announced it was closing five plants across North America as part of a global restructuring and the reorientation of its corporate strategy. The closures include the Oshawa plant, where

GM has been an employer for a century and where it has been the largest employer in the local economy, with 2592 workers at the plant alone, down from almost 23,000 at its peak in the mid-1980s (CBC News, 2018a). It is too soon to know all of the indirect local impacts resulting from this closure. GM is also reorienting its Canadian activities around technological advances in autonomous vehicles and car-sharing (see Brail and Donald, Chapter 5). In Toronto, GM is developing a mobility campus intended to house several thousand jobs, including approximately 200 high-tech jobs with GM itself (CBC News, 2018b). In Kitchener, GM has a tech-oriented team devoted to Maven, its car-sharing platform. While each of these investments will create new employment opportunities, the amount of direct job creation and the types of skills associated with these new jobs will be quite different from the losses in Oshawa.

In Kitchener, where manufacturing still accounts for 16.2 per cent of employment, the manufacturing base is more diverse. While large companies such as BlackBerry and Maple Leaf Foods no longer have manufacturing facilities in the city-region, there remains a base of large-scale manufacturing employers including Dare Foods Limited, Krug (furniture), and Rimowa (luxury luggage). In addition to these larger industrial employers, a range of smaller manufacturing companies in information and communications technology (ICT), machinery, furniture, and other manufacturing continue to operate. With the understanding that Canadian manufacturing faces challenges, business and civic leaders, universities, and government in Waterloo Region made an effort to sustain local manufacturing activity by establishing the Manufacturing Innovation Network (MIN) (Nelles, 2014). The MIN is a regional initiative intended to encourage manufacturers to adopt new business practices and technologies, become more innovative, and solve common problems through sharing information and expertise with the end goal of helping local manufacturers remain competitive in the global economy.

Manufacturing activity continues to evolve. For example, in 2017, the federal government announced an investment of $950 million in its Innovation Superclusters Initiative as a means to transform regional economies and spur innovation and job creation (see Taylor and Bradford, Chapter 3). One of the superclusters, centred in southern Ontario, focuses on advanced manufacturing and is led by NGen, a not-for-profit organization that matches manufacturing companies with new technologies. Elsewhere, new developments support artisanal and craft manufacturing. For example, Etsy and The Grommet provide digital platforms that allow for small-scale, artisanal, and craft manufacturers to reach new markets. Local organizations, such as Made In Montreal, have emerged to support local, urban manufacturing. And local makerspace initiatives, described by the National League of Cities (2016) as "new, hyperlocal manufacturing environments [that] holds potential not only for individual hobbyists but also for community-wide advances in local entrepreneurship and job creation," are providing new opportunities to access and develop manufacturing equipment, skills, and technologies (Wolf-Powers et al., 2017; Vinodrai, 2018).

Overall, it is clear that the transformation and restructuring of the manufacturing industries is far more complex than what first meets the eye. There has been a significant decline in the relative importance of the manufacturing industries and an even more acute decline in the proportion of the labour force in production-oriented occupations. Yet there remain pockets of advanced, artisanal and traditional manufacturing, as well as the emergence of new manufacturing platforms, technologies, and spaces. This raises three points regarding deindustrialization and the nature of manufacturing activity itself. First, even though employment in the manufacturing sector declined in general, the workers most affected are those in production-oriented jobs. Deindustrialization involves both the transformation of work and sectoral shifts in employment. Second, manufacturing comprises a diverse range of sub-sectors including automotive and automotive parts, steel, office furniture, machinery, food, and other goods, which are affected by different dynamics related to innovation, technological change, levels of unionization, and market demand. Some forms of manufacturing remain prominent in the Canadian landscape and some cities still have a relatively large manufacturing presence. Finally, it is too early to tell how new manufacturing spaces and platforms, such as makerspaces or Etsy, might shape urban economies.

The Rise of Knowledge-Based Activities in Canadian Cities

The previous section highlighted the substantial decline of manufacturing and its associated forms of work, as well as the place-specific evolution of this transition, given the specific histories and geographies of particular Canadian cities. Likewise, the emergence of the knowledge-based economy has been experienced differently across Canadian cities. There has been a widespread professionalization of the workforce, leading to the growth of occupations that require higher levels of formal education and specialized knowledge such as scientists, engineers, lawyers, planners, architects, writers, designers, and other artistic, cultural, social science, and technical occupations; these are the occupations associated with the creative class. Table 17.4 shows the growth of this occupational group across 30 Canadian cities between 1971 and 2016. Professionals accounted for 8.9 per cent of the labour force in 1971 but more than doubled in their relative importance by 2016, accounting for 22 per cent of the workforce. In stark contrast to the fate of processing and machining jobs (discussed above), which declined in relative terms in all 30 Canadian cities, professional occupations grew in both absolute and relative terms across all of these cities. Again, this did not occur evenly. Canada's largest cities (Toronto, Montreal, Vancouver, Ottawa) experienced some of the largest increases in the proportion of the labour force accounted for by professionals, as did several smaller regional centres such as St John's, Halifax, Victoria, and Quebec City. Recent Canadian evidence suggests that most industrial sectors, even traditional industries, are increasingly relying on knowledge-based inputs embodied in highly educated and creative workers (Wolfe, 2018).

Sudbury provides an interesting case of the shift towards knowledge-intensive economic activity. As a northern Ontario city dominated by the mining sector, including large multinational companies such as Inco (owned by the Brazilian company Vale) and Falconbridge (owned by the Swiss company Xstrata), the city has been strongly influenced by the restructuring of the mining industry (see Donald and Hall, Chapter 20). Between 1971 and 2016, the proportion of employment accounted for by Sudbury's primary industries (e.g., mining, forestry, and other extractive and natural resource-based activities) declined from 26.6 per cent to 8.8 per cent. This reflects the ongoing layoffs and outsourcing that occurred from the mid-1970s onward among local mining firms. While such downsizing has unquestionably had negative effects on Sudbury and its surrounding communities, at the same time, this has led to the emergence of a small but technology-intensive cluster of mining supply and service companies. The layoffs of highly skilled workers from the large mining companies in the city-region resulted in the emergence of a new set of firms that drew on the existing local skill base and provided services to the large mining companies. Moreover, Sudbury's high-technology mining services sector is increasingly supported through the emergence of several organizations such as the Northern Centre for Advanced Technology (NORCAT), the Centre for Excellence in Mining Innovation (CEMI), and the Sudbury Area Mining Supply and Service Association (SAMSSA); these organizations actively promote the sector's development as a means of sparking broader economic development in the city and the wider region. In addition, there is a growing role for Laurentian University as a key supplier of highly skilled labour and provider of leading-edge research (Warrian and Mulhern, 2009).

Parallel stories of how the pre-existing economic base of cities shape and condition their subsequent reinvention and transformation emerge elsewhere in Canada. Turning to Hamilton, where the steel industry has left a heavy imprint on the city's economy, the city's transition to knowledge-intensive activities has been shaped by its existing industrial base and through the activities of the local, publicly funded university. One focus of Hamilton's shift towards a more diversified,

Table 17.4 Employment in professional occupations in Canadian cities, 1971–2016

	1971		2016		1971–2016	
	Employment	**%**	**Employment**	**%**	**Change**	**% Change**
Toronto	122,055	9.8	719,150	24.1	597,095	14.3
Montreal	106,520	9.9	494,920	24.4	388,400	14.6
Vancouver	41,600	8.8	299,930	23.5	258,330	14.7
Ottawa–Gatineau	36,905	14.2	207,795	30.7	170,890	16.5
Calgary	20,035	11.3	174,355	23.6	154,320	12.3
Edmonton	22,650	10.4	145,150	20.9	122,500	10.6
Quebec City	19,480	10.9	107,825	25.8	88,345	14.9
Winnipeg	22,700	9.3	90,260	22.7	67,560	13.4
Hamilton	18,645	8.8	80,485	21.8	61,840	13.0
Kitchener	9,105	8.6	60,705	22.3	51,600	13.7
Halifax	9,945	10.4	52,725	25.2	42,780	14.8
London	12,320	9.5	51,065	21.3	38,745	11.8
Victoria	7,450	9.2	48,920	26.1	41,470	16.9
Oshawa	3,950	8.0	39,925	21.4	35,975	13.4
St Catharines–Niagara	10,140	8.1	33,485	17.8	23,345	9.7
Saskatoon	6,125	11.5	33,670	21.7	27,545	10.2
Windsor	8,255	7.8	28,850	19.3	20,595	11.5
St John's	4,540	9.4	26,250	25.5	21,710	16.0
Regina	6,080	9.9	27,140	21.5	21,060	11.7
Sherbrooke	4,000	12.4	23,630	23.5	19,630	11.1
Kingston	4,990	13.4	20,750	26.8	15,760	13.5
Guelph	3,315	11.9	19,610	23.8	16,295	11.9
Sudbury	5,390	8.7	15,375	19.7	9,985	11.0
Trois-Rivières	3,330	9.6	14,680	20.8	11,350	11.2
Thunder Bay	3,850	8.3	13,164	23.1	9,315	14.9
Saint John	3,215	7.5	12,415	21.0	9,200	13.5
Peterborough	2,980	11.0	12,550	22.5	9,570	11.5
Brantford	2,980	6.9	11,325	17.2	8,850	10.3
Sault Ste Marie	2,895	8.9	6,990	20.7	4,095	11.9
Sarnia	*3,735*	*11.4*	8,225	*19.3*	*4,490*	*7.8*
Canada	*766,550*	*8.9*	3,789,350	*22.0*	*3,022,800*	*13.1197*

Source: Statistics Canada, Census of Population, 1971 and 2016 (author's calculations).

service- and knowledge-based economy is related to health care and health sciences. While a significant part of Hamilton's growing reputation in health is due to McMaster University's School of Medicine, which offers a distinctive, experientially based form of medical education, one of the primary reasons for health services emerging in Hamilton can be attributed to the pre-existing automotive and steel industries. As Warrian (2009: 18) notes, "Hamilton citizens also tend to have generous medical benefits from unions and the government which means that specialized medical services—such as designed orthotics—are often paid for. Union health benefits have provided the financial base and market for these emergent health services firms." In other words, the presence of the Canadian Auto Workers (now part of Unifor) and United Steel Workers unions in Hamilton was critical in driving demand for and financing Hamilton's emerging health services sector, which has produced world-class innovations.

Elsewhere, the knowledge-based economies of Canadian cities have been built on long-term public investments. The emergence of Ottawa's ICT cluster specializing in photonics and telecommunications is partly the result of a longstanding investment in publicly funded government laboratories, as well as a product of particular historical events and regulatory decisions such as the founding of a key firm, Bell Northern Research (BNR), which anchored the ICT industry in Ottawa beginning in the 1950s. As Lucas, Sands, and Wolfe (2009: 194) write:

> the original decision by Northern Electric in the late-1950s to establish a research facility in the region was made after a judicial decision in the US cut off its ready access to patents from the Western Electric Co. Its purchase of a substantial tract of land on the outskirts of Ottawa as the future home of Bell Northern Research (BNR) was largely because the concentration of federal government laboratories in the nation's capital created a steady stream of industrial engineers, researchers, and managers moving into the region.

BNR (and a subsequent failed subsidiary) were the training ground for several entrepreneurs who left the company to start their own businesses in Ottawa, leading to the development of a critical mass of firms and workers specialized in ICT and related activities. Moreover, several organizations, such as Invest Ottawa (formerly the Ottawa Centre for Research and Innovation), have emerged that are dedicated to promoting networking among firms to encourage firm learning, knowledge sharing, and the solving of common problems. This local associative activity, alongside support from local and provincial government programs, the presence of national research laboratories, and the research activity of local universities, encourages the ongoing development and growth of this high-tech industry within the city.

The Cultural Economy of Canadian Cities

The creative and cultural industries are increasingly important to urban economies, especially in Canada's largest cities: Toronto, Montreal and Vancouver (Britton, Tremblay, and Smith, 2009; Grant, 2014; Wolfe and Gertler, 2016). The trajectories in each of these three cities have been shaped by their own local histories, institutions, and policies. In other words, although all three of Canada's largest urban centres have a large concentration of cultural industries, one should not simply assume their character or development paths will be identical, given different pre-existing conditions and institutions.

For example, Vancouver's development trajectory is unique compared to other Canadian cities. As Barnes and Hutton (2009: 1252) note:

> to use the Canadian economic historian Harold Innis's (1930) term, Vancouver first developed as a "local metropole" within British Columbia's staples economy and based on the extraction, processing and export of natural resources, most prominently in forestry, fishing and mining. Vancouver's role was primarily a control and distribution centre for staples goods, and only secondarily, a processing site.

Yet, the new economy has taken hold in Vancouver and the city is viewed as an appealing location for firms and workers involved in a range of creativity- and culture-oriented endeavours (Hutton, 2010). Indeed, Vancouver has emerged as a hotbed for video game production and other digital and interactive media applications, with companies like Electronic Arts—one of the largest video game developers in the world—and other gaming companies located there. Certainly, the quality-of-life factors identified by Florida (2002) related to lifestyle amenities figure heavily in accounts of why workers and firms are attracted to Vancouver (Britton, Tremblay, and Smith, 2009). However, a longer history explains the development of these industries in Vancouver. First, Vancouver's video game industry draws on resources that were already present in the local economy, including film and television, which provided the initial talent base for this emerging industry. Second, Vancouver's video game industry can be traced back to at least the early 1980s with the founding of Distinctive Software, which was acquired by Electronic Arts in the early 1990s. Similar to Vancouver's other cultural industries (such as film and television), entertainment-oriented new media firms in Vancouver generally are focused on international markets and are tied to other places through international capital flows (Barnes and Hutton, 2009; Britton, Tremblay, and Smith, 2009). Finally, several unique public and quasi-public institutions provide support to Vancouver's digital media industry, such as the University of British Columbia's Media and Graphics Interdisciplinary Centre (MAGIC), which connects researchers with other local players to enable new product development; DigiBC, a dedicated umbrella industry association, brings together various institutions, organizations, and other interested parties to promote and develop the mobile and wireless, video games, animation and special effects, social media, and online marketing industries.

In Toronto, the cultural sector is one of several key clusters that are critical to the city's economic development trajectory, and the growth of this sector rests on several important territorial assets and institutional supports. Toronto's cultural industries have benefited from public funding for the arts and culture at the local, provincial, and national levels. Moreover, a diverse set of public and private organizations support the development of these industries. For example, Toronto's film industry is supported by several city-level initiatives, including the following: (1) the Toronto Film, Television and Digital Media Office, which provides logistical and regulatory support; (2) the Toronto Film, Television and Digital Media Board made up of industry players and local councillors that creates strategies and policies to promote and support Toronto's film and television industry; and (3) Pinewood Toronto Studios (formerly Filmport), Canada's largest film and media production complex, located along the city's waterfront. In addition to these city-supported initiatives, Toronto's film industry has benefited from federal institutions such as the National Film Board (NFB) and the Canadian Broadcasting Corporation (CBC), as well as provincial programs like the Ontario Film and Television Tax Credit. Toronto's film industry exceeded $2 billion in production investment in 2016, due in part to its over 1400 productions on location per year (Toronto Film, Television and Digital Media Board, 2017). It has also established success in children's television and film production, exemplified by shows like the highly successful *Paw Patrol*, created by SpinMaster, a local firm. Toronto has also become a destination for film consumption, hosting the Toronto International Film Festival (TIFF), with its home at the TIFF Bell Lightbox in Toronto's entertainment district, as well as other events, such as HotDocs, the Canadian International Documentary Film Festival.

In addition to film and television, there is a critical mass of fashion, design, art, architecture, music, new media, and other cultural activity (Leslie and Brail, 2011; Wolfe and Bramwell, 2016). Publicly funded institutions of higher education, such as Sheridan College, Seneca College, Ontario College of Art and Design (OCAD) University, Ryerson University, and George Brown College, offer programs in animation, design, and other culture-related programs that develop the local talent so critical to the survival of these industries. And, there have been interesting developments associated with the rise of digital platforms (see Brail and Donald, Chapter 5). For example, George

Brown College hosts YouTube Space Toronto, one of nine such facilities worldwide. YouTube Space Toronto is a 3500-square-foot space that provides programming and workshops to allow local YouTube content creators and celebrities to access the necessary tools and resources to build their channels on the platform (Bogart, 2016). For other parts of the digital media and advertising industries, the presence of local demand generated by other sectors, especially financial and business services, has helped to buoy these industries.

Conclusion

This chapter has traced how the economic landscapes of Canadian cities have profoundly changed since the crisis and decline of Fordism and its associated production and employment relations, regulatory mechanisms, and other institutions. On the one hand, the occupational composition of Canadian cities reflects an increasing demand for knowledge, creativity, and formal qualifications. On the other hand, the sectoral composition of Canadian cities has slowly transformed to one where services-producing industries (including higher-order, knowledge-intensive services and lower-order services) have eclipsed goods-producing industries in terms of their relative importance to urban economies. Three key features of this economic transition were examined in detail. First, Canada's manufacturing base has declined and transformed over the past several decades. Second, there has been a shift towards knowledge-based and innovation-oriented activity. Third, creative and cultural economic activities have become increasingly important, particularly in large cities.

However, as seen throughout the chapter, economic transitions unfold slowly and urban economies evolve over time—not divorced from either broader macroeconomic, political, social, and cultural conditions or their particular local contexts. The transition towards a post-industrial, new economy has been experienced differentially across Canadian cities and their economic development trajectories are path-dependent. In other words, the specific industrial histories, individual characteristics, and territorial assets of city-regions matter to their economic development. Furthermore, it is clear that public investments and public policies at the local, provincial, and national levels have been important in shaping these outcomes. In other words, urban economies are actively created and (re)shaped by actors and institutions from the private *and* public sectors.

On a final note, the emphasis on innovation, and on creative, knowledge-based economic development and growth, points to some troubling questions about the future of work and the fortunes of individuals, communities, and cities. Scholars have raised questions about the polarizing dynamics related to the emergence of a knowledge-based, creative economy. They note that the deepening divide between professional, managerial, technical, and scientific workers (or, the creative class) and workers occupying lower-wage, routinized service jobs, is leading to greater levels of inequality as both people and places are left behind (Storper and Scott, 2009; Breau et al., 2015; Rodriguez-Pose, 2018).

Moreover, some of these same tendencies exist within the industries associated with the emerging knowledge-based economy itself, with significant implications for the urban labour markets in which they are situated. A number of emerging trends promise to spur further economic change in cities, with the potential for creating new opportunities but also erasing and transforming many current sources of employment. Technologies related to the growing use of artificial intelligence, autonomous vehicles, robotics, algorithmic and machine learning and the growth of digital platforms may further deplete job opportunities for both highly skilled workers and low-skilled workers, and leading to **winner-takes-all urbanism** (Florida, 2017; Wolfe, 2018; see also Brail and Donald, Chapter 5). The jobs that remain may become more short term and contract based, what some describe as the gig economy, leading to more people living in precarious and risky circumstances. These trends are leading to calls for cities and government to consider how to support more inclusive growth and innovation strategies. As students and scholars interested in contemporary urban and economic change, it is our task to understand the ongoing economic transitions in Canadian cities and the implications for individuals, communities, and cities.

Acknowledgements

This chapter was greatly informed by the author's participation in the Creating Digital Opportunities partnership, funded by the Social Sciences and Humanities Research Council (SSHRC) and led by David Wolfe (University of Toronto), as well as two previous SSHRC-funded studies of clusters, innovation, and economic development in Canada, and the economic and creativity performance of Canadian cities.

Review Questions

1. What economic transitions are Canadian cities undergoing?
2. How have cities adapted to changing economic circumstances?
3. What are the consequences of deindustrialization for Canadian cities?

References

Amin, A., ed. 1994. *Post-Fordism: A Reader.* Cambridge, MA: Blackwell.

Barnes, T., and T. Hutton. 2009. "Situating the new economy: Contingencies of regeneration and dislocation in Vancouver's inner city," *Urban Studies* 46: 1247–69.

Bathelt, H., A. Malmberg, and A. Maskell. 2004. "Clusters and knowledge: Local buzz, global pipelines and the process of knowledge creation," *Progress in Human Geography* 28: 31–56.

Beck, U. 2000. *The Brave New World of Work.* Cambridge: Cambridge University Press.

Bell, D. 1973. *The Coming of Post-Industrial Society: A Venture in Social Forecasting.* New York: Basic Books.

Bogart, N. 2016. *YouTube Space Opens Toronto Studio: How Budding Creators Can Turn YouTube into a Career.* At: https://globalnews.ca/news/2664824/youtube-space-opens-toronto-studio-how-budding-creators-can-turn-youtube-into-a-career/

Boyer, R. 2005. "How and why capitalisms differ," *Economy and Society* 34: 509–57.

Breau, S., D. Kogler and K. Bolton. 2015. "On the relationship between innovation and wage inequality: New evidence from Canadian cities," *Economic Geography* 90: 351–73.

Britton, J.N.H., D.-G. Tremblay, and R. Smith. 2009. "Contrasts in clustering: The example of Canadian new media," *European Planning Studies* 17: 281–301.

Bryson, J., and N. Henry. 2005. "The global production system: From Fordism to post-Fordism," in P. Daniels, M. Bradshaw, D. Shaw, and J. Sidaway, eds., *An Introduction to Human Geography: Issues for the 21st Century.* Harlow: Pearson.

CBC News. 2018a. *GM's Oshawa Plant Is Closing: Here's What You Need to Know.* At: www.cbc.ca/news/canada/gm-oshawa-closure-1.4921660

———. 2018b. *GM Canada Breaks Ground for New 7-Acre Leslieville Mobility Campus.* At: https://www.cbc.ca/news/canada/toronto/construction-begins-on-gm-leslieville-dealership-complex-1.4869512

Deslisle, F., and R. Shearmur. 2010. "Where does all the talent flow? Migration of young graduates and nongraduates, Canada 1996-2001," *The Canadian Geographer* 54: 305-23.

Florida, R. 2002. *The Rise of the Creative Class.* New York: Basic Books.

———. 2017. *The New Urban Crisis.* New York: Basic Books.

Gertler, M.S., R. Florida, G. Gates, and T. Vinodrai. 2002. *Competing on Creativity: Placing Ontario's Cities in North American Context.* Toronto: Ontario Ministry of Enterprise, Opportunity and Innovation and the Institute for Competitiveness and Prosperity.

———, and T. Vinodrai. 2009. "Life sciences and regional innovation: One path or many?" *European Planning Studies* 17: 235–61.

Glaeser, E. 2011. *Triumph of the City.* New York: Penguin.

Gouldner, A. 1979. *The Future of Intellectuals and the Rise of the New Class.* New York: Seabury.

Grant, J.L., ed. 2014. *Seeking Talent for Creative Cities: The Social Dynamics of Innovation.* Toronto: University of Toronto Press.

Harvey, D. 1990. *The Condition of Postmodernity.* Cambridge, MA: Blackwell.

Hutton, T.A. 2010. The New Economy of the Inner City: *Restructuring, Regeneration and Dislocation in the Twenty-First-Century Metropolis.* London and New York: Routledge.

Innis, H. 1930. *The Fur Trade in Canada.* Toronto: University of Toronto Press.

Jenson, J. 1990. "Representations in crisis: The roots of Canada's permeable Fordism," *Canadian Journal of Political Science* 23: 653–83.

Leslie, D., and S. Brail, 2011. "The productive role of quality of life: The case of fashion designers in Toronto," *Environment and Planning A* 43(12): 2900–17.

Lewis, N.M., and B. Donald. 2010. "A new rubric for creative city potential in Canada's smaller cities," *Urban Studies* 47: 29–54.

Ley, D. 1996. *The New Middle Class and the Remaking of the Central City*. Toronto: Oxford University Press.

Lloyd, R. 2006. *Neo-Bohemia: Art and Commerce in the Post-industrial City*. New York: Routledge.

Lowe, N., and M.S. Gertler. 2005. "Diversity and the evolution of a life-science innovation system: The Toronto region in comparative perspective," in D.A. Wolfe and M. Lucas, eds., *Global Networks and Local Linkages: The Paradox of Cluster Development in an Open Economy*. Montreal and Kingston: McGill-Queen's University Press.

——— and ———. 2009. "Building on diversity: Institutional foundations of hybrid strategies in Toronto's life sciences complex," *Regional Studies* 43: 589–603.

Lucas, M., A. Sands, and D.A. Wolfe. 2009. "Regional clusters in a global industry: ICT clusters in Canada," *European Planning Studies* 17: 189–209.

Markusen, A. 2006. "Urban development and the politics of a creative class: Evidence from the study of artists," *Environment and Planning A* 38:1921–40.

McCann, E. 2007. "Inequality and politics in the creative city-region: Questions of livability and state strategy," *International Journal of Urban and Regional Research* 31: 188–96.

Molotch, H. 2002. "Place in product," *International Journal of Urban and Regional Research* 26: 665–88.

Moos, M., and Skaburskis, A. 2007. "The characteristics and location of home workers in Montreal, Toronto and Vancouver," *Urban Studies* 44: 1781–808.

National League of Cities. 2016. *How Cities Can Grow the Maker Movement*. Washington, DC: Center for City Solutions and Applied Research, National League of Cities.

Nelles, J. 2014. "Myth making and the 'Waterloo Way': Exploring associative governance in Kitchener-Waterloo," in N. Bradford and A. Bramwell, eds., *Governing Urban Economies: Innovation and Inclusion in Canadian City-Regions*. Toronto: University of Toronto Press.

Peck, J. 2005. "Struggling with the creative class," *International Journal of Urban and Regional Research* 29: 740–70.

Piore, M.J., and C. Sabel. 1984. *The Second Industrial Divide: Possibilities for Prosperity*. New York: Basic Books.

Porter, M.E. 1998. "Clusters and the new economics of competition," *Harvard Business Review* (Nov.–Dec.): 77–90.

Rodríguez-Pose, A. 2018. "The revenge of the places that don't matter (and what to do about it)," *Cambridge Journal of Regions, Economy and Society* 11: 189–209.

Scott, A.J. 2004. "Cultural-products industries and urban economic development: Prospects for growth and market contestation in global context," *Urban Affairs Review* 39: 461–90.

———. 2007. "Capitalism and urbanization in a new key? The cognitive-cultural dimension," *Social Forces* 85: 1466–82.

Shearmur, R. 2007. "The new knowledge aristocracy: The creative class, mobility and urban growth," *Work, Organization, Labour, and Globalization* 1: 31–47.

Storper, M. 2013. *Keys to the City: How Economics, Institutions, social Interaction and Politics Shape Development*. Princeton: Princeton University Press.

———, and A.J. Scott. 2009. "Rethinking human capital, creativity and urban growth," *Journal of Economic Geography* 9: 147–67.

Tickell, A., and J. Peck. 1992. "Accumulation, regulation and the geographies of post-Fordism: Missing links in regulationist research," *Progress in Human Geography* 16: 190–218.

Toronto Film, Television and Digital Media Board. 2017. *Spotlight on Toronto: A Strategic Action Plan for the Toronto Film, Television, and Digital Media Industry*. Toronto: City of Toronto.

Vinodrai, T. 2013. "Design in a downturn? Creative work, labour market dynamics and institutions in comparative perspective," *Cambridge Journal of Regions, Economy and Society* 6: 159–76.

———. 2018. "Planning for cool: Millennials and the innovation economy of cities," in M. Moos, D. Pfeiffer and T. Vinodrai, eds., *The Millennial City: Trends, Implications and Prospects for Urban Planning and Policy*. London: Routledge.

———, and S. Keddy. 2015. "Projects and project ecologies," in C. Jones, M. Lorenzen, and J. Sapsed, eds., *The Oxford Handbook of Creative and Cultural Industries*. Oxford: Oxford University Press.

Warrian, P. 2009. "Biotech in lunch buckets: The curious knowledge networks in Steeltown," paper presented at the 10th annual meeting of the Innovation Systems Research Network, Halifax, 30 Apr.

———, and C. Mulhern. 2009. "From metal bashing to materials science and services: Advanced manufacturing and mining clusters in transition," *European Planning Studies* 17: 281–301.

Wolf-Powers, L., M. Doussard, G. Schrock, C. Heying, M. Eisenburger, and S. Marotta. 2017. "The maker movement and urban economic development," *Journal of the American Planning Association* 83: 365–76.

Wolfe, D.A. 2018. *Creating Digital Opportunity for Canada*. Toronto: Brookfield Institute for Innovation + Entrepreneurship and Innovation Policy Lab, Munk School of Global Affairs.

———, and A. Bramwell. 2016. "Toronto's fourth era: An emerging cognitive-cultural economy," in D.A. Wolfe and M.S. Gertler, eds., *Growing Urban Economies: Innovation, Creativity, and Governance in Canadian City-Regions*. Toronto: University of Toronto Press.

———, and M.S. Gertler. 2004. "Clusters from the inside and out: Local dynamics and global linkages," *Urban Studies* 41: 1071–93.

———, and M.S. Gertler. 2016. *Growing Urban Economies: Innovation, Creativity, and Governance in Canadian City-Regions*. Toronto: University of Toronto Press.

18 Innovation and Canadian Cities

Exploring Different Viewpoints

Richard Shearmur

> Cities have long been acknowledged as primary organs of cultural development . . . cities are also primary economic organs. . . . it was the fact of sustained, interdependent, creative city economies that made possible many new kinds of work, agriculture among them (Jacobs, 1969: 16, 42)

> Jacobs' model is so completely contrary to archaeological data that archaeologists have never bothered to mount a targeted attack on the notion. (Smith et al., 2014: 1525)

Introduction

It is commonplace to assume that innovation and cities are intimately linked. Newspapers regularly publish articles assessing which Canadian cities are more innovative or high-tech than others (e.g., CBC, 2018; Crane, 2016). These articles reflect a body of academic scholarship that suggests "innovative activities are the products of cities or regions" (Florida et al., 2017: 87). This assumption—that innovation emerges from cities—stands in contrast to empirical work from Canada (Shearmur and Doloreux, 2016) and elsewhere (Eder, 2018; Grillitsch and Nilsson, 2015). This work finds that "innovation occurs in all sectors of the economy, in urban and rural areas, and within micro, small, medium and large Canadian companies" (Industry Canada, 2011, Chapter 5: 11). Furthermore, innovation is not restricted to companies. It occurs in the public sector (Mazucatto, 2015), municipalities (Shearmur and Poirier, 2017), communities, and civic organizations (Klein et al., 2014).

This chapter explores the different viewpoints on the relationship between cities and innovation, and examines how the debate applies to Canadian cities. The first section provides an overview of innovation as a concept: innovation is far more than high-tech gadgets, new products, and production processes, and innovation is also difficult to measure. The second section summarizes the theory behind the idea that innovation emerges from cities, and examines patent data from Canada that supports this view. The third section returns to the various ways of defining and measuring innovation and challenges the assumption that innovation in Canada emerges only from its cities.

This does not mean that Canadian cities do not play an important role in the innovation process writ large, and there are some types of innovation that are more likely to emerge in cities. However, other forms of innovation emerge in low-density and remote areas (Doloreux et al., 2015; Petrov, 2011; Stolarick et al., 2011). For example, innovations in industries such as wine, tourism, and agriculture, as well as public sector, municipal, community and social innovation, can occur in smaller cities, towns, and rural regions. Even though this chapter is about Canadian cities, it is only by examining the overall geography of innovation that it is possible to identify what is specific about **urban innovation**.

What Is Innovation? Market and Non-Market Innovation

Innovation studies tend to consider innovation from an economic perspective. In these cases, innovation is viewed as the successful introduction of a new product, process, service, or organizational form, where "successful" is synonymous with market acceptance and profit for the innovator (Baregheh et al., 2009). This broad definition distinguishes between the *object* of innovation (i.e., the item that is new), the *process* of innovation (i.e., discovery, development, introduction, and market acceptance), and a *key actor* (i.e., the innovator, often referred to as the entrepreneur, usually a person or firm). It takes for granted that the purpose of innovation is profit or market share, and that the arbiters of innovation's value are paying clients. These assumptions also underpin Schumpeter's (1936) entrepreneurial theory of innovation. For Schumpeter, the innovation process comprises three main stages: (1) creativity and ideation; (2) development and prototyping; and (3) marketing and diffusion. These stages are coordinated by an entrepreneur who recognizes the potential of the idea, has the know-how and ability to marshal resources, and is willing to take risks during the development and marketing stages.

This view of innovation remains widely accepted. However, it has undergone refinement as it was recognized that there is feedback between stages (e.g., ideas or product improvement can be suggested by clients; prototyping can lead to reassessment of ideas) and that ideas, know-how and expertise are often gathered from external actors, including suppliers, competitors, customers, institutions or, more generally, from social interactions (Brown and Duguid, 2000). This refined view of innovation has strongly influenced and reinforced the belief that innovation more readily occurs in cities because this is where social interactions are most prevalent.

However, the assumption that innovation occurs mainly in the private sector and is market-driven is increasingly questioned (Mulgan, 2007). Indeed, the public sector has developed and introduced key technologies that underpin many of today's high-tech industries (Mazucatto, 2015). Furthermore, many innovations are introduced by government agencies (Borins, 2014), municipalities (Shearmur and Poirier, 2017), and civil society (Klein et al., 2014; Swann, 2015). The literature describes social innovation as new services, civic processes or community businesses that are introduced to address a social or community need or problem, rather than to generate profit. In many respects, the non-market innovation process resembles that of the private sector. Individuals or small organizations play the role of entrepreneur (i.e., they identify a need, find a possible solution, run pilot projects, and implement the innovation), but they do so based on a system of values that is not reducible to market value alone (Anderson, 1995; Sandel, 2013). Usually, the objectives and values that underpin social innovation are discussed and negotiated within the community, and an innovation's success is judged according to community standards. For this reason, there is no straightforward metric of success, and the innovation's widespread adoption or diffusion is often of secondary concern.

Regardless of whether an innovation is market or non-market based, it is positioned along a spectrum that runs from incremental innovation (e.g., small changes to existing products, processes, or practices) to radical innovation (e.g., major changes and world-firsts). Furthermore, innovation is inherently relative. Thus, a firm or organization can innovate by imitating what others in the industry are already doing (Godin, 2008). In the *Oslo Manual* (Organisation for Economic Co-operation and Development [OECD], 2005), which provides guidance on the definition and measurement of innovation, imitation is recognized as new-to-firm innovation. New-to-firm innovations are a critical form of innovation that enable best practices to spread and technology to diffuse. They allow firms, public agencies, and civic organizations to keep up with competition (if operating in a market context) or with evolving standards and expectations (if operating in a non-market context).

Why Do Cities Care about Innovation?

Given the nature of innovation, why link it to Canadian cities? There are three main reasons. The first is economic: cities need jobs for their citizens and tax revenues to maintain and develop the city. There is a strong assumption—driven by industry and politics (Logan and Molotch, 1987; MacLeod, 2011)—that these can only be achieved through economic growth[1] (see Hall and Donald, Chapter 20, for a counterpoint to the economic growth imperative). This aligns with economic theory, which posits the necessity of growth, and argues that growth occurs by way of innovation: better products, services, and processes replace older ones, leading to higher productivity and improved outcomes (Romer, 1994; Solow, 1994). Thus, many cities' economic policies now focus on the promotion of innovation. For example, the City of Mississauga has an *Action Plan for Innovation* (Mississauga, 2011). Montreal's economic development plan states that "to ensure its prosperity, Montreal must become a vector of innovation . . . [and of] . . . entrepreneurial economic development" (Montreal, 2018). On the west coast, the cities of Vancouver (via the Vancouver Economic Commission) and New Westminster

> . . . signed [on 16 October 2018] . . . a Memorandum of Understanding (MOU) on technology deployment and innovation. This marks the beginning of a partnership between the two parties to build a regional platform for technology deployment, demonstration, and business innovation. (Vancouver, 2018)

The hope that underpins these urban policies is that if local industries and firms are innovative—or if innovative firms can be attracted from outside—then the city's current businesses and population will benefit.

The second reason to consider innovation and cities is social. Many social and environmental problems are most concentrated in cities. For instance, income disparities are highest in cities (Chen et al., 2012);[2] cities are major contributors to climate change given their role as nodes in global energy, transport, and trade networks (World Bank, 2010); and many of the challenges associated with congestion, pollution, housing, cohabitation, and local democracy emerge in cities (see, for example, Taylor and Bradford, Chapter 3; Holden and Chang, Chapter 22; Senbel, Liem, and Lesnikowski, Chapter 24). The geographic concentration of issues and challenges means that cities offer many opportunities for experimentation and problem-solving in the social, cultural, and environmental spheres (Castells et al., 2012; Klein et al., 2014; World Bank, 2010).

The third reason for considering innovation in relation to cities is that municipalities (as administrations and organizations) manage cities, and often develop innovative ways of administering urban services, traffic, local politics, and addressing other urban issues (see Taylor and Bradford, Chapter 3). For instance, municipal governments are responsible for roads, water distribution, waste treatment plants, parks, libraries, sports centres, museums, and other public infrastructure. Therefore, they are regularly called upon to introduce new ways of designing, building, maintaining, and managing this infrastructure, often in collaboration with research institutions and private partners.[3]

Municipalities also implement bylaws and regulations that are innovative or that shape innovation. Major innovations such as town planning (Harris and Moore, 2013) were developed by local government. For example, the municipality of Westmount is the first in North America to have introduced a comprehensive zoning code—in 1909 (Fischler, 2017). More often, it is incremental regulatory changes that are introduced. For example, Surrey, BC,[4] adopted changes in waste collection to improve recycling. These innovations, whether major or minor, are created, designed, and implemented by forward-thinking municipal employees, public administrators, and politicians. They face challenges similar to entrepreneurs in the private sector when trying to bring about change (Shearmur and Poirier, 2017), but often do so for different reasons and with different objectives (Perry and Hondeghem, 2008). It is not possible to more fully explore innovation in the social, environmental, cultural, and municipal spheres in this chapter, but it is important to bear them in mind when thinking about innovation and cities.

Innovation and Canadian Cities

There are two principal theories that associate innovation and cities. First, the argument that innovation is quintessentially urban can be traced back to the work of Jane Jacobs (1969). She argues that "work that we usually consider rural has originated not in the countryside, but in cities" (13). She starts from the premise that, in Neolithic times, agricultural innovation emerged in cities and then transferred to rural areas. Jacobs then traces this through to the twentieth century, arguing that current innovations in agriculture also emerge from cities. This is then generalized into the idea that cities are crucibles of innovation because they enable a wide variety of people to meet and interact, thereby generating new ideas: creative people are thought to be especially attracted to cities (see Vinodrai, Chapter 17, on the **creative class**). Today, the idea that both a large, diverse population and high interaction potential are only available in cities sustains the notion that cities, especially large cities, are "machines of innovation" (Florida et al., 2017; Glaeser, 2011).

A second argument underpins the idea that innovation can only take place in dense urban centres. This argument focuses not on diversity but on the geographic co-location of establishments undertaking the same (Porter, 2003) or related (Frenken et al., 2007) economic activities. **Clusters** of similar or related organizations often draw on common or collective resources, including shared (or transferable) knowledge and know-how, a qualified labour force, subcontractors, financiers, infrastructure, institutions, and policy—all of which enable firms and other organizations in the cluster to quickly identify and exploit opportunities that arise in its area of expertise (or in related areas). Furthermore, such concentration leads to exchanges and collaborations between knowledgeable partners and can make clusters into leaders in their field, able to foresee opportunities and anticipate trends before less specialized or more isolated competitors (Brown and Duguid, 2000).

Many such clusters have been identified in Canadian urban areas. For example, Wolfe and Gertler (2016) introduce a number of well-documented examples from across the country (see also Vinodrai, Chapter 17). They show how these clusters differ in terms of geographic extent, governance structure, and balance between diversity and specialization. Although clusters are not *necessarily* located in highly urbanized areas, the fact that a critical mass of organizations and institutions is required means that almost all clusters are located in proximity to relatively large urban areas (Crevoisier and Camagni, 2001). Furthermore, there can be numerous clusters in a single metropolitan area (Spencer et al., 2010)—diversity and specialization can co-exist. For example, Montreal is home to clusters in artificial intelligence (High, 2017), video gaming (Pilon and Tremblay, 2013) and aeronautics.

Overall, these two theoretical approaches suggest that innovation is a predominantly urban phenomenon, since both approaches put cities forward as quintessentially innovative on the basis that density and geographic co-location enable interactions and collaborations.

Some Canadian Evidence

A collection of recent studies provide excellent discussions and evidence of the way economic innovation is deployed in Canadian cities (Wolfe, 2009; Wolfe and Gertler, 2016).[5] Their evidence, drawn from in-depth case studies from across Canada's cities, confirms that innovation occurs in cities, driven by **clusters**, mobile qualified workers, governance structures, policy, universities and market conditions. This research broadly corroborates and extends the theories outlined above, adding nuance and empirical detail.

Quantitative data also seems to confirm that innovation is principally an urban phenomenon. Patents are often used as an indicator of innovation. Figure 18.1 shows that Toronto produces about 19 per cent of all Canadian patents, followed by Montreal (12 per cent), Vancouver (11 per cent), and Ottawa (10.5 per cent). These four cities generate over 50 per cent of all Canadian patents. The 15 cities (Canada's largest cities, as well as a selection of major cities in each province) shown in Figure 18.1 comprise approximately 54 per cent of Canada's population but produce 74 per cent of Canada's patents. The data also seem to show that the largest cities—Toronto, Montreal and Vancouver—are the most innovative, but have lost ground since the late 1990s as Calgary, Waterloo, and (to a lesser extent) Ottawa-Gatineau have gained.

Looking at these data another way reveals a slightly different story. If patent production is considered on a *per capita* basis, larger cities are not necessarily the most innovative. Figure 18.2 shows that, for patents filed between 2011 and 2013 (shown as an average over this time period, referred to as "smoothed"), Waterloo is the most innovative, followed by Ottawa-Gatineau, Calgary, Vancouver, and Quebec City. Montreal and Toronto are in the middle, and some of the smaller—but locally important cities—such as Saskatoon and St. John's, produce fewer patents per capita than the rest of Canada.

How do these observations relate to the idea that innovation is quintessentially urban? They appear to confirm it. As a group, these 15 cities are disproportionately innovative, corroborating (though not proving) that size—and its associated diversity and clash of ideas—is conducive to innovation. In addition, smaller cities that are particularly innovative benefit from strong clusters (Wolfe and Gertler, 2016): Waterloo is at the heart of Canada's information and communications technologies (ICT) sector; Ottawa benefits from a diversified high-tech cluster, which itself benefits from proximity to the federal government

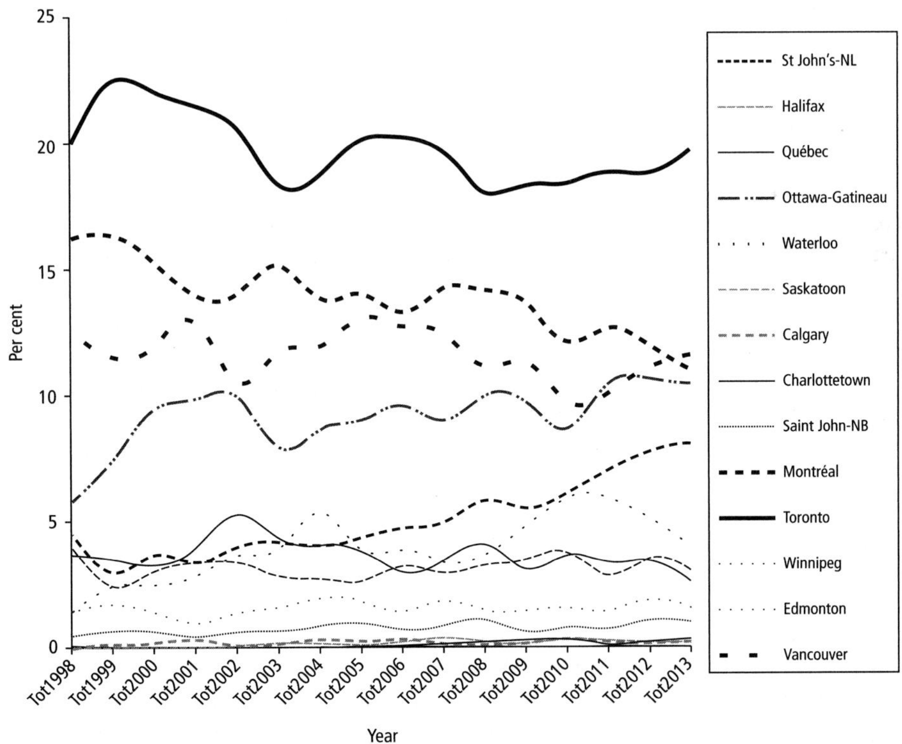

Figure 18.1 Proportion of Canadian PCT Patents, by City, 1998–2013

Note: Data are coded by place a residence (Census Division) of the inventor(s), and correspond to patents filed under the Patent Cooperation Treaty (PCT) (i.e., patents that sought international protection).

Source: OECD. At: https://stats.oecd.org/Index.aspx?DatasetCode=PATS_REGION#

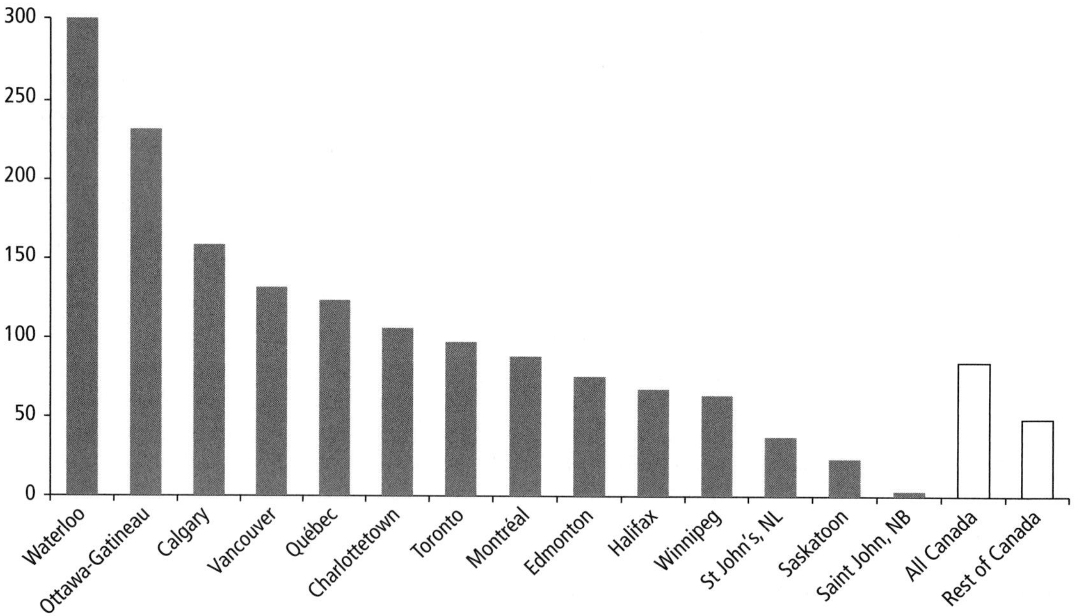

Figure 18.2 PCT Patents per One Million Inhabitants, Smoothed over 2011–2013

Note: Data are coded by place a residence (Census Division) of the inventor(s), and correspond to patents filed under the Patent Cooperation Treaty (PCT) (i.e., patents that sought international protection).

Source: Population estimates are taken from the 2016 Census at the CMA level.

(a source of finance, markets, and knowledge); Calgary is at the centre of Canada's oil industry; Vancouver has seen the rise of a digital technology and cultural cluster; and Quebec City has a strong optics and photonics cluster.

Montreal, a large city, has benefited from provincial cluster policies since the 1990s, when the government of the day introduced these policies in an attempt to boost innovation. The minister in charge of this policy later became mayor of Montreal, which reinforced the city's support for clusters. Montreal now officially houses 10 clusters: aerospace, aluminum, film and television, logistics and transport, fashion, life sciences, financial services, ICT, clean technology, and electric and smart transportation. The city provides seed money and helps firms in these sectors coordinate in view of collaborating, identifying shared interests, and lobbying. In most other cities, clusters are less formalized, but they function in similar ways—with either the city or a lead firm taking a coordinating role to strengthen an existing concentration of activity.

Clusters wax and wane. Moncton had a strong transport cluster until the mid-1980s, when the CN yards closed and trade agreements were introduced, weakening Moncton's locational advantages by reducing emphasis on east-west trade. More recently, new clusters such as a life sciences cluster and a small gaming cluster have emerged in Moncton. These clusters are focused on key companies or institutions, around which a group of suppliers and connected activities have developed.

There are many examples of clusters in Canadian cities: most emerge spontaneously—either around a large institution such as a research centre or a major company, or by virtue of the presence of many small firms in related sectors—and are then organized either through the intervention of local government (such as in Montreal) or by virtue of common interests and requirements being identified and lobbied for by the actors themselves.

Another indicator of innovative activity is venture capital—a type of finance that focuses on start-ups (often innovating in high-risk new

Table 18.1 A Geography of Venture Capital in Canada, 2018

Year 2018	Whole region		Urban		% Urban	
Region (urban)	**M $**	**Deals**	**M $**	**Deals**	**$**	**Deals**
Atlantic	37	29	-	-	-	-
Quebec (Montreal)	1298	105	1288	104	99.2	99.0
Ontario (Toronto)	1526	196	1305	160	85.5	81.6
Prairies (Calgary)	213	32	126	20	59.2	62.5
BC (Vancouver)	393	104	388	101	98.7	97.1

Note: Definitions of venture capital vary. These data provide a reasonable impression of the geography of venture capital in Canada; Please refer to the *PwC Canada and CB Insights* report for precise definitions.
Source: Adapted from PwC Canada and CB Insights, https://www.pwc.com/ca/en/industries/technology/money-tree/money-tree-q3-2018.html

technologies) with the potential to reach large and lucrative markets. These types of market are usually located in cities, and the start-up companies themselves need to be prepared for rapid growth (i.e., access to labour, real estate, accounting and legal services). Not surprisingly, the vast majority of venture capital finance in Canada is received by companies based in a few large cities (Table 18.1). In certain cases however, especially with respect to technology companies linked to resource exploration, venture capitalists also finance start-ups beyond the confines of Canada's major cities.

Is Innovation Necessarily Urban?

There is a wide body of research that presents data or case studies of the innovation process in cities that corroborates the diversity and/or cluster hypotheses, and that shows that large cities produce most innovation. These studies may also point out that cities and metropolitan areas have recently grown faster than other types of region (Proudfoot, 2016). So why isn't this sufficient to conclude that innovation occurs disproportionately in cities? There are two reasons. First, the underlying theory is not persuasive—it describes *how* innovation occurs in cities but does not demonstrate that innovation is *exclusive to* cities. Second, evidence shows that innovation occurs in low-density areas and in isolated firms; and theory is emerging that outlines how (Shearmur, 2015).

The Problem with Urban Innovation Theory

While Jacobs (1969) accurately describes knowledge exchange and innovation processes that occur in (and between) cities, her contention that innovation does not occur in rural settlements is contested by the archaeological record (Smith et al., 2014) and studies of agricultural and resource-related innovation (Perdue, 1994; Sorensen, 2008). Firms and inhabitants in rural areas possess the know-how and experience to improve techniques and innovate in areas such as agriculture. After the initial agricultural innovations that enabled cities to develop (Smith et al., 2014), it is probable that knowledge and know-how were exchanged between urban and outlying areas, leading to innovation in rural areas.[6] For example, farmers not only possess immense knowledge about techniques, soils, and weather patterns with which they work on a daily basis, but there is also no evidence that they are inherently less clever or creative than city dwellers. Farmers may not possess formal qualifications or cultural capital—but that does not make them incapable of improving techniques they are intimately familiar with. Overall, it is an urban understanding of innovation that is prioritized, and innovation by (and for) non-urban communities tends to be ignored (Shearmur, 2017).

Moving on from Jacobs, the idea that information and knowledge are difficult to obtain in smaller towns and peripheral areas may have been valid until the 1990s, but no longer fully applies.

Macpherson (2008) shows that, between 1992 and 2005, the Internet altered how manufacturers in upstate New York accessed services and external knowledge in the innovation process. By 2005 the difference between upstate and metropolitan innovators had disappeared. Shearmur and Doloreux (2015) show, for Quebec, that there is no difference in propensity to innovate between peripheral and city-based manufacturers. Not all types of innovator can operate from peripheral locations; it is often those in more technical or scientific fields who are successful (Shearmur, 2015). For some innovators, peripheral locations can be an advantage (Grabher, 2018), as less interaction decreases the likelihood of ideas (or employees) being poached (Dahlander and Gann, 2010), and it positions them closer to important knowledge, such as in the case of winemakers (Doloreux and Lord-Tarte, 2013).

As noted above, clusters are, by definition, geographic agglomerations of actors, which are viewed as hotbeds of innovation. However, this does not mean that innovation *necessarily* occurs in clusters. Geographically isolated innovators, such as those developing open-source software (Amin and Roberts, 2008), can find alternatives to permanent geographically defined clusters, through participating in online communities or trade shows and fairs, that act as temporary clusters (Bathelt and Schuldt, 2008).

Quantitative evidence—especially patents—may suffer from geographical bias, overstating innovation in cities (Shearmur, 2017). Patent data mostly capture new technologies and products, which tend to be more urban (Duranton and Puga, 2004), rather than process innovations, which tend to be more rural (Shearmur, 2011), and patents do not capture organizational innovation. Furthermore, large (often urban) corporations use patents as a defensive strategy, meaning that they patent many ideas, with no intention of commercializing them (Jaffe and Lerner, 2007). In contrast, smaller firms—which predominate in rural settings—often rely on secrecy. This is partly because they have few means of defending a patent should it be breached, and partly because secrecy can be a viable option for isolated firms.

Finally, the literature on urban innovation is generally silent on public sector and social innovation, although it assigns cultural innovation and creativity to cities (see Glaeser, 2011). However, there are many examples of policy, social and cultural innovation, and creativity in small towns and rural areas, in Canada (e.g., MacAulay, 2001; Petrov, 2011) and elsewhere (e.g., Gibson et al., 2010; Bock, 2012; Cooke, 2011).

Discussing the Empirical Evidence of Non-Urban Innovation

The above discussion casts doubt on the idea that innovation is exclusively an urban phenomenon. Arguments for, and evidence of, this have recently been summarized by Shearmur (2017), Eder (2018), Fitjar and Rodriguez-Pose (2017), and Grabher (2018). There are two main reasons why innovation in smaller towns and peripheral regions is overlooked in Canada. The first relates to data availability, measurement, and proper representation of the geographic distribution of innovative firms (Shearmur, 2013); the second relates to optics—innovation in cities is simply easier to see.

For instance, Statistics Canada runs a nationwide innovation survey every few years. This survey, geographically stratified only by province and major metropolitan area, asks firms (not each of their establishments) about their innovation activities. In principle—despite other biases that can arise when innovation is self-reported—these data should provide a more accurate geography of innovation than patents. However, since the vast majority of economic activity occurs within about 100 kilometres of major metropolitan areas (about 70 per cent of all population and jobs are within a 100 kilometres of Canada's eight largest cities; Desjardins et al., 2013), since firm headquarters tend to be more urban than their establishments, and since geographic representativeness is coarse, non-metropolitan observations are often limited or suppressed. Thus, analysis of these data reflect innovation processes that occur in and around Canada's largest cities.

As a result, in Quebec in 2005, the provincial government requested an over-sampling of the innovation survey: a *census* of manufacturers was thus obtained. Furthermore, that year the survey was conducted at the establishment (actual location), not the firm (headquarter location), level. The 2005 data are unique in Canada. These data, the only Canadian data that permit this type of geographic analysis, reveal almost no differences in propensity to innovate across Quebec. These data show that radical product innovations are (weakly) more probable in cities, and radical and incremental process innovations are (weakly) more probable in smaller cities and non-urban areas (Shearmur, 2010, 2011). These results have been broadly replicated in other Quebec surveys[7] (Shearmur and Doloreux, 2015), for services (Shearmur and Doloreux, 2009), and reflect results in other countries (e.g., Grillitsch and Nilsson, 2015; Fitjar and Rodriguez-Pose, 2011).

The second problem that occurs when innovation data are analyzed is one of optics: it can best be explained graphically. Figure 18.3 represents a stylized Canada, with two types of settlement—a large city and remote municipalities. Black dots represent innovators and white dots represent non-innovators. The geographic distribution of innovators and non-innovators resembles what is observed in Canada, and it appears to confirm that innovation is principally an urban phenomenon.

The geographic distribution in Figure 18.3 has the following properties:

- there are 14 innovators in the city and 6 innovators in remote areas—in absolute terms, the city is more innovative than peripheral jurisdictions;

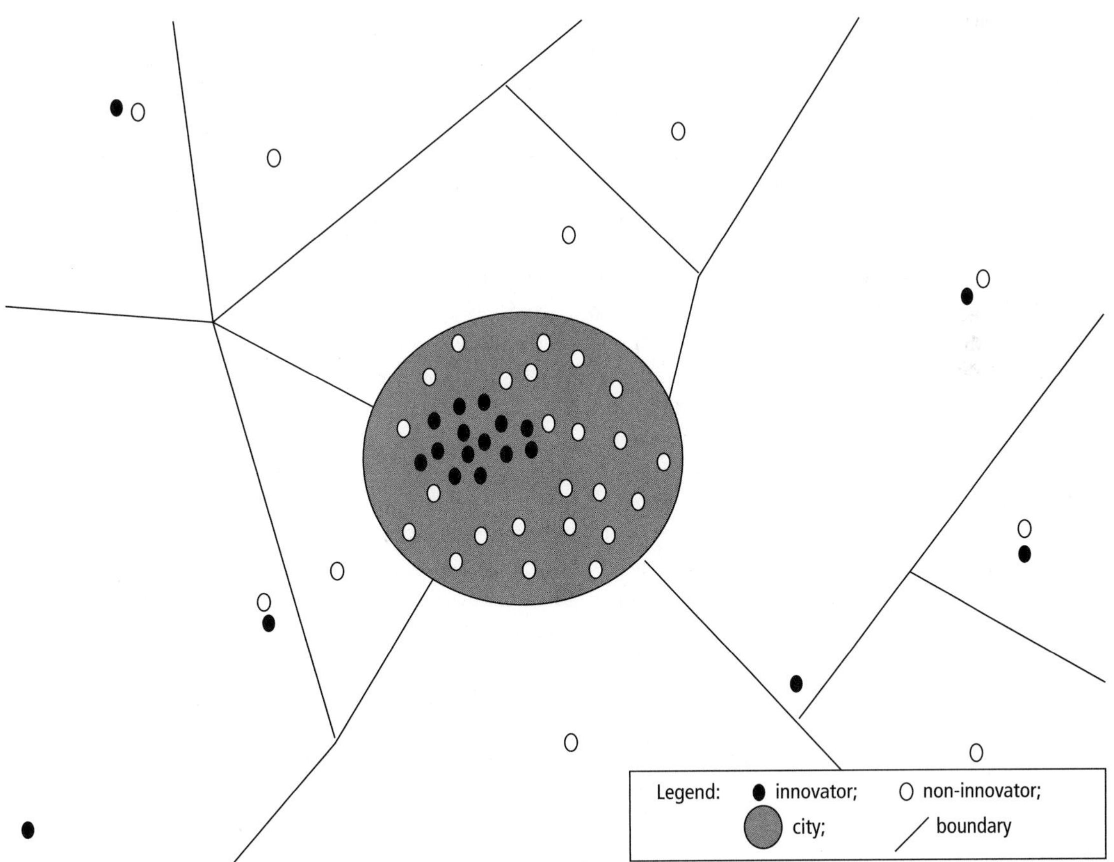

Figure 18.3 A Schematic Geography of Innovation in Canada

Source: Author's illustration.

- 100 per cent of cities (i.e., the only city) harbour innovation, whereas 50 per cent of non-urban jurisdictions have no innovators (5 out of 10); again, this corroborates the idea that innovation occurs principally in cities; and
- 36.8 per cent of city establishments innovate (14 out of 38), whereas 37.5 per cent of peripheral establishments innovate (6 out 16). This challenges conventional wisdom, theory, and the optics of Figure 18.3.

In essence, theory and analysis that insist upon innovation being purely an urban phenomenon focus on the first two points. Researchers who recognize innovation in firms located in remote areas bring the third point to attention, without, of course, denying the first two. It is empirical evidence from Quebec and from other similar geographic contexts—such as Australia (Gibson et al., 2010; Sorensen, 2008), Sweden (Grillitsch and Nilsson, 2015), Norway (Fitjar and Rodriguez-Pose, 2011)—which, combined with adequate data and/or research protocols, confirms that innovation is not necessarily less common in remote or rural environments.

What of non-market innovation? Table 18.2 presents data gathered over the period from 2005 to 2018 from the Union des Municipalités du Québec's Mérite Ovation innovation award (see Shearmur and Poirier, 2017). If one gauges municipalities' capacity to innovate by the proportion of municipalities that have won at least one prize, then larger municipalities are more innovative: 90 per cent of them have won a prize, compared to 35 per cent of the smallest municipalities. Furthermore, while larger municipalities have won an average of 3.6 prizes each, and municipalities of 20,000 to 100,000 people won an average of 1.3 prizes each, smaller municipalities have won, on average, fewer than 0.5 prizes each. These data seem to confirm that larger municipalities are more innovative.

However, if *submissions* are considered instead of *municipalities* (each municipality can submit multiple projects in a competition), then conclusions change: although smaller municipalities submit fewer projects, they are most likely to win a prize if they submit. Over 22 per cent of their submissions are recognized by an award, against only 12 to 16 per cent for larger municipalities. By this measure, it is smaller municipalities that are more innovative.

These data are exploratory; however, they demonstrate that, for municipalities too, considering data one way seems to reveal that larger municipalities are more innovative, but considering the same data another way reveals that the smallest are more innovative. Coupled with previous research that reveals the absence of metropolitan effects (Shearmur and Poirier, 2017), it can be asserted that it is not necessarily Quebec's largest municipalities that are the most innovative.

Table 18.2 Prize-Winning Municipalities and Submissions, Mérite Ovation Awards, 2005–2018

Population	Municipalities with at least one submission (2005–2018)	% of Municipalities that have won at least once (2005–2018)	Average number of prizes (2005–2018)	Total number of submissions (2005–2018)	% of winning submissions (2005–2018)
below 5000 pop.	28	35	0.42	53	22.6
5000 to 20,000	65	26	0.32	169	12.4
20,000 to 100,000	40	65	1.33	317	16.7
over 100,000	10	90	3.60	277	13.0

Source: Union des Municipalités du Québec and author.

What Particular Role Do Canadian Cities Play in Innovation?

The widely held belief that innovation in Canada is focused in cities needs to be qualified. From an absolute perspective, there is little doubt that more innovation occurs in Canada's largest cities, whether market or non-market. However, this discussion suggests that Canada's smallest and more remote areas are no less innovative than Canada's largest cities relative to their size. It is therefore misleading to claim that "innovative activities are the products of cities or regions" (Florida et al., 2017). Nonetheless, cities do play a number of key roles for innovation in Canada. This section identifies and discusses five key ways in which Canadian cities are important to innovation.

High-Impact Innovations

Over 50 per cent of the world's population now resides in urban areas, with over a third in cities of over 100,000 people. In Canada, more than 80 per cent of the population is urbanized, and 71 per cent live in cities (census metropolitan areas [CMAs]) of over 100,000 people. Innovations that respond to the needs and tastes of an urban population will more readily find large markets compared to those responding to the needs and tastes of people in smaller or more remote areas. Likewise, innovations that draw upon specific local rural knowledge will not travel as well as innovations that draw upon globalized urban culture.

Thus, notwithstanding the argument made above about innovation being no more prevalent (proportionally) in large cities than in other places, a higher proportion of lucrative innovations, and of innovations that will catch the attention of (urbanized) innovation theorists, gatekeepers, and journalists, are developed in cities. It is innovations in artificial intelligence (AI), video games, and digital technologies that make the news, potentially make money, and may have a profound influence on society. Indeed, Canadian cities regularly appear on global lists of innovative places. For example, the World Economic Forum's list of top 25 high-tech cities identifies Montreal (number 18, wearable technology, industrial design, programming), Vancouver (number 14, "Silicon Valley North"), and Toronto (number 8, information technology).[8]

Innovation in Cities vs. Remote Areas

Researchers are turning their attention to the way innovation processes differ between cities and lower-density areas. McCann (2007) presents a model that suggests that innovators further from metropolitan areas require less frequent interaction with collaborators and information sources. Shearmur and Doloreux (2015) verify this in the Quebec context, and further show that innovators in remote areas rely less on market-sourced information (e.g., clients, suppliers, personnel,[9] consultants) and more on technical information (e.g., research institutes, universities). Earlier research suggested that metropolitan-based innovators would be more experimental and radical, whereas those in smaller towns would be in more settled industries, focusing on incremental and process innovation (Duranton and Puga, 2001)—this has been confirmed (albeit weakly and with certain provisos) in Quebec (Shearmur, 2010, 2011).

Access to Resources for Firm Growth, Marketing, and Diffusion

Even if innovation occurs across the country, innovative firms that wish to grow, and innovators who wish to exploit their innovation, will often not have access to necessary advice, finance, buildings, and personnel in low-density regions (Shearmur, 2015). Thus, while it is possible to introduce an innovation in most contexts, scaling up the innovation, finding new markets and protecting intellectual property all require contacts that can only be made in large metropolitan areas. Furthermore, a fast-growing firm may need to change premises every few months, and may need to recruit large numbers of employees. These activities are most feasible in large cities with fluid labour and real estate markets.

Innovations in Urban Industries

Just as certain innovations (e.g., maple syrup harvesting, vineyard maintenance, soil drainage techniques) are likely to be developed in rural areas because of the know-how embedded in people and communities, other innovations will likely concentrate in cities because that is where the activity and know-how are located, such as high finance and cutting-edge surgery (Parr, 2002; Puga, 2010). Innovations in these industries will tend to occur in urban areas by virtue of the industries' location requirements.

Institutions and Qualified Labour

There is little reason to believe that large institutions require urban locations to function and be innovative. Cornell University is located in upstate New York, Jackson Labs is in a small village in Maine, and the atomic bomb was developed on a ranch in New Mexico. However, large institutions do tend to concentrate in cities in order to be proximate to political power and decision-makers, to have easy access to global partners and markets, and to have access to a large, qualified labour pool. Given this tendency for large institutions to locate in large cities, the innovations they produce are also qualified as "urban," although they do not necessarily rely upon the sort of interaction described by Jacobs (1969) or Porter (2003). These institutions and research labs do not locate in cities because cities are the best place for innovation—rather, cities appear to be innovative (partly) because such innovative institutions locate in them.

What Next?

Canada's cities are located in a small country (by population) at the geographic margins of the global economy. They are themselves peripheral relative to New York, Los Angeles, Beijing, London, and Tokyo. A recent report on Canadian innovation and research and development (R&D) concludes that while Canada undertakes cutting-edge R&D and initiates innovation, it is not good at retaining the intellectual property or reaping the associated beneficial impacts of these activities (Council of Canadian Academies [CCA], 2018). Innovative firms in Canada are often acquired by foreign firms, sell their patents, or move to locations that are more central to their activities, such as Silicon Valley or Hollywood. Canada introduces many innovations, but innovators migrate towards larger global centres to fully exploit their innovation's potential. On a global scale, these conclusions mirror innovation in Canadian cities. Canada's small towns and peripheral areas are just as innovative as its larger cities, but innovators tend to migrate towards its larger cities to fully exploit and develop their innovation's potential.

While the innovations and innovation processes that occur in Canada's larger cities differ from smaller regions, the basic capacity to innovate is similar across the country. What differentiates Canada's cities from its smaller regions is their capacity to provide resources (e.g., labour, finance, marketing know-how, real estate) for innovative firms to develop and grow. Moreover, larger cities are also better able to capture and internalize the benefits of innovation. As the CCA (2018) shows, this success is relative. Although Canada's large cities provide resources and markets that allow local innovations to grow into national ones, for an innovation to reach a global scale, access to global gatekeepers is required, few of whom operate in Canadian cities.

However, innovation is not only about markets. Public sector, social, and cultural innovation also occurs across Canada. While there are many examples of this, few quantitative studies exist because this type of innovation is even more difficult to measure than private sector technological innovation. Shearmur and Poirier's (2017) study, discussed in this chapter, suggests that while larger municipalities innovate more in absolute terms, smaller municipalities innovate just as much (if not more) when adjustments are made for their size. In other words, there is no metropolitan effect. Further research on non-market innovations in Canada and its cities is warranted.

Finally, a note of caution is also required about the assumed desirability of innovation. There are in fact many undesirable innovations, such as thalidomide (an anti-nausea drug, which led to the malformation of many babies), OxyContin (a painkiller that is causing the deaths of thousands of young people across Canada), and the financial innovations that led to the 2008 recession. Furthermore, the more innovative a Canadian city is, the higher its inequality (Breau et al., 2015), a relationship that is not merely correlational but causal as new technologies increasingly displace workers from middle-income jobs, accelerating income polarization as job markets segment into low-wage, often short-term and insecure, service jobs and high-income executive or technology development jobs (Tyson and Spence, 2017; see also Vinodrai, Chapter 17). Current platform-type innovations amplify this effect by instituting a winner-take-all model, whereby innovative first-movers (often tech companies employing very few people) rapidly attain dominant market positions, which they use to stifle competition (see Brail and Donald, Chapter 5). Thus, the optimistic idea that innovation's creative destruction will, in aggregate, generate net benefits may be unfounded. Innovation should be looked at critically, with far more attention to its social, environmental, and cultural externalities (which can be negative) and to its effects over time: these are unaccounted for in current approaches to evaluating innovation, which focus solely on short-term market success for the innovator. Indeed, innovation in the public sector, which can less easily ignore externalities and long-term effects, often proceeds more slowly as a result.

Although this chapter has referred to Canadian examples and data, it has also presented a wider argument: the theories that posit that innovation is quintessentially urban are flawed, not because the urban innovation processes they shed light upon are inaccurately described, but because they assume that the urban processes they theorize are the *only* processes through which innovation can occur. In particular, Jacobs's (1969) contention that innovation can *only* happen in cities—which informs much work on innovation (e.g., Florida et al., 2017; Glaeser, 2011)—is not verified by analyses that are equipped to identify innovation in non-urban places. Firms and other innovators in non-urban regions possess both unique and general knowledge, are capable of networking (albeit with less frequent face-to-face meetings), and are no less clever than their urban counterparts, though they sometimes lack the social and cultural markers that would identify them as "talent" to urban gatekeepers.

In order to understand what is specific about innovation in Canadian cities, it is first necessary to put to rest the idea that innovation *only* happens in cities. This is of particular importance to Canada since its cities are isolated, surrounded by vast tracts of low-density settlement and small towns. Recent work has effectively examined innovation *in* Canadian cities (Wolfe, 2009; Wolfe and Gertler, 2016) and clusters (Spencer et al., 2010). However, as this chapter has argued, the Canadian context also permits an exploration of how **urban innovation** can be distinguished from non-urban innovation.

Review Questions

1. What is innovation? And how is it linked to cities?
2. What aspects of cities contribute to innovation?
3. Is innovation purely an urban phenomenon in Canada? Why or why not?

Notes

1. There is a longstanding literature on steady-state (i.e., no-growth) economic processes and the current environmental crisis suggests this is a necessity (e.g., Victor, 2008): this would require fundamental changes to our current institutions and particular attention to wealth distribution. An alternative view, maybe optimistic or foolhardy, is that technological innovation will solve the environmental crisis (Krupp, 2018): from this perspective, growth-oriented policies and institutions remain unchallenged.
2. The great income disparities in cities stem from the combination of very high earnings derived from industries such as banking and entertainment (which concentrate in cities), low-wage service industries that support these high-earners, and cities' attraction of destitute people in search of opportunity or assistance.
3. Seehttps://pm.gc.ca/eng/news/2016/01/14/prime-minister-announces-support-innovative-water-technologies-southern-ontario
4. See https://www.surrey.ca/city-services/4548.aspx
5. The limitation of these books is that they do not clearly explain what distinguishes innovation in Canadian cities from innovation occurring elsewhere in Canada. They also focus upon *economic* innovation, looking at institutions only insofar as they interact with, and enable, economic innovation.
6. Jacobs recognizes this, but dismisses the possibility that this exchange can be meaningfully bidirectional: for her, it is cities that impel innovation, and rural areas that receive and apply it. She admits, "Many pre-agricultural settlements . . . may have possessed, briefly, a creative economy that flickered for a relative instant in time" (Jacobs, 1969: 42).
7. These are surveys devised and overseen by David Doloreux and myself, conducted at the establishment level.
8. See https://www.weforum.org/agenda/2017/08/these-are-the-25-most-high-tech-cities-in-the-world
9. Personnel obtain their information from social networks, visits to clients, discussions with suppliers, reading, and so forth. Apart from reading, these sources of information are easier to access in cities.

References

Amin, A. and J. Roberts, eds. 2008. *Community, Economic Creativity and Organization*. Oxford: Oxford University Press.

Anderson, E. 1995. *Values in Ethics and Economics*. Cambridge, MA: Harvard University Press.

Baregheh, A., J. Rowley, and S. Sambrook. 2009. "Towards a multidisciplinary definition of innovation," *Management Decision* 47(8): 1323–39.

Bathelt, H., and N. Schuldt. 2008. "Between Luminaires and Meat Grinders: International trade fairs as temporary clusters," *Regional Studies* 42(6): 853–68.

Bock, B. 2012. "Social sustainability and innovation: How to disentangle the buzzword and its application in the field of agriculture and rural development," *Studies in Agricultural Economics* 114: 57–63.

Borins, S. 2014. *The Persistence of Innovation in Government: A Guide for Innovative Public Servants*, Washington, DC: IBM Centre for the Business of Government.

Breau, S., D. Kogler, and K. Bolton. 2015. "On the relationship between innovation and wage inequality: New evidence from Canadian cities," *Economic Geography* 90(4): 351–73.

Brown, J., and P. Duguid. 2000. *Mysteries of the Region: Knowledge Dynamics in Silicon Valley*. At: http://citeseerx.ist.psu.edu/viewdoc/download?doi=10.1.1.500.212&rep=rep1&type=pdf

Crevoisier, O., and R. Camagni. 2001. *Les milieux urbains: Innovatiin, systèmes de production et ancrage*. Neuchâtel: Institut de Recherche Économiques et Rurales.

Castells, M., J.Caraça, and G. Cardoso, eds. 2012. *Aftermath*. Oxford: Oxford University Press.

CBC, 2018, *Move over, Waterloo and Ottawa: Calgary is now outinnovating you*. At: www.cbc.ca/news/canada/calgary/move-over-waterloo-and-ottawa-calgary-is-now-out-innovating-you-1.4676458

Council of Canadian Academies (CCA). 2018. *Competing in a Global Innovation Economy: The Current State of R&D in Canada*. Ottawa: Council of Canadian Academies. At: http://new-report.scienceadvice.ca/

Chen, W-H, J. Myles, and G. Picot. 2012. "Why have poorer neighbourhoods stagnated economically while the richer have flourished? Neighbourhood income inequality in Canadian Cities," *Urban Studies* 49(4): 877–96.

Cooke, P., 2011. "Food geography and the organic empire: Modern quests for cultural-creative related variety," in H. Bathelt, M. Feldman, and D. Kogler, eds., *Beyond Territory: Dynamic Geographies of Knowledge Creation, Diffusion and Innovation*. London: Routledge.

Crane, D. 2016. "Innovate our innovation? Think big for Canada's cities," *Globe and Mail*, May 23.

Dahlander, L., and D. Gann. 2010. "How open is innovation?" *Research Policy* 39: 699–709.

Desjardins, P-M., M. Polese, and R. Shearmur. 2013. *The Evolution of Canada's Regional Economies: Structural Patterns, Emerging Trends and Future Challenges*. Moncton: Canadian Institute for Research on Public Policy. At: https://idjs.ca/index.php/en/publications/all-publications/136-the-evolution-of

Doloreux, D., and E. Lord-Tarte. 2013. "The organisation of innovation in the wine industry: Open innovation, external sources of knowledge and proximity," *European Journal of Innovation Management* 16(2): 171–89.

Doloreux, D., R. Shearmur, and R. Guillaume. 2015. "Collaboration, (transferable and non-transferable) knowledge and innovation: A study of cool climate wine industry (Canada)," *Growth & Change* 46(1): 16–37.

Duranton, G., and D. Puga. 2001. "Nursery cities: Urban diversity, process innovation, and the life-cycle of products," *The American Economic Review* 91(5): 1454–77.

———, and ———. 2004. "Micro-foundations of urban agglomeration economies," in J.V. Henderson and J. F. Thisse, eds., *Handbook of Regional and Urban Economics*, volume 4. Amsterdam: Elsevier.

Eder, J. 2018. "Innovation in the periphery: A critical survey and research agenda," *International Review of Regional Science*. At: https://doi.org/10.1177/0160017618764279

Fischler, R. 2017. "North American zoning: Real-estate regulation—past, present and future," in A. Lehavi, eds., *One Hundred Years of Zoning and the Future of Cities*. Berlin: Springer.

Fitjar, R., and A. Rodriguez-Pose. 2011. "Innovating in the periphery: Firms, values and innovation in southwest Norway," *European Planning Studies* 19(4): 555–74.

Fitjar, R., and A. Rodriguez-Pose. 2017. "Nothing is in the air," *Growth and Change* 48(1): 22–39.

Florida, R., A. Patrick, and C. Mellander. 2017. "The city as innovation machine," *Regional Studies*, 51(1): 86–96.

Frenken, K., F. Van Oort, and T. Verburg. 2007. "Related variety, unrelated variety and regional economic growth," *Regional Studies* 41(5): 685–97.

Gibson, C., S. Luckman, and J. Willoughby-Smith. 2010. "Creativity without borders? Rethinking remoteness and proximity," *Australian Geographer* 41(1): 25–38.

Glaeser, E. 2011. *The Triumph of the City: How Our Greatest Invention Makes us Richer, Smarter, Greener, Healthier and Happier*. New York: Penguin Books.

Godin, B. 2008. *Innovation: History of a Category*, Project on the Intellectual History of Innovation, working paper 1. July 28. At: www.csiic.ca/PDF/IntellectualNo1.pdf

Grabher, G. 2018. "Marginality as strategy: Leveraging peripherality for creativity," *Environment & Planning A*. At: https://doi.org/10.1177/0308518X18784021

Grillitsch, M., and M. Nilsson. 2015. "Innovation in peripheral regions: Do collaborations compensate for a lack of local knowledge spillovers?" *Annals of Regional Science* 54(1): 299–321.

Harris, A., and S. Moore, eds. 2013. "Planning histories and practices of circulating urban knowledge," *International Journal of Urban and Regional Research* 37(5): 1499–583.

High, P. 2017. "Why Montreal has emerged as an artificial intelligence powerhouse," *Forbes*, December 17. At: https://www.forbes.com/sites/peterhigh/2017/11/06/why-montreal-has-emerged-as-an-artificial-intelligence-powerhouse/#29f5fb5923bd

Industry Canada. 2011. *Innovation Canada: A Call to Action. Expert Panel Report*. Ottawa: Industry Canada.

Jacobs, J. 1969. *The Economy of Cities*. New York: Random House.

Jaffe, A., and J. Lerner. 2007. *Innovation and Its Discontents: How Our Broken Patent System is Endangering Innovation and Progress, and What to Do About It*. Princeton: Princeton University Press.

Klein, J-L., J-L.Laville, and F.Moulaert, eds. 2014. *L'Innovation sociale*, Toulouse: Eres.

Krupp, F. 2018. *How technology is leading us to new climate change solutions*, World Economic Forum. At: https://www.weforum.org/agenda/2018/08/how-technology-is-driving-new-environmental-solutions/

Logan, J., and H. Molotch. 1987. *Urban Fortunes*. Los Angeles: University of California Press.

MacAulay, S. 2001. "The community economic development tradition in eastern Nova Scotia, Canada," *Community Development Journal* 36(2): 111–21.

MacLeod, G. 2011. Urban politics reconsidered: Growth machine to post-democratic city? *Urban Studies* 48(12): 2629–60.

MacPherson, A. 2008. "Producer service linkage and industrial innovation: Results of a twelve year tracking study of New York State manufacturers," *Growth and Change* 39(1): 1–23.

Mazucatto, M. 2015. *The Entrepreneurial State*. London: Anthem Press.

McCann, P. 2007. "Sketching out a model of innovation, face-to-face interaction and economic geography," *Spatial Economic Analysis* 2: 117–34.

Mississauga, 2011. *An Action Plan for Innovation in Mississauga*. At: www.mississauga.ca/file/COM/An_Action_Plan_for_Innovation_in_Mississauga_-_Executive_Summary.pdf

Montreal, 2018. *Accélérer Montréal*. At: http://ville.montreal.qc.ca/pls/portal/docs/PAGE/AFFAIRES_FR/MEDIA/DOCUMENTS/ACCELERER_MONTREAL.PDF

Mulgan, G. 2007. *Ready or Not: Taking Innovation in the Public Sector Seriously*. London: Nesta.

Organisation for Economic Co-operation and Development (OECD). 2005. *Oslo Manual: The Measurement of Scientific and Technological Activities*. Paris: Organisation for Economic Co-operation and Development.

Parr, J. 2002. "The location of economic activity: Central place theory and the wider urban system," in P. McCann, ed., *Industrial Location Economics*. Cheltenham: Edward Elgar.

Perdue, P. 1994. "Technological determinism in agrarian societies," in M. Smith and L. Marx, eds., *Does Technology Drive History?* Cambridge: MIT Press.

Perry, J, and A. Hondeghem, eds. 2008. *Motivation in Public Management: the Call of Public Service*, Oxford: Oxford University Press.

Petrov, A. 2011. Beyond spillovers: Interrogating innovation and creativity in the peripheries," in H. Bathelt, M. Feldman, and D. Kogler, eds., *Beyond Territory.* London: Routledge.

Pilon, S., and D-G. Tremblay. 2013. "The geography of clusters: The case of the video games clusters in Montreal and in Los Angeles," *Urban Studies Research.* At: http://dx.doi.org/10.1155/2013/957630

Porter, M. 2003. "The economic performance of regions," *Regional Studies* 37: 549–78.

Proudfoot, S. 2017. "Census 2016: A picture of a bigger, more urban Canada," *Maclean's.* At: https://www.macleans.ca/news/canada/census-2016-a-picture-of-a-bigger-more-urban-canada/ Puga, D. 2010. "The magnitude and causes of agglomeration economies," *Journal of Regional Science* 50(1): 203–19.

Romer. P. 1994. "The origins of endogenous growth," *Journal of Economic Perspectives* 8(1): 3–22.

Sandel, M. 2013. *What Money Can't Buy: The Moral Limits of Markets.* New York: Farrar Straus and Giroux.

Schumpeter, J. 1936. *The Theory of Economic Development.* Cambridge, MA: Harvard University Press (first German edition published in 1911).

Shearmur, R. 2010. "Space, place and innovation: a distance-based approach," *The Canadian Geographer* 54(1): 46–67.

———. 2011. "Innovation, regions and proximity: From neo-regionalism to spatial analysis," *Regional Studies* 45(9): 1225–43.

———. 2013. "Étudier la géographie de l'innovation: Pourquoi, comment et perspectives," in ISQ, ed., *Compendium d'Indicateurs de l'Activité Scientifique et Technologique au Québec.* Québec: Institut de la Statistique du Québec.

———. 2015. "Far from the madding crowd: Slow innovators, information value and the geography of innovation," *Growth & Change* 46(3): 424–42.

———. 2017. "Urban bias in innovation studies," in H. Bathelt, P. Cohendet, S. Henn, and L. Simon, eds., *The Elgar Companion to Innovation and Knowledge Creation.* Cheltenham: Edward Elgar.

———, and J. Poirier. 2017. "Conceptualizing nonmarket municipal entrepreneurship: Everyday municipal innovation and the roles of metropolitan context, internal resources, and learning," *Urban Affairs Review* 53(4): 718–51.

———, and D. Doloreux. 2009. "Place, space and distance: Towards a geography of knowledge intensive business services innovation," *Industry and Innovation* 16(1): 79–10.

———, and D. Doloreux. 2015. "Knowledge-intensive business services (KIBS) use and user innovation: High-order services, geographic hierarchies and internet use in Quebec's manufacturing sector," *Regional Studies* 49(10): 1654–71.

———, and D. Doloreux. 2016. "How open innovation processes vary between urban and remote environments: Slow innovators, market-sourced information and frequency of interaction," *Entrepreneurship & Regional Development* 28(5–6): 337–57.

Solow, R. 1994. "Perspectives on growth theory," *Journal of Economic Perspectives* 8(1): 45–54.

Sorensen, T. 2008. "Creativity in rural development: An Australian response to Florida (or a view from the fringe)," *Firesight and Innovation Policy* 5(1-3): 24–43.

Spencer, G., T. Vinodrai, M. Gertler, and D. Wolfe. 2010. "Do clusters make a difference? Defining and assessing their economic performance," *Regional Studies* 44(6): 697–715.

Smith, M., J. Ur, and G. Feinman. 2014. "Jane Jacobs' 'Cities First' model and archaeological reality," *International Journal of Urban and Regional Research* 38(4): 1525–35.

Stolarick, K., M. Denstedt, B. Donald, and G. Spencer. 2011. "Creativity, tourism and economic development in a rural context: The case of Prince Edward County," *Journal of Rural and Community Development* 5(1-2): 238–54.

Swann, P. 2015. *Common Innovation: How We Create the Wealth of Nations,* Cheltenham: Edward Elgar.

Tyson, L., and M. Spence. 2017. "Exploring the effects of technology on income and wealth inequality," in H. Boushey, J. DeLong, and M. Steinbaum, eds., *After Piketty: The Agenda for Economics and Inequality.* Cambridge: Harvard University Press.

Vancouver. 2018. New Westminster and Vancouver Economic Commission Sign Innovation MOU. Vancouver Economic Commission. At: www.vancouvereconomic.com/blog/media/release-new-westminster-and-vancouver-economic-commission-sign-innovation-mou/

Victor, P. 2008. *Managing Without Growth: Slower by Design Not Disaster.* Cheltenham: Edward Elgar.

Wolfe, D. 2009. *21st Century Cities in Canada: The Geography of Innovation.* Ottawa: Conference Board. At: www.mun.ca/harriscentre/Misc/21stCenturyCitiesinCanada_2009_web.pdf

———, and M. Gertler, eds. 2016. Growing Urban Economies: Innovation, Creativity and Governance in Canadian City-Regions. Toronto: University of Toronto Press.

World Bank. 2010. *Cities and Climate Change: An Urgent Agenda.* December 2010. Washington, DC: World Bank.

19 The Economic Value of Urban Land

Andrejs Skaburskis and Markus Moos

Introduction

You can fly over Canada for hours and see nothing but trees, lakes, and a few roads and in some parts even these disappear. We have lots of land. So why pay a million dollars for a very small lot in Vancouver? What determines the price of land, and how do land prices affect the development of the city? Who sets the price, and what are their interests in shaping the city? What are the social consequences of these processes, and what role does public policy play in shaping urban land markets? These questions are considered in this chapter.

Beginning with an overview of the factors that shape the value of land, we move on to discuss the characteristics of urban land markets in more detail. We distinguish between the rent and the price of land. Next, we highlight the structure of ownership and discuss the role of planning and politics in shaping land markets. We then go on to introduce in more detail some of the models economists use to understand the factors that affect urban land markets, and we discuss demographic, labour market, political, and societal shifts that are changing both real estate costs and land-use patterns. As we shall see, changing land markets have efficiency and equity implications.

The Value of Location

When we think of expensive locations, London, Paris, New York, and Tokyo might come to mind. In Canada, Vancouver and Toronto have become almost synonymous with expensive real estate, and Calgary is not far behind. High-rise condominium apartment towers dominate the Vancouver and Toronto skylines. One-bedroom condo units sell for at least $500,000, often higher. Old and modest single-family dwellings sell for well over $1 million, illustrating the high cost of land (Figure 19.1).

It may be hard to believe these figures, especially when considering that elsewhere in Canada, such as in a Quebec town on the beautiful Gaspé shoreline (Figure 19.2), a house can sell for under $50,000; one could own more than 20 homes for $1 million! But even in Canada's most expensive market, prices do not always go up: recently, newspaper articles have reported modest drops in the average house prices in major cities. Some are worried about a bubble about to burst.

What explains the geographic differences and the changes over time in the price of houses and the land they are built on? The quick answer to most questions about the value of land is the realtor's mantra, "Location, location, location." We need to know more, though, about the aspects of location that matter to people and firms and the forces that determine the attractiveness of a location. The attributes of location that are of most interest to us in this chapter are formed by the relationship a place has with other places within a city and the factors that influence this relationship. These factors vary from the local to

Andrejs Skaburskis

Figure 19.1 A Vancouver million-dollar house in the Dunbar neighbourhood

the global. The value of land and its use are determined by the conditions in the city but also by what is happening in other parts of the world. The 2008 economic crisis, for example, has led to devaluation, thus injecting risk into the valuation of property. Paying a high mortgage on a Vancouver or Toronto home suddenly looks less attractive when housing values begin to decline without an immediate rebound in sight. However, the economic cycles do not necessarily alter the factors that shape *relative* valuation of different locations.

The social and economic relationships that tie cities together into complex urban systems shape the relative value of location. For instance, Montreal's declining status as Canada's dominant economic centre helps explain why its property values increased no faster than the national average, whereas Toronto, Canada's economic powerhouse, had the highest property value appreciation in the years before the 2008 economic recession—along with the emerging regional centres of Vancouver, Calgary, and Edmonton. Historically, Montreal had an advantage in terms of the location of its port, which provided access to the West, but with the development of the railroads and the Panama Canal, western Canadian trade was increasingly

Andrejs Skaburskis

Figure 19.2 $50,000 houses on the Gaspé Peninsula, Quebec

captured by Vancouver (Germain and Rose, 2000). Partly because Toronto is closer to the US industrial belt, manufacturing industries grew more in Toronto than in Montreal after World War II. As well, the migration of the English out of Montreal during the 1970s reduced housing demand and hurt Montreal's property values considerably, thereby moving the gateway to English Canada to Toronto. And in the valuation of land, economic strength gained by an advantageous location matters.

Proximity and Accessibility

Richard Hurd (1903), regarded by some as the founder of urban land economics, tells us that the value characteristics of a location are determined by two sets of attributes: (1) those related to *proximity* and (2) those related to *accessibility*. People may be willing to pay more for locations close to parks, views, and quiet surroundings and less for locations close to municipal landfills or highway noise. Firms may also cluster together (proximity) to share infrastructure, local services, and communication opportunities among decision-makers. However, in many cases access rather than proximity is valued. *Access* refers to the ease of getting from one place to another. We might want to live by a lake but cannot on account of our need to access work, shops, and other facilities. Improvements in transport infrastructure raise the value of land in its vicinity as highways and public transit extensions enhance the accessibility of the locations connected by the system. Between 1986 and 1996, vacant land prices in the vicinity of Vancouver's SkyTrain stations increased by 251 per cent as compared to 133 per cent for prices as a whole (Landcor, 2008). Store owners want to locate near such transport hubs as well as each other to gain exposure to customers who are more likely to come to the shopping centres or main streets that let them reduce their overall travel costs. Polluting factories, on the other hand, are nuisances; most people do not want to live near them, but workers may have to live close by to access their workplace. The value of a location is a function of the advantages it offers in terms of proximity and in terms of its accessibility *relative* to other locations.

As was the case for the historic centre of cities whence urban development spread outward, proximity and accessibility to the downtown have remained most valued, and the highest per unit land prices are in the centre. The scatter plots in Figure 19.3 show how the value of urban land in 2016 in Toronto, Montreal, and Vancouver decreases with distance from the centre. The centre continues to provide primary employment and service functions in Canadian cities to this day, which explains why the price of land continues to be highest near the downtown and falls toward the periphery. There is one important exception: in Toronto, there are three tracts with higher values of residential properties (per area) than those in the downtown. These are located near the intersection of Sheppard Avenue and Yonge Street, a developing employment and amenity node just north of the downtown.

Increases in business-related travel have also made airport locations more important to many firms than downtown locations, and subcentres have developed around the major airports that partially explain the other secondary peaks in land values (Figure 19.3). Land values also increase around "suburban downtowns" and emerging employment nodes in suburban locations, where firms move for lower rents as well as to locate closer to an increasingly suburban labour force as cities continue to spread outward (Filion and Gad, 2006; Moos and Mendez, 2013).

External Factors Affecting the Value of Land

Factors external to the city also affect the value of land. For example, the Vancouver land market heated up in the late 1980s as offshore investors and immigrants bought real estate largely due to worries about the return of Hong Kong to Chinese rule. Moreover, the **globalization** of financial markets has made it easier for foreign investors to buy real estate in distant places. Immigration has dramatically increased in recent years, thereby expanding the demand for land and housing in the large Canadian cities (Ley and Tutchener, 2001; see Harris, Chapter 11). Labour markets are closely related to housing demand, the latter being a function of household formation, income, wealth, and preferences for housing space. Our demand for housing and land is also related to how much income we earn currently (i.e., monetary income) and how much we expect to earn over a longer period of time (i.e., permanent income). And, due to changes in immigration policies that evaluate migrants on a points system, recent immigrants to Canada tend to have higher permanent incomes, which increase their housing consumption (Moos and Skaburskis, 2010). Moreover, if immigrants hold strong preferences for particular locations on account of their ethnic makeup, the growth in demand for these neighbourhoods increases their land values. In this context, therefore, it would be fair to say that the federal government's immigration policy has become an important factor in shaping land markets in our largest cities.

The Influence of Demography, Planning, and Societal Factors

Changes in the population's age composition and in household and family formation rates also affect the demand for housing; smaller households buy smaller houses or, more often, condominiums. Changing lifestyles among the young have increased the relative value of inner-city locations near restaurants, entertainment, and people with potentially similar interests (Skaburskis, 1999; 2012; see Moos, Chapter 6). Similarly, changes in marriage arrangements can affect the relative value of neighbourhoods: the increasing propensity of well-educated women to marry well-educated men since 1980 was observed by Picot, Miles, and Wen-Hao (2011). This propensity resulted in the increased inequality in household incomes and, therefore, in neighbourhood incomes and property values (see Walks, Chapter 10). Some firms have responded to this shift in residential location of young professionals by moving back downtown. And some developers are targeting condominiums to young, single professional women who prefer inner-city locations (Kern, 2010).

Land markets are affected by planning efforts and municipal infrastructure projects that

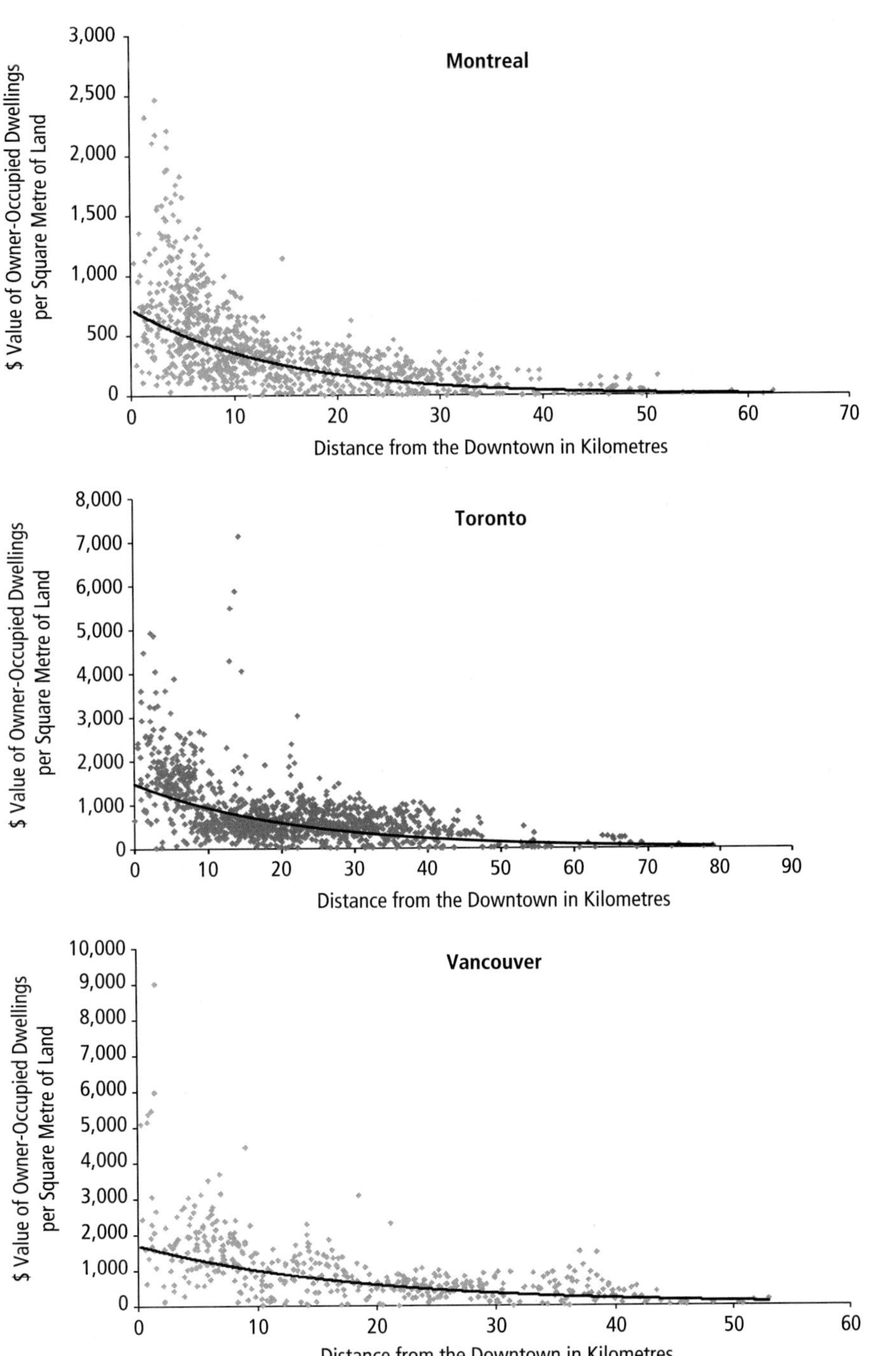

Figure 19.3 Property Value Gradients in Canada's Three Largest Cities

Source: Authors' calculations from 2016 census data with research assistance from Margaret Ellis-Young.

determine where public facilities are built, which waterfronts are beautified, what parts of the city are connected with roads and transit systems, and what parts of the city receive the landfills and other nuisances. Zoning, in particular, is a highly political process that sets out the pattern of land-use activities and builds expectations over the potential for land value appreciation. Decisions about zoning changes are often released only after being finalized to avoid speculative behaviour in land markets created by uncertainty. The role of government is highlighted by Charles Schultze's (1977: 30) bold assertion that the free market is "made by government": land ownership and its transactions are possible only within the protected environment formed by government.

The economic models that explain urban land values show that the land-use patterns that would be formed by a perfectly competitive market place reflect the "highest and best use" of the land. In other words, in the absence of "third-party effects," land markets promote the efficient use of land by ensuring that the potential users of a parcel of land are those who value and can pay the most for its location attributes. Zoning helps reduce or eliminate the negative "third-party effects" one property owner can impose on a neighbour by building too high or too near the edge of the property. Fairness issues are raised when the market brings changes that hurt the more vulnerable populations by forcing them to move or by reducing the supply of lower-priced housing. Nick Blomley (2004) argues that our system of property ownership can seem definitive and even natural whereas, in fact, it is made possible by a regulatory system that favours property owners. Renters often have little to no claim to their space and are potentially harmed by changes in land prices that bring windfalls to the owners. The societal value of land is determined, therefore, not just by the price someone is willing and able to pay but also by the occupants' emotional, cultural, and affective ties to place.

The Characteristics of Urban Land Markets

Markets facilitate the selling and buying of goods and services, and create incentives for firms to provide the goods and services consumers most want, and to do so efficiently. Changes in prices tell producers how demand is changing. The price of a good or service is determined in the market by the interplay of demand and supply. A rise in demand unmatched by an increase in supply raises prices. Land markets are unique in that the supply of land at any location is fixed. This means that any increase in demand for land at a location will increase its price. While the supply of urban land can increase as more rural and agricultural lands are converted to urban uses, this expansion makes the inner-city locations even more desirable *relative* to the retreating peripheral locations; and, due to the fixed supply, inner-city land prices increase with city growth. As land becomes more expensive, developers use less in creating real estate, and city densities increase. The demand for land affects its price and, therefore, how it is used as well as how the city looks and functions. Understanding land markets is a key to understanding urban geography. Knowledge of property values and understanding of the dynamics that drive them are also prerequisites to informed land-use planning.

Rent and Prices

In examining land and property markets, economists distinguish between the value of *using* the property from the value of *owning* the property. In this sense, there are two markets in land: one for its use and another for its ownership. When considering the value of the use of a property, we use the concept of a "rent." The rent is the amount a household or a firm pays or would be willing to pay for the *use* of a property for a period of time. Homeowners are regarded as paying "rent" to themselves for the use of the property they themselves own. When the property is sold, a "price" is paid. The price reflects the value of owning the property and usually is determined by the expected value of the use of the property in the future or, in other words, by the rent it is expected to generate in the future. Changes in the expectation of future uses and rents will change the market price of the land much as change in expectations change the value of investments in the stock market.

The distinction between these two markets is important because the factors determining price can be different from those determining current rents, but the two measures of value are linked. In a world in which the future is known and nothing changes, the highest price someone would offer would be the present value of the expected future rents—determined by the capitalization process. If a property would yield rents starting next year, for all years into the future, the price of the land would be the rent divided by the discount rate:

(1) Price = annual rent/discount rate

If we estimate the discount rate at 6 per cent and the rent on a property is \$1200 per month, then the price for that property as determined by equation (1) would be as follows:

(2) Price = (\$1200 per month × 12 months)/0.06 = \$240,000

When we look at a growing city, we see much higher prices: one cannot buy a condominium unit that would rent for \$1200 a month in any of our major cities for as little as \$240,000. The price is much higher because the rents are expected to increase in the future with continued city growth. The formula needs to capitalize the future increase in rents by using the "net of growth" discount rate; the simplified formula then becomes this:

(3) Price = annual rent/(discount rate – expected growth rate in rents)

Using the above numerical example but adding an expected growth in rents of 2 per cent a year changes the price calculation as follows:

(4) Price = (\$1200 × 12)/(0.06 – 0.02) = \$360,000

The growth rate is not known but is guessed at by the prospective buyers. The difference in the price of the property—the difference between \$240,000 and \$360,000—is due to the *expectation* of a 2 per cent annual growth in rent. Equation 3 can help us illustrate two important features of the land market that affect the way cities change. First, since the growth rate is not known but is guessed at, people's expectations of their city's future affect growth rates and subsequently **urban form**. If people expect the city to keep growing, then this expectation will drive up land prices and create the incentive for developers and builders to use less land in producing real estate. Anticipated future changes in land use affect the market's view of future rents and, therefore, affect its current price. Announcing a transportation improvement, for example, offers to increase the value of the location in the future and therefore the current price. New plans that will change the use of land affect its price at the time they become known. For example, when Ontario announced plans for a greenbelt around Toronto, the price of the farmland in the greenbelt dropped because this growth boundary eliminated the possibility of gaining future urban rents.

The second feature revealed by Equation 3 relates to the stability of the market. The amount people will be willing to pay for land increases as their expectations of future growth rates increase. When they base their expectations on recent past trends, they may be induced to buy larger houses with the hope of increasing the amount they can gain from future price increases. As others follow, prices rise across the city and fuel expectations of an even higher future growth rate and prices increase. As the denominator in Equation 3 decreases in size, the price bubble forms until people recognize that there is no longer a connection between prices and realistic future rents. The bubble bursts and prices tumble. People who had taken mortgages they can no longer afford and households who had refinanced their homes to make other purchases find themselves underwater with a mortgage debt that is greater than the value of their homes. As in the US housing market meltdown that triggered the recent economic crisis, the viability of the banks and other financial institutions holding far too many worthless mortgages is threatened and the crisis broadens to affect the whole economy of the country and the world.

Of course, the 2008 economic crisis was not solely triggered by misguided expectations on the borrowers' part. The downturn has been the worst since the Great Depression of the 1930s and has drawn worldwide attention to the US sub-prime mortgage sector. Lending institutions actively pursued households who would not traditionally qualify for a mortgage in order to make money on the initial sale of the mortgage. Loans were made to working-class and racially marginalized communities, with hidden costs in schemes commonly called "predatory" (Wyly et al., 2009). Most of these loans were grouped together, and shares in these bundles were sold to investors. While housing prices kept climbing, the system remained stable. A household about to default on its mortgage could sell the house at a higher price and cover all costs. However, once prices stabilized and houses were being taken back by the mortgage lenders, people began to realize that mortgages exceeded the value of their properties; as a result, the financial system began to collapse. The increase in foreclosures caused house prices to drop further, making the problem even worse by bringing bankruptcies in other sectors, loss of life savings, and widespread unemployment. Land, property markets, and the institutions that govern them can have very far-reaching consequences.

The Structure of Land Ownership

There are broad consequences to the way society determines who owns land and how to gain access to it. An understanding of the structure of ownership is therefore important for understanding the consequences of how land markets operate. "A man without land is nothing," Duddy Kravitz is told by his grandfather in Mordecai Richler's (1959) novel *The Apprenticeship of Duddy Kravitz* about the young Duddy growing up on Montreal's St. Urbain Street. Duddy becomes obsessed with the idea of attaining property, doing anything regardless of its legality or morality to attain it. Since the mid-twentieth century, the majority of young households in North America have shared his goal. Our lending institutions have engineered financial instruments to permit and encourage households to achieve home ownership. The notion that property ownership comes with elevated rights, privileges, and wealth potential has pervaded scholarly and popular Western thought for many years (Ronald, 2008); indeed, the first concept of "universal" suffrage in Western democratic societies limited voting to male property owners. Today we know that households at risk of homelessness or coping with other forms of poverty are more likely to be renters than owners (Hulchanski and Shapcott, 2004; Skaburskis, 2004). The advocates for a tax on rent, sometimes called a *Ricardian land tax*, argue that the value of a parcel of land at a particular location, and thus its rent, is due to its relationship with other parcels of land and the overall growth and development of the city. Hence, the value is socially constructed, raising questions as to whether the value of increases in land prices should accrue only to the owners of the land. In the late nineteenth century, Henry George was among the most vocal advocates of a land tax. Not only would such a tax be fair but also efficient since owners could do nothing to avoid paying the tax.

Clearly, property rights and the ability to gain profit from one's possessions are essential characteristics of any efficiently operating market. In order to engage in the selling and buying of goods and services, we must know who owns the good or service in question. But if the ability to attain property is unequally distributed and the status of owning property comes with elevated rights and privileges, we can begin to ask questions about the role of public policy in addressing these potential inequalities. Democratic rights are, of course, no longer defined by property ownership as they once were, but observers continue to argue that due to differential tax treatments, to wealth gain potential, and to the social status of owning, renters are at an unjust disadvantage simply because of their tenure.

Conceptualizing the Spatial Structure of Urban Land Markets

The conceptual frameworks we use to understand land markets today have evolved from classic economic theory derived from the thinking of the eighteenth-century economist Adam Smith. Examining the rents charged to tenant farmers, Smith (1970 [1776]) views rent as the residual value: the difference between the cost of growing produce and the price paid for the produce at the market. He observes that, regardless of what was grown, land rents near the market are higher than rents further from the

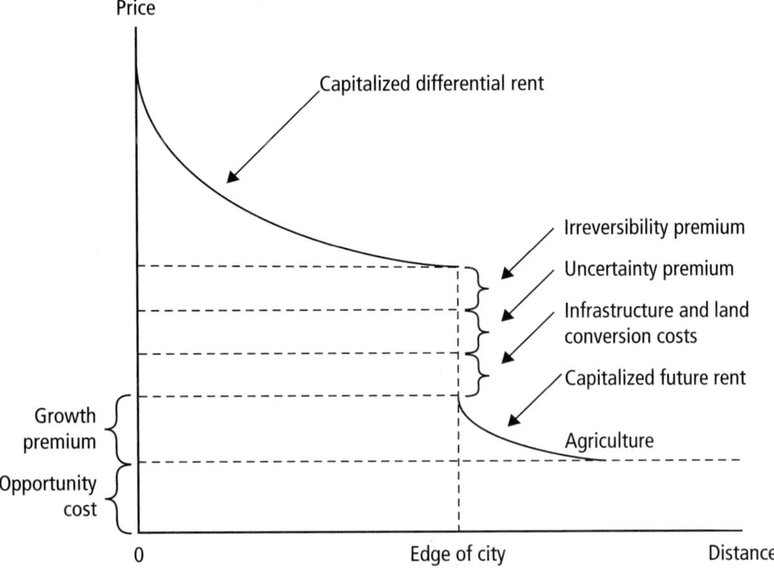

Figure 19.4 Hypothetical Land Price Gradient

Note: The price of an inner-city lot is set by the present value of the agricultural rents at the periphery plus the growth, uncertainty, and irreversibility premiums. It is set by the developer's cost of providing the on-site infrastructure plus the development cost charges for the public facilities and off-site infrastructure. Prices inside the city also are determined by the capitalized differential rents due to the relative increase in the attractiveness of locations closer to the centre.

market on account of the added cost of transportation. David Ricardo, recognized as the founder of rent theory, defines **land rent** as "that portion of the produce of the earth, which is paid to the landlord for the use of the original and indestructible powers of the soil" (Ricardo, 1969 [1817]: 33). The price of produce is set by the cost of growing it on the least productive land. This land yields no residual value for the landowner. If we assume all produce fetches the same price in the market regardless of where it is grown (an assumption that may require revision due to growing interest in locally grown produce), the owners of more fertile land can extract a rent from their tenant farmers equal to the difference between costs and price. "Ricardian rent" levels throughout a city are set by the differences in the relative attractiveness of sites and neighbourhoods. "Fertility" can also be transformed into the "accessibility" attribute—the more accessible sites are similar to the more "fertile" plots in Ricardo's model. The base rent for urban uses is set by the least attractive land that has to be used to house the population. From a commuting point of view, these locations are at the periphery of the city. From here the urban land rent profile starts to rise from the agricultural rent level, increasing toward the centre to reflect the households' valuation of the reduced commute, as illustrated in Figure 19.4.

This very simple model still pertains to the way we view the city today. Putting aside other conditions affecting the commute (an admittedly restrictive assumption) and assuming employment were concentrated in the centre of the city, as it was historically until mid-twentieth century in most Canadian and American cities, then some households would have shorter commutes than others; this difference would affect the amount they would pay for housing (see Harris, Chapter 11). Land prices near the centre are driven up by people who want to reduce their commute costs. The neo-classical models of land use and urban density are built on the classical ideas. William Alonso (1964), for instance, develops the concept of a "bid-rent" map with contours that trace the amount that a household would be willing to pay at each location while keeping its satisfaction constant (Figure 19.5). The **bid-rent curves** depict the

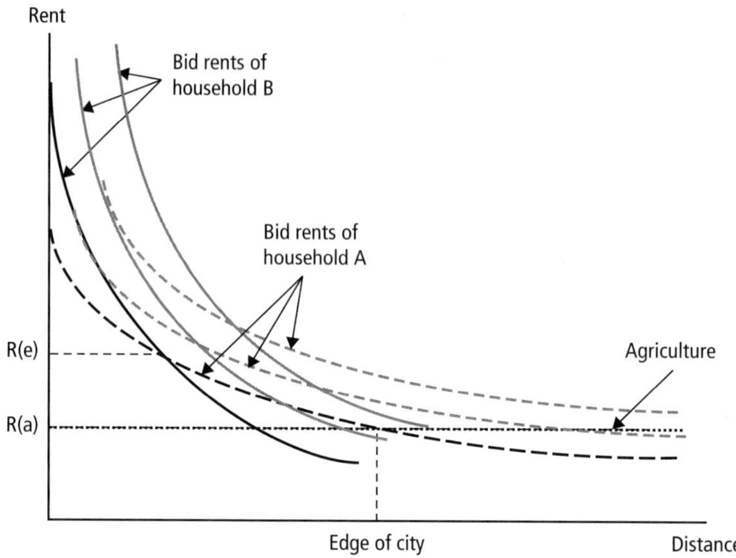

Figure 19.5 Alonso's Bid-Rent Map

Note: The figure assumes there are two households, A and B, each trying to bid for one of the two sites from agriculture for their own use. The solid lines are the bid rents of household B, and the broken lines are those of household A. Higher bid-rent curves, as indicated in grey, give lower welfare to a household because the household pays more rent for the site. Lines in black represent the highest household welfare, but are the minimum rent to landowners. Land rent from agriculture is denoted by the dotted line.

rent/distance trade-off that would make the householders indifferent to their location. Households maximize their well-being by making a trade-off between better access to the city centre and lower-priced land. Since higher-income households tend to want larger lots, they are drawn toward the periphery by the lower land costs. Lower-income households occupy the more expensive inner-city land at higher densities to reduce their commute costs. Since they buy less housing and occupy less land, they would not benefit from the lower price of suburban land as much as higher-income households. Alonso's (1964) model explains the growth of the suburbs and the spread of the urban region as a function of increasing incomes and the preference for large houses. Richard Muth (1969) and Edwin Mills (1969) expand the theory and explicitly introduce housing markets.

The "New" Urban Economics

The "new" urban economics introduces the time dimension to the classical and neo-classical economic models and helps answer such questions as the following: Why do we find parking lots on very expensive downtown land when they clearly cannot generate the land value of a condominium tower? Why is suburban development discontiguous in places and leapfrogging into the countryside to the chagrin of most city planners? Why is the market price of agricultural land at the edge of the city so high compared to the capitalized agricultural rents? We start examining these questions through the capitalization process that links the expected future land rents with land prices and then discuss the costs and benefits of delaying development.

The market price of land at the edge of the city is set by its value in agricultural use and by the present value of the expected future growth in rents at that location. In a growing city, the boundary that sets the base rent for the rest of the city is continuously being pushed further from the centre, making the land within the built-up part of the city *relatively* more attractive. Since buyers recognize that the land at the city's current edge will command higher rents in the future, they are willing to pay a premium now to gain the future increases, as described by Equation 3 earlier in this chapter. Another complication is introduced by the possibility that development options can change over time. If a vacant lot became trapped within an expanding city, its eventual development would be different from the development that would take place when the land is at the very edge of the city. The intensity with which vacant land inside a growing city can be developed increases over time when zoning permits; and this intensity, in turn, increases the residual value of the land. It can, therefore, benefit an owner to hold the land vacant for an extended period of time before converting it to more intense use.

Hotelling's (1931) model builds the basis for understanding the process of land valuation over time. His model starts by recognizing that equilibrium in investment markets requires that assets, such as land, appreciate in value at a rate equal to the relevant interest rate (Shoup, 1970). If the rate at which the asset's price increases is lower than the rate available in other equally risky investments, then its price drops as investors sell to buy more profitable assets. The reduction in the price of an asset increases the *rate* of return created by a given increase in price. Wicksell (1935) explains the profit-maximizing of timed decision-making with the example of a landowner waiting for trees to grow before cutting them down to sell their lumber. Waiting allows the trees to grow larger and yield more lumber when they are cut, but waiting precludes the use of the funds that would be gained by the sale of the lumber. The most profitable time to harvest the trees is when the rate of growth in their value drops to equal the rate of return on the alternative investment, as predicted by Hotelling's model. Thus, in the context of city development the models predict that the owner of a downtown lot may delay its redevelopment until demand for space in an office tower has grown enough to justify the new building. While a smaller condominium on the site might be profitable now, waiting for the site to "ripen" and allow the construction of the office tower might offer even larger profits.

Shoup (1970: 40) specifically applies the profit-maximizing timing model to urban land use and dispels the notion that "development or redevelopment would or should occur as soon as the development value of a site, net of clearance costs, exceeds the value of the existing improved property, as is sometimes stated." The most profitable development time is when the rate of change in the value of the development that can take place on a site is equal to the interest rate on equally risky alternative investments (Wicksell, 1935). Holding land vacant, or using it as a parking lot, is worthwhile when the rate of increase in the present value of the most profitable project that can be placed on the land exceeds the rate of return on alternative investments. The change in the profitability of development is due to the change in what would be built on the land and due to the fact that development freezes that use for a long time. If the development potential is not changing over time, then a site will be developed now or never.

Arnott and Lewis (1979) show that waiting for development changes the density of development. As owners hold vacant land and the suburbs grow around them, land prices increase and less is used in developing real estate: higher land prices encourage higher-density development. Keeping some land vacant can help curb "sprawl" in the long run by leaving room for a new wave of much higher-density development (Peiser, 1989). Had room for infill development not been left, the new development would be at the periphery of the urban region and the total time and effort spent on commuting to the centre would be higher.

The development of land fixes its use for a long period and, thereby, prevents the property owner from taking advantage of possible future changes that could increase his or her options. The presence of uncertainty increases the value of having options, and owners will delay development until prices rise to cover their perceived cost of the uncertainty. Owners have many different reasons for holding land, as pointed out by Brown, Phillips, and Roberts (1981). Some are keeping the land for their children and have no interest in making a profit. Some may place intrinsic value on the ownership itself regardless of the returns they can get for its sale or development. Others may have different discount rates or expectations of growth rates that affect their timing of the sale or of development. The variety of reasons for owning land implies that owners will sell or develop their land at different times. The result is the discontiguous pattern of suburban development that most city planners deplore. Keeping unused land within the city's perimeter increases infrastructure costs, but it also leaves room for more intense development at a later time.

Since the development decision is irreversible, owners will wait until prices rise to cover their perceived cost of the irreversibility of the development decision (Capozza and Helsley, 1989); the resulting land price gradient is illustrated in Figure 19.4. The base price is set by the residual value of the land just outside the periphery. The growth, uncertainty, and irreversibility premiums are added along with the cost of infrastructure to determine the market price of buildable lots at the periphery. The capitalized differential rent is added only inside the city. Constraints on development or increases in development costs at the edge of the city increase the price of all land inside the city. Zoning bylaws and differences in neighbourhood attributes create discontinuities in the rent and price profiles, resulting in a gradient that is not as smooth as the one depicted in Figure 19.4.

The Changing Profile of Cities

Canadian metropolitan areas, historically characterized by monocentric development patterns and sharply falling density gradients, have dramatically changed over the past 30 years (see Grant and Filion, Chapter 12). Not only has their reach expanded, but an amalgam of demographic, labour market, and socio-economic shifts has resulted in what many believe to be fundamental changes in the spatial and social structure of Canadian cities (see Townshend and Walker, Chapter 7; Walks, Chapter 10; and Vinodrai, Chapter 17). The return of higher-income households to the inner city is perhaps the most important change in the structure of cities in the past half-century (see Bain and Mark, Chapter 15). The changes are spectacularly visible in Vancouver and Toronto, where high-rise condominiums have taken the place of industrial lands and rail yards (Figure 19.6). Redevelopment of the inner city, through a process called **gentrification,** is attributed to the growth of smaller, non-family households, the emergence of a "new middle class" of highly educated, quaternary-sector workers—employed in professional service sector occupations—who reject the suburban lifestyle, and **deindustrialization** that left many inner cities with undervalued properties (Ley, 1996). The incoming population tends to have specific preferences and values that translate into political lobbying for infrastructure and cultural facilities, which, when provided, further increase the value of inner-city land (as shown in Figure 19.7). In gentrification, the lower-income populations are displaced

Andrejs Skaburskis

Figure 19.6 Vancouver's redeveloped inner city

Andrejs Skaburskis

Figure 19.7 Use of nudity by the "new middle class" to protest car culture. The ideologies of politically astute residents can shape transport and land-use policies with implications for property markets

by the incoming, wealthier households. The lobbying and policies that facilitate inner-city investment, and support its upper- and middle-class residents, have even been called "revanchist" for what some describe as blatant attempts to exclude vulnerable populations (Smith, 1996).

To be sure, most population growth is still in the suburbs. Ever since the introduction of streetcars in the early 1900s, transportation improvements have continued to permit suburban development (see Addie, Fiedler, and Keil, Chapter 16). The rate of expansion increased dramatically in the 1950s with growth in prosperity, the development of mortgage financial institutions and instruments, and the construction of intra-urban highways. The neo-classical economic models, and their focus on the trade-offs between commuting and housing costs, help explain the outward movement of the well-off and the filtering-down of inner-city housing from higher- to lower-income households. The increasing income levels allow people to buy larger houses on larger lots, making them more sensitive to the per-square-metre price of lots and more willing to accept the longer commutes from the suburbs. Technological change and rising incomes increased automobile ownership, making jobs accessible from distant locations. One of the most important changes in the last half of the twentieth century was women's changing employment prospects, which in part contributed to **suburbanization**. Not only did this increase the income of households with two income earners, but it also increased their commuting costs. Some companies moved their clerical offices out into the suburbs to be closer to their predominantly female workforce, helping to intensify the emerging centres that would define the new polycentric urban region (see Vinodrai, Chapter 17).

In turn, the spread of employment across the region increased the spread of the city. Surprising to many observers, however, was the resurgence of interest in the inner city by the well-off, leading to "the return of the café society" as some have called this trend. The filtering process in housing used to pass housing down to lower-income people as it aged and as the higher-income households moved out to newer housing of higher quality. This process has reversed in most Canadian metropolitan areas since the early 1980s (see Harris, Chapter 11). Without exception, in all Canadian metropolitan areas—small and large, growing or staying still—the average rents in the older stock, which is located in the inner city, are higher than in the newer buildings (Skaburskis, 2006). This reversal reflects a revaluation of inner-city locations by higher-income households. This revaluation can be explained by at least two viewpoints, discussed below: the neo-classical and the structuralist. Many believe, too, that the pursuit of more entrepreneurial urban policies is also partly responsible for the changes (see Taylor and Bradford, Chapter 3).

Neo-Classical Viewpoint

Wheaton (1977) shows that the perceived cost of commute time has been increasing with household income, enough to change the way the commute-cost/housing-price trade-off is made and to make some higher-income people want to move back to the inner city. Demographic changes also contribute to the displacement of lower-income households by middle- and higher-income earners when gentrification occurs. Average household sizes are declining in North America and Europe, and smaller households occupy less land and can benefit more by paying a higher per-square-foot price for land to reduce their commute costs. Employment insecurity makes central locations more attractive by reducing commute costs over the long run as job locations change. The increase in the proportion of one- and two-person households helps account for the increase in the demand for central locations. Two-worker households are more likely to locate near the urban centre to minimize joint commute costs when they work in different parts of the city. An inner-city location helps solve the co-location problem of well-educated, dual-career couples and helps "preserve the marriages of dual career households" (Costa and Kahn, 2000: 1289). Women professionals are more likely to find work in the downtown (Rose and Villeneuve, 1998).

The higher density of the inner city increases the options that single people have in the marriage market. The delay in family formation since the 1950s has reduced young people's need for space and increased the time and money spent on leisure activities. Increases in the education of young people translate into a greater demand for the services and amenities generally found in the downtown, such as theatres, museums, cafés, and sushi bars. By observing that rents in cities increased faster than wages, Glaeser, Kolko, and Saiz (2001) conclude that demand for urban amenities increased faster than urban productivity. They suggest that people live in downtowns increasingly to consume, not to work. Increasing traffic congestion and gasoline costs also make inner-city locations more attractive.

Structuralist Argument

A number of geographers see the changes in inner-city land markets as linked to shifts in the social structure brought about by global economic restructuring (Badcock, 1992; Maher, 1994). The restructuring of the global economy—often labelled as a transition from **Fordism** to **post-Fordism**—has resulted in the growth of both high- and low-level service sector occupations. This two-tiered employment structure and a declining **welfare state** under **neo-liberalism** have produced an hour glass–shaped income distribution and an ever-growing income gap between high- and low-income earners, referred to as "social polarization" (Castells, 2002; Esping-Anderson, Assimakopoulou, and van Kersbergen, 1993; Hamnett and Cross, 1998; Sassen, 1990, 1991). It is this growing inequality that spurred in part the recent Occupy Wall Street movement. **Polarization** materializes in changes in the social geography of the city through segregation and neighbourhood inequalities (Marcuse and van Kempen, 2000; Walks, 2001). Those at the top end of the income structure can out-compete others in the housing market, resulting in ever-escalating markets that fewer people can attain. The trends can help to explain the declining willingness of private developers to construct rental buildings that tend to be occupied by lower-income populations. Some observers have called this the "dual" or the aforementioned "revanchist" city (Mollenkopf and Castells, 1992; Smith, 1996). Some explanations of inner-city reinvestment draw on a Marxist perspective of investment cycles (Harvey, 1985; Smith, 1986). During Fordism, capital left downtown in favour of suburban expansion, contributing to inner-city decline. In the post-Fordist context, **revitalization/regeneration** and redevelopment through renovations and new investments returned value to inner-city land.

Left with empty manufacturing buildings and polluted industrial sites, localities are devising new policies to attract and expand their workforce in the growing sectors, such as information and communications technology and cultural industries (Figure 19.8). These policies focus on securing private and public investment in urban amenities, such as beautified streetscapes and waterfront revitalization, and holding festivals and farmers' markets. It is believed that these kinds of amenities and events will be attractive to the so-called "knowledge workers" or, to borrow Richard Florida's (2002) now much-used phrase, the **creative class** of the **new economy** (Hall, 2006). A larger share of public resources is directed toward urban amenities than in the past through **urban renewal** projects and **place-making** and "place-marketing" strategies (Hackworth, 2007; Kipfer and Keil, 2002; see Bain and Mark, Chapter 15).

Markus Moos

Figure 19.8 Construction of cultural amenities in place of manufacturing in Montreal

Urban Entrepreneurial Policies

The focus on urban amenity provision as a public policy has been labelled as "entrepreneurial" and differs in important ways from the "managerial" policies of the past, when a greater share of resources was dedicated to public services, such as housing (Harvey, 1989). The entrepreneurial policies change the amenity attributes and attractiveness of specific locations, shaping land markets differently than did the managerial policies that preceded them. Instead of investing in social infrastructure that promoted the development of stable neighbourhoods, the new policies harness the growth-generating potential of real estate markets by providing the type of infrastructure that enhances the value of land and location. Existing owners, investors, and the local tax base benefit from appreciating markets, while those wanting to enter the market or needing to pay escalating rents find it increasingly difficult to do so. Arguably, the new entrepreneurial policies cater to the consumption preferences of the workers in professional and managerial occupations, preferences not necessarily shared by those at the bottom of the income spectrum. The issue gets at fundamental ideological foundations that separate entrepreneurial from managerial urban policies and at questions of whether government should help facilitate the operations of the market or whether it should help smooth its inequities.

Conclusion

In a market framework, land values necessarily differ by location, and the land's uses and value are determined by those who have the means to pay for it. Valuing location in that manner may be efficient, but not necessarily equitable. Governments in Canada have traditionally intervened in land markets for reasons of equity and social justice when the inability to afford land and housing leads to a seriously disadvantaged social position. Whether recent increases in affordability problems and homelessness warrant renewed government programs is a question outside the scope of this chapter (see Walks, Chapter 10; Harris, Chapter 11), but the economics of property markets clearly have broad consequences that warrant discussion about the relative role that markets and public policies ought to play in shaping our cities.

Review Questions

1. What factors—both internal and external to the city—determine the relative value of one location versus other possible locations for a home or business?
2. Why do we find expensive downtown land being held vacant or used for parking lots instead of being developed immediately at its highest and best use?

References

Alonso, W. 1964. *Location and Land Use: Toward a General Theory of Land Rent*. Cambridge, MA: Harvard University Press.

Arnott, R.J., and F. Lewis. 1979. "The transition of land to urban use," *Journal of Political Economy* 87, 4: 161–9.

Badcock, B.A. 1992. "Adelaide's heart transplant, 1970–88: 1. Creation, transfer, and capture of 'value' within the built environment," *Environment and Planning A* 24: 215–41.

Blomley, N. 2004. *Unsettling the City: Urban Land and the Politics of Property*. New York: Routledge.

Brown, J.H., R.S. Phillips, and N. Roberts. 1981. "Land markets at the urban fringe: New insights for policy makers," *Journal of the American Planning Association* 47: 131–44.

Capozza, D., and R. Helsley. 1989. "The fundamentals of land prices and urban growth," *Journal of Urban Economics* 26: 295–306.

Castells, M. 2002. *The Information Age: Economy, Society and Culture, Volume 1: The Rise of the Network Society*. Malden, MA: Blackwell.

Costa, D., and M. Kahn. 2000. "Power couples: Changes in the locational choice of the college educated, 1940–1990," *Quarterly Journal of Economics* 115: 1287–1315.

Esping-Anderson, G., Z. Assimakopoulou, and K. van Kersbergen. 1993. "Trends in contemporary class structuration: A six-nation comparison," in G. Esping-Anderson, ed., *Changing Classes: Stratification and Mobility in Post-Industrial Societies*. London: Sage.

Filion, P., and G. Gad. 2006. "Urban and suburban downtowns: Trajectories of growth and decline," in T. Bunting and P. Filion, eds., *Canadian Cities in Transition: Local Through Global Perspectives*. Toronto: Oxford University Press.

Florida, R. 2002. *The Rise of the Creative Class*. New York: Basic Books.

Germain, A., and D. Rose. 2000. *Montreal: The Quest for a Metropolis*. New York: John Wiley & Sons Inc.

Glaeser, E., J. Kolko, and A. Saiz. 2001. "Consumer city," *Journal of Economic Geography* 1: 27–50.

Hackworth, J. 2007. *The Neoliberal City: Governance, Ideology, and Development in American Urbanism*. Ithaca, NY: Cornell University Press.

Hall, C.M. 2006. "Urban entrepreneurship, corporate interests and sports mega-events: The thin policies of competitiveness within the hard outcomes of neoliberalism," *Sociological Review* 54: 59–70.

Hamnett, C., and D. Cross. 1998. "Social polarisation and inequality in London: The earnings evidence 1979–1995," *Environment and Planning C* 16: 659–80.

Harvey, D. 1985. *The Urbanization of Capital*. Baltimore: Johns Hopkins University Press.

———. 1989. "From managerialism to entrepreneurialism: The transformation in urban governance in late capitalism," *Geografiska Annaler* 71B: 3–17.

Hotelling, H. 1931. "The economics of exhaustible resources," *Journal of Political Economy* 39: 137–75.

Hulchanski, D., and M. Shapcott, eds. 2004. *Finding Room: Policy Options for a Canadian Rental Housing Strategy.* Toronto: Centre for Urban and Community Studies Press.

Hurd, R. 1903. *Principles of City Land Values.* New York: Record and Guide.

Kern, L. 2010. "Gendering re-urbanisation: Women and new-build gentrification in Toronto," *Population, Space and Place* 16: 363–79.

Kipfer, S., and R. Keil. 2002. "Toronto Inc? Planning the competitive city in the new Toronto," *Antipode* 34: 227–64.

Landcor. 2008. *Lessons from Expo 86 for the 2010 Winter Games.* The Landcor Report, Special Edition, 19 Aug. Vancouver: Landcor Data Corporation.

Ley, D. 1996. *The New Middle Class and the Remaking of the Central City.* Oxford: Oxford University Press.

———, and J. Tutchener. 2001. "Immigration, globalization and house prices in Canada's gateway cities," *Housing Studies* 16: 199–223.

Maher, C. 1994. "Housing prices and geographical scale: Australian cities in the 1980s," *Urban Studies* 31: 5–27.

Marcuse, P., and R. van Kempen, eds. 2000. *Globalizing Cities: A New Spatial Order?* Oxford: Blackwell.

Mills, E. 1969. "The value of urban land," in H. Perloff, ed., *The Quality of Urban Environment.* Baltimore, MD: Resources for the Future, John Hopkins University Press.

Mollenkopf, J.H., and M. Castells, eds. 1991. *Dual City: Restructuring New York.* New York: Russell Sage Foundation.

Moos, M., and P. Mendez. 2019. "Suburbanization and the remaking of metropolitan Canada," in R. Keil, ed., *Suburban Constellations: Governance, Land and Infrastructure in the 21st Century.* Berlin: Jovis.

———, and A. Skaburskis. 2010. "The globalization of urban housing markets: Immigration and changing housing demand in Vancouver," *Urban Geography* 31: 724–49.

Muth, R. 1969. *Cities and Housing.* Chicago, IL: University of Chicago Press.

Peiser, R.B. 1989. "Density and urban sprawl," *Land Economics* 65: 194–204.

Picot, G., J. Myles, and C. Wen-Hao. 2011. "Why have poorer neighbourhoods stagnated economically while the richer have flourished? Neighbourhood income inequality in Canadian cities," *Urban Studies* 49: 877–96.

Ricardo, D. 1969 [1817]. *The Principles of Political Economy and Taxation.* London: Everyman's Library, J.M. Dent and Sons.

Richler, M. 1959. *The Apprenticeship of Duddy Kravitz.* Don Mills, ON: A. Deutsch.

Ronald, R. 2008. *The Ideology of Home Ownership: Homeowner Societies and the Role of Housing.* New York: Palgrave Macmillan.

Rose, D., and P. Villeneuve. 1998. "Engendering class in the metropolitan city: Occupational pairings and income disparities among two-earner couples," *Urban Geography* 19: 123–59.

Sassen, S. 1990. "Economic restructuring and the American city," *Annual Review of Sociology* 16: 465–90.

———. 1991. *The Global City: New York, London, Tokyo.* Princeton, NJ: Princeton University Press.

Schultze, C. 1977. *The Public Use of Private Interest.* Washington, DC: Brookings Institution.

Shoup, D. 1970. "The optimal timing of urban land development," *Papers of the Regional Science Association* 75: 33–44.

Skaburskis, A. 1999. "Modelling the Choice of Tenure and Building Type," *Urban Studies* 36: 2199–2215.

———. 2004. "Decomposing Canada's growing housing affordability problem: Do city differences matter?" *Urban Studies* 41: 117–49.

———. 2006. "Filtering, city change and the supply of low-priced housing in Canada," *Urban Studies* 43: 533–58.

———. 2012. "Gentrification and Toronto's changing household characteristics and income distribution," *Journal of Planning Education and Research* 32: 191–203.

Smith, A. 1970 [1776]. *The Wealth of Nations.* London: Everyman's Library, J.M. Dent and Sons.

Smith, N. 1986. "Gentrification, the frontier, and the restructuring of urban space," in N. Smith and P. Williams, eds., *Gentrification of the City.* Boston, MA: Allen and Unwin.

———. 1996. *The New Urban Frontier: Gentrification and the Revanchist City.* New York: Routledge.

Walks, A. 2001. "The social ecology of the post-Fordist/global city? Economic restructuring and socio-spatial polarisation in the Toronto urban region," *Urban Studies* 38: 407–47.

Wheaton, W.C. 1977. "Income and urban residence: An analysis of consumer demand for location," *American Economic Review* 67: 620–31.

Wicksell, K. 1935. *Lectures on Political Economy.* London: Routledge.

Wyly, E., M. Moos, D. Hammel, and E. Kabahizi. 2009. "Cartographies of race and class: Mapping the class-monopoly rents of American sub-prime mortgage capital," *International Journal of Urban and Regional Research* 33: 332–54.

20 Slow Growth and Decline in Canadian Cities

Heather M. Hall and Betsy Donald

Introduction

One notable finding from the 2016 Canadian census is that over 57 per cent of the Canadian urban system is either growing slowly (40.1 per cent) or declining (17.1 per cent). Geographers, planners, and policy-makers have been involved in planning for decline *within* the city for several decades. This is not new. Research on intra-urban declivity has examined downtown and inner-city deterioration along with derelict **brownfields sites**, waterfronts, and industrial lands. What is new, however, is a heightened awareness, and growing reality, of rising disparities *between* cities—those that are rapidly growing and those that are not. This uneven inter-urban geography is confronting all developed countries to a greater or lesser degree; it is increasingly attracting attention across Canada (Bourne and Simmons, 2003; Hall, 2009; Hall and Hall, 2008), Germany (Oswalt, 2005; Lötscher, 2005), the United States (Popper and Popper, 2002), and the United Kingdom (Cocks and Couch, 2012). This literature shares a common view that if current demographic, economic, and policy trends persist, then the gap between cities that are rapidly growing and those that are not will become increasingly accentuated. The uneven inter-urban geography among Canadian cities will challenge the way we think about urban development issues because the dominant discourse in Canada is about planning for growth, not decline. How we manage decline in many of our Canadian cities will become one of the most pressing urban policy issues of the twenty-first century.

This chapter is organized into four sections. The first section provides a typology of urban decline and includes a description of the broader demographic, economic, and policy trends producing this new and accentuated pattern of uneven growth. In the next section, we focus on the negative perceptions and "psychology of failure" that often accompany academic or political labelling of a "slow-growth" or "declining city." A discussion of the challenges and opportunities associated with different growth trajectories constitutes the third section. In the fourth section, we focus on the geographic, planning, and policy implications for Canadian cities experiencing slow growth and decline. We conclude with a discussion of further challenges confronting urban areas in Canada.

Terminology, Trends, and Discourse

What Are "Slow-Growth" and "Declining" Cities?

In the Canadian urban geography literature, cities experiencing slower or declining rates of growth are referred to as "slow-growth" or "declining" (Bourne and Simmons, 2003; Bunting and Filion, 2001; Hall, 2009; Hall and Hall, 2008; Leo and

Anderson, 2006; Simmons, 2003). In Europe and the United States, the term "shrinking cities" is commonly used, along with "stagnating cities" or "urban areas in difficulty" (Atkinson, 2001; Bontje, 2004; ECOTEC Research and Consulting, 2007; Lötscher, 2005; Rybczynski and Linneman, 1999). While these different terms incorporate a number of trends, the common measurement for slow growth and decline is population change over time. No concise method exists for measuring this change, however, which produces a variety of results. For example, American policy analyst Anthony Downs (1994) classifies cities based on population change in a 10-year period as rapidly declining if they lost more than 4.9 per cent of their population, slowly declining if they lost between 0.1 and 4.9 per cent, slowly growing if they gained less than 10 per cent, and rapidly growing if they gained more than 10 per cent. In Canada, five-year periods are frequently used, although this method only considers short-term trends and may not provide an accurate depiction of long-term trajectories of change (Bourne and Simmons, 2003; Bunting and Filion, 2001; Hall and Hall, 2008; Leo and Anderson, 2006; Simmons, 2003).

Where Are Slow-Growth and Declining Cities?

Canada is not alone in having declining cities. Similar patterns of slow growth and decline are emerging across Europe, Australia, South America, and the United States. Prior to the 2007 American sub-prime mortgage crisis and the 2008 global economic slowdown, Oswalt and Rienitz (2006) estimated that 25 per cent of world cities with populations over 100,000 were in decline (see also Hollander and Németh, 2011). This trend has intensified since the global economic slowdown. In the United States, for example, land abandonment has once again become a major issue in cities like Detroit, Michigan, and Gary, Indiana. In the most severe cases, residential and commercial property owners abandon their property, leading to drastic decreases in land prices, home values, and municipal revenue (Hackworth, 2012).

Meanwhile, in Canada, 87 of 152 census agglomerations (CAS) and **census metropolitan areas (CMAS)** experienced slow growth or decline between 2006 and 2016 (Table 20.1). As seen in Figure 20.1, this uneven pattern of growth in Canada is more pronounced in small and mid-size urban areas. Campellton, NB, for example, had the largest decline, at -9.3 per cent, followed by Elliot Lake, ON (-7.0 per cent), and Cape Breton, NS (-6.8 per cent). Meanwhile, most large city-regions, such as Toronto, Calgary, and Vancouver (with the exception of Montreal and Hamilton), as well as some small and mid-size cities in oil- and mineral-rich regions, grew rapidly.

In most developed countries, this uneven inter-urban geography is largely a result of several powerful demographic, economic, and policy trends (Bourne and Simmons, 2003; Hall and Hall, 2008; Hollander and Németh, 2011; Hospers, 2014; Lötscher, 2005; Martinez-Fernandez 2012a, 2012b; Müller, 2004; Hart, 2018). In Canada, we are in the final stages of the demographic transition in which lower fertility rates are resulting in the decreased role of natural increase for population growth (see Townshend and Walker, Chapter 7). As a result, immigration is the major source of demographic change; it is well documented that the majority of domestic and international migrants locate in large city-regions like Toronto and Vancouver, further accentuating the uneven population distribution between cities (Bourne and Rose, 2001; Bourne and Simmons, 2003; see Kobayashi and Preston, Chapter 8). In some cities, especially on the Prairies, another major source of demographic change is the rising Indigenous population (see Fawcett and Walker, Chapter 4). These demographic trends are expected to continue.

Economic restructuring and technological changes in Canada have also led to fewer jobs in the resource and manufacturing sectors, while job growth related to the knowledge economy has concentrated primarily in large city-regions as well as areas within their zone of influence (Barnes et al., 2000; Bourne and Simmons, 2003; Gertler, 2001; Norcliffe, 1994; see Vinodrai, Chapter 17). In their work on Atlantic Canada and Quebec, Polèse and Shearmur (2002) argue that having a resource-dependent economic base is another precondition

Table 20.1 Trends in the Canadian Urban System, 2006–2016

	Large urban areas (>500,000)	Mid-size urban areas (50,000–500,000)	Small urban areas (10,000–50,000)	All urban areas
Number of urban areas	10	51	91	152
Total pop. 2006	16,808,861	6.956,910	2,086,263	27,059,824*
Total pop. 2016	19,278,422	7,616,500	2,238,005	30,619,033*
Overall % pop. change, 2006–2016	15.3	9.5	7.3	13.2*
Number of slowly growing areas**	2	18	41	61
% slowly growing	20.0	35.3	45.1	40.1
Number of declining urban areas**	0	7	19	26
% declining	0	13.7	20.9	17.1
Lowest % pop. change	Hamilton (7.9)	Cape Breton, NS (–6.8)	Campbellton, NB (–9.3)	Campbellton, NB (–9.3)
Highest % pop. change	Calgary (29.0)	Wood Buffalo Alta (39.3)	Okotoks, Alta (68.4)	Okotoks, Alta (68.4)

*The Canadian population grew by 11.2 per cent between 2006 and 2016, from 31,612,897 to 35,151,728

**Slowly growing cities are defined as cities that grew by less than 10 per cent while declining cities are those that lost population between 2006 and 2016.

Source: Statistics Canada.

for decline. Resource dependency is an issue because the economies of many smaller peripheral cities often lack the economic diversification to counteract the volatile boom-bust cycle of a natural resources–based economy. More importantly, these cities have little local control over resource decision-making, often receive little direct revenue or royalties from these resources, and often generate few gains from value-added development (e.g., through manufacturing products from raw materials).

Other conditions contributing to decline relate to changing state policies and geography. In Canada, shifting national capital flows to the United States along with the changing role of government through trade liberalization and increased provincial responsibilities have contributed to what Bourne and Simmons (2003) refer to as "new fault lines" between cities that are rapidly growing and those that are not. Added to this is the shift away from state redistribution policies aimed at reducing spatial inequalities and toward policies that stress competitiveness, innovation, and development from within urban areas (see Vinodrai, Chapter 17; Shearmur, Chapter 18). In Canada, a peripheral location in the national or continental space economy or a location away from a major transportation axis or trade route are also cited as preconditions for decline (Polèse and Shearmur, 2006: 42). Thus, all of these macro social, economic, and political factors have accumulated to produce a new pattern of uneven growth, which has further accentuated existing patterns between a small number of cities and regions that are rapidly growing and a larger number of those that are not (Table 20.2).

Who Cares about "Declining" Cities?

The literature on slow growth and decline is growing in Europe, Australia, and the United States, especially after the Great Recession, which led to foreclosures and abandoned properties. Articles on decline have appeared in the *Journal of the American Planning Association*, the *International*

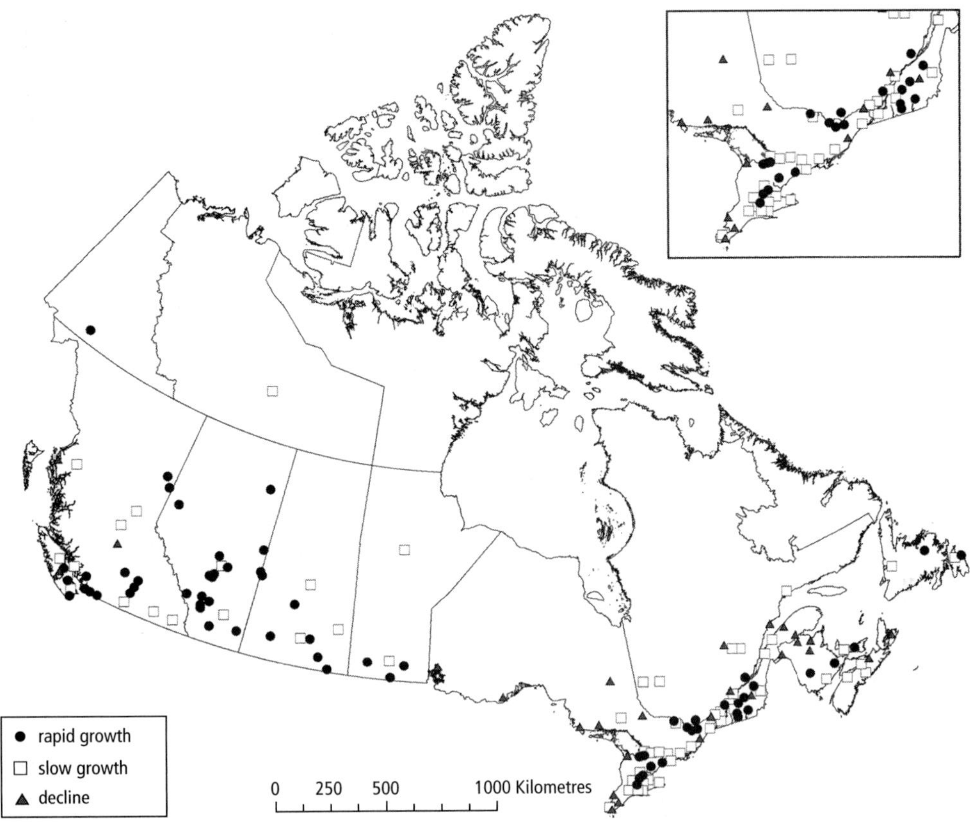

Figure 20.1 Pattern of Growth and Decline in the Canadian Urban System, 2001–11

Journal of Urban and Regional Research, *International Planning Studies*, *European Planning Studies* and *Plan Canada*. A number of research networks and projects have also emerged, including the German government-funded Shrinking Cities research project (www.shrinkingcities.com) and Shrink Smart—a European Union Framework VII-funded research project (www.shrinksmart.eu) (Cocks and Couch, 2012; Hollander and Németh, 2011; Hospers, 2014).

Much of the early writing on decline was focused on describing the causes, characteristics, and consequences of decline or slow growth (Bourne and Rose, 2001; Bourne and Simmons, 2003; Bunting and Filion, 2001; Leo and Anderson, 2006; Leo and Brown, 2000; Lötscher, 2005; Lötscher, Howest, and Basten, 2004; Martinez-Fernandez et al., 2012a; Müller, 2004; Müller and Siedentop, 2004; Oswalt, 2005; Polèse and Shearmur, 2006; Reckien and Martinez-Fernandez, 2011; Simmons, 2003). These scholars initiated the discussion by providing detailed national accounts of the causes and by depicting the pattern of growth and decline. More recent research has examined the social, fiscal, economic, spatial, planning, and policy challenges and responses associated with shrinkage or decline (Bontje, 2004; Cocks and Couch, 2012; Franz, 2004; Hackworth, 2012; Hall, 2009; Hollander and Németh, 2011; Hospers, 2014; Mäding, 2004; Müller and Siedentop, 2004; Simmons and Bourne, 2007; Wiechmann, 2008; Wiechmann and Pallagst, 2012). Research has also explored the psychological dimension of open public discussion of the topic due to the growth-centric mentality that permeates civic politics and planning (Hall, 2009; Hollander and Németh, 2011; Hospers, 2014; Leo and Anderson, 2006; Oswalt, 2005; Popper and Popper,

Table 20.2 Summary of Macro Trends Contributing to the Pattern of Uneven Growth

Trend	Description
Demographic	• Canada in the final stages of the demographic transition • Lower fertility rates and decreasing natural increase • Immigration to and migration within Canada major source of demographic growth • Demographic concentration occurring in large Canadian city-regions
Economic	• Economic restructuring and technological changes in manufacturing and resource sectors • Shift to a more knowledge-intensive economy • Resource dependency and limits of (profitable) resource exploitation • Economic concentration in large Canadian city-regions
State policy	• Shifting national trade to the United States • Shifting Canadian capital flows • National trade liberalization and devolved responsibilities to lower levels of government • Shift away from state redistribution policies aimed at reducing spatial inequalities
Locational	• Peripheral location in the Canadian urban system • Location away from a major transportation axis or trade route • Outside the influence zone of a large city-region • Size: small and mid-size cities

2002; Seasons, 2007). A common thread running throughout all of this research is that the current demographic, economic, and policy trends most likely will persist and the spatial fallout of these trends will deepen the uneven inter-urban geography across Canada and internationally.

Psychology of Failure and Political Sensitivity

How is the topic of decline perceived in the conduct of urban affairs? In a study on the social dynamics of economic performance in Kingston, Ontario, a slowly growing city, a key informant explained that the city has a "psychology of failure" that plays a strong role in its economic **governance** (Lewis and Donald, 2010: 36; see also Taylor and Bradford, Chapter 3).

> If a place has a personality, it can also have a psychology. Just as individual psychologies reflect their experiences, their traumas, and upbringing and raising and so on, Kingston has had, one can argue, a psychology of loss, of failure. It didn't grow, it wasn't industrialized, it was bypassed by commerce, the military left, the capital left.

This psychology of failure and government malaise that often accompanies these slow-growth and declining cities is a pressing issue that influences all aspects of decision-making from land-use planning to economic development. Likewise, Hospers (2014) describes the "urban mindware" or image of a city and the impacts of being labelled a shrinking city. He notes that this negatively loaded term can discourage local empowerment and make people feel inferior. Moreover, Audirac (2018) discusses the territorial or spatial stigma that is associated with being labelled a "shrinking city," which can impact residents and their sense of self, business development, the quality of public services, the image of the community, and the decisions of policy-makers.

In the local policy arena, the issue of decline is rarely broached unless the discussion is focused on reversing the trend (Müller and Siedentop, 2004; Seasons, 2007). Seasons (2007: 6) refers to decline as the "policy elephant in the living room," while Hall (2009) found that planners, economic developers, and policy-makers were uncomfortable discussing the topic at the local level. This discomfort is a direct result of the negative perceptions and psychology that accompany decline (Robinson, 1981). For example, Leo and Anderson (2006: 393), in their research on Winnipeg, argue that there is a perception that "any city that is not growing rapidly is being 'left behind.'" Hall (2007) discovered that local officials in Greater Sudbury associated decline with death, producing a negative image that would counteract investment and local economic development. One key informant suggested that "if you're not growing, you're dying." Put simply, cities view growth as expected, desirable, and necessary, while any trajectory other than growth is seen as failure. Moreover, anticipating decline is equivalent to accepting this failure.

This obsession with urban growth is firmly rooted in the politics of local economic development. From the early days of North American cities, demographic and economic growth and new urban development have been perceived essentially as ingredients for maintaining and improving urban quality of life. Harvey Molotch (1976) emphasized the entrenched attachment to growth in his now classic framing of the city as a "growth machine." Molotch (1976: 310) further stated that "the clearest indication of success at growth is a constantly rising urban area population." Growth is depicted as the magic potion that creates jobs, reinforces the tax base, and provides resources to solve social problems (Logan and Molotch, 1987). In the current economic context, a positive urban image is perceived to be a vital component for urban competitiveness; no city wants to be labelled as declining (Avraham, 2004).

Decline creates a policy dilemma for those cities experiencing it. Yet, if current trends persist, some cities will be unable to reverse their demographic situation and may have to make the best of what is perceived to be a non-ideal situation. Incessantly planning for future growth in its absence, however, may prove to be more costly in the long term. We argue that approaches reflecting place-based realities are essential because slow growth and no growth do not necessarily have to be viewed as problems if they are managed and planned for. As an example, Greater Sudbury has changed and improved dramatically over the last few decades without population growth, suggesting that city decline is different, but not deadly. Over the 45-year period between 1971 and 2016, the population of Sudbury has grown and declined cyclically, from about 170,000 to just over 164,000. Despite this trend, the economic base has expanded so that Sudbury has become a regional service centre for northeastern Ontario for medical care, retail, tourism, government, and education. In 2005, the Northern Ontario School of Medicine was opened with a campus in Sudbury. This is the first medical school to be opened in Canada in over 30 years. In 2013, Laurentian University's McEwen School of Architecture opened in downtown Sudbury. Like the medical school, it is the first school of architecture to open in Canada in over 40 years (Laurentian University, 2013). When the city was rapidly growing during the 1960s, the natural environment was barren and polluted due to smelting techniques used by the mining industry. Indeed, a widespread rumour claimed that US astronauts, in preparation for landing on the moon, trained on the supposedly similar barren terrain that Sudbury provided. While this rumour has since been dispelled, it gives some indication of the former state of the landscape. Since that time, key agents in the city have planted over eight million trees, earning a United Nations commendation, among other distinctions. The Sudbury case depicts how communities can change and improve without demographic growth. As we discuss further in the following section, both declining *and* growing cities have unique opportunities and challenges.

Growth, Slow Growth, and Decline: Challenges and Opportunities

A common perception exists that rapid growth is the desideratum of urban development. There are challenges, however, associated with this type of development trajectory. Hall and Hall (2008) discuss how urban growth can lead to environmental concerns, such as rural fringe development (e.g., sprawl) and the loss of prime agricultural land. Furthermore, challenges arise related to the provision and maintenance costs of infrastructure and municipal services as a city grows. Conflicts may arise between urban and rural land uses or other community social tensions, such as has occurred in Fort McMurray, located within the Regional Municipality of Wood Buffalo, Alberta. Between 2001 and 2011, the population of this region increased by 57 per cent, from 42,580 to 66,896 people (Statistics Canada, 2013). This rapid growth was a direct result of the region's proximity to the Athabasca oil sands and the booming oil and gas industry. Hoernig et al. (2005) discuss the social impacts of this rapid growth. High wages due to labour shortages in the oil sands development industry lead to shortages in other sectors that are unable to match the high oil wages. Other issues related to the quick demographic changes include a serious shortage of affordable housing, a high cost of living, homelessness, social and individual stresses, and, of course, environmental degradation on a massive scale. Thus, rapid growth is not a panacea but has challenges and opportunities that require realistic planning solutions.

On the positive side, large, growing cities often offer demographic and cultural diversity as a result of their large immigrant populations, economic diversity in terms of employment opportunities and choices, and urban diversity, which includes more selection in dwelling types, such as lofts and condominiums (Gertler et al., 2002; see Bain and Mark, Chapter 15; Vinodrai, Chapter 17). In slow-growth or declining cities, on the other hand, the populations tend to be relatively homogeneous, with a higher proportion of aging residents due to limited economic opportunities, which lead to youth out-migration and low immigrant settlement (Hanlon and Halseth, 2005; Simard and Simard, 2005). Declining cities can also become increasingly dependent on grants from provincial and federal governments due to their shrinking tax base. However, slow-growth and declining cities in peripheral locations often boast an abundance of landscapes rich in natural amenities: lakes and forests, better air quality, and less traffic congestion (Hall, 2009; Simard and Simard, 2005).

Leo and Anderson (2006) compared Vancouver, a fast-growing urban area, with Winnipeg, a slow-growing urban area. They found that housing costs are cheaper in Winnipeg, yet Vancouver has more ability to pay for its infrastructure and services. Vancouver also has sharper contrasts between the rich and poor. A remarkable characteristic of some **slow-growth cities** is evident in Winnipeg, where dispersion has occurred despite its slow growth. This contrasts with Vancouver, a fast-growing city, which has had more success in limiting the region's outward growth. This dispersion pattern in Winnipeg is related to decisions made by city officials who are more accepting of pro-development proposals in their desire to grow at any cost. Often these developments are unsuitable with regard to location, densities, and land uses and may place further pressure on services and infrastructure. In Greater Sudbury, Hall (2007) identifies a similar planning practice to "take anything anywhere," in which decline is used as an opportunity to push through proposals. This trend has also occurred in Kingston, where a sprawl-like spatial pattern has emerged over the last 10 years without a significant population increase. This "any growth is good growth" mentality becomes a problem, however, when city officials fail to see that more development also produces more pressure on municipal services and increased expenditures (Siegel, 2002).

Growth is an important component in municipal budgeting for infrastructure and services, presenting a challenge for cities in decline. Indeed, growth brings money into municipal accounts. In Canadian municipal budgeting, the three main sources of revenue are user fees; unconditional or conditional grants; and taxes on the assessed value of property. Complicating municipal financing, moreover, has been the downloading in recent years of responsibilities from senior levels of government to municipalities, without augmenting municipalities' capacity to secure new revenue streams (Bradford, 2002, 2004; Federation of Canadian Municipalities [FCM], 2006; Kitchen, 2002; Mintz and Roberts, 2006; Sancton, 2000, 2006; Vojnovic and Poel, 2000; see Taylor and Bradford, Chapter 3). For example, Mintz and Roberts (2006) describe how the Ontario government downloaded various social welfare programs, including social assistance, child-care, immigration services, **social housing**, and homelessness. Kitchen (2002: 5) argues that this downloading was the "largest redistribution of services between a province and its municipalities ever witnessed in Canada," thereby further fuelling a growth mentality because new property development increases the tax base and, ultimately, municipal revenue. Without substantial new growth, the purse strings on the municipal budget become tighter and the challenge becomes how to deliver up-to-date and quality services and infrastructure without considerable new growth occurring.

One last opportunity associated with slow growth and decline is the option to manage land resources in a more efficient and forward-looking manner (Robinson, 1981). In rapidly growing cities, policies can be reactive, whereas in slow-growth and declining cities, the potential exists to become more proactive. Strategies like impact analyses and strategic assessments can be employed before making development decisions to determine how foreseeable changes will affect the city presently *and* over time. In reality, however, declining cities are not necessarily using that time to practice proactive planning. Thus, a variety of place-based opportunities and challenges are associated with rapid-growth, slow-growth, and declining communities (Table 20.3).

Implications for Urban Geography, Planning, and Policy

The 1950s and 1960s were a time of rapid economic and demographic growth, and during these decades "most Canadian planning standards, norms and zoning practices were developed" (Wolfe, 1995: 55). Historically, then, planning has been tied to growth, and the current framework does not provide the proper tools for urban areas experiencing anything other than rapid growth at the urban scale.[1] Nevertheless, planners do have experience managing areas in decline, such as inner cities, brownfields, and core areas, to increase livability and attract investment. Most planning concepts, policies, and statutes, however, are devoid of the word *decline*, with common approaches like **smart growth** and the *Places to Grow Act* 2005 in Ontario focused on growth. For example, the *Growth Plan for Northern Ontario 2011* includes no reference to demographic decline facing many communities across the region. A similar tale is told for the United States and Germany. Popper and Popper (2002) argue that US planning tools are designed to manage a rising economy and

Table 20.3 Summary of Challenges and Opportunities

	Rapid-Growth Cities	Slow-Growth and Declining Cities
Opportunities	• Demographic and cultural diversity • Employment diversity • Variety in housing options and the built form • Higher tax base	• Lower cost of living • Housing affordability • Less traffic congestion • Better air quality • Natural amenities • Opportunities to be more proactive about decision-making
Challenges	• Sharper contrasts between rich and poor • Environmental concerns related to increased • Land-use conflicts • Social conflicts • Extreme pressure on infrastructure and services • Pollution (air quality) • Congestion	• Aging population • Homogeneous population dispersion • Youth out-migration • Limited economic opportunities • "Any growth is good growth" mentality for urban development • Grant dependency • Less ability to pay for infrastructure and services • Land abandonment

Source: Compiled by authors. For similar summary tables, see Bourne and Simmons, 2003 and Robinson, 1981.

population. In Germany, municipal, regional, and state planning approaches are unprepared to meet the challenges of demographic decline (Bontje, 2004; Müller, 2004; Müller and Siedentop, 2004). More fundamentally, growth-centred solutions to no-growth scenarios are unrealistic and prevent the acceptance and management of decline as an acceptable trajectory. In the following sections we offer some solutions to ignite the search for alternatives.

Creating a New Decline Discourse

Canadian urbanists and policy-makers must turn their attention to the certainty that decline or slower growth will continue in some places and begin in others. Hall and Hall (2008: 14) argue that the Canadian urban literature "repeats, and arguably reinforces, the message that quantitative [population] growth is the only legitimate response to decline" and that "[c]onsumers of this research are implicitly dared to concede defeat if they disagree with this perspective." If we expect places to realistically plan for decline, then the topic can no longer be avoided (Bunting and Filion, 2001). Focus must now turn to how cities in decline can maintain quality of life when revenue sources depend on growth. This may require alternative financial tools and policies for slow-growth and declining cities. For example, many European cities rely much less on the property tax base. In fact, Canadian municipalities have one of the highest dependencies on property taxes in the world. Alternative revenue sources more prominent in other jurisdictions include user fees, sales taxes, and income-based taxes. These alternative sources—especially the income tax—may diminish the pressure on cities to continuously seek land-development-related growth (see Slack, 2005, and Courchene, 2005, for discussions of fiscal constraints in Canadian cities). At a more fundamental level, the perceived notion that slow-growing or declining places are "failing" in the metropolitan growth competition needs to change if we expect the political sensitivity of discussing the issue to diminish and become a more solutions-oriented conversation.

New Decline-Oriented Techniques: "Smart Decline" and "Qualitative Development"

Hospers (2014) identifies four policy responses to decline in Europe: (1) trivializing shrinkage, (2) countering shrinkage, (3) accepting shrinkage, and (4) utilizing shrinkage. Likewise, Pallagst, Fleschurz and Said (2017: 16) discuss a number of policy and strategy stages associated with shrinking cities, including Farke's four stages (ignoring, observing without acceptance, certain acceptance, and acceptance), as well as Danielzyk et al.'s typology (decline as a vicious circle, expansive strategy, maintenance strategy, and planning for decline).

However, Canadian planning legislation and policy tools have ignored decline, and are largely ill equipped to manage decline at the urban scale. Existing concepts like growth management, growth controls, and smart growth all imply the existence of substantial growth to manage. Müller and Siedentop (2004) call for a decline paradigm to establish more decline-oriented planning tools. Their suggestions include, but are not limited to, focusing on redevelopment, recycling land and buildings, adapting infrastructure, and planning strategically. Likewise, Popper and Popper (2002: 22) argue that we need a "smart decline" strategy requiring us to think "about who and what remains," and that this "may entail reorganizing or eliminating some services and providing different ones." In Germany, Hospers (2014) identifies *Gesundschrumpfen*, or healthy shrinking, while in the Netherlands he identifies *slimpen,* or smart shrinking. Hollander et al. (2009) point out that the corporate expression of "right-sizing" is being touted in the United States, while Pallagst (2013) discusses the notion of "palliative planning."

Hollander et al. (2009) identify the more corporate expression of "right-sizing" that is being used in rustbelt cities, like Detroit, where land abandonment in the inner city has become a significant concern for policy-makers. Németh and Hollander (2016) describe the following three actions for right-sizing strategies: (1) preparing a strategy that understands the "hot spots of decline," including vacant land, and includes technical experts (e.g., planners, urban designers, landscape architects) and the public; (2) investing in these areas using land banks, for example; and (3) regulation to encourage a desired private or public market response (e.g., temporary zoning to create urban gardens on vacant land).

Based on previous research, we offer a number of alternative planning approaches to assist urban geographers, planners, and city officials (Hall, 2009). Our hope is to kick-start conversations on how to accept, manage, and plan for decline. We start with **qualitative development**, a concept used by Sudbury in the 1980s, which centres on the existing built form, promoting redevelopment, infilling, and conservation. These concepts are simple to implement and already essential to good planning. The challenge, however, is to forget about the growth mentality associated with quantitative development or expanding growth, services, and infrastructure in the city. An expansive urban boundary is *not* a metaphor for success. Other decline-oriented approaches involve realigned development, where settlement boundaries and services are scaled back or reduced to reflect current expectations, while maintaining and improving the existing built form. Controlled development would mean halting municipal infrastructure or service expansions to facilitate new growth, with developers paying the costs associated with development and identifying spaces for redevelopment and infilling. A final decline-oriented technique that we offer is internal zoning, whereby a hierarchy of limited settlement areas are predetermined for development. Each zone or settlement area slated for urbanization or re-urbanization would be developed before another zone could be selected. Essentially, the key is not to overdevelop but to think strategically about where growth should occur within the city, which is something all cities should be doing. A summary of these approaches is provided in Table 20.4.

Hollander and Németh (2011: 358–62) make five recommendations for smart decline planning processes:

1. Include and recognize multiple voices.
2. Be political and deliberative.
3. Be aware of various communication techniques, and provide information to mobilize citizens.
4. Be transparent and utilize different types and sources of information.
5. Be regional in scope but ensure local control and implementation.

Focusing upon the complexity of the process, Hospers (2014) argues that urban shrinkage is a complex urban governance issue that requires the involvement of a wide variety of urban actors (e.g., local government, corporations, schools, business networks, local associations and citizens).

Table 20.4 Summary of Decline-Oriented Techniques

Decline-oriented technique	Description
Decline-oriented planning/ Smart decline	• Focus on redevelopment, recycling land and buildings, adapting infrastructure, and decline paradigm strategic planning • Plan for who and what remains • Might involve reorganizing or eliminating some services and providing alternatives
Qualitative development	• A concept used by Sudbury, ON, in the 1980s • Focus on the existing built form, promoting redevelopment, infilling, and conservation
Realigned development	• Settlement boundaries and services are scaled back to reflect current expectations • Focus on maintaining and improving the existing built form
Controlled development	• No infrastructure or services expansion to facilitate new growth • Developers pay the costs associated with development • Identify spaces for redevelopment and infilling
Internal zoning	• Hierarchy of limited settlement areas are predetermined for development

Source: Compiled by authors from Müller and Siedentop, 2004; Popper and Popper, 2002; and Hall, 2009.

Fischler and Wolfe (2006: 348) describe efforts of the Montreal Metropolitan Community (MMC), which is planning for a smaller population by the year 2050.[2] They explain that the MMC approach is to "consolidate already urbanized areas, to rationalize regional infrastructure investment, and to arbitrate the inter-municipal competition for households and businesses by restricting development on greenfield sites." Arguably, little risk befalls a city that plans for decline but experiences growth, other than negative image. If a city plans for growth and experiences decline, however, the risk is overbuilding and extending services and infrastructure that will be too costly to manage and upgrade in the future.

Moving Away from One-Size-Fits-All

In recent decades, government has become more concerned with large city-regions and their competition on a global scale (Donald, 2005). As Bürkner (2005) explains, state intervention policies that once aimed to *balance* spatial disparities between cities have shifted to *targeting* cities and regions that are already economically competitive or that are perceived to have potential. However, Atkinson (2001) cautions that this focus on competition may actually accentuate disparities and create more place-based winners and losers. In Ontario, most government policy has paid attention to the Greater Toronto Area and its surrounding region, exemplified through the City of Toronto Act, 2006, and the 2006 Growth Plan for the Greater Golden Horseshoe. The first provides the city with powers and legislative freedoms that other municipalities do not have, while the latter was designed to guide where and how growth should occur in order to promote healthy, strong, and prosperous communities in this large conurbation (Ontario Ministry of Municipal Affairs and Housing, 2009; Ontario Ministry of Public Infrastructure Renewal, 2006). More recently, a report published for the Ontario government by the Martin Prosperity Institute promotes the Greater Toronto Area as pivotal to a secure future for Ontario, centred on the creative economy (Florida and Martin, 2009).

We are not suggesting that Toronto and the broader Golden Horseshoe region are undeserving of attention. We are suggesting, however, that as the rift between have and have-not cities in the Canadian urban system widens, a new way of thinking will be needed to balance the needs of large, growing city-regions and their slow-growing and declining counterparts. This will require place-based, locally or regionally sensitive policies. It will require us to move beyond supply-side policies aimed at competitiveness that do little to manage spatial disparities between cities.

A similar spatial penchant for large city-regions is seen in much of the contemporary economic geography literature. The creative economy literature is biased by design to favour large cities like Toronto and Vancouver, while much of the literature on regional innovation systems and **clusters** is focused on economically advanced regions like Silicon Valley, Boston, Baden-Württemberg, and Toronto (Hall and Donald, 2009; see Vinodrai, Chapter 17; Shearmur, Chapter 18). Unfortunately, this spatial bias creates a dualistic depiction that shows large cities, on the one hand, as creative and cosmopolitan economic drivers, and, on the other hand, many smaller peripheral places as economic failures. Hayter, Barnes, and Bradshaw (2003: 18) remind us that "[t]here is a whole world out there and not just a few cores or clusters."

Determining If Population Growth Matters

As described earlier in the chapter, growth is commonly measured in terms of population change. But is population growth everything? Is there a better way to measure a city's prosperity? Given the current economic realities and the expected demographic situation in many Canadian cities, we contend that these questions will require future attention. Population growth is important for municipal revenue generation but little is known about savings that could be made by planning for a smaller population. Rybczynski and Linneman (1999) believe that cities experiencing decline should not be trying to grow big again but, rather, should find out how to prosper as great, smaller cities. Other suggestions include planning for sustainability, livability, and quality of life (Gonzáles, 2006). Furthermore, Molotch (1976: 328) foresees opportunities in the city that asks "what it can do for its people rather than what it can do to attract more people."

In Germany, Oswalt (2005: 13) describes a new slogan, "shrinking as a new potential," to suggest that decline can result in a new compact urban core, which is an ideal European **urban form**. Meanwhile, the concept of Slow Cities or Cittaslow evolved out of the **slow food movement** in Italy. These two movements are responses to the fast-paced growth and cultural standardization most often associated with the broader trend of **globalization**. The slow cities movement encourages "local, traditional cultures, a relaxed pace of life, and conviviality" (Knox, 2005: 6). The founding principles of Cittaslow include "working towards calmer and less polluted physical environments, conserving local aesthetic traditions and fostering local crafts, produce, and cuisine" (Knox, 2005: 6). At the heart of the movement are the concepts of health, sustainability, and the environment (see Rees, Chapter 21). Critics of this model raise concerns about whether this may reinforce or generate divisive politics and exclusionary dynamics observed in many places left behind (Rodriguez-Pose, 2018). Interestingly, participating cities are required to have populations under 50,000 (Knox, 2005). This model moves away from the obsession with rapid growth and instead encourages slow growth as the ideal trajectory for cities. While not a panacea, declining and slow-growing cities in Canada could draw on the Cittaslow movement for inspiring a new reality in which quality of life, quality of place, quality of employment opportunities, and sustainability become central to any policy-making, regardless of size. Thus, emphasizing local strengths and a *better* place, rather than a bigger place, could provide the opportunity for an alternative vision: one that ignores the obsession with growth and promotes urban quality (see Walker and Blakely, Chapter 13).

Conclusions: Moving Forward on a New Urban Agenda

Population growth remains concentrated in major urban centres across Canada, from Toronto to Calgary to Vancouver and Saskatoon. Many of the reasons behind this growth are linked more fundamentally to broader changes in the global economy and to increased flows of people, capital, and ideas from around the world to Canada. As a consequence, researchers and policy-makers interested in the Canadian urban system have become fascinated with documenting and promoting the dynamic, diverse, and creative Canadian urban centres. As detailed in this chapter and elsewhere in this book, new research projects have explored the social dynamics of economic performance in these fast-growing cities, and politicians have called for more explicit urban-based policies at the national level (see Taylor and Bradford, Chapter 3; Vinodrai, Chapter 17). However, this spatial bias accentuates inter-urban differences that are increasing across Canada.

As we have demonstrated, 57.2 per cent of the Canadian urban system is either growing slowly (40.1 per cent) or declining (17.1 per cent). The questions we must ask are as follows: How are we going to plan for Canadian cities in decline? What are the realistic urban scenarios and what are the effective policy tools at our disposal? Now is the time for urban scholars to turn more actively to insights from the environmental, green, sustainable, and quality-of-life literatures for ideas on how to plan for slow growth and decline (see Rees, Chapter 21; Holden and Chang, Chapter 22). Before doing this, however, we must first permit ourselves to face this reality and find new ways of constructively conversing about preparing for slow growth and decline in our Canadian cities. Only then can we enhance our collective capacity to respond to policy problems that are increasingly defined by the complex interdependence and spatially concentrated expression of the dynamic Canadian urban system. Coming to the aid of Canada's slow-growing or declining cities does not necessarily have to be a "zero-sum" view of power relations between Canada's have and have-not places. Rather, doing so can be aligned with our ability and history as a nation to embed **place-based public policies** that are sensitive to the realities of different regions across space and with our reputation as a nation to deal with deeper economic unevenness with broader multi-scaled policy effects.

It is worth closing this chapter with a cautionary message about the importance of recognizing the needs of slow growth and declining cities. As Audriac (2018) argues, "shrinking cities may represent the canary in the coal mine of global systemic transformations. This includes the Trump presidency delivered, among other things, by votes of discontent with long-term urban decline, de-industrialization and free trade." Put simply, slow growth and declining cities are increasingly the "places left behind" by globalization, breeding a politics of discontent. As Rodriguez-Pose (2018) argues, these perceived "places that don't matter" are waging their revenge at the ballot box via populist politics. For example, many industrial cities and rural areas experiencing decline in the United Kingdom voted for Brexit, while many rust-belt states in the United States voted for Donald Trump. As the number of slow-growth and declining cities increase in Canada, and around the world, better policies are needed to address this territorial inequality or the politics of discontent will create more turmoil and divisions.

Acknowledgements

We would like to thank Laurent Martel from Statistics Canada for their statistical assistance as well as Cameron Miller from the University of Waterloo for his mapping assistance.

Review Questions

1. What is a "slow-growth" or "declining" city, and where are they located in the Canadian urban system?
2. Is a "qualitative development" strategy a good idea for municipalities in order to foster growth, or is it strictly a strategy for managing decline?

Notes

1. Our definition of the Canadian urban system includes all census agglomerations (CAs) and census metropolitan areas (CMAs) (see also Simmons and Bourne, 2013).
2. The MMC is a planning and funding body that represents 82 municipalities (Communauté métropolitaine de Montréal, 2013).

References

Atkinson, R. 2001. "The emerging 'urban agenda' and the European spatial development perspective: Towards an EU urban policy?" *European Planning Studies* 9: 385–406.

Audirac, I. 2018. "Shrinking cities: An unfit term for American urban policy?" *Cities* 75: 12-19.

Avraham, E. 2004. "Media strategies for improving an unfavorable city image," *Cities* 21: 471–9.

Barnes, T.J., W.J. Coffey, D.W. Edgington, M.S. Gertler, and G. Norcliffe. 2000. "Canadian economic geography at the millennium," *Canadian Geographer* 44: 4–24.

Bontje, M. 2004. "Facing the challenge of shrinking cities in East Germany: The case of Leipzig," *GeoJournal* 61: 13–21.

Bourne, L.S., and D. Rose. 2001. "The changing face of Canada: The uneven geographies of population and social change," *Canadian Geographer* 45: 105–19.

———, and J. Simmons. 2003. "New fault lines? Recent trends in the Canadian urban system and their implications for planning and public policy," *Canadian Journal of Urban Research* 12, 1 (suppl.): 22–47.

Bradford, N. 2002. *Why Cities Matter: Policy Research Perspectives for Canada*. Ottawa, ON: Canadian Policy Research Networks. At: www.cprn.org/documents/29277_en.pdf

———. 2004. *Place Matters and Multi-level Governance: Perspectives on a New Urban Policy Paradigm*. Ottawa, ON: Canadian Policy Research Networks. At: www.cprn.org/documents/26856_fr.pdf

Bunting, T., and P. Filion. 2001. "Uneven cities: Addressing rising inequality in the twenty-first century," *Canadian Geographer* 45: 126–31.

Bürkner, H.J. 2005. "Polarization and Peripherization," in P. Oswalt, ed., *Shrinking Cities: Volume 1—International Research*. Ostfildern-Ruit, Germany: Hatje Cantz.

Cocks, M., and C. Couch. 2012. "The governance of a shrinking city: Housing renewal in the Liverpool conurbation, UK," *International Planning Studies* 17: 277–301.

Communauté métropolitaine de Montréal. 2013. *Who Are We—Establishment*. At: http://cmm.qc.ca/who-are-we/establishment/

Courchene, T.J. 2005. *Citistates and the State of Cities: Political-Economy and Fiscal-Federalism Dimensions*. Montreal: Institute for Research in Public Policy (IRPP Working Paper Series no. 2005–03).

Donald, B. 2005. "The politics of local economic development in Canada's city-regions: New dependencies, new deals, and a new politics of scale," *Space and Polity* 9: 261–81.

Downs, A. 1994. *New Visions for Metropolitan America*. Washington, DC: Brookings Institution.

ECOTEC Research and Consulting Ltd. 2007. *State of European Cities Report: Adding Value to the European Urban Audit*. Birmingham: ECOTEC Research and Consulting Ltd. At: http://ec.europa.eu/regional_ policy/sources/docgener/studies/pdf/urban/stateof cities_2007.pdf

Federation of Canadian Municipalities (FCM). 2006. *Building Prosperity from the Ground Up: Restoring Municipal Fiscal Imbalance*. Ottawa: FCM. At: www.fpeim.ca/2006_fcm_Building_Prosperity_from_the_Ground_Up.pdf

Fischler, R., and J.M. Wolfe. 2006. "Contemporary planning," in T. Bunting and P. Filion, eds., *Canadian Cities in Transition: Local through Global Perspectives*, 3rd edn. Toronto: Oxford University Press.

Florida, R., and R. Martin. 2009. *Ontario in the Creative Age*. Toronto: Martin Prosperity Institute. At: http://martinprosperity.org/media/pdfs/MPI%20Ontario%20Report%202009%20v3.pdf

Franz, P. 2004. "Shrinking cities–shrinking economy? The case of East Germany," *German Journal of Urban Studies* 44, 1: online journal.

Gertler, M.S. 2001. "Urban economy and society in Canada: Flows of people, capital and ideas," ISUMA: *The Canadian Journal of Policy Research* 2(3): 119–30.

———, R. Florida, G. Gates, and T. Vinodrai. 2002. *Competing on Creativity: Placing Ontario's Cities in North American Context*. Toronto: Ontario Ministry of Enterprise, Opportunity and Innovation and the Institute for Competitiveness and Prosperity. At: www.competeprosper.ca/research/CompetingOnCreativity_061202.pdf

Gonzáles, S. 2006. *The Northern Way: A Celebration or a Victim of the New City-Regional Government Policy?* Swindon: Economic and Social Research Council Postgraduate Research Programme, Working Paper 28.

Hackworth, J. 2012. *A Critical Examination of Market-Centred Land Abandonment Policies in Shrinking American Cities*. Toronto: University of Toronto (Working Paper Geography & Program in Planning).

Hall, H., and P. Hall. 2008. "Decline and no growth: Canada's forgotten urban interior," *Canadian Journal of Regional Studies* 31: 1–18.

Hall, H.M. 2007. *Being Realistic about Planning in No Growth Communities: Challenges, Opportunities, and Foundations for a New Agenda*. Waterloo, ON: University of Waterloo, MA thesis.

———. 2009. "Slow growth and decline in Greater Sudbury: Challenges, opportunities, and foundations for a new planning agenda," *Canadian Journal of Urban Research* 18: 1–26.

———, and B. Donald. 2009. *Innovation and Creativity on the Periphery: Challenges and Opportunities in Northern Ontario*. Toronto: Martin Prosperity Institute Working Paper. At: http://martinprosperity.org/media/pdfs/Innovation_and_creativity_on_the_Periphery-H_Hall-B _Donald.pdf

Hanlon, N., and G. Halseth. 2005. "The greying of resource communities in northern British Columbia: Implications for health care delivery in already-underserviced communities," *Canadian Geographer* 49: 1–24.

Hartt, M.D. 2018. "How cities shrink: Complex pathways to population decline," *Cities* 75: 38–49.

Hayter, R., T.J. Barnes, and M.J. Bradshaw. 2003. "Relocating resource peripheries to the core of economic geography's theorizing: Rationale and agenda," *Area* 25, 1: 15–23.

Hoernig, H., D. Leahy, Z.X. Zhuang, R. Early, L. Randall, and G. Whitelaw. 2005. "Planning for people: Integrating social issues and processes into planning practice," *Berkeley Planning Journal* 18: 35–55.

Hollander, J.B., K.M. Pallagst, T. Schwarz, and F.J. Popper. 2009. "Chapter 4: Planning shrinking cities," *Progress in Planning* 72(4): 223–32.

———, and Németh, J. 2011. "The bounds of smart decline: A foundational theory for planning shrinking cities," *Housing Policy Debate* 21: 349–67.

Hospers, G. 2014. "Policy responses to urban shrinkage: From growth thinking to civic engagement," *European Planning Studies* 22(7): 1507–23.

Kitchen, H.M. 2002. *Municipal Revenue and Expenditure Issues in Canada* (Canadian Tax Paper No. 107). Toronto: Canadian Tax Foundation.

Knox, P.L. 2005. "Creating ordinary places: Slow cities in a fast world," *Journal of Urban Design* 10: 1–11.

Laurentian University. 2013. School of Architecture. Sudbury, ON: Laurentian University. At: laurentian.ca/content/school-architecture-0

Leo, C., and K. Anderson. 2006. "Being realistic about urban growth," in T. Bunting and P. Filion, eds., *Canadian Cities in Transition: Local through Global Perspectives*, 3rd edn. Toronto: Oxford University Press.

———, and W. Brown. 2000. "Slow growth and urban development policy," *Journal of Urban Affairs* 22: 193–213.

Lewis, N., and B. Donald. 2010. "A new rubric for 'creative city' potential in Canada's smaller cities," *Urban Studies* 47: 29–54.

Logan, W., and H. Molotch. 1987. *Urban Fortunes: The Political Economy of Place*. Berkeley, CA: University of California Press.

Lötscher, L. 2005. "Shrinking East German cities?" *Geographia Polonica* 78: 79–98.

———, F. Howest, and L. Basten. 2004. "Eisenhüttenstadt: Monitoring a shrinking German city," *Dela* 21: 361–70.

Mäding, H. 2004. "Demographic change and local government finance—Trends and expectations," *German Journal of Urban Studies* 44, 1: online journal.

Martinez-Fernandez, C., I. Audirac, S. Fol, and E. Cunningham-Sabot. 2012a. "Shrinking cities: Urban challenges of globalization," *International Journal of Urban and Regional Research*, 36: 213–25.

———, N. Kubo, A. Noya, and T. Weyman. 2012b. *Demographic Change and Local Development: Shrinkage, Regeneration and Social Dynamics*. Paris: OECD (Working Paper from the OECD Local Economic and Employment Development [LEED] Programme).

Mintz, J.M., and T. Roberts. 2006. *Running on Empty: A Proposal to Improve City Finances*. Toronto: C.D. Howe Institute.

Molotch, H.L. 1976. "The city as a growth machine: Toward a political economy of place," *American Journal of Sociology* 82: 309–32.

Müller, B. 2004. "Demographic change and its consequences for cities—Introduction and overview," *German Journal of Urban Studies* 44: online journal.

———, and S. Siedentop. 2004. "Growth and shrinkage in Germany—Trends, perspectives and challenges for spatial planning and development," *German Journal of Urban Studies* 44, 1: online journal.

Németh, J., and J. Hollander. 2016. "Right-sizing shrinking cities: A landscape and design strategy for abandoned properties," *Journal of Landscape Architecture* 11(2): 90-100.

Norcliffe, G. 1994. "Regional labour market adjustments in a period of structural transformation: An assessment of the Canadian case," *Canadian Geographer* 38: 2–17.

Ontario Ministry of Municipal Affairs and Housing. 2009. City of Toronto Act, 2006. At: www.mah.gov.on.ca/Page343.aspx

Ontario Ministry of Public Infrastructure Renewal. 2006. *Growth Plan for the Greater Golden Horseshoe*. Toronto: Ontario Ministry of Public Infrastructure Renewal. At: www.placestogrow.ca/images/pdfs/FPLAN-ENG-WEB- ALL.pdf

Oswalt, P., ed. 2005. *Shrinking Cities: Volume 1—International Research*. Ostfildern-Ruit, Germany: Hatje Cantz.

———. 2005. "Introduction," in P. Oswalt, ed., *Shrinking Cities: Volume 1—International Research*. Ostfildern-Ruit, Germany: Hatje Cantz.

———, and T. Rienitz, ed. 2006. *Shrinking Cities, Volume 2: Interventions*. Ostfildern, Germany: Hatje Cantz Verlag.

Pallagst, K. 2013. "A Review of 'Design after Decline: How America Rebuilds Shrinking Cities,'" *Journal of the American Planning Association* 79(2): 11.

Polèse, M., and R. Shearmur, with P.M. Desjardins and M. Johnson. 2002. *The Periphery in the Knowledge Economy: The Spatial Dynamics of the Canadian Economy and the Future of Non-Metropolitan Regions in Quebec and the Atlantic Provinces*. Montreal and Moncton: Institut national de la recherche scientifique and the Canadian Institute for Research on Regional Development.

———, and ———. 2006. "Why some regions will decline: A Canadian case study with thoughts on local development strategies," *Papers in Regional Science* 85: 23–46.

Popper, D.E., and F.J. Popper. 2002. "Small can be beautiful: Coming to terms with decline—Americans tend to think that places, once settled, stay settled," *Planning* 68, 7: 20–3.

Reckien, D., and C. Martinez-Fernandez. 2011. "Why do cities shrink?" *European Planning Studies*, 19: 1375–97.

Robinson, I.M. 1981. *Canadian Urban Growth Trends: Implications for a National Settlement Policy*. Vancouver, BC: University of British Columbia Press.

Rodríguez-Pose, A. 2018. "The revenge of the places that don't matter (and what to do about it)," *Cambridge Journal of Regions, Economy and Society* 11(1): 189–209.

Rybczynski, W., and P.D. Linneman. 1999. "How to save our shrinking cities," *The Public Interest* 135: 30–44.

Sancton, A. 2000. "Amalgamations, service realignment, and property taxes: Did the Harris government have a plan for Ontario's municipalities?" *Canadian Journal of Regional Science* 23: 135–56.

———. 2006. "City politics: Municipalities and multi-level governance," in T. Bunting and P. Filion, eds., *Canadian Cities in Transition: Local through Global Perspectives*, 3rd edn. Toronto: Oxford University Press.

Seasons, M. 2007. "Planning for uneven growth," *Plan Canada* 47(2): 6.

Siegel, D. 2002. "Urban finance at the turn of the century: Be careful what you wish for," in E.P. Fowler and D. Siegel, eds., *Urban Policy Issues: Canadian Perspectives*, 2nd edn. Toronto: Oxford University Press.

Simard, M., and C. Simard. 2005. "Toward a culturalist city: A planning agenda for peripheral mid-size cities," *Canadian Journal of Urban Research* 14: 38–56.

Simmons, J. 2003. *Cities in Decline: The Future of Urban Canada*. Toronto: Ryerson University, Centre for the Study of Commercial Activity.

———, and L.S. Bourne. 2007. "Living with population growth and decline," *Plan Canada* 47, 2: 13–21.

———, and L.S. Bourne. 2013. *The Canadian Urban System in 2011: Looking Back and Projecting Forward*. Research Paper 228. Toronto: Cities Centre University of Toronto.

Slack, E. 2005. "Easing the fiscal restraints on Canadian cities," *Dialogues: Canada West Foundation* (Spring): 19–20.

Statistics Canada. 2013a. Based on one dataset showing 2001 CA populations in 2011 geography and another showing 2001 CMA populations in 2011 geography.

Vojnovic, I., and D. Poel. 2000. "Provincial and municipal restructuring in Canada: Assessing expectations and outcomes," *Canadian Journal of Regional Science* 23: 1–6.

Wiechmann, T. 2008. "Errors expected-aligning urban strategy with demographic uncertainty in shrinking cities," *International Planning Studies* 13: 431–46.

———, and K.M. Pallagst. 2012. "Urban shrinkage in Germany and the USA: A comparison of transformation patterns and local strategies," *International Journal of Urban and Regional Research* 36: 261–80.

Wolfe, J.M. 1995. "Canada," in D. Lyddon, ed., *International Manual of Planning Practice*, 2nd edn. The Hague: International Society of City and Regional Planners.

V The Environmental Imperative

The final part of this book is dedicated to the natural systems that all cities depend on for survival. While environmental issues have been of concern for a long time, recent pressing issues such as climate change challenge our current ways of urban living. Chapter 21 considers cities from a biophysical perspective. Readers learn about the dire state of our natural environment, and the threat this poses to human survival. The chapter introduces the concept of the ecological footprint to demonstrate the impact that urbanization has on the planet. Readers also learn how to utilize certain aspects of cities to lessen our impact on the natural environment.

Chapter 22 focuses on specific ways that cities tackle environmental issues through the lens of sustainability. The chapter outlines sprawl as an environmental issue that planners in particular have been trying to reverse. Chapter 22 considers the pros and cons of sprawl, and its commonly promoted antidote, density. Readers learn how climate change and other sustainability issues are being addressed in cities but also are urged to consider the implications for affordability. Drawing on an array of illuminating examples, the authors also examine resilience as a way to address climate change in cities.

Chapter 23 is about food systems. The authors describe the characteristics of our urban food system and its linkages to sustainability. Readers learn about the growing integration of agricultural activities into cities, and about planning and policy regarding food provision in cities. The authors offer rich case studies on community gardens, urban farms, rooftop production, food deserts, and vertical/indoor/container farming.

Finally, Chapter 24 is about (re)designing cities to address climate change. Readers learn about the implications of greenhouse gases on climate and the impacts of climate change. The emphasis of the chapter is on how to reduce greenhouse gas production through urban design. Compact and mixed-use developments are highlighted as examples of ways to reduce greenhouse gas emissions. The chapter details approaches to implementing green infrastructure and environmentally friendly building/neighbourhood designs. The authors outline how climate action in cities is being undertaken and the challenges that such efforts confront.

After reading the chapters in Part V, readers will have understanding of

- the urgency for action to address environmental issues in cities;
- the environmental impact of cities and how the ecological footprint can be used to measure this impact;

- the approaches cities have pursued to address sustainability issues, particularly relating to sprawl and resilience;
- the characteristics of our food systems and attempts to integrate food production into cities; and
- the role of greenhouse gas emissions in contributing to climate change and the role of urban design in mitigating these emissions.

21 Getting Serious about Urban Sustainability

Eco-Footprints and the Vulnerability of Twenty-First-Century Cities

William E. Rees

> *At the heart of this assessment is a stark warning. Human activity is putting such a strain on the natural functions of the Earth that the ability of the planet's ecosystems to sustain future generations can no longer be taken for granted.* (MEA, 2005: 5)

Introduction: Framing the Analysis

This chapter is concerned with the long-term sustainability of cities and urban civilization. My starting premise is that, because of accelerating global ecological change, cities everywhere are facing unprecedented challenges to their functional integrity, even survival. Unprecedented challenges require unprecedented solutions. In keeping with ecological reality, I depart from most urban scholarship, which assumes a humanities and social-science perspective. Instead, I approach the "urban question" from a mainly biophysical point of view. Accelerating global change makes clear that society will not be able to ensure the sustainability of cities without a much fuller understanding of cities as ecological entities subject to biophysical laws.

With this in mind, the chapter begins with a brief consideration of the organic origins of cities and their subsequent evolution. Permanent settlements became possible as a result of a technology-induced shift in the ecological "niche" of humans from hunter-gathering to farming about 10 millennia ago. However, the subsequent alienation of urban techno-industrial society from nature has produced modern cities that are not only incomplete as human ecosystems but that are aggressively hostile to the natural ecosystems that sustain them.

The subsequent section uses ecological footprint analysis to illustrate the ecological "load" that inhabitants of modern cities impose on the natural world. The key question is this: How much of the surface of the earth do we occupy *ecologically* to sustain our material lifestyles? Pay attention, urban planners (also see Moos et al., 2006)! The eco-footprints of typical cities are hundreds of times larger than their political or built-up areas. In any functionally meaningful sense, does not this productive "hinterland" constitute urban land as much as does a sterile parking lot within the city limits?

I then consider the increasing vulnerability of modern cities to global ecological change. Urbanization represents the greatest mass migration of people ever. More people will be added to the world's cities in the first three or four decades of the twenty-first century, mostly through immigration, than had accumulated on the entire planet by the mid-twentieth century. But this trend implicitly assumes climate stability, adequate resources, and geopolitical calm. Just how secure will the world's six billion urbanites be if cities are besieged by climate change, rising sea levels, energy and food shortages, and violent conflict by mid- to late-century?

The final section examines the ecological leverage that cities can exercise in society's general quest for sustainability. What can cities do to decrease their eco-footprints and enhance their own prospects? How might rethinking the "city-as-ecosystem" help humanity to live sustainably within the carrying capacity of the earth?

Setting the Ecological Stage

While few people think of them as such, cities are biophysical entities. The fundamentally *organic* nature of cities is underscored by the fact that permanent settlements are actually a product of a change in human ecological circumstances. "The city" is an emergent phenomenon made possible by people's adoption of agriculture 10,000 years ago. Humanity's slowly developing ability to produce regular food surpluses triggered a truly "autocatalytic process—one that catalyses itself in a positive feedback cycle, going faster and faster once it has started" (Diamond, 1997). More food made higher population densities possible, enabled large permanent settlements with the specialized skills and inventiveness this implies, and shortened the time-spacing between children. This, in turn, enabled the higher populations to produce still more people, which increased both the demand for food and the technical and organizational capacity to produce it.

The first small, more-or-less permanent human settlements appeared barely 9000 years ago and another 3500 years passed before the first definable cities, with socially stratified societies and marked division of labour, emerged in southwest Asia (today's Middle East) around 5500 BCE. In short, while we tend today to take the existence of cities for granted, they actually have a remarkably short history. They have been part of human reality for merely 3 per cent of the time since "modern" humans—*Homo sapiens*—stumbled onto the world stage about 250,000 years ago.

But there is more to this story than "surplus food leads to urban civilization." The shift from the nomadic hunter-gatherer lifestyle to a more agriculture-based, settlement-centred way of life represents a major transformation of human ecological reality and may well constitute the most critical branch-point to date in the evolution of *Homo sapiens*. First, consider that with large-scale agriculture, people switched from merely taking what wild nature had to offer to manipulating entire landscapes in order to redirect as much as possible of nature's productivity to strictly human ends. In this way, humans became the most significant "patch-disturbance" species on earth (Rees, 2000). Indeed, agriculture and agriculture-induced urbanization constitute a great leap forward in an accelerating process that has gradually seen humans become the most important geological force changing the face of the planet.

Second—and, regrettably, given the enormous ecological impacts of industrial cities—the very process of urbanization insulates city dwellers from the negative consequences of their own eco-dysfunction. The migration of people to cities distances them physically from the ecosystems that support them and from the direct negative consequences of landscape degradation. In short, the separation of people's lives and livelihoods *from* the land diminishes urbanites' sense of felt connectedness *to* the land—humanity's apparent abandonment of the countryside critically reshapes people's spatial relationships and psychological sensitivities to nature. Thus doubly blinded, many urbanites, particularly in high-income developed countries, are blissfully unaware that they remain ecological actors and fail to see the growing threat their consumer lifestyles pose to the distant ecosystems on which they remain dependent.

Modern humans' failure to appreciate themselves as ecological beings reflects a deep cognitive bias. Our evolving techno-scientific paradigm has erected such a powerful perceptual barrier between humanity and the rest of the natural world that "human exceptionalism" has become a defining characteristic of industrial society. Certainly, modern urbanites do not generally think of "the city" in biophysical terms. Even urban scholars have only recently acknowledged and begun to study the human ecological dimensions of urbanization and cities. Most discussions of urbanization still view the process mainly as a demographic, economic, and technological phenomenon made possible by the intensification of agriculture, increased resource productivity, and various technological advances. The "smart cities" movement, characterized by the integration of multiple information and communication technologies to better coordinate urban activities and manage cities' resources for sustainability and improved well-being, is a recent example. In short, cities are perceived as concentrations of people; as areas dominated by the built environment; as places of intense social interaction; as the seats of government; as hotbeds of political conflict; as the nexus of national transportation and communication systems; and as the engines of national economic growth—but rarely as ecological entities embedded within the ecosphere. Some observers actually—and falsely—interpret urbanization as evidence that humanity is *transcending* nature, that the human enterprise is "decoupling" from "the environment."

This chapter is intended to address this perceptual gap. A major purpose is to show that, while urbanization represents a dramatic shift in urbanites' spatial/psychological relationships to the land, *there is no corresponding change in eco-functional relationships*. Indeed, far from reducing people's dependence on productive ecosystems, urbanization generally implies an increase in our per capita **ecological footprints**. From this perspective, urbanization and the modern city remain bio-ecological phenomena fully explicable only in light of human evolutionary history and fundamental biophysical laws. Failure to understand ourselves and our cities as ecological entities will doom our quest for global sustainability and expose an increasingly vulnerable global urban civilization to the spectre of collapse.

Cities and the Human Ecosystem

By now almost everyone interested in cities is familiar with the term **urban ecosystem**. Prominent urban analysts have long recognized that the city could be conceived as an ecosystem (e.g., Douglas, 1981), and today there is even a scientific journal called *Urban Ecosystems*. Nevertheless, the concept itself remains ambiguous. For example, a majority of the papers in *Urban Ecosystems* focus on the impacts of urbanization on plants and animals or on remnant "natural" ecosystems within the city. This shows that most natural scientists who study urban ecosystems cast the city as a somewhat unnatural habitat for *other* species.[1] To ecologists, the urban ecosystem consists of the assemblage of non-human species in the city, and the purpose of their inquiries is to determine how these species have adapted to the structural and chemical characteristics of the "built environment" (Rees, 2003). Remarkably, humans are excluded from the analysis except as their actions affect these other species. This conception of urban ecosystems is a clear reflection of the human exceptionalism, which sees people as separate from nature.

On the other hand, those who do acknowledge humans as the major ecological actors in the city err if they see "the city" per se as the modern human ecosystem. To qualify as a complete human ecosystem, a city would have to contain a sufficient complement of producer organisms (green plants), macro-consumers (animals, including humans), micro-consumers (bacteria and fungi), and abiotic factors to support its human population indefinitely. Any *complete* ecosystem consists of a self-organizing, self-sustaining assemblage of living species existing in complementary relationships with each other and the physical environment. Ecosystems are energized by the unidirectional cascade of solar energy and maintained in perpetuity by the continuous recycling of chemical nutrients.

Clearly, from this perspective modern cities are, at best, functionally incomplete *heterotrophic* (consuming) ecosystems. Some essential defining parts are missing altogether (e.g., virtually the entire relevant producer complex), and others (micro-consumers) are insufficiently abundant for functional integrity. Just as significantly, the spatial separation of people from the rest of their supportive ecosystems (e.g., agricultural and forest lands) inhibits the on-site application of organic matter and the recycling of phosphorus, nitrogen, and other nutrients contained in human wastes.

In effect, urbanization transforms local, integrated, cyclical ecological production systems into global, horizontally disintegrated, unidirectional, throughput systems (Rees, 1997). Ironically, the resultant continuous "leakage" of nutrients from farmland in shipments of food to cities (and ultimately the sea) threatens to undermine organic agriculture even as it gains ground in the urban marketplace.

It is instructive to think of "the city" as the human analogue of a livestock feedlot (Rees, 2003). Feedlots, like cities, are densely populated almost entirely by a single macro-consumer species—for example, cattle (or pigs, or chickens, which today are raised using even more constrained industrial methods). However, the grain fields that produce the feed for feedlot animals may be located hundreds of kilometres away. Also missing are adequate populations of micro-consuming decomposers. Having separated the functionally inseparable, industrial feedlots short-circuit even the possibility of within-system decomposition and nutrient recycling. As a result, vast quantities of manure containing vital nutrients are rarely re-deposited on range- or cropland for nutrient recycling but, rather, are disposed of inappropriately, contaminating soils and surface and subsurface waters at a distance and over large areas.[2]

Cities are obviously more ecologically complex than feedlots but the analogy holds in structural terms. The largest and functionally most important components of urbanites' ecosystems—the assemblage of producer organisms that feed city dwellers and provide them with oxygen; most of the micro-consumers that complete their nutrient cycles; and the various sub-systems that perform myriad other vital life-support functions—are all located externally, in rural areas scattered all over the planet. Also, like feedlots, cities generate enormous quantities of waste that they cannot assimilate internally, making cities the major source of pollution of the global commons. In short, cities are nodes of intense resource consumption and waste generation entirely dependent for their survival on the productive and assimilative capacities of distant hinterland ecosystems. In both ecological and spatial terms, "the city" constitutes only a small fraction of the total urban-centred human ecosystem.

The Human Ecological Footprint

The next questions are these: Just how extensive is the human ecosystem? How much of the productive capacity of the ecosphere do humans need to sustain urban industrial society? One way to approach these questions is through ecological footprint analysis (Rees, 1992, 2013; Wackernagel and Rees, 1996). A variant and extension of energy and material flows assessment (Haberl et al., 2004), eco-footprint analysis starts from a comprehensive inventory of the annualized energy and material flows required to support any specified population—an individual, a city, a nation, or the entire world. We also quantify the flow of waste carbon dioxide generated in the production/consumption process (carbon dioxide is the greatest waste by weight of industrial countries, hence the "carbon footprint"). Eco-footprinting is further based on the fact that many of these material and energy flows can be converted into a corresponding area of productive land and water ecosystems. Thus, we formally define the ecological footprint of a specified population as follows:

> The area of land and water ecosystems required, on a continuous basis, to produce the bioresources that the population consumes and to assimilate its (carbon-dioxide) wastes, wherever on Earth the relevant land/water is located.[3]

A population's eco-footprint (EF) depends on four factors: the population size; the average material standard of living; the productivity of the land/water base (whether local or "imported" in trade goods); and the efficiency of resource harvesting, processing, and use. Regardless of the relative importance of these factors and how they interact, every population has an ecological footprint. The productive land and water area captured by eco-footprint analysis represents much of the "natural capital" or *biocapacity* required to meet the study population's energy and material demands.[4]

Significantly, population EFs constitute mutually exclusive appropriations of nature. The biocapacity used by one population is not available for use by another. True, the grain grown in a particular region may wind up in export shipments to several countries, but the total area of cropland involved is the sum of the areas required by the individual populations. In the final analysis, *all human populations are competing for the limited biocapacity of the earth.*

Table 21.1 shows the equivalence-adjusted[5] per capita EFs and domestic biocapacities for a selection of countries from among the richest to among the poorest using 2008 data from the *Living Planet Report 2012* (World Wildlife Fund [WWF], 2012; see also WWF, 2014, 2016). Note the considerably larger load imposed on the ecosphere by wealthy, mainly urban consumers compared to that imposed by mainly rural peasants. The citizens of wasteful, high-income countries like Canada and the United States have average EFs of 6 to 10 global hectares (gha), or up to 20 times larger than the EFs of the citizens of the world's poorest countries, such as Bangladesh and Afghanistan. European countries and Japan typically have per capita EFs in the 4 to 6 gha range. China is

Table 21.1 The Eco-Footprints and Biocapacities of Selected Nations

Country	Per capita eco-footprint (gha)	Per capita domestic biocapacity (gha)	Overshoot factor
World	**2.7**	**1.8**	**1.5**
United States	7.2	3.9	1.8
Australia	6.7	14.6	0.5
Canada	6.4	15.0	0.4
Netherlands	6.3	1.0	6.3
Greece	4.9	1.6	3.1
France	4.9	3.0	1.6
United Kingdom	4.7	1.5	3.1
Germany	4.6	1.9	2.4
Japan	4.2	0.6	7.0
Malaysia	3.9	2.4	1.6
Hungary	3.5	2.8	1.3
Mexico	3.3	1.5	2.2
Brazil	2.9	9.6	0.3
Thailand	2.5	1.1	2.3
China	2.1	0.9	2.3
Peru	2.0	3.9	0.5
Nigeria	1.4	0.9	1.6
Ethiopia	1.1	0.7	1.6
Indonesia	1.1	1.4	0.8
India	0.9	0.4	2.3
Bangladesh	0.7	0.3	2.3
Afghanistan	0.5	0.4	1.3

Source: Data extracted from WWF, 2012.

fairly representative of the emerging economies, which show rapidly growing EFs of 1.5 to 3 gha. These data reflect the growing global income gap: 82 per cent of the wealth generated in 2017 went to the richest 1 per cent of the world's population while the poorest half of humanity—3.7 billion people—saw no increase (Oxfam, 2018).

The final column of Table 21.1 shows each country's "overshoot factor." This is a simple ratio of the national average eco-footprint compared to per capita domestic biocapacity. Countries with overshoot factors larger than one impose a greater burden on the ecosphere than could be supported by their domestic ecosystems. These countries are at least partially dependent on trade and on exploitation of the global commons to maintain their current average per capita consumption levels. The Netherlands, for example, uses six times as much productive land/water outside its borders as is found within the country. Japan's demand for biocapacity is seven times its domestic supply. Both the United States and China are also significantly in overshoot. All such countries are running "ecological deficits" with the rest of the world.

A few countries with overshoot ratios less than one are living within their "natural incomes" and thus seem to have ecological surpluses. They only *seem* to have surpluses, however, because the extra biocapacity in most cases is being traded away or otherwise employed to cover the ecological deficits of other countries. The agricultural, forestry, and fisheries surpluses of Canada, for example, serve a large export market. Trade, therefore, contributes proportionately to the ongoing degradation of the nation's soils, forests, and fish stocks (Kissinger and Rees, 2009).

Ominously, the world as a whole is in a state of overshoot (first line of Table 21.1). Human demand exceeded the earth's regenerative/assimilative capacity by about 50 per cent in 2008. With the erosion of biocapacity and rising per capita consumption, this number reached 65 per cent by 2012 (WWF, 2016). We are living and growing, in part, by depleting and dissipating as waste the fossil carbon fuels and even the "renewable" natural capital (fish, forests, soils, etc.) that have accumulated in ecosystems over millennia.

The Global Reach of Cities

Cities, of course, virtually all have in an ecological deficits. Urban populations are almost totally dependent on rural people, ecosystems, and life-support processes that are increasingly scattered all over the planet (Girardet, 2004; Newman and Jennings, 2008; Rees, 1992; 2003). In some respects, this relationship is a two-way, mutualistic one—rural areas benefit from urban markets, from the products of urban factories, from urban-based services, from technology transfers from urban areas, and so forth. However, while rural populations have survived historically without cities, the ecological dependence of urbanites on "the **hinterland**" is almost absolute. *There can be no urban sustainability without rural sustainability.* Understanding the nature of rural–urban interdependence is essential to understanding the total human ecosystem and to understanding **urban sustainability**.

In theory, estimating a city's eco-footprint is no different from estimating that of an entire country. In practice, however, matters are more complicated because of a lack of local data—no statistical or planning agencies monitor the flow of biophysically significant goods and services across municipal boundaries. While some urban EF studies do attempt to compile local data (Moore and Rees, 2013), others use "quick and dirty" extrapolation from national eco-footprint estimates, sometimes with adjustments for local conditions, income differences, and so forth. (e.g., Federation of Canadian Municipalities [FCM], 2005). The latter method produces more accurate city footprint numbers for highly urbanized, high-income countries than for less-urbanized, poorer countries.

So, just how great is a typical modern city's debt to the global countryside? Despite methodological and data-quality differences, urban eco-footprint studies invariably show that the EFs of typical modern high-income cities exceed their geographic or political areas by two to three orders of magnitude. Following are examples:

- Based on locally adjusted per capita EF estimates, the people of Toronto and Vancouver "occupy" an ecosystem area outside their municipal boundaries 292 and 390 times larger, respectively, than the cities themselves (FCM, 2005). Even the lower-density metropolitan areas of these cities have EFs 57 times bigger than the respective urban regions (Table 21.2). The citizens of Toronto and Vancouver might want to contemplate the implications of this growing extra-territorial dependence as they sprawl out over Canada's most productive farmland in an era of global change. Where will they turn when they can no longer import essential foods from distant elsewheres?

- Under varying management assumptions of cities' ability to cope with regional waste-management issues, Folke et al. (1997) estimated that the 29 largest cities of the Baltic region require, for resources and certain categories of waste assimilation, an area of forest, agricultural, marine, and wetland ecosystems 565–1130 times larger than the area of the cities themselves.
- With a population of 33 million and a per capita EF of about 4.2 gha, metropolitan Tokyo's total eco-footprint is 138,600,000 gha. However, the entire domestic biocapacity of Japan is only about 76,200,000 gha. In short, Tokyo, with only 26 per cent of Japan's population, lives on an area of productive ecosystems 1.8 times larger than the nation's entire terrestrial biocapacity.[6] Clearly, if Japan were required by changing global circumstances to subsist on its domestic biocapacity, the country would have difficulty supporting even the population of its capital city.
- Warren-Rhodes and Koenig (2001) estimated that Hong Kong, with almost seven million people, has a total eco-footprint of 332,150 km^2 to 478,300 km^2 (5.0–7.2 ha per capita) (the range reflects two estimates of carbon sink land requirements). Hong Kong's eco-footprint is at least 303 times the total land area of the Hong Kong Special Administrative Region (1097 km^2) and 3020 times the built-up area of the city (110 km^2).

These data show clearly that, in material terms, "sustainable city" is an oxymoron (Rees, 1997). Modern cities are ecological black holes sweeping up the productivity of a vastly larger and increasingly global resource hinterland and spewing an equivalent quantity of waste back into it. They are compact nodes of consumption living parasitically on the productivity and assimilative capacity of a vastly larger "undeveloped" area, much of which may be half a planet away.

While some have interpreted the consumptive and polluting power of cities as an anti-urban argument, it is nothing of the sort. Cities do have enormous ecological footprints; however, as we shall see, cities actually offer several advantages over more dispersed settlement patterns in the quest for sustainability (see Holden and Chang, Chapter 22; Senbel, Liem, and Lesnikowski, Chapter 24). Eco-footprinting also suggests several other paradoxes about current perceptions of cities. For example, why is the lifeless asphalt of the mall parking lot considered to be urban land, while cropland vital to the survival of the city is not? What does it mean for urban planning if 99.5 per cent of the de facto urban (eco)system lies outside the municipal boundaries, out of sight and beyond the control of those it supports? Is it not time to redefine the city-as-system?

Table 21.2 The Eco-Footprints of Toronto and Vancouver

City or region	Population (2006)	Per capita eco-footprint (gha)	Area (hectares)	Total eco-footprint (gha)	Ratio of EF to actual area
Vancouver	578,041	7.71	11,400	4,456,696	390
Metro Vancouver	2,116,580	7.71	278,736	16,318,832	57
Toronto	2,503,281	7.36	63,000	18,424,148	292
Greater Toronto	5,555,912	7.36	712,500	40,891,512	57

Source: EF data from FCM, 2005.

Finally, eco-footprinting underscores a material reality that is all but ignored in the sustainability literature—no individual, no city, and no country can achieve sustainability if the system of which it is a part is unsustainable. Vancouver, Toronto, or Montreal might become exemplars of sustainable urban design and lifestyles (Vancouver is trying with its "Greenest City" initiative), but if the global system of which they are a part remains on an unsustainable path, then even our model cities would be taken down by, for example, severe climate change, depleted resources, and resultant geopolitical instability. Given such interdependence, the best any sub-global system can attain independently is a state of quasi-sustainability. "Quasi-sustainable" describes that level of energy/material consumption per capita that, if extended to the whole world, would result in global sustainability (Rees, 2009). By 2012, quasi-sustainability implied a per capita eco-footprint of less than 1.7 gha (the equitable per capita share of global biocapacity [WWF, 2016]). Since Canadians' average eco-footprint exceeded 7 gha per capita in 2012, we would have to reduce consumption by 76 per cent to meet the quasi-sustainability standard (also known as "one-planet living" [Moore and Rees, 2013])!

The Vulnerability of Modern Cities

Increasing global interdependence obviously has enormous implications for the security of urban populations in an era of global change. Cities have grown so large and have such enormous eco-footprints, not because size necessarily confers great advantage, but simply because they could. Historically, **globalization** and trade have ensured the abundant supplies and uninterrupted flows of the energy and other material resources required to grow the modern metropolis. But this raises an increasingly awkward question in an era of global change: just how secure is any city of millions, or even a relative "town" of 100,000, if resource scarcity, shifting climate, or geopolitical unrest threatens to cut it off from vital sources of supply? There are several interrelated reasons to believe this is not an idle question. For example:

1. Reliable food supplies should be of increasing concern to urbanizing populations (Kissinger and Rees, 2009). The Food and Agriculture Organization (FAO) (2009) projections suggest that feeding a world population of 9.1 billion people in 2050 would require raising overall food production by some 70 per cent over 2007 levels by 2050. Developing countries would have to almost double production.[7] Achieving increases of this magnitude may be difficult. Assessing land degradation is notoriously difficult—estimates of the amount of degraded productive land vary from less than one to over six billion hectares (Gibbs and Salmon, 2014). Nevertheless, by some estimates, 24 billion tonnes of fertile soil or 10 million hectares of topsoil are lost every year and as much as a third of the earth's productive land is already degraded (Nkonya et al., 2016); enough once-arable land has been lost to production to feed 1.5 billion people (more than twice Europe's 2016 population). Meanwhile, depending on the climate and agricultural practices, topsoil is being dissipated 16 to 300 times as fast as it is regenerated. (FAO officials suggest that we have only about 60 years of harvests left [FAO, 2009, 2015]). So far, the impact has been masked because we have managed to substitute fossil fuel for depleted soils and landscape degradation, but that may be about to change.
2. Modern cities are made possible by abundant, cheap fossil fuel. No other resource has changed the structure of economies, the nature of technologies, the balance of geopolitics, and the quality of human life as much as petroleum (Duncan and Youngquist, 1999). Fossil fuels, especially oil, still supply about 80 per cent of humanity's total energy demand and are essential in agriculture, heavy transportation, construction and much electricity generation. Oil is a major factor in the so-called green revolution. Mechanization, diesel-powered irrigation, the capacity to double-crop, and agrochemicals (fertilizers and pesticides) made from oil and natural gas account for 79–96

per cent of the increased yields of wheat, rice, and maize production since 1967 (Cassman, 1999; Conforti and Giampietro, 1997). "Peak oil," including "peak tight oil" extraction, may therefore represent a singular event in modern history, posing a unique challenge to food production, transportation, geopolitical stability and, with them, urban security (Campbell, 1999; Duncan and Youngquist, 1999; Laherrere, 2003).[8]

3. To complicate matters, the world may actually have to abandon fossil fuels before we run out. Some scientists believe that climate change is the greatest threat to civilization and could even bring the world to the edge of anarchy (e.g., Center for Strategic and International Studies [CSIS], 2007; Schwartz and Randall, 2003). This is a conundrum for cities—they remain dependent on fossil fuels but continuing to use them risks climate catastrophe. Even modest shifts in weather patterns could disrupt historic precipitation regimes, thus undermining both agricultural production and urban water supplies. In *The Age of Consequences*, Washington's Center for Strategic and International Studies suggests that human-induced warming could end peaceful global integration as various nations contract inwardly to conserve what they need—or expand outwardly to *take* what they need—for survival. In the event of "severe climate change," corresponding to an average increase in global temperature of 2.6°C (now deemed increasingly likely), non-linear changes in biophysical systems will give rise to non-linear socio-political events. People will abandon regions where food and water are scarce and flee areas devastated by increasingly frequent severe storms, droughts, rising seas, and floods. Combined with food, energy, and water shortages, cross-border migration will impose great pressure on the cohesion of nations. War—even nuclear war—is possible (CSIS, 2007).

Such dismal scenarios seem increasingly likely in light of recent climate studies (e.g., Anderson and Bows, 2008; Hansen et al., 2008; World Bank, 2012). The 2015 Paris climate accord aims to keep the global temperature increase this century to well below 2°C, ideally to only 1.5°C, which would require maintaining atmospheric carbon dioxide (CO_2) levels below 450 and 400 parts per million by volume (ppmv) respectively. (The pre-industrial level was 280 ppmv). However, by early 2018 atmospheric CO_2 was already pressing 410 ppmv and rising along with other greenhouse gases. The combined climate forcing effect of all greenhouse gases had already reached the equivalent of 445 ppmv CO_2 by 2015 and was rising at 4 ppm per year (meaning that we are already 50 per cent likely to exceed the 1.5°C warming target [European Environment Agency [EEA], 2018]). Worse, according to Anderson and Bows (2008), "an optimistic interpretation of the current framing of climate change implies that stabilization of greenhouse gases much below 650 ppmv CO_2e [carbon dioxide equivalents] is improbable." To stabilize at 650 ppmv CO_2e, the majority of OECD nations will soon have to begin decarbonizing at rates in excess of 6 per cent per year, which might require a planned economic recession (Anderson and Bows, 2008). We should note that 650 ppmv CO_2e implies a catastrophic 4°C increase in mean global temperature, compared to the "mere" 2.6°C increase assumed in CSIS's already horrific "severe climate change" scenario.[9] Models suggest that 4°C warming would be sufficient to convert much of the United States, Southern Europe, China, India, Africa, and South America into uninhabitable wastelands (see Vince, 2009), displacing billions of people and jeopardizing prospects for maintaining global civilization.

Toward the "One Planet" City

> Industrialized world reductions in material consumption, energy use, and environmental degradation of over 90 per cent will be required by 2040 to meet the needs of a growing world population fairly within the planet's ecological means. (Business Council for Sustainable Development [BCSD], 1993)

This is a world in overshoot—we have exceeded long-term global carrying capacity—yet both population and per capita consumption are increasing and material expectations are rising all over the world. This is a fundamentally unsustainable situation—to raise just the present world population to North American material standards sustainably would require the biocapacity of three additional earth-like planets (Rees, 2013). The *really* inconvenient truth is that, to achieve sustainability, global energy and material throughput must decrease, not grow.

Techno-industrial society is a self-proclaimed, science-based society but we are not acting consistently with our best science. The risks associated with anticipated global change certainly demand major restructuring and may well require economic contraction. For sustainability with greater equity, wealthy countries will have to free up the ecological space necessary for needed growth in the developing world (Rees, 2008). High-income nations should therefore work to limit the energy and material throughput required to sustain urban life (Lenzen, Dey, and Barney, 2004; Newman and Jennings, 2008; Rees, 2009). For example, to achieve one-planet living, Canadians should be taking steps *now* to reduce their eco-footprints by the necessary 72 per cent from 6.4 gha to their "equitable earth-share" of 1.8 gha per capita (Table 21.1).

Clearly, achieving such targets will require a dramatic shift in prevailing economic beliefs, values, and consumer lifestyles—60–70 per cent of the material flows through cities are attributable to personal consumption. Although much of the technology needed to make the transition relatively painless is already available (von Weizsäcker et al., 2009), and "managing without growth" is both economically possible and could *improve* quality of life (Victor, 2008), there is scant evidence that the necessary cultural shift is underway. Certainly no national government, the United Nations, or any other official international organization has begun to articulate in public the social, regulatory, trade, tax, and related economic policies required to induce the needed reductions in energy and material throughput. Despite repeated warnings that staying our present course spells catastrophe for billions of people (MEA, 2005; UCS, 1992), the modern world remains mired in a swamp of cognitive dissonance and collective denial (Rees, 2009).

Mainstream responses to date not only fail to address the problem of ecological overshoot, but seem designed, instead, to reproduce the status quo by other means. Such "innovations" as electric cars, green buildings, **smart growth**, the **New Urbanism**, green consumerism, and even much of the eco-cities movement generally assume that we can achieve sustainability through technological innovation and greater material and economic efficiency alone. This is a conceptual error—historically, efficiency has actually *increased* consumption by, for example, raising incomes and lowering prices. With more money chasing cheaper goods and services, throughput rises; improved efficiency simply makes industrial society more efficiently unsustainable. By contrast, some eco-villages are inspired by social goals, such as sharing and community co-operation, that can result in reductions in the ecological footprint (Moos et al., 2006).

The Urban Sustainability Multiplier

> The climate crisis won't be solved by changing light bulbs and inflating your tires more, planting a tree and driving a little less. It's going to require a truly fundamental shift in how we build our cities and live in them. (Register, 2009)

Getting serious about urban sustainability obviously requires more determined action than society has yet been willing to contemplate. Fortunately, the very factors that make wealthy cities weigh so heavily on the ecosphere—the concentration of people and the localized intensity of energy/material consumption and waste generation—also give them considerable economic and technical leverage in shrinking their eco-footprints.

To enable cities to take full advantage of this leverage, provincial and municipal governments must create the land-use legislation and zoning bylaws that urban planners need to eliminate sprawl and consolidate and densify existing built-up areas. Compact cities have the potential to be vastly less energy- and material-intensive than today's sprawling suburban cities.[10] The economies of scale and the agglomeration economies associated with high-density settlements confer a substantial "urban sustainability multiplier" on cities (Rees, 1999). For example, Walker and Rees (1997) show that the increased density and consequent energy and material savings associated with condos and high-rises, compared to single-family houses, can reduce that part of the per capita urban ecological footprint associated with housing type and related transportation needs by about 40 per cent.

As noted, however, efficiency gains alone will not enable society to achieve "one-planet living" (Moore and Rees, 2013). Sustainability and security demand that cities everywhere become less consumption-driven and more materially self-reliant. Indeed, cities may be forced down this unfamiliar path by either or both the rising cost of oil-based transportation or the rapid phase-out of fossil fuels needed to avoid severe climate change (target: 80 per cent decarbonization by 2050). Certainly, there is no place for the fossil-fuelled automobile—or any cars at all—in the eco-cities of the future (Register, 2009). For these reasons, urban designers must begin now to rethink cities so they function as complete ecosystems. This is the ultimate form of bio-mimicry.

Bio-mimicry at the city level requires re-localizing many ecological and economic functions. The least vulnerable and most resilient urban ecosystem might be a new form of regional eco-city state (or bioregion) designed so that a densely built-up core is surrounded by its supportive ecosystems (see Grant and Filion, Chapter 12; Blay-Palmer and Landman, Chapter 23).[11] The central idea is to consolidate as much as possible of the city's productive hinterland (its eco-footprint) within close proximity to its consumptive centre. In effect, without preventing *essential* trade, this would enable our city-centred eco-regions to function as more nearly complete human ecosystems. Such a transformed home place, "rather than being merely the site of consumption, [would], through its very design, produce some of its own food and energy, as well as become the locus of work for its residents" (Van der Ryn and Calthorpe, 1986). Eco-city states would be less a burden on, and more a contributor to, the life-support functions of the ecosphere than contemporary cities.

Importantly, the bioregionalized city would reconnect urban populations both physically and psychologically to "the land." Because inhabitants would be, and would see themselves to be, more directly dependent on local ecosystems, they would have a powerful incentive—currently negated by imports—to manage their land and water resources sustainably in the face of global change. (Ideally, political control over the productive land and resource base of the consolidated region would pass from the provinces to new eco-city state governments.) Less reliant on imports, their populations would be partially insulated from at least external climate vagaries, resource shortages, and distant violent conflicts. Future urban bioregions could enhance their resilience in the face of rapid change by adopting the following policies:

- Re-localize local food production/processing, conserve regional farmland, and encourage local food co-ops.
- Re-localize production of clothing, furniture, housewares and other essential goods to minimize unnecessary travel and shipping while maintaining essential inter-regional linkages.
- Densify the urban core to increase efficiency of shared infrastructure—transportation, water, sewage, electrical and recycling systems—and to reduce demand for land.
- Otherwise reduce the per capita biophysical and economic costs of municipal services by capitalizing on economies of scale with neighbourhood-level infrastructure—electricity co-generation, heat pumps for

district heating/cooling, heat extraction from industrial waste heat systems, and so forth.
- Increase opportunities/incentives for housing co-ops, car-sharing, tool libraries, and other co-operative relationships that have lower capital requirements (lower consumption) per household.
- Discourage (and eventually eliminate) private auto use by investing in walking, cycling, and regional public transportation systems.
- Otherwise reduce demand for energy and eventually eliminating fossil energy consumption.
- Implement multiple redundant energy systems (wind, solar, hydro, etc.) as appropriate to specific location to buffer the consequences of a single system's failure.
- Encourage "industrial ecology"—low throughput and closed-loop industries—through taxes, licensing, and related positive incentives.
- Build to last—for example, penalize planned obsolescence of physical structures and manufactured goods.
- Encourage the adoption of one-earth lifestyles—"social contagion" can help reduce per capita demand for energy, material, and water resources by facilitating the spread of sustainable lifestyle choices.

Note that if the entire world were organized into an interconnected system of self-reliant bioregions, each managed in a way to conserve adequate per capita stocks of natural capital, the aggregate effect would be global sustainability.

Canada is one of the few countries with sufficient space and resources for many of its cities to be reorganized along bioregional lines; with political commitment, the bioregional ideal could also guide ongoing urbanization in developing countries. Regrettably, it is no longer an option for many large consumer cities, particularly in small countries. Remember Tokyo with its eco-footprint spanning two Japans? The best such megacities can do in the face of global change is to reduce their material demands as much as possible and hope their national governments can negotiate reliable supplies of vital resources from areas that still have surplus capacity. (Otherwise, some cities may have to disperse.) Clearly, not all urban regions can be self-sustaining, but in a crowded world every urban agglomeration has an obligation to contribute to global sustainability by taking appropriate action within its own jurisdiction (McGranahan and Satterthwaite, 2003).

Of course, "appropriate action" will vary greatly. This discussion has focused mainly on the sustainability implications of global change for wealthy cities. But many millions of people live under deplorable conditions in the barrios, **slums**, and squatter settlements of Third World cities. The greatest environmental problems facing these people are poor diets and the lack of potable water and sanitary sewage. Unlike the already wealthy, the world's urban poor would benefit from *increased* incomes and *greater* consumption. In this light, an appropriate sustainability strategy for rich and poor cities alike would be to manage gross consumption so that average per capita eco-footprints converge from above and below toward 1.8 gha, each person's equitable share of global biocapacity.[12]

Conclusion

Historically, environmental concerns about cities were confined to the local public health effects of air, water, and land pollution. Today, the fourfold expansion of human numbers and the order of magnitude increase in economic activity during the twentieth century alone has raised ecological concerns to the global level.

Paradoxically, however, the world community remains addicted to material growth. Many studies of urban metabolism (e.g., Brunner, 2007) seem concerned mainly with alternative technologies and better resource management that would enable cities to maintain their growth trajectories. Some do recognize the accelerating degradation of earth systems, but imply a fairly smooth succession to the point where "modern megacities will . . . begin to climax when global fossil fuel reserves are exhausted and global water and food resources are maximally utilized" (Decker et al., 2000). Few acknowledge the possibility of implosion or urge the kind of dramatic response to prevent it that now seems justified by global change science.

This chapter has attempted to make that leap. I argue that we are witnessing the dissipative destruction of globally essential ecosystems and vital life-support functions and that the process is accelerating with population growth and rising material expectations. Given the increasing probability of severe climate change, resource shortages, large-scale population displacements, and resultant geopolitical chaos, the world should be focused on corrective action *now*. The transition to sustainability will be anything but smooth and predictable if there is further delay.

Environmental scientists are sometimes dismissed as purveyors of gloom and doom. However, if global change science is correct, it is defenders of unfettered individualism and the growth-based status quo that are leading us toward the abyss. Theoretically, we can still turn "spaceship Earth" around—humans are an intelligent species uniquely capable of rational analysis, extensive co-operation, and planning ahead to shape a desirable future. However, to exercise these qualities, global society must soon acknowledge that we are in an unprecedented crisis. We must also recognize it as a collective problem requiring collective solutions. No person, city, or nation can be sustainable on its own. We are all part of the same ecosystemic hierarchy in which human subsystems are utterly dependent on maintaining the operational integrity of the whole (the ecosphere itself).

Individual action helps, of course; ordinary people should abandon frenetic consumerism and embrace the personal and community benefits of living in sufficiency (FOE, 2018). However, the really heavy lifting will require government intervention in the economy *for the common good*. Individuals cannot implement the ecological tax programs, trade regimes, public transit projects, and so forth, required for sustainability. Even mainstream economists should agree that government has a major role—after all, human-induced global change is indicative of gross market failure.

Indeed, achieving global sustainability will require a concerted multi-layered *inter*-governmental program of sustainability planning that incorporates population reduction, articulation of less material-intense but more satisfying lifestyles, and experiments in reshaping cities in the image of natural ecosystems.

Certainly, business as usual is not an option. Rich countries will have to abandon the idea of continuous economic growth and learn to share more equitably the economic and ecological output of the planet. The immediate material objective of the great restructuring should be to reduce gross consumption and waste production to match the regenerative capacity of the ecosphere. The ultimate goal is to create a positive future for all, characterized by ecological stability, greater social equity, and enhanced economic security.

Transition toward an ecologically sustainable urban society is a 100-year project and there is no excuse not to begin immediately. Certainly, the problems of both developed and developing world cities are well documented, and many partial solutions have been proposed (e.g., Marcotullio and McGranahan, 2007; Martine et al., 2008; Satterthwaite, 1999). Some of the best and most accessible handbooks for urban sustainability are explicitly based on treating cities as true ecosystems (Newman and Jennings, 2008; Register, 2006). The only question is whether our growth-addicted global culture is capable of acknowledging and responding effectively to the challenge. Our species is facing its greatest test: whether high-intelligence, reasoned analysis, and common interests can trump base emotions, instinctive intransigence, and myopic self-interest. Humanity can theoretically continue to thrive but only if urban civilization is able to adapt purposefully to living more equitably within the regenerative capacity of the only planet most of us will ever know.

Review Questions

1. Why are cities not functioning as true ecosystems?
2. What are the implications of the ecological footprint perspective in terms of the global environmental impacts of cities, and in terms of cities' environmental vulnerability?
3. What can be done to reduce cities' ecological footprint?
4. What are the main environmental threats on the horizon for cities?

Notes

1. One paper in the March 2009 issue of the journal even struggles with the question of whether "the ecosystem concept [is] relevant when humans are part of the system" (Pickett and Grove, 2009). The remaining eight articles focus on other aspects of non-human ecology.
2. Since livestock feedlots are sub-systems of the human urban industrial system, it is not surprising that they are similar to cities in eco-structure.
3. For fuller details of the method, including inclusions, exceptions, and limitations, see Rees (2006, 2013), WWF (2012, 2014, 2016), and links at https://www.footprintnetwork.org/
4. Eco-footprint analysis (EFA) underestimates the total human impact on the earth because it captures only those demands that can readily be converted into corresponding, mutually exclusive ecosystem areas (e.g., food and fibre production, carbon sequestration, built-up areas). No single indicator can represent all relevant variables; biodiversity loss, most pollutants (including the hundreds of toxic chemical residues accumulating in our food chain), and other sources, are not reflected in EFA.
5. To enable fair comparisons among countries, the national EF and biocapacity data in Table 21.1 are presented in terms of standardized or "global" hectares (gha)—that is, the equivalent area of ecosystems of global average productivity.
6. The area of Japan is only about 37,770,000 ha, but Japan's terrestrial ecosystems are considerably more productive than the world average, which increases the country's biocapacity to over 76,000,000 gha.
7. This situation is complicated by the diversion of grain, especially maize, to biofuel production.
8. Conventional crude extraction peaked in 2006, and it is doubtful whether the output of tar sands, shale oil, and other forms of unconventional petroleum will keep pace with the accelerating depletion of conventional oil fields (see Hughes, 2013, 2018). Despite the hype over wind and solar electricity, renewable alternatives are not yet major contributors to the energy budget and are often poor substitutes for fossil fuels.
9. Even the generally conservative World Bank (2012) agrees that the world is on track for 4°C warming and that this "must be avoided."
10. Many North Americans fear density, but the City of Vancouver illustrates that excellent urban design can actually draw families from the suburbs to live in high-density urban communities. In 2008, the city actually had to restrict further residential high-rise construction in its high-amenity downtown core because it was displacing office development.
11. For a history and philosophy of the bioregional movement, see Carr (2005).
12. Note that as the human population increases and productive ecosystems are degraded, this "equitable earth-share" declines.

References

Anderson, K., and A. Bows. 2008. "Reframing the climate change challenge in light of post-2000 emission trends," *Philosophical Transactions of the Royal Society A* 266: 3863–82. At: http://rsta.royalsocietypublishing.org/content/366/1882/3863.abstract

Business Council for Sustainable Development (BCSD). 1993. *Getting Eco-Efficient* (Report of the BCSD First Antwerp Eco-Efficiency Workshop, Nov. 1993). Geneva: BCSD.

Brunner, P.H. 2007. "Reshaping urban metabolism," *Journal of Industrial Ecology* 11(2): 11–13.

Campbell, C.C. 1999. *The Imminent Peak of World Oil Production*. At: www.hubbertpeak.com/campbell/commons.htm

Carr, M. 2005. *Bioregionalism and Civil Society: Democratic Challenges to Corporate Globalism*. Vancouver, BC: University of British Columbia Press.

Cassman, K.G. 1999. "Ecological intensification of cereal production systems: Yield potential, soil quality, and precision agriculture," *Proceedings of the National Academy of Science* 96: 5952–9.

Center for Strategic and International Studies (CSIS). 2007. *The Age of Consequences: The Foreign Policy and National Security Implications of Climate Change.* Washington, DC: CSIS. At: www.csis.org/media/csis/pubs/071105_ageof consequences.pdf

Conforti, P., and M. Giampietro. 1997. "Fossil energy use in agriculture: An international comparison," *Agriculture, Ecosystems and Environment* 65: 231–43.

Decker, E.H., S. Elliott, F.A. Smith, D.R. Blake, and F.S. Rowland. 2000. "Energy and material flow through the urban ecosystem," *Annual Review, Energy and Environment* 25: 685–740.

Diamond, J. 1997. *Guns, Germs, and Steel: The Fates of Human Societies.* New York: Norton.

Douglas, I. 1981. "The city as an ecosystem," *Progress in Physical Geography* 5: 315–67.

Duncan, R.C., and W. Youngquist. 1999. "Encircling the peak of world oil production," *Natural Resources Research* 8: 219–32.

European Environment Agency (EEA). 2018. *Atmospheric greenhouse gas concentrations.* At: https://www.eea.europa.eu/data-and-maps/indicators/atmospheric-greenhouse-gas-concentrations-10/assessment

Federation of Canadian Municipalities (FCM). 2005. *Ecological Footprints of Canadian Municipalities and Regions.* Edmonton, Alberta: Report for the Federation of Canadian Municipalities prepared by Anielski Management. At: www.anielski.com/Documents/EFA%20Report%20FINAL%20Feb%202.pdf

Folke, C., A. Jansson, J. Larsson, and R. Costanza. 1997. "Ecosystem appropriation by cities," *Ambio* 26: 167–72.

Food and Agriculture Organization (FAO). 2009. *Global Agriculture towards 2050*, Rome: Food and Agriculture Organization of the United Nations. At: www.fao.org/fileadmin/templates/wsfs/docs/Issues_papers/HLEF2050_Global_Agriculture.pdf

———. 2015. *Status of the World's Soil Resources*, Rome: Food and Agriculture Organization of the United Nations.

Friends of the Earth Europe (FOE). 2018. *Sufficiency—Moving beyond the gospel of eco-efficiency.* At: www.foeeurope.org/sufficiency

Gibbs, H.K. and J.M. Salmon. 2014. "Mapping the world's degraded lands," *Applied Geography* 57: 12–21.

Girardet, H. 2004. "The metabolism of cities," in S. Wheeler and T. Beatley, eds., *Sustainable Urban Development Reader.* London: Routledge.

Haberl, H., M. Fischer-Kowalski, J. Krausmann, H. Weisz, and V. Winiwarter. 2004. "Progress toward sustainability? What the conceptual framework of material and energy flow accounting (MEFA) can offer," *Land Use Policy* 21: 199–213.

Hansen, J., M. Sato, P. Kharecha, D. Beerling, R. Berner, V. Masson-Delmotte, M. Pagani, M. Raymo, D.L. Royer, and J.C. Zachos. 2008. "Target atmospheric CO_2: Where should humanity aim?" *The Open Atmospheric Science Journal* 2: 217–31.

Hughes, J.D. 2013. *Drill Baby, Drill—Can Unconventional Fuels Usher in a New Era of Energy Abundance?* Santa Rosa, CA: Post Carbon Institute.

———. 2018. Shale Reality Check—Drilling Into the U.S. Government's Rosy Projections for Shale Gas & Tight Oil Production Through 2050. Santa Rosa, CA: Post Carbon Institute.

Kissinger, M., and W.E. Rees. 2009. "Footprints on the prairies: Degradation and sustainability of Canadian agricultural land in a globalizing world," *Ecological Economics* 68: 2309–15.

Laherrere J. 2003. "Forecast of Oil and Gas Supply to 2050," paper presented to Petrotech 2003, New Delhi. At: www.hubbertpeak.com/laherrere/Petrotech 090103.pdf

Lenzen, M., C. Dey, and F. Barney. 2004. "Energy requirements of Sydney households," *Ecological Economics* 49: 375–99.

McGranahan, G., and D. Satterthwaite. 2003 "Urban centres: An assessment of urban sustainability," *Annual Review of Environmental Resources* 28: 243–74.

Marcotullio, P., and G. McGranahan, eds. 2007. *Scaling Urban Environmental Challenges—From Local to Global and Back.* London: Earthscan.

Martine, G., G. McGranahan, M. Montgomery, and R. Fernández-Castilla, eds. 2008. *The New Global Frontier—Urbanization, Poverty and Environment in the 21st Century.* London: Earthscan.

Millennium Ecosystem Assessment (MEA). 2005. "Living beyond our means: Natural assets and human well-being (Statement from the Board)," *Millennium Ecosystem Assessment.* At: www.millenniumassessment.org/documents/document.429.aspx.pdf

Moore, J., and W.E. Rees. 2013. "Getting to one planet living," in *State of the World—Is Sustainability Still Possible?* Washington, DC: Island Press.

Moos, M., J. Whitfield, L. Johnson, and J. Andrey. 2006. "Does design matter? The ecological footprint as a planning tool at the local level," *Journal of Urban Design* 11: 195–224.

Newman, P., and I. Jennings. 2008. *Cities as Sustainable Ecosystems.* Washington, DC: Island Press.

Nkonya E., Mirzabaev A., von Braun J. 2016. "Economics of Land Degradation and Improvement: An Introduction and Overview," in Nkonya E., Mirzabaev A., von Braun J., eds., *Economics of Land Degradation and Improvement—A Global Assessment for Sustainable Development.* Springer, Cham. doi.org/10.1007/978-3-319-19168-3_1

Oxfam. 2018. *Reward Income, Not Wealth.* Oxfam briefing paper. Oxford: Oxfam UK. At: https://d1tn3vj7xz9fdh.cloudfront.net/s3fs-public/file_attachments/bp-reward-work-not-wealth-220118-en.pdf

Pickett, S.T.A., and J.M. Grove. 2009. "Urban ecosystems: What would Tansley do?" *Urban Ecosystems* 12: 1–8.

Rees, W.E. 1992. “Ecological footprints and appropriated carrying capacity: What urban economics leaves out,” *Environment and Urbanization* 4: 120–30.

———. 1997. “Is ‘sustainable city’ an oxymoron?” *Local Environment* 2: 303–10.

———. 1999. “The built environment and the ecosphere: A global perspective,” *Building Research and Information* 27: 206–20.

———. 2000. “Patch disturbance, eco-footprints, and biological integrity: Revisiting the Limits to Growth,” in D. Pimentel, L. Westra, and R. Noss, eds., *Ecological Integrity: Integrating Environment, Conservation and Health*. Washington, DC: Island Press.

———. 2003. “Understanding urban ecosystems: An ecological economics perspective,” in A. Berkowitz et al., eds., *Understanding Urban Ecosystems*. New York: Springer-Verlag.

———. 2006. “Ecological footprints and biocapacity: Essential elements in sustainability assessment,” in J. Dewulf and H. Van Langenhove, eds., *Renewables-Based Technology: Sustainability Assessment*. Chichester: John Wiley and Sons.

———. 2008. “Human nature, eco-footprints and environmental injustice,” *Local Environment: The International Journal of Justice and Sustainability* 13: 685–701.

———. 2009. “The ecological crisis and self-delusion: Implications for the building sector,” *Building Research and Information* 37: 300–11.

———. 2013. “Ecological footprint, concept of,” in S.A. Levin, ed., *Encylopedia of Biodiversity*, 2nd edn., Vol. 2. Waltham, MA: Academic Press.

Register, R. 2006. *EcoCities: Rebuilding Cities in Balance with Nature*, rev. edn. Gabriola Island, BC: New Society. See also: “An interview with Richard Register, author of *Ecocities: Building Cities in Balance with Nature*.” At: www.sustainablecityblog.com/2009/03/richard-register-interview/

———. 2009. “Cities can save the earth,” *Foreign Policy in Focus*, 12 May. At: www.fpif.org/fpiftxt/6113

Satterthwaite, D., ed. 1999. *Sustainable Cities*. London: Earthscan.

Schwartz, P., and D. Randall. 2003. *An Abrupt Climate Change Scenario and Its Implications for United States National Security*. Washington, DC: A report commissioned by the US Defense Department.

Union of Concerned Scientists (UCS). 1992. *World Scientists' Warning to Humanity*. At: www.ucsusa.org/about/1992-world-scientists.html

Van der Ryn, S., and P. Calthorpe. 1986. *Sustainable Communities: A New Synthesis for Cities and Towns*. San Francisco, CA: Sierra Club Books.

Victor, P. 2008. *Managing without Growth: Slower by Design, Not Disaster*. Cheltenham: Edward Elgar.

Vince, G. 2009. “Surviving in a warmer world,” *New Scientist* 201, 2697: 29–33.

von Weizsäcker, E., K. Hargroves, M. Smith, C. Desha, and P. Stasinopoulos. 2009. *Factor 5: Transforming the Global Economy through 80% Increase in Resource Productivity*. London and Droemer, Germany: Earthscan.

Wackernagel, M., and W.E. Rees. 1996. *Our Ecological Footprint: Reducing Human Impact on the Earth*. Gabriola Island, BC: New Society.

Walker, L., and W.E. Rees. 1997. “Urban density and ecological footprints: An analysis of Canadian households,” in M. Roseland, ed., *Ecocity Dimensions*. Gabriola Island, BC: New Society.

Warren-Rhodes, K., and A. Koenig. 2001. “Ecosystem appropriation by Hong Kong and its implications for sustainable development,” *Ecological Economics* 39: 347–59.

World Bank. 2012. *Turn Down the Heat: Why a 4°C Warmer World Must Be Avoided*. Washington DC: World Bank. At: http://documents.worldbank.org/curated/en/2012/11/17097815/turn-down-heat-4%C2%B0c-warmer-world-must-avoided

World Wildlife Fund (WWF). 2012. *Living Planet Report 2012*. Gland, Switzerland: World Wide Fund for Nature.

———. 2014. *Living Planet Report 2014*. Gland, Switzerland: World Wide Fund for Nature.

———. 2016. *Living Planet Report 2016*. Gland, Switzerland: World Wide Fund for Nature.

22 The Ups and Downs of a Sustainable and Climate Resilient Development Path in Canadian Cities

Meg Holden and Robin Chang

Introduction

Canadian cities sit in a catch-22 position when it comes to planning to meet sustainability and climate **resilience** goals. On the one hand, there is a long-lived backlash, amongst planners and urban professionals, against sprawling suburban landscapes. It is taken as a duty of Canadian urbanists to fight against their spread. On the other hand, the cities that have most successfully reversed the suburbanization trend by creating "livable" downtowns and compact, complete communities are witnessing serious declines in housing affordability, compromising sustainable cities in other ways. This chapter examines both sides of this contemporary **urban sustainability** dilemma in Canadian cities.

The trouble with these conflicting urban climate change and sustainability action dynamics becomes clear when considered in terms of how Canadian cities have tried to improve their sustainability and climate performance. In this chapter, we examine the ups and downs of suburban and more compact development forms through Canadian cases, including the Metro Vancouver Housing + Transportation Affordability Study and Transit-Oriented Affordable Housing Study, and studies of greenhouse gas (GHG) emissions profiles. We consider the intended and perverse consequences of living-first and family-first downtown policies in Vancouver and Toronto. We also introduce efforts to take bold moves toward urban density beyond livable downtowns in Vancouver's Making Room initiative, and to achieve and certify an integrated model of sustainable neighbourhood solutions across Canada and around the world.

Addressing climate change in Canadian cities involves its own ups and downs that only partially overlap with sustainable cities work. In the second half of this chapter, we consider the particular demands that climate change makes of urbanists, from mitigation goals in Edmonton to the Coastal Flood Adaptation Strategy in Surrey. In closing, we introduce the concept of urban resilience to Canadian cities to provide opportunities and dynamics of success rather than particular end states—incomplete rather than complete streets, for example, acupunctural land-use activation, and inclusive citizen engagement in "depaving."

Sprawl as the Sustainable City's Nemesis

The classical Canadian suburban form carries the label "sprawl." Sprawl has held the status of sinister villain amongst urbanists and planners in Canada for several generations. Picture single detached homes arranged in rows on spacious, grassy lots—set back from the neighbours, from the street, and from other land uses by wide roadways designed for private automobiles. This is the image of **urban sprawl**. The exclusionary zoning and planning practices that produce sprawl have been blamed for enabling environments that create enclaves of middle- or upper-class socio-economic status, discouraging social and economic mixing and local economic development. Sprawl is also blamed for limiting opportunities for resident health and well-being by prioritizing time for long-distance automobile commuting over other activities, like physical exercise and social time together with family or neighbours. These long commutes and automobile-dependent transportation patterns are considered disproportionately to harm children, the elderly, new immigrants, and women.

By encroaching on farmland and natural areas, sprawl also reduces biodiversity and ecosystem integrity, and worsens environmental risks through high rates of paving that exacerbate water-borne pollution and flooding risks. It is a development pattern that demands excessive water and energy to maintain large homes, gas-powered vehicles, and mono-cultured, chemical-dependent landscaping. It destroys and constrains wildlife habitat. Sprawl often also has negative implications for social capital among neighbours, who are seldom at home in bedroom communities, who have limited common spaces for neighbourhood interaction, and whose values in their neighbourhood are dominated by private property rather than a shared neighbourhood mentality. Sprawling suburbs are costly for municipalities to maintain because of the length and quantity of infrastructure that must be provided for them to approach urban levels of service for necessities like roads, sewers and sanitation, water, energy, and emergency services. Many of these costs imposed by sprawl are hidden from the people making decisions to live or work in those environments (Blais, 2010; Downs, 2005; Soule, 2006; Thompson, 2013; Woodcock et al., 2011). Let's examine two hidden costs of sprawl more closely: transportation costs and GHG emissions costs.

Sprawl Costs: Metro Vancouver Housing + Transportation (H+T) Cost Burden Study

The cost of housing and the cost of transportation are inversely linked in the majority of Canadian cities. In past decades, the cost of housing in urban central cores has risen more than has been the case in suburban areas. Central core areas also hold the bulk of jobs and educational opportunities, as well as the bulk of alternatives to private cars in getting from A to B. Homes in the suburbs are more affordable than homes in the central city, but are associated with increased transportation costs. There is not sufficient data on how much more suburban households spend on daily transportation compared to urban households. In practice, the only realistic option for many house-seekers across the country has been to "drive (away from the central city) until you qualify (for a mortgage)." This excludes the additional costs of driving itself that ought to be factored into the affordability equation.

For decades, policy-makers have relied on the 30 per cent rule of thumb for defining housing affordability—namely, that housing is affordable if it costs no more than 30 per cent of gross household income. While useful for understanding what kinds of rents households can reasonably afford, this rule of thumb is silent on the second major component of cost of living: transportation. In 2015, Metro Vancouver undertook a new kind of cost analysis of the affordability of living in the region's different municipalities. Contrary to typical approaches that compare only the cost of housing as a proxy for cost of living, the H+T study takes into account both housing and commuting costs in relation to work or school. Commuting costs include the amount spent on public transit as well as automobile insurance, use, maintenance, fuel, and parking. In summary, the study found that, when both the cost of housing and the cost of transportation are factored into household costs, it is the far-flung suburban city of Langley that is the costliest place for renters to live in the Vancouver region, not the City of Vancouver proper. This study begins to bring some of the costs of sprawl out of hiding, so that people making locational choices can better account for them (Metro Vancouver, 2015).

If studies like the H+T study are to make a difference, urban policy change is needed to support better locational decision-making by households. In the case of Metro Vancouver, the regional government has teamed up with the provincial government, the regional transportation authority TransLink, and other partners to create a Transit-Oriented Affordable Housing Study. The ultimate goal of this study is to expand the availability of housing affordable to households earning less than $50,000 annually in transit-oriented locations, region-wide. This effort is expected to create more opportunities for people with lower incomes to live in transit-efficient places,[1] both reducing the transportation expenses of housing for these households, and cutting down on the larger package of "sprawl-related costs" for everyone in the region (Metro Vancouver, 2018).

Sprawl Costs: Greenhouse Gas (GHG) Emissions

Hoornweg et al. (2011) grapple with the toll of sprawl in terms of calculating how sprawl exacerbates GHG emissions. They find that compact urban form can cut GHG emissions in half when compared with a sprawling built form. They provide evidence from different national contexts, that big cities are better at reducing per capita GHG emissions. By their measures, average per capita emissions in New York City (10.5 tCO_2e/capita[2]) are half those in Denver (21.5 tCO_2e/capita), per capita emissions in Toronto (11.6 tCO_2e/capita) are just over half those of Calgary (17.7 tCO_2e/capita), and those in Seoul (4.1 tCO_2e/capita) are less than half those of the Republic of Korea as a whole (11.46 tCO_2e/capita). This research group has also found this relationship to hold for central cities compared to the suburban fringe. Residents in the Toronto city core produced 6.42 tCO_2e per capita on average, compared to 7.74 tCO_2e for suburban Torontonians (VandeWeghe and Kennedy, 2007).[3] Examining the policy approaches that have shown the greatest impact on reducing GHG emissions profiles for cities, Hoornweg et al. (2011) document the dominant impact of land-use planning policies that encourage compact city development. The biggest impacts of land use decisions on reducing GHG emissions are in reducing local travel demand, increasing the use of public and active transportation, discouraging private automobile use, and zoning to promote multi-family and connected housing types.

Transportation and GHG emissions are two significant examples of how "sprawl," in almost every respect, acts as a stand-in for unsustainable urban development. The concept of sustainable urban development articulates a goal toward which all of our work in cities should be oriented—namely, providing for the needs of present generations such that future generations will have the ability to meet their own needs (United Nations Conference on Environment and Development [UNCED], 1992). An underlying assumption of the idea of sustainable development is that efforts on this path can create a "triple bottom line," where economic, social, and environmental sustainability all positively correlate. And, even beyond a correlation, sustainable development thinking assumes that seeking something better for cities than sprawl in economic, social, and environmental terms can lead to "synergistic" or "holistic" solutions that go beyond the outcomes that could be attained by addressing any one objective by itself. In this way, moving toward thinking and acting for sustainable development in cities cannot be encapsulated in any particular list of policies or initiatives, but is best thought of as a "development path." The concept of a development path was defined by the 2007 Intergovernmental Panel on Climate. Change (IPCC) Fourth Assessment Report as

> a complex array of technological, economic, social, institutional, cultural, and biophysical characteristics that determine the interactions between human and natural systems, including consumption and production patterns in all countries, over time at a particular scale (Sathaye et al., 2007: 696).

The costs of sprawl are like a series of wrong turns that have oriented cities down an unsustainable development path. It is by rethinking the direction of this development path entirely that we can move cities toward sustainability.

Charting a Sustainable Development Path: The Up Side of Sprawl?

The notion of changing cities' development paths to move toward compactness, sustainability, and climate resilience is theoretically sound. But does it fit with Canadian values? Canada's major cities have opted over the past generation for compact development (Neptis Foundation, 2014). This has been an effective strategy, in the suburbs and in the urban cores alike, in the sense that compact or complete communities have become home to an increasing share of Canadians. At the same time, a substantial proportion of Canadians value their suburban lifestyles and resist the introduction of urbanization strategies in their neighbourhood. In fact, efforts to change the development path of many suburban Canadian communities toward compactness and higher density have met increasing public resistance in recent years. In a survey conducted by the Real Estate Foundation of BC, for example, the majority of respondents across age groups and genders, in rural and urban communities alike, expressed dissatisfaction with public consultation in the development process. Just over half of British Columbians felt that ordinary people do not have enough of a say in decisions about their neighbourhoods. Almost 40 per cent of British Columbians *living in cities* thought that little or no future development should consist of compact, high-density, high-rise buildings (Real Estate Foundation of BC, 2015).[4] In numerous instances, this resistance to densification by existing residents has delayed or prevented the development of new multi-family housing units (Quastel et al., 2012). When they reach a point at which scarce redevelopment opportunities remain in the city's central core, Canadian cities face new kinds of challenges in advancing policy and planning initiatives to densify the remainder of the city.

Across Canada, suburban places from Don Mills, Ontario, to Garrison Woods in Calgary have continuous and ongoing popular appeal to Canadians. In fact, although Canada refers to itself as one of the most urbanized nations in the world, David Gordon proposes that Canada became a suburban nation in 1981 and has continued to grow more suburban ever since (Gordon and Shirokoff, 2014). Meanwhile, other Canadian urban scholars have taken a different tack to understanding the suburban phenomenon in Canada. This different tack has entailed research to demonstrate how the places we call suburbs have been themselves steadily urbanizing since the 1970s, based upon different measures of urbanization. So, for example, many Canadian suburbs have undergone land-use intensification, the introduction of mixed-use nodes, rapid transit and secondary town centres, and have introduced complete communities and sustainability strategies. An outcome of the new brownfield and greyfield redevelopments are new suburban-urban hybrid places of medium- and high-density communities where industrial uses or shopping malls once stood (see Grant and Filion, Chapter 12).

So where does the truth lie? In fact, it lies in both suburban and urban answers—depending on how these are defined. Gordon and Shirokoff (2014) define suburban places based on automobile-dependent commuting patterns, whereas Grant and Filion (see Chapter 12) look at housing density. In terms of density, Canadians are urbanizing; in terms of mobility choices, the development path is proving more resistant to alternatives to automobile dependency.

Is There a Downside to Density?

Since the 1970s, the City of Vancouver and a growing suite of other Canadian cities have promoted housing intensification. These policies, initiatives, and processes propelled the central area of Vancouver from a state of industrial and commercial decline to its current renown near the apex of global rankings for livability (Holden and Scerri, 2013). Whereas the default for planning downtowns had been to encourage the use of central cores primarily as economic engines, Vancouver's "living-first downtown" policies set the stage for complete community development in all Canadian urban downtowns. A "living-first" or livability-focused development path for central cities integrates a rich social mix of residents, along with institutional, commercial, park, and livability-related land uses, at a level of intensity to make non-automobile transportation economical (Punter, 2003; Berelowitz, 2005) (Figure 22.1). Non-automobile transportation has accounted for the majority of trips in the City of Vancouver since 2014.

A trend that now appears not just in Vancouver but in all major Canadian cities, this livability-focused development path has a darker side (Peck, 2005; White and Punter, 2017). In fact, as much as the shift toward "living-first downtowns" may at one time have appeared as a triumph of the anti-sprawl position and the value of promoting human community over the pursuit of capital growth, it seems now that the shift we have experienced is toward new forms of capital growth. In Vancouver and Toronto in particular, the cost of land and housing in the central cities has skyrocketed, leaving a very different picture of "living-first downtowns" than what can been seen in a typical suburban Canadian grocery store checkout or schoolyard. Living-first planning, as an approach, has no difficulty holding up its success in attracting a rich social mix. The

Michael Wakely

Figure 22.1 This neighbourhood landscape in Vancouver depicts some of the other sustainability features that become possible as housing density increases, including active transportation, ecological landscaping and "depaving," and more opportunities for neighbourly interactions.

problem is that the mix of people in the everyday spaces of these core areas is increasingly stratified, socio-economically. Fewer and fewer can afford to live, work, and play full-time. Increasing shares of the social mix come from the people who can only afford to stay for coffee, people who can only afford to work there serving the coffee, and people who sleep outside in the high-quality public places.

At the same time as they grow in popularity, support and enthusiasm for higher-density neighbourhoods and attached housing are tempered by growing negative associations including a sense of lifestyle unaffordability, polarizing class dynamics, crowding, loneliness and social isolation, and lack of neighbourliness and community life. Bundled together, these trends can be thought of as the dark side of living-first cities. These all factor into a shifting sense of Canadians' support for and hesitancy about the value of compact urban living, sprawling suburban living, and everything in between.

In an extensive review of housing markets in Canadian cities, the Canada Mortgage and Housing Corporation (2018) uncovered that, particularly for the cities where growth in housing demand has been strongest, Vancouver and Toronto, the supply of new single detached houses is decreasing.[5] New construction of condos, on the other hand, has increased. The scale of this trend is larger than what can be explained by housing affordability alone. So what is going on? Most of the country's highest-paying jobs are located in dense cities, and residents of these cities have the most disposable income, which they are willing to spend on housing, while being less willing to sacrifice in distance from the centre. As a result, single detached homes simply are no longer an option for a growing share of urban professionals.[6] This shift in preference for housing that is locationally efficient over housing that is bigger and ground oriented demands new thinking and action by urban professionals to make sure high-density areas are designed and equipped to meet the needs of their populations. In Toronto, upon the realization that 32 per cent of households with children in the Toronto area live in mid- and high-rise buildings (whereas only 3.8 per cent of units in buildings built between 1996 and 2011 had three or more bedrooms), the city launched a set of "Growing Up Guidelines" to ensure that multi-unit housing in high-density communities was better designed to accommodate the needs of households with children and youth (City of Toronto, 2016).

What If Creating Compact, Livable Downtown Cores Was the Easy Part?

If our goal is to guide Canadian cities toward a more sustainable development path, the achievements of compact downtowns may, in fact, have been the "easy part." The harder parts of tackling automobile dependency, inequity, and distinguishing the economic from the livability value of homes are still in need of solutions. Canada's cities and those fortunate enough to own real estate in them have realized the economic value of greater housing density. But, what happens as this drive for density moves into more traditional neighbourhoods, outside of the downtown? One impact that is now widespread in the Canadian urban landscape is the trend of secondary suites as an increasingly common feature of single-family-zoned neighbourhoods. Secondary suites, also called laneway or coach houses, "granny" or "nanny" flats, "mortgage helpers" or (typically in the United States) "accessory dwelling units," are small homes located on the same lot as a primary home, sometimes as a garage conversion or basement suite. Looking at the Metro Vancouver region, the regional government estimates that there were between 85,340 and 93,620 secondary suites in the region in 2014, representing 26–29 per cent of the rental households in the region. Numbers of secondary suites are growing across Canada. The Canada Mortgage and Housing Corporation (CMHC, 2018) found that for every 10 single detached dwellings started in 2014, approximately 8 "mortgage helpers" started alongside. Historically, these units have typically been illegal. However, many municipalities now permit and even encourage this trend. For many growing cities, this type of housing fits the bill for the lamented "missing middle" housing that is needed to better meet households' demands for housing that is in the middle of density and cost ranges. "Missing middle" is a term used to describe housing that sits in between apartment towers and single detached dwellings in terms of density and massing, as well as the sense that this middle range of housing types is underprovided in contemporary Canadian cities.

Taking a step that would be considered anathema to the notion of a single-family-zoned neighbourhood a decade ago, the City of Vancouver introduced the Making Room initiative in summer 2018. Making Room permits duplexes as well as laneway houses and basement suites on all single-family-zoned parcels city-wide (City of Vancouver, 2018).[7] In certain neighbourhoods that are closer to commercial areas and rapid transit hubs, the Making Room initiative goes further to allow triplexes, quadruplexes, and low-rise apartment buildings where only single-family houses were historically permitted.

This initiative is being advanced to address affordable housing shortages through "gentle density," or density that is advanced in forms and at rates that are incremental to the height, massing, and number of residential units per building, compared to neighbouring lots. It holds a range of implications for sustainability. The argument in favour of "gentle density" takes the anti-sprawl discourse into account by emphasizing green space within existing neighbourhoods, that building heights should not be excessive, and that road width and speed should be minimized. On the other hand, as quoted in *The Globe and Mail*, one citizen speaking at a Vancouver public hearing asserted that this move "will destroy what makes Vancouver such a green and charming city" (Bula, 2018). Beyond the question of divergent values, this example elevates the need for urban professionals to ask, "What is an acceptable rate of change in Canadian communities?" This, in turn, implies a need for effective public engagement. Effective public engagement provides an opportunity to understand a given local context and improve our collective understanding of pressing land-use issues and acceptable solutions. Public participation experiences are also key opportunities to increase citizens' sense of responsibility for and ownership of their cities and neighbourhoods. An engaged public, invested with a sense of civic responsibility and mutual aid, is in an ideal position to contribute to long-term planning for decisions about their communities that affect long-term sustainability. Urban planning and policy professionals in most Canadian cities are learning to engage with the public as allies to effectively leverage the potential of deliberation. Their intention is to guide outcomes in order to increase the perceived value of the ingredients for sustainability and climate responsibility that cities offer.

Model Sustainable Neighbourhoods in Canadian Cities

The neighbourhood scale has long made sense as the right scale for people looking to build, plan, orient, and organize their lives differently from what is offered in mainstream culture. The idea of designing and building an urban neighbourhood as a model of sustainable living in the city has origins in the thought of sustainable urbanists like Richard Register (2009), Manuel Ruano (1999), and Paolo Soleri (2001). The idea has intuitive appeal: rather than attempt a skinny street here, a parklet there, a community garden in that sector, and a green building across the city—a combination of ecological, economic development, social, governance, and cultural components of sustainability are sited together in one manageable piece of the city. Doing so, we may create the kind of holistic result that would allow people in that neighbourhood to feel the difference in a visceral way, and be more willing to change their behaviour than we are when told to do something that is "better for the planet."

Some countries, like France and China, are pursuing model sustainable neighbourhood developments in a top-down, centralized way. Others, such as Denmark and India, pursue these neighbourhoods as demonstration projects, usually demonstrating advances in energy and smart systems technologies. A few countries, including Sweden and Singapore, have designed sustainable neighbourhood-development approaches for export to other countries. Canada, like the United States, Australia, and other countries, is pursuing the development of model sustainable neighbourhoods at the voluntary will of private sector urban-development companies and city leaders, but with little support from central government, so far.[8]

After about 15 years of catching on as a niche practice in urban sustainability planning, policy, architecture, engineering, and design, Canada has a limited number of models to show for its efforts. Victoria's Dockside Green (Songhees and Esquimalt First Nations territory) (see Figure 22.2), and Vancouver's Southeast False Creek (Musqueam, Squamish, and Tsleil-Waututh territory), are two of the most established and best-known models, whose first phase opened to residents in 2008 and 2010, respectively. Other neighbourhoods are earlier on in their construction, such as Zibi on Chaudière and Albert Islands in Ottawa-Gatineau (Algonquin and Mohawk territory), and Blatchford in Edmonton (Enoch Cree territory). Beyond the models that these neighbourhoods offer of higher-density, mixed-use urban living, they showcase different advances in energy efficiency and neighbourhood energy systems, water and waste-management systems, priorities for active transportation, and efforts in local economic development, higher-quality public and common spaces, and social mixing. Building-specific energy efficiency technologies employed include Passive House and Net Zero—the Canadian government has committed to instituting a "net-zero-ready" building code by 2032, as part of our Paris Agreement commitments. Further

Meg Holden

Figure 22.2 This building is a cornerstone of the Dockside Green model sustainable neighbourhood in Victoria, BC. In addition to its green building features, such as its green roof and solar and wind infrastructure, it also offers social features, such as the neighbourhood café and neighbourhood greenspace, as well as access to the Galloping Goose bike trail.

attention and evaluations are needed in order to understand the impacts that these and other comparable efforts are having on their cities and city-life possibilities.

While some actors in this realm of sustainable building and development emphasize the context-dependency of any model sustainable neighbourhood, efforts to articulate generalizable principles to guide and evaluate what a model sustainable neighbourhood is and does are on the rise. From the Ahwahnee Principles (Local Government Commission, 1991) to LEED (US Green Building Council, 2013) or One Planet Living (Bioregional, 2017) certification, such systems can be helpful to orient theorists and practitioners to what it takes to make and maintain a model sustainable neighbourhood in a contemporary city. The trouble is, the number and variety of these heuristic tools is multiplying faster than the efforts to demonstrate the work in practice! In order to understand the nature of the potential and realized contribution that model sustainable neighbourhoods make to our cities at home and abroad, we need to understand the values and motivations of those who are undertaking the work of designing, building, financing, and then living, working and playing in model sustainable neighbourhoods (Sturgeon et al., 2016).

Approaching Resilience in Canadian Cities

Is planning for urban sustainability the same as planning for climate safety and climate change reduction? Urbanists differ in how they answer this question. The difference often depends upon the distinction between *mitigation* and *adaptation*—reducing the risk of change versus getting ready for impending change, respectively. For those who take a climate change mitigation view, the only worthy sustainability work in Canadian cities should be measured in terms of the GHG emissions reductions that the work achieves. For example, the City of Edmonton, through its *The Way We Green* policy, takes a mitigation approach to climate change policy. It sets a goal of becoming "a carbon-neutral city"; this is consistent with climate change planning because it is a plan to reduce GHG emissions. Specifically, Edmonton plans to reduce its GHG emissions from city operations by 50 per cent by 2020 (based upon a 2008 baseline), and by 100 per cent to reach "neutrality" at some point in the long-term future (City of Edmonton, 2012). Such an approach supports global climate action plans and agreements, like the Paris Agreement,[9] and signals political will to change business as usual in order to "fight" climate change. It has also triggered a steep learning curve in Canadian cities related to how to measure and account for invisible GHG emissions as a liability (and sometimes also as a tradeable commodity).

Climate Resilience and Urban Sustainability Are Not Always Commensurate Goals

While accounting for GHG emissions is an important and necessary step in recognizing the need for a stable and predictable climate system, Canadian cities increasingly recognize that this cannot be where urban climate policy ends. Reducing GHG emissions may advance certain urban sustainability goals, but its approach to mitigation is not necessarily consistent with all the urban sustainability goals discussed earlier in this chapter. For example, to reduce GHG emissions in the urban transportation system, we could reduce private vehicle use, by changing our land-use patterns and increasing the viability of public, active, and shared transportation systems. By changing the organization of space so that people are less inclined to drive rather than opt for other, more climate-friendly means of mobility, we are doing climate change mitigation work that also helps us adapt over time to a way of life that is less GHG-intensive, healthier, and better for sustainability overall. Another way to reduce the GHG intensity of urban transportation could be to incentivize the use of electric and biofuel-powered private vehicles. This latter approach may have the same mitigation potential as the first in the short term, but over the long term, does not share the same kinds of sustainability co-benefits as the first approach. It would not lead to any efforts to change land-use patterns, or the private automobile mobility preference in general, and would therefore have questionable results for urban sustainability. Higher-level examples of how the Government of British Columbia has classified the fossil fuel natural gas as "clean energy," in the same category as solar or wind (Bailey and Stueck, 2012), or how the Government of Canada insists that oil pipeline expansions are needed in order to meet national GHG reduction targets (Jaccard, 2018), illustrate the complex politics generated by trying to reconcile climate change action with sustainability goals.

Climate Adaptation and Urban Resilience Planning

Along the continuum of policies that Canadian cities are developing to respond to the realities of climate change is climate adaptation planning. Such strategies do not target GHG emissions reductions but instead focus on developing plans to adapt the city's physical, institutional, and social infrastructure such that the city is able to function well and protect lives and property in the face of climate-induced crisis (Figure 22.3). The case of the City of Surrey's Coastal Flood Adaptation Strategy (CFAS) represents one leading urban climate adaptation policy approach.

The City of Surrey's land area includes about 20 per cent coastal floodplain. Historically, the city has managed this space with sea dykes, sea dams, ditches, pumps, and spillways (Figure 22.4). These measures have worked effectively to open up this floodplain area to residential development, agricultural development worth over $100 million annually in farm revenue, popular parks, beaches, and recreation areas, and many kilometres of important transportation, water and energy infrastructure. Currently, over 1500 people live in Surrey's coastal floodplain, many in the Crescent Beach neighbourhood and others in the Semiahmoo First Nation. However, the mitigation measures that have served these areas well to date are not expected to hold up in a climate-changed future, with rising sea levels, increased frequency

Chris Boulton

Figure 22.3 An example of climate adaptation infrastructure, this dry pond in a park in Edmonton is designed to also serve a stormwater management function.

Robin Chang

Figure 22.4 This older breakwater infrastructure at Crescent Beach, Surrey, will not be sufficient to hold back the rising sea level of the next 20 years. This realization prompted Surrey to undertake the Coastal Flood Adaptation Strategy (CFAS).

of storm surges, precipitation, and flooding. Surrey anticipates that it will face multiplying threats of water ingress: higher tides from an elevated sea level along with storm surges, higher river levels from increased precipitation, reduced field drainage from increased flooding, and even higher river levels in spring from increased glacial melting in the mountains that feed the Serpentine and Nicomekl Rivers. Along the British Columbia coast, sea levels are expected to rise at least one metre by 2100.

Surrey's answer was to launch the CFAS. Between 2016 and 2018, the CFAS included five phases: (1) city-wide education and awareness-building around what is at risk and what the community values, (2) exploring the adaptation options, (3) developing adaptation strategies, (4) detailing the preferred strategies, and (5) reporting back to the community. The CFAS is innovative not only in how it tackles climate change adaptation and resilience, but also because of its inclusive engagement process. Whereas protecting residents from risk and danger might in some strategic approaches imply a government that is working to reassure residents about the strength and capacity of engineered solutions, the CFAS takes strides toward transparency with residents even when conventional protections are insufficient and risk failure. For Surrey, this new transparency has meant not only making projections and models available to residents, but also translating these data into visual models and timelines, maps and scenarios, and community meetings at which residents are asked to contribute their values-based assessments, ideas, and alternatives, along with council reports. The City of Surrey has gone even farther in its public engagement through initiatives that raise the popular visibility of the risks at hand in the coastal floodplain—for example, a photo contest for nature, storm, and activity photos in the affected areas.

The technical risks to public and private property associated with climate change disasters are severe. Nevertheless, risks are just as relevant to the life and health of all communities, and even more so for already designated at-risk populations. While the risks are greatest in the coastal area, Surrey expects dyke infrastructure nearly 10 kilometres inland to be vulnerable to being overtopped by 2040. Risks consist of

- a narrowing of the intertidal zone, as sea levels rise on one side and concrete dykes on the other, reducing the habitat available in this critical migratory route for birds;
- habitat risks to other fish, animals, and plants;
- reductions in salmon spawning area;
- loss of beloved public spaces, including Surrey's only sandy public beach;
- loss of access to trails and recreation areas during storm events;
- threats to key transportation routes;
- threats to key infrastructure corridors serviced by utilities, railways, and the freight and passengers that travel the 10 kilometres of affected provincial highways;
- threats to evacuation routes;
- potential loss of vehicles that could be flooded off of roadways;
- potential floodwater damage to the ground floors of homes; and
- potential failures to electric infrastructure.

Surrey's CFAS is being examined as a model by neighbouring municipalities, including the City of Vancouver, as they realize that they have too much at stake not to plan differently to achieve urban resilience, which is discussed next.

Urban Resilience Planning

Increasingly, climate strategies are referred to as neither mitigation nor adaptation-focused, but under the common banner of resilience. The notion of resilience has multiple origins. As an ecological concept, it originated as a measure of persistence in a dynamic of instability. The ecological approach to resilience was defined by Holling (1973: 14) as "the ability of a system to return to an equilibrium state after a temporary disturbance." The adaptability of a system to a host of threats and challenges over time would provide the best measure of resilience in this ecological understanding.

Within the discipline of engineering, resilience is defined and used somewhat differently to describe systems that are "fail safe" in the sense that crises and extreme events have been forecasted, modelled, and incorporated into system designs that are able to maintain efficient and safe functioning even when things fall apart. Based on this understanding, the measure of resilience success would be less related to adaptability and more related to the sheer speed at which a system could return to "business as usual" under a worst case scenario disruption.

Davoudi and others (2012) introduced an essential addition to the ecological or engineering dichotomy in thinking about urban resilience. By introducing resilience thinking to cities, the idea is to help provide new tools for intentional and transformational change. The argument is that resilience framing could prove useful in planning if it assists planners in their work to make change more palatable for change-averse people, and that transformative change opportunities might be sought where disruption has already been thrust on a city through stress or crisis. This has sometimes been referred to as an evolutionary approach to resilience. A key aspect of evolutionary resilience thinking is that, in an urban planning context, the resilience dynamic should not be limited to rising to a challenge to "bounce back" from a disaster event to a pre-disaster state. Instead, resilience should be thought of as actually "bouncing forward" toward a new state, perhaps never before achieved. According to an evolutionary way of thinking about resilience, crisis can break down pre-existing structures, institutions, and relationships, opening up space for new pioneers and structures and opportunistic relationships, the likes of which could scarcely have been imagined previously. Whereas in ecology this process may tend toward re-establishment of a similar structure as the one that existed before, and in structural engineering the system would be considered a failure if it did not revert to its previous functional state, an evolutionary process in a city can represent a breakthrough to a new state, particularly if demographic, infrastructural, financial, soil, energy, water, or climate conditions are altered drastically by the disaster.

The evolutionary approach to resilience also holds more weight when it comes to considering the social dimensions of resilience. Working from a social learning perspective, Lakey (2015) promotes the use of resilience as a way to describe effectiveness in a social movement. This effectiveness becomes a condition in which individuals and groups exhibit mental and social strength in the face of crisis that facilitates the maintenance of well-being under adverse conditions.

A socially useful plan for community resilience may help articulate the injustices that often seem as unmovable as mountains or mature forests. It may also help recognize in advance where certain kinds of system shocks or drastic changes might propel opportunities to improve the standing or life opportunities of disadvantaged people in a community, or close the social gap between the "haves" and the "have-nots." This pursuit of a social justice frame of resilience would work toward an improved state of social justice and community well-being compared to that which existed prior to a shock. In considering the work that the concept of resilience may yet do in Canadian cities, it is worth keeping in mind that nothing specifically ties this concept to a response to climate change. Indeed, "resilience" is a term urban professionals may also use in relation to an economic transition, industrial shift, social health, or response to disease.

Based upon careful comparative consideration and synthesis of the growing host of definitions of urban resilience, Meerow et al. (2016: 39) arrive at this definition:

> Urban resilience refers to the ability of an urban system—and all its constituent socio-ecological and socio-technical networks across temporal and spatial scales—to maintain or rapidly return to desired functions in the face of a disturbance, to adapt to change, and to quickly transform systems that limit current or future adaptive capacity.

Urban Resilience as Adaptive Management

As the concept of urban resilience has entered into policy practice in Canadian cities, it has been treated as an exercise in articulating the desired outcome of community planning with an emphasis on aversion of particular disasters. Cities are constantly exposed to natural disasters, but climate models as well as recent memories of the 2013 Southern Alberta flood and the 2016 Fort McMurray wildfire suggest that catastrophes are becoming more normal. The June 2013 flood in southern Alberta, for example, was caused by the concurrence of heavy rainfall an snowpack melting in the Bow and Elbow Rivers, forcing the declaration of 32 states of local emergency and the evacuation of 175,000 residents; downtown Calgary was inaccessible for a week, and the cost of recovery was in the billions of dollars (Haney and McDonald-Hunter, 2016). Resilience thinking, applied to cities, permits and gives structure to planning for a diverse range of possible futures, such that these options can be considered more clearly by residents. In turn, the community can be positioned to determine the kinds of future social, economic, and built configurations that are most desirable, and perhaps also the best means to move toward a preferable future in spite of shocks and surprises. This iterative process of envisioning the future, taking steps toward that vision, taking stock and changing course along the way, is known as adaptive management (Folke, 2006; Wilkinson et al., 2010; Wardekker et al., 2010). The ability of a community to engage in adaptive management is known as its adaptive capacity.

Put another way, adaptive management toward urban resilience may begin with an assessment of the risk of certain historical disasters or disturbances, followed by planning for the best techniques and resources needed to respond effectively to each of these. This kind of urban resilience planning puts a premium on consideration of dynamics of change and uncertainty, promoting "resilience in cities" as opposed to an imaginary fail-safe "resilient city" (Ernstson et al., 2010). Strategies of adaptive management consider structures and systems throughout the city as being in a state of flux, whether this flux is detectable or not, and as offering an opportunity to learn about optimizing these flows through practice (Ahern, 2011). Urban resilience thinking also sometimes resists the tendencies of planning toward comprehensiveness, instead valuing principles of self-organization, bottom-up action, and experimental and collaborative approaches taken by diverse groups with diverse expertise (Chang, 2018; Folke et al., 2003; Olsson et al., 2004; Wilkinson et al., 2010). In this way, we can think of cities pursuing a resilience-oriented planning and development strategy as adaptively managing their development path. What defines a city's development path is a mostly uncoordinated assemblage of choices made at a range of formal and informal levels of decision-making, and a range of scales from the individual and ultra-local to the influences of global markets and migrations. Complicating matters further, the locus of decision-making when it comes to development paths is fragmented:

> Decisions about the development of the most significant sectors that shape emission profiles—energy, industry, transportation and land use—are made by ministries and companies that do not regularly attend to climate risks. The same is true for even more indirect influences on these sectoral pathways, including financial, macro-economic, and trade practices and policies (Sathaye et al., 2007: 696).

Using the concept and logic of the development path, the political, social, and cultural case for adaptive shifts toward sustainability, climate resilience and risk reduction need to be made on multiple fronts in order for the plans and policies we make to guide our cities to be consistent and progressive. On their own, environmental, sustainability, climate, and resilience plans and initiatives will prove futile if Canadian cities keep treading the same development path from the past generation.

Vancouver: Adaptive Experimentalism in False Creek Flats

Impromptu, voluntary, and citizen-led initiatives generate enthusiasm around planning initiatives that, while entirely consistent with an adaptive management approach, could not be replicated by a city government acting in a top-down way. The sense of tactic and the incomplete nature of planning that goes into such initiatives are key to their appeal, and may be important to the integration of resilience values into the planning and development of Canadian cities. We present two examples here: one of an attempt by the City of Vancouver to integrate adaptive experimentalism into planning for the False Creek Flats, and one a citizen-driven effort operating in cities across Canada.

Since at least 2012, the City of Vancouver has pursued strategies of adaptive and temporary uses of public space (Lydon and Garcia, 2015; Oswalt et al., 2013). Improving the pedestrian experience has inspired other cities like Winnipeg to try something similar (City of Vancouver, 2017a; Wohl, 2017; Downtown Winnipeg BIZ, 2016). On top of an improved pedestrian experience, Vancouver can now point to these same interventions as a form of adaptive management toward climate resilience. Perceived as an industrial area in decline, the False Creek Flats area is projected to be partially under water within the next century, given the expected sea level rise in False Creek. Always a low-lying area, this increased risk of flooding has driven new design guidelines anticipating higher street elevations (City of Vancouver, 2017b). The area is unsuitable for long-term residential use but remains central to the city's intent for more innovative and strategic employment and industrial space planning and programming, which will include live-work spaces. In preparing the redevelopment plan for this area, the city has consulted with business stakeholders and neighbouring communities, via an ideas competition and several rounds of interviews, surveys, workshops, and open houses (City of Vancouver, 2017c, p. 3). The City's engagement with local communities has also been counterpointed by international engagement through its membership in the C40 Cities network and the inclusion of the neighbourhood's proposed Innovation Hub as a component of its entry in the C40 *Reinventing Cities* competition (C40 Cities, 2019). Planning for the Flats has to contribute to the city's sustainability goals, and specifically to its goal to increase access to green jobs while securing industrial land for new and creative uses (City of Vancouver, 2019). The specific target is to diversify employment options and triple the job density between 2017 and 2047 on the 450 acres of the site from roughly 10,000 to over 30,000 jobs (Vancouver Economic Commission, 2017). This change represents a doubling of total employment floor area from 5.4 million square feet to over 11 million square feet (City of Vancouver, 2017d).

The new Flats planning framework strives to intensify innovative forms of land-use activation through flexible and stacked land reuse and conversion that spatially supports core and "back-of-house" industrial functions (City of Vancouver, 2017d). Intended for new mixed-use developments on the city-owned lots in the area, new licensing processes have been established to allow small to mid-size innovative and creative initiatives access to new "mobile," "orphan," and "temporarily underutilized" spaces. The range of policy initiatives proposed will support amenities including but not limited to co-location, resource sharing, and community finance mechanisms that take advantage of the Business Improvement Area status as well as management bodies such as a non-profit industrial development corporation (Vancouver Economic Commission, 2017) to facilitate affordable workspaces and rentable live-work options that cater to artistic and entrepreneurial lifestyles. As for the 36 acres of city park land in the area, the plan is for multiple initiatives including urban forestry, stormwater management, flood-risk protection, and biofiltration. This and existing residential areas aim to "thoughtfully" connect and will frame and integrate the False Creek Flats area into the established urban landscape (City of Vancouver, 2017d).

Depave Paradise: Adapting Ecologies for (In)complete Streets

Another facilitator of adaptive land use for urban climate resilience is the community organization Depave Paradise, which helps convert land and aims to revalue both ecosystem qualities and social qualities in public space. The civic organization works to support voluntary community projects that renew and reclaim neglected urban spaces by depaving impermeable surfaces. This opens up public space to greener, more ecosystem-friendly landscape treatments and social uses, decreases pollution from asphalt and concrete runoff, allows better soil permeability, and lightens the load on municipal stormwater systems. Depave Paradise encourages the increase of biodiversity through native landscaping and planting diverse flora, natural stormwater management and water infiltration through natural drainage, lessening urban heat islands, and GHG emissions reductions (Depave Paradise, 2018b). Sponsored by Green Communities Canada, Depave Paradise has already established projects in 29 locations in 15 cities and 5 provinces. The growth in popularity and spread of this initiative suggests a social and cultural readiness across Canadian cities to engage differently with public spaces and to create new stories about the values of public spaces. Depave Paradise statistics show that from 2012 to 2018, the work of the initiative has diverted roughly 4700 m^3 of stormwater annually (about the volume of water held by two Olympic-sized swimming pools), while almost 800 kg (the weight of a Volkswagen Beetle) of pollution has been kept out of waterways (Depave Paradise, 2018a). Across all projects, thousands of volunteers come out to lend a hand in depaving, embedding in these volunteers a sense of responsibility, contribution to sustainability and resilience goals, and the value of such engagement.

Whereas a previous generation of sustainability planners may have heralded such an initiative as leading toward "complete streets," a critical discourse is emerging around the corollary value of "incomplete streets." In their work on incomplete streets, Zavestoski and Agyeman (2015) argue that paved public spaces are relics of an auto-centric paradigm as promoted by the "complete streets" and complete communities ideals. They argue for a critical examination of the efforts to green and activate transportation planning in cities in the past 15 years, and for more emphasis on promoting inclusive and citizen-engaged work to replan the uses of our streets and other paved areas. While the transportation planners and engineers who have long been entrusted with planning our cities' paved areas and networks have a preference for "complete" streets planning, an adaptive management and social-evolutionary approach to urban resilience calls for putting more of the work in the hands of the public. Doing this will certainly result in "incomplete" streets, but from the evolutionary or transformative standpoint of resilience, we should not see this as a negative outcome. While the underlying logic of complete communities has been one of comprehensive self-sufficiency, the logic underlying incomplete streets is one of adaptiveness, equity, and inclusion. At the same time, incomplete street design is inspired by landscape architecture concepts that create working green, urban landscapes as they re-establish hydrological balance, healthy wildlife habitat, and can "harness natural processes in the service of the urban environment" (Girling and Kellett, 2005: 58).

Depave Paradise's self-initiated urban greening projects illustrate societal acceptance for more adaptive, integrated, sustainable, and green models that rebalance urban design. A clear willingness to reduce barriers between natural and urban systems and strengthen opportunities to integrate quality urban design and ecosystem services is emerging (Girling and Kellett, 2005: 14) through more adaptive and civic-led "green" models.

Conclusion: Toward an Urban Development Path at the Nexus of Sustainability and Resilience

This chapter has presented an understanding of the challenges and opportunities faced by Canadian cities related to aspirations for sustainability and crises of climate change. Throughout, we have talked about how different trends, values, patterns, and policies in cities can be understood as relating to the need for action on these fronts in different ways—that is, that lead toward a new development path for Canadian cities.

Cities in Canada and around the world generate the lion's share of GHG emissions—perhaps as much as 80 per cent. The explanation offered for this is that while the concentration of development, residents, and workers in space through compact urban planning provides possibilities for energy and emissions efficiencies, it equally provides the possibility of multiplier effects of all kinds. On the one hand, cities amplify income, wealth, purchasing power, consumption, and production; but on the other hand, most of this amplified activity generates GHG emissions. Was Bill Rees (1992; also see Rees, Chapter 21) right when he said a generation ago that "However bright its economic star, every city is an ecological black hole"? How does this calculus account for the political and cultural work that cities do today to raise awareness and political will for moves toward different lifestyle modes that are less GHG intensive?

Common to the three cases of urban resilience planning discussed here—Surrey's Coastal Flood

Adaptation Strategy, False Creek Flats planning for a future flood zone, and the Depave Paradise organization supporting citizens in their desire to rip up the pavement in their local public spaces—are elements of building urban resilience capacity, focused specifically on addressing climate change and sustainability challenges. The environmental implications of such models clearly demonstrate potential improvement in management of urban landscapes, better biodiversity outcomes, and improved water quality and flow. The reuse and repurposing of urban land in this way provides proof that environmental and infrastructural change can be mobilized to benefit human and ecological communities (Lokman, 2017: 5). At different scales, in different institutional structures, with different priorities and opportunity spaces, these initiatives all work to advance a new development path for Canadian cities in a context of sustainability threats and climate change risks. All of them experimental, with uncertainty embedded from point of initiation forward, these examples demonstrate samples of progressive work in Canadian cities. A critical perspective is needed to see these projects for their consequences along the way, so that the pursuit of resilience and sustainability in our cities remains a goal with social, ecological, and economic justice values built in. After all, despite the strong indication of an urban resilience approach to sustainability goals, it remains to be seen how and to what extent urban resilience work will fully contribute to social and ecological sustainability. A lesson exists in the complete shift in understanding of the social justice implications of a "living-first downtown" considered in this chapter, from a planning and development project worth fighting for in the 1980s, to a social justice nightmare that now haunts planners in Canada's gentrifying urban cores.

These examples also demonstrate a recalibration of sustainability aims away from equilibrium thinking and toward a dynamic "safe-to-fail" resilience perspective (Ahern, 2011: 341–42). The difference that this new perspective makes to practice is that it aims for strategies of value even under stressors and strains that have not yet been experienced, as opposed to strategies that are expected never to fail under conditions that hold constant. In other words, flexibility in planning and design can encourage transdisciplinary approaches to research and practice, and the attendant capacity building associated with this. This shift in perception of the strategies that constitute sustainable urban development reflects new potential for innovations across fields such as planning, urban design, ecology, engineering, natural resource management, and sociology.

There are numerous points of commonality between the pursuit of urban resilience and the pursuit of urban sustainability. Both framings explicitly recognize the value of cross-disciplinary work and thought in order to improve the ability of actors with different perspectives, skills, and capacities to solve problems together that they could not solve alone. Both carry an aspiration for transformation of our cities toward a higher level of well-being and civilization than we have seen in our past. At the same time, the pursuit of urban resilience is silent on a number of fronts that are considered essential to progress on sustainability, such as biodiversity, waste, and intergenerational equity. And, not to be missed, the notion of resilience is silent on the idea of "limits to growth," long a clarion call of the environmental sustainability movement, and a major sticking point for the implementation of limits-oriented sustainability solutions within a capitalist growth machine. It is accepted that effectively addressing climate resilience in cities will entail large investments, and that a sizeable proportion of these investments will be spent to pour concrete, itself a major source of GHG emissions. Stronger engineered systems and reinforcements of dykes, seawalls, evacuation routes, and critical urban institutions such as hospitals and evacuation centres may produce cities that are in a better position to withstand severe storms and other unexpected crises, but whether they also create the conditions for more long-term social, economic, and environmental sustainability is another question.

Review Questions

1. How has the perception of "sprawl" changed over time in Canadian cities?
2. Can urban resilience strategies effectively reduce the impact and address the consequences of climate change in Canadian cities? What are some of the other consequences of taking a resilience approach in urban planning and policy?
3. Characterize the "dark side" of planning for livability in some Canadian cities. What, if any, initiatives could be put in place to make livability an approach with fewer negative trade-offs?

Notes

1. Transit-efficient locations are locations in close proximity to means of getting around without a gas-powered vehicle. They may include bus or rapid transit, bicycle routes or stations, walking routes, and other active transportation modes.
2. Greenhouse gas emissions are often measured in tonnes of carbon dioxide equivalent per person (tCO_2e per capita). The calculation of CO_2 equivalent takes different greenhouse gases such as methane, water, and ozone into account by rendering their GHG intensity equivalent to that of CO_2. A GHG measurement expressed per capita indicates the level of emissions of an area, such as a city, as if each resident of that city were responsible for an equal share.
3. Countervailing this story about urban GHG efficiencies, this research group has by contrast shown that cities amplify waste generation (Hoornweg, Phada-Tata, and Kennedy, 2013).
4. The survey was conducted by McAllister Opinion Research and included a representative sample of 1701 British Columbians.
5. The CMHC study analyzed data for the period 2010–2016.
6. The report also uncovers an interesting, and perhaps new dynamic in this drive toward density: many condo buyers do not stop at investing their income in one home: about a quarter of condos in Vancouver and a third in Toronto are owned as investment properties and occupied by renters.
7. The possible permutations of proposed allowable dwellings on affected single-family-zoned lots include one principal residence (fee simple ownership) plus one basement suite (rental) plus one laneway house (rental); or two duplex residences (strata ownership) with one basement suite each (rental). In all variations, the maximum buildable area remains what was already permitted for a single-family home.
8. To learn more about trends in model sustainable neighbourhoods and their assessment systems, check out these websites: www.transformativetools.org/ and https://ecourbanismworldwide.wordpress.com/
9. The Paris Climate Agreement is named after the host city of the 2015 agreement struck by the nations that are party to the United Nations Framework Convention on Climate Change. There are 197 parties to have ratified the agreement, committing these nations to voluntary and independent actions that will strengthen efforts to address climate change and keep global temperature rise in this century below 2°Celsius above pre-industrial averages. See https://unfccc.int/process-and-meetings/the-paris-agreement/the-paris-agreement

References

Ahern, J. 2011. "From fail-safe to safe-to-fail. Sustainability and resilience in the new urban world," *Landscape and Urban Planning* 100(4): 341–43.

Bailey, I., and W. Stueck. 2012. "B.C. Liberals declare natural gas a clean energy source," *Globe and Mail*, 21 June. At: www.theglobeandmail.com/news/british-columbia/bc-liberals-declare-natural-gas-a-clean-energy-source/article4362331/

Berelowitz, L. 2005. *Dream City. Vancouver and the Global Imagination*. Vancouver: Douglas & MacIntyre.

Bioregional. 2017: *One Planet Principles*. At: www.bioregional.com/guide-one-planet-principles/

Blais, P. 2010. Perverse Cities. Hidden Subsidies, Wonky Policy, and Urban Sprawl. Vancouver: UBC Press.

Bula, F. 2018. "Vancouver to allow duplexes on most single-family lots as city seeks to unlock more housing options," *Globe and Mail*, 20 September. At: www.theglobeandmail.com/canada/british-columbia/article-vancouver-to-allow-duplexes-on-most-single-family-lots-as-city-seeks/

C40 Cities. 2019. *Vancouver Innovation Hub*. At: https://www.c40reinventingcities.org/en/sites/vancouver-innovation-hub-1314.html

Canada Mortgage and Housing Corporation (CMHC). 2018. *Examining Escalating House Prices in Large Canadian Metropolitan Centres*. Ottawa: CMHC.

Chang, R.A. 2018. "Temporary use & collective action: How urban planning practices contribute to adaptive capacity building for economic resilience," *plaNext* 7: 82–99.

City of Edmonton. 2012. *City Operations Greenhouse Gas Management Plan. Supporting The Way We Green*. With assistance of Office of Environment. At: www.edmonton.ca/city_government/documents/PDF/CityOperationsGHGManagementPlan.pdf

City of Toronto. 2016. *Growing Up. Planning for Children in New Vertical Communities. Study Update*. With assistance of City Planning Division. Toronto: City of Toronto.

City of Vancouver. 2017a. *Report—2017 VIVA Vancouver Program Update*. Administrative Report. Vancouver: City of Vancouver.

———. 2017b. "False Creek Flats Urban Design Policies and Guidelines for IC-3," in *Land Use and Development Policies and Guidelines*. Vancouver: City of Vancouver.

———. 2017c. "Urban Design. Part 2—Urban design and plan implementation." *False Creek Flats Info Boards*. At: http://vancouver.ca/files/cov/false-creek-flats-info-boards-jan-2017-part-2.pdf

———. 2017d. *False Creek Flats Area Plan*. Approved by Vancouver City Council 17 May 2017. At: https://vancouver.ca/files/cov/false-creek-flats-plan-2017-05-17.pdf

———. 2018. *City Endorses Making Room Program to Increase Housing Choices in Vancouver. Vancouver*. At: https://vancouver.ca/news-calendar/city-endorses-making-room-program-to-increase-housing-choices-in-vancouver.aspx.

———. 2019. *False Creek Flats Plan*. At: https://vancouver.ca/home-property-development/false-creek-flats.aspx

Davoudi, S., K. Shaw, L.J. Haider,, A.E. Quinlan, G.D. Peterson, C. Wilkinson, et al. 2012. *Resilience. A Bridging Concept or a Dead End? "Reframing" Resilience: Challenges for Planning Theory and Practice Interacting Traps: Resilience Assessment of a Pasture Management System in Northern Afghanistan Urban Resilience: What Does It Mean in Planning Practice? Resilience as a Useful Concept for Climate Change Adaptation? The Politics of Resilience for Planning: A Cautionary Note*. At: www.tandfonline.com/doi/pdf/10.1080/14649357.2012.677124

Depave Paradise. 2018a. *Where?* At: www.depaveparadise.ca/where.html

———. 2018b. *Why?* At: www.depaveparadise.ca/why.html

Downs, A. 2005. "Smart growth: Why we discuss it more than we do it," *Journal of the American Planning Association* 71(4): 367–78.

Downtown Winnipeg BIZ. 2016. *Imagine Portage & Main*. At: http://downtownwinnipegbiz.com/portageandmain/

Ernstson, H., S.E. van der Leeuw, C.L. Redman, D.J. Meffert, G. Davis, C. Alfsen, and T. Elmqvist. 2010. "Urban transitions. On urban resilience and human-dominated ecosystems," *AMBIO: A Journal of the Human Environment* 39(8): 531–45.

Folke, C. 2006. "Resilience: The emergence of a perspective for social–ecological systems analyses," *Global Environmental Change* 16(3): 253–67.

———, F. Berkes, and J. Colding, eds. 2003. *Navigating Social-Ecological Systems. Building Resilience for Complexity and Change*. Cambridge, NY: Cambridge University Press.

Girling, C.L., and R. Kellett. 2005. *Skinny Streets and Green Neighborhoods. Design for Environment and Community*. Washington DC: Island Press.

Gordon, D.L.A., and I. Shirokoff. 2014. *Suburban Nation? Population Growth in Canadian Suburbs, 2006–2011*. Council for Canadian Urbanism. Working Paper #1. Council for Canadian Urbanism. At: www.canadianurbanism.ca/wp-content/uploads/2014/07/CanU%20WP1%20Suburban%20Nation%202006-2011%20Text%20and%20Atlas%20comp.pdf

Haney, T., and C. McDonald-Harker. 2016. "The river is not the same anymore: Environmental risk and uncertainty in the aftermath of the High River, Alberta, Flood," *Social Currents* 4(6): 1–19.

Holden, M., and A. Scerri. 2013. "More than this: Liveable Melbourne meets liveable Vancouver," *Cities* 31: 444–53.

Holling, C.S. 1973. "Resilience and stability of ecological systems," *Annual Review of Ecology, Evolution, and Systematics* 4(1): 1–23.

Hoornweg, D., P. Bhada-Tata, C. Kennedy. 2013. "Environment: Waste production must peak this century," *Nature* 502(7473): 615–17.

———, L. Sugar, and C.L. Trejos Gómez. 2011. "Cities and greenhouse gas emissions: Moving forward," *Environment and Urbanization* 23(1): 207–27.

Jaccard, M. 2018. "Trudeau's Orwellian logic. We reduce emissions by increasing them," *Globe and Mail*, 20 February. At: www.theglobeandmail.com/opinion/trudeaus-orwellian-logic-reduce-emissions-by-increasing-them/article38021585/

Lakey, G. 2015. "When the next crisis comes, which movements will seize the opportunity?" *Resilience*. At: www.resilience.org/stories/2015-08-15/when-the-next-crisis-comes-which-movements-will-seize-the-opportunity/

Local Government Commission. 1991. *The Ahwahnee Principles for Resource-Efficient Communities*. At: www.lgc.org/who-we-are/ahwahnee/principles/

Lokman, K. 2017. "Vacancy as a laboratory. Design criteria for reimagining social-ecological systems on vacant urban lands," *Landscape Research* 42(7): 1–19.

Lydon, M., and A. Garcia, eds. 2015. *Tactical Urbanism. Short-Term Action for Long-Term Change*. Washington, DC: Island Press.

Meerow, S., J.P. Newell, and M. Stults. 2016. "Defining urban resilience. A review," *Landscape and Urban Planning* 147: 38–49.

Metro Vancouver. 2015. *The Metro Vancouver Housing and Transportation Cost Burden Study. A New Way of Looking at Affordability*. At: www.metrovancouver

.org/services/regional-planning/PlanningPublications/HousingAndTransportCostBurdenReport2015.pdf.

———. 2018. *Transit-Oriented Affordable Housing Study.* At: www.metrovancouver.org/services/regional-planning/housing-affordability/transit-oriented/Pages/default.aspx

Neptis Foundation. 2014. *Rethinking Sprawl.* At: www.neptis.org/publications/rethinking-sprawlhttp://www.neptis.org/publications/rethinking-sprawl

Olsson, P., C. Folke, and F. Berkes. 2004. "Adaptive comanagement for building resilience in social–ecological systems," *Environmental Management* 34(1): 75–90.

Oswalt, P., K. Overmeyer, and P. Misselwitz, eds. 2013. *Urban Catalyst. The Power of Temporary Use.* Berlin: Dom Pub.

Peck, J. 2005. "Struggling with the creative class," *International Journal of Urban and Regional Research* 29(4): 740–70.

Punter, J.V. 2003. *The Vancouver Achievement. Urban Planning and Design.* Vancouver: UBC Press.

Quastel, N., M. Moos, and N. Lynch. 2012. "Sustainability-as-density and the return of the social: The case of Vancouver, British Columbia," *Urban Geography* 33(7): 1055–84.

Real Estate Foundation of BC. 2015. *Public Views on Sustainability and the Built Environment.* Highlights of Finding. With assistance of McAllister Opinion Research. Vancouver: Real Estate Foundation of BC.

Rees, W.E. 1992. "Ecological footprints and appropriated carrying capacity: What urban economics leaves out," *Environment and Urbanization* 4(2): 121–30.

Register, R. 2009. Ecocities. *Rebuilding Cities in Balance with Nature,* rev. edn. Gabriola Island, BC: New Society Publishers.

Ruano, M. 1999. *Ecourbanism. Sustainable Human Settlements. 60 Case Studies.* Barcelona: Gustavo Gili (AD+E, Arquitectura y diseño + ecología).

Sathaye, J., A. Najam, C. Cocklin, T. Heller, F. Lecocq, J. Llanes-Regueiro, et al. 2007. "Sustainable development and mitigation," in B. Metz, O.R. Davidson, P.R. Bosch, R. Dave, L A. Meyer, eds., *Climate Change 2007. Mitigation. Contribution of Working Group III to the Fourth Assessment Report of the Intergovernmental Panel on Climate Change.* Cambridge, NY: Cambridge University Press.

Soleri, P., and J. Strohmeier. 2001. *The Urban Ideal. Conversations with Paolo Soleri.* Berkeley, CA: Berkeley Hills Books.

Soule, D.C. 2006. "The cost of sprawl," in D.C. Soule, ed., *Urban Sprawl. A Comprehensive Reference Guide.* Westport, CT, and London: Greenwood.

Sturgeon, D., M. Holden, and A. Molina. 2016. "What does neighborhood theory mean for ecourbanism? Introduction to the themed issue on 'Ecourbanism Worldwide,'" *Articulo—Journal of Urban Research* (14). At: http://journals.openedition.org/articulo/pdf/3128

Thompson, D. 2013. *Suburban Sprawl. Exposing Hidden Costs, Identifying Innovations. Sustainable Prosperity.* At: http://thecostofsprawl.com/report/SP_SuburbanSprawl_Oct2013_opt.pdf

United Nations Conference on Environment and Development (UNCED). 1992. *Agenda 21.* Rio de Janeiro. At: https://sustainabledevelopment.un.org/content/documents/Agenda21.pdf

US Green Building Council (USGBC). 2013. *Better Buildings Are Our Legacy.* LEED *Is Green Building.* At: https://new.usgbc.org/leed

Vancouver Economic Commission. 2017. *The Flats Economic Development Strategy.* At: www.vancouvereconomic.com/wp-content/uploads/2017/05/Flats-Strategy_Final_May8.compressed-1.pdf

VandeWeghe, J.R., and C. Kennedy. 2007. "A spatial analysis of residential greenhouse gas emissions in the Toronto census metropolitan area," *Journal of Industrial Ecology* 11(2): 133–44.

Wardekker, J.A., A. de Jong, J.M. Knoop, and J.P. van der Sluijs. 2010. "Operationalising a resilience approach to adapting an urban delta to uncertain climate changes," *Technological Forecasting and Social Change* 77(6): 987–98.

White, J.T., and J. Punter. 2017. "Toronto's 'Vancouverism': Developer adaptation, planning responses and the challenge of design quality," *Town Planning Review* 88(2): 173–200. Wilkinson, C., L. Porter, and J. Colding. 2010. "Metropolitan planning and resilience thinking: A practitioner's perspective," *Critical Planning* 17: 25–44.

Wohl, S. 2017. "Tactical urbanism as a means of testing relational processes in space. A complex systems perspective," *Planning Theory* 17(4): 472-93.

Woodcock, J., O.H. Franco, N. Orsini, and I. Roberts. 2011. "Non-vigorous physical activity and all-cause mortality: Systematic review and meta-analysis of cohort studies," *International Journal of Epidemiology* 40(1): 121–38.

Zavestoski, S., and J. Agyeman, eds. 2015. *Incomplete Streets. Processes, Practices, and Possibilities.* London: Routledge.

23 Food Systems and the City

Alison Blay-Palmer and Karen Landman

Food is a sustaining and enduring necessity. Yet among the basic essentials for life—air, water, shelter, and food—food has only recently become a focus of serious professional planning interest. This has been a puzzling omission because, as a discipline, planning marks its distinctiveness by being comprehensive in scope and attentive to the temporal dimensions and spatial interconnections among important facets of community life (American Planning Association, 2007: 1).

Introduction

Food offers a prism for understanding more about the challenges faced in cities and regions (Born and Purcell, 2006; Morgan, 2008). Food can help us envision more sustainable, livable urban spaces that are better integrated with their adjacent landscapes. Urban geography, planning, and sociology contribute to rethinking the role of food as we move to more sustainable communities. By placing food at the centre of thinking about quality and the (re)imagining of producer–consumer linkages, we are able to conceive of space and relationships in new ways (Blay-Palmer et al., 2018; Hinrichs and Lyson, 2008; Marsden, 2008; Morgan, Marsden, and Murdoch, 2006). As planners and others clamour for new approaches and tools, food has surfaced as a missing link in the urban (re)creation project. While the evolution of agriculture and urban life are linked (Steel, 2008), in the past century in North America agriculture has traditionally been considered a rural preoccupation, separate from urban issues. Engaging with food as an integral part of the urban system is still considered by some to be unusual. Based on the recognition of the potential for a more integrated approach to food, this chapter addresses how our cities could and should be fed; as such, it also delivers a strong rationale for using food as a lens to frame policy and develop even more relevant perspectives on urban growth and development.

The chapter unfolds in three parts. First, to provide a meaningful context, the issue of food and planning in Canada is tackled. Next, we explore innovative moves by planning and policy groups that signal a radical change in the role of food as a theoretical and policy framing tool. We then turn to case studies that offer examples of different initiatives that can assist academics and practitioners interested in putting food on the agenda. In many ways, academic and policy concerns about how a city feeds itself reflect the most profound changes within the nascent transition from modern to coherent postmodern world views. The evolution of land-use practices, community well-being, and the rise of science and technology are all well represented in the Canadian food story. These shifts provide us with a context for the evolution of planning for **food systems** as a consideration in Canadian cities.

Food and Urban Spaces in Canada

The history of food in Canada encapsulates many changes in our society since the mid-1800s. The place of rural, urban, and food communities in our planning traditions also reflect these shifts. In a broader societal context, the association of food with rural landscapes is founded in our farm community roots. At the turn of the twentieth century, some 63 per cent of Canadians lived in rural areas (Statistics Canada, 2007). As modern society took hold, however, there was a gradual shift in population from rural to urban communities. As the economy became much more deeply entrenched in manufacturing, farm size expanded, partially in response to rural depopulation. Planning and food intersected peripherally through land preservation initiatives as people attempted to protect fertile rural land from development. This was also the time when increased mechanization and the application of nitrogen-based fertilizers and chemical pesticides meant that fewer people were needed to farm bigger tracts of land. In 1921, Canada had 711,090 farms with an average size of 198 acres. By 2016, Canada had 193,492 farms remaining, with an average of 820 acres per farm, more than a fourfold increase in average acreage per farm, through land consolidation, as compared to the early twentieth century (Statistics Canada, 2016). With fewer links to food producers, places of food production and food consumption became increasingly disconnected.

In addition to demographic changes, food systems also reflect technological changes and help us understand the challenges that innovation poses to planners. Since the mid-nineteenth century, the transformation of food production from a local, small-scale, manual activity to a predominately technologically complex, industrial, monoculture production system mirrors key changes in our relationship to nature. The "green revolution" that began in the 1950s marked a technological shift, with increased reliance on irrigation, chemicals, and sophisticated seed technology packages for food production, plus distribution via new, long-distance, refrigerated modes of transport. The stated goal was to eliminate famines by mass-producing food. Coincident with this green revolution was the reorientation of large chemical companies away from wartime production and toward more domestic markets, a shift dominated by the manufacture and distribution of agricultural chemicals, such as pesticides and fertilizers. As food production became more industrialized and corporatized under this food production regime (Friedmann and McMichael, 1989; Goodman and Redclift, 1989), urban and rural spaces became increasingly separate, and people became physically and psychologically distanced from their food sources (Kneen, 1995). This has accelerated so that much of Canada's food now comes from a global industrial food system, travelling long distances and producing 3.3 million tonnes of CO_2 (Kissinger 2012).

In the past, planners have been challenged to reconcile conflicting needs across jurisdictions and across interest groups. On the one hand, there is a need to preserve valuable farmland, often in the shadow of urban development; on the other, many urbanites leave cities for suburban, peri-urban, and rural areas, bringing along their idealized visions of rural landscapes. These idyllic imaginaries must be reconciled with the realities of farming, which includes odours; access to roadways for the movement of machinery, livestock, and crops; and the conservation of agriculturally productive lands (Bryant, Russwurm, and McLennan, 1982). In the face of pressures from "city folk" to gentrify the countryside and turn more land from food production to housing development, golf courses, and other amenities, the primary provincial response has been to preserve agricultural land. The 1976 Ontario Agricultural Code of Practice (Caldwell, 2000), the Agricultural Land Reserve (ALR) enacted in British Columbia in 1973, and the Greenbelt legislation enacted in Ontario in 2005 (Wilson, 2008) all reflect the desire to protect agricultural land for food production. The closer one gets to urban centres, the more intense the conflict—as of 2001, half of Canadian urban centres were built on agriculturally viable land (Statistics Canada, 2001, cited in Campbell, 2006). However, despite provincial initiatives to protect land, urban pressures for outward expansion and development frequently supersede the need to protect quality land for food production. The case of the British Columbia ALR is instructive. Despite the provincial government's having set aside land for agriculture in the 1970s, the reality is that poor-quality land in remote parts of the province has been swapped by developers for high-quality land closer to urban centres for the purposes of development, such as second homes in "cottage" communities in Invermere (Campbell, 2006). In this way, development trumps the protection of fertile farmland.

Our lifestyles also affect how we eat. In recent times in Canada the pace of life has accelerated—families, often out of economic necessity, moved from one to two incomes, exerting pressure for food preparation to be faster and more efficient. This has produced tension in Canadians' attitudes to food, which is captured through the debate about affordable access to quality food, such as organically grown foods. This is contrasted against the trade-offs linked to large-scale retail developments, our reliance on a suburban car culture, and the apparent efficiency offered by large retailers that regularly opt for quantity over quality. Locally based food production, consumption linkages, new health and fitness awareness, and environmental concerns are reflections of individual attempts to redress some of these challenges. As part of the move to healthier living, in the last decade we have witnessed an increased desire on the part of some consumers to reconnect with and understand what they are putting into their bodies (Whatmore, 2002). This desire has created opportunities for many innovative relationships between people and their food. For example, the creation of community-supported agriculture (CSA), farmers' markets, the **slow food movement**, and the **100-mile diet** (Smith and MacKinnon, 2006) all demand that spaces exist in or near urban centres for consumers and farmers to connect. In a move towards emphasizing healthful food, the Guiding Principles for the new *Canada Food Guide* released in 2018 include "Regular intake of vegetables, fruit, whole grains, and protein-rich foods—especially plant-based sources of protein. . . . Limited intake of processed or prepared foods high in sodium, sugars or saturated fats" (Government of Canada, 2017).

Other healthy food initiatives are linked to urban food production. These include a suite of garden types, including rooftops, backyards, and **community gardens**, as well as the emergence of urban farms. For this change to be significant and lasting, a reconceptualization of how people get access to food will ultimately be required. This, in turn, is connected to how we conceive of city spaces.

As the archetypical big-box store, Walmart provides interesting insights into food and retail (Donald, 2008). Walmart's motto is "Save money. Live better." The company honours this commitment by being fiercely competitive through its "lean retailing" strategy that (1) concentrates purchasing power; (2) vertically integrates global production–delivery chains; and (3) maintains low-wage operations (Christopherson, 2007: 453). When Walmart entered the organic food market in 2006, it used its global connections to access fresh produce and encouraged its multinational corporate suppliers, such as Kellogg's, to develop lines of organic processed food. The move by Walmart into organics provoked widespread controversy about alternative food systems, specifically organics (Donald and Blay-Palmer, 2006). Organic supporters' vision is founded on small-scale, trust-based, locally rooted ecological production. Walmart, on the other hand, sees organic food as a new niche opportunity and has revved up its suppliers to produce more of what Guthman (2003, 2007) has dubbed "yuppie chow." While organic production methods lower the use of chemicals, other core organic/sustainability principles, including ecologically sound production methods, fair labour standards, and animal welfare, are not on Walmart's table (see Rees, Chapter 21). As well, to gain access to Walmart's organics, consumers need to be part of the sprawling suburban car culture (Donald and Blay-Palmer, 2006; Dupuis, 2000; Guthman, 2003).

Clearly, then, food represents a complex and nuanced challenge. But nothing hits closer to home than food. Given the growing advocacy on the part of some consumer groups,[1] food could shape how we see the future of cities. In the case of food in the cityscape, there is considerable public stakeholder support for locally produced foods but not a lot of knowledge. Food offers a unique lens for rethinking city spaces as it integrates land-use planning, transportation, water management, and environmental quality, as well as regulations and regulatory regimes concerned with human and environmental health. Given the environmental impact of food production, **urban sprawl**, the need for green infrastructure, and heightened consumer awareness about food safety and food-related health issues, relevant public policy is pressing. In addition, ground-up initiatives remain disjointed, fragmented, and small-scale. The recent City-Region Food System projects piloted in seven cities (Lusaka and Kitwe in Zambia; Medellin, Colombia; Colombo, Sri Lanka; Toronto, Canada; Utrecht, The Netherlands; and Quito, Ecuador) point to the need for and potential of multi-stakeholder engagement and policy coherence across city-regions as a way to address pressing issues, including waste management, food access, and local economic development such as opportunities for women through multi-scaled policy (Blay-Palmer et al., 2018). Integrated models such as **smart growth** that could empower municipalities and encourage city officials to take steps in new directions may also have traction (see Grant and Filion, Chapter 12). As a group, planners may well be in a unique position to offer this direction.

To understand how food can help make linkages to more sustainable cities, we need to understand how food and cities are connected. If we adopt a food perspective to frame urban challenges, food accounts for a substantial level of a city's **ecological footprint** (see Rees, Chapter 21). For example, food is the largest component of the ecological footprint with an average of 1.2 global hectares per capita per annum (Goldstein et al., 2017). As we have now crossed the threshold where, globally, more people live in urban areas than in rural areas, planning for food as part of the urban environment promises to make cities greener in many ways. In recent years there has been concern about the following:

- environmental impacts of cities and their growing greenhouse gas footprints;
- the escalating cost of transporting food, particularly in the context of the food-for-fuel debate; and
- the need to recycle waste/nutrients within the urban system.

As experts at the Global Footprint Network explain:

> The global effort for sustainability will be won, or lost, in the world's cities, where urban design may influence over 70 per cent of people's Ecological Footprint. High-Footprint cities can reduce this demand on nature greatly with existing technology. Many of these savings also cut costs and make cities more livable. (Wackernagel et al., 2006)

Research on food issues clarifies changing relationships between people, food, and the city, particularly in the context of creating alternative food networks through shorter, increasingly trust-based food chains. Central to these discussions are farmers' markets (Connell, Smithers, and A. Joseph, 2008; Feagan, 2008), alternative supply chains (Goodman, 2003; Maye, Holloway, and Kneafsey, 2007; Morgan, Marsden, and Murdoch, 2006), and the role of retailers as supply chains shift in the existing food system. Food planning and policy can help address all of these challenges.

Food, Planning, and Policy

Various bodies within North America and the European Union (EU) have incorporated food into their planning agendas. A notable early accomplishment at the organizational level is that of the American Planning Association (APA), which in 2007 produced the highly influential Policy Guide on Community and Regional Food Planning. The inclusion of food in planning is not exclusive to the United States, however. In 2008, the Association of European Schools of Planning (AESOP) established a special thematic group to study food and planning, marking another pivotal point. Food became another critical planning issue along with property rights, new technologies, urban design, ethics, transportation, and risk mitigation. Food is an essential addition to planning, as it offers a lens to see and understand the entire urban-to-rural landscape as a system. As such, this new focus can contribute to the creation of healthy communities. Thus, the importance of getting planners to respond immediately to food in broad-based policy initiatives cannot be underestimated.

In Canada, planning for food is increasingly evident in policy formation. Organizations, most notably the Toronto Food Policy Council (TFPC), have helped pave the way for planners to include food policy in their toolboxes. Early initiatives opened up opportunities and inspired others to use food as a vehicle to deal with a range of challenges, from community beautification to improving youth self-esteem. As TFPC manager Wayne Roberts explained,

> food security is good for Toronto's integrity, cohesion and reputation as "the city that works" which in turn attracts tourists and business. Everyone benefits when fresh, local food is available at vibrant farmers' markets, lively main street grocery stores and picturesque community gardens. (Roberts, 2013)

Recent initiatives reflect this multifaceted policy approach to food. For example, the City of Toronto recently won a Milan Urban Food Policy Pact Award for its work with new Canadian communities through its program "Community Food Works for Newcomer Settlement: Using Food as Tool for Settlement and Interaction" that trains and provides certification in food handling, food literacy, and employment skills to newcomers to the area (Miller and Blay-Palmer, 2018). The Food by Ward project documents food assets and gaps within Toronto and links this to city councillors as an advocacy tool that is being used to locate food resources such as mobile food markets in low-income underserved areas.

With this holistic approach in mind, planners at the Region of Waterloo Public Health Unit take a systems approach to food at a regional scale. They have produced groundbreaking work on food miles, the distance travelled from farm to fork, as they connect the dots between what the region could produce for local consumption, what food the region exported, and the resulting unnecessary food miles due to a mismatch between the two. Using data about existing production capacity, Xuereb (2005) estimated that if the region ate what it could produce, it would reduce its **greenhouse gas emissions** by the equivalent of over 16,000 cars every year. Waterloo's public health planners have also been leaders in producing local food maps, showing where nearby producers are located as part of their broader undertaking to understand the linkages between health, food prices, local sustainability, and rural–urban relationships.

Increasingly, municipal governments have raised the profile of food by taking on various aspects of the food cause. For example, in June 2008, the mayor of Markham, Ontario, announced a local food procurement initiative that would "help support Ontario's farm economy, address climate change, reduce greenhouse gas emissions and pesticide use, curb urban sprawl, protect Ontario farm lands and promote sustainable farming practices" (Scarpitti, 2008). The City of Toronto (2009) announced a similar plan in the fall of 2008 for 37 daycares, with the potential to expand the program to senior citizen homes and homeless shelters. The directive to purchase up to 50 per cent of food from local sources was unanimously approved by Toronto City Council. Local environmental groups claimed that "[i]ncreasing the amount of local produce will support farmers in the Greenbelt and Southern Ontario and the decrease in greenhouse gases emitted from food being transported to Toronto from the other side of the globe will improve air quality" (Hanes, 2008). Other sustainable food systems initiatives in Toronto include the 2015–2017 City-Region Food Systems project in the Greater Golden Horseshoe, which included a food flow analysis for carrots, eggs and poultry, apples, potatoes, and beef. In Toronto this project was led through the Centre for Sustainable Food Systems with the RUAF Foundation. RUAF and the UN-FAO undertook research in six other cities including Kitwe and Lusaka in Zambia; Colombo, Lanka; Medellin, Colombia; Quito, Ecuador; and Utrecht, Netherlands. Key policy recommendations from the Toronto reports are provided in Box 23.1.

Box 23.1 Opportunities for Strengthening the City-Region Food System (CRFS)

Stakeholder input identified eight key Policy Recommendations (PR):

1. Develop and support transition to increased mid-scale infrastructure (regional processing, distribution, marketing), in order to reduce resource inefficiency from redundant trade, including traffic congestion and GHG emissions.
2. Establish financial resources that support a range of scales and stages, including small and mid-scale.
3. Establish scale-appropriate safety and operational regulations and feasibility assessments for mid-scale infrastructure such as regional food hubs.
4. Increase research and educational opportunities directed at regional agriculture and regional infrastructure needs linked to shorter supply chains.
5. Provide sufficient social assistance, through a guaranteed income or other measures, to ensure that everyone can afford to eat locally produced healthy food.
6. Establish a national food policy and a national school food policy.
7. Ensure widespread formalization and implementation of public procurement policies for local and sustainable food (with percentages and budgets to meet policy goals).
8. Revise the labour practices, government support and subsidy programs to ensure the necessary skilled labour for all food system areas with tenure security and fair compensation.

Source: Miller and Blay-Palmer (2018: 85).

In this regard, the intersection of policy and food reflects the importance of good **governance** (see Taylor and Bradford, Chapter 3). The hierarchical, silo approach that dominates much of current policy and planning results in disconnected outcomes that serve the needs of a few and often create more problems than solutions. A systemic approach founded on networks and informed by a broad grasp of challenges that include economic, social, environmental, policy, and design considerations could improve the urban environment.

In Canada, we have begun the work to achieve planning for this broad-based and integrative city. In 2007, Duany Plater-Zyberk and Company developed an agrarian urban masterplan for Southlands, a 200-hectare site outside Vancouver. Four years later, Duany's book *Garden Cities: Theory & Practice of Agrarian Urbanism* (Duany, 2011) captured this vision for the integration of agriculture at all levels of development. Also in 2011, de la Salle and Holland published *Agricultural Urbanism*, which proposes that all aspects of the food system be made visible in our cities. In June 2011, the Ontario Professional Planners Institute (OPPI) released *Healthy Communities and Planning for Food Systems in Ontario—A Call to Action* to raise awareness of the need to consciously plan for healthy food systems at the community level. We are beginning to make progress on a number of urban food fronts, as illustrated by the case studies in the following section.

Case Studies

Numerous projects in Canada currently capture the reality and/or potential offered by strengthening local food systems. The few examples offered here are from multiple perspectives—production, distribution, and consumption—and illustrate how more sustainable food initiatives could fit into a systematic urban **regeneration** context.

Community Gardens

Community gardens are not new in North America. During World War I and especially World War II, households and neighbourhoods were encouraged to plant Victory Gardens; communal plots were set aside so that food could be produced locally. The success of **community gardens** lies in the ability to meet a range of community needs, including the provision of beautiful and safe community spaces, informal meeting places, and personal well-being. Community gardens can be found across Canada. In Montreal, the first community garden was created in 1936. Today the city has 97 community gardens with over 8200 plots. The gardens are multi-functional, meeting recreational, food, accessibility, and community needs of Montreal citizens. The city provides seeds, soil, flowers, tool storage facilities, and access to water to support these initiatives (Montreal, 2008).

In Winnipeg, community gardens can be rented from the city for a nominal amount every year ($28.35 for an unserviced plot and $42.20 for a serviced plot). Plots are managed by the Public Works Department, which stipulates strict conditions of use (Winnipeg, 2013). Vancouver also offers community gardening opportunities, primarily on its own parkland. In this case the city prepares the garden space—breaking ground, applying compost, and setting out individual plots—and the community is expected to manage the land in an open and inclusive way (Vancouver, 2009).

In Toronto, community gardens are viewed as places to build safe community spaces where people can interact, create beauty, and get some physical activity and mental peace. As part of its network, Toronto has an eight-acre urban farm, Black Creek Community Farm, that operates in one of the most economically disadvantaged areas of the city. This farm provides youth with leadership opportunities. It also aims to break down barriers between gangs as youth get to know their peers from other parts of the city.[2] In Ottawa, Just Food provides support to community gardens through education including "How to Start a Community Garden" and food preservation workshops; start-up funds through the Community Garden Development Fund; lobbying for more land; and awareness raising and network building (Just Food, 2018).

Despite the value of these community gardens, there are typically two problems. First, the costs associated with infrastructure and maintenance include the need to insure gardens and the people who work in them; second, pressures from land development are unavoidable. In the first case, basic infrastructure must be provided to gardeners, including reliable access to water, compost, mulch, and general maintenance. In some instances, gardens are made available to communities at no charge and municipalities absorb operational expenses. The most onerous cost is insurance, which is in fact such a burden that it has prevented some gardens from getting off the ground. In Kingston, Ontario, urban agriculture and community gardening activities are being proposed as part of the Official Plan. As part of this initiative, the designation of urban agriculture (UA) as a "community facility" is being recommended so that gardens could be covered under city insurance policies.

With respect to development threats, the benefits and success of community gardens can also be their downfall. The classic case in this regard comes from New York City. Starting in the mid-1990s with the bulldozing of 20 community gardens, the city moved to sell land used by communities as gardens since 1918 as part of efforts to revitalize communities. The rationale to sell garden land was that the lots had previously been building sites, so they were only temporary gardens. Ironically, once "blighted" neighbourhoods were revived in part by community gardens and witnessed increasing land prices and **gentrification**, these lands came under pressure for housing and retail space. A second issue was the need for affordable housing lands. In 2002, a settlement was reached between the city and the state, protecting over 200 gardens, some of which ended up in land trusts, while the rest were designated as parkland (Elder, 2005).

Urban Farms

Positive economic considerations also are emerging from urban farms. A fascinating initiative is SPIN (small plot intensive) farming, a new farming model that is taking hold in cities across North America. The first SPIN farm was developed in Saskatoon on half an acre of land pieced together from a number of arable residential plots, accessed via a handshake agreement or through payment in vegetables, with each plot ranging in size from 45 to 275 square metres. SPIN farmers report benefits from reduced wind and pest damage, more frequent harvests of fast-growing produce (e.g., lettuce and radishes) due to the urban microclimate, and a market literally at the doorstep. A government-funded pilot SPIN project in Philadelphia documented that gross revenue of over $50,000 can be earned from half an acre of urban land using SPIN techniques. SPIN farming has also been reported in British Columbia, Ontario, and Iowa (SPIN Farming, 2009). The most-cited benefits of SPIN farming are the low start-up costs, profitability, and the maintenance of urban green spaces. New forms of urban farming such as SPIN offer economic alternatives as well, providing a positive and empowering way to move from urban blight to beautification.

Non-profit farms, such as that of Just Food in the Ottawa region, work on community-based initiatives for social justice. In the Ottawa Greenbelt, on the former grounds of the National Capital Commission's tree nursery, the Just Food Farm offers access to land, training, and shared infrastructure and equipment to help foster viable small-scale agriculture. Just Food also offers support for other initiatives, such as the Karen Community Farm Project, Permaculture Ottawa's Community Urban Food Forest (cuff), and the Ottawa Food Bank Community Harvest Program including the community garden "Plant-a-row, Donate-a-row" program (Just Food, 2013, 2018). At the University of Guelph, the Guelph Centre for Urban Organic Farming (GCUOF) has the primary purpose to be a learning and research platform. However, its involvement with and outreach to the community has numerous added benefits, including strengthened relationships between the university and the surrounding community through volunteerism and other initiatives (Nourishing Communities, 2015).

Rooftop Production

Rooftops offer another opportunity for cities to foster urban food production. In 2011, Montreal's Lufa Farms built a 3000-square-metre commercial rooftop greenhouse with the goal to recycle water, reduce energy consumption, and grow food without using synthetic pesticides. Produce is sold through a weekly-box subscription model, with delivery to over 100 drop-off points in and around the city. Lufa Farms partners with 25 other food producers to offer their subscribers an extensive product line beyond their own 40 varieties of vegetables. From the Quebec landscape, partners provide such items as maple syrup, jams and jellies, mustards and oils—even wild foods, such as cattail hearts and marinated milkweed pods (Lufa Farms, 2013). In spite of having to overcome zoning, building code, engineering, and lease and taxation issues, Lufa Farms has built new greenhouses in Laval and Boston, and plans to scout for possibilities in Toronto, Chicago, and New York (Meeting of Minds, 2013).

The Carrot Common (2013) in Toronto developed from a partnership between a workers co-operative and a private developer interested in community development. Profits go to social justice food initiatives and organic agriculture projects. The Big Carrot, an organic food store, is on the main floor of the Carrot Common, along with 17 other stores. The second floor is a holistic health centre, and on the roof is the Carrot Green Roof, a food-producing garden where many community events take place and workshops on urban agriculture are held. While not focused on rooftop food production as a primary goal, the rooftop offers a space for the diffusion of food production knowledge and methods throughout the city.

The YWCA Metro Vancouver, in partnership with the Society Promoting Environmental Conservation, offers urban agriculture courses at the YWCA Rooftop Food Garden. Their goal is to provide training for sustainable urban agriculture, on topics such as urban soil management, season extension, growing in small spaces, and small-business planning. At the same time, volunteers have experimented with 40 different varieties of fruits, herbs, and vegetables since the Rooftop Food Garden began in the summer of 2006. The food is provided to low-income single mothers and their children through the YWCA programs in Vancouver (YWCA Metro Vancouver, 2013).

Elimination of the Food Desert

Important work has been going on in North America and the UK in the last decades to identify and find solutions to problems associated with **food deserts** (for the UK, see Wrigley, 2002; for Chicago, see Block and Kouba, 2006; for Alberta, see Smoyer-Tomic, Spence, and Amrhein, 2006). According to work done in the UK, a food desert exists in "areas of relative exclusion where people experience physical and economic barriers to accessing healthy food" (Reisig and Hobbiss, 2000: 138). Food deserts are linked with poor diet due to inadequate access to enough healthy and culturally appropriate food, which in turn is linked to compromised health. In North America, food deserts are often related to the advent of big-box stores that require car access and can result in diminished food access in urban areas. Food swamps are a version of food deserts where communities only have access to unhealthy, ultra-processed foods (Luan and Quick, 2015). Planners can address food access challenges through better design. The move to walkable communities, including higher-density urban use, is part of the solution. A study in Saskatoon highlights useful strategies on the potential of zoning to encourage multi-purpose neighbourhoods of high population density with incentives for retailers to establish food outlets in "food desert" areas (Peters and MacCreary, 2008). Food desert studies have been conducted in other Canadian cities and point to the links between food access and income (Páez et al., 2010, in Montreal; Wang et al., 2014, in Edmonton; Larsen and Gilliland, 2009, in London, Ontario).

While radical shifts to more **livable cities** are being established, interim measures can be adopted. Small farmers' markets, for example, can be located where commuters congregate and/or in underused spaces in low-income areas. The Region of Waterloo Public Health Department is experimenting with this option. Weekly summer markets were set up in parking lots around the City of Waterloo. Community centres, hospitals, and other public venues were included. Disadvantaged communities were targeted as a way to provide ready access to fresh, healthy food. While the project is still being assessed, preliminary results indicate that people who had access to the markets increased the quantity and variety of food they ate. The markets also became a meeting place and helped to build a stronger sense of community.

Another example involves taking food to people, or people to food. Both solutions offer ways to reconnect people with their food. In Toronto, the FoodShare Good Food Market is a fruit and vegetable store that operates out of a converted school bus. In 2016, the mobile market served 10 underserved, low-income locations reaching more than 6600 community members, delivering nearly 73,000 pounds (approximately 33,2000 kg) of fresh fruits and vegetables (FoodShare, 2018). A pilot project funded by the New York State Health Department also illustrates this approach (Figures 23.1 and 23.2). In this case, Veggie Mobiles provide low-income communities with much-needed fresh produce. The concept is simple and allows people such as diabetics, who need fresh, low-sodium food, access to appropriate produce. They also provide food education by offering a "Taste and Take" event on Wednesdays when people can sample and take home a selected produce item.

Santropol Roulant (2018) in Montreal is another example of creative food delivery. This project also targets disadvantaged citizens, in this case people who are elderly and/or have mobility challenges. Affiliated with McGill University, the project provides "Meals on Wheels" by delivering

Capital Roots

Figure 23.1 The Veggie Mobile, a travelling produce market operated by Capital Roots in New York's Capital Region

hot meals using bicycles and bringing together youth and senior citizens. In addition to operating as a food hub, Santropol Roulant (2018) grows its own produce as much as possible on rooftops and in small urban spaces, such as balconies. By using intense growing techniques that combine hydroponics with biodynamic and permaculture growing strategies, project members are able to produce up to six times more food than normally would be expected from the small areas under "cultivation."

Other jurisdictions are addressing poverty through food access programs. In this vein, the Toronto FoodShare program is among the most recognized of its kind in Canada. Established in 1985, FoodShare improves access to food through school meal programs, the Good Food Box, incubator kitchen facilities, youth programs, and food education programs, as well as the Good Food Mobile Markets mentioned earlier (FoodShare, 2018). The Child Hunger Eradication Program (CHEP) in Saskatoon is another Canadian success story. By focusing on food security and child poverty issues, the program mobilizes parents, school groups, volunteers, and the local business community to improve food education and access to children living in poverty (CHEP, 2009). Both FoodShare and CHEP work to build linkages with local farmers as part of their food security strategies. However, while these programs are remarkable initiatives that reflect people's ability to work with limited resources, they are also small in scale and exist largely outside the main operations of cities; as such, they reflect fundamental problems in urban environments and highlight the pressing need to make food access an integrated consideration for urban development.

Capital Roots

Figure 23.2 Interior of The Veggie Mobile, operated by Capital Roots

Vertical, Indoor, and Container "Farming"

While the idea of a sustainable vertical "farm" is futuristic at the moment, due to the challenges of producing food in vertical conditions, there are interesting ideas being generated on this type of production. Dickson Despommiers and his students at Columbia University have worked to conceptualize **vertical farms** (Despommiers, 2009). While still on the drawing board, they combine principles of urban agriculture and hydroponics in multi-storey indoor farms (Figure 23.3).

Vertical farms, they suggest, would be able to provide food year-round while using space more efficiently. Depending on the crop, a vertical hectare could be four to six times as productive as a hectare of outdoor land, while for crops such as strawberries, indoor cultivation would be 30 times as productive per hectare. Growing indoors also addresses issues of drought and other weather-related challenges faced by

Courtesy of Blake Kurasek

Figure 23.3 Prototypical model of a vertical farm

farmers. Proposed water recirculation systems would make vertical farms closed-loop food production systems as waste and water could be captured and reflowed into the system. Vertical farms could also offer urban employment opportunities as this type of food production would be labour intensive. The proposal for vertical farms offers interesting possible scenarios for the future of healthy food production within more sustainable cities.

Container and indoor farms are an offshoot of vertical farms. There has been an explosion of these "farms" in recent years. These models grow greens in shipping containers or indoors by combining a variety of soil-based growing, aquaponics, and hydroponics innovations. As one example, an economically competitive proof of concept from the United States was announced in late 2017. Local Roots Farms uses a shipping container and aquaponics to grow the equivalent of 5 acres of greens with no chemical inputs and 95 per cent less water than field-grown greens. As the organization's promotional material explains,

> We design, manufacture and operate the world's most productive indoor farming solutions. We believe the key to a more sustainable future requires eliminating supply-chain risks and undoing the commoditization of the food industry. That's why our Local Roots Family is building a distributed network of cutting-edge farms throughout the world to grow the freshest, healthiest food possible. (Local Roots Farms, 2018)

Typically, these models are indoor sites that grow food in cities, significantly reducing food miles. To date, while some models incorporate fish, adding a protein dimension, and/or strawberries, the majority of these models focus on greens. While an important part of a healthy diet, we need more than greens, so it will be interesting to see what the next innovations create in the years to come. And, given the need for food and space for food production, there are interesting opportunities for planning.

Conclusion

Clearly, food offers an integrative lens to urban geographers, planners, policy-makers, and other practitioners who seek to move cities in more sustainable directions. This chapter outlined the trajectory Canada and its cities have taken in addressing food issues. The overview makes it clear that planners need to focus on food systems in the urban context and that food connections to peri-urban and rural communities can be improved. While the case studies point to places where food is making its way onto the urban planning, policy, and design agenda, much more is needed.

Through food we can develop progressive dialogues about, and address, issues facing urban centres. A food prism allows us to examine challenges, such as production, distribution, and waste management, in terms that are relevant to many members in a community. The need for food is something we all have in common. Planning for food therefore provides a language and a space for uniting seemingly disparate interests. Food provides a place from which to appreciate the extent of connectivity within and beyond city systems and to situate urban areas within their foodsheds (Kloppenberg et al., 1996). This connects consumers to producers, and urban landscapes to rural ones. Food offers the chance to break down the silos between health, social justice, environmental well-being, and economic development. Recalling the emphasis that the American Planning Association (APA) has placed on the positive role food can play for planners, we can begin to understand how food can help us to address the most pressing issues for cities—deteriorating urban environments (see Rees, Chapter 21), crumbling and declining economies (see Hall and Donald, Chapter 20), and social decay (see Walks, Chapter 10)—to create more robust and resilient living spaces. While it is essential to recognize the pressures on urban resources and capacities, integrated food systems offer solutions to many related and pressing problems.

For this to take place, large-scale integration and incorporation with policies such as sustainability and smart-growth plans are necessary. Specific references to food and land use that protects food production are needed in official plans to give food-based initiatives traction. Planners and municipal officials can help move food onto the agenda. In a newspaper interview, Paula Jones, a San Francisco food activist, explains,

> All the individual efforts are super important, but we need policymakers and business at the table, too. . . . Government can bring in not only the policy but also the funding and technical expertise that it takes to drive large-scale, systemic changes. (Rich, 2008)

Food, then, can provide the basis for partnerships between public and private interests as we conceive of and make real our cities of the future. Bringing projects such as rooftop greenhouses from the drawing board into urban centres requires many partners at the table. To facilitate this process, we will need to legitimize, integrate, and standardize the role of food in urban planning, design, and policy. As recognized by the APA, food needs to be accorded a policy home so citizens, policy-makers, and bureaucrats can be given a food-centric mandate. As this chapter has demonstrated, this is happening in some jurisdictions.

However, a permanent set of champions could bring food onto the agenda and ensure that hard dollars are committed to food-based initiatives so that we avoid relying largely or entirely on soft money and volunteers. This task calls for professional engagement at all scales—from the local to the international—to take ownership of food and to provide mandates that need to be entrenched in law. With these points in mind, it appears that the realm of planning may offer the best place for food. Planning could offer the integrated perspective now missing for food. While food will continue to be connected to issues of rurality, communities, and health, urban planning may be a way to bring food forward as a focal point for a new understanding of planning, design, environmental quality, and quality of life in urban areas.

Food provides a rallying point to re-imagine cities so they are truly "just cities" (Fainstein, 2010; MacLaren and Agyeman, 2015). Planning for food can give us the chance to redefine our cities in sustainable terms and to empower citizens to be engaged in everyday acts of change, lifting common, daily acts to ones of transformation (Lefebvre, 1991). Indeed, food can provide a platform in the creation of the "new city."

Review Questions

1. Why should professional planners be concerned with urban food supplies and distribution?
2. How have urban food systems changed over the past 100 years?
3. How is the urban food system changing today?

Notes

1. Sales of organic food increased an average of 20 per cent annually from the early 1990s into the twenty-first century. Knudson (2007) reported that organic sales in the US were set to increase by 71 per cent between 2006 and 2011.
2. See also the award-winning work by Will Allen, "Growing Power in Milwaukee and Chicago," at www.macfound.org/site/c.lkLXJ8MQKrH/b.4537249/k.29CA/Will_Allen.htm

References

American Planning Association. 2007. *Policy Guide on Community and Regional Food Planning.* At: www.planning.org/policy/guides/adopted/food.htm

Blay-Palmer, A., G. Santini, M. Dubbelling, H. Renting, M. Taguchi, and T. Giordano. 2018. "Validating the city region food system approach: Enacting inclusive, transformational city region food systems," *Sustainability* 10, 1680: 1-23.

Block, D., and J. Kouba. 2006. "A comparison of the affordability of a market basket in two communities in the Chicago area," *Public Health and Nutrition* 9: 837–45.

Born, B., and M. Purcell. 2006. "Avoiding the local food trap: Scale and food systems in planning research," *Journal of Planning Education and Research* 26: 195–207.

Bryant, C.R., L.H. Russwurm, and A.G. McLennan. 1982. *The City's Countryside: Land and Its Management in the Rural–Urban Fringe.* London: Longman.

Caldwell, W. 2000. *Rural Non-Farm Development: Its Impact on the Viability and Sustainability of Agricultural and Rural Communities.* Toronto: Government of Ontario, Ministry of Agriculture, Food and Rural Affairs, Research and Corporate Services Division Research Branch.

Campbell, C. 2006. *Forever Farmland: Reshaping the Agricultural Land Reserve for the 21st Century.* Vancouver, BC: David Suzuki Foundation.

Carrot Common. 2013. Carrot Common—Home of the Big Carrot! At: www.carrotcommon.com

Child Hunger Eradication Program (CHEP). 2009. "Good food incorporated." At: www.chep.org/index.html

Christopherson, S. 2007. "Barriers to 'US Style' lean retailing: The case of Wal-Mart's failure in Germany," *Journal of Economic Geography* 7: 451–69.

Connell, D., J. Smithers, and A. Joseph. 2008. "Farmers' markets and the 'good food' value chain: A preliminary study," *Local Environment* 13: 169–85.

de la Salle, J., and M. Holland, eds. 2011. *Agricultural Urbanism: Handbook for Building Sustainable Food Systems in 21st Century Cities.* Winnipeg, MB: Green Frigate Books.

Despommiers, D. 2009. "Vertical farms: Home." At: www.verticalfarm.com/

Donald, B. 2008. "Food systems planning and sustainable cities and regions: The role of the firm in sustainable food capitalism," *Regional Studies* 42: 1251–62.

———, and A. Blay-Palmer. 2006. "The urban food economy: Cultural consumption for the urban elite or social inclusion opportunity?" *Environment and Planning* A 38: 1901–20.

Duany, A. 2011. *Garden Cities: Theory & Practice of Agrarian Urbanism.* London: The Prince's Foundation for the Built Environment, Duany Plater-Zyberk & Company.

Dupuis, M. 2000. "Not in my body: BGH and the rise of organic milk," *Agriculture and Human Values* 17: 285–95.

Elder, R.F. 2005. "Protecting New York City's community gardens," *New York University Environmental Law Journal* 13: 769–800. At: www3.law.nyu.edu/journals/envtllaw/issues/vol13/3/v13_n3_elder.pdf

Fainstein, S. 2010. *The Just City.* Ithaca, NY: Cornell University Press.

Feagan, R. 2008. "Direct marketing: Towards sustainable local food systems?" *Local Environment* 13: 161–7.

FoodShare. 2018. Good Food Market. At: https://foodshare.net/program/mobile/

Friedmann, H., and P. McMichael. 1989. "Agriculture and the state system: The rise and decline of national agricultures, 1870 to the present," *Sociologia Ruralis* 29: 93–117.

Goldstein, B., M. Birkved, J. Fernández, and M. Hauschild. 2017. "Surveying the environmental footprint of urban food consumption," *Journal of Industrial Ecology* 21(1): 151-65.

Goodman, D. 2003. "The quality 'turn' and alternative food practices: Reflections and agenda," *Journal of Rural Studies* 19: 1–7.

———, and M. Redclift. 1989. *From Peasant to Proletarian: Capitalist Development and Agrarian Transitions.* New York: Blackwell.

Government of Canada. 2017. "Guiding principles," At: https://www.foodguideconsultation.ca/guiding-principles-detailed

Guthman, J. 2003. "Fast food/organic food: Reflexive tastes and the making of 'yuppie chow,'" *Social and Cultural Geography* 4: 45–58.

———. 2007. "The Polanyian Way? Voluntary Food Labels as Neoliberal Governance," *Antipode* 39: 456–78.

Hanes, A. 2008. "Toronto's target: 50% local food," *National Post*, 31 Oct. At: http://network.nationalpost.com/np/blogs/toronto/archive/tags/Local+food/default.aspx

Hinrichs, C., and T. Lyson, eds. 2008. *Remaking the North American Food System: Strategies for Sustainability.* Lincoln, NE: University of Nebraska Press.

Just Food. 2013. The Just Food Farm: A Community Food and Sustainable Agriculture Hub. At: www.ca/just-food-farm/

———. 2018. Community Gardening. At: http://justfood.ca/community-gardening-network/

Kissinger, M. 2012. "International trade related food miles–The case of Canada," *Food Policy* 37(2): 171-78.

Kloppenberg, J., et al. 1996. "Coming into the foodshed," *Agriculture and Human Values* 13(3): 33–41.

Kneen, B. 1995. *From Land to Mouth: Understanding the Food System*, 2nd edn. Toronto: NC Press.

Knudson, W. 2007. *The Organic Food Market. East Lansing,* MI: Michigan State University, Strategic Marketing Institute Working Paper 01-0407.

Larsen, K., and J. Gilliland. 2009. "A farmers' market in a food desert: Evaluating impacts on the price and availability of healthy food," *Health & Place* 15(4): 1158-62.

Lefebvre, H. 1991. *The Production of Space.* London: Blackwell.

Local Roots Farms. 2018. *Home. Welcome to the future of farming.* At: https://www.localrootsfarms.com/

Luan, H., J. Law, and M. Quick. 2015. "Identifying food deserts and swamps based on relative healthy food access: A spatio-temporal Bayesian approach," *International Journal of Health Geographics* 14(1): 37.

Lufa Farms. 2013. Lufa Farms: Farming for a Better Future. At: http://lufa.com/en/

McLaren, D., and J. Agyeman 2015. *Sharing Cities: A Case for Truly Smart and Sustainable Cities*. MIT Press.

Marsden, T., ed. 2008. *Sustainable Communities: New Spaces for Planning, Participation and Engagement*. London: Elsevier.

Maye, D., L. Holloway, and M. Kneafsey, eds. 2007. *Alternative Food Geographies: Representation and Practice*. Oxford: Elsevier

Meeting of Minds. 2013. Beyond the Green Roof: Greenhouses Grow Possibilities for Urban Building Owners. At: http://cityminded.org/greenhouses-grow-possibilities-for-urban-building-owners-5809

Miller, S. and A. Blay-Palmer. 2018. *Assessment and Planning of the Toronto City Region Food System*. RUAF Foundation, Wilfrid Laurier University, and the Food and Agriculture Organization of the United Nations. At: www.fao.org/3/CA1111EN/ca1111en.pdf

Montreal, City of. 2008. "Montreal, living in Montreal, community gardens." At: http://ville.montreal.qc.ca/pls/portal/docs/page/plan_urbanisme_en/media/documents/061030_2_1_en.pdf

Morgan, K. 2008. "Greening the realm: Sustainable food chains and the public plate," *Regional Studies* 42: 1237–50.

———, T. Marsden, and J. Murdoch. 2006. *Worlds of Food: Place, Power and Provenance in the Food Chain*. London: Oxford University Press.

Nourishing Communities. 2015. At: http://nourishingontario.ca/the-social-economy-of-food/the-guelph-centre-for-urban-organic-farming/

Páez, A., R. Gertes Mercado, S. Farber, C. Morency, and M. Roorda. 2010. "Relative accessibility deprivation indicators for urban settings: Definitions and application to food deserts in Montreal," *Urban Studies* 47(7): 1415–38.

Peters, E., and T. McCreary. 2008. "Poor neighbourhoods and the changing geography of food retailing in Saskatoon, Saskatchewan, 1984–2004," *Canadian Journal of Urban Research* 17: 78–106.

Reisig, V., and A. Hobbiss. 2000. "Food deserts and how to tackle item: A study of one city's approach," *Health Education Journal* 59: 137–49.

Rich, D. 2008. "Farming the city: Planners start thinking of how to feed us," *San Francisco Chronicle*, 22 Mar., F-1.

Roberts, W. 2013. *No Nonsense Guide to World Food*. Toronto: Between the Lines.

Santropol Roulant. 2018. *What Is the Roulant?* At: https://santropolroulant.org/en/

Scarpitti, F. 2008. "Announcement on Markham's agreement with Local Food Plus to adopt LFP standards and zero waste program for municipal catering and food services," Markham, ON: Municipality of Markham, 4 June, 2008 Smog Summit, Toronto. At: www.organicfoodcouncil.org/files/downloads/Announcement%20on%20Markham.pdf

Smith, A., and J. MacKinnon. 2006. *The 100-Mile Diet: A Year of Local Eating*. Toronto: Vintage.

Smoyer-Tomic, K., J. Spence, and C. Amrhein. 2006. "Food deserts in the prairies? Supermarket accessibility and neighbourhood need in Edmonton, Canada," *Professional Geographer* 58: 307–26.

spin Farming. 2009. "spin makes agriculture accessible to anyone, anywhere." At: www.spinfarming.com/whatsSpin/

Statistics Canada. 2007. "A statistical portrait of agriculture, Canada and provinces: Census years 1921 to 2006." At: www.statcan.gc.ca/pub/95-632-x/2007000/t/4185571-eng.htm

———. 2016. "Census of agriculture, number and area of farms and farmland area by tenure, Canada and provinces." At: www5.statcan.gc.ca/cansim/a47

Steel, Carolyn. 2008. *Hungry City: How Food Shapes Our Lives*. London: Chatto & Windus.

Toronto, City of. 2009. "Living in Toronto: Green Bin Program." At: www.toronto.ca/greenbin/index.htm

Vancouver, City of. 2009. Vancouver's Community Gardens. "Parks and recreation." At: http://vancouver.ca/parks/parks/comgardn.htm

Wackernagel, M., J. Kitzes, D. Moran, S. Goldfinger, and M. Thomas. 2006. "The ecological footprint of cities and regions: Comparing resource availability with resource demand," *Environment and Urbanization* 18: 103–12. At: www.footprintnetwork.org/en/index.php/GFN/page/footprint_for_cities/

Wang, H., F. Qiu, and B. Swallow. 2014. "Can community gardens and farmers' markets relieve food desert problems? A study of Edmonton, Canada," *Applied Geography 55*: 127–37.

Whatmore, S. 2002. *Hybrid Geographies: Natures, Cultures, Spaces*. London: Sage.

Wilson, S.J. 2008. *Ontario's Wealth, Canada's Future: Appreciating the Value of the Greenbelt's Eco-Services*. Vancouver, BC: David Suzuki Foundation.

Winnipeg, City of. 2013. "Allotment gardens/Community gardens." At: www.winnipeg.ca/publicworks/parksandfields/CommunityGardens/communitygardens.asp

Wrigley, N. 2002. "'Food deserts' in British cities: Policy context and research priorities," *Urban Studies* 39: 2029–40. At: www.informaworld.com/smpp/content~db=all~content=a713707798

Xuereb, M. 2005. Food Miles: Environmental Implications of Food Imports to Waterloo Region. Waterloo, ON: Region of Waterloo Public Health. At: http://chd.region.waterloo.on.ca/web/health.nsf/0/54ED787F44A-CA44C852571410056AEB0/$file/FOOD_MILES_REPORT.pdf

YWCA Metro Vancouver. 2013. YWCA Metro Vancouver Rooftop Food Garden. At: www.ywcavan.org/content/ywca_Metro_Vancouver_Rooftop_Food_Garden_/605

24 Climate Change Threats and Urban Design Responses

Maged Senbel, Simon Liem, and Alexandra Lesnikowski

Introduction

City-building professionals and politicians confront significant challenges in the face of climate change. They must make difficult decisions in the design of cities in order to accommodate urban lifestyles and to manage the material and energy flows needed to sustain them. Their goal is to provide communities with a high quality of life using lower levels of **greenhouse gas (GHG) emissions** while simultaneously building resilience against the threats of a changing climate. The responses to these challenges, as this chapter argues, are interrelated.

The chapter includes eight sections. The first two sections ("Linking Cities and Greenhouse Gas Emissions" and "Impacts of Climate Change in Canada") connect the urban phenomenon and the production of GHG and describe consequences of these gases on climate. The next four sections ("Neighbourhood Design and Climate Change," "Health Consequences of Neighbourhood Design," "Building Typology, Energy, and Emissions," and "Green Infrastructure and Adaptive Design") concentrate on different aspects of neighbourhood design as an instrument to reduce the emission of GHG. Finally, the two last sections ("Implementing Climate Action" and "Challenges and Opportunities for Building Healthy, Low-Carbon Communities") concern the implementation of GHG reduction measures pertaining to neighbourhood design as well as circumstances affecting the effectiveness of these measures.

Linking Cities and Greenhouse Gas (GHG) Emissions

Climate change is possibly the most serious challenge facing modern civilization (National Research Council, 2009) and is arguably only a symptom of a much larger overshoot of the planet's carrying capacity (Rees, 2012; see Rees, Chapter 21). Average global temperatures are increasing as GHG emission levels continue to rise. The best estimates of temperature rise produced by the Intergovernmental Panel on Climate Change project a rise in global average temperature between 1°C and 3.7°C by the end of this century (Stocker, Qin, and Plattner, 2013). In 2011, the National Round Table on the Environment and the Economy (NRTEE) estimated that even if reductions in emissions are successful in limiting the global average temperature rise to under 2°C, the costs of climate change in Canada would continue to rise from $5 billion per year in 2020 to between $21 billion and $43 billion per year by the 2050s (NRTEE, 2011). Costs of flooding alone could be between $1 billion and $8 billion per year by the 2050s. Between 16,000 and 28,000 dwelling units will be at risk of permanent flooding from sea level rise and temporary flooding by storm surges. If the objective of limiting the global average temperature rise to under 2°C is *not* met, even adaptive efforts that are many times more expensive will be unable to safeguard human settlements from the devastating effects of global warming and climate change.

While Canada produces only 2 per cent of total global emissions, it has one of the highest per capita emissions rates in the world. In 2014, Canada emitted 15 tonnes of CO_2 per person compared to the world average of 5 tonnes, making it the third largest emitter among Western nations (World Bank, 2018). The building and transportation sectors together produce 37 per cent of emissions—12 per cent from buildings, and 25 per cent from transportation. Relative to 1990 levels, emissions from these two sectors have grown by about 58 megatonnes (58 million tonnes) (Environment Canada, 2018). Despite the higher per capita emissions of rural residents (Dodman, 2009), more than 80 per cent of Canadians live in urban areas, so the configurations of buildings, land use, and transportation in cities are vital to GHG emissions reduction efforts.

While there is a complex relationship between **urban form** and GHG emissions, many believe that compact urban land-use patterns and an emphasis on public transportation are important for climate change mitigation. Transit-oriented development, which densifies residential, commercial, and other land uses around public transit corridors, is one method that has had some success around the world at improving the sustainability of cities (Cervero, 2016) and lowering GHG emissions (National Academies of Sciences, Engineering, and Medicine, 2015).

A 2014 study of American households' energy use found that an increase of population-weighted density of 10 per cent was correlated with a reduction of CO_2 emissions from travel and buildings by 4.8 per cent and 3.5 per cent, respectively, highlighting the impact low-density **urban sprawl** can have on consumption habits (S. Lee and Lee, 2014). Along with compact development, pricing policies on gasoline, parking, and road use can also help support behavioural changes away from automobile use toward public transit, cycling, and walking in the short and medium terms to lessen carbon emissions from urban areas (Heres-Del-Valle and Niemeier, 2011; B. Lee and Lee, 2013).

Impacts of Climate Change in Canada

As planners and policy-makers work on mitigating climate change through policies that would reduce emissions, they must simultaneously adapt to the effects of climate change that are already occurring in Canada. The frequency of hot and cold temperature extremes, forest fires, and floods are increasing, and we may soon experience an ice-free summer in the Arctic Ocean. The sea around the Atlantic provinces has been rising at a rate of three millimetres per year, which has necessitated the adaptation of shorelines and infrastructure to resist more flooding. The southern Prairies, in particular, are facing more frequent droughts, severe rainfalls, and forest fires (Warren and Lemmen, 2014). In Canada's North, where temperatures are warming at quicker paces than the rest of the globe, Indigenous peoples have been forced to adapt their hunting practices, traditional lifestyles, and travel patterns in response to climate-induced permafrost thawing, sea level rise, changes in sea ice cover, and coastal erosion (Champalle et al., 2013; Ford, McDowell, and Jones, 2014; Pearce et al., 2011).

Extreme weather events have already had significant effects in Canada. A heat wave that struck Ontario in 2012 disrupted fruit production, causing a loss estimated at $100 million (Warren and Lemmen, 2014). In 2010, a Quebec heat wave was estimated to have caused an excess 280 deaths (Bustinza et al., 2013). As temperature events such as these become more common, the risk of heat-related illnesses and death, particularly for

Figure 24.1 Adapting to Reduce the Urban Heat Island Effect in Montreal, QC

The Rosemont–La Petite-Patrie borough in Montreal has adopted a number of measures to reduce the urban heat island effect. The borough requires property owners who are replacing or constructing roofs to build with highly reflective materials or use vegetative roofs to reduce the heat retention of buildings. The neighbourhood also runs a "Ruelles Vertes" program to cover alleyways with vegetation that can retain more moisture than impermeable surfaces, resulting in a cooling effect.

individuals with respiratory and/or cardiovascular conditions, will be heightened and can be exacerbated by the urban form (Belanger et al., 2008). In highly urbanized areas, the predominance of asphalt for roads and parking lots, the concentration of heat-emitting mechanical systems for buildings, and the paucity of plants and permeable surfaces that hold moisture all contribute to higher temperatures. This is known as urban heat island effect (UHI). Warmer summers and declining air quality in cities will increase mortality and illness, and raise the costs of the health-care system. When it occurs alongside heavy air pollution, the UHI can be especially dangerous for children, people with allergy symptoms, and those suffering from pulmonary and cardiovascular diseases (Belanger et al., 2008).

Several Canadian cities have implemented policies or are in the process of implementing policies to reduce the impacts of UHI. Figure 24.1 shows an example from Montreal that typifies the kind of urban design responses that would both minimize heat gain and maximize moisture retention in urban areas. Responding to UHI is a form of climate change adaptation, which is different to the emissions reduction strategies of climate change mitigation. In addition to implementing strategies to reduce emissions to mitigate the onset of climate change, adaptation prepares cities for changes as they come. Cities are starting to implement policies and design infrastructure that would guard against extreme weather events. By doing so they are building **resilience** in the face of predictable threats like rising sea levels (which are mentioned later in this chapter), more frequent and more severe river flooding, and more intense storms (Childers et al., 2015).

Neighbourhood Design and Climate Change

The urban design response to climate change is an evolution of a long tradition of attempting to work within the limits and constraints of nature and natural systems. The modern application of this tradition began with Ian McHarg's (1969) ideas of creating a built environment that is more sensitive to, and respectful of, ecological systems. Many others followed and by the beginning of the twenty-first century, **urban sustainability**,

New Urbanism, and **smart growth** became popular themes among planners concerned about climate change (Farr, 2008). The New Urbanism and smart growth movements emerged in response to the proliferation of sprawling auto-oriented suburban developments throughout Canada and the United States. New Urbanism advocates for a return to traditional neighbourhood design principles, including the following: town centres that anchor the social and economic life of neighbourhoods; fine-grain and walkable streets that encourage active transportation and interaction among neighbours; architectural diversity with small building setbacks that contribute to a neighbourhood's vibrancy; and access to reliable and convenient transit that connects to other neighbourhood centres (see Grant and Filion, Chapter 12). Smart growth is similarly concerned with integrating social, economic, and residential activity through compact **mixed-use development** supported by transit, bicycle, and pedestrian networks. Douglass Farr defines sustainable urbanism as "walkable and transit-served urbanism integrated with high-performance buildings and high-performance infrastructure" (Farr, 2008: 42). At the heart of these efforts is the goal of improving the quality of life of urban residents while reducing the energy and resources dedicated to the automobile.

Informing smart growth and New Urbanism is growing evidence that neighbourhood design has an impact on transportation mode choice, and in particular on how much people choose to drive. Researchers have found that a number of neighbourhood design elements help reduce driving and encourage reliance on walking, on cycling, or on transit. These include access to high-quality transit stops, wide sidewalks, well-connected pedestrian and cycling paths, available bike parking, having a mixture of land uses, a higher density of street networks, smaller block sizes, engaging street frontages, good lighting, and usable street furniture (Aziz et al., 2017; Ameli et al., 2015; Berrigan, Pickle, and Dill, 2010; Boulange et al., 2017; Crowley, Shalaby, and Zarei, 2009; Ewing and Cervero, 2010; Guo, 2009; C. Lee and Moudon, 2008; Mitra, Siva and Kehler, 2015). Designing streets to include features that are human scaled and memorable and that have a clear definition of public space, along with visual complexity, navigability, coherence, and interconnectivity between the physical elements of the street, all help to increase walking as a travel choice (Ewing and Bartholomew, 2013). Streets that are treated as social spaces, in which people interact, help to improve safety, comfort, and accessibility for persons of all ages and abilities, regardless of travel mode (see El-Geneidy and Grisé, Chapter 14; Walker and Blakley, Chapter 13).

Practitioners cannot rely on neighbourhood design alone to reduce driving and emissions. One study found a weak correlation between compact development and reduced automobile travel (Stevens, 2016). Older studies found that self-selection or demographic profiles are more significant determinants than the design of different locales. People move to walkable neighbourhoods because they are already interested in walking, cycling, and taking transit; they do not become walkers or cyclists in reaction to the design of their new neighbourhood (Cao, Mokhtarian, and Handy, 2007; Schwanen and Mokhtarian, 2005). Even so, the potential for more walking and transit use increases with compact, mixed-use neighbourhoods, and some causal linkages have been detected between physical environments and travel behaviour (Cerin et al., 2017; Ewing and Cervero, 2010).

Health Consequences of Neighbourhood Design

At the same time that planners and geographers have been working to promote walkable communities, and researchers have been gathering evidence on their merits, poor physical health has emerged as a major priority for changing neighbourhood design. Auto-oriented development patterns are a significant contributor to respiratory ailments, pedestrian injuries and fatalities, low levels of physical activity, and rising obesity (Maibach, Steg, and Anable, 2009). Conversely, children living in walkable neighbourhoods with access to park space and a nearby supermarket are less likely to be obese (Saelens et al., 2012).

In Vancouver, evidence shows that individuals living in the most walkable neighbourhoods are half as likely to be overweight as those living in the least walkable neighbourhoods (Frank et al., 2009). In addition, proximity and regular exposure to nature can help urban residents cope with stress and can even contribute to boosting their immune system (Selhub and Logan, 2012).

Large cities in Canada are recognizing the critical importance of creating pedestrian-friendly neighbourhoods for public health and safety, and of encouraging lively, sociable communities. The City of Vancouver's master transportation plan, *Transportation 2040*, prioritizes pedestrian access above all other modes of transportation and aims to increase the share of trips taken on foot, by bike, and by transit to two-thirds of all trips by 2040. Online tools and applications like Walk Score and RateMyStreet are gaining popularity, and walking audits such as the Pedestrian Environment Review System help assess the walking accessibility of different destinations.

Building Typology, Energy, and Emissions

Another co-benefit of compact development, in addition to reduced auto-dependence and increased physical activity, is the opportunity for greater efficiencies in building energy consumption. Building emissions come from two sources: "embodied" energy consumed in the materials and construction of the building itself, and the emissions resulting from building operations (e.g., heating, cooling, and lighting). Energy requirements in residential buildings differ widely from detached single-family dwellings, to low- and mid-rise multi-family structures, to high-rise apartment buildings. The construction of compact, multi-family residential buildings is an important component of low-carbon urban development. In communities with a low existing share of compact housing, there is significant potential for energy and GHG savings (Pitt, 2013).

Life cycle assessments (LCA) track energy consumption through the construction, occupation, and decommissioning of buildings and, thereby, provide a complete and comparable profile of energy consumption across buildings. LCA and other studies that combine the embodied energy of construction with the ongoing operating energy of inhabiting the building reveal that high-rise developments are the least energy consumptive per capita (Senbel et al., 2014). Even though relative to wood-frame construction, high-rise apartment buildings carry a much higher embodied energy from the steel and concrete and mechanical systems, and additional operating energy loads from lighting, heating, and cooling common rooms and hallways (Norman et al., 2006; Perkins et al., 2009; VandeWeghe and Kennedy, 2007), high-rises still perform better than New Urbanism suburban communities and much better than suburbs made exclusively of large single-family homes (Brown and Southworth, 2008; Senbel et al., 2014). As shown in Figure 24.2, mechanisms can be used to increase the environmental performance of relatively low-density residential areas.

Compactness increases the number of shared walls and floors, thereby reducing the surface area of individual dwelling units exposed to the elements. Higher densities also increase the potential for **district energy systems**, which centralize heating and cooling equipment for several buildings or even entire urban sectors and can recycle waste heat or use waste biomass for fuel. Because district energy systems require a minimum level of density to be viable, they are more commonly found in dense downtowns like Toronto, Montreal, and Vancouver, and on university campuses like that of the University of British Columbia (Andrews, 2008).

Green Infrastructure and Adaptive Design

Green urbanism seeks to go beyond New Urbanism and smart growth to achieve zero-carbon and zero-waste urban development. It emphasizes renewable resources, adaptive reuse of building stock, shared energy production, and densification within urban growth boundaries

"Aerial photo of Drake Landing Solar Community, 2007." Natural Resources Canada, 2007. Reproduced with the permission of the Minister of Natural Resources of Canada, 2014

Figure 24.2 Drake Landing Solar Community, Okotoks, BC

Drake Landing is the first subdivision of single-family detached dwellings in North America to derive more than 90 per cent of heating needs from solar thermal energy. A district heating system generates solar thermal energy from 800 rooftop collectors and stores the energy underground during summer months for distribution to each of the 52 homes during winter. The heating system reduces annual GHG emissions from each house by an estimated five tonnes.

(Lehmann, 2010). Green urbanism also draws from E.O. Wilson's idea of **biophilia** (the existence of a bond between humans and other forms of life) to ensure that urban environments are integrated with natural systems, and that urban dwellers have access to green spaces, such as parks, urban forests, and waterfronts. The integration of high-quality and well-connected patches of habitat for native plants and animals throughout neighbourhoods allows functioning ecosystems to live alongside urban residents. The psychological and educational benefits of urban nature are well documented. Green spaces in the fabric of urban life promote both ecosystem functioning and the health and well-being of urban residents (Selhub and Logan, 2012).

Parc du Mont-Royal in Montreal, Stanley Park in Vancouver, and High Park in Toronto provide both recreational space and habitat for local wildlife and plant species. Cities across Canada have more recently begun integrating urban forests and "wild areas" in parks to create ecological corridors that integrate native species into the urban fabric (Figure 24.3). Framing urban systems as embedded within natural systems also helps to facilitate urban design that minimizes the adverse impacts of the natural hazards that come with climate change. For example, landscape elements like **bioswales** (the use of vegetation, compost, and stones to purify runoff water) and permeable surface materials are now considered essential to stormwater management efforts. Unlike pipes and concrete culverts, bioswales and permeable surfaces work to naturally reduce runoff during high precipitation

Figure 24.3 Urban Forest Management Plan, Edmonton, AB

Edmonton is home to about 300,000 publicly owned trees. Since 2000, the city has lost over 30,000 trees due to drought and pests, problems that are expected to increase under changing climatic conditions. Recognition of the importance of the urban forest in keeping neighbourhoods cool, improving air quality, sequestering CO_2 emissions, and reducing storm water runoff has encouraged the city to plan for tree replacement, fire and disaster management, and planting strategies. The urban forest plan has also set a target to double the city's tree canopy from 10.3 to 20 per cent.

events; they also filter pollutants while allowing stormwater to percolate into the ground. These stormwater management systems reduce flood damage and improve the environmental performance of urban infrastructure.

Regenerative design theory argues for a more fundamental restructuring of the city's relationship with nature so that our cities are designed to benefit native ecosystems. Rather than simply doing less harm, regenerative design suggests that if we were to build neighbourhoods based on patterns identified in native ecosystems, we would begin to erase the artificial boundary between the built and natural environments. Urban development would seek to restore rather than destroy ecosystem habitats. This has implications for watershed management, urban agriculture, urban ecology, and energy and water resource management.

In Vancouver, the conversion of former industrial lands along False Creek to residential and commercial space has culminated in increasing levels of habitat restoration. With the beginning of high-rise residential construction in the 1980s, remediation of shoreline habitat included the gradual phasing out of the combined (storm and sanitary) sewer overflow system that drains into the creek. The development of the Olympic Village in Southeast False Creek saw more ambitious restoration with the creation of Habitat Island (Figure 24.4). Using a design that emphasizes native vegetation and tree snags, the island attracts waterfowl and bald eagles. In the fall of 2008, herring returned to spawn along one kilometre of the

Maged Senbel

Figure 24.4 Olympic Village, Vancouver, BC

Southeast False Creek in Vancouver was selected as the site for the Olympic Village for the 2010 Olympic and Paralympic Winter Games. A former industrial district comprising sawmills, port operations, and shipping, the area was redeveloped into a mixed-use, mixed-income neighbourhood designed for 11,000–13,000 residents. The neighbourhood is connected with the False Creek seawall pedestrian and cycling network, is served by two light rapid transit stations, and features a naturalized shoreline called Habitat Island.

shoreline for the first time in decades. However, industrial waste trapped in the ground around the creek remains a problem. Toxic chemicals continue to leak into the creek bed, preventing marine life from utilizing the bottom of the creek. Difficulties in dealing with these chemicals underscore complexities inherent in the regenerative process.

On land, restoring habitat helps to revive tree canopies, which improves rainfall retention and in turn helps reduce the urban heat island effect. Surface temperatures beneath individual and clustered trees and over grassy surfaces are typically cooler than over concrete or asphalt. Parks, for example, have been shown to be on average 0.95°C cooler during the day than non-green sites (Bowler et al., 2010). Integration of green roofs and living walls into buildings also provides an opportunity to increase urban micro ecosystems while improving buildings' energy performance (Smith and Levermore, 2008).

Restoration of natural ecosystems when combined with urban design responses can help cities adapt to climate change conditions, including sea level rise and flooding. Adaptive building designs that accommodate flooding (for example, by locating all vital systems and functioning to higher floors) and floating residential developments are important design opportunities for vulnerable coastal and floodplain communities. Additionally, restoration and preservation of natural habitats along shorelines is a common strategy to protect communities from sea level rise, erosion, storms, and flooding. Living shorelines, as they are called, are flexible and cost-effective alternatives to hard adaptations like dykes, dams, or locks that some communities are pursuing in tandem with environmental protection. In some cases, communities are resorting to managed retreat from vulnerable land and relocating homes, businesses, and recreational facilities away from areas at risk of flood and erosion.

Implementing Climate Action

A number of factors impact the success of local climate action. Leadership from municipal staff and elected officials, jurisdictional powers, organizational capacity, and adequate financing are critical factors in local climate action (Robinson and Gore, 2005). Mandates from senior levels of government tend to motivate local jurisdictions to address climate change (Senbel, Fergusson, and Stevens, 2013), especially where municipalities previously considered climate change to be a regional rather than a local problem (Tang et al., 2010). In 2008, for example, the province of British Columbia passed Bill 27, requiring local governments to set targets to reduce GHG emissions and identify implementable policies and actions in their official community plans. Despite the absence of any enforcement mechanism in the legislation, BC municipalities registered high rates of compliance by setting emission reduction targets (Senbel, Fergusson, and Stevens, 2013; Stevens and Senbel, 2012).

Every Canadian province has created and implemented a climate action plan, with Saskatchewan being the last to adopt a plan in late 2017 (Government of Saskatchewan, 2017; Holmes et al., 2012). Alberta was the first Canadian province to pass legislation on GHG emissions reduction, but targets focus on large industrial emitters and the province's climate change action plan allows emissions to continue growing until 2020. The strategy focuses largely on achieving emissions reductions through energy efficiency gains, greening energy production, and carbon sequestration. *Go Green: Ontario's Action Plan on Climate Change* set an emissions reduction target of 15 per cent below 1990 levels by 2020, and it launched a $17.5 billion rapid transit investment in the Greater Toronto and Hamilton Area. The Ontario Greenbelt Act was passed to protect undeveloped land around Toronto by limiting urban sprawl. Ontario has also mandated the closure of all coal-fired power plants. The province of Quebec has set an emissions reduction target of 20 per cent below 1990 levels by 2020, the highest emissions reduction target of any Canadian province or territory. The Quebec Climate Change Action Plan emphasizes transitions to sustainable land-use planning, including densification around public transit corridors, mixed-use planning, and enhancement of existing town cores. Quebec also plans to develop a sustainable building strategy (Harper et al., 2016).

Community environmental activism affects how proactive local governments are on climate planning (Millard-Ball, 2012; Pitt, 2010). Having a history of disasters, proximity to coastal areas, and population concentration all influence perceptions of risk and willingness to engage in climate change planning (Tang et al., 2010). **Non-governmental organizations (NGOs)** also play an important role in mobilizing and supporting Canadian municipalities on climate change planning. The Federation of Canadian Municipalities and ICLEI Canada—Local Governments for Sustainability (formerly the International Council for Local Environmental Initiatives)—formed the Partners for Climate Protection (PCP) network for municipalities that have committed to act on climate change. Municipalities are supported through a five-milestone framework toward developing a local action plan to reduce GHG emissions. In 2018, more than 450 Canadian municipalities were members of the PCP network.

Local governments face a number of challenges in their efforts to mitigate, and adapt to, climate change. A major impediment confronting mitigation is the difficulty of reducing total emissions through higher-density developments that reduce *per capita* emissions but that, because of the growing population, fail to reduce *total* emissions or prevent an increase of total emissions (Senbel et al., 2013). An absence of reliable data and forecasting tools for the local scale, methodological uncertainties in tracking emissions and projecting impacts, and resource constraints further complicate local efforts to design low-carbon neighbourhoods (Measham et al., 2011; Pitt and Randolph, 2009). Planning for adaptation is made difficult by uncertainty about climate change futures and their impact on human systems. The complexity and global nature of climate change science require efforts to raise awareness of climate change and to stoke interest in mediation and adaptation. Public engagement strategies that employ visualizations of neighbourhoods under different scenarios can help simplify the complexity of climate change issues, facilitating understanding and encouraging effective climate action (Sheppard et al., 2011).

Challenges and Opportunities for Building Healthy, Low-Carbon Communities

Exacerbating the climate change challenge is the potential for conflict in implementing both adaptation and mitigation. Hamin and Gurran (2009) point to a "density conundrum" in climate planning, where emissions reduction requires high-density urban environments that reduce driving and building energy use, and adaptation requires open spaces for stormwater management, migration of species, and cooling in urban areas. Optimal urban forms must therefore minimize conflicts between mitigation and adaptation by setting urban growth boundaries, designing open spaces to achieve multiple goals—such as urban agriculture and floodplain protection—and building green transportation routes that enable public transit (Hamin and Gurran, 2009). A strong bias in civic agendas toward mitigation over adaptation means that neighbourhood design approaches have generally not considered intersections between these two approaches to climate change (Measham et al., 2011). A notable exception is Halifax, which has a mitigation plan that complements a robust sea level rise adaptation plan (Figure 24.5).

A common challenge with increased density is the public perception of a dichotomy between sustainability and livability. High costs of housing in transit-rich neighbourhoods; perceptions of poor environmental quality in urban centres; and increased noise, traffic, crowding and crime: these all make it difficult to promote density (Ancell and Thompson-Fawcett, 2008; Howley, Scott, and Redmond, 2009). Higher-density neighbourhoods must therefore achieve a high quality of urban design to appeal to residents. Neighbourhood design that emphasizes green space and the adaptation of space to the presence of children can help to cast a positive light on compact neighbourhoods. However, while quality materials and amenities might help make density more desirable, they exacerbate the affordability problem.

Involving community residents in climate-change planning is important to address resistance to higher-density developments and large investments in public transit infrastructure. Visualization media, such as film, along with digital 3-D and energy consumption modelling, may also make relationships between climate change and urban form more accessible to community residents. Senbel and Church (2011) found that these approaches must be embedded in a broader effort to build trust and dialogue between communities and decision-makers if they are to contribute to abating community opposition to changes in neighbourhood form.

Emphasizing the co-benefits of urban planning for climate change whereby compact development also provides other benefits for municipalities and their residents helps to make the case more compelling. For example, higher densities result in lower per capita costs for public infrastructure and services. The same goes for health promotion through transportation, land use, and building-design decisions, which generate a range of positive side-effects: improvements in public safety, air quality, physical activity,

Figure 24.5 Adaptation Planning for Sea Level Rise, Halifax, NS

The Halifax Regional Municipality has undertaken an extensive adaptation planning process to address risks of sea level rise in Halifax Harbour. Using LiDAR mapping for Digital Elevation Models, future flood hazard maps, vulnerability analysis, and community consultations, Halifax is drafting a comprehensive adaptation strategy. In the meantime, the Municipal Planning Strategy and Land Use Bylaw requires coastal developments to be at least 2.5 metres above the current high-water mark, subject to ongoing sea level rise monitoring and analysis.

accessibility, local food production, preservation of natural resources, urban greening, and contact with nature (Younger et al., 2008). For example, urban design practices that increase shading and green space in neighbourhoods, enhance **walkability** and cycling infrastructure, and improve access to public transportation lead to increases in physical activity, improved social connectivity and mental health outcomes, and decreased cardiovascular and respiratory diseases (Cheng and Berry, 2013).

Conclusion

Structural changes in the way we design our neighbourhoods can improve public health outcomes, reduce GHG emissions, and mitigate dangerous climate change impacts like the urban heat island effect. Changing the dominant pattern of sprawling urban development in Canadian cities may go a long way toward addressing some of the major challenges that urban areas face. Urban design that mixes land uses alongside safe and enjoyable pedestrian and cycling pathways can help. Building compact, mixed-use, transit-oriented communities with networks of streets that are safe for walking and cycling; and creating networks of parks that provide habitat for native flora and fauna and have the capacity to filter, store, and absorb storm water; are practical approaches for making low-carbon, resilient, and healthy Canadian cities.

Review Questions

1. Why is it important to take measures that reduce the environmental impact of urban areas?
2. What approaches are most likely to minimize the environmental damage caused by cities?
3. What are the likely impacts of climate change on Canadian cities and what are some cities doing to plan for climate change?
4. What is your opinion on the different examples presented in the chapter? Which are most likely to achieve their environmental objectives?

References

Ameli, S.H., S. Hamidi, A. Garfinkel-Castro, and R. Ewing. 2015. "Do better urban design qualities lead to more walking in Salt Lake City, Utah?" *Journal of Urban Design* 20: 393–410.

Aziz, H.M.A., N.N. Nagle, A.M. Morton, et al. 2017. "Exploring the impact of walk–bike infrastructure, safety perception, and built-environment on active transportation mode choice: a random parameter model using New York City commuter data" *Transportation* 2017. https://doi.org/10.1007/s11116-017-9760-8.

Ancell, S., and M. Thompson-Fawcet. 2008. "The social sustainability of medium density housing: A conceptual model and Christchurch case study," *Housing Studies* 23: 423–41.

Andrews, C.J. 2008. "Greenhouse gas emissions along the rural-urban gradient," *Journal of Environmental Planning and Management* 51: 847–70.

Belanger, D., P. Berry, V. Bouchet, D. Charron, K-L. Clarke, B. Doyon, and M. Fleury. 2008. *Human Health in a Changing Climate: A Canadian Assessment of Vulnerabilities and Adaptive Capacity*. Ottawa: Health Canada.

Berrigan, D., L.W. Pickle, and J. Dill. 2010. "Associations between street connectivity and active transportation," *International Journal of Health Geographics* 9. At: www.ij-healthgeographics.com/content/9/1/20

Boulange, C., L. Gunn, B. Giles-Corti, S. Mavoa, C. Pettit, and H. Badland. 2017. "Examining associations between urban design attributes and transport mode choice for walking, cycling, public transport and private motor vehicle trips," *Journal of Transport & Health* 6: 155–66.

Bowler, D.E., L. Buyung-Ali., T.M. Knight, and A.S. Pullin. 2010. "Urban greening to cool towns and cities: A systematic review of the empirical evidence," *Landscape and Urban Planning* 97: 147–55.

Brown, M., and F. Southworth. 2008. "Mitigating climate change through green buildings and smart growth," *Environment and Planning A* 40: 653–75.

Bustinza, R., G. Lebel, P. Gosselin, D. Bélanger, and F. Chebana. 2013. "Health impacts of the July 2010 heat wave in Québec, Canada," *BMC Public Health* 13: 56.

Cao, X., P.L. Mokhtarian, and S.L. Handy. 2007. "Do changes in neighborhood characteristics lead to changes in travel behavior ? A structural equations modeling approach," *Transportation* 34: 535–56.

Cerin, E., A. Nathan, J. van Cauwenberg, D.W. Barnett, and A. Barnett. 2017. "The neighbourhood physical environment and active travel in older adults: A systematic review and meta-analysis," *International Journal of Behavioral Nutrition and Physical Activity* 14: 2–23.

Cervero, R. 2016. "Public transport and sustainable urbanism: Global lessons," in C. Curtis, J.L. Renne, L. Bertolini, eds., *Transit Oriented Development: Making It Happen*. New York, NY: Routledge.

Champalle, C., P. Tudge, E. Sparling, R. Riedlsperger, J. Ford, T. Bell, and M. Tremblay. 2013. *Adapting the Built Environment in a Changing Northern Climate: A Review of Climate Hazard-Related Mapping and Vulnerability Assessments of the Built Environment in Canada's North to Inform Climate Change Adaptation*. Ottawa: Natural Resources Canada.

Cheng, J.J., and P. Berry. 2013. "Health co-benefits and risks of public health adaptation strategies to climate change: A review of current literature," *International Journal of Public Health* 58: 305–11.

Childers, L.D., L.M. Cadenasso, M.J. Grove, V. Marshall, B. McGrath, and T.S. Pickett. 2015. "An ecology for cities: A transformational nexus of design and ecology to advance climate change resilience and urban sustainability," *Sustainability* 7: 3774–91.

Crowley, D.F., A.S. Shalaby, and H. Zarei. 2009. "Access walking distance, transit use, and transit-oriented development in North York City Center, Toronto, Canada," *Transportation Research Record* 2110: 96–105.

Dodman, D. 2009. "Blaming cities for climate change? An analysis of urban greenhouse gas emissions inventories," *Environment and Urbanization* 21: 185–201.

Environment Canada. 2018. *National Inventory Report 1990–2016—Greenhouse Gas Sources and Sinks in Canada: Executive Summary*. At: https://www.canada.ca/content/dam/eccc/documents/pdf/climate-change/emissions-inventories-reporting/nir-executive-summary/National Inventory Report Executive Summary 2018.pdf

Ewing, R., and K. Bartholomew. 2013. *Pedestrian- and Transit-Oriented Design*. Washington, DC: Urban Land Institute and American Planning Association.

———, and R. Cervero. 2010. "Travel and the built environment," *Journal of the American Planning Association* 76: 265–94.

Farr, D. 2008. *Sustainable Urbanism: Urban Design with Nature.* Hoboken, NJ: John Wiley & Sons.

Ford, J.D., G. McDowell, and J. Jones, J. 2014. "The state of climate change adaptation in the Arctic," *Environmental Research Letters*, 9(10).

Frank, L.D., M. Winters, B. Patterson, and C.L. Craig. 2009. *Promoting Physical Activity through Health Community Design.* Vancouver, BC: Faculty of Medicine, University of British Columbia.

Government of Saskatchewan. 2017. *Saskatchewan Climate Change Strategy Focuses on Readiness and Resilience* [Press release], 4 December.

Guo, Z. 2009. "Does the pedestrian environment affect the utility of walking? A case of path choice in downtown Boston," *Transportation Research Part D* 14: 343–52.

Hamin, E.M., and N. Gurran. 2009. "Urban form and climate change: Balancing adaptation and mitigation in the U.S. and Australia," *Habitat International* 33: 238–45.

Harper, D., S. Wong, A. Drost, T. Crossman, D. Taylor, N. Bakker, N. Chernawsky, and M. Nolin. 2016. "An overview of various provincial climate change policies across Canada and their impact on renewable energy generation," *Energy Regulation Quarterly* 4(4).

Heres-Del-Valle, D., and D. Niemeier. 2011. "CO_2 emissions: Are land-use changes enough for California to reduce VMT? Specification of a two-part model with instrumental variables," *Transportation Research Part B: Methodological* 45: 150–61.

Holmes, M., D.M. Lingl, I. Bruce, M. Carter, and F. Moola. 2012. *All Over the Map 2012.* Vancouver, BC: David Suzuki Foundation.

Howley, P., M. Scott, and D. Redmond. 2009. "Sustainability versus liveability: An investigation of neighbourhood satisfaction," *Journal of Environmental Planning and Management* 52: 847–64.

Lee, B., and Y. Lee. 2013. "Complementary pricing and land use policies: Does it lead to higher transit use?" *Journal of the American Planning Association*, 79(4): 314–28.

Lee, C., and A.V. Moudon. 2008. "Neighbourhood design and physical activity," *Building Research & Information* 36: 395–411.

Lee, S., and B. Lee. 2014. "The influence of urban form on GHG emissions in the U.S. household sector," *Energy Policy* 68: 534–549.

Lehmann, S. 2010. *The Principles of Green Urbanism: Transforming the City for Sustainability.* London: Earthscan.

Maibach, E., L. Steg, and J. Anable. 2009. "Promoting physical activity and reducing climate change: Opportunities to replace short car trips with active transportation," *Preventive Medicine* 49: 326–7.

McHarg, I. 1969. *Design with Nature.* New York: The Natural History Press.

Measham, T.G., B.L. Preston, T.F. Smith, C. Brooke, R. Gorddard, G. Withycombe and C. Morrison. 2011. "Adapting to climate change through local municipal planning: Barriers and challenges," *Mitigation and Adaptation Strategies for Global Change* 16: 889–909.

Millard-Ball, A. 2012. "Do city climate plans reduce emissions?" *Journal of Urban Economics* 71: 289–311.

Mitra, R., H. Siva, and M. Kehler. 2015. "Walk-friendly suburbs for older adults? Exploring the enablers and barriers to walking in a large suburban municipality in Canada," *Journal of Aging Studies* 35: 10–19.

National Academies of Sciences, Engineering, and Medicine. 2015. *Quantifying Transit's Impact on GHG Emissions and Energy Use—The Land Use Component.* Washington, DC: The National Academies Press.

National Research Council. 2009. *Restructuring Federal Climate Research to Meet the Challenges of Climate Change.* Washington, DC: The National Academies Press.

Norman, J., H.L. MacLean, and C.A. Kennedy. 2006. "Comparing high and low residential density: Life-cycle analysis of energy use and greenhouse gas emissions," *Journal of Urban Planning and Development* 132: 10–21.

National Round Table on the Environment and the Economy (NRTEE). 2011. *Paying the Price: The Economic Impacts of Climate Change for Canada.* Ottawa: NRTEE.

Pearce, T., J.D. Ford, F. Duerden, B. Smit, M. Andrachuk, L. Berrang-Ford, and T. Smith. 2011. "Advancing adaptation planning for climate change in the Inuvialuit Settlement Region (ISR): A review and critique," *Regional Environmental Change*, 11: 1–17.

Perkins, A., S. Hamnett, S. Pullen, R. Zito, and D. Trebilcock. 2009. "Transport, housing and urban form: The life cycle energy consumption and emissions of city centre apartments compared with suburban dwellings," *Urban Policy and Research* 27: 377–96.

Pitt, D. 2010. "The impact of internal and external characteristics on the adoption of climate mitigation policies by US municipalities," *Environment and Planning* C 28: 851–71.

———. 2013. "Evaluating the greenhouse gas reduction benefits of compact housing development," *Journal of Environmental Planning and Management* 56: 588–606.

———, and J. Randolph. 2009. "Identifying obstacles to community climate protection planning," *Environment and Planning C* 27: 841–57.

Rees, W. 2012. "The way forward: Survival 2100," *Solutions for a Sustainable and Desirable Future* 3(2). At: www.the solutionsjournal.com/node/1113

Robinson, P.J., and C.D. Gore. 2005. "Barriers to Canadian municipal response to climate change," *Canadian Journal of Urban Research* 14: 102–20.

Saelens, B.E., J.F. Sallis, L.D. Frank, S.C. Couch, C. Zhou, T. Colburn, and K.L. Cain. 2012. "Obesogenic neighborhood environments, child and parent obesity: The Neighborhood Impact on Kids study," *American Journal of Preventive Medicine* 42: 57–64.

Schwanen, T., and P.L. Mokhtarian. 2005. "What affects commute mode choice: Neighborhood physical structure or preferences toward neighborhoods?" *Journal of Transport Geography* 13: 83–99.

Selhub, E.M., and A.C. Logan. 2012. *Your Brain on Nature.* New York: John Wiley & Sons.

Senbel, M., and S.P. Church. 2011. "Design empowerment: The limits of accessible visualization media in neighborhood densification," *Journal of Planning Education and Research* 31: 423–37.

———, D. Fergusson, and M. Stevens. 2013. "Local responses to regional mandates: Assessing municipal greenhouse gas emissions reduction targets in British Columbia," *Sustainability: Science, Practice, & Policy* 9: 28–41.

———, W. Giratallah, K. Zhang, and M. Kissinger. 2014. "Compact development without transit: GHG emissions from three versions of high-density residential neighbourhoods in Vancouver," *Environment and Planning A* 46(5):1226–43.

Sheppard, S.R.J., A. Shaw, D. Flanders, S. Burch, A. Wiek, J. Carmichael, and J. Robinson. 2011. "Future visioning of local climate change: A framework for community engagement and planning with scenarios and visualisation," *Futures* 43: 400–12.

Smith, C., and G. Levermore. 2008. "Designing urban spaces and buildings to improve sustainability and quality of life in a warmer world," *Energy Policy* 36: 4558–62.

Stevens, M. 2016. "Does compact development make people drive less?" *Journal of the American Planning Association* 83: 1, 7–18.

———, and M. Senbel. 2012. "Examining municipal response to a provincial climate action planning mandate in British Columbia, Canada," *Local Environment* 17: 837–61.

Stocker, T.F., D. Qin, G.-K. Plattner. 2013. "Technical summary," in T.F. Stocker, D. Qin, G.-K. Plattner, M. Tignor, S.K. Allen, J. Boschung, A. Nauels, Y. Xia, V. Bex and P.M. Midgley, eds., *Climate Change 2013: The Physical Science Basis. Contribution of Working Group I to the Fifth Assessment Report of the Intergovernmental Panel on Climate Change*. Cambridge: Cambridge University Press.

Tang, Z., S.D. Brody, C. Quinn, L. Chang, and T. Wei. 2010. "Moving from agenda to action: Evaluating local climate change action plans," *Journal of Environmental Planning and Management* 53: 41–62.

VandeWeghe, J.R., and C. Kennedy. 2007. "A spatial analysis of residential greenhouse gas emissions in the Toronto census metropolitan area," *Journal of Industrial Ecology* 11: 133–44.

Warren, F.J., and D.S. Lemmen. 2014. "Synthesis," in F.J. Warren and D.S. Lemmen, eds., *Canada in a Changing Climate: Sector Perspectives on Impacts and Adaptation.* Ottawa: Natural Resources Canada.

World Bank. 2018. *World Development Indicators Databank.* At: https://data.worldbank.org/indicator/EN.ATM.CO2E.PC

Younger, M., H.R. Morrow-Almeida, S.M. Vindigni, and A.L. Dannenberg. 2008. "The built environment, climate change, and health: Opportunities for co-benefits," *American Journal of Preventive Medicine* 35: 517–26.

Glossary

100-mile diet A local food concept, popularized by Canadian authors Alisa Smith and J.B. MacKinnon in a book by the same name, that encourages people to consume only food produced within 100 miles of their home.

age-friendly city/community A concept advanced by public agencies that asks local governments to take into account the particular needs of older people, who make up an increasing proportion of the population. This vision often draws on a three-part framework: (1) participation; (2) health; and (3) security and independence of the older population.

automobility Refers to the heavy reliance on the automobile as well as the influence of high levels of car use on urban form and lifestyle.

bid-rent curves A modelling concept in economic geography used to understand and depict the trade-offs made by economic agents (e.g., a household, a firm) between rent and distance. At any point along one bid-rent curve, the economic agent is equally satisfied with the combination of location and the rent cost to occupy that location. Any negative change in the desired distance from the urban centre, for example, along one bid-rent curve, is compensated for by an equally desirable change in rent cost, such that the economic agent remains indifferent.

biophilia A hypothesis that posits an instinctive bond between humans and other forms of life. It assumes that humans are naturally attracted to nature.

bioswales Provide a natural solution to the removal of pollution and silt from surface runoff water. They consist of slightly sloped channels filled with vegetation, compost, and stones.

brownfield sites Former industrial locations that can become the object of redevelopment efforts and may require decontamination; see *greyfield sites*.

built environment. Refers to the physical landscape, usually urban, that has been created by humans. Examples include cities, buildings, infrastructure systems, and human-made open spaces. In contrast to the natural environment.

business improvement areas (BIA) Districts within a city, composed primarily of retail and consumer-oriented establishments, where business owners have banded together, agreeing to pay costs (usually through an added municipal tax) to support renovations to make the area more attractive, functionally up to date, and competitive (e.g., street furniture and planting, parking, pedestrian amenities). BIA dues also serve to finance publicity campaigns and promotional events.

census metropolitan area (CMA) Statistics Canada defines CMA as follows: "Area consisting of one or more neighbouring municipalities situated around a core. A census metropolitan area must have a total population of at least 100,000 of which 50,000 or more live in the core." At: www12.statcan.gc.ca/census-recensement/2011/ref/dict/geo009-eng.cfm

central business district (CBD) District generally close to the historical centre of a metropolitan region hosting a concentration of employment, retailing, and institutions. CBDs represent a portion of downtowns. Downtowns cover a wider area, including high-density housing.

central city The core municipality of a metropolitan area containing multiple municipalities.

citizenship Formal legal rights and responsibilities conferred automatically upon the citizens of a state, as well as rights (e.g., voting) from which non-citizens are "excluded."

city charter Special legislation that specifies the structure and authority of a single municipality. Most Canadian municipalities derive their authority from general legislation—"municipal acts" of general application. General municipal legislation does not apply to cities with charters, which in Canada include Toronto, Vancouver, Saint John, Winnipeg, among others.

cluster (economic) A dense network of geographically proximate companies and institutions that do business with one another or are related to one another in some way.

cognitive-cultural capitalism (or **cognitive-cultural economy**) A term used to describe the contemporary economy. It is associated with the increased use of digital technologies, alongside the increasing prevalence of cultural and knowledge-intensive (cognitive) work leading to the growth of technology-intensive, service, financial, craft, and cultural industries.

cohort Made up of people born in the same time period.

cohort approach A common research method used to compare people of the same age in different time periods, or to follow a given birth cohort over the members' lifetime.

commodification Making a commodity of some intangible attribute of urban space. Commodification of the core, for example, would entail the notion that one can purchase (or own) some of the ambience that is attributed to a core area.

common interest development (CID) This housing formula entails common ownership and management of shared facilities such as swimming pools, recreation halls and security systems, which can include gates and guards. CIDs can be targeted at certain groups, mostly the elderly.

community gardens Land provided to individuals and/or groups who contract to actively use and maintain vegetation they have planted. Such gardens are believed to be a step toward municipal food self-sufficiency.

competitive city A city that competes, economically and culturally, with other cities on a national, continental, and especially global scale. Today, competition to gain *world city* status, or to strengthen a city's position in the global network, is believed to be a primary factor underlying a city's urban agenda, especially larger, fast-growing cities.

complete street design An approach to designing and apportioning street right-of-way to prioritize safe use by all users, including pedestrians, bicyclists, transit riders, and private automobile users. This often involves some combination of cycling lanes, improved pedestrian crossings and sidewalks, transit lanes, and private vehicle lanes. Complete streets are also designed to account for users' different ages and abilities.

consolidated local government Metropolitan areas are understood to be more consolidated when they contain relatively few municipalities, or a higher proportion of their population residing within the central city, or both.

core housing need A measure of the housing circumstances of Canadians that combines three standards for housing: (1) adequacy, such that it does not require major repair; (2) suitability, as defined by the National Occupancy Standards for number and type of household members per room; and (3) affordability, as defined by the shelter cost-to-income ratio of 30 per cent of gross household income. A household that fails to meet any of the three standards and is unable to access alternative local housing is said to be in core housing need.

creative class A term coined by Richard Florida to describe a segment of the labour force that he argues is responsible for driving economic growth and prosperity in the twenty-first century. The creative class is composed of professional and knowledge-intensive occupations (jobs) where people create new ideas, new technologies, and new creative content.

deindustrialization The reversal of industrialization whereby former industrial facilities close down due to international outsourcing, competition from newer forms of production, or the obsolescence of the goods they produced.

digital city City in which digital and mobile technologies are adopted and used—often to enhance the performance of urban services and provide data to firms, governments, and citizens.

district energy systems (DES) These systems centralize heating and cooling facilities servicing areas generally containing several buildings. The purpose of DES is to improve the energy efficiency of these facilities.

ecological footprint The resource requirements of an urban area measured in terms of the surface of the earth needed to produce these resources. *Ecological footprint* can also refer to the surface of the planet needed to absorb (neutralize) the pollution generated by an urban area.

edge cities Concentrations of employment and retail located at the fringe of large metropolitan areas in an otherwise traditionally lower-density suburban area.

ethnic enclaves Spatial concentrations formed by residents' preference to live near others from the same ethno-cultural group rather than by the processes of exclusion, as is the case in *ghettos*.

ethnoburb A suburban area hosting a concentration of minority ethnic residents and businesses. Such areas have become points of entry for new immigrants.

express powers A legal doctrine that holds that municipalities can only perform tasks or functions that are authorized by provincial legislation, and nothing more. In the United States, this is called "Dillon's Rule," after judge John Forrest Dillon's description of state pre-eminence over local government in Clinton v Cedar Rapids and the Missouri River Railroad (24 Iowa 455; 1868).

exurban Refers to portions of metropolitan regions located beyond the suburbs in environments that are still predominantly rural.

financialization Refers to the situation in which the financial sector increases in importance in relation to other sectors of the economy. In this situation, some individuals are able to reap profits purely from their involvement in the financial sector instead of the production of goods and services.

food deserts Areas of a city, usually low-income areas, without accessible outlets that provide healthy and affordable food for household consumption.

food systems The areas and agents that constitute the supply end of the food chain along with all the components of food distribution and consumption in cities.

Fordism A period of economic development that lasted roughly from the 1920s until the late 1970s, when growth rested on a correspondence between rising consumption and increasing mass production. Fordism required ongoing *Keynesian*-type government interventions to stimulate consumption.

Fordist-Keynesian Economic development and economic and social policy-making that relied on government intervention in the form of various welfare-state and demand stimulation measures. The period lasted from the end of World War II until the late 1970s; see *Fordism*; *Keynesianism*.

Friendship Centres Emerged in Canadian cities starting in the 1950s (but really took off in the 1960–70s) to provide a place for service referrals, advocacy, and social, cultural, and recreational programs for urban Indigenous peoples.

General-purpose municipal governments In Canada, these refer to public corporations, created by provincial governments, with powers of self-government and delegated jurisdiction over multiple fields. The latter distinguishes them from special-purpose bodies.

gentrification The process whereby high-income households purchase and upgrade central city housing that once was occupied by residents of a significantly lower income. Today, some would consider other kinds of residential upgrading, such as condominium development, as gentrification.

ghettos Spaces in cities that segregate low-income and/or minority households who lack the freedom, as a consequence of income and/or prejudice, to move into residential zones elsewhere in the city. Originally used in the eighteenth, nineteenth, and early twentieth centuries to refer to neighbourhoods that housed segregated Jewish populations.

global city Very large cities that interact as much (if not more)—in terms of the flows of information, finances, goods, and people—with other places globally as with cities in their own country, and where growth is propelled by global rather than local factors. Various typologies rank different cities on a global hierarchy.

globalization The growing tendency for economic, political, and cultural exchanges to take place at a world scale.

governance In contrast to *government*, which refers to formal state institutions, *governance* also involves non-governmental actors. Governmental and non-governmental actors work together to address complex problems or achieve collective projects through joint mobilization of their respective resources and knowledge. Governance thus provides a much broader perspective on the political process than the concept of government does.

green urbanism The creation or transformation of cities in a fashion that respects the environment. Green urbanism concerns the form cities take along with the behaviour of organizations and people therein.

greenhouse gas emissions A host of gases, such as carbon dioxide and methane, emitted when burning fossil fuels, such as coal, oil, and natural gas. There is concern among scientists and society that greenhouse gas emissions are contributing to global climate change.

greyfield sites Abandoned retail locations; see *brownfield sites*.

gross domestic product (GDP) The sum of all the value added by individuals and organizations engaged in the production of goods and services in a given country.

indigeneity A term to describe the basis for Indigenous right claims and mobilization for embedding Indigenous self-determination in society's institutions at all scales.

Indigenous density A term attributed to the work of Chris Andersen, where he proposes it as a corrective alternative to the focus on how Indigenous peoples are "different" from a societal mainstream. The focus upon "density" rather than difference relates to understanding Indigenous self-determination in a layered, contextual, and complex way, such a way that it will sometimes incorporate strategies consistent with the settler mainstream, when suitable to the circumstances, and other times will take remarkably different approaches emanating from Indigenous worldviews, protocols, and processes.

Indigenous urbanism The participation in, production, and re-territorialization of urban space by Indigenous peoples using their own approaches in the pursuit of enjoying a good urban life.

induced demand An increase in the supply of transportation infrastructure (e.g., adding an additional lane for traffic) that leads to an increase in use/consumption of that infrastructure (e.g., more drivers on roads). This is because when more roads are built, new demand for roads is cultivated because the cost of driving (for example, as measured by time spent in congested traffic) is reduced. As the time-cost of driving goes down, people consume more of it, filling up the new road capacity and, in the medium to long run, reaching the prior level of congestion, though now with more vehicles on the road overall.

infrastructure deficit The gap between the amount of spending on the maintenance of infrastructure systems actually occurring versus the amount required to keep systems in good repair. Some define it broadly to include the amount of investment that would be required, but has not been made, to expand infrastructure systems in proportion to population growth.

intensification An increase in the density of development on a given parcel of land or in a neighbourhood or sector of a city. Intensification leads to higher dwelling, and generally also population, densities.

Keynesianism An economic approach formulated by John Maynard Keynes, according to which the market economy benefits from countercyclical government spending.

Keynesianism has been associated with public sector economic development and social programs.

knowledge-based economy Perspective by which economic development increasingly depends on the presence of an educated workforce. The importance of knowledge in the economy is related to *deindustrialization*, automation, and the growth of the high-order tertiary sector.

knowledge-intensive economic activity That part of the economy based on ideas and higher-order services, as opposed to manufacturing and primary (resource) production.

land rent A value derived within a land market for the use of land, affected by site characteristics such as location. An economic agent (e.g., firm or household) is willing to pay a certain rent to the landowner for the use of the owner's property for a period of time. For comparability of land values across an urban area, it is common to think of landowners who use that land themselves (e.g., for their private home), instead of renting it to others, as effectively paying "rent" to themselves for use of their property.

life course A concept recognizing that individuals move through stages in life defined in part by their personal biographies but also converging around transitional events that are roughly in common throughout a population (e.g., leaving school, leaving the parental home, entering a conjugal relationship). Life course transitions can be examined schematically by grouping key transitional events into meaningful life stages.

livable cities Cities generally considered "good" places to live. Often, livability is assessed using clearly defined indicators. Canadian cities have generally ranked high in published statistical reports that claim to measure urban quality of life or livability.

low income cutoffs (LICOs) Statistics Canada defines LICO as "income threshold below which a family will likely devote a larger share of its income on the necessities of food, shelter and clothing than the average family." At: www.statcan.gc.ca/pub/75f0002m/2012002/lico-sfr-eng.htm

massing The combined effect of building height, distribution of bulk, and silhouette. The massing of a building or group of buildings is often considered in relationship to the surrounding streetscape and the overall urban design qualities being promoted for the street or precinct as a whole.

mixed-use development Forms of urban development that comprise different types of activities. Mixed-use developments are often proposed as an instrument to reduce the dependence on the automobile.

multiculturalism The official policy of the Canadian government that minority groups participate fully in Canadian society while also maintaining distinctively different cultural values, practices, and institutions, provided these adhere to the Canadian Charter of Rights and Freedoms and to provincial human rights legislation.

multi-level governance Governance relationships that involve multiple levels of government in policy and planning.

municipal autonomy. The degree to which municipalities can make choices independent of the priorities of upper-level governments or of businesses. With respect to the former, the distinction is often made been fiscal autonomy (having the financial resources to pay for policy choices) and jurisdictional autonomy (having the legal authority to make policy choices).

neo-liberalism Tendency for a withdrawal of governments from the economic and social scene, so as to increase reliance on the private sector and market processes. Neo-liberalism was meant to reverse *Keynesian* policies.

new-build gentrification Wherein the local state primes the redevelopment of former industrial sites, including large tracts adjacent to downtown cores, by inviting and offering incentives to private developers to create new residential zones comprised essentially of high-rise or low-rise condominium buildings.

New Deal for Cities and Communities A policy initiative of the Liberal government led by Paul Martin (2003–6) that sought to directly involve the federal government in urban affairs. More generally, the "New Deal" refers to the federal and provincial response in the 1998–2006 period to local demands for more authority, autonomy, and resources.

new economy An economy that reflects recent economic changes stemming from deindustrialization, the rise of high-order tertiary activities, and globalization.

New Urbanism An urban design movement proposing a return to pre–World War II urban development features and vernacular architecture. New urbanism designs emphasize the orientation of buildings toward the street, a grid street pattern, a diversity of housing types and uses, human-scale built form, garages in back lanes, and urban infrastructure facilitating a variety of transportation modes besides the car. Architects Andres Duany and Elizabeth Plater-Zyberk have been two of the most vocal and prolific champions of the New Urbanism movement.

NIMBY (not in my back yard) Reactions against changes happening around one's residence. NIMBY movements are usually targeted at intensification of land use, infrastructure developments, and uses and activities that local residents do not want near them, such as strip clubs, halfway houses, group homes, and landfills. These movements can be locally based or consist of federations of local groups.

non-governmental organizations (NGOs) Organizations that provide goods or services that might normally be delivered by a government agency—for example, homeless shelters. During neo-liberal times, Canadian cities have relied more heavily on NGOs to provide important municipal services that would have been provided by an arm of government during the era of the welfare state.

own-source revenues Revenue raised by a government under its own authority.

path dependence A perspective by which certain patterns and behaviours are long-lasting and difficult to alter because they are supported by existing institutional arrangements and processes.

place-based public policy Policy initiatives that address the distinct contextual characteristics of local problems, typically informed by the knowledge of municipalities and local stakeholders.

place-making Planning efforts to insert physical/architectural features and events into the urban environment to help make a city or a particular part of a city more appealing, and thus more competitive locally and globally.

platform economy The growing predominance of firms using digital technologies and algorithms to match buyers and suppliers of a good or service. Examples of firms leading the platform economy are Airbnb and Uber.

polarization A distribution that is skewed toward the extreme ends of the attribute that is being measured—akin to an hourglass. Under conditions of the new economy, income is said to be polarized because major segments in the population fall into either relatively high- or low-income groups.

post-Fordism The period succeeding *Fordism*, characterized by a dismantling of Fordist mechanisms and their replacement by more market-oriented (*neo-liberal*) processes.

power centres Clustering of specialized stores of different size along with discount department stores in an automobile-oriented environment. In contrast with shopping malls, there is little common space in power centres, notwithstanding large parking areas.

producer services Services contracted out that cater to producers of final demand goods or services—for example, contracted legal work, accounting, maintenance, and cleaning.

qualitative development An approach to urban development that departs from a fixation on urban expansion and population growth (i.e., quantitative development), focusing instead on the existing built environment, infilling and redeveloping, and conserving or adapting existing buildings for reuse, with attention to preserving and accentuating a sense of place and urban quality, often at a pedestrian scale.

regenerative design The expression refers to forms of design that rely on natural processes to minimize impact on the environment. Regenerative design typically makes use of closed loops to prevent the generation of waste.

regional planning The making of land-use, infrastructure, social, or economic policies for broad territories that span multiple municipal boundaries. Canadian urban examples include Metro Vancouver's Livable Region Strategic Plan and the Government of Ontario's policy framework for the broader Toronto region, the Growth Plan for the Greater Golden Horseshoe.

resilience The capacity of a community to bounce back from a disaster or shock, either to the previously existing state, or progressively, to a better state of operations than that preceding the disaster.

revitalization/regeneration Renewal or regrowth of an obsolete sector of the economy or area of the city, such as the reinvigoration of the core and inner city in Canadian metropolitan areas in the twenty-first century.

row plexes Form of housing consisting of two (duplex) or three (triplex) units each occupying a full floor, one on top of another. In their Montreal variant, each unit in a plex tends to have its direct outdoor access.

second-wave feminism Originated in the early 1960s and widened the concern of first-wave feminism for legal obstacles to gender equality (e.g., suffrage and property rights) to a much broader range of issues, including those relating to reproduction and daily-life gender relations as well as an end to all forms of discrimination or unequal treatment of women in employment based on marital or parental status.

settler A term used in reference to a person, community, institution, or state descended from colonial traditions in countries like Canada, such as those derived from British or French traditions, whose power is privileged and superimposed onto Indigenous territory and community structures, often by force and violence.

shared street design An approach to designing streets that prioritizes the most vulnerable user, the pedestrian. Other travel modes (e.g., bicycles, private automobiles) move through the street at a pace and priority that is respectful of the pedestrian pace, safety, and right-of-way. Curbs and lane markings are either absent or are very subtle. Paving materials are chosen to privilege the needs and urban quality considerations of pedestrians, not the rapid movement of vehicles.

sharing economy An economic system in which assets or services are shared between private individuals, either free or for a fee, typically mediated through the Internet.

shelter poverty The expression refers to low-income households' difficulty in finding adequate housing. It also pertains to the consequences of these households' poor-quality housing and the allocation of a high proportion of their income to housing.

slow food movement A global grassroots movement that celebrates healthy and accessible food produced in local communities through sustainably and environmentally friendly production methods.

slow-growth cities Cities where population growth over a 10-year period is less than 10 per cent. Given the high proportion of the Canadian urban system on slow-growth trajectories or in decline (losing population), urbanists are calling for more sophisticated and realistic approaches to urban development that are not centrally focused on unrealistic expectations of continuous growth; see *qualitative development*.

slums Neighbourhoods in which poverty is concentrated, often due to the spatial concentration of poor-quality and under-maintained rental housing discrimination.

smart city An urban area that uses different types of electronic data collection techniques (e.g., video cameras, sensors, apps, meters) to supply information that can then be used to manage and monitor city-wide assets and resources efficiently.

smart growth A perspective on urban development that promotes forms of growth that respect the environment, show concern for quality of life, and attempt to reduce infrastructure expenses.

social exclusion The multi-dimensional process that prevents certain individuals and groups from securing the recognition, rights, and resources to participate fully in mainstream society through adequate employment, housing, health, and education. The spatial dynamics of social exclusion are often most visible in inner-city neighbourhoods with limited services and networks, concentrating poverty and compounding its negative effects.

social housing Government-funded housing provided to low-income households whose housing needs are not adequately met by the private real estate industry. Rent is subsidized such that the household does not pay more than 30 per cent of its gross income.

special-purpose bodies (SPBs) Local government bodies that are authorized to perform a single function. *Dependent* local SPBs, such as local transit commissions and library boards, are accountable to, and often mostly funded by, general-purpose municipalities. *Independent* local SPBs are separately constituted from general-purpose municipalities. In Canada, the latter include Ontario's conservation authorities, Vancouver's directly elected parks board, and directly elected school boards.

spheres of jurisdiction A legislative framework for delegating provincial authority to local governments that defines general areas of jurisdiction in which they may act with broad discretion. This is distinct from the *express powers* doctrine.

suburbanism A diversity of ways of living, generally occurring at the urban periphery, distinct from the ways people generally live their lives in the city or in the countryside. In a North American context, suburban ways of living are often, but not always, associated with car use and single-family-dwelling ownership. Suburban ways of living can also be found in some central locations.

suburbanization Non-central population and economic growth occurring in tandem with the spatial expansion of an urban area.

surveillance city A term used to take a more critical perspective on the adoption and use of smart city strategies that identifies the growing ability of governments (and firms) to collect data on individuals and raises questions about privacy.

survival curve Depiction of the proportion of a population surviving at a particular age in life. Given the very low infant mortality rate and significantly reduced mortality at older ages, demographers and human health experts raise the possibility of nearly all humans living to a genetically fixed age limit as mortality at earlier ages becomes less common, creating a rectangular survival curve.

symbolic capital A concept attributed to the work of Pierre Bourdieu, whereby a level of symbolic capital would be attached to the prominence and acceptability of a particular set of tastes, meanings, and values, translating into positive (or negative) cultural recognition.

technoburbs Concentrations of high-tech industries in a suburban setting, the outcome of the movement of employment away from central cities permitted by advances in information and communications technology.

temporary foreign workers (TFWs) Workers allowed into a country for a prescribed period in specified employment. As such, most rights of citizenship are not available to TFWs.

topophilia Love of place; a term coined by geographer Y.-F. Tuan. The term pertains to the growing interest for place in planning and an awareness of the importance of place for many people. The opposite term, *topophobia*, denotes fear of place.

Tower in the Park Model of urban development conceived by Le Corbusier, which consists of high-rise buildings set in a park-like environment. The model has been popular all over the world and has been criticized by Jane Jacobs, among many others.

transnational Refers to people, activities, and organizations that are based in more than one country. The term can be used in reference to an immigrant who attains citizenship in one country but keeps up ties with his or her place of origin and/or former residence.

travel demand management (TDM) A recent strategy used by transportation planners. In the past traffic was simply forecast and accommodated, but TDM attempts to change the demand itself rather than simply accommodate demand—for example, shifting hours of work in one or more large employment sectors in order to reduce congestion during periods of rush hour or peak load.

urban dynamics Human behaviour taking place in cities; also, journey patterns within urban areas.

urban ecosystem How natural systems function within the built environments of cities.

urban form The configuration of urban areas. Urban form can pertain to the distribution and density of activities within metropolitan regions or to design features of specific places within cities.

urban innovation An approach to transformational change that privileges cities as twenty-first-century drivers of innovation, uniquely bringing together creative individuals and diverse networks to generate and test new ideas or unproven solutions. Urban-scale innovations tackle major societal challenges ranging from sustainable growth and the digital economy to affordable housing and inter-cultural accommodation.

urban renewal Strategic reuse of an area of the city that is underused and often run down due to forces of change and transition. Urban renewal schemes are usually planned comprehensively under the direction of professional planners and at least partially funded by one or more levels of government.

urban sprawl Describes patterns of low-density development outward from the city centre, transforming often rural—*greenfield*—land into new suburban areas. New outward urban development does not necessarily constitute urban sprawl. *Sprawl* refers specifically to the features of the pattern of development, such as its low density of residents and jobs, and the separation of land uses across different zoning districts, at large distances from one another.

urban sustainability Conditions required to ensure the long-term availability of the natural resources (including pure water and air) required for the existence of urban settlements. Urban sustainability is increasingly perceived in a global context, such as the contribution of cities to planetary environmental degradation (e.g., global warming). *Sustainability* can also be defined in more narrow economic terms.

vertical farms The use of high-density urban space for purposes of food cultivation. The term spans a spectrum of practices, from roof gardens to "factory farms."

walkability Configurations of urban space that are pedestrian-friendly and so promote walking from place to place within walkable sub-areas. A major goal of twenty-first-century land-use planning is to increase the walkability of Canadian cities.

weak-mayor system A system of local government in which the mayor has few formal powers in addition to those of councillors, and so can further a policy agenda only through persuasion and compromise. Weak-mayor systems are criticized for inhibiting innovative city-wide policy-making. In contrast with strong-mayor systems, in which the mayor can veto council decisions, set the city's budget, and directly control the municipal public service.

welfare state Strong state/government involvement in the provision of basic needs, such as health care, housing, and old age security, as well as government intervention in matters more typically dealt with by the private sector, such as wage rates. In Canada the term is most often associated with the Fordist period of urban economic growth.

winner-takes-all urbanism Condition that allows a very small number of "superstar" cities—usually large cities with a high proportion of highly educated workers and dynamic clusters of economic activity—to accrue the wealth and benefits of innovation and creativity-oriented economic activity. This situation is posited to lead to the growth of political unrest, the rise of populism, and growing levels of inter- and intra-urban inequality.

Index

Aboriginal peoples. *see* Indigenous peoples
absolute homelessness, 186
accessibility: definition and concept, 255, 256, 264; development in Canada, 257; and jobs, 265–8; land value and location, 349–50; as measure, 256, 264–9; *vs.* mobility, 255–6, 273; planning for, 273; and public transportation, 257, 266; and travel demand, 261; and walkability, 269
activity-based model of transportation, 263
adaptive management and design, 405, 406–8, 409–12, 436–40, 441
administration, arrangements, 10–11
adults. *see* seniors; young adults
"affective resistance," 65
affordable housing: availability and access, 189, 206, 207; cost, 206, 207, 398; and density of housing, 402, 403; and gentrification, 282, 286–7; in suburbs, 305; and sustainability, 397; young adults, 102–3
Affordable Housing Initiative (AHI), 206
age: biological and social meanings, 89; and car commuting, 161; as change factor, 17; cohort and generations, 89–91; four "ages" life course, 115, 129; geographies of and segregation, 95–100; and housing, 91–5, 125–6; inequalities, 98–104; labour and employment, 117–18; and life course, 91, 115–18, 129; and lifestyle, 123; living arrangements, 120; residential decisions, 91–5; residential geography, 96–8, 105; and retirement, 117–18; social aspects, 91, 95–6; strata in society, 90
age-friendly communities, 104, 105
Age Friendly Communities: A Call to Action, 104
ageism, 98–9
aging, 90, 104–5; *see also* seniors
Agricultural Land Reserve (ALR), 419
agriculture: cases studies in urban areas, 423–5, 427–8; development and land preservation, 418–19; and fossil fuels, 388–9; green revolution, 388, 418; and innovation, 334, 337; and land prices, 355, 356; in origin of cities, 382; and sustainability of cities, 388; in urban and rural systems, 417, 418; *see also* food production
Airbnb, 81–2
Alberta, 40, 409, 440
Albert Street (Regina), 232
Alexander, C., and colleagues, 240–1, 249, 255
Alliance Against Displacement, 304
Alonso, William, 355–6
Alphabet Inc., 82
Amazon, HQ2 and facilities, 76–7
amenity provision, 361
American Planning Association (APA), 420, 428, 429
Andersen, C., 55
Anderson, K., 389
Arar, Maher, 145
art, on streets, 244–8
Assiniboine Avenue (Winnipeg), 236
Association of European Schools of Planning (AESOP), 420
attachment to places, 9, 10
Attainable Homes (Calgary), 128–9
Audriac, I., 376
automated data, 71, 72
automated vehicles (AVs), 272–3
automobiles. *see* cars
automobility, 160
automotive industry, 322–3, 326
auto-oriented pods, 226–7

baby boom, impact, 22, 111
baby boomers, 104–5, 111, 125, 198
Barnes, T., 326
Belanger, Y., 65
beliefs, and change, 18, 22, 27, 217
Bell Canada, 72
Bell Northern Research (BNR), 326
bicycle lanes, 236, 237
bid-rent curves and map, 355–6
big data, 71–2
big four tech firms, 76
"The Big Move" plan, 267
Bill C-45, 66
biocapacity, 384–5, 386
bio-mimicry for cities, 391
bioregionalism, 391, 392
bioswales, 437–8
birth rate, Canada, 7
BIXI system, 236
Blackfoot people, 61–3
Bonds, A., 52
Borrows, J., 53
Boushie, Colten, 232
Bows, A., 389
Bridget Jones's Diary, 124
British Columbia, 41, 46–7, 400, 419, 440
broadband Internet, 71
Broadway Avenue (Saskatoon), 244
brownfield sites, 279
builders and building industry, 195–8, 206–7, 216
building facades, 243–4
built environment, 8–9, 15–16, 125–7
Burbaby, BC, 304
business improvement areas, 247
business improvement associations (BIAs), 285
business parks, 227

Calgary, AB: change in transportation, 19; complete communities, 224–5; condos, 282; homelessness, 189, 191, 282; Indigenous urbanism, 61–3; millennials, 123–4; residential communities, 127, 128–9; residential geography, 96–7; suburbs and commute, 294

Canada Food Guide, 419
Canada Mortgage and Housing Corporation (CMHC), 197, 205, 206–7, 216, 402
Canada-Ontario Immigration Agreement (COIA), 44
capitalism, 314
capitalization, 3, 8–9
Carrot Common, 424
cars: and accessibility, 257; and age, 161; automated vehicles (AVs), 272–3;
auto-oriented pods, 226–7; congestion, 259–61; dissatisfaction with, 25; electric vehicles (EVs), 272; GHG emissions, 271–2; impact, 20, 22–3; importance and role, 257, 258, 273; and mobility, 255–6; and proximity, 6; in recentralization model, 29; and ride-hailing, 81; in shared streets, 238, 239; on streets, 236; and suburbs, 217, 220–1; and urban forms, 18, 215, 216, 217, 220–1, 228; and urban sprawl, 257–8; use and driver's licences, 93; and women, 160, 161; *see also* transportation
Carter, T., 140
CBD (central business district), 23, 221–2
census families, 121, 122
census metropolitan areas (CMAs). *see* CMAs
census tracts, 182
central business district (CBD), 23, 221–2
C40 Cities network, 411
change, urban: built environment, 15–16; in Canada, 19–29; and capitalization, 8–9; and consumption, 22; culture and values, 217; and demography, 17, 27, 87, 109, 110–22, 129, 216–17; and economy, 17, 19, 26, 27, 216, 313, 314–15, 318–22, 328; in education, 116; and employment, 313, 328; and environment, 381, 393; evidence and signs, 15–16; factors of, 15, 16–19, 26, 27, 215–17; and food production, 418; and generations, 90–1; and governments, 25–6; in households, 118–22, 200; and housing, 215–16; and inertia, 15–16; inner city, 23, 277–8, 358–9; and land value and markets, 358–61; life course and lifestyles, 115–22; in neighbourhoods, 199–200, 278–9; and planning, 18–19, 27; and recentralization model, 27, 29; and transportation, 16, 18, 19, 20; and urban development, 16, 216, 442; and urban forms, 16, 26–7, 29, 215–17; and urban structures, 19–29; values and beliefs, 18, 22, 27, 217; *see also* transition
Child Hunger Eradication Program (CHEP), 426
child poverty, 201
Chinatowns, 139, 241, 242
cities. *see* specific topic
citizenship, and public space, 232, 233, 252
Cittaslow, 375
city charters, 40–1
City-Region Food System (CRFS), 420, 422
climate change: action implementation, 440–1; adaptive management and design, 406–8, 436–40, 441; and buildings, 436; and GHG emissions, 432–3; impact and costs, 11–12, 389, 432–4; mitigation, 405, 441; and neighbourhoods, 434–6, 441–2; and resilience, 405–8, 412; and risk, 408, 409, 410; urban forms and planning, 441–2
clusters, and innovation, 334, 336, 338
CMAs (census metropolitan areas): age segregation, 98, 100; commuting, 162, 258–60; condos, 125–6, 204, 218; congestion, 259–61; homelessness, 189–90; home ownership, 204; immigrants, 137, 141, 147; income, 162; income inequality, 175–8, 179, 182–3; Indigenous population, 51, 52; land value and markets, 351; LGBTQ+ and same-sex couples, 159; living arrangements, 116; mode share, 258–60; nodal development, 220; patents, 334–6; population, 34, 36, 159, 222, 293; public transportation, 258–61; rental vacancies, 203; residential geography, 96–8; slow-growth and declining cities, 365, 366; Walk Score, 269–71; *see also* specific cities
CO_2 emissions, 271–2, 389, 433
Coastal Flood Adaptation Strategy (CFAS), 406–8
cohort approach, 89–91
collaborative multi-level urban governance, 33–4, 39, 42–5, 47–8
colonization and colonialism, 53–4
commerce. *see* economy
common interest developments (CID), 205
common-law couples, 121, 122
Community Futures Organizations (CFOs), 44
community gardens, 423
commuting: age and cars, 161; in CMAs, 162, 258–60; costs, 398; and gender, 160–2; and housing, 200; and land prices, 355–6, 360; and mode share, 258; public transportation, 258–60; and suburbs, 20, 23, 294, 359
compact development, 400–3, 436
complete communities urban forms, 224–6
complete streets, 236–8, 412
condominiums: description, 217–18; drawbacks, 280; and gentrification, 166–7; as market, 198; as mixed-use form, 222, 223–4; policy and development, 204, 217–18, 279, 280; popularity, 125–6, 279, 402; on waterfront, 280, 281–2
congestion, in transportation, 256, 258, 259–61
construction techniques, 17
consumption, 7, 22
container farming, 428
co-op housing, 205–6
core housing need, 165, 207
Cornell community, 224, 225, 300
covenants, 205
creative cities, 169
creative class: and income inequality, 178; knowledge economy, 5, 279; new economy, 316–17; and place, 10, 234, 316, 361; and sexualities, 169; theory and criticisms, 316–17
crisis of 2008, 353–4
Culhane, D., 65
cultural industry, development, 326–8
culture: and gentrification, 283, 284, 286; in inner city, 279; and new economy, 317, 326–8; on streets, 244–8
cumulative opportunity measure, 264–5, 266
cyber-attacks, 73
cycling, and public space, 234–5, 236, 237

Daniels decision, 65
Dartmouth, NS, 226–7

death, in Canada, 112, 113
debt of households, 180, 181, 207
decentralization of cities, 6
declining cities: concept, 364–5; decline-oriented approaches, 373; economy, 365–6, 368, 369; location, 365–6, 368; new discourse and techniques, 372–4; opportunities and challenges, 370–1, 374–6; patterns and trends, 364, 366, 367, 368; planning and management, 364, 369, 370, 371–4, 375; and population, 365, 369; psychology of failure, 368–9; research on, 366–8; *see also* slow-growth cities
decolonization, 66
deindustrialization, 5, 177, 299; and new economy, 313, 315, 317
demographic transition in Canada, 365
demography: built environment, 125–7; in Canada, 110–15; and change, 17, 27, 87, 109, 110–22, 129, 216–17; description and impact, 17; households and families, 119–22; housing and lifestyle, 125–7, 216–17; and immigration, 134–7; and land value and markets, 350; and life course, 115–18; slow-growth and declining cities, 365, 368; *see also* population
Denny, Dave, 247
density and densification, 279–80, 305, 400–3, 433, 436, 441
depaving and Depave Paradise, 411–12
design review committees (DRC), 244
developers, 197, 216, 218
development, urban: and agricultural land, 418–19; and change, 16, 216, 442; compact development, 400–3; dissatisfaction with, 23, 25; and economy, 17; flagship strategies, 278; home building, 197; and homelessness, 282; and industrialization, 20; and knowledge economy, 278, 279, 324–6; and land market, 356–8; livability-focused development, 400–2; models, 19, 22; postwar period, 22; public participation, 287–8; and resilience, 410, 412; social aspects, 282, 288; as spectacle, 278; of suburbs, 22, 23, 293–8, 301, 303, 359; and sustainability, 399–400; in transition, 277–8
development path, 410
Diamond, Jared, 12
digital cities: big data, 71–2; case studies, 77–83; city-wide applications and data collection, 72–3, 77, 81; definitions and concepts, 70–1; disruptive technologies, 73–5, 84; and economy, 75–7, 84; federal program, 78; and governance, 72, 73, 83–4; and public space, 75; ride-hailing and regulations, 78–81; security and hacking, 73; short-term rentals, 81–2; and surveillance, 73
digital divide, 71
digital technology, adoption, 71
directed data, 71–2
direct observation, of public space, 234–5
disability-free life expectancy, 118
dispersed urban environment, 20, 22–3, 25, 26
divorce, 121
DIY work on homes, 197, 202
Dockside Green (Victoria, BC), 404
domestic work, 157–8, 300
Downs, Anthony, 365
downtown: growth and resurgence, 20, 26–7; livability-focused development, 401; population in CMAs, 222; residential areas, 221–2; revitalization zones, 220, 221–4; in suburbs, 26, 301; *see also* inner city
Downtown Eastside (Vancouver, BC), 65, 281–2
Downtown Yonge Business Improvement Area, 247
Drake Landing, BC, solar community, 437
DrinkleFest, 247, 248
driver's licences, 93
Duany Plater-Zyberk and Company, 422

East Clayton community, 224
"echo" generation, 198
eco-cities, 391–2
ecological footprint: of cities, 386–8; as concept, 384–5, 387–8; and food, 420, 421; of humans, 384–6, 390; of nations, 385–6; overshoot, 385–6, 390; reduction, 390
ecology, 381, 382–4, 391–3
economy: and accessibility, 257; big four tech firms, 76; and capitalism, 314; and capitalization, 9; and change, 17, 19, 26, 27, 216, 313, 314–15, 318–22, 328; crises and bailouts, 180–1; in digital cities, 75–7, 84; and environment, 4; financialization, 180; and gentrification, 286; and housing, 9; and immigration, 138, 139–40, 147; importance to cities, 4–5; and inequality, 180; and infrastructure, 8; and innovation, 332–3; integration with US, 178, 180; land value and markets, 354–8; and manufacturing, 317–24; platform economy and firms, 74–5, 315; and policy-making, 38; in post-1975 cities, 25; and proximity, 6; and public space, 247, 250–1; and reproduction, 7–8; restructuring and inner city, 360–1; service economy, 5; sharing economy, 74; slow-growth and declining cities, 365–6, 368, 369; and suburbs, 26; and sustainability, 390, 391, 392–3; and transformation of cities, 317–28; and urban development, 17; and urban dispersion model, 25; *see also* knowledge economy; new economy
edge cities, 301
Edmonton, AB, 96–7, 246–7, 405, 406, 438
education, 8, 101–2, 116
elderly. *see* seniors
elections, and suburbs, 304
electric vehicles (EVs), 272
employment. *see* labour and employment
energy, 208, 388–9, 391, 433, 436
entrepreneurial policies, 361
environment: awareness, 11–12, 217; and change, 381, 393; definition and role for cities, 3; degradation and damage, 11, 392–3; and economy, 4; and housing, 208; importance, 11; as property of cities, 11–12; and transportation, 271–2; vulnerability of cities, 388–9
environment, built, 8–9, 15–16, 125–7
environmentalism, 217
ethnic covenants, 205
ethnic enclaves, 144, 184–5
ethnoburbs, 140, 185, 298, 304
Europe, 258–9, 260, 261, 372
exports, 4–5

extreme weather events, 433–4
exurban sprawl, 292

facades of buildings, 243–4
False Creek (Vancouver, BC), 410–11, 438–9
families: with children, 121; in cities, 93–5; configuration, 120–2, 181; income, 163–4, 181; living arrangements, 120–1, 122; *see also* lone-parent families
farmers' markets, 425
farming. *see* agriculture
Farr, Douglass, 435
federal government: housing policy and programs, 205–7, 208; immigration policy, 146–8; manufacturing policy and programs, 323; in multi-level governance, 33, 39, 42–5, 47–8; Smart Cities Challenge, 78
federal-local partnerships with municipalities, communities, and Indigenous peoples, 43, 44–5
federal-provincial/territorial agreement with municipal participation, 43–4, 45
Federation of Canadian Municipalities (FCM), 43
feminist perspective: on sexuality, 155–6; on suburbs, 300; on transportation, 160
"feral" suburbs, 295
fertility, in Canada, 110–12
film industry, 327
filtering, and neighbourhoods, 198–9
financialization, 180
financial sector, and housing, 197
fire escapes, 246
First Age, 115, 116
First Nations, 52, 55, 57–9; *see also* Indigenous peoples
Fishman, Robert, 295
flagship strategies of development, 278
floodplains and climate change, 406–8
Florida, Richard, 169, 316
food deserts, 425–6
food production, 388, 418, 422–5, 427–8; *see also* agriculture
FoodShare Good Food Market, 425, 426
food swamps, 425
food systems: case studies, 422–8; change and challenges, 418–20, 428–9; City-Region Food System (CRFS), 422; community gardens, 423; eco-footprint and sustainability, 420, 421; food access, 425–6; food deserts, 425–6; and healthier living, 419; planning and policy, 417, 418–19, 420–2, 428–9; rooftop production, 424–5; urban farms, 424; vertical, indoor and container farming, 427–8
Ford, Henry, 314
Ford, Rob, 45
Fordist period and Fordism, 20, 22–3, 314, 361
forms, urban: and capitalization, 9; and cars, 18, 215, 216, 217, 220–1, 228; and change, 16, 26–7, 29, 215–17; climate change, 441–2; and IT, 25; mixed-use forms, 218–26, 228–9, 303; new forms, 215, 218–26, 228–9; and place, 9; planning, 28; and production, 4; public policy, 217–18; recentralization model, 29; segregated-use forms, 218–19, 226–8, 229; in suburbs, 219–21, 301–3; and transportation, 16, 20; types, 218–19; and walking, 222, 227, 228
Fort McMurray, AB, 369–70
fossil fuels, and sustainability, 388–9
four-step demand model of transportation, 261–4
Fourth Age, 115, 118
Free Flow Dance Theatre Company, 246
freeways and highways, 23, 25, 257
freight transport, 263
Frost, H., 277, 278
funding, 8, 11

Garden City, 19, 22
gardens, 288
Garrison Woods community, 225
Gas Tax Fund, 39
gated communities, 126, 227–8, 229
gateways, 241
gay bashing, 168
gay people. *see* LGBTQ+ people
gay villages, 167–9
Gehl, Jan, 280
gender: and commuting, 160–2; concept, 155; and education, 116; and gentrification, 165–7, 168; housework and work at home, 201–2; and housing, 200–1; LGBTQ+ in CMAs, 159; life expectancy and survival patterns, 112, 113; and lone parenthood, 163–5; and marriage and relationships, 120; and mobility, 157, 160–2; "normality" and models, 156; and population, 112; retirement age, 117; and suburbs, 300; and transportation, 160–2
General Motors (GM), 322–3
generational wage gap, 101–2
generations: comparisons between, 89–91; and inequalities, 100–4; residential decisions, 91–5; social aspects, 91
gentrification: and culture, 283, 284, 286; description and impact, 183, 199, 200, 282–3, 286–7, 358–9; and economy, 286; and gender, 165–7, 168; and hipsters, 285–6; inner cities, 26–7, 183, 189, 199, 282–5; resistance and alternatives, 286–8; and retail, 284–6; scholarly literature, 283–4; stages, 283
geography. *see* place
geography of age(s), 95–100
Germany, 372, 375
Gertler, M.S., and colleagues, 316–17, 322, 334
ghettos and ghettoization, 184–5, 300
GHG (greenhouse gases) emissions: in Canada, 432–3; and cities, 12, 399, 412, 432–3; and climate change, 432–3; in houses, 208; legislation, 440; reduction and neutrality, 405, 440; and transportation, 271–2, 433; and urban sprawl, 399–400
gig economy, 328
Gini coefficient, 182
Glass, Ruth, 283
Global Age Friendly Cities Project (WHO), 105
Global Footprint Network, 420

globalization, 4, 177–8, 189
Goheen, Peter, 75, 233
governance: arrangements and budgets, 11; and big data, 73; and change, 18, 27, 33; constraints, 45–8; corporatization, 73; definition and manifestation, 1, 3, 10; and digital cities, 72, 73, 83–4; and food systems, 422; and immigration, 43–4; local government, 34; multi-level in Canada, 33–4, 39, 42–5, 47–8; municipalities (*see* municipalities and governance); overview, 33–4; and policy-making, 38–9; as property of cities, 10–11; suburbs, 303; transportation, 46–7
Granville Island (Vancouver, BC), 281
gravity-based opportunity measure, 265–6
Greater Golden Horseshoe (GGH), 6, 224, 279
Greater Toronto-Hamilton Area (GTHA), 267, 268
greenhouse gases. *see* GHG
green infrastructure, 436–40
greening of cities, 280
green revolution, 388, 418
green spaces, in suburbs, 26
Green Urbanism, 436–7
growth, urban: importance, 368–9, 370–1, 373; measurement, 365; and planning, 371–2, 374; and production, 4–5; size bias, 374; smart growth, 224, 435; *see also* rapid-growth cities; slow-growth cities
Guelph, ON, 424

habitat integration and restoration, 437–40
Habitat Island (Vancouver, BC), 438–9
Halifax, NS, 226–7, 244, 441, 442
Hamilton, ON, 324, 326
health, 258, 419, 435–6, 441–2
health care and industry, 7, 17–18, 104–5, 326
Hérouxville, QC, 146
heteronormativity, 155, 156–9
high income earners, 176–7, 178
highways and freeways, 23, 25, 257
hipsters, 285–6
history of cities, importance, 1
Hollander, J.B., 372–3
Holling, C.S., 408
home, meaning and uses, 201–2
homelessness: causes and factors, 187–9, 206; in CMAs, 189–90; continuum, 187; counts and alleviation, 189–91; definitions and description, 186–7; funding and costs, 8; and housing, 206, 282; public policy, 189, 191; and urban development, 282
home ownership: financing and mortgages, 197; as goal, 354; immigrants, 140–4, 201, 217; trends and rates, 197, 200, 204; women, 166–7; by young adults, 103–4
homes, builders and industry, 195–8, 206–7, 216
Hong Kong, eco-footprint, 387
Hoornweg, D., and colleagues, 399
Hospers, G., 372
Hotelling's model, 357
household debt, 180, 181, 207
households: changes to, 118–22, 200; description and formation, 198; income inequality, 175–7, 181–4; living arrangements, 121–2, 181; size and number of, 119–20
housework, 201–2
housing: affordability (*see* affordable housing); and age, 91–5, 125–6; and change, 215–16; and cohorts, 90, 91, 198; and commuting, 200; core housing need, 165, 207; costs, 398; demography and lifestyle, 125–7, 216–17; in dense cities, 402; development, 197; and economy, 9; and employment, 200; and energy, 208; filtering process, 195, 198–9; financing and mortgages, 197; and gentrification, 183; geographical variation, 202–4; and immigration, 140–4, 201, 217; and income, 128–9, 185–6, 195; Indigenous peoples, 207; inequalities, 181–4, 191–2; land value and markets, 204, 350; and lone parenthood, 165; as market, 198, 203, 204–5, 207, 215–16; mobility, 199–200; occupancy factors, 198–201; policy and government, 205–7, 208; prices, 195, 198, 203, 207; problems in Canada, 207–8; production and building, 195–8, 206–7, 216; renovations and repairs, 197–8, 202; rural *vs.* urban, 204; secondary suites and "missing middle" housing, 402–3; short-term rentals, 82; social aspects, 200–1; trends, 402; and urban sprawl, 398–9; young adults, 102–4
Housing First, 206
H+T study (Housing + Transportation), 398–9
Hubbard, P., 284
human capital, 316–17
human exceptionalism, 382, 383
human settlements, origin, 382
human vision, 242–3
Hutton, T., 326

Idle No More, 66
image of city, 278–9, 285
immigrants: border crossings from US, 148; Chinese and Japanese, 135, 139, 140, 144, 298; in CMAs, 137, 141, 147; concentration, 143–4; country of birth, 136; geographical origins, 114; housing and home ownership, 140–4, 201, 217; income inequalities, 178, 179; and population growth, 113–14, 135, 365; and religion, 146; settlement patterns, 141, 143; in suburbs, 143, 144, 298–9, 304, 305–6; temporary migrants and residents, 134, 137, 148; transportation and mobility, 161
immigration: as demographic process, 134–7; discrimination, 144–6; and economy, 138, 139–40, 147; and employment, 148, 149; ghettos and enclaves, 143–4; and governance, 43–4; history, 134–6; and inequality, 144–6, 178; and land value and markets, 350; as life course, 137; MTV phenomenon, 135–6, 137; points system, 147; as process, 149–50; and provinces, 148–9; racial minorities, 136, 140, 145–6, 147; and reproduction, 7; settlement services, 149; as social process, 137–8; transnational aspects, 138, 144; trends and policy, 134, 146–9, 150
improvisational opportunities, 288

income: families, 163–4, 181; generational wage gap, 101–2; of high earners, 176–7, 178; and housing, 128–9, 185–6, 195; Indigenous peoples, 56; and transportation of women, 160, 161–2
income inequalities: in CMAs, 175–8, 179, 182–3; and economy, 180; in households, 175–7, 181–4; and immigrants, 178, 179; low *vs.* high incomes, 176–7; and polarization, 175, 177–8, 216; racial aspects, 184–5
incomplete streets, 412
Indian Act (1876), 52, 54, 55
Indigenous density, 57
Indigenous peoples: in cities, 51–2, 54–5, 66–7; homelessness, 206; housing, 207; labour and employment, 56; legal aspects and cases, 53, 64–5; life expectancy, 113; local government, 34; in multi-level governance, 44–5; and municipalities, 62–4; organizations in cities, 57; population, 51, 52, 54, 55, 112; rights and resurgence, 53, 63–6; Saskatoon Public Library, 60–1; settler colonialism, 51, 53–4, 66; socio-economic characteristics, 55–6; and suburbs, 293, 295; symbolic capital, 57; terminology and identity, 51–2; TRC principles and reconciliation, 64; urbanism and examples, 57–63; urbanization, 54–6, 57; urban reserves, 54, 57–9; women, 65
indoor farms, 428
induced demand, 263
industrial areas and cities, 10, 20
inequality: and age, 98–104; and economy, 180; and globalization, 177–8; home ownership, 185–6, 187; and housing, 181–4, 191–2; and immigration, 144–6, 178; in income (*see* income inequalities); increase, 175; in neighbourhoods, 182–5; and neo-liberalism, 178, 179–81; as process, 191–2; response by government, 192; social factors, 177–82
inertia, and change, 15–16
infant mortality, in Canada, 113
information technology (IT), 18, 25, 326; *see also* digital cities
infrastructure, funding, 8
inner city: change in, 23, 277–8, 358–9; and culture, 279; and density, 279–80; employment, 200; gentrification, 26–7, 183, 189, 199, 282–5; land value and markets, 360–1; neighbourhood typology, 277–8; as place to live, 26–7; redevelopment and revitalization, 278, 279; residential decisions, 93–5; *see also* downtown
innovation: in Canada, 333–7, 338–9, 342; in cities, 331, 333–4, 337–40, 341–2; concept and stages, 332; critical view, 343; future of, 342–3; market and non-market, 332–3, 340, 342; and new economy, 315–16, 322, 323, 328; new-to-firm, 332; non-urban and remote areas, 331, 337–40, 341–2; patents and venture capital, 334–7, 338; social aspects, 332, 333; urban innovation, 337–8
Innovation Superclusters Initiative (ISI), 44, 323
intensification, 279, 280, 303, 304
Internet, 71, 76
interprovincial migration, 114–15
Inuit, 52, 55; *see also* Indigenous peoples
Inwood, J., 52
Ishikawa, Sara, 240–1, 249, 255
islamophobia, 145
IT. *see* information technology (IT)

Jacobs, Allan, 239–40, 241
Jacobs, Jane: on innovation, 334, 337, 343; and mixed-use forms, 219; streets and public spaces, 75, 235, 240, 241, 244
Jacques, P.B., 278
John Hirsch Place (Winnipeg), 239
Jones, Paula, 429
Just Food Farm, 424
"Justice for our Stolen Children" tipi camp, 66

Keil, R., 178, 279, 303, 304, 306
Kenney, M., 75
Keynesianism, 314
Kingston, ON, 72, 232, 368, 423
Kitchener, ON, 245, 323
Kitchin, R., 71, 73
knowledge-based activities, rise, 324–6
knowledge economy: development, 278, 279, 324–6; employment, 324–6, 328; importance, 5; and place, 10; and reproduction, 8
knowledge workers, 316

labour and employment: accessibility to jobs, 265–8; and age, 117–18; automotive industry, 322–3; and change, 313, 328; Fordism, 314; at home, 201–2; and housing, 200; and immigration, 148, 149; Indigenous peoples, 56; by industry in Canada, 318; in knowledge economy, 324–6, 328; in manufacturing, 318–23; in new economy, 316; professional occupations, 324, 325; and reproduction, 7, 8; and suburbs, 300; and technology, 75–7; transportation, 160–2; women, 7, 156, 157, 158, 160–2, 200, 359
landmarks, 244
land rent and prices, 352–7
land tax, 354
land use: and change, 16; and GHG emissions, 399; and land value and markets, 350, 352, 353; by municipalities, 205; and transportation, 255–6, 263
land value and markets: and change, 358–61; characteristics, 352–4; CMAs, 351; and demography, 350; and entrepreneurial policies, 361; factors, 347–9, 350; and housing, 204, 350; and immigration, 350; land development and profit, 356–8; land ownership structure, 354; and location, 347–9, 350; markets structure, 354–6; planning and land use, 350, 352, 353; polarization, 360–1; proximity and accessibility, 349–50; rent and prices, 352–7; tax on rent, 354
laneway houses, 402
laneways, 246–7
Lewis, John, 105
Ley, David, 234, 277, 278, 283
LGBTQ+ people: in cities, 158–9; discrimination, 156; "villages" and neighbourhoods, 167–9
life course: age and demography, 91, 115–18; and change, 115–22; and immigration, 137
life cycle assessments (LCA), 436

life expectancy, 112–13, 118
life sciences industries, 322
lifestyles, 98, 122–5
livability and livable cities, 280
livability-focused development, 400–2
living alone, 119–20, 166
local autonomy track, 37, 39–42, 46–7
local government: climate action, 440–1; description and types, 34; financial resources and funding, 11; and governance, 34; and provinces, 37; reorganization, 47; *see also* municipalities
Local Roots Farms, 428
location, and growing up, 91–2
lofts, 284
lone-parent families/lone parenthood, 121, 122, 162–5
lone-person households, 119–20
Lowe, N., 322
low-income cut-offs (LICO), 176
Lucas, M., 326
Lufa Farms, 424
Lui, C.-W., and colleagues, 105
Lynch, Nicholas, 234

machining occupations, employment, 319, 321–2
McKenzie Towne community, 224–5
Making Room initiative, 403
manufacturing: change in Canada, 5, 318–24; and economy, 317–24; employment, 318–23; Fordism, 314
Manufacturing Innovation Network (MIN), 323
Market Square (Kingston), 232
Markham, ON, 224, 225, 421
marriage and married couples, 120, 121, 122
mayors, 46
"Meals on Wheels," 425–6
Medicine Hill, AB (Moh'kins tsis), 61–3
Meerow, S., and colleagues, 409
Meewasin Skating Rink (Saskatoon), 250
Meier, Richard, 6
men, 158–9; *vs.* women (*see* gender)
Mercury Opera's "104 Underground," 246, 247
Mérite Ovation innovation award, 340
Métis, 52, 54, 55; urbanism and Saskatoon Public Library, 59–61; *see also* Indigenous peoples
Metrotown (Burnaby, BC), 305
migration, 113–15
millennials, 100–1, 123–4
mining industry and employment, 324
Misquadis v Canada, 64–5
"missing middle" housing, 402–3
mixed nodes, 219–20
mixed-use urban forms, 218–26, 228–9, 303
mobility: *vs.* accessibility, 255–6, 273; and congestion, 256; definition, 255; and gender, 157, 160–2; in housing, 199–200; and transportation, 255–6
mode share in transportation, 258–61
Moh'kins tsis, AB (Medicine Hill), 61–3
Molotch, Harvey, 369
Moncton, NB, 336
Montreal, QC: accessibility, 256, 267; cycling network, 234–5, 236, 237; food access, 425–6; gay village, 168; gentrification, 165, 166; immigrants, 134–6, 137, 139, 143; innovation, 336; life sciences industries, 322; lone parenthood, 163–4, 165; nodal development, 220; parklets, 248; planning, 374; property value, 348, 351; public art, 245–6; public spaces and streets, 237, 242, 245–6, 251; residential geography, 96–7; short-term rentals, 82; suburbs and commute, 294; urban farming, 423, 424; urban heat measures, 434; Walk Score, 269, 270, 271; wheelchairs accessibility, 266, 267; young adults' housing, 102–3
mortgages, 197, 203, 205, 207, 353–4
multiculturalism, 138, 150
multigenerational/multi-family households, 299
Mumford, Lewis, 6
municipalities: autonomy, 36–8, 46–7; and big data, 72–3; climate action, 440–1; description and tier systems, 34, 35; downloading of responsibilities on, 370–1; fiscal resources, 41–2; housing and land use, 205; and Indigenous peoples, 62–4; and innovation, 333, 340; mayors, 46; politics, 36; population, 34–5, 36; by province and territory, 35; recognition and authority, 40–1, 45–6; revenue and budget, 11, 370–1, 372; ride-hailing and regulations, 78–81; role, 35–6; urban forms policy, 217–18; zoning and commerce, 285; *see also* local government
municipalities and governance, 18, 34–8; constraints, 45–8; multi-level, 33, 39, 42–5, 47–8; and provinces, 37, 39–42, 46–7
Muskeg Lake Cree Nation, 58

National Housing Strategy (NHS), 43, 206
National League of Cities, 323
national urban policy, 38–9, 48
neighbourhoods: change, 199–200, 278–9; and climate change, 434–6, 441–2; community land trusts, 287–8; filtering, 198–9; inequalities, 182–5; racial aspects, 184–5; residential decisions, 93–5; and sustainability, 401, 403–5; typology, 277–8
Németh, J., 372–3
neo-classical view of inner city, 360
neo-liberalism, 178, 179–81, 218
new-build gentrification, 166
new economy: of cities, 315–17; clustering and agglomeration, 315–16; and culture, 317, 326–8; description, 6, 313, 315; and innovation, 315–16, 322, 323, 328; labour and employment, 316; polarization, 328; transition to, 314–15
new-to-firm innovation, 332
New Urbanism, 19, 26, 126–7, 224–5, 435
New York City, food systems, 423, 425
nodes, 219–20
non-market innovation, 332–3, 340, 342
numbered treaties, 53

Occupy protests, 100
office parks, 227

Okotoks, BC, 437
Olympic Village (Vancouver, BC), 438–9
Ontario: climate action, 440; complete communities, 224; downloading of responsibilities, 370–1; food policy and production, 421, 422; growth and planning, 371, 374; house builders, 195–6; immigration and governance, 44; manufacturing, 319, 320, 322–3; municipalities and local government, 40, 41, 47; ride-hailing and regulations, 79, 80
Ontario Professional Planners Institute (OPPI), 104, 422
organic food, 419
Oshawa, ON, 322–3
Ossington Avenue (Toronto), 285–6
Ottawa, ON, 189, 190, 326, 423, 424
Ottawa-Gatineau, ON/QC, 96–7
over 65s. *see* seniors
ownership. *see* home ownership

paid and unpaid work, 201–2, 300
Parkdale Neighbourhood Land Trust (PNLT), 287–8
parking lots, 241–2
parklets, 248
parks and urban forests, 437, 438, 439
Partners for Climate Protection (PCP) network, 440
Paskapoo Slopes (AB), 61–3
patents, and innovation, 334–6, 338
path dependence, 8, 16, 19
Peck, J., 178, 303
pedestrians. *see* walking and walkability
"permeable Fordism," 314
pharmaceuticals, 322
place: attachment to, 9, 10; definition and manifestation, 3; and growing up, 91–2; and housing, 203–4; importance, 9; as property of cities, 9–10; and public policy, 38–9, 48; and public space, 234; sense of, 9–10, 234
Place du Musée project (Montreal), 245–6
place-making, 234, 278
Places to Grow Act (Ontario, 2005), 279, 371
planning: for aging society, 104–5; and change, 18–19, 27; for climate change, 441–2; food and food systems, 417, 418–19, 420–2, 428–9; influence, 18–19; and land value and markets, 350, 352, 353; and livability, 280; models, 18–19, 22; public participation, 287–8, 403; and resilience, 408–10, 412; slow-growth and declining cities, 364, 369, 370, 371–4, 375; urban forms, 28
platform economy and firms, 74–5, 315
polarization: and income inequality, 175, 177–8, 216; land value and markets, 360–1; and new economy, 328; and nodal development, 223; socio-spatial, 182–5
policy: and economy, 38; food and food systems, 421–2, 429; "urban lens" tool, 42; *see also* public policy
pollution, by cities, 11
population: in Canada, 110, 112, 135, 375; in CMAs, 34, 36, 159, 222, 293; and immigration, 113–14, 135, 365; Indigenous peoples, 51, 52, 54, 55, 112; as measure of growth, 365, 369, 374–5; municipalities, 34–5, 36; pyramids, 111, 112; slow-growth and declining cities, 365, 369; urban and rural, 92
Portlandia, 124, 125
post-Fordism, 314, 361
post-suburbia, 302–4
poverty: child poverty, 201; in neighbourhoods, 184–5; racialization, 178, 184–5; in suburbs, 299–300
power centres, 226–7
prices: of housing, 195, 198, 203, 207; land and commuting, 355–6, 360; and rent in market, 352–7
"priority neighbourhoods," 299–300
private communities, 227–8
private sector, innovation, 332
processing occupations, employment, 319, 321–2
production, 3, 4–5, 7
professional occupations, employment, 324, 325
Project for Public Spaces (PPS), 234
protest movements, 25
provinces: agricultural land, 418–19; climate action, 440; condos, 217–18; governance and municipalities, 37, 39–42, 46–7; and housing, 206; immigration policy, 148–9; local government system, 34; migration rates, 114–15; in multi-level governance, 43–4; neo-liberal reforms, 180; population in municipalities, 34–5; recognition of municipalities, 40; urban reserves, 58–9; *see also* territories; specific provinces
Provincial Nominee Program, 149
proximity, 3, 5–6, 10, 349–50
psychology of failure, 368–9
public engagement, 287–8, 403, 441
public housing, 205
public policy: amenity provision, 361; homelessness, 189, 191; housing, 205–7, 208; immigration, 146–9; place-based, 38–9, 48; for sustainability, 393; urban forms, 217–18
public sector, 7–8, 332
public services, and big data, 72
public space: and citizenship, 232, 233, 252; and commerce, 247, 250–1; contest, 251–2; cycling, 234–5, 236, 237; depaving, 411–12; description, 75; and digital cities, 75; knowledge and study, 234–5, 252; parklets, 248; and place, 234; role, 233–4; squares, 249–51; streets as, 232–3; symbolism and state control, 233–4; three Ms, 232, 236; youth in, 95
public surveys, of public space, 235
public transportation: and accessibility, 257, 266; CMAs, 258–61; commuting, 258–60; development, 23; and gender, 160–1; history, 256–7; in mixed-used forms, 219, 225–6; mode share, 258–61; and recentralization model, 29; and ride-hailing, 81; and streets, 236; suburban town centres, 220
Pulido, Laura, 295

qualitative development, 373
Quarry Park (Calgary), 127
quasi-sustainability, 388
Quebec City, QC, 96–7, 233

Quebec province: climate action, 440; immigrants, 146, 148–9; innovation, 338–9, 340; lone parenthood, 164–5; municipalities, 40; short-term rentals, 82; women and car ownership, 160
queer neighbourhoods, 167–9

race and poverty, 178, 184–5
racial minorities, 136, 140, 145–6, 147
racial profiling, 145–6
rail and railways, 20
ransoms to hackers, 73
rapid-growth cities, 364, 367, 369–70, 371
"Reaching Home" program, 43
real estate market, 198
real property, 9, 352–3; *see also* land value and markets
recentralization model, 29
recession of 2008, 353–4
reconciliation with Indigenous peoples, 64
redevelopment, 16, 278, 279
refugees, 147–8, 201
regenerative design, 438
Regina, SK, 66, 232
registered Indians, population, 55
Réinventer Montréal: Transportation Plan 2008, 236
relative homelessness, 186
religion, and immigration, 146
Relph, E., 10
renovations and repairs, 197–8, 202
rentals/rental market, 81–2, 189, 195, 203
rent and prices, in land market, 352–7
rent gap thesis, 283
reproduction, 3, 6–8
Réseau Blanc, 236
residential communities, 126–9
residential decisions, 91–5
residential development and forms, 218, 219, 220–4, 279–80
residential geography, 96–8
residential intensification, 26–7
resilience: and adaptive management, 406–8, 409–12, 413; and climate adaptation, 405–8, 412; concept and goals, 408–9, 412; planning, 408–10, 412; and sustainability, 405–6, 412
resource communities, 4, 365–6
retail, and gentrification, 284–6
retirement, and life course, 116–18
Ricardian rent and land tax, 354, 355
Ricardio, David, 355
ride-hailing and regulations, 78–81
Riley, M., 90
River Landing, Saskatoon, 245
roads, 23, 25; *see also* streets
Roberts, Wayne, 421
rooftop food production, 424–5
rooming houses, 287–8
Rosemont–La Petite-Patrie, QC, 434
Rosenberg, M., 95
Round Prairie, SK, 59–61
Royal Proclamation, 1763, 53
rural areas: agriculture, 418; digital divide, 71; housing, 204; innovation, 331, 337–40, 341–2; population, 92
Ryder, N., 91

Saanich, BC, 105
safety, 240, 300
same-sex couples, 121–2, 156, 158–9; *see also* LGBTQ+
Sands, A., 326
San Francisco, 248
Santropol Roulant, 425–6
Saskatchewan, 112
Saskatoon, SK: cycling, 236; food access, 425, 426; Indigenous peoples, 56, 58; and Métis, 59–60; parking lots, 241–2; public art and performance, 246; public spaces and streets, 232, 235, 240, 243, 244, 245, 246, 248, 250; skating rink, 250; streets and culture, 246, 247; transparency, 243; urban farm, 424; urban reserve, 58
Saskatoon Public Library, 60–1
Schumpeter, J., 332
Scott, A.J., 314
sea level rise adaptation, 406–7, 439, 442
Second Age, 115, 116–17, 123
secondary suites, 402
segregated-use urban forms, 218–19, 226–8, 229
seniors: age segregation, 98, 100; and life course, 117–18; lifestyle, 125; planning for, 104–5; residential communities, 126; residential geography, 96–8; *see also* aging
sense of place, 9–10, 234
service economy, 5
settler colonialism, 51, 53–4, 66
seven properties of cities, 2–12
Sex and the City, 124
sexuality, 155–6; and social change, 156–9
sexual minorities. *see* LGBTQ+ people
shared streets, 237, 238–9
sharing economy, 74
shelter poverty, 186
Shields, Rob, 295
shopping streets, gentrification, 284–6
shorelines, 439–40
short-term rentals, 81–2
Shoup, D., 357
shrinking cities. *see* slow-growth cities
Sidewalk Labs proposal, 82–3
sidewalks, 240
Siemiatycki, E., 303
Silverstein, Murray, 240–1, 249, 255
single parents. *see* lone-parent families
single-person households, 119–20
single young female (SYF), 124–5
site survey, of public space, 234
size of urban regions, 19, 27
skating rinks, 250
Slow Cities, 375

slow-growth cities: concept, 364–5; economy, 365–6, 368, 369; location, 365–6, 368; opportunities and challenges, 370–1, 374–6; patterns and trends, 364, 366, 367, 368; planning, 369, 370, 371–4, 375; and population, 365, 369; psychology of failure, 368–9; *vs.* rapid-growth cities, 364; *see also* declining cities
slums, 184
Smart Cities Challenge (SCC), 45, 78
smart city, 70; *see also* digital cities
smart decline, 373
smart growth, 224, 435
Smith, Adam, 354–5
Smith, Neil, 283–4
social aspects and processes: and age, 89, 95–6; for generations, 91; and housing, 200–1; and immigration, 137–8; and inequality, 177–82; and innovation, 332, 333; polarization, 182–5; and resilience, 409; in suburbs, 299–300; in urban development, 282, 288
social housing, 165, 200, 205–6
solar community, 437
space, urban. *see* urban structures and space
Spadina Crescent East (Saskatoon, SK), 232
speculative builders, 197, 198
"spheres of jurisdiction" framework, 40
SPIN (small plot intensive) farming, 424
sprawl: anti-sprawl initiatives, 401, 403; and cars, 257–8; description and impact, 397–8, 399; and GHG emissions, 399–400; reduction, 397; and sustainability, 397–405; and transportation, 398–9
squares, design and programming, 249–51
Square Saint-Louis (Montreal), 251
Stanley, Gerald, 232
stormwater management, 406, 437–8
streetcars, 20, 257
streets: adaptive management, 411–12; and cars, 236, 238, 239; complementarity, 244; complete streets, 236–8, 412; culture and art, 244–8; definition features, 240–2; depaving, 411–12; design, 235–44, 435; furniture, 245; ground floor frontage, 243; as outdoor space, 241; programming, 244–8; as public space, 232–3; qualities of great streets, 239–44; and safety, 240; shared streets, 237, 238–9; transparency, 242–3
structural view of inner city, 360–1
structures, urban. *see* urban structures and space
suburbanisms, 292, 306
suburbs and suburbanization: appearance, 20; attachment to, 10, 26; and capitalization, 9; cars, 217, 220–1; commuting, 20, 23, 294, 359; compact development, 400; costs to households, 398; definition and characteristics, 292–3, 294, 296–7, 306, 400; development, 22, 23, 293–8, 301, 303, 359; diversity in, 297–8, 306; downtowns, 26, 301; and elections, 304; "feral" peripheries, 295; function, 302–3; and gender, 300; governance, 303; home ownership, 204; and immigrants, 143, 144, 298–9, 304, 305–6; on Indigenous peoples' land, 293, 295; innovations in, 26; labour and work, 300; as living places, 292; move to, 199–200; politics, 303–6; population, 293; post-suburbia, 302–4; post-WWII, 26, 199; and poverty, 299–300; power centres, 226–7; as process, 302, 306; public transportation, 220; residential geography, 96–8; social aspects, 299–300; sprawl, 397–8; stereotypes and representation, 295–7, 298, 306; town centres, 219–21; transportation, 160–2, 257, 398–9; urban forms, 219–21, 301–3; urbanization, 400; urban structures, 23, 24; and women, 160–2, 300
Sudbury, ON, 324, 369, 370, 373
Summit of the Americas, 233
Supreme Court of Canada, Indigenous rights and title, 53, 64
surface parking, 241–2
Surrey, BC, 224, 299, 406–8
surveillance capitalism, 83
surveillance city, 73
survival patterns and curve, in Canada, 113
sustainability: actions needed, 390–2, 393; of cities, 386, 387–8; compact and dense cities, 400–3; conflicting dynamics, 397; "development path," 399–400; economy and consumption, 390, 391, 392–3; example, 404; food systems and production, 388, 420, 421; as goal, 280, 399; in neighbourhoods, 401, 403–5; overshoot, 390; public policy, 393; quasi-sustainability, 388; and resilience, 405–6, 412; and transportation, 258, 269, 273, 435; and urban sprawl, 397–405
sustainable development, importance, 12

taxation, 45, 46–7, 82, 180, 354
taxis, 79, 80–1
technoburbs, 301
technology: and change, 17–18, 27; disruptive technologies, 73–5, 84; and economy, 75–7, 84; and employment, 75–7; impact, 70; *see also* digital cities
teleworking, 202
temporary migrants and residents, 134, 137, 148
10-Year Plan to End Homelessness (Calgary, AB), 282
terra nullius, 53
territories, 35, 149; *see also* provinces
Third Age, 115, 117–18, 123, 125
Third Avenue (Saskatoon), 248
3T's (technology, talent, and tolerance), 316–17
three M's (movement, market, meeting), 232, 236
Tickell, A., 178
time periods, cohort approach, 90–1
tiny homes, 128
Tokyo, eco-footprint, 387, 392
topophilia, 9, 10
Toronto, ON: accessibility to jobs, 268; Amazon's HQ2, 76; boundaries and proximity, 6; community land trust, 287–8; condos, 223; cultural industry, 327–8; as digital city, 76, 78–80, 82–3; eco-footprint, 386–7; ethnoburbs, 298, 299; families in, 94; food access, 425, 426; food policy and production, 421; gentrification, 284–6; governance, 41, 45, 47; growth and planning, 374; homelessness, 189; immigrants, 135–6, 137, 139–40, 141, 143, 145–6; income inequalities and housing, 183–4; life sciences industries, 322; lone parenthood, 163–4, 165; mixed-use forms, 220, 221–2; power centre, 227; property value, 348–9, 350, 351; reorganization, 47; residential geography, 97–8, 99;

ride-hailing and regulations, 78–80; Sidewalk Labs proposal, 82–3; streets and culture, 247; urban farming, 423, 424; Walk Score, 269, 270–1; young adults, 97–8, 99, 102–3
Toronto Food Policy Council (TFPC), 421
Tory, John, 47
Tower-in-the-Park housing, 19, 22
traffic, 23, 25; *see also* commuting
trains, 20
transgender people, 156, 169
transition, 26–7, 91, 115; *see also* change, urban
transit-oriented development, 220
TransLink and SkyTrain, 46–7, 93
transport analysis zones (TAZs), 262
transportation: and accessibility (*see* accessibility); active modes, 258; automated vehicles (AVs), 272–3; Calgary, AB, 19; and change, 16, 18, 19, 20; and congestion, 256, 258, 259–61; costs, 398; current systems, 258–61; environmental impact, 271–2; future of, 271–3; gender and employment, 160–2; and GHG emissions, 271–2, 433; history in Canada, 256–7; induced demand, 263; and land use, 255–6, 263; municipal governance, 46–7; and proximity, 6; and size of urban region, 19; and suburbs, 160–2, 257, 398–9; and sustainability, 258, 269, 273, 435; as technologies, 18; travel demand models, 261–4; urban forms, 16, 20; and urban sprawl, 398–9; and women, 160–2; *see also* commuting; public transportation
travel demand management, 264
travel demand models, 261–4
treaties with Indigenous peoples, 53
Trinity Development Group and Trinity Hills Project, 62–3
Trudeau, Justin, and government, 42, 43, 44, 47
Truth and Reconciliation Commission (TRC) report and principles, 64
Tuan, Yi-Fu, 9
20th Street West Streetscape Improvement Project (Saskatoon), 240
Twitter, 72

Uber, 78–9
United Nations Declaration on the Rights of Indigenous Peoples (UNDRIP), 53, 65
United States: accessibility, 256; economic integration, 178, 180; energy use, 433; gated communities, 126; growth and planning, 371–2; immigration, 148; poverty and ghettos, 184–5; recession and property prices, 353–4; slow-growth and declining cities, 365; transportation, 258–9, 260, 261; urban development, 23
universal properties of cities, 2–12
unpaid and paid work, 201–2, 300
Urban Aboriginal Strategy (UAS), 44
urban areas: population, 92; urban forms, 218–20
urban change. *see* change, urban
urban core. *see* inner city
urban development. *see* development, urban
urban dispersion, 20, 22–3, 25, 26
urban ecosystem, 383–4
urban farms, 424
urban forms. *see* forms, urban
urban growth, 4–5
urban heat island effect (UHI), 434
urbanism: concept, 56–7, 217; Indigenous peoples, 57–63
urbanization: and ecology, 381–4; and housing, 204; Indigenous peoples, 54–6, 57; and sexuality, 156–9; of suburbs, 400
"urban lens" policy tool, 42
urban planning. *see* planning
Urban Programming for Indigenous Peoples (UPIP), 44–5
urban renewal, 22
urban reserves, 54, 57–9
urban sprawl. *see* sprawl
urban structures and space: pre-1945, 20, 21; 1945-1975, 20, 22–3, 24; 1975 to present, 23, 25–7, 28; change in, 19–29; dissatisfaction with, 23, 25; suburbs, 23, 24

vacant land, value, 356–7
values, and change, 18, 22, 27, 217
Vancouver, BC: accessibility to jobs, 265; adaptive management, 410–11, 438–9; "affective resistance" by Indigenous peoples, 65; condos, 222, 223; cultural industry, 326–7; dense and compact housing, 400–1, 402, 403; eco-footprint, 386–7; ethnoburbs, 298, 299; food planning and policy, 422; governance, 41, 46–7; immigrants, 135, 137, 139, 142, 143, 144; income and wage gap, 101; Indigenous peoples, 54; livability as city, 280, 282; livability-focused development, 400–1; mixed-use forms, 220, 222–3; property value, 348, 349, 351; as rapid-growth city, 370; residential geography, 96–7; ride-hailing and public transportation, 81; urban farming, 423, 424–5; urban planning and design, 281; urban sprawl and affordability, 398–9; Walk Score and walking, 269, 270, 271, 436; waterfront, 281; young adults, 94, 95, 102–3
"Vancouverism," 281
Vaz, L.F., 278
Veggie Mobile, 426, 427
venture capital, 336–7
Veracini, Lorenzo, 294
vertical farming, 427–8
Victoria, BC, 404
video game industry, 327
visible minorities, 114, 179, 185
vision, of humans, 242–3
Vitiello, D., 140
volunteered data, 71–2
voters, and suburbs, 304
vulnerability of cities, 388–9

Wachsmuth, D., 74
Walker, R., 65
walking and walkability: and health, 435–6; pre-1945, 20; shared streets, 237; streets and public space, 240; and urban forms, 222, 227, 228; Walk Score and CMAs, 269–71
Walk Score, 269–71
Walmart, and food, 419
Warrian, P., 326
waste from cities, 384

waterfront, use and condos, 280–2
Waterfront Toronto agency, 82, 83
Waterloo, ON, 105, 421, 425
weight issues, 435–6
Weisler, A., 74
welfare state, 8, 178, 180, 314
wheelchairs, 266, 267
white privilege, 295
Wilson, D., 95, 279
Winnipeg, MB, 55–6, 236, 256, 370, 423
Wolfe, D.A., 326, 334
women: cars, 160, 161; condos and home ownership, 166–7; employment, 7, 156, 157, 158, 160–2, 200, 359; and gentrification, 165–6; housework, 201, 202; Indigenous women, 65; "individualization," 156–7; lifestyle, 124–5; as lone parents, 163, 165; marriage and domestic work, 157–8; *vs.* men (*see* gender); mobility as immigrants, 161; reproduction and production in cities, 7; same-sex partnerships, 159; in suburbs, 160–2, 300; and transportation, 160–2; *see also* gender
Wong-Tam, Kristyn, 247
woonerf, 239
work and workforce. *see* labour and employment
work at home, 201–2
World Health Organization (WHO), 105
Wyly, E., 303

Yellow Quill First Nation, 59
Yonge Street (Toronto), 247
Young, D., 304
young adults: age segregation, 98, 100; housing and ownership, 102–4; income and wage gap, 101–2; inequalities for, 100–4; life course, 116–17; lifestyle, 123–5; living arrangements, 116, 120–1; in parental home, 102, 116; residential decisions and geography, 92–5, 96–8; in Toronto, 97–8, 99, 102–3; in Vancouver, 94, 95, 102–3
YWCA Metro Vancouver, 424–5

zoning, 205, 285, 352
Zysman, J., 75